Finite Mathematics

Finite Mathematics

Sixth Edition, Hybrid

Stefan Waner
Hofstra University

Steven R. Costenoble
Hofstra University

Australia • Brazil • Japan • Korea • Mexico • Singapore • Spain • United Kingdom • United States

Finite Mathematics, Sixth Edition, Hybrid
Stefan Waner, Steven R. Costenoble

Publisher: Richard Stratton

Development Editor: Jay Campbell

Editorial Assistant: Danielle Hallock

Media Editor: Andrew Coppola

Brand Manager: Gordon Lee

Marketing Communications Manager: Linda Yip

Content Project Manager: Alison Eigel Zade

Senior Art Director: Linda May

Manufacturing Planner: Doug Bertke

Rights Acquisition Specialist: Shalice Shah-Caldwell

Production Service: MPS Limited

Text Designer: RHDG Design

Cover Designer: Chris Miller

Cover Image: RHDG/Ben Hopfer

Compositor: MPS Limited

Library of Congress Control Number: 2012947705

Student Edition:
ISBN-13: 978-1-285-05631-9
ISBN-10: 1-285-05631-0

Brooks/Cole
20 Channel Center Street
Boston, MA 02210
USA

Cengage Learning is a leading provider of customized learning solutions with office locations around the globe, including Singapore, the United Kingdom, Australia, Mexico, Brazil and Japan. Locate your local office at **international.cengage.com/region**

Cengage Learning products are represented in Canada by Nelson Education, Ltd.

For your course and learning solutions, visit **www.cengage.com.**

Purchase any of our products at your local college store or at our preferred online store **www.cengagebrain.com.**

Instructors: Please visit **login.cengage.com** and log in to access instructor-specific resources.

Printed in Canada
1 2 3 4 5 6 7 16 15 14 13 12

Brief Contents

Contents

CHAPTER 3 Systems of Linear Equations and Matrices 143

CHAPTER 4 Matrix Algebra and Applications 187

CHAPTER 5 Linear Programming 249

CHAPTER 6 Sets and Counting 311

Preface

Finite Mathematics, sixth edition, hybrid, is intended for a one- or two-term course for students majoring in business, the social sciences, or the liberal arts. Like the earlier editions, the sixth edition, hybrid of *Finite Mathematics* is designed to address the challenge of generating enthusiasm and mathematical sophistication in an audience that is often underprepared and lacks motivation for traditional mathematics courses. We meet this challenge by focusing on real-life applications and topics of current interest that students can relate to, by presenting mathematical concepts intuitively and thoroughly, and by employing a writing style that is informal, engaging, and occasionally even humorous.

The sixth edition, hybrid goes further than earlier editions in implementing support for a wide range of instructional paradigms: from traditional face-to-face courses to online distance learning courses, from settings incorporating little or no technology to courses taught in computerized classrooms, and from classes in which a single form of technology is used exclusively to those incorporating several technologies. We fully support three forms of technology in this text: TI-83/84 Plus graphing calculators, spreadsheets, and powerful online utilities we have created for the book. In particular, our comprehensive support for spreadsheet technology, both in the text and online, is highly relevant for students who are studying business and economics, where skill with spreadsheets may be vital to their future careers.

Key Features of the Hybrid Edition

Many mathematics courses are evolving into lecture-lab courses or into courses in which all assignments—homework and even tests—are delivered online. Furthermore, distance learning is growing rapidly. Instructors are increasingly faced with the challenge of integrating technology and teaching the concepts and skills of statistics.

In this hybrid text, as with the hardcover text, the goal is to provide instructors and students with the tools they can use to meet this challenge.

What does hybrid mean? A hybrid text involves an integration of several products and can be used best in a course with a strong blend of lecture time and online course work.

For this hybrid edition, the end-of-section exercises have been removed from the text and are available exclusively in Enhanced WebAssign, an easy-to-use online homework system. This slightly smaller book will be more manageable for the student who is spending his or her homework time on a computer.

Although this text has been tailored for instructors and students who use online homework, the focus of the expositions remains unchanged.

Our Approach to Pedagogy

Real-World Orientation We are confident that you will appreciate the diversity, breadth, and abundance of examples and exercises included in this edition. A large number of these are based on real, referenced data from business, economics, the life sciences, and the social sciences. Examples and exercises based on dated information have generally been replaced by more current versions; applications based on unique or historically interesting data have been kept.

Adapting real data for pedagogical use can be tricky; available data can be numerically complex, intimidating for students, or incomplete. We have modified and streamlined many of the real-world applications, rendering them as tractable as any "made-up" application. At the same time, we have been careful to strike a pedagogically sound balance between applications based on real data and more traditional "generic" applications. Thus, the density and selection of real data–based applications has been tailored to the pedagogical goals and appropriate difficulty level for each section.

Readability We would like students to read this book. We would like students to *enjoy* reading this book. Thus, we have written the book in a conversational and student-oriented style, and have made frequent use of question-and-answer dialogues to encourage the development of the student's mathematical curiosity and intuition. We hope that this text will give the student insight into how a mathematician develops and thinks about mathematical ideas and their applications.

Rigor We feel that mathematical rigor need not be antithetical to the kind of applied focus and conceptual approach that are earmarks of this book. We have worked hard to ensure that we are always mathematically honest without being unnecessarily formal. Sometimes we do this through the question-and-answer dialogues and sometimes through the "Before we go on . . ." discussions that follow examples, but always in manner designed to provoke the interest of the student.

Five Elements of Mathematical Pedagogy to Address Different Learning Styles The "Rule of Four" is a common theme in many texts. Implementing this approach, we discuss many of the central concepts **numerically**, **graphically**, and **algebraically** and clearly delineate these distinctions. The fourth element, **verbal communication** of mathematical concepts, is emphasized through our discussions on translating English sentences into mathematical statements and in our extensive Communication and Reasoning exercises at the end of each section. A fifth element, **interactivity**, is implemented through expanded use of question-and-answer dialogues but is seen most dramatically in the student Website. Using this resource, students can interact with the material in several ways: through interactive tutorials in the form of games, chapter summaries, and chapter review exercises, all in reference to concepts and examples covered in sections and with online utilities that automate a variety of tasks, from graphing to regression and matrix algebra.

Exercise Sets Our comprehensive collection of exercises provides a wealth of material that can be used to challenge students at almost every level of preparation and includes everything from straightforward drill exercises to interesting and rather challenging applications. The exercise sets have been carefully graded to move from straightforward basic exercises and exercises that are similar to examples in the text to more interesting and advanced ones, marked as "more advanced" for easy reference. There are also several much more difficult exercises, designated as "challenging." We have also included, in virtually every section of every chapter, interesting applications based on real data, Communication and Reasoning exercises that help students articulate mathematical concepts and recognize common errors, and exercises ideal for the use of technology.

Many of the scenarios used in application examples and exercises are revisited several times throughout the book. Thus, for instance, students will find themselves using a variety of techniques, from solving systems of equations to linear programming to analyze the same application. Reusing scenarios and important functions provides unifying threads and shows students the complex texture of real-life problems.

New to This Edition

Content

- Chapter 1 (page 35): We now include, in Section 1.1, careful discussion of the common practice of representing functions as equations and vice versa; for instance, a cost equation like $C = 10x + 50$ can be thought of as defining a cost *function* $C(x) = 10x + 50$. Instead of rejecting this practice, we encourage the student to see this connection between functions and equations and to be able to switch from one interpretation to the other.

  Our discussion of functions and models in Section 1.2 now includes a careful discussion of the algebra of functions presented through the context of important applications rather than as an abstract concept. Thus, the student will see from the outset *why* we want to talk about sums, products, etc. of functions rather than simply *how* to manipulate them.

- Chapter 2 (page 101): The Mathematics of Finance is now Chapter 2 of the text because the discussion of many important topics in finance relates directly to the first discussions of compound interest and other mathematical models in Chapter 1. Note that our discussion of the Mathematics of Finance does not require the use of logarithmic functions to solve for exponents analytically but instead focuses on numerical solution using the technologies we discuss. However, the use of logarithms is presented as an option for students and instructors who prefer to use them.
- **Case Studies:** A number of the Case Studies at the ends of the chapters have been extensively revised, using updated real data, and continue to reflect topics of current interest, such as subprime mortgages, hybrid car production, and the diet problem (in linear programming).

Current Topics in the Applications

- We have added and updated numerous real data exercises and examples based on topics that are either of intense current interest or of general interest to contemporary students, including Facebook, XBoxes, iPhones, iPads, foreclosure rates, the housing crisis, subprime mortgages, stock market gyrations, shorting the stock market, and even travel to Cancun. (Also see the list, in the inside back cover, of the corporations we reference in the applications.)

Exercises

- We have expanded the chapter review exercise sets to be more representative of the material within the chapter. Note that all the applications in the chapter review exercises revolve around the fictitious online bookseller, *OHaganBooks.com,* and the various—often amusing—travails of *OHaganBooks.com* CEO John O'Hagan and his business associate Marjory Duffin.
- We have added many new conceptual Communication and Reasoning exercises, including many dealing with common student errors and misconceptions.

End-of-Chapter Technology Guides

- Our end-of-chapter detailed Technology Guides now discuss the use of spreadsheets in general rather than focusing exclusively on Microsoft® Excel, thus enabling readers to use any of the several alternatives now available, such as Google's online Google Sheets®, Open Office®, and Apple's Numbers®.

Continuing Features

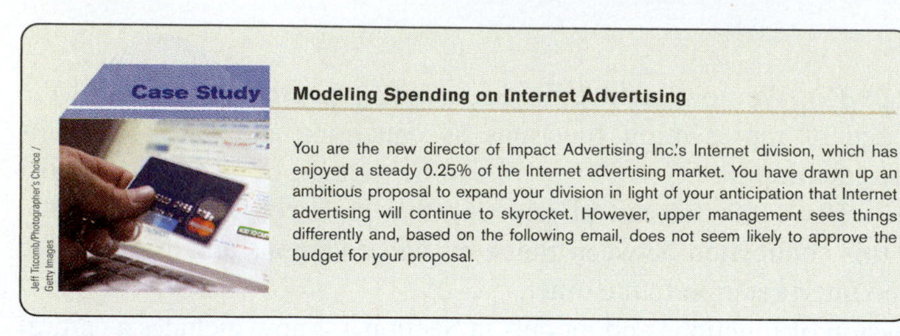
Case Study

Modeling Spending on Internet Advertising

You are the new director of Impact Advertising Inc.'s Internet division, which has enjoyed a steady 0.25% of the Internet advertising market. You have drawn up an ambitious proposal to expand your division in light of your anticipation that Internet advertising will continue to skyrocket. However, upper management sees things differently and, based on the following email, does not seem likely to approve the budget for your proposal.

Jeff Titcomb/Photographer's Choice/Getty Images

- **Case Studies** Each chapter ends with a section entitled "Case Study," an extended application that uses and illustrates the central ideas of the chapter, focusing on the development of mathematical models appropriate to the topics. These applications are ideal for assignment as projects, and to this end we have included groups of exercises at the end of each.
- **Before We Go On** Most examples are followed by supplementary discussions, which may include a check on the answer, a discussion of the feasibility and significance of a solution, or an in-depth look at what the solution means.
- **Quick Examples** Most definition boxes include quick, straightforward examples that a student can use to solidify each new concept.
- **Question-and-Answer Dialogue** We frequently use informal question-and-answer dialogues that anticipate the kinds of questions that may occur to the student and also guide the student through the development of new concepts.

Q: *It seems clear from the figure that the second model in Example 1 gives a better fit. Why bother to compute SSE to tell me this?*

A: The difference between the two models we chose is so great that it is clear from the graphs which is the better fit. However, if we used a third model with $m = 0.25$ and $b = 9.1$, then its graph would be almost indistinguishable from that of the second, but a slightly better fit as measured by SSE = 1.68.

- **Marginal Technology Notes** We give brief marginal technology notes to outline the use of graphing calculator, spreadsheet, and Website technology in appropriate examples. When necessary, the reader is referred to more detailed discussion in the end-of-chapter Technology Guides.
- **End-of-Chapter Technology Guides** We continue to include detailed TI-83/84 Plus and Spreadsheet Guides at the end of each chapter. These Guides are referenced liberally in marginal technology notes at appropriate points in the chapter, so instructors and students can easily use this material or not, as they prefer. Groups of exercises for which the use of technology is suggested or required appear throughout the exercise sets.

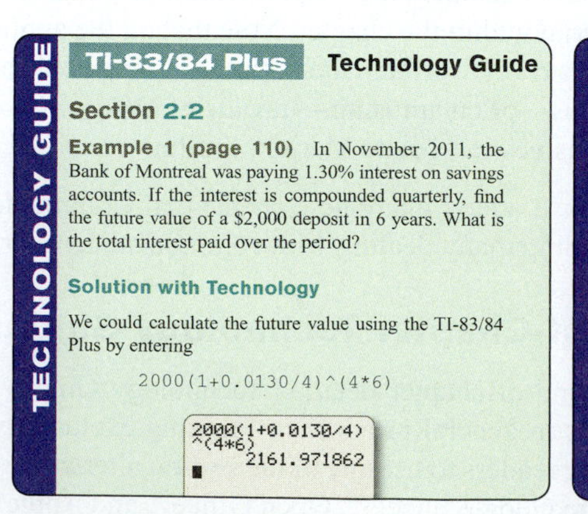
TECHNOLOGY GUIDE

TI-83/84 Plus Technology Guide

Section 2.2

Example 1 (page 110) In November 2011, the Bank of Montreal was paying 1.30% interest on savings accounts. If the interest is compounded quarterly, find the future value of a \$2,000 deposit in 6 years. What is the total interest paid over the period?

Solution with Technology

We could calculate the future value using the TI-83/84 Plus by entering

```
2000(1+0.0130/4)^(4*6)
```


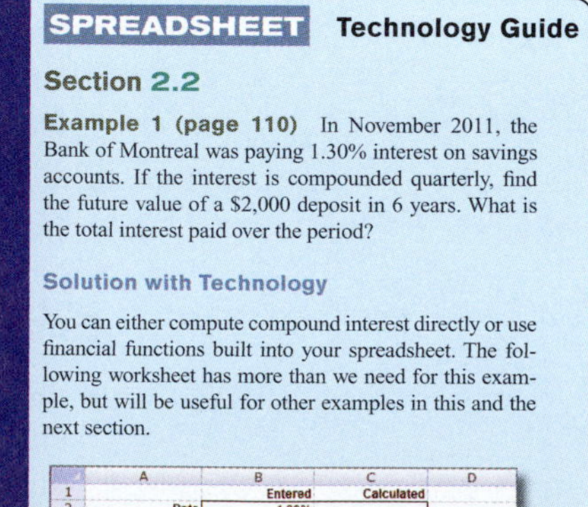
SPREADSHEET Technology Guide

Section 2.2

Example 1 (page 110) In November 2011, the Bank of Montreal was paying 1.30% interest on savings accounts. If the interest is compounded quarterly, find the future value of a \$2,000 deposit in 6 years. What is the total interest paid over the period?

Solution with Technology

You can either compute compound interest directly or use financial functions built into your spreadsheet. The following worksheet has more than we need for this example, but will be useful for other examples in this and the next section.

- **Communication and Reasoning Exercises for Writing and Discussion** These are exercises designed to broaden the student's grasp of the mathematical concepts and develop modeling skills. They include exercises in which the student is asked to provide his or her own examples to illustrate a point or design an application with a given solution. They also include "fill in the blank" type exercises, exercises that invite discussion and debate, and exercises in which the student must identify common errors. These exercises often have no single correct answer.

Supplemental Material

For Instructors and Students

Enhanced WebAssign®

Content

Exclusively from Cengage Learning, Enhanced WebAssign® combines the exceptional mathematics content in Waner and Costenoble's text with the most powerful online homework solution, WebAssign. Enhanced WebAssign engages students with immediate feedback, rich tutorial content, videos, animations, and an interactive eBook, helping students to develop a deeper conceptual understanding of the subject matter. The interactive eBook contains helpful search, highlighting, and note-taking features.

Instructors can build online assignments by selecting from thousands of text-specific problems, supplemented if desired with problems from any Cengage Learning textbook. Flexible assignment options give instructors the ability to choose how feedback and tutorial content is released to students as well as the ability to release assignments conditionally based on students' prerequisite assignment scores. Increase student engagement, improve course outcomes, and experience the superior service offered through CourseCare. Visit us at http://webassign.net/cengage or www.cengage.com/ewa to learn more.

Service

Your adoption of Enhanced WebAssign® includes CourseCare, Cengage Learning's industry leading service and training program designed to ensure that you have everything that you need to make the most of your use of Enhanced WebAssign. CourseCare provides one-on-one service, from finding the right solutions for your course to training and support. A team of Cengage representatives, including Digital Solutions Managers and Coordinators as well as Service and Training Consultants, assists you every step of the way. For additional information about CourseCare, please visit www.cengage.com/coursecare.

Our Enhanced WebAssign training program provides a comprehensive curriculum of beginner, intermediate, and advanced sessions, designed to get you started and effectively integrate Enhanced WebAssign into your course. We offer a flexible online and recorded training program designed to accommodate your busy schedule. Whether you are using Enhanced WebAssign for the first time or an experienced user, there is a training option to meet your needs.

www.WanerMath.com

The authors' Website, accessible through www.WanerMath.com and linked within Enhanced WebAssign, has been evolving for more than a decade and has been receiving increasingly more recognition. Students, raised in an environment in which computers permeate both work and play, now use the Internet to engage with the material in an active way. The following features of the authors' Website are fully integrated with the text and can be used as a personalized study resource as well as a valuable teaching aid for instructors:

- *Interactive tutorials* on almost all topics, with guided exercises, which also can be used in classroom instruction or in distance learning courses
- *More challenging game versions of tutorials* with randomized questions that complement the traditional interactive tutorials and can be used as in-class quizzes
- *Detailed interactive chapter summaries* that review basic definitions and problem-solving techniques and can act as pre-test study tools
- *Downloadable Excel tutorials* keyed to examples given in the text
- *Online utilities for use in solving many of the technology-based application exercises*. The utilities, for instructor use in class and student use out of class, include a function grapher and evaluator, regression tools, a matrix algebra tool, linear programming tools, and a line entry calculator that calculates permutations and combinations and expands multinomial expressions.
- *Chapter true-false quizzes* with feedback for many incorrect answers
- *Supplemental topics* including interactive text and exercise sets for selected topics not found in the printed texts
- *Spanish versions* of chapter summaries, tutorials, game tutorials, and utilities

For Students

Student Solutions Manual *by Waner and Costenoble*
ISBN: 9781285085586
The student solutions manual provides worked-out solutions to the odd-numbered exercises in the text, plus problem-solving strategies and additional algebra steps and review for selected problems.

To access this and other course materials and companion resources, please visit **www.cengagebrain.com**. At the CengageBrain.com home page, search for the ISBN of your title (from the back cover of your book) using the search box at the top of the page. This will take you to the product page where free companion resources can be found.

For Instructors

Complete Solution Manual *by Waner and Costenoble*
ISBN: 9781285085593
The instructor's solutions manual provides worked-out solutions to all of the exercises in the text.

Solution Builder *by Waner and Costenoble*
ISBN: 9781285085609
This time-saving resource offers fully worked instructor solutions to all exercises in the text in customizable online format. Adopting instructors can sign up for access at www.cengage.com/solutionbuilder.

PowerLecture™ with ExamView® *by Waner and Costenoble*
ISBN: 9781285085654
This CD-ROM provides the instructor with dynamic media tools for teaching, including Microsoft® PowerPoint® lecture slides, figures from the book, and the Test Bank. You can create, deliver, and customize tests (both print and online) in minutes with ExamView® Computerized Testing, which includes Test Bank items in electronic format. In addition, you can easily build solution sets for homework or exams by linking to Solution Builder's online solutions manual.

Instructor's Edition
ISBN: 9781285056296

www.WanerMath.com
The Instructor's Resource Page at www.WanerMath.com features an expanded collection of instructor resources, including an updated corrections page, an expanding set of author-created teaching videos for use in distance learning courses, and a utility that automatically updates homework exercise sets from the fifth edition to the sixth.

Acknowledgments

This project would not have been possible without the contributions and suggestions of numerous colleagues, students, and friends. We are particularly grateful to our colleagues at Hofstra and elsewhere who used and gave us useful feedback on previous editions. We are also grateful to everyone at Cengage Learning for their encouragement and guidance throughout the project. Specifically, we would like to thank Richard Stratton and Jay Campbell for their unflagging enthusiasm and Alison Eigel Zade for whipping the book into shape.

We would also like to thank our accuracy checker, Jerrold Grossman, and the numerous reviewers who provided many helpful suggestions that have shaped the development of this book.

Igor Fulman, *Arizona State University*
Tom Rosenwinkel, *Concordia University Texas*
Kim Ricketts, *Northwest-Shoals Community College*
Elaine Fitt, *Bucks County Community College*
Xian Wu, *University of South Carolina*
Catherine Remus, *University of Tennessee*
Hugh Cornell, *University of North Florida*
Upasana Kashyap, *The Citadel*
Laura Urbanski, *Northern Kentucky University*

Stefan Waner
Steven R. Costenoble

0

Precalculus Review

DreamPictures/Taxi/Getty Images

Website
www.WanerMath.com

- At the Website you will find section-by-section interactive tutorials for further study and practice.

Introduction

In this chapter we review some topics from algebra that you need to know to get the most out of this book. This chapter can be used either as a refresher course or as a reference.

There is one crucial fact you must always keep in mind: The letters used in algebraic expressions stand for numbers. All the rules of algebra are just facts about the arithmetic of numbers. If you are not sure whether some algebraic manipulation you are about to do is legitimate, try it first with numbers. If it doesn't work with numbers, it doesn't work.

0.1 Real Numbers

The **real numbers** are the numbers that can be written in decimal notation, including those that require an infinite decimal expansion. The set of real numbers includes all integers, positive and negative; all fractions; and the irrational numbers, those with decimal expansions that never repeat. Examples of irrational numbers are

$$\sqrt{2} = 1.414213562373\ldots$$

and

$$\pi = 3.141592653589\ldots$$

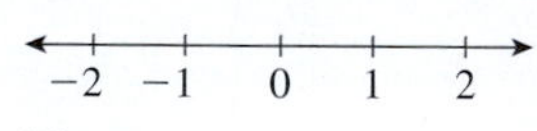

Figure 1

It is very useful to picture the real numbers as points on a line. As shown in Figure 1, larger numbers appear to the right, in the sense that if $a < b$ then the point corresponding to b is to the right of the one corresponding to a.

Intervals

Some subsets of the set of real numbers, called **intervals**, show up quite often and so we have a compact notation for them.

Interval Notation

Here is a list of types of intervals along with examples.

	Interval	*Description*	*Picture*	*Example*
Closed	$[a, b]$	Set of numbers x with $a \le x \le b$	a b (includes end points)	$[0, 10]$
Open	(a, b)	Set of numbers x with $a < x < b$	a b (excludes end points)	$(-1, 5)$
Half-Open	$(a, b]$	Set of numbers x with $a < x \le b$	a b	$(-3, 1]$
	$[a, b)$	Set of numbers x with $a \le x < b$	a b	$[0, 5)$

Infinite	$[a, +\infty)$	Set of numbers x with $a \le x$	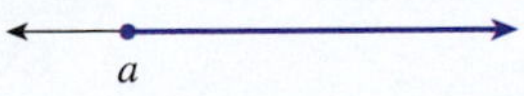	$[10, +\infty)$
	$(a, +\infty)$	Set of numbers x with $a < x$		$(-3, +\infty)$
	$(-\infty, b]$	Set of numbers x with $x \le b$		$(-\infty, -3]$
	$(-\infty, b)$	Set of numbers x with $x < b$		$(-\infty, 10)$
	$(-\infty, +\infty)$	Set of all real numbers		$(-\infty, +\infty)$

Operations

There are five important operations on real numbers: addition, subtraction, multiplication, division, and exponentiation. "Exponentiation" means raising a real number to a power; for instance, $3^2 = 3 \cdot 3 = 9$; $2^3 = 2 \cdot 2 \cdot 2 = 8$.

A note on technology: Most graphing calculators and spreadsheets use an asterisk * for multiplication and a caret sign ^ for exponentiation. Thus, for instance, 3×5 is entered as `3*5`, $3x$ as `3*x`, and 3^2 as `3^2`.

When we write an expression involving two or more operations, like

$$2 \cdot 3 + 4$$

or

$$\frac{2 \cdot 3^2 - 5}{4 - (-1)}$$

we need to agree on the order in which to do the operations. Does $2 \cdot 3 + 4$ mean $(2 \cdot 3) + 4 = 10$ or $2 \cdot (3 + 4) = 14$? We all agree to use the following rules for the order in which we do the operations.

Standard Order of Operations

Parentheses and Fraction Bars First, calculate the values of all expressions inside parentheses or brackets, working from the innermost parentheses out, before using them in other operations. In a fraction, calculate the numerator and denominator separately before doing the division.

Quick Examples

1. $6(2 + [3 - 5] - 4) = 6(2 + (-2) - 4) = 6(-4) = -24$
2. $\dfrac{(4 - 2)}{3(-2 + 1)} = \dfrac{2}{3(-1)} = \dfrac{2}{-3} = -\dfrac{2}{3}$
3. $3/(2 + 4) = \dfrac{3}{2 + 4} = \dfrac{3}{6} = \dfrac{1}{2}$
4. $(x + 4x)/(y + 3y) = 5x/(4y)$

Exponents Next, perform exponentiation.

Quick Examples

1. $2 + 4^2 = 2 + 16 = 18$
2. $(2 + 4)^2 = 6^2 = 36$

Note the difference.

3. $2\left(\frac{3}{4-5}\right)^2 = 2\left(\frac{3}{-1}\right)^2 = 2(-3)^2 = 2 \times 9 = 18$
4. $2(1 + 1/10)^2 = 2(1.1)^2 = 2 \times 1.21 = 2.42$

Multiplication and Division Next, do all multiplications and divisions, from left to right.

Quick Examples

1. $2(3 - 5)/4 \cdot 2 = 2(-2)/4 \cdot 2$ — Parentheses first
   $= -4/4 \cdot 2$ — Left-most product
   $= -1 \cdot 2 = -2$ — Multiplications and divisions, left to right
2. $2(1 + 1/10)^2 \times 2/10 = 2(1.1)^2 \times 2/10$ — Parentheses first
   $= 2 \times 1.21 \times 2/10$ — Exponent
   $= 4.84/10 = 0.484$ — Multiplications and divisions, left to right
3. $4\frac{2(4-2)}{3(-2 \cdot 5)} = 4\frac{2(2)}{3(-10)} = 4\frac{4}{-30} = \frac{16}{-30} = -\frac{8}{15}$

Addition and Subtraction Last, do all additions and subtractions, from left to right.

Quick Examples

1. $2(3 - 5)^2 + 6 - 1 = 2(-2)^2 + 6 - 1 = 2(4) + 6 - 1$
   $= 8 + 6 - 1 = 13$
2. $\left(\frac{1}{2}\right)^2 - (-1)^2 + 4 = \frac{1}{4} - 1 + 4 = -\frac{3}{4} + 4 = \frac{13}{4}$
3. $3/2 + 4 = 1.5 + 4 = 5.5$
4. $3/(2 + 4) = 3/6 = 1/2 = 0.5$

Note the difference.

5. $4/2^2 + (4/2)^2 = 4/2^2 + 2^2 = 4/4 + 4 = 1 + 4 = 5$

Entering Formulas

Any good calculator or spreadsheet will respect the standard order of operations. However, we must be careful with division and exponentiation and use parentheses as necessary. The following table gives some examples of simple mathematical expressions and their equivalents in the functional format used in most graphing calculators, spreadsheets, and computer programs.

Mathematical Expression	*Formula*	*Comments*
$\frac{2}{3-x}$	`2/(3-x)`	Note the use of parentheses instead of the fraction bar. If we omit the parentheses, we get the expression shown next.
$\frac{2}{3}-x$	`2/3-x`	The calculator follows the usual order of operations.
$\frac{2}{3\times 5}$	`2/(3*5)`	Putting the denominator in parentheses ensures that the multiplication is carried out first. The asterisk is usually used for multiplication in graphing calculators and computers.
$\frac{2}{x}\times 5$	`(2/x)*5`	Putting the fraction in parentheses ensures that it is calculated first. Some calculators will interpret `2/3*5` as $\frac{2}{3\times 5}$, but `2/3(5)` as $\frac{2}{3}\times 5$.
$\frac{2-3}{4+5}$	`(2-3)/(4+5)`	Note once again the use of parentheses in place of the fraction bar.
2^3	`2^3`	The caret ^ is commonly used to denote exponentiation.
2^{3-x}	`2^(3-x)`	Be careful to use parentheses to tell the calculator where the exponent ends. Enclose the *entire exponent* in parentheses.
2^3-x	`2^3-x`	Without parentheses, the calculator will follow the usual order of operations: exponentiation and then subtraction.
3×2^{-4}	`3*2^(-4)`	On some calculators, the negation key is separate from the minus key.
$2^{-4\times 3}\times 5$	`2^(-4*3)*5`	Note once again how parentheses enclose the entire exponent.
$100\left(1+\frac{0.05}{12}\right)^{60}$	`100*(1+0.05/12)^60`	This is a typical calculation for compound interest.
$PV\left(1+\frac{r}{m}\right)^{mt}$	`PV*(1+r/m)^(m*t)`	This is the compound interest formula. PV is understood to be a single number (present value) and not the product of P and V (or else we would have used `P*V`).
$\frac{2^{3-2}\times 5}{y-x}$	`2^(3-2)*5/(y-x)` or `(2^(3-2)*5)/(y -x)`	Notice again the use of parentheses to hold the denominator together. We could also have enclosed the numerator in parentheses, although this is optional. (Why?)
$\frac{2^y+1}{2-4^{3x}}$	`(2^y+1)/(2-4^(3*x))`	Here, it is necessary to enclose both the numerator and the denominator in parentheses.
$2^y+\frac{1}{2}-4^{3x}$	`2^y+1/2-4^(3*x)`	This is the effect of leaving out the parentheses around the numerator and denominator in the previous expression.

Accuracy and Rounding

When we use a calculator or computer, the results of our calculations are often given to far more decimal places than are useful. For example, suppose we are told that a square has an area of 2.0 square feet and we are asked how long its sides are. Each side is the square root of the area, which the calculator tells us is

$$\sqrt{2} \approx 1.414213562$$

However, the measurement of 2.0 square feet is probably accurate to only two digits, so our estimate of the lengths of the sides can be no more accurate than that. Therefore, we round the answer to two digits:

$$\text{Length of one side} \approx 1.4 \text{ feet}$$

The digits that follow 1.4 are meaningless. The following guide makes these ideas more precise.

Significant Digits, Decimal Places, and Rounding

The number of **significant digits** in a decimal representation of a number is the number of digits that are not leading zeros after the decimal point (as in .0005) or trailing zeros before the decimal point (as in 5,400,000). We say that a value is **accurate to *n* significant digits** if only the first n significant digits are meaningful.

When to Round

After doing a computation in which all the quantities are accurate to no more than n significant digits, round the final result to n significant digits.

Quick Examples

1. 0.00067 has two significant digits. *The 000 before 67 are leading zeros.*
2. 0.000670 has three significant digits. *The 0 after 67 is significant.*
3. 5,400,000 has two or more significant digits. *We can't say how many of the zeros are trailing.*[1]
4. 5,400,001 has 7 significant digits. *The string of zeros is not trailing.*
5. Rounding 63,918 to three significant digits gives 63,900.
6. Rounding 63,958 to three significant digits gives 64,000.
7. $\pi = 3.141592653...$ $\frac{22}{7} = 3.142857142...$ Therefore, $\frac{22}{7}$ is an approximation of π that is accurate to only three significant digits (3.14).
8. $4.02(1 + 0.02)^{1.4} \approx 4.13$ *We rounded to three significant digits.*

[1] If we obtained 5,400,000 by rounding 5,401,011, then it has three significant digits because the zero after the 4 is significant. On the other hand, if we obtained it by rounding 5,411,234, then it has only two significant digits. The use of scientific notation avoids this ambiguity: 5.40×10^6 (or 5.40 E6 on a calculator or computer) is accurate to three digits and 5.4×10^6 is accurate to two.

One more point, though: If, in a long calculation, you round the intermediate results, your final answer may be even less accurate than you think. As a general rule,

When calculating, don't round intermediate results. Rather, use the most accurate results obtainable or have your calculator or computer store them for you.

When you are done with the calculation, *then* round your answer to the appropriate number of digits of accuracy.

0.1 EXERCISES

Access end-of-section exercises online at **www.webassign.net**

0.2 Exponents and Radicals

In Section 1 we discussed exponentiation, or "raising to a power"; for example, $2^3 = 2 \cdot 2 \cdot 2$. In this section we discuss the algebra of exponentials more fully. First, we look at *integer* exponents: cases in which the powers are positive or negative whole numbers.

Integer Exponents

Positive Integer Exponents

If a is any real number and n is any positive integer, then by a^n we mean the quantity $a \cdot a \cdot \cdots \cdot a$ (n times); thus, $a^1 = a$, $a^2 = a \cdot a$, $a^5 = a \cdot a \cdot a \cdot a \cdot a$. In the expression a^n the number n is called the **exponent**, and the number a is called the **base**.

Quick Examples

$3^2 = 9$ $\qquad 2^3 = 8$

$0^{34} = 0$ $\qquad (-1)^5 = -1$

$10^3 = 1{,}000$ $\qquad 10^5 = 100{,}000$

Negative Integer Exponents

If a is any real number *other than zero* and n is any positive integer, then we define

$$a^{-n} = \frac{1}{a^n} = \frac{1}{a \cdot a \cdot \cdots \cdot a} \quad (n \text{ times})$$

Quick Examples

$2^{-3} = \frac{1}{2^3} = \frac{1}{8}$ $1^{-27} = \frac{1}{1^{27}} = 1$

$x^{-1} = \frac{1}{x^1} = \frac{1}{x}$ $(-3)^{-2} = \frac{1}{(-3)^2} = \frac{1}{9}$

$y^7 y^{-2} = y^7 \frac{1}{y^2} = y^5$ 0^{-2} is not defined

Zero Exponent

If a is any real number other than zero, then we define

$$a^0 = 1$$

Quick Examples

$3^0 = 1$ $1{,}000{,}000^0 = 1$

0^0 is not defined

When combining exponential expressions, we use the following identities.

Exponent Identity	Quick Examples
1. $a^m a^n = a^{m+n}$	$2^3 2^2 = 2^{3+2} = 2^5 = 32$ $x^3 x^{-4} = x^{3-4} = x^{-1} = \frac{1}{x}$ $\frac{x^3}{x^{-2}} = x^3 \frac{1}{x^{-2}} = x^3 x^2 = x^5$
2. $\frac{a^m}{a^n} = a^{m-n}$ if $a \neq 0$	$\frac{4^3}{4^2} = 4^{3-2} = 4^1 = 4$ $\frac{x^3}{x^{-2}} = x^{3-(-2)} = x^5$ $\frac{3^2}{3^4} = 3^{2-4} = 3^{-2} = \frac{1}{9}$
3. $(a^n)^m = a^{nm}$	$(3^2)^2 = 3^4 = 81$ $(2^x)^2 = 2^{2x}$
4. $(ab)^n = a^n b^n$	$(4 \cdot 2)^2 = 4^2 2^2 = 64$ $(-2y)^4 = (-2)^4 y^4 = 16y^4$
5. $\left(\frac{a}{b}\right)^n = \frac{a^n}{b^n}$ if $b \neq 0$	$\left(\frac{4}{3}\right)^2 = \frac{4^2}{3^2} = \frac{16}{9}$ $\left(\frac{x}{-y}\right)^3 = \frac{x^3}{(-y)^3} = -\frac{x^3}{y^3}$

Caution

- In the first two identities, the bases of the expressions must be the same. For example, the first gives $3^2 3^4 = 3^6$, but does *not* apply to $3^2 4^2$.
- People sometimes invent their own identities, such as $a^m + a^n = a^{m+n}$, which is wrong! (Try it with $a = m = n = 1$.) If you wind up with something like $2^3 + 2^4$, you are stuck with it; there are no identities around to simplify it further. (You can factor out 2^3, but whether or not that is a simplification depends on what you are going to do with the expression next.)

EXAMPLE 1 Combining the Identities

$$\frac{(x^2)^3}{x^3} = \frac{x^6}{x^3} \qquad \text{By (3)}$$
$$= x^{6-3} \qquad \text{By (2)}$$
$$= x^3$$

$$\frac{(x^4y)^3}{y} = \frac{(x^4)^3y^3}{y} \qquad \text{By (4)}$$
$$= \frac{x^{12}y^3}{y} \qquad \text{By (3)}$$
$$= x^{12}y^{3-1} \qquad \text{By (2)}$$
$$= x^{12}y^2$$

EXAMPLE 2 Eliminating Negative Exponents

Simplify the following and express the answer using no negative exponents.

a. $\dfrac{x^4y^{-3}}{x^5y^2}$ **b.** $\left(\dfrac{x^{-1}}{x^2y}\right)^5$

Solution

a. $\dfrac{x^4y^{-3}}{x^5y^2} = x^{4-5}y^{-3-2} = x^{-1}y^{-5} = \dfrac{1}{xy^5}$

b. $\left(\dfrac{x^{-1}}{x^2y}\right)^5 = \dfrac{(x^{-1})^5}{(x^2y)^5} = \dfrac{x^{-5}}{x^{10}y^5} = \dfrac{1}{x^{15}y^5}$

Radicals

If a is any non-negative real number, then its **square root** is the non-negative number whose square is a. For example, the square root of 16 is 4, because $4^2 = 16$. We write the square root of n as $\sqrt{n}$. (Roots are also referred to as **radicals**.) It is important to remember that $\sqrt{n}$ is never negative. Thus, for instance, $\sqrt{9}$ is 3, and not -3, even though $(-3)^2 = 9$. If we want to speak of the "negative square root" of 9, we write it as $-\sqrt{9} = -3$. If we want to write both square roots at once, we write $\pm\sqrt{9} = \pm 3$.

The **cube root** of a real number a is the number whose cube is a. The cube root of a is written as $\sqrt[3]{a}$ so that, for example, $\sqrt[3]{8} = 2$ (because $2^3 = 8$). Note that we can take the cube root of any number, positive, negative, or zero. For instance, the cube root of -8 is $\sqrt[3]{-8} = -2$ because $(-2)^3 = -8$. Unlike square roots, the cube root of a number may be negative. In fact, the cube root of a always has the same sign as a.

Higher roots are defined similarly. The **fourth root** of the *non-negative* number a is defined as the non-negative number whose fourth power is a, and written $\sqrt[4]{a}$. The **fifth root** of any number a is the number whose fifth power is a, and so on.

Note We cannot take an even-numbered root of a negative number, but we can take an odd-numbered root of any number. Even roots are always positive, whereas odd roots have the same sign as the number we start with. ■

EXAMPLE 3 *n*th Roots

$\sqrt{4} = 2$	Because $2^2 = 4$
$\sqrt{16} = 4$	Because $4^2 = 16$
$\sqrt{1} = 1$	Because $1^2 = 1$
If $x \geq 0$, then $\sqrt{x^2} = x$	Because $x^2 = x^2$
$\sqrt{2} \approx 1.414213562$	$\sqrt{2}$ is not a whole number.
$\sqrt{1+1} = \sqrt{2} \approx 1.414213562$	First add, then take the square root.[2]
$\sqrt{9+16} = \sqrt{25} = 5$	Contrast with $\sqrt{9} + \sqrt{16} = 3 + 4 = 7$.
$\frac{1}{\sqrt{2}} = \frac{\sqrt{2}}{2}$	Multiply top and bottom by $\sqrt{2}$.
$\sqrt[3]{27} = 3$	Because $3^3 = 27$
$\sqrt[3]{-64} = -4$	Because $(-4)^3 = -64$
$\sqrt[4]{16} = 2$	Because $2^4 = 16$
$\sqrt[4]{-16}$ is not defined	Even-numbered root of a negative number
$\sqrt[5]{-1} = -1$, since $(-1)^5 = -1$	Odd-numbered root of a negative number
$\sqrt[n]{-1} = -1$ if n is any odd number	

Q: *In the example we saw that $\sqrt{x^2} = x$ if x is non-negative. What happens if x is negative?*

A: If x is negative, then x^2 is positive, and so $\sqrt{x^2}$ is still defined as the non-negative number whose square is x^2. This number must be $|x|$, the **absolute value of *x***, which is the non-negative number with the same size as x. For instance, $|-3| = 3$, while $|3| = 3$, and $|0| = 0$. It follows that

$$\sqrt{x^2} = |x|$$

for every real number x, positive or negative. For instance,

$$\sqrt{(-3)^2} = \sqrt{9} = 3 = |-3|$$

and $$\sqrt{3^2} = \sqrt{9} = 3 = |3|.$$

In general, we find that

$$\sqrt[n]{x^n} = x \text{ if } n \text{ is odd, and } \sqrt[n]{x^n} = |x| \text{ if } n \text{ is even.}$$

[2] In general, $\sqrt{a+b}$ means the square root of the *quantity* $(a + b)$. The radical sign acts as a pair of parentheses or a fraction bar, telling us to evaluate what is inside before taking the root. (See the Caution on the next page.)

We use the following identities to evaluate radicals of products and quotients.

Radicals of Products and Quotients

If a and b are any real numbers (non-negative in the case of even-numbered roots), then

$$\sqrt[n]{ab} = \sqrt[n]{a}\sqrt[n]{b}$$ Radical of a product = Product of radicals

$$\sqrt[n]{\frac{a}{b}} = \frac{\sqrt[n]{a}}{\sqrt[n]{b}} \quad \text{if } b \neq 0$$ Radical of a quotient = Quotient of radicals

Notes

- The first rule is similar to the rule $(a \cdot b)^2 = a^2b^2$ for the square of a product, and the second rule is similar to the rule $\left(\frac{a}{b}\right)^2 = \frac{a^2}{b^2}$ for the square of a quotient.
- ***Caution*** There is no corresponding identity for addition:

  $\sqrt{a+b}$ is *not* equal to $\sqrt{a} + \sqrt{b}$

(Consider $a = b = 1$, for example.) Equating these expressions is a common error, so be careful! ■

Quick Examples

1. $\sqrt{9 \cdot 4} = \sqrt{9}\sqrt{4} = 3 \times 2 = 6$ — Alternatively, $\sqrt{9 \cdot 4} = \sqrt{36} = 6$
2. $\sqrt{\frac{9}{4}} = \frac{\sqrt{9}}{\sqrt{4}} = \frac{3}{2}$
3. $\frac{\sqrt{2}}{\sqrt{5}} = \frac{\sqrt{2}\sqrt{5}}{\sqrt{5}\sqrt{5}} = \frac{\sqrt{10}}{5}$
4. $\sqrt{4(3+13)} = \sqrt{4(16)} = \sqrt{4}\sqrt{16} = 2 \times 4 = 8$
5. $\sqrt[3]{-216} = \sqrt[3]{(-27)8} = \sqrt[3]{-27}\sqrt[3]{8} = (-3)2 = -6$
6. $\sqrt{x^3} = \sqrt{x^2 \cdot x} = \sqrt{x^2}\sqrt{x} = x\sqrt{x} \quad \text{if } x \geq 0$
7. $\sqrt{\frac{x^2+y^2}{z^2}} = \frac{\sqrt{x^2+y^2}}{\sqrt{z^2}} = \frac{\sqrt{x^2+y^2}}{|z|}$ — We can't simplify the numerator any further.

Rational Exponents

We already know what we mean by expressions such as x^4 and a^{-6}. The next step is to make sense of *rational* exponents: exponents of the form p/q with p and q integers as in $a^{1/2}$ and $3^{-2/3}$.

Q: *What should we mean by $a^{1/2}$?*

A: The overriding concern here is that all the exponent identities should remain true. In this case the identity to look at is the one that says that $(a^m)^n = a^{mn}$. This identity tells us that

$$(a^{1/2})^2 = a^1 = a.$$

That is, $a^{1/2}$, when squared, gives us a. But that must mean that $a^{1/2}$ is the *square root* of a, or

$$a^{1/2} = \sqrt{a}.$$

A similar argument tells us that, if q is any positive whole number, then

$$a^{1/q} = \sqrt[q]{a}, \text{ the } q\text{th root of } a.$$

Notice that if a is negative, this makes sense only for q odd. To avoid this problem, we usually stick to positive a.

Q: *If p and q are integers (q positive), what should we mean by* $a^{p/q}$*?*

A: By the exponent identities, $a^{p/q}$ should equal both $(a^p)^{1/q}$ and $(a^{1/q})^p$. The first is the qth root of a^p, and the second is the pth power of $a^{1/q}$, which gives us the following.

Conversion Between Rational Exponents and Radicals

If a is any non-negative number, then

$$a^{p/q} = \sqrt[q]{a^p} = \left(\sqrt[q]{a}\right)^p.$$

↑ Using exponents ↑ ↑ Using radicals

In particular,

$$a^{1/q} = \sqrt[q]{a}, \text{ the } q\text{th root of } a.$$

Notes

- If a is negative, all of this makes sense only if q is odd.
- All of the exponent identities continue to work when we allow rational exponents p/q. In other words, we are free to use all the exponent identities even though the exponents are not integers. ■

Quick Examples

1. $4^{3/2} = (\sqrt{4})^3 = 2^3 = 8$
2. $8^{2/3} = (\sqrt[3]{8})^2 = 2^2 = 4$
3. $9^{-3/2} = \dfrac{1}{9^{3/2}} = \dfrac{1}{(\sqrt{9})^3} = \dfrac{1}{3^3} = \dfrac{1}{27}$
4. $\dfrac{\sqrt{3}}{\sqrt[3]{3}} = \dfrac{3^{1/2}}{3^{1/3}} = 3^{1/2-1/3} = 3^{1/6} = \sqrt[6]{3}$
5. $2^2 2^{7/2} = 2^2 2^{3+1/2} = 2^2 2^3 2^{1/2} = 2^5 2^{1/2} = 2^5\sqrt{2}$

EXAMPLE 4 Simplifying Algebraic Expressions

Simplify the following.

a. $\dfrac{(x^3)^{5/3}}{x^3}$ **b.** $\sqrt[4]{a^6}$ **c.** $\dfrac{(xy)^{-3}y^{-3/2}}{x^{-2}\sqrt{y}}$

Solution

a. $\dfrac{(x^3)^{5/3}}{x^3} = \dfrac{x^5}{x^3} = x^2$

b. $\sqrt[4]{a^6} = a^{6/4} = a^{3/2} = a \cdot a^{1/2} = a\sqrt{a}$

c. $\dfrac{(xy)^{-3}y^{-3/2}}{x^{-2}\sqrt{y}} = \dfrac{x^{-3}y^{-3}y^{-3/2}}{x^{-2}y^{1/2}} = \dfrac{1}{x^{-2+3}y^{1/2+3+3/2}} = \dfrac{1}{xy^5}$

Converting Between Rational, Radical, and Exponent Form

In calculus we must often convert algebraic expressions involving powers of x, such as $\dfrac{3}{2x^2}$, into expressions in which x does not appear in the denominator, such as $\dfrac{3}{2}x^{-2}$. Also, we must often convert expressions with radicals, such as $\dfrac{1}{\sqrt{1+x^2}}$, into expressions with no radicals and all powers in the numerator, such as $(1+x^2)^{-1/2}$. In these cases, we are converting from **rational form** or **radical form** to **exponent form.**

Rational Form

An expression is in **rational form** if it is written with positive exponents only.

Quick Examples

1. $\dfrac{2}{3x^2}$ is in rational form.
2. $\dfrac{2x^{-1}}{3}$ is not in rational form because the exponent of x is negative.
3. $\dfrac{x}{6} + \dfrac{6}{x}$ is in rational form.

Radical Form

An expression is in **radical form** if it is written with integer powers and roots only.

Quick Examples

1. $\dfrac{2}{5\sqrt[3]{x}} + \dfrac{2}{x}$ is in radical form.
2. $\dfrac{2x^{-1/3}}{5} + 2x^{-1}$ is not in radical form because $x^{-1/3}$ appears.
3. $\dfrac{1}{\sqrt{1+x^2}}$ is in radical form, but $(1+x^2)^{-1/2}$ is not.

Exponent Form

An expression is in **exponent form** if there are no radicals and all powers of unknowns occur in the numerator. We write such expressions as sums or differences of terms of the form

Constant × (Expression with x)^p As in $\frac{1}{3}x^{-3/2}$

Quick Examples

1. $\frac{2}{3}x^4 - 3x^{-1/3}$ is in exponent form.
2. $\frac{x}{6} + \frac{6}{x}$ is not in exponent form because the second expression has x in the denominator.
3. $\sqrt[3]{x}$ is not in exponent form because it has a radical.
4. $(1+x^2)^{-1/2}$ is in exponent form, but $\frac{1}{\sqrt{1+x^2}}$ is not.

EXAMPLE 5 Converting from One Form to Another

Convert the following to rational form:

a. $\frac{1}{2}x^{-2} + \frac{4}{3}x^{-5}$ **b.** $\frac{2}{\sqrt{x}} - \frac{2}{x^{-4}}$

Convert the following to radical form:

c. $\frac{1}{2}x^{-1/2} + \frac{4}{3}x^{-5/4}$ **d.** $\frac{(3+x)^{-1/3}}{5}$

Convert the following to exponent form:

e. $\frac{3}{4x^2} - \frac{x}{6} + \frac{6}{x} + \frac{4}{3\sqrt{x}}$ **f.** $\frac{2}{(x+1)^2} - \frac{3}{4\sqrt[5]{2x-1}}$

Solution For (a) and (b), we eliminate negative exponents as we did in Example 2:

a. $\frac{1}{2}x^{-2} + \frac{4}{3}x^{-5} = \frac{1}{2} \cdot \frac{1}{x^2} + \frac{4}{3} \cdot \frac{1}{x^5} = \frac{1}{2x^2} + \frac{4}{3x^5}$

b. $\frac{2}{\sqrt{x}} - \frac{2}{x^{-4}} = \frac{2}{\sqrt{x}} - 2x^4$

For (c) and (d), we rewrite all terms with fractional exponents as radicals:

c. $\frac{1}{2}x^{-1/2} + \frac{4}{3}x^{-5/4} = \frac{1}{2} \cdot \frac{1}{x^{1/2}} + \frac{4}{3} \cdot \frac{1}{x^{5/4}}$

$$= \frac{1}{2} \cdot \frac{1}{\sqrt{x}} + \frac{4}{3} \cdot \frac{1}{\sqrt[4]{x^5}} = \frac{1}{2\sqrt{x}} + \frac{4}{3\sqrt[4]{x^5}}$$

d. $\frac{(3+x)^{-1/3}}{5} = \frac{1}{5(3+x)^{1/3}} = \frac{1}{5\sqrt[3]{3+x}}$

For (e) and (f), we eliminate any radicals and move all expressions involving x to the numerator:

e. $\frac{3}{4x^2} - \frac{x}{6} + \frac{6}{x} + \frac{4}{3\sqrt{x}} = \frac{3}{4}x^{-2} - \frac{1}{6}x + 6x^{-1} + \frac{4}{3x^{1/2}}$

$$= \frac{3}{4}x^{-2} - \frac{1}{6}x + 6x^{-1} + \frac{4}{3}x^{-1/2}$$

f. $$\frac{2}{(x+1)^2} - \frac{3}{4\sqrt[5]{2x-1}} = 2(x+1)^{-2} - \frac{3}{4(2x-1)^{1/5}}$$
$$= 2(x+1)^{-2} - \frac{3}{4}(2x-1)^{-1/5}$$

Solving Equations with Exponents

EXAMPLE 6 Solving Equations

Solve the following equations:

a. $x^3 + 8 = 0$ **b.** $x^2 - \frac{1}{2} = 0$ **c.** $x^{3/2} - 64 = 0$

Solution

a. Subtracting 8 from both sides gives $x^3 = -8$. Taking the cube root of both sides gives $x = -2$.

b. Adding $\frac{1}{2}$ to both sides gives $x^2 = \frac{1}{2}$. Thus, $x = \pm\sqrt{\frac{1}{2}} = \pm\frac{1}{\sqrt{2}}$.

c. Adding 64 to both sides gives $x^{3/2} = 64$. Taking the reciprocal (2/3) power of both sides gives

$$(x^{3/2})^{2/3} = 64^{2/3}$$
$$x^1 = \left(\sqrt[3]{64}\right)^2 = 4^2 = 16$$

so $$x = 16.$$

0.2 EXERCISES

Access end-of-section exercises online at **www.webassign.net** ENHANCED WebAssign

0.3 Multiplying and Factoring Algebraic Expressions

Multiplying Algebraic Expressions

Distributive Law

The **distributive law** for real numbers states that

$$a(b \pm c) = ab \pm ac$$
$$(a \pm b)c = ac \pm bc$$

for any real numbers a, b, and c.

Quick Examples

1. $2(x-3)$ is *not* equal to $2x-3$ but is equal to $2x-2(3)=2x-6$.
2. $x(x+1)=x^2+x$
3. $2x(3x-4)=6x^2-8x$
4. $(x-4)x^2=x^3-4x^2$
5. $(x+2)(x+3)=(x+2)x+(x+2)3$
$=(x^2+2x)+(3x+6)=x^2+5x+6$
6. $(x+2)(x-3)=(x+2)x-(x+2)3$
$=(x^2+2x)-(3x+6)=x^2-x-6$

There is a quicker way of expanding expressions like the last two, called the "FOIL" method (First, Outer, Inner, Last). Consider, for instance, the expression $(x+1)(x-2)$. The FOIL method says: Take the product of the first terms: $x\cdot x=x^2$, the product of the outer terms: $x\cdot(-2)=-2x$, the product of the inner terms: $1\cdot x=x$, and the product of the last terms: $1\cdot(-2)=-2$, and then add them all up, getting $x^2-2x+x-2=x^2-x-2$.

EXAMPLE 1 FOIL

a. $(x-2)(2x+5)=\underset{\text{First}}{2x^2}+\underset{\text{Outer}}{5x}-\underset{\text{Inner}}{4x}-\underset{\text{Last}}{10}=2x^2+x-10$

b. $(x^2+1)(x-4)=x^3-4x^2+x-4$

c. $(a-b)(a+b)=a^2+ab-ab-b^2=a^2-b^2$

d. $(a+b)^2=(a+b)(a+b)=a^2+ab+ab+b^2=a^2+2ab+b^2$

e. $(a-b)^2=(a-b)(a-b)=a^2-ab-ab+b^2=a^2-2ab+b^2$

The last three are particularly important and are worth memorizing.

Special Formulas

$(a-b)(a+b)=a^2-b^2$ Difference of two squares

$(a+b)^2=a^2+2ab+b^2$ Square of a sum

$(a-b)^2=a^2-2ab+b^2$ Square of a difference

Quick Examples

1. $(2-x)(2+x)=4-x^2$
2. $(1+a)(1-a)=1-a^2$
3. $(x+3)^2=x^2+6x+9$
4. $(4-x)^2=16-8x+x^2$

Here are some longer examples that require the distributive law.

EXAMPLE 2 Multiplying Algebraic Expressions

a. $(x+1)(x^2+3x-4) = (x+1)x^2 + (x+1)3x - (x+1)4$
$= (x^3+x^2) + (3x^2+3x) - (4x+4)$
$= x^3 + 4x^2 - x - 4$

b. $\left(x^2 - \frac{1}{x} + 1\right)(2x+5) = \left(x^2 - \frac{1}{x} + 1\right)2x + \left(x^2 - \frac{1}{x} + 1\right)5$
$= (2x^3 - 2 + 2x) + \left(5x^2 - \frac{5}{x} + 5\right)$
$= 2x^3 + 5x^2 + 2x + 3 - \frac{5}{x}$

c. $(x-y)(x-y)(x-y) = (x^2 - 2xy + y^2)(x-y)$
$= (x^2 - 2xy + y^2)x - (x^2 - 2xy + y^2)y$
$= (x^3 - 2x^2y + xy^2) - (x^2y - 2xy^2 + y^3)$
$= x^3 - 3x^2y + 3xy^2 - y^3$

Factoring Algebraic Expressions

We can think of factoring as applying the distributive law in reverse—for example,

$$2x^2 + x = x(2x+1),$$

which can be checked by using the distributive law. Factoring is an art that you will learn with experience and the help of a few useful techniques.

Factoring Using a Common Factor

To use this technique, locate a **common factor**—a term that occurs as a factor in each of the expressions being added or subtracted (for example, x is a common factor in $2x^2 + x$, because it is a factor of both $2x^2$ and x). Once you have located a common factor, "factor it out" by applying the distributive law.

Quick Examples

1. $2x^3 - x^2 + x$ has x as a common factor, so
$2x^3 - x^2 + x = x(2x^2 - x + 1)$

2. $2x^2 + 4x$ has $2x$ as a common factor, so
$2x^2 + 4x = 2x(x+2)$

3. $2x^2y + xy^2 - x^2y^2$ has xy as a common factor, so
$2x^2y + xy^2 - x^2y^2 = xy(2x + y - xy)$

4. $(x^2+1)(x+2) - (x^2+1)(x+3)$ has x^2+1 as a common factor, so
$(x^2+1)(x+2) - (x^2+1)(x+3) = (x^2+1)[(x+2) - (x+3)]$
$= (x^2+1)(x+2-x-3)$
$= (x^2+1)(-1) = -(x^2+1)$

5. $12x(x^2-1)^5(x^3+1)^6 + 18x^2(x^2-1)^6(x^3+1)^5$ has $6x(x^2-1)^5(x^3+1)^5$ as a common factor, so

$$\begin{aligned}&12x(x^2-1)^5(x^3+1)^6 + 18x^2(x^2-1)^6(x^3+1)^5\\&= 6x(x^2-1)^5(x^3+1)^5[2(x^3+1)+3x(x^2-1)]\\&= 6x(x^2-1)^5(x^3+1)^5(2x^3+2+3x^3-3x)\\&= 6x(x^2-1)^5(x^3+1)^5(5x^3-3x+2)\end{aligned}$$

We would also like to be able to reverse calculations such as $(x+2)(2x-5) = 2x^2-x-10$. That is, starting with the expression $2x^2-x-10$, we would like to **factor** it to get the expression $(x+2)(2x-5)$. An expression of the form ax^2+bx+c, where a, b, and c are real numbers, is called a **quadratic** expression in x. Thus, given a quadratic expression ax^2+bx+c, we would like to write it in the form $(dx+e)(fx+g)$ for some real numbers d, e, f, and g. There are some quadratics, such as x^2+x+1, that cannot be factored in this form at all. Here, we consider only quadratics that do factor, and in such a way that the numbers d, e, f, and g are integers (whole numbers; other cases are discussed in Section 5). The usual technique of factoring such quadratics is a "trial and error" approach.

Factoring Quadratics by Trial and Error

To factor the quadratic ax^2+bx+c, factor ax^2 as $(a_1x)(a_2x)$ (with a_1 positive) and c as c_1c_2, and then check whether or not $ax^2+bx+c = (a_1x \pm c_1)(a_2x \pm c_2)$. If not, try other factorizations of ax^2 and c.

Quick Examples

1. To factor x^2-6x+5, first factor x^2 as $(x)(x)$, and 5 as (5)(1):

$(x+5)(x+1) = x^2+6x+5.$ No good

$(x-5)(x-1) = x^2-6x+5.$ Desired factorization

2. To factor $x^2-4x-12$, first factor x^2 as $(x)(x)$, and -12 as $(1)(-12)$, $(2)(-6)$, or $(3)(-4)$. Trying them one by one gives

$(x+1)(x-12) = x^2-11x-12.$ No good

$(x-1)(x+12) = x^2+11x-12.$ No good

$(x+2)(x-6) = x^2-4x-12.$ Desired factorization

3. To factor $4x^2-25$, we can follow the above procedure, or recognize $4x^2-25$ as the difference of two squares:

$$4x^2-25 = (2x)^2-5^2 = (2x-5)(2x+5).$$

Note: Not all quadratic expressions factor. In Section 5 we look at a test that tells us whether or not a given quadratic factors.

Here are examples requiring either a little more work or a little more thought.

EXAMPLE 3 Factoring Quadratics

Factor the following: **a.** $4x^2 - 5x - 6$ **b.** $x^4 - 5x^2 + 6$

Solution

a. Possible factorizations of $4x^2$ are $(2x)(2x)$ or $(x)(4x)$. Possible factorizations of -6 are $(1)(-6)$, $(2)(-3)$. We now systematically try out all the possibilities until we come up with the correct one.

$(2x)(2x)$ and $(1)(-6)$:	$(2x + 1)(2x - 6) = 4x^2 - 10x - 6$	No good
$(2x)(2x)$ and $(2)(-3)$:	$(2x + 2)(2x - 3) = 4x^2 - 2x - 6$	No good
$(x)(4x)$ and $(1)(-6)$:	$(x + 1)(4x - 6) = 4x^2 - 2x - 6$	No good
$(x)(4x)$ and $(2)(-3)$:	$(x + 2)(4x - 3) = 4x^2 + 5x - 6$	Almost!
Change signs:	$(x - 2)(4x + 3) = 4x^2 - 5x - 6$	Correct

b. The expression $x^4 - 5x^2 + 6$ is not a quadratic, you say? Correct. It's a quartic (a fourth degree expression). However, it looks rather like a quadratic. In fact, it is quadratic *in* x^2, meaning that it is

$$(x^2)^2 - 5(x^2) + 6 = y^2 - 5y + 6$$

where $y = x^2$. The quadratic $y^2 - 5y + 6$ factors as

$$y^2 - 5y + 6 = (y - 3)(y - 2)$$

so

$$x^4 - 5x^2 + 6 = (x^2 - 3)(x^2 - 2)$$

This is a sometimes useful technique.

Our last example is here to remind you why we should want to factor polynomials in the first place. We shall return to this in Section 5.

EXAMPLE 4 Solving a Quadratic Equation by Factoring

Solve the equation $3x^2 + 4x - 4 = 0$.

Solution We first factor the left-hand side to get

$$(3x - 2)(x + 2) = 0.$$

Thus, the product of the two quantities $(3x - 2)$ and $(x + 2)$ is zero. Now, if a product of two numbers is zero, one of the two must be zero. In other words, either $3x - 2 = 0$, giving $x = \frac{2}{3}$, or $x + 2 = 0$, giving $x = -2$. Thus, there are two solutions: $x = \frac{2}{3}$ and $x = -2$.

0.3 EXERCISES

Access end-of-section exercises online at **www.webassign.net**

0.4 Rational Expressions

Rational Expression

A **rational expression** is an algebraic expression of the form $\frac{P}{Q}$, where P and Q are simpler expressions (usually polynomials) and the denominator Q is not zero.

Quick Examples

1. $\frac{x^2 - 3x}{x}$ — $P = x^2 - 3x,\ Q = x$
2. $\frac{x + \frac{1}{x} + 1}{2x^2y + 1}$ — $P = x + \frac{1}{x} + 1,\ Q = 2x^2y + 1$
3. $3xy - x^2$ — $P = 3xy - x^2,\ Q = 1$

Algebra of Rational Expressions

We manipulate rational expressions in the same way that we manipulate fractions, using the following rules:

	Algebraic Rule	Quick Example
Product:	$\frac{P}{Q} \cdot \frac{R}{S} = \frac{PR}{QS}$	$\frac{x+1}{x} \cdot \frac{x-1}{2x+1} = \frac{(x+1)(x-1)}{x(2x+1)} = \frac{x^2-1}{2x^2+x}$
Sum:	$\frac{P}{Q} + \frac{R}{S} = \frac{PS+RQ}{QS}$	$\frac{2x-1}{3x+2} + \frac{1}{x} = \frac{(2x-1)x + 1(3x+2)}{x(3x+2)} = \frac{2x^2+2x+2}{3x^2+2x}$
Difference:	$\frac{P}{Q} - \frac{R}{S} = \frac{PS-RQ}{QS}$	$\frac{x}{3x+2} - \frac{x-4}{x} = \frac{x^2-(x-4)(3x+2)}{x(3x+2)} = \frac{-2x^2+10x+8}{3x^2+2x}$
Reciprocal:	$\frac{1}{\left(\frac{P}{Q}\right)} = \frac{Q}{P}$	$\frac{1}{\left(\frac{2xy}{3x-1}\right)} = \frac{3x-1}{2xy}$
Quotient:	$\frac{\left(\frac{P}{Q}\right)}{\left(\frac{R}{S}\right)} = \frac{P}{Q} \cdot \frac{S}{R} = \frac{PS}{QR}$	$\frac{\left(\frac{x}{x-1}\right)}{\left(\frac{y-1}{y}\right)} = \frac{xy}{(x-1)(y-1)} = \frac{xy}{xy-x-y+1}$
Cancellation:	$\frac{P\cancel{R}}{Q\cancel{R}} = \frac{P}{Q}$	$\frac{(x-1)(xy+4)}{(x^2y-8)(x-1)} = \frac{xy+4}{x^2y-8}$

Caution Cancellation of summands is *invalid*. For instance,

$$\frac{x + (2xy^2 - y)}{x + 4y} = \frac{(2xy^2 - y)}{4y}$$ ✗ *WRONG!* Do *not* cancel a summand.

$$\frac{x(2xy^2 - y)}{4xy} = \frac{(2xy^2 - y)}{4y}$$ ✔ *CORRECT* Do cancel a factor.

Here are some examples that require several algebraic operations.

EXAMPLE 1 Simplifying Rational Expressions

a. $$\frac{\left(\frac{1}{x+y} - \frac{1}{x}\right)}{y} = \frac{\left(\frac{x - (x+y)}{x(x+y)}\right)}{y} = \frac{\left(\frac{-y}{x(x+y)}\right)}{y} = \frac{-y}{xy(x+y)} = -\frac{1}{x(x+y)}$$

b. $$\frac{(x+1)(x+2)^2 - (x+1)^2(x+2)}{(x+2)^4} = \frac{(x+1)(x+2)[(x+2) - (x+1)]}{(x+2)^4}$$

$$= \frac{(x+1)(x+2)(x+2-x-1)}{(x+2)^4} = \frac{(x+1)(x+2)}{(x+2)^4} = \frac{x+1}{(x+2)^3}$$

c. $$\frac{2x\sqrt{x+1} - \frac{x^2}{\sqrt{x+1}}}{x+1} = \frac{\left(\frac{2x(\sqrt{x+1})^2 - x^2}{\sqrt{x+1}}\right)}{x+1} = \frac{2x(x+1) - x^2}{(x+1)\sqrt{x+1}}$$

$$= \frac{2x^2 + 2x - x^2}{(x+1)\sqrt{x+1}} = \frac{x^2 + 2x}{\sqrt{(x+1)^3}} = \frac{x(x+2)}{\sqrt{(x+1)^3}}$$

0.4 EXERCISES

Access end-of-section exercises online at **www.webassign.net** ENHANCED WebAssign

0.5 Solving Polynomial Equations

Polynomial Equation

A **polynomial equation** in one unknown is an equation that can be written in the form

$$ax^n + bx^{n-1} + \cdots + rx + s = 0$$

where $a, b, \ldots, r$, and s are constants.

We call the largest exponent of x appearing in a nonzero term of a polynomial the **degree** of that polynomial.

Quick Examples

1. $3x + 1 = 0$ has degree 1 because the largest power of x that occurs is $x = x^1$. Degree 1 equations are called **linear** equations.
2. $x^2 - x - 1 = 0$ has degree 2 because the largest power of x that occurs is x^2. Degree 2 equations are also called **quadratic equations**, or just **quadratics**.
3. $x^3 = 2x^2 + 1$ is a degree 3 polynomial (or **cubic**) in disguise. It can be rewritten as $x^3 - 2x^2 - 1 = 0$, which is in the standard form for a degree 3 equation.
4. $x^4 - x = 0$ has degree 4. It is called a **quartic**.

Now comes the question: How do we solve these equations for x? This question was asked by mathematicians as early as 1600 BCE. Let's look at these equations one degree at a time.

Solution of Linear Equations

By definition, a linear equation can be written in the form

$$ax + b = 0. \qquad a \text{ and } b \text{ are fixed numbers with } a \neq 0.$$

Solving this is a nice mental exercise: Subtract b from both sides and then divide by a, getting $x = -b/a$. Don't bother memorizing this formula; just go ahead and solve linear equations as they arise. If you feel you need practice, complete the exercises for this section at www.webassign.net. (You can access WebAssign with the access code that came with this book.)

Solution of Quadratic Equations

By definition, a quadratic equation has the form

$$ax^2 + bx + c = 0. \qquad a, b, \text{ and } c \text{ are fixed numbers and } a \neq 0.$$ [3]

The solutions of this equation are also called the **roots** of $ax^2 + bx + c$. We're assuming that you saw quadratic equations somewhere in high school but may be a little hazy about the details of their solution. There are two ways of solving these equations—one works sometimes, and the other works every time.

Solving Quadratic Equations by Factoring (works sometimes)

If we can factor[4] a quadratic equation $ax^2 + bx + c = 0$, we can solve the equation by setting each factor equal to zero.

[3] What happens if $a = 0$?

[4] See the section on factoring for a review of how to factor quadratics.

Quick Examples

1. $x^2 + 7x + 10 = 0$

   $(x + 5)(x + 2) = 0$ Factor the left-hand side.

   $x + 5 = 0$ or $x + 2 = 0$ If a product is zero, one or both factors is zero.

   Solutions: $x = -5$ and $x = -2$

2. $2x^2 - 5x - 12 = 0$

   $(2x + 3)(x - 4) = 0$ Factor the left-hand side.

   $2x + 3 = 0$ or $x - 4 = 0$

   Solutions: $x = -3/2$ and $x = 4$

Test for Factoring

The quadratic $ax^2 + bx + c$, with a, b, and c being integers (whole numbers), factors into an expression of the form $(rx + s)(tx + u)$ with r, s, t, and u integers precisely when the quantity $b^2 - 4ac$ is a perfect square. (That is, it is the square of an integer.) If this happens, we say that the quadratic **factors over the integers**.

Quick Examples

1. $x^2 + x + 1$ has $a = 1$, $b = 1$, and $c = 1$, so $b^2 - 4ac = -3$, which is not a perfect square. Therefore, this quadratic does not factor over the integers.
2. $2x^2 - 5x - 12$ has $a = 2$, $b = -5$, and $c = -12$, so $b^2 - 4ac = 121$. Because $121 = 11^2$, this quadratic does factor over the integers. (We factored it above.)

Solving Quadratic Equations with the Quadratic Formula (works every time)

The solutions of the general quadratic $ax^2 + bx + c = 0$ ($a \neq 0$) are given by

$$x = \frac{-b \pm \sqrt{b^2 - 4ac}}{2a}.$$

We call the quantity $\Delta = b^2 - 4ac$ the **discriminant** of the quadratic (Δ is the Greek letter delta), and we have the following general rules:

- If Δ is positive, there are two distinct real solutions.
- If Δ is zero, there is only one real solution: $x = -\dfrac{b}{2a}$. (Why?)
- If Δ is negative, there are no real solutions.

Quick Examples

1. $2x^2 - 5x - 12 = 0$ has $a = 2$, $b = -5$, and $c = -12$.

$$x = \frac{-b \pm \sqrt{b^2 - 4ac}}{2a} = \frac{5 \pm \sqrt{25 + 96}}{4} = \frac{5 \pm \sqrt{121}}{4} = \frac{5 \pm 11}{4}$$

$$= \frac{16}{4} \text{ or } -\frac{6}{4} = 4 \text{ or } -3/2$$

Δ is positive in this example.

2. $4x^2 = 12x - 9$ can be rewritten as $4x^2 - 12x + 9 = 0$, which has $a = 4$, $b = -12$, and $c = 9$.

$$x = \frac{-b \pm \sqrt{b^2 - 4ac}}{2a} = \frac{12 \pm \sqrt{144 - 144}}{8} = \frac{12 \pm 0}{8} = \frac{12}{8} = \frac{3}{2}$$

Δ is zero in this example.

3. $x^2 + 2x - 1 = 0$ has $a = 1$, $b = 2$, and $c = -1$.

$$x = \frac{-b \pm \sqrt{b^2 - 4ac}}{2a} = \frac{-2 \pm \sqrt{8}}{2} = \frac{-2 \pm 2\sqrt{2}}{2} = -1 \pm \sqrt{2}$$

The two solutions are $x = -1 + \sqrt{2} = 0.414\ldots$ and $x = -1 - \sqrt{2} = -2.414\ldots$

Δ is positive in this example.

4. $x^2 + x + 1 = 0$ has $a = 1$, $b = 1$, and $c = 1$. Because $\Delta = -3$ is negative, there are no real solutions.

Δ is negative in this example.

Q: *This is all very useful, but where does the quadratic formula come from?*

A: To see where it comes from, we will solve a general quadratic equation using "brute force." Start with the general quadratic equation.

$$ax^2 + bx + c = 0.$$

First, divide out the nonzero number a to get

$$x^2 + \frac{bx}{a} + \frac{c}{a} = 0.$$

Now we **complete the square:** Add and subtract the quantity $\frac{b^2}{4a^2}$ to get

$$x^2 + \frac{bx}{a} + \frac{b^2}{4a^2} - \frac{b^2}{4a^2} + \frac{c}{a} = 0.$$

We do this to get the first three terms to factor as a perfect square:

$$\left(x + \frac{b}{2a}\right)^2 - \frac{b^2}{4a^2} + \frac{c}{a} = 0.$$

(Check this by multiplying out.) Adding $\frac{b^2}{4a^2} - \frac{c}{a}$ to both sides gives:

$$\left(x + \frac{b}{2a}\right)^2 = \frac{b^2}{4a^2} - \frac{c}{a} = \frac{b^2 - 4ac}{4a^2}.$$

Taking square roots gives

$$x + \frac{b}{2a} = \frac{\pm\sqrt{b^2 - 4ac}}{2a}.$$

Finally, adding $-\frac{b}{2a}$ to both sides yields the result:

$$x = -\frac{b}{2a} + \frac{\pm\sqrt{b^2 - 4ac}}{2a}$$

or

$$x = \frac{-b \pm \sqrt{b^2 - 4ac}}{2a}.$$

Solution of Cubic Equations

By definition, a cubic equation can be written in the form

$$ax^3 + bx^2 + cx + d = 0.$$ *a, b, c, and d are fixed numbers and $a \neq 0$.*

Now we get into something of a bind. Although there is a perfectly respectable formula for the solutions, it is very complicated and involves the use of complex numbers rather heavily.[5] So we discuss instead a much simpler method that *sometimes* works nicely. Here is the method in a nutshell.

Solving Cubics by Finding One Factor

Start with a given cubic equation $ax^3 + bx^2 + cx + d = 0$.

Step 1 By trial and error, find one solution $x = s$. If a, b, c, and d are integers, the only possible *rational* solutions[6] are those of the form $s = \pm$(factor of d)/(factor of a).

Step 2 It will now be possible to factor the cubic as

$$ax^3 + bx^2 + cx + d = (x - s)(ax^2 + ex + f) = 0$$

To find $ax^2 + ex + f$, divide the cubic by $x - s$, using long division.[7]

Step 3 The factored equation says that either $x - s = 0$ or $ax^2 + ex + f = 0$. We already know that s is a solution, and now we see that the other solutions are the roots of the quadratic. Note that this quadratic may or may not have any real solutions, as usual.

Quick Example

To solve the cubic $x^3 - x^2 + x - 1 = 0$, we first find a single solution. Here, $a = 1$ and $d = -1$. Because the only factors of ± 1 are ± 1, the only possible rational solutions are $x = \pm 1$. By substitution, we see that $x = 1$ is a solution. Thus, $(x - 1)$ is a factor. Dividing by $(x - 1)$ yields the quotient $(x^2 + 1)$. Thus,

$$x^3 - x^2 + x - 1 = (x - 1)(x^2 + 1) = 0$$

so that either $x - 1 = 0$ or $x^2 + 1 = 0$.

Because the discriminant of the quadratic $x^2 + 1$ is negative, we don't get any real solutions from $x^2 + 1 = 0$, so the only real solution is $x = 1$.

[5]It was when this formula was discovered in the 16th century that complex numbers were first taken seriously. Although we would like to show you the formula, it is too large to fit in this footnote.

[6]There may be *irrational* solutions, however; for example, $x^3 - 2 = 0$ has the single solution $x = \sqrt[3]{2}$.

[7]Alternatively, use "synthetic division," a shortcut that would take us too far afield to describe.

Possible Outcomes When Solving a Cubic Equation

If you consider all the cases, there are three possible outcomes when solving a cubic equation:

1. One real solution (as in the Quick Example on page 25)
2. Two real solutions (try, for example, $x^3 + x^2 - x - 1 = 0$)
3. Three real solutions (see the next example)

EXAMPLE 1 Solving a Cubic

Solve the cubic $2x^3 - 3x^2 - 17x + 30 = 0$.

Solution First we look for a single solution. Here, $a = 2$ and $d = 30$. The factors of a are ± 1 and ± 2, and the factors of d are $\pm 1, \pm 2, \pm 3, \pm 5, \pm 6, \pm 10, \pm 15$, and ± 30. This gives us a large number of possible ratios: $\pm 1, \pm 2, \pm 3, \pm 5, \pm 6, \pm 10, \pm 15, \pm 30, \pm 1/2, \pm 3/2, \pm 5/2, \pm 15/2$. Undaunted, we first try $x = 1$ and $x = -1$, getting nowhere. So we move on to $x = 2$, and we hit the jackpot, because substituting $x = 2$ gives $16 - 12 - 34 + 30 = 0$. Thus, $(x - 2)$ is a factor. Dividing yields the quotient $2x^2 + x - 15$. Here is the calculation:

$$
\begin{array}{r|l}
 & 2x^2 + x - 15 \\
\hline
x - 2 & 2x^3 - 3x^2 - 17x + 30 \\
 & \underline{2x^3 - 4x^2} \\
 & \qquad\quad x^2 - 17x \\
 & \qquad\quad \underline{x^2 - 2x} \\
 & \qquad\qquad -15x + 30 \\
 & \qquad\qquad \underline{-15x + 30} \\
 & \qquad\qquad\qquad\quad 0.
\end{array}
$$

Thus,

$$2x^3 - 3x^2 - 17x + 30 = (x - 2)(2x^2 + x - 15) = 0.$$

Setting the factors equal to zero gives either $x - 2 = 0$ or $2x^2 + x - 15 = 0$. We could solve the quadratic using the quadratic formula, but, luckily, we notice that it factors as

$$2x^2 + x - 15 = (x + 3)(2x - 5).$$

Thus, the solutions are $x = 2$, $x = -3$ and $x = 5/2$.

Solution of Higher-Order Polynomial Equations

Logically speaking, our next step should be a discussion of quartics, then quintics (fifth degree equations), and so on forever. Well, we've got to stop somewhere, and cubics may be as good a place as any. On the other hand, since we've gotten so far, we ought to at least tell you what is known about higher order polynomials.

Quartics Just as in the case of cubics, there is a formula to find the solutions of quartics.[8]

---

[8] See, for example, *First Course in the Theory of Equations* by L. E. Dickson (New York: Wiley, 1922), or *Modern Algebra* by B. L. van der Waerden (New York: Frederick Ungar, 1953).

Quintics and Beyond All good things must come to an end, we're afraid. It turns out that there is no "quintic formula." In other words, there is no single algebraic formula or collection of algebraic formulas that gives the solutions to all quintics. This question was settled by the Norwegian mathematician Niels Henrik Abel in 1824 after almost 300 years of controversy about this question. (In fact, several notable mathematicians had previously claimed to have devised formulas for solving the quintic, but these were all shot down by other mathematicians—this being one of the favorite pastimes of practitioners of our art.) The same negative answer applies to polynomial equations of degree 6 and higher. It's not that these equations don't have solutions; it's just that they can't be found using algebraic formulas.[9] However, there are certain special classes of polynomial equations that can be solved with algebraic methods. The way of identifying such equations was discovered around 1829 by the French mathematician Évariste Galois.[10]

[9]What we mean by an "algebraic formula" is a formula in the coefficients using the operations of addition, subtraction, multiplication, division, and the taking of radicals. Mathematicians call the use of such formulas in solving polynomial equations "solution by radicals." If you were a math major, you would eventually go on to study this under the heading of Galois theory.

[10]Both Abel (1802–1829) and Galois (1811–1832) died young. Abel died of tuberculosis at the age of 26, while Galois was killed in a duel at the age of 20.

0.5 EXERCISES

Access end-of-section exercises online at **www.webassign.net**

0.6 Solving Miscellaneous Equations

Equations often arise in calculus that are not polynomial equations of low degree. Many of these complicated-looking equations can be solved easily if you remember the following, which we used in the previous section:

Solving an Equation of the Form $P \cdot Q = 0$

If a product is equal to 0, then at least one of the factors must be 0. That is, if $P \cdot Q = 0$, then either $P = 0$ or $Q = 0$.

Quick Examples

1. $x^5 - 4x^3 = 0$

$x^3(x^2 - 4) = 0$ — Factor the left-hand side.

Either $x^3 = 0$ or $x^2 - 4 = 0$ — Either $P = 0$ or $Q = 0$.

$x = 0, 2$ or -2. — Solve the individual equations.

2. $(x^2 - 1)(x + 2) + (x^2 - 1)(x + 4) = 0$

$(x^2 - 1)[(x + 2) + (x + 4)] = 0$ — Factor the left-hand side.

$(x^2 - 1)(2x + 6) = 0$

Either $x^2 - 1 = 0$ or $2x + 6 = 0$ — Either $P = 0$ or $Q = 0$.

$x = -3, -1,$ or 1. — Solve the individual equations.

EXAMPLE 1 Solving by Factoring

Solve $12x(x^2 - 4)^5(x^2 + 2)^6 + 12x(x^2 - 4)^6(x^2 + 2)^5 = 0$.

Solution
Again, we start by factoring the left-hand side:

$$\begin{aligned}12x(x^2 - 4)^5(x^2 + 2)^6 &+ 12x(x^2 - 4)^6(x^2 + 2)^5 \\ &= 12x(x^2 - 4)^5(x^2 + 2)^5[(x^2 + 2) + (x^2 - 4)] \\ &= 12x(x^2 - 4)^5(x^2 + 2)^5(2x^2 - 2) \\ &= 24x(x^2 - 4)^5(x^2 + 2)^5(x^2 - 1).\end{aligned}$$

Setting this equal to 0, we get:

$$24x(x^2 - 4)^5(x^2 + 2)^5(x^2 - 1) = 0,$$

which means that at least one of the factors of this product must be zero. Now it certainly cannot be the 24, but it could be the x: $x = 0$ is one solution. It could also be that

$$(x^2 - 4)^5 = 0$$

or

$$x^2 - 4 = 0,$$

which has solutions $x = \pm 2$. Could it be that $(x^2 + 2)^5 = 0$? If so, then $x^2 + 2 = 0$, but this is impossible because $x^2 + 2 \geq 2$, no matter what x is. Finally, it could be that $x^2 - 1 = 0$, which has solutions $x = \pm 1$. This gives us five solutions to the original equation:

$$x = -2, -1, 0, 1, \text{ or } 2.$$

EXAMPLE 2 Solving by Factoring

Solve $(x^2 - 1)(x^2 - 4) = 10$.

Solution Watch out! You may be tempted to say that $x^2 - 1 = 10$ or $x^2 - 4 = 10$, but this does not follow. If two numbers multiply to give you 10, what must they be? There are lots of possibilities: 2 and 5, 1 and 10, $-500{,}000$ and -0.00002 are just a few. The fact that the left-hand side is factored is nearly useless to us if we want to solve this equation. What we will have to do is multiply out, bring the 10 over to the left, and hope that we can factor what we get. Here goes:

$$\begin{aligned}x^4 - 5x^2 + 4 &= 10 \\ x^4 - 5x^2 - 6 &= 0 \\ (x^2 - 6)(x^2 + 1) &= 0\end{aligned}$$

(Here we used a sometimes useful trick that we mentioned in Section 3: We treated x^2 like x and x^4 like x^2, so factoring $x^4 - 5x^2 - 6$ is essentially the same as factoring $x^2 - 5x - 6$.) *Now* we are allowed to say that one of the factors must be 0: $x^2 - 6 = 0$ has solutions $x = \pm\sqrt{6} = \pm 2.449\ldots$ and $x^2 + 1 = 0$ has no real solutions. Therefore, we get exactly two solutions, $x = \pm\sqrt{6} = \pm 2.449\ldots$.

To solve equations involving rational expressions, the following rule is very useful.

Solving an Equation of the Form $P/Q = 0$

$$\text{If } \frac{P}{Q} = 0, \text{ then } P = 0.$$

How else could a fraction equal 0? If that is not convincing, multiply both sides by Q (which cannot be 0 if the quotient is defined).

Quick Example

$$\frac{(x+1)(x+2)^2 - (x+1)^2(x+2)}{(x+2)^4} = 0$$

$$(x+1)(x+2)^2 - (x+1)^2(x+2) = 0 \qquad \text{If } \tfrac{P}{Q} = 0, \text{ then } P = 0.$$

$$(x+1)(x+2)[(x+2) - (x+1)] = 0 \qquad \text{Factor.}$$

$$(x+1)(x+2)(1) = 0$$

Either $x + 1 = 0$ or $x + 2 = 0$,

$x = -1$ or $x = -2$

$x = -1$ $\qquad$ $x = -2$ does not make sense in the original equation: it makes the denominator 0. So it is not a solution and $x = -1$ is the only solution.

EXAMPLE 3 Solving a Rational Equation

Solve $1 - \dfrac{1}{x^2} = 0$.

Solution Write 1 as $\frac{1}{1}$, so that we now have a difference of two rational expressions:

$$\frac{1}{1} - \frac{1}{x^2} = 0.$$

To combine these we can put both over a common denominator of x^2, which gives

$$\frac{x^2 - 1}{x^2} = 0.$$

Now we can set the numerator, $x^2 - 1$, equal to zero. Thus,

$$x^2 - 1 = 0$$

so

$$(x-1)(x+1) = 0,$$

giving $x = \pm 1$.

Before we go on... This equation could also have been solved by writing

$$1 = \frac{1}{x^2}$$

and then multiplying both sides by x^2. ■

EXAMPLE 4 Another Rational Equation

Solve $\frac{2x-1}{x} + \frac{3}{x-2} = 0$.

Solution We *could* first perform the addition on the left and then set the top equal to 0, but here is another approach. Subtracting the second expression from both sides gives

$$\frac{2x-1}{x} = \frac{-3}{x-2}$$

Cross-multiplying [multiplying both sides by both denominators—that is, by $x(x-2)$] now gives

$$(2x-1)(x-2) = -3x$$

so

$$2x^2 - 5x + 2 = -3x.$$

Adding $3x$ to both sides gives the quadratic equation

$$2x^2 - 2x + 1 = 0.$$

The discriminant is $(-2)^2 - 4 \cdot 2 \cdot 1 = -4 < 0$, so we conclude that there is no real solution.

Before we go on... Notice that when we said that $(2x-1)(x-2) = -3x$, we were *not* allowed to conclude that $2x - 1 = -3x$ or $x - 2 = -3x$. ■

EXAMPLE 5 A Rational Equation with Radicals

Solve $\frac{\left(2x\sqrt{x+1} - \frac{x^2}{\sqrt{x+1}}\right)}{x+1} = 0$.

Solution Setting the top equal to 0 gives

$$2x\sqrt{x+1} - \frac{x^2}{\sqrt{x+1}} = 0.$$

This still involves fractions. To get rid of the fractions, we could put everything over a common denominator ($\sqrt{x+1}$) and then set the top equal to 0, or we could multiply the whole equation by that common denominator in the first place to clear fractions. If we do the second, we get

$$2x(x+1) - x^2 = 0$$
$$2x^2 + 2x - x^2 = 0$$
$$x^2 + 2x = 0.$$

Factoring,

$$x(x+2) = 0$$

so either $x = 0$ or $x + 2 = 0$, giving us $x = 0$ or $x = -2$. Again, one of these is not really a solution. The problem is that $x = -2$ cannot be substituted into $\sqrt{x+1}$, because we would then have to take the square root of -1, and we are not allowing ourselves to do that. Therefore, $x = 0$ is the only solution.

0.6 EXERCISES

Access end-of-section exercises online at **www.webassign.net**

0.7 The Coordinate Plane

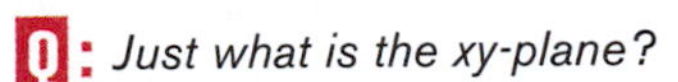

Q: *Just what is the xy-plane?*

A: The *xy*-plane is an infinite flat surface with two perpendicular lines, usually labeled the ***x*-axis** and ***y*-axis**. These axes are calibrated as shown in Figure 2. (Notice also how the plane is divided into four **quadrants**.)

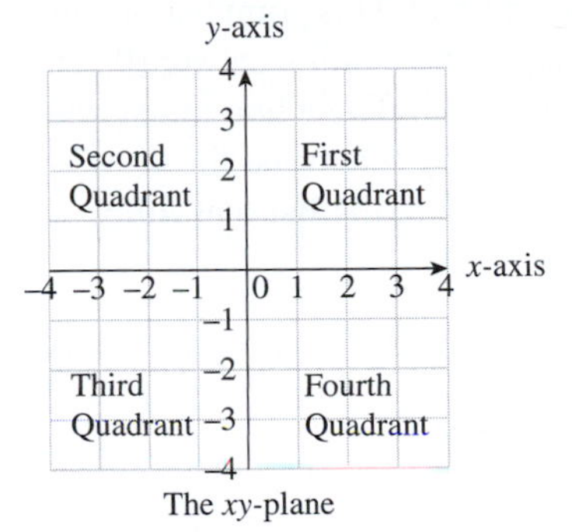

The *xy*-plane

Figure 2

Thus, the *xy*-plane is nothing more than a very large—in fact, infinitely large—flat surface. The purpose of the axes is to allow us to locate specific positions, or **points**, on the plane, with the use of **coordinates**. (If Captain Picard wants to have himself beamed to a specific location, he must supply its coordinates, or he's in trouble.)

Q: *So how do we use coordinates to locate points?*

A: The rule is simple. Each point in the plane has two coordinates, an ***x*-coordinate** and a ***y*-coordinate**. These can be determined in two ways:

1. The x-coordinate measures a point's distance to the right or left of the y-axis. It is positive if the point is to the right of the axis, negative if it is to the left of the axis, and 0 if it is on the axis. The y-coordinate measures a point's distance above or below the x-axis. It is positive if the point is above the axis, negative if it is below the axis, and 0 if it is on the axis. Briefly, the x-coordinate tells us the *horizontal* position (distance left or right), and the y-coordinate tells us the *vertical* position (height).
2. Given a point P, we get its x-coordinate by drawing a vertical line from P and seeing where it intersects the x-axis. Similarly, we get the y-coordinate by extending a horizontal line from P and seeing where it intersects the y-axis.

This way of assigning coordinates to points in the plane is often called the system of **Cartesian** coordinates, in honor of the mathematician and philosopher René Descartes (1596–1650), who was the first to use them extensively.

Here are a few examples to help you review coordinates.

EXAMPLE 1 Coordinates of Points

a. Find the coordinates of the indicated points. (See Figure 3. The grid lines are placed at intervals of one unit.)

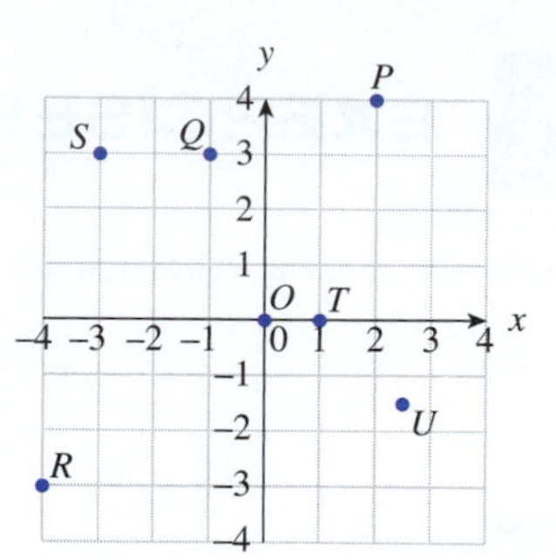

Figure 3

b. Locate the following points in the xy-plane.

$$A(2, 3),\ B(-4, 2),\ C(3, -2.5),\ D(0, -3),\ E(3.5, 0),\ F(-2.5, -1.5)$$

Solution

a. Taking them in alphabetical order, we start with the origin O. This point has height zero and is also zero units to the right of the y-axis, so its coordinates are (0, 0). Turning to P, dropping a vertical line gives $x = 2$ and extending a horizontal line gives $y = 4$. Thus, P has coordinates (2, 4). For practice, determine the coordinates of the remaining points, and check your work against the list that follows:

$$Q(-1, 3),\ R(-4, -3),\ S(-3, 3),\ T(1, 0),\ U(2.5, -1.5)$$

b. In order to locate the given points, we start at the origin (0, 0), and proceed as follows. (See Figure 4.)

To locate A, we move 2 units to the right and 3 up, as shown.

To locate B, we move -4 units to the right (that is, 4 to the *left*) and 2 up, as shown.

To locate C, we move 3 units right and 2.5 down.

We locate the remaining points in a similar way.

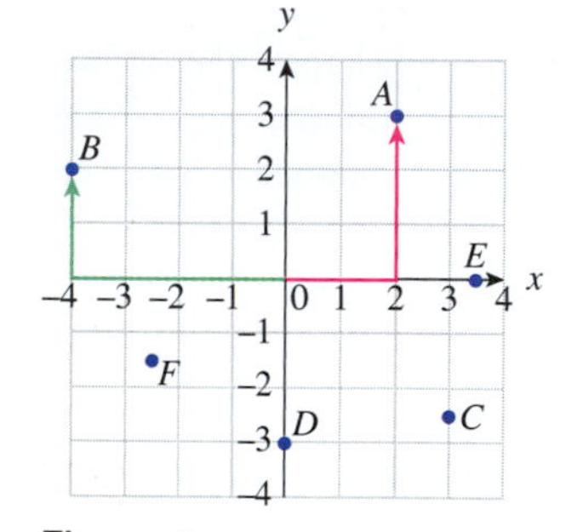

Figure 4

The Graph of an Equation

One of the more surprising developments of mathematics was the realization that equations, which are algebraic objects, can be represented by graphs, which are geometric objects. The kinds of equations that we have in mind are equations in x and y, such as

$$y = 4x - 1, \quad 2x^2 - y = 0, \quad y = 3x^2 + 1, \quad y = \sqrt{x - 1}.$$

The **graph** of an equation in the two variables x and y consists of all points (x, y) in the plane whose coordinates are solutions of the equation.

EXAMPLE 2 Graph of an Equation

Obtain the graph of the equation $y - x^2 = 0$.

Solution We can solve the equation for y to obtain $y = x^2$. Solutions can then be obtained by choosing values for x and then computing y by squaring the value of x, as shown in the following table:

x	−3	−2	−1	0	1	2	3
$y = x^2$	9	4	1	0	1	4	9

Plotting these points (x, y) gives the following picture (left side of Figure 5), suggesting the graph on the right in Figure 5.

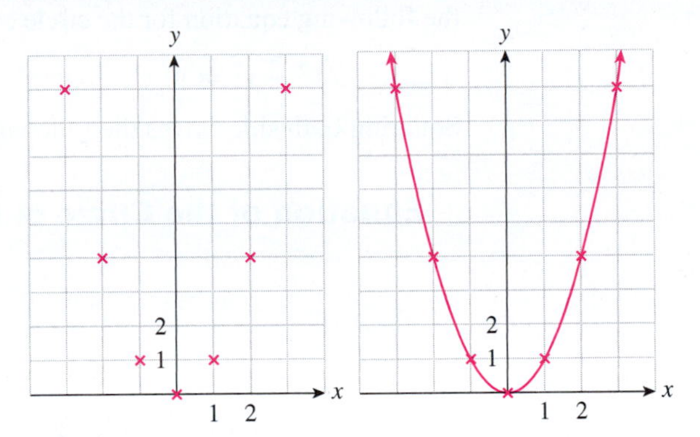

Figure 5

Distance

The distance between two points in the xy-plane can be expressed as a function of their coordinates, as follows:

Distance Formula

The distance between the points $P(x_1, y_1)$ and $Q(x_2, y_2)$ is

$$d = \sqrt{(x_2 - x_1)^2 + (y_2 - y_1)^2} = \sqrt{(\Delta x)^2 + (\Delta y)^2}.$$

Derivation

The distance d is shown in the figure below.

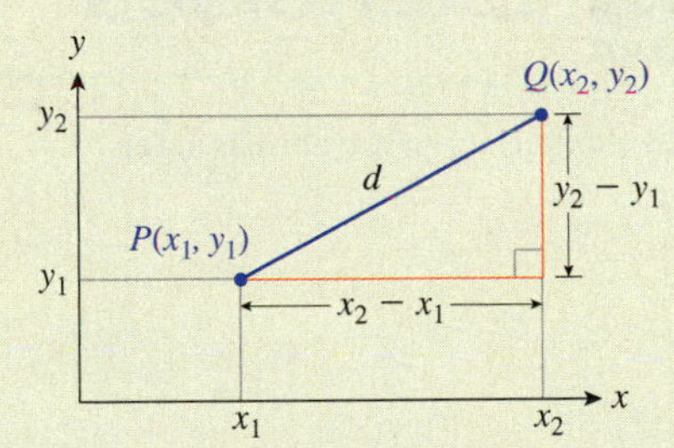

By the Pythagorean theorem applied to the right triangle shown, we get

$$d^2 = (x_2 - x_1)^2 + (y_2 - y_1)^2.$$

Taking square roots (d is a distance, so we take the positive square root), we get the distance formula. Notice that if we switch x_1 with x_2 or y_1 with y_2, we get the same result.

Quick Examples

1. The distance between the points $(3, -2)$ and $(-1, 1)$ is
$$d = \sqrt{(-1 - 3)^2 + (1 + 2)^2} = \sqrt{25} = 5.$$

2. The distance from (x, y) to the origin $(0, 0)$ is
$$d = \sqrt{(x - 0)^2 + (y - 0)^2} = \sqrt{x^2 + y^2}.$$ Distance to the origin

The set of all points (x, y) whose distance from the origin $(0, 0)$ is a fixed quantity r is a circle centered at the origin with radius r. From the second Quick Example, we get the following equation for the circle centered at the origin with radius r:

$$\sqrt{x^2 + y^2} = r. \qquad \text{Distance from the origin} = r.$$

Squaring both sides gives the following equation:

Equation of the Circle of Radius *r* Centered at the Origin

$$x^2 + y^2 = r^2$$

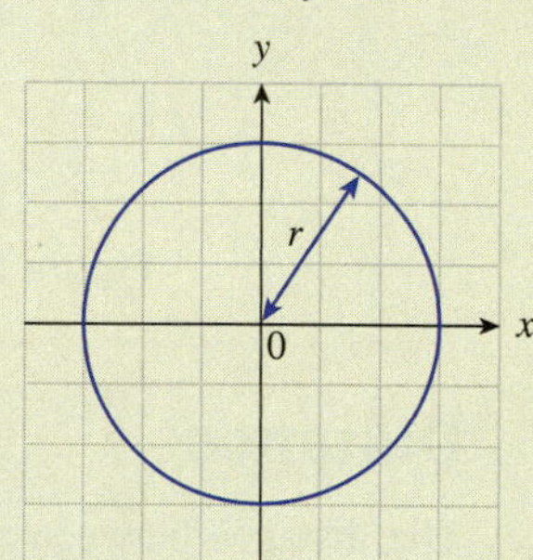

Quick Examples

1. The circle of radius 1 centered at the origin has equation $x^2 + y^2 = 1$.
2. The circle of radius 2 centered at the origin has equation $x^2 + y^2 = 4$.

0.7 EXERCISES

Access end-of-section exercises online at **www.webassign.net** ENHANCED WebAssign

1

Functions and Applications

Website

www.WanerMath.com

At the Website you will find:

- Section-by-section tutorials, including game tutorials with randomized quizzes
- A detailed chapter summary
- A true/false quiz
- Additional review exercises
- Graphers, Excel tutorials, and other resources
- The following extra topic:

  New Functions from Old: Scaled and Shifted Functions

Case Study Modeling Spending on Internet Advertising

You are the new director of *Impact Advertising Inc.'s* Internet division, which has enjoyed a steady 0.25% of the Internet advertising market. You have drawn up an ambitious proposal to expand your division in light of your anticipation that Internet advertising will continue to skyrocket. The VP in charge of Financial Affairs feels that current projections (based on a linear model) do not warrant the level of expansion you propose. **How can you persuade the VP that those projections do not fit the data convincingly?**

Jeff Titcomb/Photographer's Choice / Getty Images

Introduction

To analyze recent trends in spending on Internet advertising and to make reasonable projections, we need a mathematical model of this spending. Where do we start? To apply mathematics to real-world situations like this, we need a good understanding of basic mathematical concepts. Perhaps the most fundamental of these concepts is that of a function: a relationship that shows how one quantity depends on another. Functions may be described numerically and, often, algebraically. They can also be described graphically—a viewpoint that is extremely useful.

The simplest functions—the ones with the simplest formulas and the simplest graphs—are linear functions. Because of their simplicity, they are also among the most useful functions and can often be used to model real-world situations, at least over short periods of time. In discussing linear functions, we will meet the concepts of slope and rate of change, which are the starting point of the mathematics of change.

algebra Review
For this chapter, you should be familiar with real numbers and intervals. To review this material, see **Chapter 0.**

In the last section of this chapter, we discuss *simple linear regression*: construction of linear functions that best fit given collections of data. Regression is used extensively in applied mathematics, statistics, and quantitative methods in business. The inclusion of regression utilities in computer spreadsheets like Excel® makes this powerful mathematical tool readily available for anyone to use.

1.1 Functions from the Numerical, Algebraic, and Graphical Viewpoints

The following table gives the approximate number of Facebook users at various times since its establishment early in 2004.[1]

Year t (Since start of 2004)	0	1	2	3	4	5	6
Facebook Members n (Millions)	0	1	5.5	12	58	150	450

Let's write $n(0)$ for the number of members (in millions) at time $t = 0$, $n(1)$ for the number at time $t = 1$, and so on (we read $n(0)$ as "n of 0"). Thus, $n(0) = 0$, $n(1) = 1$, $n(2) = 5.5, \ldots, n(6) = 450$. In general, we write $n(t)$ for the number of members (in millions) at time t. We call n a **function** of the variable t, meaning that for each value of t between 0 and 6, n gives us a single corresponding number $n(t)$ (the number of members at that time).

In general, we think of a function as a way of producing new objects from old ones. The functions we deal with in this text produce new numbers from old numbers. The numbers we have in mind are the *real* numbers, including not only positive and negative integers and fractions but also numbers like $\sqrt{2}$ or π. (See Chapter 0 for more on real numbers.) For this reason, the functions we use are called **real-valued functions of a real variable**. For example, the function n takes the year since the start of 2004 as input and returns the number of Facebook members as output (Figure 1).

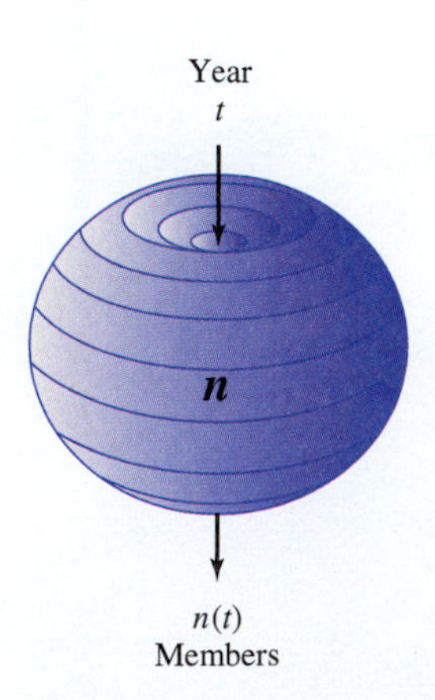

Figure 1

[1]Sources: www.facebook.com, www.insidefacebook.com.

The variable t is called the **independent variable**, while n is called the **dependent variable** as its value depends on t. A function may be specified in several different ways. Here, we have specified the function n **numerically** by giving the values of the function for a number of values of the independent variable, as in the preceding table.

Q: *For which values of t does it make sense to ask for n(t)? In other words, for which years t is the function n defined?*

A: Because $n(t)$ refers to the number of members from the start of 2004 to the start of 2010, $n(t)$ is defined when t is any number between 0 and 6, that is, when $0 \leq t \leq 6$. Using interval notation (see Chapter 0), we can say that $n(t)$ is defined when t is in the interval $[0, 6]$.

The set of values of the independent variable for which a function is defined is called its **domain** and is a necessary part of the definition of the function. Notice that the preceding table gives the value of $n(t)$ at only some of the infinitely many possible values in the domain $[0, 6]$. The domain of a function is not always specified explicitly; if no domain is specified for the function f, we take the domain to be the largest set of numbers x for which $f(x)$ makes sense. This "largest possible domain" is sometimes called the **natural domain**.

The previous Facebook data can also be represented on a graph by plotting the given pairs of numbers $(t, n(t))$ in the xy-plane. (See Figure 2. We have connected successive points by line segments.) In general, the **graph** of a function f consists of all points $(x, f(x))$ in the plane with x in the domain of f.

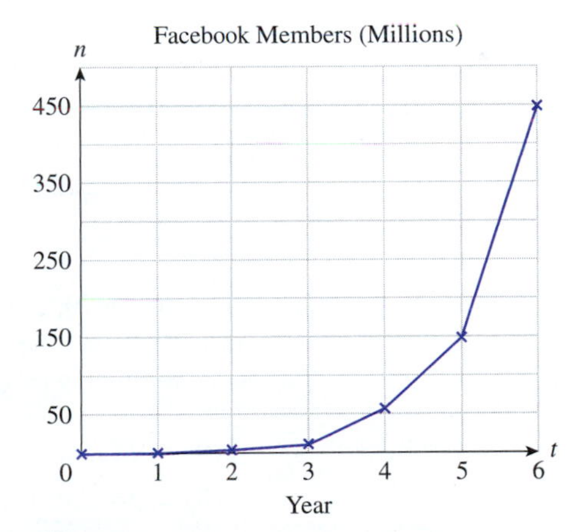

Figure 2

In Figure 2 we specified the function n **graphically** by using a graph to display its values. Suppose now that we had only the graph without the table of data. We could use the graph to find approximate values of n. For instance, to find $n(5)$ from the graph, we do the following:

1. Find the desired value of t at the bottom of the graph ($t = 5$ in this case).
2. Estimate the height (n-coordinate) of the corresponding point on the graph (around 150 in this case).

Thus, $n(5) \approx 150$ million members.*

* In a graphically defined function, we can never know the y-coordinates of points exactly; no matter how accurately a graph is drawn, we can obtain only *approximate* values of the coordinates of points. That is why we have been using the word *estimate* rather than *calculate* and why we say $n(5) \approx 150$ rather than $n(5) = 150$.

In some cases we may be able to use an algebraic formula to calculate the function, and we say that the function is specified **algebraically**. These are not the only ways in which a function can be specified; for instance, it could also be specified **verbally**, as in "Let $n(t)$ be the number of Facebook members, in millions, t years since the start of 2004."* Notice that any function can be represented graphically by plotting the points $(x, f(x))$ for a number of values of x in its domain.

* Specifying a function verbally in this way is useful for understanding what the function is doing, but it gives no numerical information.

Here is a summary of the terms we have just introduced.

Functions

A **real-valued function f of a real-valued variable x** assigns to each real number x in a specified set of numbers, called the **domain** of f, a unique real number $f(x)$, read "f of x." The variable x is called the **independent variable**, and f is called the **dependent variable**. A function is usually specified **numerically** using a table of values, **graphically** using a graph, or **algebraically** using a formula. The **graph of a function** consists of all points $(x, f(x))$ in the plane with x in the domain of f.

Quick Examples

1. **A function specified numerically:** Take $c(t)$ to be the world emission of carbon dioxide in year t since 2000, represented by the following table:[2]

t (Year Since 2000)	$c(t)$ (Billion Metric Tons of CO_2)
0	24
5	28
10	31
15	33
20	36
25	38
30	41

The domain of c is $[0, 30]$, the independent variable is t, the number of years since 2000, and the dependent variable is c, the world production of carbon dioxide in a given year. Some values of c are:

$c(0) = 24$ — 24 billion metric tons of CO_2 were produced in 2000.

$c(10) = 31$ — 31 billion metric tons of CO_2 were produced in 2010.

$c(30) = 41$ — 41 billion metric tons of CO_2 were projected to be produced in 2030.

[2]Figures for 2015 and later are projections. Source: Energy Information Administration (EIA) (www.eia.doe.gov)

Graph of *c*: Plotting the pairs $(t, c(t))$ gives the following graph:

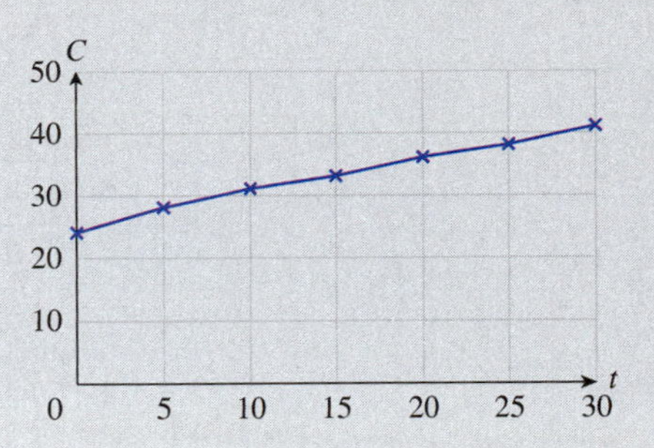

2. **A function specified graphically:** Take $m(t)$ to be the median U.S. home price in thousands of dollars, t years since 2000, as represented by the following graph:[3]

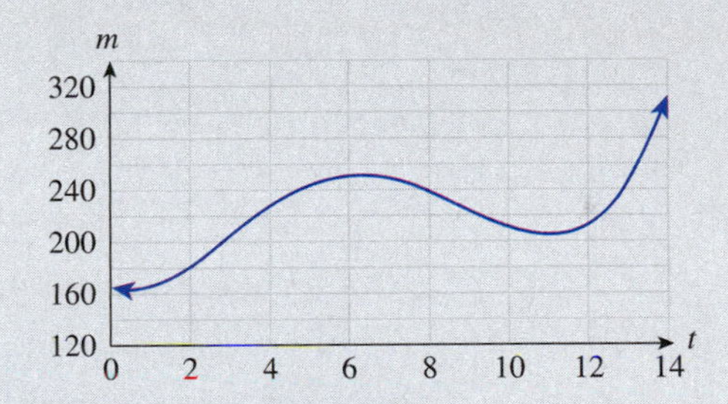

The domain of m is $[0, 14]$, the independent variable is t, the number of years since 2000, and the dependent variable is m, the median U.S. home price in thousands of dollars. Some values of m are:

$m(2) \approx 180$ The median home price in 2002 was about \$180,000.

$m(10) \approx 210.$ The median home price in 2010 was about \$210,000.

3. **A function specified algebraically:** Let $f(x) = \frac{1}{x}$. The function f is specified algebraically. The independent variable is x and the dependent variable is f. The natural domain of f consists of all real numbers except zero because $f(x)$ makes sense for all values of x other than $x = 0$. Some specific values of f are

$$f(2) = \frac{1}{2} \qquad f(3) = \frac{1}{3} \qquad f(-1) = \frac{1}{-1} = -1$$

$f(0)$ is not defined because 0 is not in the domain of f.

[3]Source for data through end of 2010: www.zillow.com/local-info.

4. The graph of a function: Let $f(x) = x^2$, with domain the set of all real numbers. To draw the graph of f, first choose some convenient values of x in the domain and compute the corresponding y-coordinates $f(x)$:

x	-3	-2	-1	0	1	2	3
$f(x) = x^2$	9	4	1	0	1	4	9

Plotting these points $(x, f(x))$ gives the picture on the left, suggesting the graph on the right.*

* If you plot more points, you will find that they lie on a smooth curve as shown. That is why we did not use line segments to connect the points.

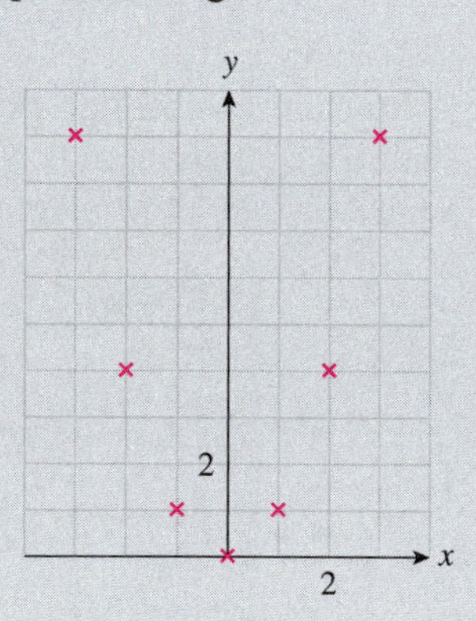

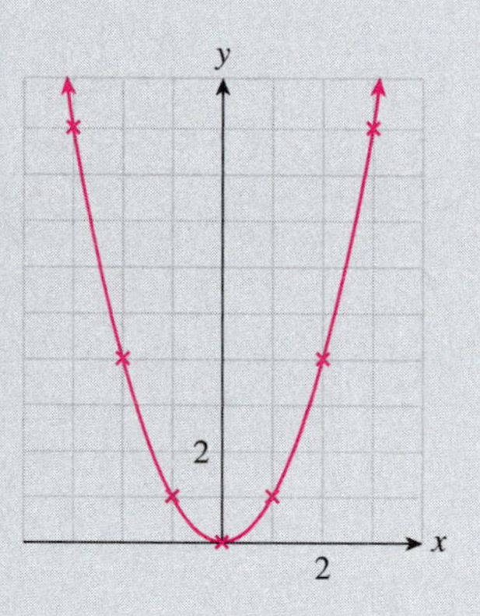

(This particular curve happens to be called a **parabola**, and its lowest point, at the origin, is called its **vertex**.)

EXAMPLE 1 iPod Sales

The total number of iPods sold by Apple up to the end of year x can be approximated by

$$f(x) = 4x^2 + 16x + 2 \text{ million iPods } (0 \le x \le 6),$$

where $x = 0$ represents 2003.[4]

a. What is the domain of f? Compute $f(0)$, $f(2)$, $f(4)$, and $f(6)$. What do these answers tell you about iPod sales? Is $f(-1)$ defined?

b. Compute $f(a)$, $f(-b)$, $f(a+h)$, and $f(a)+h$ assuming that the quantities a, $-b$, and $a+h$ are in the domain of f.

c. Sketch the graph of f. Does the shape of the curve suggest that iPod sales were accelerating or decelerating?

Solution

a. The domain of f is the set of numbers x with $0 \le x \le 6$—that is, the interval $[0, 6]$. If we substitute 0 for x in the formula for $f(x)$, we get

$$f(0) = 4(0)^2 + 16(0) + 2 = 2.$$

By the end of 2003 approximately 2 million iPods had been sold.

[4]Source for data: Apple quarterly earnings reports at www.apple.com/investor/.

Similarly,

$$f(2) = 4(2)^2 + 16(2) + 2 = 50$$ By the end of 2005 approximately 50 million iPods had been sold.

$$f(4) = 4(4)^2 + 16(4) + 2 = 130$$ By the end of 2007 approximately 130 million iPods had been sold.

$$f(6) = 4(6)^2 + 16(6) + 2 = 242.$$ By the end of 2009 approximately 242 million iPods had been sold.

As -1 is not in the domain of f, $f(-1)$ is not defined.

b. To find $f(a)$ we substitute a for x in the formula for $f(x)$ to get

$$f(a) = 4a^2 + 16a + 2.$$ Substitute a for x.

Similarly,

$$f(-b) = 4(-b)^2 + 16(-b) + 2$$ Substitute $-b$ for x.

$$= 4b^2 - 16b + 2$$ $(-b)^2 = b^2$

$$f(a+h) = 4(a+h)^2 + 16(a+h) + 2$$ Substitute $(a+h)$ for x.

$$= 4(a^2 + 2ah + h^2) + 16a + 16h + 2$$ Expand.

$$= 4a^2 + 8ah + 4h^2 + 16a + 16h + 2$$

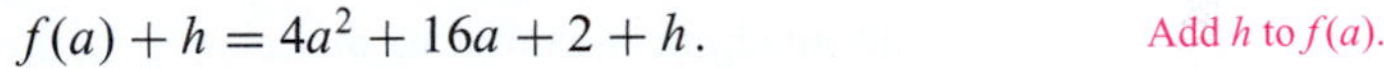

$$f(a) + h = 4a^2 + 16a + 2 + h.$$ Add h to $f(a)$.

Note how we placed parentheses around the quantities at which we evaluated the function. If we tried to do without any of these parentheses we would likely get an error:

Correct expression: $f(a+h) = 4(a+h)^2 + 16(a+h) + 2.$ ✓

NOT $4a + h^2 + 16a + h + 2x$

Also notice the distinction between $f(a+h)$ and $f(a) + h$: To find $f(a+h)$, we replace x by the quantity $(a+h)$; to find $f(a) + h$ we add h to $f(a)$.

c. To draw the graph of f we plot points of the form $(x, f(x))$ for several values of x in the domain of f. Let us use the values we computed in part (a):

x	0	2	4	6
$f(x) = 4x^2 + 16x + 2$	2	50	130	242

Graphing these points gives the graph shown in Figure 3, suggesting the curve shown on the right.

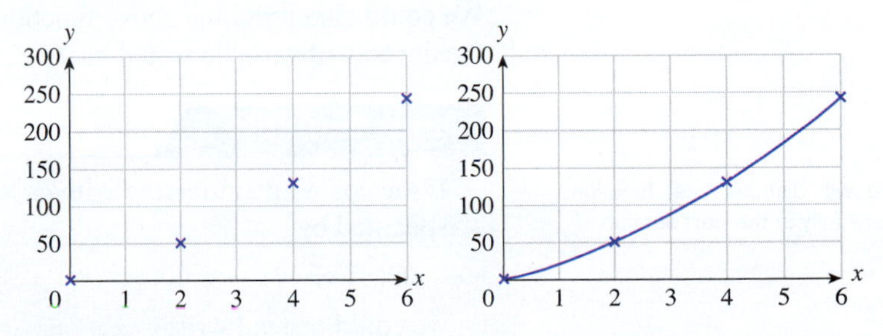

Figure 3

The graph becomes more steep as we move from left to right, suggesting that iPod sales were accelerating.

using Technology

See the Technology Guides at the end of the chapter for detailed instructions on how to obtain the table of values and graph in Example 1 using a TI-83/84 Plus or Excel. Here is an outline:

TI-83/84 Plus

Table of values:

`Y1=4X^2+16X+2`

[2ND] [TABLE].

Graph: [WINDOW];

Xmin = 0, Xmax = 6

[ZOOM] [0].

[More details on page 90.]

Spreadsheet

Table of values: Headings x and $f(x)$ in A1–B1; x-values 0, 2, 4, 6 in A2–A5.

`=4*A2^2+16*A2+2`

in B2; copy down through B5.

Graph: Highlight A1 through B5 and insert a Scatter chart. [More details on page 95.]

Website

www.WanerMath.com

Go to the Function Evaluator and Grapher under Online Utilities, and enter

`4x^2+16x+2`

for y_1. To obtain a table of values, enter the x-values 0, 1, 2, 3 in the Evaluator box, and press "Evaluate" at the top of the box. Graph: Set Xmin = 0, Xmax = 6, and press "Plot Graphs".

Before we go on... The following table compares the value of f in Example 1 with the actual sales figures:

x	0	2	4	6
$f(x) = 4x^2 + 16x + 2$	2	50	130	242
Actual iPod Sales (Millions)	2	32	141	240

The actual figures are only stated here for (some) integer values of x; for instance, $x = 4$ gives the total sales up to the end of 2007. But what were, for instance, the sales through June of 2008 ($x = 4.5$)? This is where our formula comes in handy: We can use the formula for f to **interpolate**—that is, to find sales at values of x other than those between values that are stated:

$$f(4.5) = 4(4.5)^2 + 16(4.5) + 2 = 155 \text{ million iPods.}$$

We can also use the formula to **extrapolate**—that is, to predict sales at values of x *outside* the domain—say, for $x = 6.5$ (that is, sales through June 2009):

$$f(6.5) = 4(6.5)^2 + 16(6.5) + 2 = 275 \text{ million iPods.}$$

As a general rule, extrapolation is far less reliable than interpolation: Predicting the future from current data is difficult, especially given the vagaries of the marketplace.

We call the algebraic function f an **algebraic model** of iPod sales because it uses an algebraic formula to model—or mathematically represent (approximately)—the annual sales. The particular kind of algebraic model we used is called a **quadratic model**. (See the end of this section for the names of some commonly used models.) ■

Functions and Equations

Instead of using the usual "function notation" to specify a function, as in, say,

$$f(x) = 4x^2 + 16x + 2, \qquad \text{Function notation}$$

we could have specified it by an equation by replacing $f(x)$ by y:

$$y = 4x^2 + 16x + 2 \qquad \text{Equation notation}$$

(the choice of the letter y is a convention, but any letter will do).

Technically, $y = 4x^2 + 16x + 2$ is an equation and not a function. However, an equation of this type, $y = \textit{Expression in } x$, can be thought of as "specifying y as a function of x." When we specify a function in this way, the variable x is the independent variable and y is the dependent variable.

We could also write the above function as $f = 4x^2 + 16x + 2$, in which case the dependent variable would be f.

Quick Example

If the cost to manufacture x items is given by the "cost function"* C specified by

$$C(x) = 40x + 2{,}000, \qquad \text{Cost function}$$

we could instead write

$$C = 40x + 2{,}000 \qquad \text{Cost equation}$$

and think of C, the cost, as a function of x.

* We will discuss cost functions more fully in the next section.

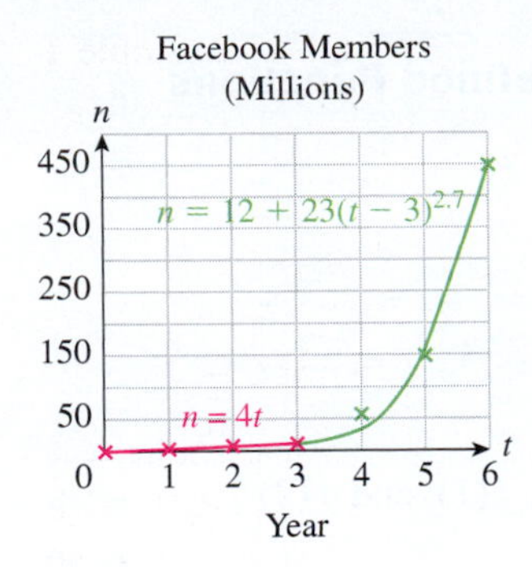

Figure 4

Function notation and equation notation, sometimes using the same letter for the function name and the dependent variable, are often used interchangeably. It is important to be able to switch back and forth between function notation and equation notation, and we shall do so when it is convenient.

Look again at the graph of the number of Facebook users in Figure 2. From year 0 through year 3 the membership appears to increase more-or-less linearly (that is, the graph is almost a straight line), but then curves upward quite sharply from year 3 to year 6. This behavior can be modeled by using two different functions: one for the interval $[0, 3]$ and another for the interval $[3, 6]$ (see Figure 4).

A function specified by two or more different formulas like this is called a **piecewise-defined function**.

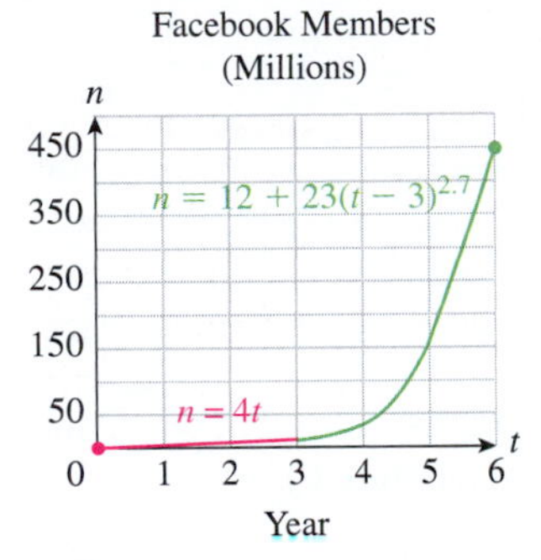

Figure 5

EXAMPLE 2 A Piecewise-Defined Function: Facebook Membership

The number $n(t)$ of Facebook members can be approximated by the following function of time t in years ($t = 0$ represents January 2004):

$$n(t) = \begin{cases} 4t & \text{if } 0 \le t \le 3 \\ 12 + 23(t-3)^{2.7} & \text{if } 3 < t \le 6 \end{cases} \quad \text{million members.}$$

What was the approximate membership of Facebook in January 2005, January 2007, and June 2009? Sketch the graph of n by plotting several points.

Solution We evaluate the given function at the corresponding values of t:

Jan. 2005 ($t = 1$): $n(1) = 4(1) = 4$ — Use the first formula because $0 \le t \le 3$.

Jan. 2007 ($t = 3$): $n(3) = 4(3) = 12$ — Use the first formula because $0 \le t \le 3$.

June 2009 ($t = 5.5$): $n(5.5) = 12 + 23(5.5 - 3)^{2.7} \approx 285.$ — Use the second formula because $3 < t \le 6$.

Thus, the number of Facebook members was approximately 4 million in January 2005, 12 million in January 2007, and 285 million in June 2009.

To sketch the graph of n we use a table of rounded values of $n(t)$ (some of which we have already calculated above), plot the points, and connect them to sketch the graph:

t	0	1	2	3	4	5	6
$n(t)$	0	4	8	12	35	161	459

First Formula (t = 0 to 3) — Second Formula (t = 4 to 6)

The graph (Figure 5) has the following features:

1. The first formula (the line) is used for $0 \le t \le 3$.
2. The second formula (ascending curve) is used for $3 < t \le 6$.
3. The domain is $[0, 6]$, so the graph is cut off at $t = 0$ and $t = 6$.
4. The heavy solid dots at the ends indicate the endpoints of the domain.

using Technology

See the Technology Guides at the end of the chapter for detailed instructions on how to obtain the table of values and graph in Example 2 using a TI-83/84 Plus or Excel. Here is an outline:

TI-83/84 Plus

Table of values:

```
Y1=(X≤3)*(4X)+(X>3)*
(12+23*abs(X-3)^2.7)
```

[2ND] [TABLE].

Graph: [WINDOW]; Xmin = 0, Xmax = 6; [ZOOM] [0].

[More details on page 90.]

Spreadsheet

Table of values: Headings t and $n(t)$ in A1–B1; t-values 0, 1, . . . , 6 in A2–A8.

```
=(A2<=3)*(4*A2)+(A2>3)*
(12+23*abs(A2-3)^2.7)
```

in B2; copy down through B8.

Graph: Highlight A1 through B8 and insert a Scatter chart. [More details on page 96.]

 **Website**

www.WanerMath.com

Go to the Function Evaluator and Grapher under Online Utilities, and enter

```
(x≤3)*(4x)+(x>3)*
  (12+23*abs(x-3)^2.7)
```

for y_1. To obtain a table of values, enter the x-values 0, 1, . . . , 6 in the Evaluator box, and press "Evaluate." at the top of the box. Graph: Set Xmin = 0, Xmax = 6, and press "Plot Graphs."

EXAMPLE 3 More Complicated Piecewise-Defined Functions

Let f be the function specified by

$$f(x) = \begin{cases} -1 & \text{if } -4 \leq x < -1 \\ x & \text{if } -1 \leq x \leq 1 \\ x^2 - 1 & \text{if } 1 < x \leq 2 \end{cases}.$$

a. What is the domain of f? Find $f(-2)$, $f(-1)$, $f(0)$, $f(1)$, and $f(2)$.

b. Sketch the graph of f.

Solution

a. The domain of f is $[-4, 2]$, because $f(x)$ is specified only when $-4 \leq x \leq 2$.

$f(-2) = -1$ We used the first formula because $-4 \leq x < -1$.

$f(-1) = -1$ We used the second formula because $-1 \leq x \leq 1$.

$f(0) = 0$ We used the second formula because $-1 \leq x \leq 1$.

$f(1) = 1$ We used the second formula because $-1 \leq x \leq 1$.

$f(2) = 2^2 - 1 = 3$ We used the third formula because $1 < x \leq 2$.

b. To sketch the graph by hand, we first sketch the three graphs $y = -1$, $y = x$, and $y = x^2 - 1$, and then use the appropriate portion of each (Figure 6).

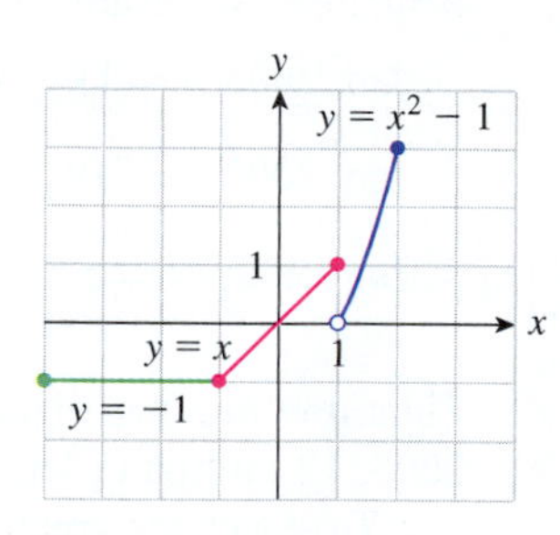

Figure 6

Note that solid dots indicate points on the graph, whereas the open dots indicate points not on the graph. For example, when $x = 1$, the inequalities in the formula tell us that we are to use the middle formula (x) rather than the bottom one ($x^2 - 1$). Thus, $f(1) = 1$, not 0, so we place a solid dot at (1, 1) and an open dot at (1, 0).

using Technology

For the function in Example 3, use the following technology formula (see the technology discussion for Example 2):

```
(X<-1)*(-1)
 +(-1≤X)*(X≤1)*X
 +(1<X)*(X^2-1)
```

Vertical Line Test

Every point in the graph of a function has the form $(x, f(x))$ for some x in the domain of f. Because f assigns a *single* value $f(x)$ to each value of x in the domain, it follows that, in the graph of f, there should be only one y corresponding to any such value of x—namely, $y = f(x)$. In other words, *the graph of a function cannot contain two or more points with the same x-coordinate—that is, two or more points on the same vertical line.* On the other hand, a vertical line at a value of x not in the domain will not contain any points in the graph. This gives us the following rule.

Vertical-Line Test

For a graph to be the graph of a function, every vertical line must intersect the graph in *at most* one point.

Quick Examples

As illustrated below, only graph B passes the vertical line test, so only graph B is the graph of a function.

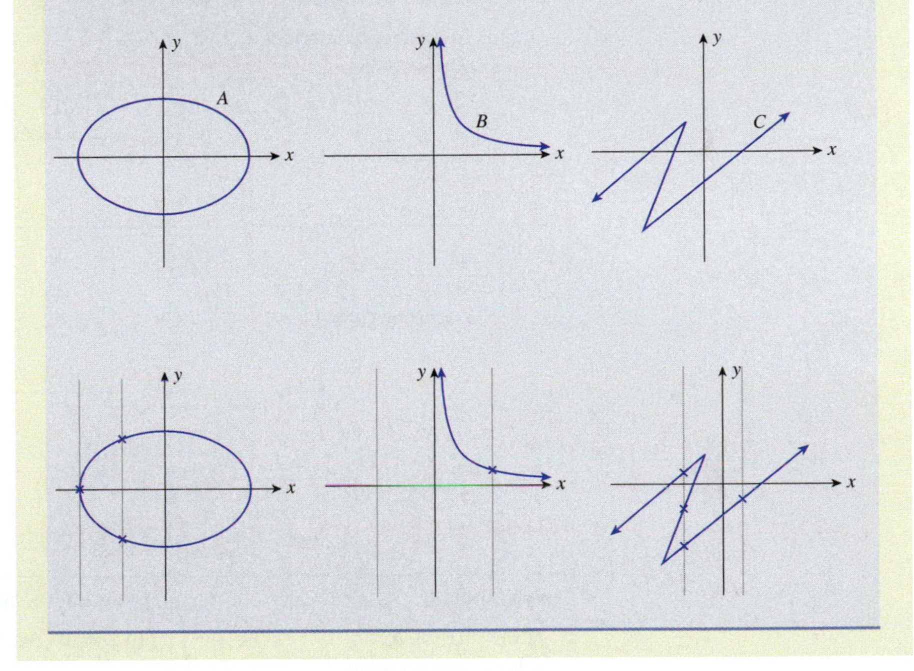

Table 1 lists some common types of functions that are often used to model real world situations.

Table 1 A Compendium of Functions and Their Graphs

Type of Function	*Examples*	
Linear $f(x) = mx + b$ m, b constant Graphs of linear functions are straight lines. The quantity m is the **slope** of the line; the quantity b is the ***y*-intercept** of the line. [See Section 1.3.]	$y = x$ 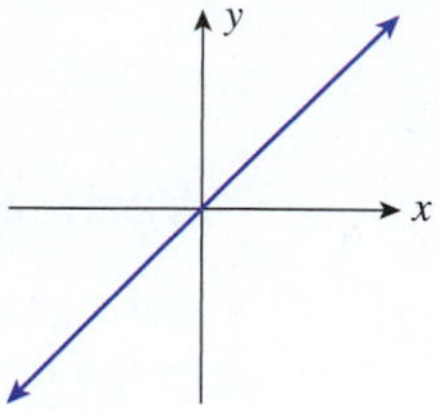	$y = -2x + 2$ 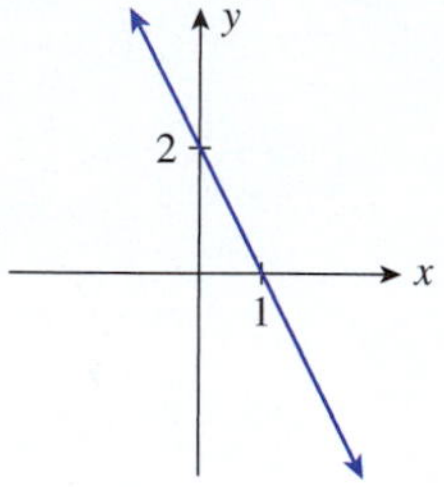
Technology formulas:	`x`	`-2*x+2`
Quadratic $f(x) = ax^2 + bx + c$ a, b, c constant $(a \neq 0)$ Graphs of quadratic functions are called **parabolas**.	$y = x^2$ 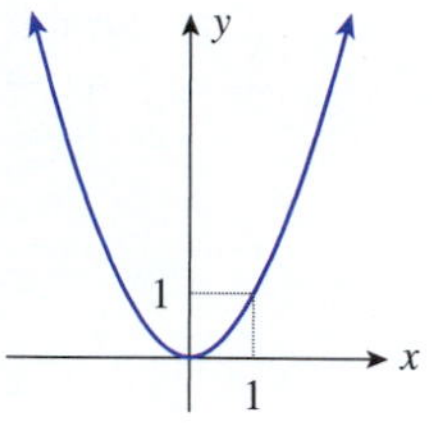	$y = -2x^2 + 2x + 4$ 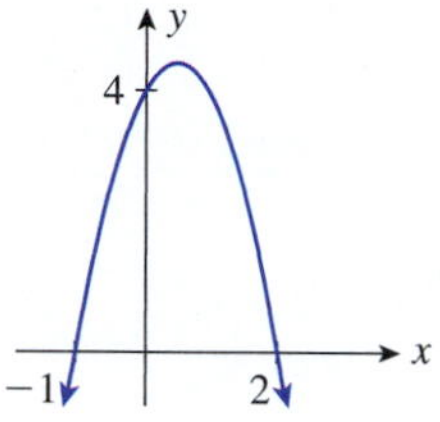
Technology formulas:	`x^2`	`-2*x^2+2*x+4`
Cubic $f(x) = ax^3 + bx^2 + cx + d$ a, b, c, d constant $(a \neq 0)$	$y = x^3$ 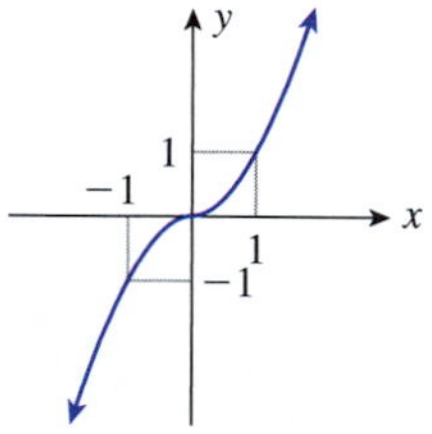	$y = -x^3 + 3x^2 + 1$ 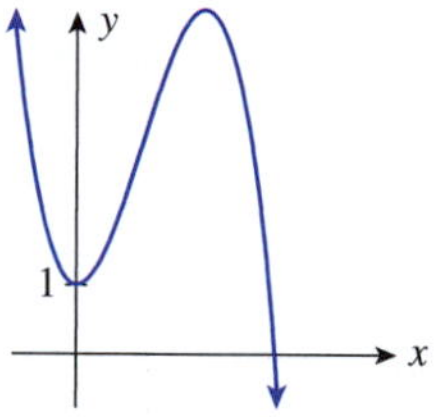
Technology formulas:	`x^3`	`-x^3+3*x^2+1`
Polynomial $f(x) = ax^n + bx^{n-1} + \ldots + rx + s$ $a, b, \ldots, r, s$ constant (includes all of the above functions)	All the above, and $f(x) = x^6 - 2x^5 - 2x^4 + 4x^2$ 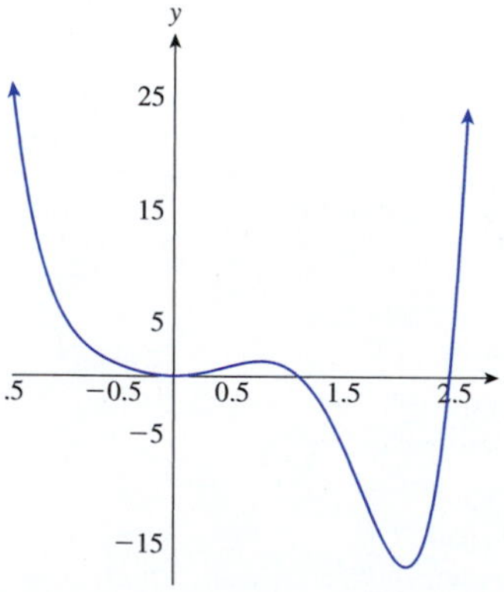	
Technology formula:	`x^6-2x^5-2x^4+4x^2`	

Table 1 *(Continued)*

Type of Function	*Examples*	
Exponential $f(x) = Ab^x$ A, b constant ($b > 0$ and $b \neq 1$) The y-coordinate is multiplied by b every time x increases by 1.	$y = 2^x$ 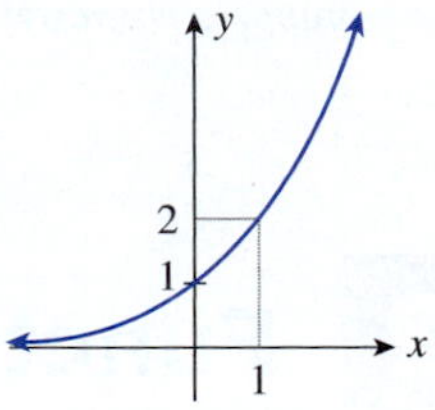 y is doubled every time x increases by 1.	$y = 4(0.5)^x$ 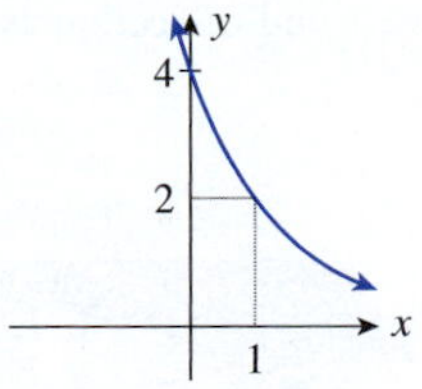 y is halved every time x increases by 1.
Technology formulas:	`2^x`	`4*0.5^x`
Rational $f(x) = \frac{P(x)}{Q(x)}$; $P(x)$ and $Q(x)$ polynomials The graph of $y = 1/x$ is a **hyperbola**. The domain excludes zero because $1/0$ is not defined.	$y = \frac{1}{x}$ 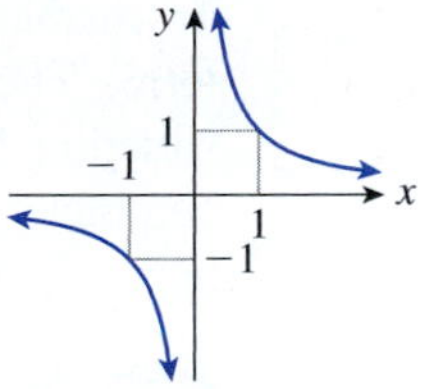	$y = \frac{x}{x-1}$ 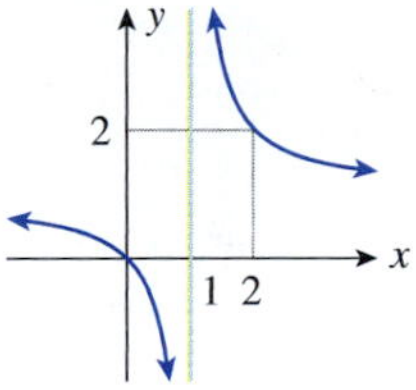
Technology formulas:	`1/x`	`x/(x-1)`
Absolute value For x positive or zero, the graph of $y = \lvert x \rvert$ is the same as that of $y = x$. For x negative or zero, it is the same as that of $y = -x$.	$y = \lvert x \rvert$ 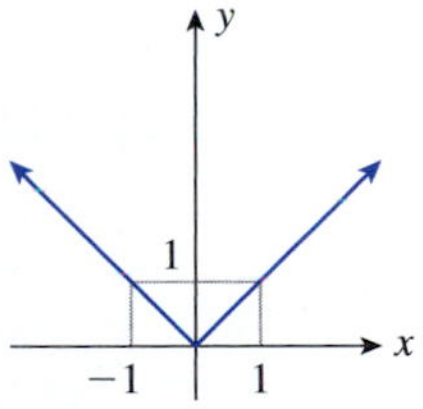	$y = \lvert 2x + 2 \rvert$ 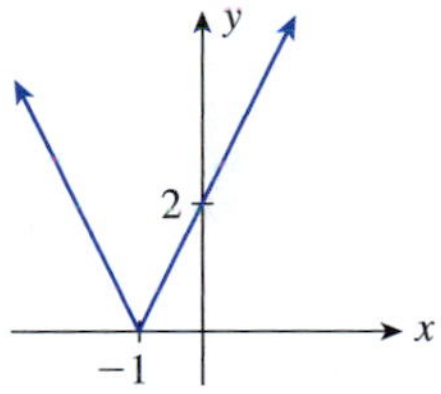
Technology formulas:	`abs(x)`	`abs(2*x+2)`
Square Root The domain of $y = \sqrt{x}$ must be restricted to the nonnegative numbers, because the square root of a negative number is not real. Its graph is the top half of a horizontally oriented parabola.	$y = \sqrt{x}$ 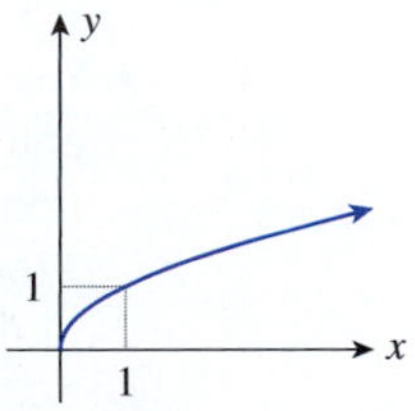	$y = \sqrt{4x - 2}$ 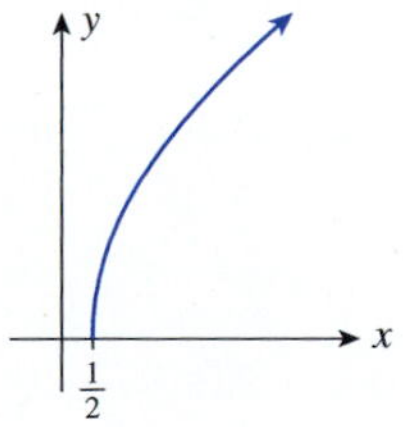
Technology formulas:	`x^0.5` or `√(x)`	`(4*x-2)^0.5` or `√(4*x-2)`

Go to the Website and follow the path

Online Text

→ New Functions from Old: Scaled and Shifted Functions

where you will find complete online interactive text, examples, and exercises on scaling and translating the graph of a function by changing the formula.

Functions and models other than linear ones are called **nonlinear**.

1.1 EXERCISES

Access end-of-section exercises online at **www.webassign.net** ENHANCED WebAssign

1.2 Functions and Models

The functions we used in Examples 1 and 2 in Section 1.1 are **mathematical models** of real-life situations, because they model, or represent, situations in mathematical terms.

Mathematical Modeling

To mathematically model a situation means to represent it in mathematical terms. The particular representation used is called a **mathematical model** of the situation. Mathematical models do not always represent a situation perfectly or completely. Some (like Example 1 of Section 1.1) represent a situation only approximately, whereas others represent only some aspects of the situation.

Quick Examples

1. The temperature is now 10°F and increasing by 20° per hour.
   Model: $T(t) = 10 + 20t$ (t = time in hours, T = temperature)
2. I invest \$1,000 at 5% interest compounded quarterly. Find the value of the investment after t years.
   Model: $A(t) = 1{,}000\left(1 + \frac{0.05}{4}\right)^{4t}$ (This is the compound interest formula we will study in Example 6.)
3. I am fencing a rectangular area whose perimeter is 100 ft. Find the area as a function of the width x.
   Model: Take y to be the length, so the perimeter is
   $$100 = x + y + x + y = 2(x + y).$$
   This gives
   $$x + y = 50.$$
   Thus the length is $y = 50 - x$, and the area is
   $$A = xy = x(50 - x).$$
4. You work 8 hours a day Monday through Friday, 5 hours on Saturday, and have Sunday off. Model the number of hours you work as a function of the day of the week n, with $n = 1$ being Sunday.
   Model: Take $f(n)$ to be the number of hours you work on the nth day of the week, so
   $$f(n) = \begin{cases} 0 & \text{if } n = 1 \\ 8 & \text{if } 2 \le n \le 6 \\ 5 & \text{if } n = 7 \end{cases}.$$
   Note that the domain of f is $\{1, 2, 3, 4, 5, 6, 7\}$—a discrete set rather than a continuous interval of the real line.

5. The function

$$f(x) = 4x^2 + 16x + 2 \text{ million iPods sold } (x = \text{years since } 2003)$$

in Example 1 of Section 1.1 is a model of iPod sales.

6. The function

$$n(t) = \begin{cases} 4t & \text{if } 0 \le t \le 3 \\ 12 + 23(t-3)^{2.7} & \text{if } 3 < t \le 6 \end{cases} \text{ million members}$$

(t = years since January 2004) in Example 2 of Section 1.1 is a model of Facebook membership.

Types of Models

Quick Examples 1–4 are **analytical models**, obtained by analyzing the situation being modeled, whereas Quick Examples 5 and 6 are **curve-fitting models**, obtained by finding mathematical formulas that approximate observed data. All the models except for Quick Example 4 are **continuous models**, defined by functions whose domains are intervals of the real line, whereas Quick Example 4 is a **discrete model** as its domain is a discrete set, as mentioned above. Discrete models are used extensively in probability and statistics.

Cost, Revenue, and Profit Models

EXAMPLE 1 Modeling Cost: Cost Function

As of August 2010, Yellow Cab Chicago's rates amounted to \$2.05 on entering the cab plus \$1.80 for each mile.[5]

a. Find the cost C of an x-mile trip.

b. Use your answer to calculate the cost of a 40-mile trip.

c. What is the cost of the second mile? What is the cost of the tenth mile?

d. Graph C as a function of x.

Solution

a. We are being asked to find how the cost C depends on the length x of the trip, or to find C as a function of x. Here is the cost in a few cases:

Cost of a 1-mile trip: $C = 1.80(1) + 2.05 = 3.85$ — 1 mile at \$1.80 per mile plus \$2.05

Cost of a 2-mile trip: $C = 1.80(2) + 2.05 = 5.65$ — 2 miles at \$1.80 per mile plus \$2.05

Cost of a 3-mile trip: $C = 1.80(3) + 2.05 = 7.45$ — 3 miles at \$1.80 per mile plus \$2.05

Do you see the pattern? The cost of an x-mile trip is given by the linear function

$$C(x) = 1.80x + 2.05.$$

[5]According to their Web site at www.yellowcabchicago.com.

Notice that the cost function is a sum of two terms: The **variable cost** $1.80x$, which depends on x, and the **fixed cost** 2.05, which is independent of x:

$$\text{Cost} = \text{Variable Cost} + \text{Fixed Cost}.$$

The quantity 1.80 by itself is the incremental cost per mile; you might recognize it as the *slope* of the given linear function. In this context we call 1.80 the **marginal cost**. You might recognize the fixed cost 2.05 as the *C-intercept* of the given linear function.

b. We can use the formula for the cost function to calculate the cost of a 40-mile trip as

$$C(40) = 1.80(40) + 2.05 = \$74.05.$$

c. To calculate the cost of the second mile, we *could* proceed as follows:

Find the cost of a 1-mile trip: $C(1) = 1.80(1) + 2.05 = \$3.85$.

Find the cost of a 2-mile trip: $C(2) = 1.80(2) + 2.05 = \$5.65$.

Therefore, the cost of the second mile is $\$5.65 - \$3.85 = \$1.80$.

But notice that this is just the marginal cost. In fact, the marginal cost is the cost of each additional mile, so we could have done this more simply:

Cost of second mile = Cost of tenth mile = Marginal cost = \$1.80.

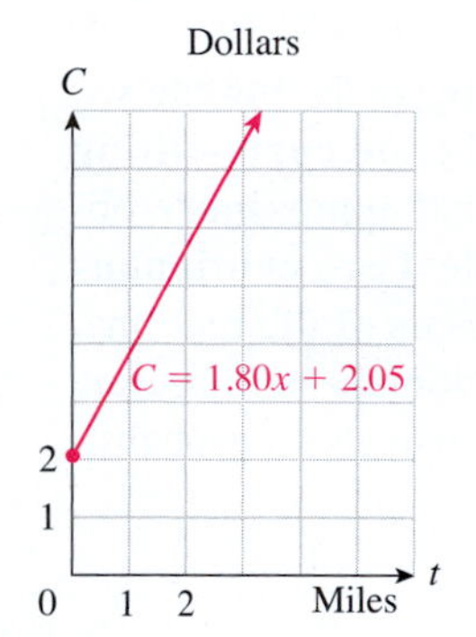

Figure 7

d. Figure 7 shows the graph of the cost function, which we can interpret as a *cost vs. miles* graph. The fixed cost is the starting height on the left, while the marginal cost is the slope of the line: It rises 1.80 units per unit of x. (See Section 1.3 for a discussion of properties of straight lines.)

Before we go on... The cost function in Example 1 is an example of an *analytical model:* We derived the form of the cost function from a knowledge of the cost per mile and the fixed cost.

As we discussed in Section 1.1, we can specify the cost function in Example 1 using equation notation:

$$C = 1.80x + 2.05. \quad \text{Equation notation}$$

Here, the independent variable is x, and the dependent variable is C. (This is the notation we have used in Figure 7. Remember that we will often switch between function and equation notation when it is convenient to do so.) ■

Here is a summary of some terms we used in Example 1, along with an introduction to some new terms:

Cost, Revenue, and Profit Functions

A **cost function** specifies the cost C as a function of the number of items x. Thus, $C(x)$ is the cost of x items, and has the form

$$\text{Cost} = \text{Variable cost} + \text{Fixed cost}$$

where the variable cost is a function of x and the fixed cost is a constant. A cost function of the form

$$C(x) = mx + b$$

is called a **linear cost function**; the variable cost is mx and the fixed cost is b. The slope m, the **marginal cost**, measures the incremental cost per item.

The **revenue**, or **net sales**, resulting from one or more business transactions is the total income received. If $R(x)$ is the revenue from selling x items at a price of m each, then R is the linear function $R(x) = mx$ and the selling price m can also be called the **marginal revenue**.

The **profit**, or **net income**, on the other hand, is what remains of the revenue when costs are subtracted. If the profit depends linearly on the number of items, the slope m is called the **marginal profit**. Profit, revenue, and cost are related by the following formula.

$$\text{Profit} = \text{Revenue} - \text{Cost}$$
$$P(x) = R(x) - C(x).^*$$

* We say that the profit function P is the **difference** between the revenue and cost functions, and express this fact as a formula about functions: $P = R - C$. (We will discuss this further when we talk about the algebra of functions at the end of this section.)

If the profit is negative, say $-\$500$, we refer to a **loss** (of \$500 in this case). To **break even** means to make neither a profit nor a loss. Thus, breakeven occurs when $P = 0$, or

$$R(x) = C(x). \qquad \text{Breakeven}$$

The **break-even point** is the number of items x at which breakeven occurs.

Quick Example

If the daily cost (including operating costs) of manufacturing x T-shirts is $C(x) = 8x + 100$, and the revenue obtained by selling x T-shirts is $R(x) = 10x$, then the daily profit resulting from the manufacture and sale of x T-shirts is

$$P(x) = R(x) - C(x) = 10x - (8x + 100) = 2x - 100.$$

Breakeven occurs when $P(x) = 0$, or $x = 50$.

EXAMPLE 2 Cost, Revenue, and Profit

The annual operating cost of *YSport* Fitness gym is estimated to be

$$C(x) = 100{,}000 + 160x - 0.2x^2 \text{ dollars} \qquad (0 \le x \le 400),$$

where x is the number of members. Annual revenue from membership averages \$800 per member. What is the variable cost? What is the fixed cost? What is the profit function? How many members must *YSport* have to make a profit? What will happen if it has fewer members? If it has more?

Solution The variable cost is the part of the cost function that depends on x:

$$\text{Variable cost} = 160x - 0.2x^2.$$

The fixed cost is the constant term:

$$\text{Fixed cost} = 100{,}000.$$

The annual revenue *YSport* obtains from a single member is \$800. So, if it has x members, it earns an annual revenue of

$$R(x) = 800x.$$

For the profit, we use the formula

$$\begin{aligned} P(x) &= R(x) - C(x) && \text{Formula for profit} \\ &= 800x - (100{,}000 + 160x - 0.2x^2) && \text{Substitute } R(x) \text{ and } C(x). \\ &= -100{,}000 + 640x + 0.2x^2. \end{aligned}$$

To make a profit, *YSport* needs to do better than break even, so let us find the break-even point: the value of x such that $P(x) = 0$. All we have to do is set $P(x) = 0$ and solve for x:

$$-100{,}000 + 640x + 0.2x^2 = 0.$$

Notice that we have a quadratic equation $ax^2 + bx + c = 0$ with $a = 0.2$, $b = 640$, and $c = -100{,}000$. Its solution is given by the quadratic formula:

$$\begin{aligned} x = \frac{-b \pm \sqrt{b^2 - 4ac}}{2a} &= \frac{-640 \pm \sqrt{640^2 + 4(0.2)(100{,}000)}}{2(0.2)} \\ &\approx \frac{-640 \pm 699.71}{2(0.2)} \\ &\approx 149.3 \text{ or } -3{,}349.3. \end{aligned}$$

using Technology

Excel has a feature called "Goal Seek," which can be used to find the point of intersection of the cost and revenue graphs numerically rather than graphically. See the downloadable Excel tutorial for this section at the Website.

We reject the negative solution (as the domain is $[0, 400]$) and conclude that $x \approx 149.3$ members. To make a profit, should *YSport* have 149 members or 150 members? To decide, take a look at Figure 8, which shows two graphs: On the left we see the graph of revenue and cost, and on the right we see the graph of the profit function.

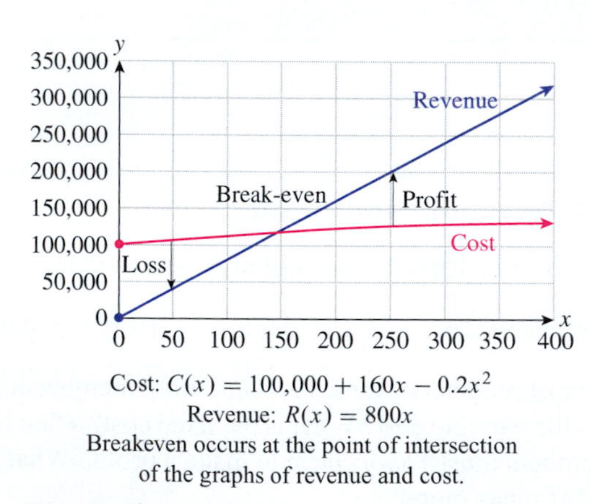

Cost: $C(x) = 100{,}000 + 160x - 0.2x^2$
Revenue: $R(x) = 800x$
Breakeven occurs at the point of intersection of the graphs of revenue and cost.

Profit: $P(x) = -100{,}000 + 640x + 0.2x^2$
Breakeven occurs when $P(x) = 0$

Figure 8

For values of x less than the break-even point of 149.3, $P(x)$ is negative, so the company will have a loss. For values of x greater than the break-even point, $P(x)$ is positive, so the company will make a profit. Thus, *YSport Fitness* needs at least 150 members to make a profit. (Note that we rounded 149.3 up to 150 in this case.)

Demand and Supply Models

The demand for a commodity usually goes down as its price goes up. It is traditional to use the letter q for the (quantity of) demand, as measured, for example, in sales. Consider the following example.

EXAMPLE 3 Demand: Private Schools

The demand for private schools in Michigan depends on the tuition cost and can be approximated by

$$q = 77.8p^{-0.11} \text{ thousand students} \qquad (200 \le p \le 2{,}200), \qquad \text{Demand curve}$$

where p is the net tuition cost in dollars.[6]

a. Use technology to plot the demand function.

b. What is the effect on demand if the tuition cost is increased from \$1,000 to \$1,500?

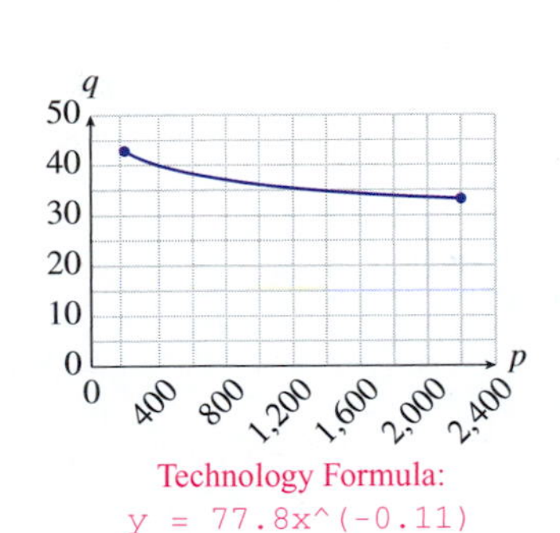

Technology Formula:
`y = 77.8x^(-0.11)`

Figure 9

Solution

a. The demand function is given by $q(p) = 77.8p^{-0.11}$. Its graph is known as a **demand curve** (Figure 9).

b. The demand at tuition costs of \$1,000 and \$1,500 is

$$q(1{,}000) = 77.8(1{,}000)^{-0.11} \approx 36.4 \text{ thousand students}$$
$$q(1{,}500) = 77.8(1{,}500)^{-0.11} \approx 34.8 \text{ thousand students.}$$

The change in demand is therefore

$$q(1{,}500) - q(1{,}000) \approx 34.8 - 36.4 = -1.6 \text{ thousand students.}$$

We have seen that a demand function gives the number of items consumers are willing to buy at a given price, and a higher price generally results in a lower demand. However, as the price rises, suppliers will be more inclined to produce these items (as opposed to spending their time and money on other products), so supply will generally rise. A **supply function** gives q, the number of items suppliers are willing to make available for sale*, as a function of p, the price per item.

* Although a bit confusing at first, it is traditional to use the same letter q for the quantity of supply and the quantity of demand, particularly when we want to compare them, as in the next example.

Demand, Supply, and Equilibrium Price

A **demand equation** or **demand function** expresses demand q (the number of items demanded) as a function of the unit price p (the price per item). A **supply equation** or **supply function** expresses supply q (the number of items a supplier is willing to make available) as a function of the unit price p (the price per item). It is usually the case that demand decreases and supply increases as the unit price increases.

[6] The tuition cost is net cost: tuition minus tax credit. The model is based on data in "The Universal Tuition Tax Credit: A Proposal to Advance Personal Choice in Education," Patrick L. Anderson, Richard McLellan, J.D., Joseph P. Overton, J.D., Gary Wolfram, Ph.D., Mackinac Center for Public Policy, www.mackinac.org/

Demand and supply are said to be in **equilibrium** when demand equals supply. The corresponding values of p and q are called the **equilibrium price** and **equilibrium demand**. To find the equilibrium price, determine the unit price p where the demand and supply curves cross (sometimes we can determine this value analytically by setting demand equal to supply and solving for p). To find the equilibrium demand, evaluate the demand (or supply) function at the equilibrium price.

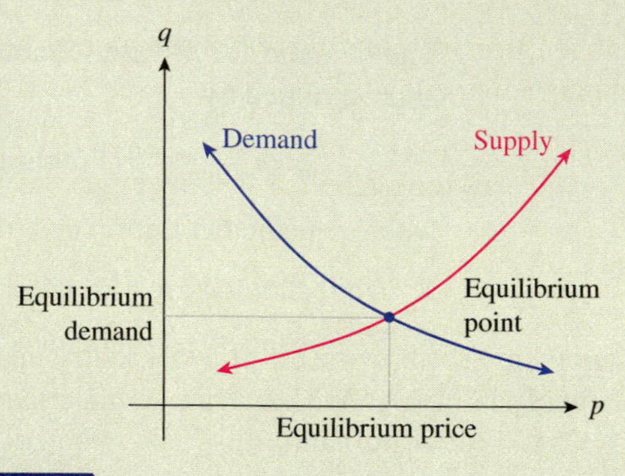

Quick Example

If the demand for your exclusive T-shirts is $q = -20p + 800$ shirts sold per day and the supply is $q = 10p - 100$ shirts per day, then the equilibrium point is obtained when demand = supply:

$$-20p + 800 = 10p - 100$$
$$30p = 900, \text{ giving } p = \$30.$$

The equilibrium price is therefore \$30 and the equilibrium demand is $q = -20(30) + 800 = 200$ shirts per day. What happens at prices other than the equilibrium price is discussed in Example 4.

Note In economics it is customary to plot the independent variable (price) on the vertical axis and the dependent variable (demand or supply) on the horizontal axis, but in this book we follow the usual mathematical convention for all graphs and plot the independent variable on the horizontal axis.

EXAMPLE 4 Demand, Supply, and Equilibrium Price

Continuing with Example 3, suppose that private school institutions are willing to create private schools to accommodate

$$q = 30.4 + 0.006p \text{ thousand students} \qquad (200 \le p \le 2{,}200) \qquad \text{Supply curve}$$

who pay a net tuition of p dollars.

a. Graph the demand curve of Example 3 and the supply curve given here on the same set of axes. Use your graph to estimate, to the nearest \$100, the tuition at which the demand equals the supply. Approximately how many students will be accommodated at that price, known as the **equilibrium price**?

b. What happens if the tuition is higher than the equilibrium price? What happens if it is lower?

c. Estimate the shortage or surplus of openings at private schools if tuition is set at \$1,200.

Solution

a. Figure 10 shows the graphs of demand $q = 77.8p^{-0.11}$ and supply $q = 30.4 + 0.006p$. (See the margin note for a brief description of how to plot them.)

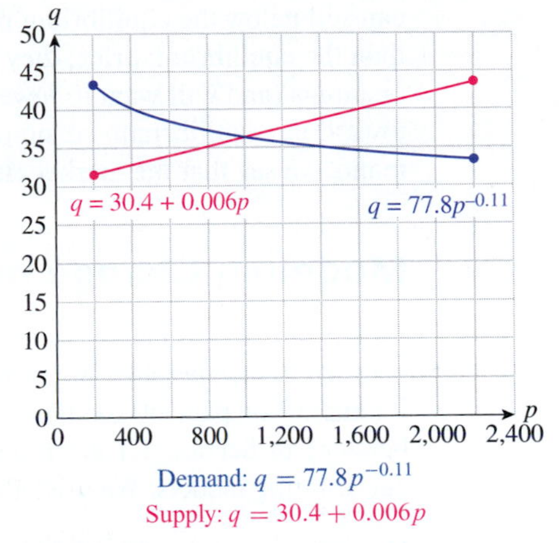

Demand: $q = 77.8p^{-0.11}$
Supply: $q = 30.4 + 0.006p$

Figure 10

The lines cross close to $p = \$1,000$, so we conclude that demand = supply when $p \approx \$1,000$ (to the nearest \$100). This is the (approximate) equilibrium tuition price. At that price, we can estimate the demand or supply at around

Demand: $q = 77.8(1,000)^{-0.11} \approx 36.4$

Supply: $q = 30.4 + 0.006(1,000) = 36.4$ Demand = Supply at equilibrium

or 36,400 students.

b. Take a look at Figure 11, which shows what happens if schools charge more or less than the equilibrium price.

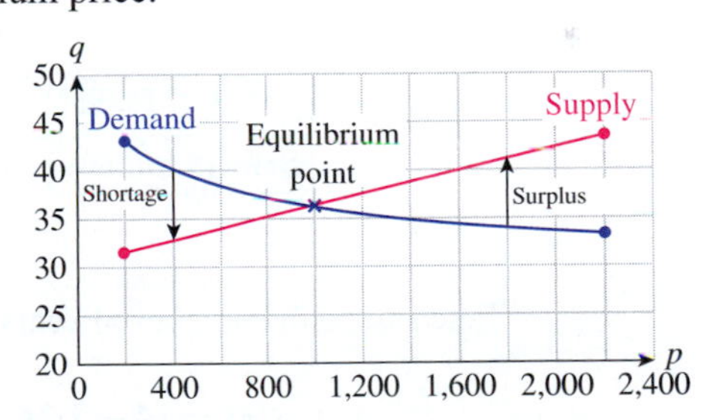

Figure 11

If tuition is, say, \$1,800, then the supply will be larger than demand and there will be a surplus of available openings at private schools. Similarly, if tuition is less—say \$400—then the supply will be less than the demand, and there will be a shortage of available openings.

c. The discussion in part (b) shows that if tuition is set at \$1,200 there will be a surplus of available openings. To estimate that number, we calculate the projected demand and supply when $p = \$1,200$:

Demand: $q = 77.8(1,200)^{-0.11} \approx 35.7$ thousand seats

Supply: $q = 30.4 + 0.006(1,200) = 37.6$ thousand seats

Surplus = Supply − Demand $\approx 37.6 - 35.7 = 1.9$ thousand seats.

So, there would be a surplus of around 1,900 available seats.

 **using Technology**

See the Technology Guides at the end of the chapter for detailed instructions on how to obtain the table of values and graph in Example 4 using a TI-83/84 Plus or Excel. Here is an outline:

TI-83/84 Plus

Graphs:
`Y1=77.8*X^(-0.11)`
`Y2=30.4+0.006*X`
[2ND] [TABLE] Graph:
Xmin = 200, Xmax = 2200;
[ZOOM] [0] [More details on page 91.]

Spreadsheet

Headings p, Demand, Supply in A1–C1; p-values 200, 300, ..., 2200 in A2-A22.
`=77.8*A2^(-0.11)` in B2
`=30.4+0.006*A2` in C2
Copy down through C22.
Highlight A1–C22; insert Scatter chart. [More details on page 96.]

 **Website**

www.WanerMath.com
Go to the Function Evaluator and Grapher under Online Utilities, and enter
`77.8*x^(-0.11)` for y_1 and
`30.4+0.006*x` for y_2.
Graph: Set Xmin = 200, Xmax = 2200, and press "Plot Graphs".

➡ **Before we go on...** We just saw in Example 4 that if tuition is less than the equilibrium price there will be a shortage. If schools were to raise their tuition toward the equilibrium, they would create and fill more openings and increase revenue, because it is the supply equation—and not the demand equation—that determines what one can sell below the equilibrium price. On the other hand, if they were to charge more than the equilibrium price, they will be left with a possibly costly surplus of unused openings (and will want to lower tuition to reduce the surplus). Prices tend to move toward the equilibrium, so supply tends to equal demand. When supply equals demand, we say that the market **clears.** ■

Modeling Change over Time

Things around us change with time. Thus, there are many quantities, such as your income or the temperature in Honolulu, that are natural to think of as functions of time. Example 1 on page 40 (on iPod sales) and Example 2 on page 43 (on Facebook membership) in Section 1.1 are models of change over time. Both of those models are curve-fitting models: We used algebraic functions to approximate observed data.

Note We usually use the independent variable t to denote time (in seconds, hours, days, years, etc.). If a quantity q changes with time, then we can regard q as a function of t. ■

In the next example we are asked to select from among several curve-fitting models for given data.

T EXAMPLE 5 Model Selection: Sales

The following table shows annual sales, in billions of dollars, by Nike from 2005 through 2010:[7]

Year	2005	2006	2007	2008	2009	2010
Sales ($ billion)	13.5	15	16.5	18.5	19	19

Take t to be the number of years since 2005, and consider the following four models:

(1) $s(t) = 14 + 1.2t$ — Linear model

(2) $s(t) = 13 + 2.2t - 0.2t^2$ — Quadratic model

(3) $s(t) = 14(1.07^t)$ — Exponential model

(4) $s(t) = \dfrac{19.5}{1 + 0.48(1.8^{-t})}$ — Logistic model

a. Which models fit the data significantly better than the rest?

b. Of the models you selected in part (a), which gives the most reasonable prediction for 2013?

[7]Figures are rounded. Source: http://invest.nike.com.

Solution

a. The following table shows the original data together with the values, rounded to the nearest 0.5, for all four models:

t	0	1	2	3	4	5
Sales ($billion)	13.5	15	16.5	18.5	19	19
Linear: $s(t) = 14 + 1.2t$ Technology: `14+1.2*x`	14	15	16.5	17.5	19	20
Quadratic: $s(t) = 13 + 2.2t - 0.2t^2$ Technology: `13+2.2*x-0.2*x^2`	13	15	16.5	18	18.5	19
Exponential: $s(t) = 14(1.07^t)$ Technology: `14*1.07^x`	14	15	16	17	18.5	19.5
Logistic: $s(t) = \dfrac{19.5}{1 + 0.48(1.8^{-t})}$ Technology: `19.5/(1+0.48*1.8^(-x))`	13	15.5	17	18	18.5	19

Notice that all the models give values that seem reasonably close to the actual sales values. However, the quadratic and logistic curves seem to model their behavior more accurately than the others (see Figure 12).

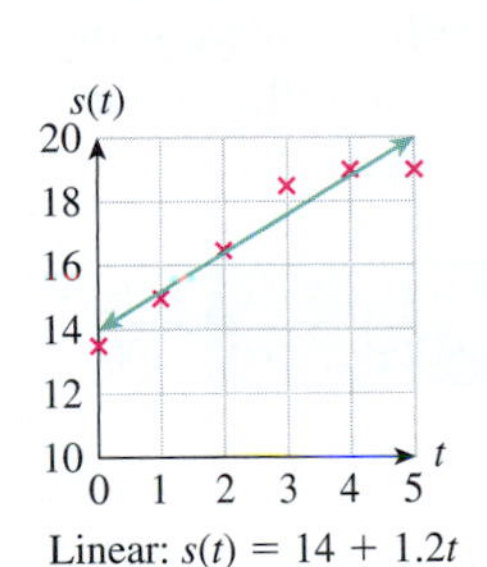

Linear: $s(t) = 14 + 1.2t$

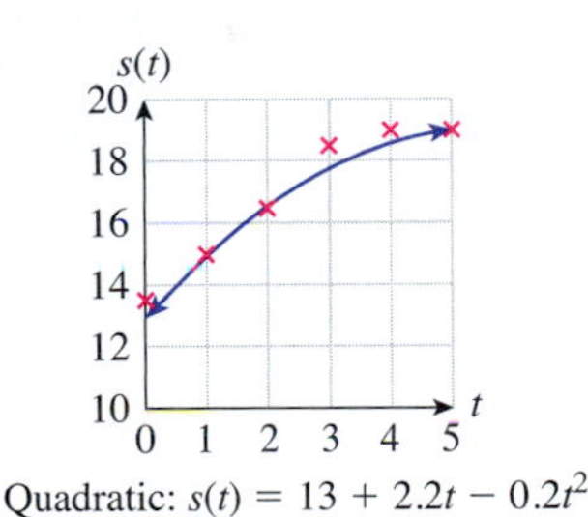

Quadratic: $s(t) = 13 + 2.2t - 0.2t^2$

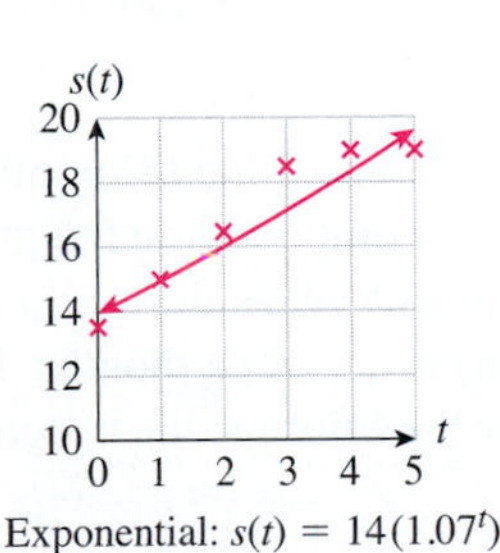

Exponential: $s(t) = 14(1.07^t)$

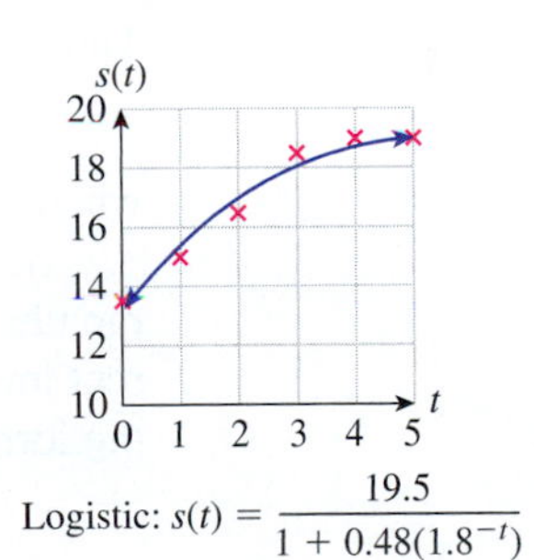

Logistic: $s(t) = \dfrac{19.5}{1 + 0.48(1.8^{-t})}$

Figure 12

We therefore conclude that the quadratic and logistic models fit the data significantly better than the others.

b. Although the quadratic and logistic models both appear to fit the data well, they do not both extrapolate to give reasonable predictions for 2013:

$$\text{Quadratic Model: } s(8) = 13 + 2.2(8) - 0.2(8)^2 = 17.8$$

$$\text{Logistic Model: } s(8) = \frac{19.5}{1 + 0.48(1.8^{-8})} \approx 19.4.$$

using Technology

See the Technology Guides at the end of the chapter for detailed instructions on how to obtain the table and graphs in Example 5 using a TI-83/84 Plus or Excel. Here is an outline:

TI-83/84 Plus
[STAT] EDIT; enter the values of t in L_1, and $r(t)$ in L_2.
Plotting the points: [ZOOM] [9]
Adding a curve:
Turn on Plot1 in Y= screen. Then (for the second curve)
`Y1=13+2.2*X-0.2*X^2`
Press [GRAPH] [More details on page 91.]

Spreadsheet
Table of values: Headings t, Sales, $s(t)$ in A1–C1;
t-values in A2-A7; sales values in B2–B7. Formula in C2:
`=13+2.2*A2-0.2*A2^2`
(second model)
Copy down through C7
Graph: Highlight A1–C7; insert Scatter chart. [More details on page 97.]

Website
www.WanerMath.com
In the Function Evaluator and Grapher utility, enter the data and model(s) as shown below. Set xMin = 0, xMax = 5 and press "Plot Graphs".

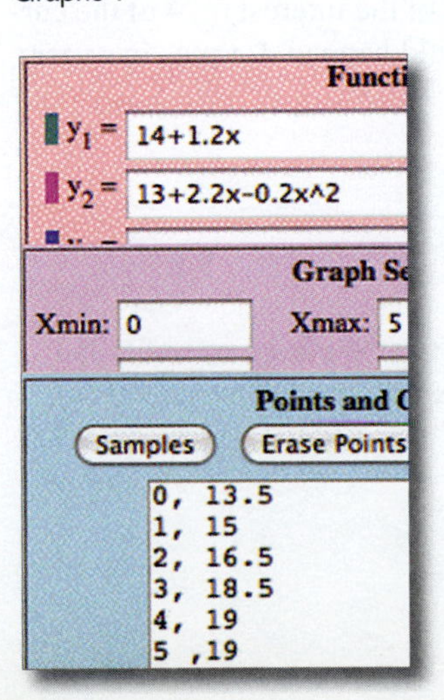

Notice that the quadratic model predicts a significant *decline* in sales whereas the logistic model predicts a more reasonable modest increase. This discrepancy can be seen quite dramatically in Figure 13.

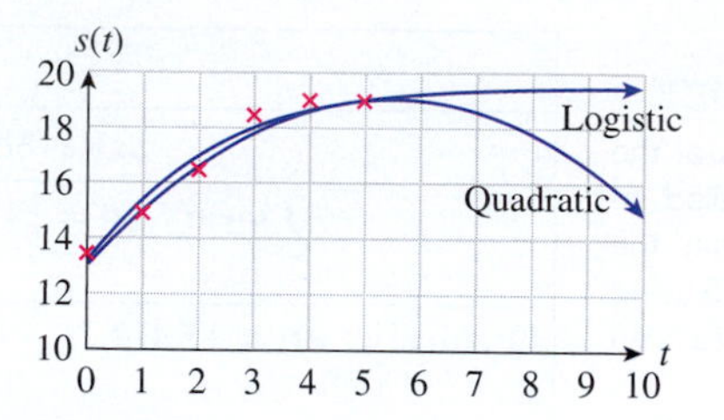

Figure 13

We now derive an analytical model of change over time based on the idea of **compound interest**. Suppose you invest \$500 (the **present value**) in an investment account with an annual yield of 15%, and the interest is reinvested at the end of every year (we say that the interest is **compounded** or **reinvested** once a year). Let t represent the number of years since you made the initial \$500 investment. Each year, the investment is worth 115% (or 1.15 times) of its value the previous year. The **future value** A of your investment changes over time t, so we think of A as a function of t. The following table illustrates how we can calculate the future value for several values of t:

t	0	1	2	3
Future Value $A(t)$	500	575	661.25	760.44
	A	$500(1.15)$	$500(1.15)^2$	$500(1.15)^3$

×1.15 ×1.15 ×1.15

Thus, $A(t) = 500(1.15)^t$. A traditional way to write this formula is

$$A(t) = P(1 + r)^t,$$

where P is the present value ($P = 500$) and r is the annual interest rate ($r = 0.15$).

If, instead of compounding the interest once a year, we compound it every three months (four times a year), we would earn one quarter of the interest ($r/4$ of the current investment) every three months. Because this would happen $4t$ times in t years, the formula for the future value becomes

$$A(t) = P\left(1 + \frac{r}{4}\right)^{4t}.$$

Compound Interest

If an amount (**present value**) P is invested for t years at an annual rate of r, and if the interest is compounded (reinvested) m times per year, then the **future value** A is

$$A(t) = P\left(1 + \frac{r}{m}\right)^{mt}.$$

A special case is **interest compounded once a year:**

$$A(t) = P(1 + r)^t.$$

Quick Example

If \$2,000 is invested for two and a half years in a mutual fund with an annual yield of 12.6% and the earnings are reinvested each month, then $P = 2{,}000$, $r = 0.126$, $m = 12$, and $t = 2.5$, which gives

$$A(2.5) = 2{,}000\left(1 + \frac{0.126}{12}\right)^{12\times 2.5} \qquad \texttt{2000*(1+0.126/12)^(12*2.5)}$$

$$= 2{,}000(1.0105)^{30} = \$2{,}736.02.$$

EXAMPLE 6 Compound Interest: Investments

Consider the scenario in the preceding Quick Example: You invest \$2,000 in a mutual fund with an annual yield of 12.6% and the interest is reinvested each month.

a. Find the associated exponential model.

b. T Use a table of values to estimate the year during which the value of your investment reaches \$5,000.

c. Use a graph to confirm your answer in part (b).

Solution

a. Apply the formula

$$A(t) = P\left(1 + \frac{r}{m}\right)^{mt}$$

with $P = 2{,}000$, $r = 0.126$, and $m = 12$. We get

$$A(t) = 2{,}000\left(1 + \frac{0.126}{12}\right)^{12t}$$

$$= 2{,}000(1.0105)^{12t}. \qquad \texttt{2000*(1+0.126/12)^(12*t)}$$

This is the exponential model. (What would happen if we left out the last set of parentheses in the technology formula?)

b. We need to find the value of t for which $A(t) = \$5{,}000$, so we need to solve the equation

$$5{,}000 = 2{,}000(1.0105)^{12t}.$$

Logarithms can be used to solve this equation algebraically, but we can answer the question numerically using a graphing calculator, a spreadsheet, or the Function Evaluator and Grapher utility at the Website. Just enter the model and compute the balance at the end of several years. Here are examples of tables obtained using three forms of technology:

 **using Technology**

TI-83/84 Plus

`Y1=2000(1+0.126/12)^(12X)`

[2ND] [TABLE]

Spreadsheet

Headings t and A in A1–B1; t-values 0–11 in A2–A13.

`=2000*(1+0.126/12)^(12*A2)`

in B2; copy down through B13.

 **Website**

www.WanerMath.com

Go to the Function Evaluator and Grapher under Online Utilities, and enter

`2000(1+0.126/12)^(12x)`

for y_1. Scroll down to the Evaluator, enter the values 0–11 under x-values and press "Evaluate."

X	Y_1
5	3742.9
6	4242.7
7	4809.3
8	5451.5
9	6179.5
10	7004.7
11	7940.1

Y_1=5451.50618802

TI-83/84 Plus

	A	B
1	t	A
2	0	$ 2,000.00
3	1	$ 2,267.07
4	2	$ 2,569.81
5	3	$ 2,912.98
6	4	$ 3,301.97
7	5	$ 3,742.91
8	6	$ 4,242.72
9	7	$ 4,809.29
10	8	$ 5,451.51
11	9	$ 6,179.49

Excel

x-Values	y_1-Values
3	2912.98
4	3301.97
5	3742.91
6	4242.72
7	4809.29
8	5451.51

Website

Because the balance first exceeds $5,000 at $t = 8$ (the end of year 8), your investment has reached $5,000 during year 8.

c. Figure 14 shows the graph of $A(t) = 2{,}000(1.0105)^{12t}$ together with the horizontal line $y = 5{,}000$. The graphs cross between $t = 7$ and $t = 8$, confirming that year 8 is the first year during which the value of the investment reaches $5,000.

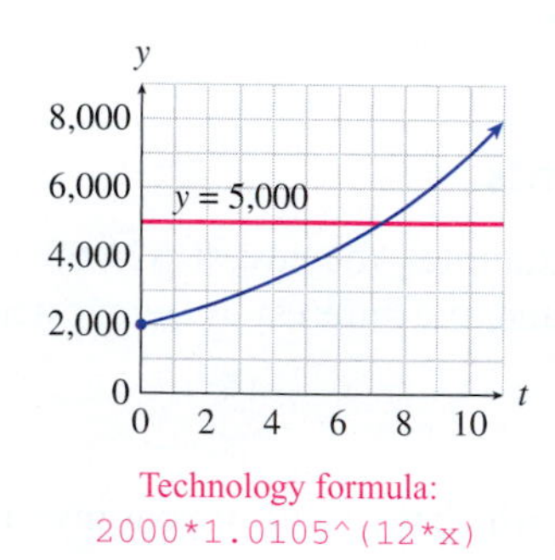

Technology formula:
`2000*1.0105^(12*x)`

Figure 14

The compound interest examples we saw above are instances of **exponential growth:** a quantity whose magnitude is an increasing exponential function of time. The decay of unstable radioactive isotopes provides instances of **exponential decay:** a quantity whose magnitude is a *decreasing* exponential function of time. For example, carbon 14, an unstable isotope of carbon, decays exponentially to nitrogen. Because carbon 14 decay is extremely slow, it has important applications in the dating of fossils.

EXAMPLE 7 Exponential Decay: Carbon Dating

The amount of carbon 14 remaining in a sample that originally contained A grams is approximately

$$C(t) = A(0.999879)^t,$$

where t is time in years.

a. What percentage of the original amount remains after one year? After two years?

b. Graph the function C for a sample originally containing 50 g of carbon 14, and use your graph to estimate how long, to the nearest 1,000 years, it takes for half the original carbon 14 to decay.

c. A fossilized plant unearthed in an archaeological dig contains 0.50 g of carbon 14 and is known to be 50,000 years old. How much carbon 14 did the plant originally contain?

Solution

Notice that the given model is exponential as it has the form $f(t) = Ab^t$. (See page 47.)

a. At the start of the first year, $t = 0$, so there are

$$C(0) = A(0.999879)^0 = A \text{ grams.}$$

At the end of the first year, $t = 1$, so there are

$$C(1) = A(0.999879)^1 = 0.999879A \text{ grams;}$$

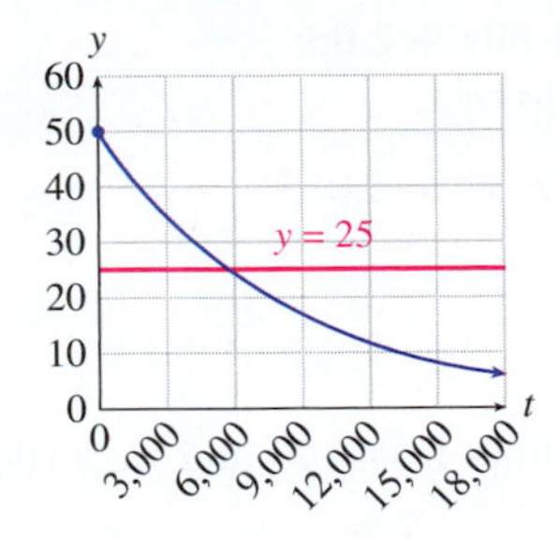

Technology formula:
`50*0.999879^x`

Figure 15

that is, 99.9879% of the original amount remains. After the second year, the amount remaining is

$$C(2) = A(0.999879)^2 \approx 0.999758A \text{ grams},$$

or about 99.9758% of the original sample.

b. For a sample originally containing 50 g of carbon 14, $A = 50$, so $C(t) = 50(0.999879)^t$. Its graph is shown in Figure 15. We have also plotted the line $y = 25$ on the same graph. The graphs intersect at the point where the original sample has decayed to 25 g: about $t = 6{,}000$ years.

c. We are given the following information: $C = 0.50$, $A =$ the unknown, and $t = 50{,}000$. Substituting gives

$$0.50 = A(0.999879)^{50{,}000}.$$

Solving for A gives

$$A = \frac{0.5}{0.999879^{50{,}000}} \approx 212 \text{ grams}.$$

Thus, the plant originally contained 212 g of carbon 14.

Before we go on...

The formula we used for A in Example 7(c) has the form

$$A(t) = \frac{C}{0.999879^t},$$

which gives the original amount of carbon 14 t years ago in terms of the amount C that is left now. A similar formula can be used in finance to find the present value, given the future value. ■

Algebra of Functions

If you look back at some of the functions considered in this section, you will notice that we frequently constructed them by combining simpler or previously constructed functions. For instance:

Quick Example 3 on page 48:	Area = Width × Length: $A(x) = x(50 - x)$
Example 1:	Cost = Variable Cost + Fixed Cost: $C(x) = 1.80x + 2.05$
Quick Example on page 51:	Profit = Revenue − Cost: $P(x) = 10x - (8x + 100)$.

Let us look a little more deeply at each of the above examples:

Area Example: $A(x) = \text{Width} \times \text{Length} = x(50 - x)$:
Think of the width and length as separate functions of x:

$$\text{Width: } W(x) = x; \qquad \text{Length: } L(x) = 50 - x$$

so that

$$A(x) = W(x)L(x). \qquad \text{Area = Width × Length}$$

We say that the area function A is the **product of the functions** W and L, and we write

$$A = WL. \qquad A \text{ is the product of the functions } W \text{ and } L.$$

To calculate $A(x)$, we multiply $W(x)$ by $L(x)$.

Cost Example: $C(x) = \text{Variable Cost} + \text{Fixed Cost} = 1.80x + 2.05$:
Think of the variable and fixed costs as separate functions of x:

* F is called a constant function as its value, 2.05, is the same for every value of x.

$$\text{Variable Cost: } V(x) = 1.80x; \qquad \text{Fixed Cost: } F(x) = 2.05^{*}$$

so that

$$C(x) = V(x) + F(x). \qquad \text{Cost} = \text{Variable Cost} + \text{Fixed Cost}$$

We say that the cost function C is the **sum of the functions V and F**, and we write

$$C = V + F. \qquad C \text{ is the sum of the functions } V \text{ and } F.$$

To calculate $C(x)$, we add $V(x)$ to $F(x)$.

Profit Example: $P(x) = \text{Revenue} - \text{Cost} = 10x - (8x + 100)$:
Think of the revenue and cost as separate functions of x:

$$\text{Revenue: } R(x) = 10x; \qquad \text{Cost: } C(x) = 8x + 100$$

so that

$$P(x) = R(x) - C(x). \qquad \text{Profit} = \text{Revenue} - \text{Cost}$$

We say that the profit function P is the **difference between the functions R and C**, and we write

$$P = R - C. \qquad P \text{ is the difference of the functions } R \text{ and } C.$$

To calculate $P(x)$, we subtract $C(x)$ from $R(x)$.

Algebra of Functions

If f and g are real-valued functions of the real variable x, then we define their **sum s, difference d, product p,** and **quotient q** as follows:

$s = f + g$ is the function specified by $s(x) = f(x) + g(x)$.

$d = f - g$ is the function specified by $d(x) = f(x) - g(x)$.

$p = fg$ is the function specified by $p(x) = f(x)g(x)$.

$q = \dfrac{f}{g}$ is the function specified by $q(x) = \dfrac{f(x)}{g(x)}$.

Also, if f is as above and c is a constant (real number), then we define the associated **constant multiple m of f** by

$m = cf$ is the function specified by $m(x) = cf(x)$.

Note on Domains

In order for any of the expressions $f(x) + g(x)$, $f(x) - g(x)$, $f(x)g(x)$, or $f(x)/g(x)$ to make sense, x must be simultaneously in the domains of both f and g. Further, for the quotient, the denominator $g(x)$ cannot be zero. Thus, we specify the domains of these functions as follows:

Domain of $f + g$, $f - g$, and fg: All real numbers x simultaneously in the domains of f and g

Domain of f/g: All real numbers x simultaneously in the domains of f and g such that $g(x) \neq 0$

Domain of cf: Same as the domain of f

Quick Examples

1. If $f(x) = x^2 - 1$ and $g(x) = \sqrt{x}$ with domain $[0, +\infty)$, then the sum s of f and g has domain $[0, +\infty)$ and is specified by $s(x) = f(x) + g(x) = x^2 - 1 + \sqrt{x}$.
2. If $f(x) = x^2 - 1$ and $c = 3$, then the associated constant multiple m of f is specified by $m(x) = 3f(x) = 3(x^2 - 1)$.
3. If there are $N = 1{,}000t$ Mars shuttle passengers in year t who pay a total cost of $C = 40{,}000 + 800t$ million dollars, then the cost per passenger is given by the quotient of the two functions,

$$\text{Cost per passenger} = q(t) = \frac{C(t)}{N(t)}$$
$$= \frac{40{,}000 + 800t}{1{,}000t} \text{ million dollars per passenger.}$$

The largest possible domain of C/N is $(0, +\infty)$, as the quotient is not defined if $t = 0$.

1.2 EXERCISES

Access end-of-section exercises online at **www.webassign.net** 

1.3 Linear Functions and Models

Linear functions are among the simplest functions and are perhaps the most useful of all mathematical functions.

Linear Function

A **linear function** is one that can be written in the form

$f(x) = mx + b$ Function form

or

$y = mx + b$ Equation form

Quick Example

$f(x) = 3x - 1$

$y = 3x - 1$

where m and b are fixed numbers. (The names m and b are traditional.*)

* Actually, c is sometimes used instead of b. As for m, there has even been some research into the question of its origin, but no one knows exactly why the letter m is used.

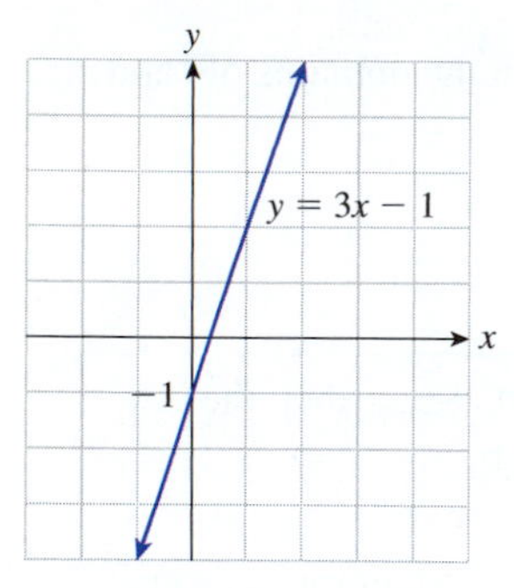

Figure 16

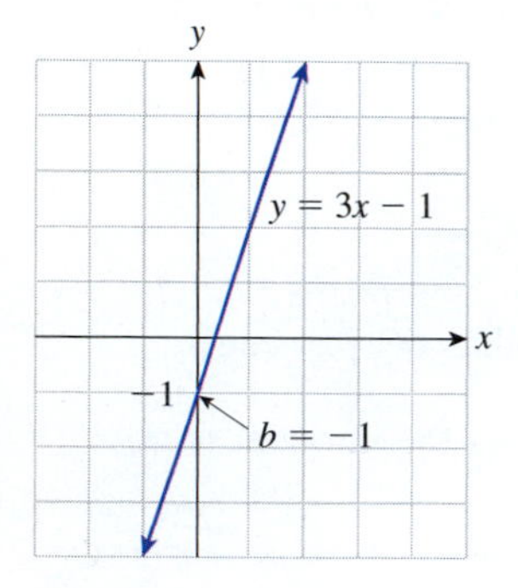

y-intercept $= b = -1$
Graphically, b is the y-intercept of the graph.

Figure 17

Linear Functions from the Numerical and Graphical Point of View

The following table shows values of $y = 3x - 1$ $(m = 3, b = -1)$ for some values of x:

x	−4	−3	−2	−1	0	1	2	3	4
y	−13	−10	−7	−4	−1	2	5	8	11

Its graph is shown in Figure 16.

Looking first at the table, notice that setting $x = 0$ gives $y = -1$, the value of b.

Numerically, b is the value of y when $x = 0$.

On the graph, the corresponding point $(0, -1)$ is the point where the graph crosses the y-axis, and we say that $b = -1$ is the ***y*-intercept** of the graph (Figure 17).

What about m? Looking once again at the table, notice that y increases by $m = 3$ units for every increase of 1 unit in x. This is caused by the term $3x$ in the formula: for every increase of 1 in x we get an increase of $3 \times 1 = 3$ in y.

Numerically, y increases by m units for every 1-unit increase of x.

Likewise, for every increase of 2 in x we get an increase of $3 \times 2 = 6$ in y. In general, if x increases by some amount, y will increase by three times that amount. We write:

$$\text{Change in } y = 3 \times \text{Change in } x.$$

The Change in a Quantity: Delta Notation

If a quantity q changes from q_1 to q_2, the **change in *q*** is just the difference:

$$\text{Change in } q = \text{Second value} - \text{First value}$$
$$= q_2 - q_1.$$

Mathematicians traditionally use Δ (delta, the Greek equivalent of the Roman letter D) to stand for change, and write the change in q as Δq.

$$\Delta q = \text{Change in } q = q_2 - q_1$$

Quick Examples

1. If x is changed from 1 to 3, we write

$$\Delta x = \text{Second value} - \text{First value} = 3 - 1 = 2.$$

2. Looking at our linear function, we see that, when x changes from 1 to 3, y changes from 2 to 8. So,

$$\Delta y = \text{Second value} - \text{First value} = 8 - 2 = 6.$$

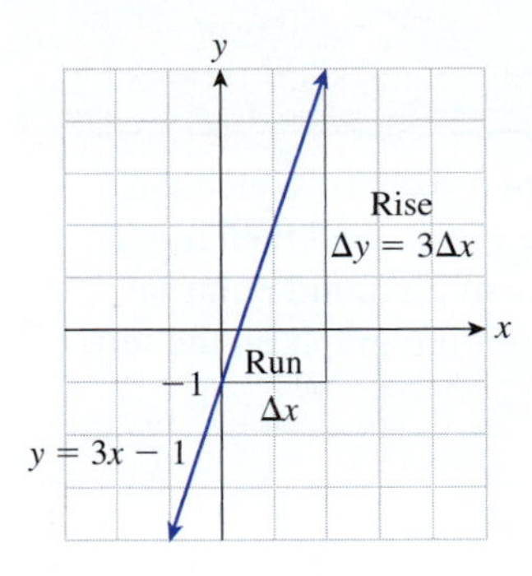

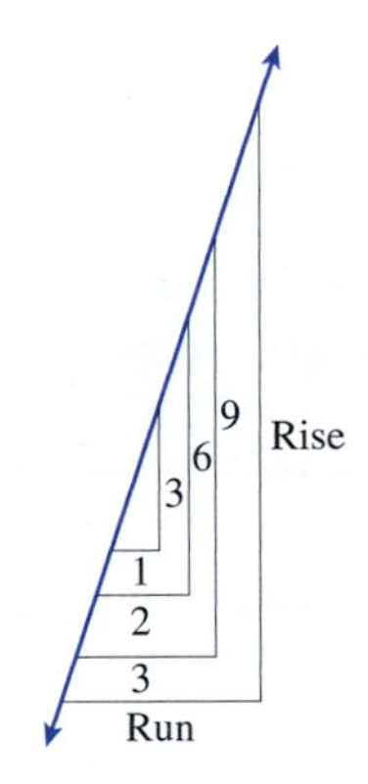

Slope $= m = 3$
Graphically, m is the slope of the graph.

Figure 18

Using delta notation, we can now write, for our linear function,

$$\Delta y = 3\Delta x \qquad \text{Change in } y = 3 \times \text{Change in } x.$$

or

$$\frac{\Delta y}{\Delta x} = 3.$$

Because the value of y increases by exactly 3 units for every increase of 1 unit in x, the graph is a straight line rising by 3 units for every 1 unit we go to the right. We say that we have a **rise** of 3 units for each **run** of 1 unit. Because the value of y changes by $\Delta y = 3\Delta x$ units for every change of Δx units in x, in general we have a rise of $\Delta y = 3\Delta x$ units for each run of Δx units (Figure 18). Thus, we have a rise of 6 for a run of 2, a rise of 9 for a run of 3, and so on. So, $m = 3$ is a measure of the steepness of the line; we call m the **slope of the line**:

$$\text{Slope} = m = \frac{\Delta y}{\Delta x} = \frac{\text{Rise}}{\text{Run}}.$$

In general (replace the number 3 by a general number m), we can say the following.

The Roles of *m* and *b* in the Linear Function *f*(*x*) = *mx* + *b*

Role of *m*

Numerically If $y = mx + b$, then y changes by m units for every 1-unit change in x. A change of Δx units in x results in a change of $\Delta y = m\Delta x$ units in y. Thus,

$$m = \frac{\Delta y}{\Delta x} = \frac{\text{Change in } y}{\text{Change in } x}.$$

Graphically m is the slope of the line $y = mx + b$:

$$m = \frac{\Delta y}{\Delta x} = \frac{\text{Rise}}{\text{Run}} = \text{Slope}.$$

For positive m, the graph rises m units for every 1-unit move to the right, and rises $\Delta y = m\Delta x$ units for every Δx units moved to the right. For negative m, the graph drops $|m|$ units for every 1-unit move to the right, and drops $|m|\Delta x$ units for every Δx units moved to the right.

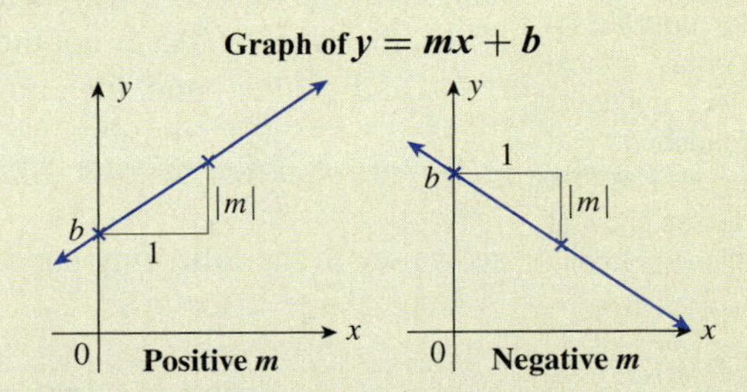

Role of *b*

Numerically When $x = 0$, $y = b$.

Graphically b is the y-intercept of the line $y = mx + b$.

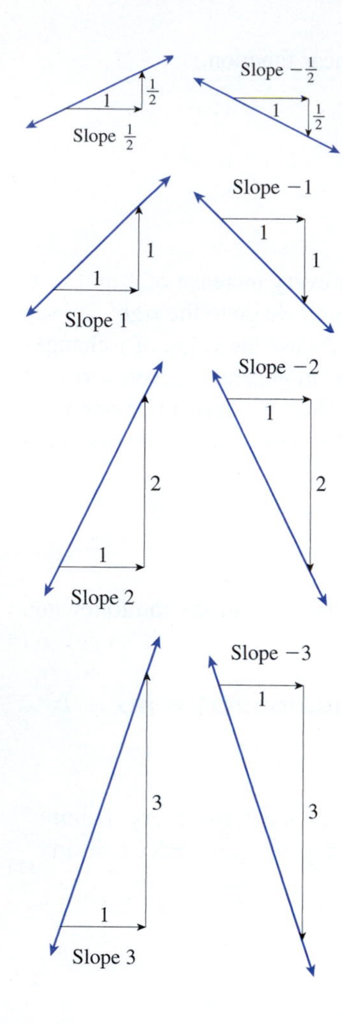

Figure 19

Quick Examples

1. $f(x) = 2x + 1$ has slope $m = 2$ and y-intercept $b = 1$. To sketch the graph, we start at the y-intercept $b = 1$ on the y-axis, and then move 1 unit to the right and up $m = 2$ units to arrive at a second point on the graph. Now connect the two points to obtain the graph on the left.

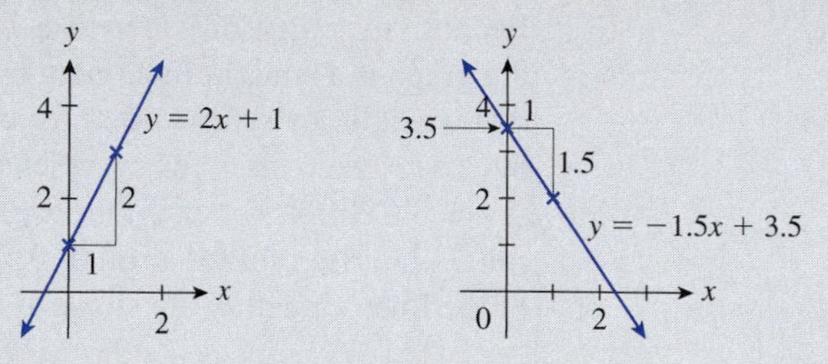

2. The line $y = -1.5x + 3.5$ has slope $m = -1.5$ and y-intercept $b = 3.5$. Because the slope is negative, the graph (above right) goes *down* 1.5 units for every 1 unit it moves to the right.

It helps to be able to picture what different slopes look like, as in Figure 19. Notice that the larger the absolute value of the slope, the steeper is the line.

EXAMPLE 1 Recognizing Linear Data Numerically and Graphically

Which of the following two tables gives the values of a linear function? What is the formula for that function?

x	0	2	4	6	8	10	12
$f(x)$	3	−1	−3	−6	−8	−13	−15

x	0	2	4	6	8	10	12
$g(x)$	3	−1	−5	−9	−13	−17	−21

Solution The function f cannot be linear: If it were, we would have $\Delta f = m\Delta x$ for some fixed number m. However, although the change in x between successive entries in the table is $\Delta x = 2$ each time, the change in f is not the same each time. Thus, the ratio $\Delta f/\Delta x$ is not the same for every successive pair of points.

On the other hand, the ratio $\Delta g/\Delta x$ is the same each time, namely,

$$\frac{\Delta g}{\Delta x} = \frac{-4}{2} = -2$$

as we see in the following table:

Δx		$2-0=2$	$4-2=2$	$6-4=2$	$8-6=2$	$10-8=2$	$12-10=2$
x	0	2	4	6	8	10	12
$g(x)$	3	−1	−5	−9	−13	−17	−21
Δg		$-1-3 = -4$	$-5-(-1) = -4$	$-9-(-5) = -4$	$-13-(-9) = -4$	$-17-(-13) = -4$	$-21-(-17) = -4$

using Technology

See the Technology Guides at the end of the chapter for detailed instructions on how to obtain a table with the successive quotients $m = \Delta y/\Delta x$ for the functions f and g in Example 1 using a TI-83/84 Plus or Excel. These tables show at a glance that f is not linear. Here is an outline:

TI-83/84 Plus

STAT EDIT; Enter values of x and $f(x)$ in lists L_1 and L_2.
Highlight the heading L_3 then enter the following formula (including the quotes)

```
"ΔList(L2)/ΔList(L1)"
```

[More details on page 92.]

Spreadsheet
Enter headings x, $f(x)$, Df/Dx in cells A1–C1, and the corresponding values from one of the tables in cells A2–B8. Enter

```
=(B3-B2)/(A3-A2)
```

in cell C2, and copy down through C8.
[More details on page 98.]

Thus, g is linear with slope $m = -2$. By the table, $g(0) = 3$, hence $b = 3$. Thus,

$$g(x) = -2x + 3. \qquad \text{Check that this formula gives the values in the table.}$$

If you graph the points in the tables defining f and g above, it becomes easy to see that g is linear and f is not; the points of g lie on a straight line (with slope -2), whereas the points of f do not lie on a straight line (Figure 20).

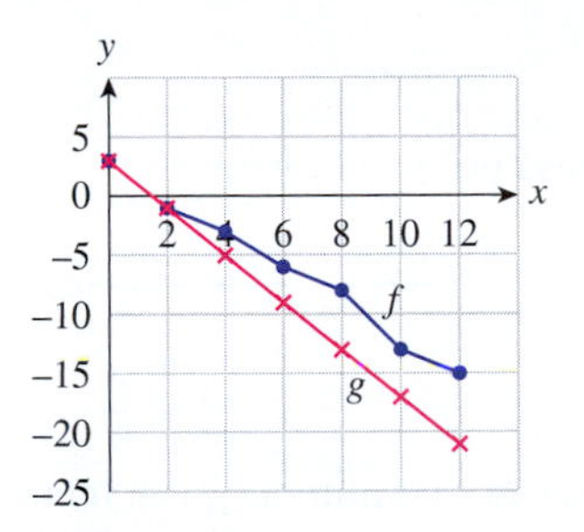

Figure 20

Finding a Linear Equation from Data

If we happen to know the slope and y-intercept of a line, writing down its equation is straightforward. For example, if we know that the slope is 3 and the y-intercept is -1, then the equation is $y = 3x - 1$. Sadly, the information we are given is seldom so convenient. For instance, we may know the slope and a point other than the y-intercept, two points on the line, or other information. We therefore need to know how to use the information we are given to obtain the slope and the intercept.

Computing the Slope

We can always determine the slope of a line if we are given two (or more) points on the line, because any two points—say (x_1, y_1) and (x_2, y_2)—determine the line, and hence its slope. To compute the slope when given two points, recall the formula

$$\text{Slope} = m = \frac{\text{Rise}}{\text{Run}} = \frac{\Delta y}{\Delta x}.$$

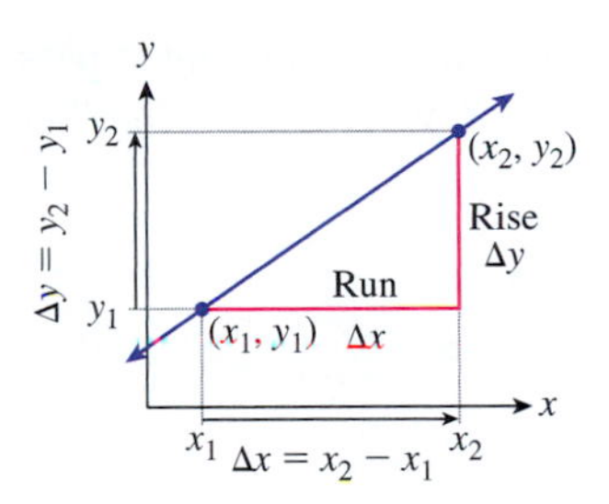

Figure 21

To find its slope, we need a run Δx and corresponding rise Δy. In Figure 21, we see that we can use $\Delta x = x_2 - x_1$, the change in the x-coordinate from the first point to the second, as our run, and $\Delta y = y_2 - y_1$, the change in the y-coordinate, as our rise. The resulting formula for computing the slope is given in the box.

Computing the Slope of a Line

We can compute the slope m of the line through the points (x_1, y_1) and (x_2, y_2) using

$$m = \frac{\Delta y}{\Delta x} = \frac{y_2 - y_1}{x_2 - x_1}.$$

Quick Examples

1. The slope of the line through $(x_1, y_1) = (1, 3)$ and $(x_2, y_2) = (5, 11)$ is

$$m = \frac{\Delta y}{\Delta x} = \frac{y_2 - y_1}{x_2 - x_1} = \frac{11 - 3}{5 - 1} = \frac{8}{4} = 2.$$

Notice that we can use the points in the reverse order: If we take $(x_1, y_1) = (5, 11)$ and $(x_2, y_2) = (1, 3)$, we obtain the same answer:

$$m = \frac{\Delta y}{\Delta x} = \frac{y_2 - y_1}{x_2 - x_1} = \frac{3 - 11}{1 - 5} = \frac{-8}{-4} = 2.$$

2. The slope of the line through $(x_1, y_1) = (1, 2)$ and $(x_2, y_2) = (2, 1)$ is

$$m = \frac{\Delta y}{\Delta x} = \frac{y_2 - y_1}{x_2 - x_1} = \frac{1 - 2}{2 - 1} = \frac{-1}{1} = -1.$$

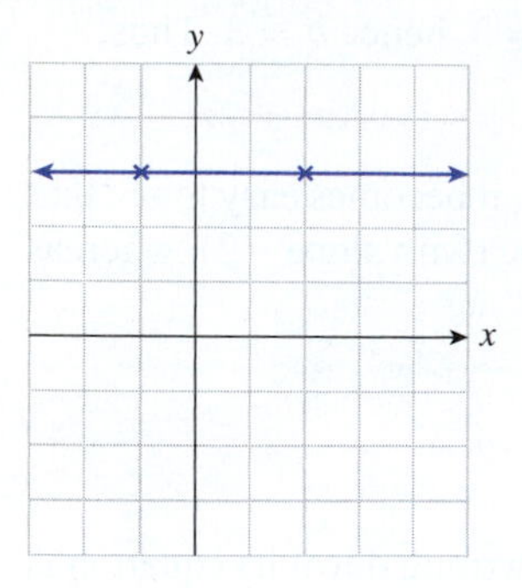

Figure 22

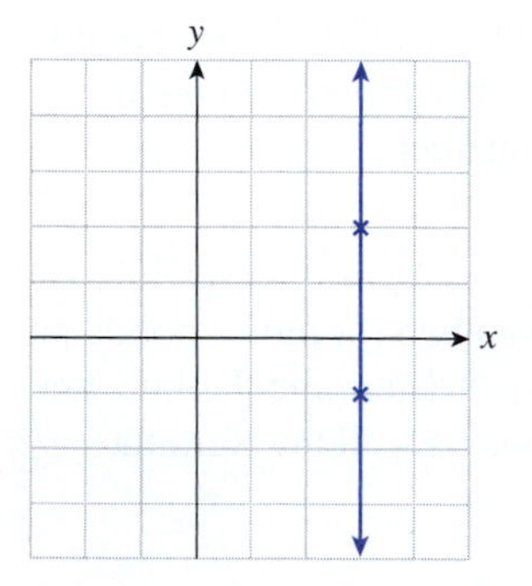

Vertical lines have undefined slope.

Figure 23

3. The slope of the line through (2, 3) and (−1, 3) is

$$m = \frac{\Delta y}{\Delta x} = \frac{y_2 - y_1}{x_2 - x_1} = \frac{3 - 3}{-1 - 2} = \frac{0}{-3} = 0.$$

A line of slope 0 has zero rise, so is a *horizontal* line, as shown in Figure 22.

4. The line through (3, 2) and (3, −1) has slope

$$m = \frac{\Delta y}{\Delta x} = \frac{y_2 - y_1}{x_2 - x_1} = \frac{-1 - 2}{3 - 3} = \frac{-3}{0},$$

which is undefined. The line passing through these points is *vertical*, as shown in Figure 23.

Computing the *y*-Intercept

Once we know the slope m of a line, and also the coordinates of a point (x_1, y_1), then we can calculate its y-intercept b as follows: The equation of the line must be

$$y = mx + b,$$

where b is as yet unknown. To determine b we use the fact that the line must pass through the point (x_1, y_1), so that (x_1, y_1) satisfies the equation $y = mx + b$. In other words,

$$y_1 = mx_1 + b.$$

Solving for b gives

$$b = y_1 - mx_1.$$

In summary:

Computing the *y*-Intercept of a Line

The y-intercept of the line passing through (x_1, y_1) with slope m is

$$b = y_1 - mx_1.$$

Quick Example

The line through (2, 3) with slope 4 has

$$b = y_1 - mx_1 = 3 - (4)(2) = -5.$$

Its equation is therefore

$$y = mx + b = 4x - 5.$$

EXAMPLE 2 Finding Linear Equations

Find equations for the following straight lines.

a. Through the points (1, 2) and (3, −1)

b. Through (2, −2) and parallel to the line $3x + 4y = 5$

c. Horizontal and through (−9, 5)

d. Vertical and through (−9, 5)

 **using Technology**

See the Technology Guides at the end of the chapter for detailed instructions on how to obtain the slope and intercept in Example 2(a) using a TI-83/84 Plus or a spreadsheet. Here is an outline:

TI-83/84 Plus

STAT EDIT; Enter values of x and y in lists L_1 and L_2.
Slope: Home screen
`(L2(2)-L2(1))/(L1(2)-L1(1))→M`
Intercept: Home screen
`L2(1)-M*L1(1)`
[More details on page 93.]

Spreadsheet

Enter headings x, y, m, b, in cells A1–D1, and the values (x, y) in cells A2–B3. Enter
`=(B3-B2)/(A3-A2)`
in cell C2, and
`=B2-C2*A2`
in cell D2.
[More details on page 98.]

Solution

a. To write down the equation of the line, we need the slope m and the y-intercept b.

- ***Slope*** Because we are given two points on the line, we can use the slope formula:

$$m = \frac{y_2 - y_1}{x_2 - x_1} = \frac{-1-2}{3-1} = -\frac{3}{2}.$$

- ***Intercept*** We now have the slope of the line, $m = -3/2$, and also a point—we have two to choose from, so let us choose $(x_1, y_1) = (1, 2)$. We can now use the formula for the y-intercept:

$$b = y_1 - mx_1 = 2 - \left(-\frac{3}{2}\right)(1) = \frac{7}{2}.$$

Thus, the equation of the line is

$$y = -\frac{3}{2}x + \frac{7}{2}. \qquad y = mx + b$$

b. Proceeding as before,

- ***Slope*** We are not given two points on the line, but we are given a parallel line. We use the fact that *parallel lines have the same slope*. (Why?) We can find the slope of $3x + 4y = 5$ by solving for y and then looking at the coefficient of x:

$$y = -\frac{3}{4}x + \frac{5}{4} \qquad \text{To find the slope, solve for } y.$$

so the slope is $-3/4$.

- ***Intercept*** We now have the slope of the line, $m = -3/4$, and also a point $(x_1, y_1) = (2, -2)$. We can now use the formula for the y-intercept:

$$b = y_1 - mx_1 = -2 - \left(-\frac{3}{4}\right)(2) = -\frac{1}{2}.$$

Thus, the equation of the line is

$$y = -\frac{3}{4}x - \frac{1}{2}. \qquad y = mx + b$$

c. We are given a point: $(-9, 5)$. Furthermore, we are told that the line is horizontal, which tells us that the slope is $m = 0$. Therefore, all that remains is the calculation of the y-intercept:

$$b = y_1 - mx_1 = 5 - (0)(-9) = 5$$

so the equation of the line is

$$y = 5. \qquad y = mx + b$$

d. We are given a point: $(-9, 5)$. This time, we are told that the line is vertical, which means that the slope is undefined. Thus, we can't express the equation of the line in the form $y = mx + b$. (This formula makes sense only when the slope m of the line is defined.) What can we do? Well, here are some points on the desired line:

$$(-9, 1), (-9, 2), (-9, 3), \ldots$$

so $x = -9$ and $y = \textit{anything}$. If we simply say that $x = -9$, then these points are all solutions, so the equation is $x = -9$.

Applications: Linear Models

Using linear functions to describe or approximate relationships in the real world is called **linear modeling**.

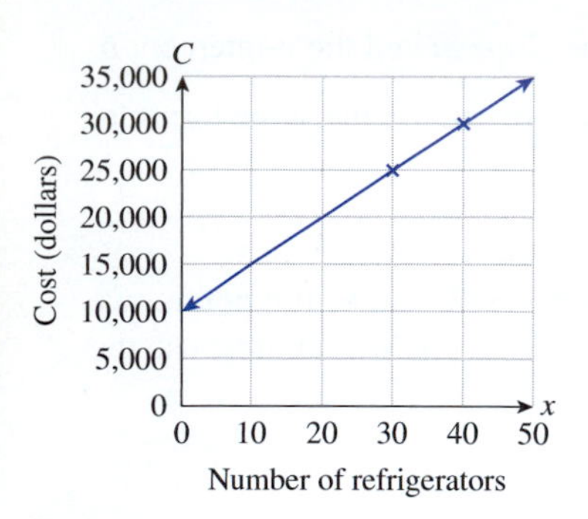

Figure 24

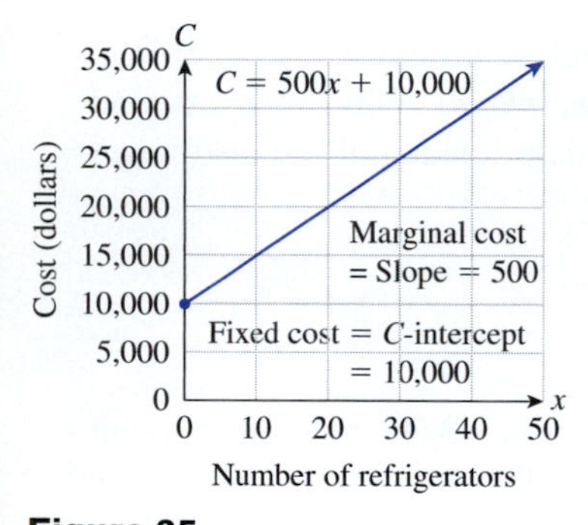

Figure 25

Recall from Section 1.2 that a **cost function** specifies the cost C as a function of the number of items x.

EXAMPLE 3 Linear Cost Function from Data

The manager of the FrozenAir Refrigerator factory notices that on Monday it cost the company a total of \$25,000 to build 30 refrigerators and on Tuesday it cost \$30,000 to build 40 refrigerators. Find a linear cost function based on this information. What is the daily fixed cost, and what is the marginal cost?

Solution We are seeking the cost C as a linear function of x, the number of refrigerators sold:

$$C = mx + b.$$

We are told that $C = 25{,}000$ when $x = 30$, and this amounts to being told that (30, 25,000) is a point on the graph of the cost function. Similarly, (40, 30,000) is another point on the line (Figure 24).

We can use the two points on the line to construct the linear cost equation:

- ***Slope*** $m = \dfrac{C_2 - C_1}{x_2 - x_1} = \dfrac{30{,}000 - 25{,}000}{40 - 30} = 500$ — C plays the role of y.
- ***Intercept*** $b = C_1 - mx_1 = 25{,}000 - (500)(30) = 10{,}000.$ — We used the point $(x_1, C_1) = (30, 25{,}000)$.

The linear cost function is therefore

$$C(x) = 500x + 10{,}000.$$

Because $m = 500$ and $b = 10{,}000$ the factory's fixed cost is \$10,000 each day, and its marginal cost is \$500 per refrigerator. (See page 51 in Section 1.2.) These are illustrated in Figure 25.

Before we go on... Recall that, in general, the slope m measures the number of units of change in y per 1-unit change in x, so it is measured in units of y per unit of x:

Units of Slope = Units of y per unit of x.

In Example 3, y is the cost C, measured in dollars, and x is the number of items, measured in refrigerators. Hence,

Units of Slope = Units of y per Unit of x = Dollars per refrigerator.

The y-intercept b, being a value of y, is measured in the same units as y. In Example 3, b is measured in dollars. ■

using Technology

To obtain the cost equation for Example 3 with technology, apply the Technology note for Example 2(a) to the given points (30, 25,000) and (40, 30,000) on the graph of the cost equation.

In Section 1.2 we saw that a **demand function** specifies the demand q as a function of the price p per item.

EXAMPLE 4 Linear Demand Function from Data

You run a small supermarket and must determine how much to charge for Hot'n'Spicy brand baked beans. The following chart shows weekly sales figures (the demand) for Hot'n'Spicy at two different prices.

Price/Can	\$0.50	\$0.75
Demand (cans sold/week)	400	350

a. Model these data with a linear demand function. (See Example 4 in Section 1.2.)

b. How do we interpret the slope and q-intercept of the demand function?

Solution

a. Recall that a demand equation—or demand function—expresses demand q (in this case, the number of cans of beans sold per week) as a function of the unit price p (in this case, price per can). We model the demand using the two points we are given: (0.50, 400) and (0.75, 350).

using Technology

To obtain the demand equation for Example 4 with technology, apply the Technology note for Example 2(a) to the given points (0.50, 400) and (0.75, 350) on the graph of the demand equation.

$$\textbf{\textit{Slope:}} \quad m = \frac{q_2 - q_1}{p_2 - p_1} = \frac{350 - 400}{0.75 - 0.50} = \frac{-50}{0.25} = -200$$

$$\textbf{\textit{Intercept:}} \; b = q_1 - mp_1 = 400 - (-200)(0.50) = 500$$

So, the demand equation is

$$q = -200p + 500. \qquad q = mp + b$$

b. The key to interpreting the slope in a demand equation is to recall (see the "Before we go on" note at the end of Example 3) that we measure the slope in *units of y per unit of x*. Here, $m = -200$, and the units of m are units of q per unit of p, or the number of cans sold per dollar change in the price. Because m is negative, we see that the number of cans sold decreases as the price increases. We conclude that the weekly sales will drop by 200 cans per $1 increase in the price.

To interpret the q-intercept, recall that it gives the q-coordinate when $p = 0$. Hence it is the number of cans the supermarket can "sell" every week if it were to give them away.*

* Does this seem realistic? Demand is not always unlimited if items were given away. For instance, campus newspapers are sometimes given away, and yet piles of them are often left untaken. Also see the "Before we go on" discussion at the end of this example.

Before we go on...

Q: *Just how reliable is the linear model used in Example 4?*

A: The *actual* demand graph could in principle be obtained by tabulating demand figures for a large number of different prices. If the resulting points were plotted on the *pq* plane, they would probably suggest a curve and not a straight line. However, if you looked at a small enough portion of any curve, you could closely *approximate* it by a straight line. In other words, *over a small range of values of p, a linear model is accurate*. Linear models of real-world situations are generally reliable only for small ranges of the variables. (This point will come up again in some of the exercises.)

■

The next example illustrates modeling change over time t with a linear function of t.

EXAMPLE 5 Modeling Change Over Time: Growth of Sales

The worldwide market for portable navigation devices was expected to grow from 50 million units in 2007 to around 530 million units in 2015.[8]

[8] Sales were expected to grow to more than 500 million in 2015 according to a January 2008 press release by Telematics Research Group. Source: www.telematicsresearch.com.

a. Use this information to model annual worldwide sales of portable navigation devices as a linear function of time t in years since 2007. What is the significance of the slope?

b. Use the model to predict when annual sales of mobile navigation devices will reach 440 million units.

Solution

a. Since we are interested in worldwide sales s of portable navigation devices as a function of time, we take time t to be the independent coordinate (playing the role of x) and the annual sales s, in million of units, to be the dependent coordinate (in the role of y). Notice that 2007 corresponds to $t = 0$ and 2015 corresponds to $t = 8$, so we are given the coordinates of two points on the graph of sales s as a function of time t: (0, 50) and (8, 530). We model the sales using these two points:

$$m = \frac{s_2 - s_1}{t_2 - t_1} = \frac{530 - 50}{8 - 0} = \frac{480}{8} = 60$$

$$b = s_1 - mt_1 = 50 - (60)(0) = 50$$

So, $$s = 60t + 50 \text{ million units.} \qquad s = mt + b$$

The slope m is measured in units of s per unit of t; that is, millions of devices per year, and is thus the *rate of change of annual sales*. To say that $m = 60$ is to say that annual sales are increasing at a rate of 60 million devices per year.

b. Our model of annual sales as a function of time is

$$s = 60t + 50 \text{ million units.}$$

Annual sales of mobile portable devices will reach 440 million when $s = 440$, or

$$440 = 60t + 50$$

Solving for t, $$60t = 440 - 50 = 390$$

$$t = \frac{390}{60} = 6.5 \text{ years},$$

which is midway through 2013. Thus annual sales are expected to reach 440 million midway through 2013.

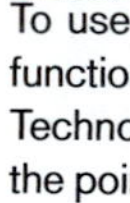

using Technology

To use technology to obtain s as a function of t in Example 5, apply the Technology note for Example 2(a) to the points (0, 50) and (8, 530) on its graph.

EXAMPLE 6 Velocity

You are driving down the Ohio Turnpike, watching the mileage markers to stay awake. Measuring time in hours after you see the 20-mile marker, you see the following markers each half hour:

Time (h)	0	0.5	1	1.5	2
Marker (mi)	20	47	74	101	128

Find your location s as a function of t, the number of hours you have been driving. (The number s is also called your **position** or **displacement**.)

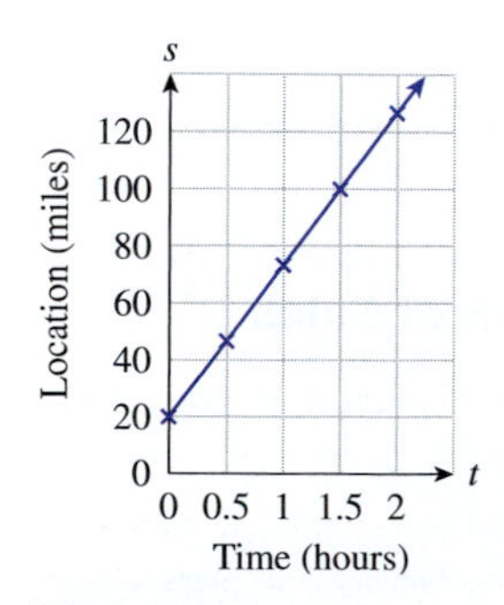

Figure 26

Solution

If we plot the location s versus the time t, the five markers listed give us the graph in Figure 26. These points appear to lie along a straight line. We can verify this by calculating how far you traveled in each half hour. In the first half hour, you traveled $47 - 20 = 27$ miles. In the second half hour you traveled $74 - 47 = 27$ miles also. In fact, you traveled exactly 27 miles each half hour. The points we plotted lie on a straight line that rises 27 units for every 0.5 unit we go to the right, for a slope of $27/0.5 = 54$.

To get the equation of that line, notice that we have the s-intercept, which is the starting marker of 20. Thus, the equation of s as a function of time t is

$$s(t) = 54t + 20. \quad \text{We used } s \text{ in place of } y \text{ and } t \text{ in place of } x.$$

using Technology

To use technology to obtain s as a function of t in Example 6, apply the Technology note for Example 2(a) to the points (0, 20) and (1, 74) on its graph.

Notice the significance of the slope: For every hour you travel, you drive a distance of 54 miles. In other words, you are traveling at a constant velocity of 54 mph. We have uncovered a very important principle:

In the graph of displacement versus time, velocity is given by the slope.

Linear Change over Time

If a quantity q is a linear function of time t,

$$q = mt + b,$$

then the slope m measures the **rate of change** of q, and b is the quantity at time $t = 0$, the **initial quantity**. If q represents the position of a moving object, then the rate of change is also called the **velocity**.

Units of *m* and *b*

The units of measurement of m are units of q per unit of time; for instance, if q is income in dollars and t is time in years, then the rate of change m is measured in dollars per year.

The units of b are units of q; for instance, if q is income in dollars and t is time in years, then b is measured in dollars.

Quick Example

If the accumulated revenue from sales of your video game software is given by $R = 2{,}000t + 500$ dollars, where t is time in years from now, then you have earned \$500 in revenue so far, and the accumulated revenue is increasing at a rate of \$2,000 per year.

Examples 3–6 share the following common theme.

General Linear Models

If $y = mx + b$ is a linear model of changing quantities x and y, then the slope m is the rate at which y is increasing per unit increase in x, and the y-intercept b is the value of y that corresponds to $x = 0$.

Units of *m* and *b*

The slope m is measured in units of y per unit of x, and the intercept b is measured in units of y.

Quick Example

If the number n of spectators at a soccer game is related to the number g of goals your team has scored so far by the equation $n = 20g + 4$, then you can expect 4 spectators if no goals have been scored and 20 additional spectators per additional goal scored.

FAQs

What to Use as *x* and *y*, and How to Interpret a Linear Model

Q: *In a problem where I must find a linear relationship between two quantities, which quantity do I use as x and which do I use as y?*

A: The key is to decide which of the two quantities is the independent variable, and which is the dependent variable. Then use the independent variable as x and the dependent variable as y. In other words, *y depends on x.*

Here are examples of phrases that convey this information, usually of the form *Find y [dependent variable] in terms of x [independent variable]:*

- Find the cost in terms of the number of items. $y = \text{Cost}, x = \#\ \text{Items}$
- How does color depend on wavelength? $y = \text{Color}, x = \text{Wavelength}$

If no information is conveyed about which variable is intended to be independent, then you can use whichever is convenient.

Q: *How do I interpret a general linear model $y = mx + b$?*

A: The key to interpreting a linear model is to remember the units we use to measure m and b:

> *The slope m is measured in units of y per unit of x; the intercept b is measured in units of y.*

For instance, if $y = 4.3x + 8.1$ and you know that x is measured in feet and y in kilograms, then you can already say, "y is 8.1 kilograms when $x = 0$ feet, and increases at a rate of 4.3 kilograms per foot" without knowing anything more about the situation!

1.3 EXERCISES

Access end-of-section exercises online at **www.webassign.net**

1.4 Linear Regression

We have seen how to find a linear model given two data points: We find the equation of the line that passes through them. However, we often have more than two data points, and they will rarely all lie on a single straight line, but may often come close to doing so. The problem is to find the line coming *closest* to passing through all of the points.

Suppose, for example, that we are conducting research for a company interested in expanding into Mexico. Of interest to us would be current and projected growth in that country's economy. The following table shows past and projected per capita gross domestic product (GDP)[9] of Mexico for 2000–2014.[10]

Year t ($t = 0$ represents 2000)	0	2	4	6	8	10	12	14
Per Capita GDP y ($1,000)	9	9	10	11	11	12	13	13

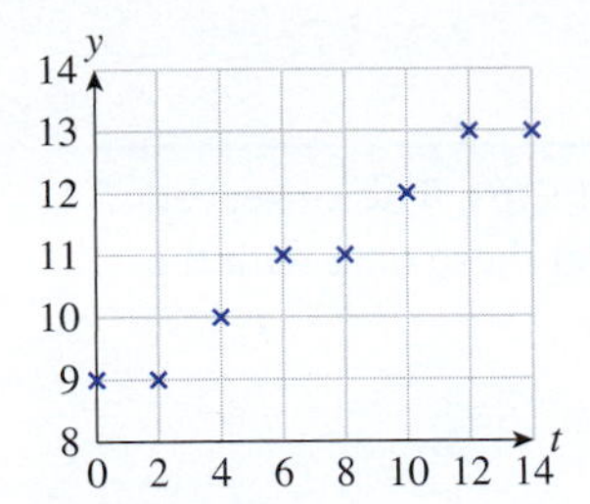

Figure 27(a)

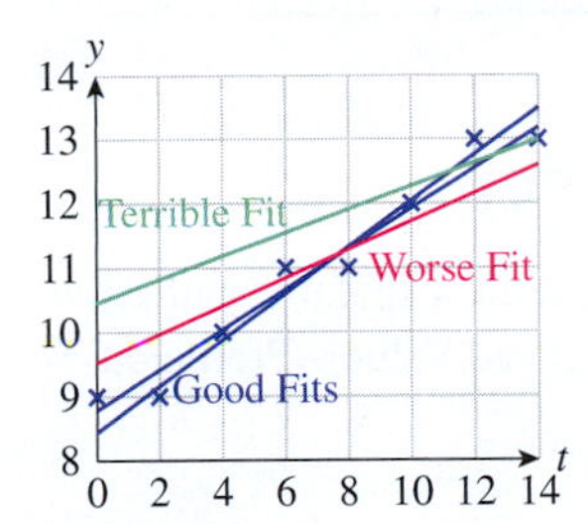

Figure 27(b)

A plot of these data suggests a roughly linear growth of the GDP (Figure 27(a)). These points suggest a roughly linear relationship between t and y, although they clearly do not all lie on a single straight line. Figure 27(b) shows the points together with several lines, some fitting better than others. Can we precisely measure which lines fit better than others? For instance, which of the two lines labeled as "good" fits in Figure 27(b) models the data more accurately? We begin by considering, for each value of t, the difference between the actual GDP (the **observed value**) and the GDP predicted by a linear equation (the **predicted value**). The difference between the predicted value and the observed value is called the **residual**.

$$\text{Residual} = \text{Observed Value} - \text{Predicted Value}$$

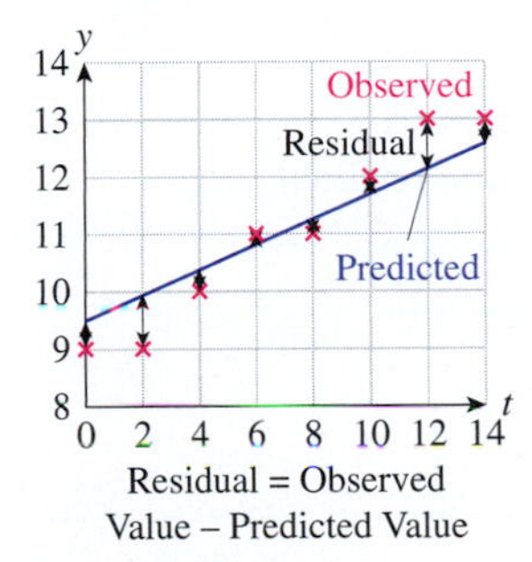

Residual = Observed Value – Predicted Value

Figure 28

On the graph, the residuals measure the vertical distances between the (observed) data points and the line (Figure 28) and they tell us how far the linear model is from predicting the actual GDP.

The more accurate our model, the smaller the residuals should be. We can combine all the residuals into a single measure of accuracy by adding their *squares*. (We square the residuals in part to make them all positive.*) The sum of the squares of the residuals is called the **sum-of-squares error, SSE**. Smaller values of SSE indicate more accurate models.

Here are some definitions and formulas for what we have been discussing.

* Why not add the absolute values of the residuals instead? Mathematically, using the squares rather than the absolute values results in a simpler and more elegant solution. Further, using the squares always results in a *single* best-fit line in cases where the x-coordinates are all different, whereas this is not the case if we use absolute values.

Observed and Predicted Values

Suppose we are given a collection of data points $(x_1, y_1), \ldots, (x_n, y_n)$. The n quantities $y_1, y_2, \ldots, y_n$ are called the **observed y-values**. If we model these data with a linear equation

$$\hat{y} = mx + b, \qquad \hat{y} \text{ stands for "estimated } y\text{" or "predicted } y\text{."}$$

then the y-values we get by substituting the given x-values into the equation are called the **predicted y-values**:

$$\hat{y}_1 = mx_1 + b \qquad \text{Substitute } x_1 \text{ for } x.$$
$$\hat{y}_2 = mx_2 + b \qquad \text{Substitute } x_2 \text{ for } x.$$
$$\ldots$$
$$\hat{y}_n = mx_n + b. \qquad \text{Substitute } x_n \text{ for } x.$$

[9] The GDP is a measure of the total market value of all goods and services produced within a country.

[10] Data are approximate and/or projected. Sources: CIA World Factbook/www.indexmundi.com, www.economist.com.

Quick Example

Consider the three data points (0, 2), (2, 5), and (3, 6). The observed y-values are $y_1 = 2$, $y_2 = 5$, and $y_3 = 6$. If we model these data with the equation $\hat{y} = x + 2.5$, then the predicted values are:

$$\hat{y}_1 = x_1 + 2.5 = 0 + 2.5 = 2.5$$
$$\hat{y}_2 = x_2 + 2.5 = 2 + 2.5 = 4.5$$
$$\hat{y}_3 = x_3 + 2.5 = 3 + 2.5 = 5.5.$$

Residuals and Sum-of-Squares Error (SSE)

If we model a collection of data $(x_1, y_1), \ldots, (x_n, y_n)$ with a linear equation $\hat{y} = mx + b$, then the **residuals** are the n quantities (Observed Value − Predicted Value):

$$(y_1 - \hat{y}_1), (y_2 - \hat{y}_2), \ldots, (y_n - \hat{y}_n).$$

The **sum-of-squares error (SSE)** is the sum of the squares of the residuals:

$$\text{SSE} = (y_1 - \hat{y}_1)^2 + (y_2 - \hat{y}_2)^2 + \cdots + (y_n - \hat{y}_n)^2.$$

Quick Example

For the data and linear approximation given above, the residuals are:

$$y_1 - \hat{y}_1 = 2 - 2.5 = -0.5$$
$$y_2 - \hat{y}_2 = 5 - 4.5 = 0.5$$
$$y_3 - \hat{y}_3 = 6 - 5.5 = 0.5$$

and so SSE $= (-0.5)^2 + (0.5)^2 + (0.5)^2 = 0.75$.

using Technology

See the Technology Guides at the end of the chapter for detailed instructions on how to obtain the tables and graphs in Example 1 using a TI-83/84 Plus or a spreadsheet. Here is an outline:

TI-83/84 Plus

[STAT] EDIT
Values of t in L_1, and y in L_2.
Predicted y: Highlight L_3. Enter `0.5*L1+8`
Squares of residuals: Highlight L_4. Enter `(L2-L3)^2`
SSE: Home screen `sum(L4)`
Graph: `Y1=0.5X+8`
Y = screen: Turn on Plot 1 [ZOOM] (STAT) [More details on page 93.]

Spreadsheet

Headings t, y, y-hat, Residual^2, m, b, and SSE in A1–F1.
t-values in A2–A9, y-values in B2–B9; 0.25 for m and 9 for b in E2–F2
Predicted y: `=$E$2*A2+$F$2` in C2 and copy down to C9.
Squares of residuals: `=(B2-C2)^2` in D2 and copy down to D9.
SSE: `=SUM(D2:D9)` in G2
Graph: Highlight A1–C9. Insert a Scatter chart.
[More details on page 98.]

EXAMPLE 1 Computing SSE

Using the data above on the GDP in Mexico, compute SSE for the linear models $y = 0.5t + 8$ and $y = 0.25t + 9$. Which model is the better fit?

Solution We begin by creating a table showing the values of t, the observed (given) values of y, and the values predicted by the first model.

Year t	Observed y	Predicted $\hat{y} = 0.5t + 8$
0	9	8
2	9	9
4	10	10
6	11	11
8	11	12
10	12	13
12	13	14
14	13	15

We now add two new columns for the residuals and their squares.

Year t	Observed y	Predicted $\hat{y} = 0.5t + 8$	Residual $y - \hat{y}$	Residual^2 $(y - \hat{y})^2$
0	9	8	$9 - 8 = 1$	$1^2 = 1$
2	9	9	$9 - 9 = 0$	$0^2 = 0$
4	10	10	$10 - 10 = 0$	$0^2 = 0$
6	11	11	$11 - 11 = 0$	$0^2 = 0$
8	11	12	$11 - 12 = -1$	$(-1)^2 = 1$
10	12	13	$12 - 13 = -1$	$(-1)^2 = 1$
12	13	14	$13 - 14 = -1$	$(-1)^2 = 1$
14	13	15	$13 - 15 = -2$	$(-2)^2 = 4$

SSE, the sum of the squares of the residuals, is then the sum of the entries in the last column,

$$\text{SSE} = 8.$$

Repeating the process using the second model, $0.25t + 9$, yields the following table:

Year t	Observed y	Predicted $\hat{y} = 0.25t + 9$	Residual $y - \hat{y}$	Residual^2 $(y - \hat{y})^2$
0	9	9	$9 - 9 = 0$	$0^2 = 0$
2	9	9.5	$9 - 9.5 = -0.5$	$(-0.5)^2 = 0.25$
4	10	10	$10 - 10 = 0$	$0^2 = 0$
6	11	10.5	$11 - 10.5 = 0.5$	$0.5^2 = 0.25$
8	11	11	$11 - 11 = 0$	$0^2 = 0$
10	12	11.5	$12 - 11.5 = 0.5$	$0.5^2 = 0.25$
12	13	12	$13 - 12 = 1$	$1^2 = 1$
14	13	12.5	$13 - 12.5 = 0.5$	$0.5^2 = 0.25$

This time, SSE = 2 and so the second model is a better fit.

Figure 29 shows the data points and the two linear models in question.

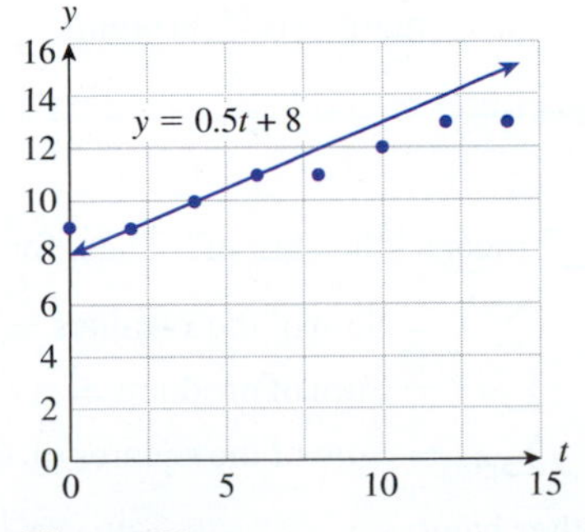

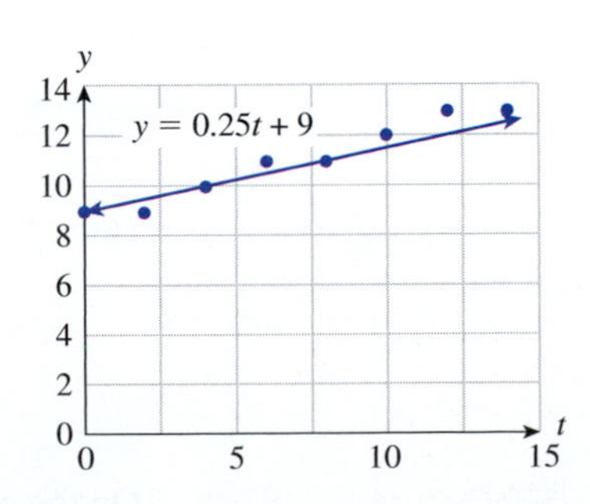

Figure 29

➡ Before we go on...

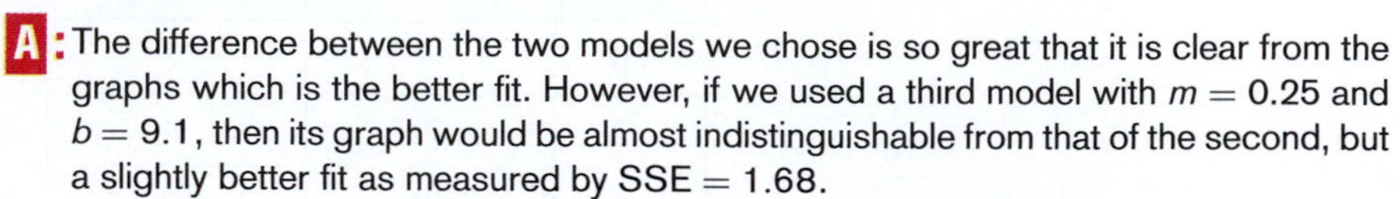

Q: *It seems clear from the figure that the second model in Example 1 gives a better fit. Why bother to compute SSE to tell me this?*

A: The difference between the two models we chose is so great that it is clear from the graphs which is the better fit. However, if we used a third model with $m = 0.25$ and $b = 9.1$, then its graph would be almost indistinguishable from that of the second, but a slightly better fit as measured by SSE = 1.68.

■

Among all possible lines, there ought to be one with the least possible value of SSE—that is, the greatest possible accuracy as a model. The line (and there is only one such line) that minimizes the sum of the squares of the residuals is called the **regression line**, the **least-squares line**, or the **best-fit line**.

To find the regression line, we need a way to find values of m and b that give the smallest possible value of SSE. As an example, let us take the second linear model in the example above. We said in the "Before we go on" discussion that increasing b from 9 to 9.1 had the desirable effect of decreasing SSE from 2 to 1.68. We could then increase m to 0.26, further reducing SSE to 1.328. Imagine this as a kind of game: Alternately alter the values of m and b by small amounts until SSE is as small as you can make it. This works, but is extremely tedious and time-consuming.

Fortunately, there is an algebraic way to find the regression line. Here is the calculation. To justify it rigorously requires calculus of several variables or linear algebra.

Regression Line

The **regression line (least squares line, best-fit line)** associated with the points $(x_1, y_1), (x_2, y_2), \ldots, (x_n, y_n)$ is the line that gives the minimum SSE. The regression line is

$$y = mx + b,$$

where m and b are computed as follows:

$$m = \frac{n(\sum xy) - (\sum x)(\sum y)}{n(\sum x^2) - (\sum x)^2}$$

$$b = \frac{\sum y - m(\sum x)}{n}$$

$$n = \text{number of data points.}$$

The quantities m and b are called the **regression coefficients**.

Here, "$\sum$" means "the sum of." Thus, for example,

$$\sum x = \text{Sum of the } x\text{-values} = x_1 + x_2 + \cdots + x_n$$

$$\sum xy = \text{Sum of products} = x_1y_1 + x_2y_2 + \cdots + x_ny_n$$

$$\sum x^2 = \text{Sum of the squares of the } x\text{-values} = x_1{}^2 + x_2{}^2 + \cdots + x_n{}^2.$$

On the other hand,

$$\left(\sum x\right)^2 = \text{Square of } \sum x = \text{Square of the sum of the } x\text{-values.}$$

EXAMPLE 2 Per Capita Gross Domestic Product in Mexico

In Example 1 we considered the following data on the per capita gross domestic product (GDP) of Mexico:

Year x ($x = 0$ represents 2000)	0	2	4	6	8	10	12	14
Per Capita GDP y (\$1,000)	9	9	10	11	11	12	13	13

Find the best-fit linear model for these data and use the model to predict the per capita GDP in Mexico in 2016.

using Technology

See the Technology Guides at the end of the chapter for detailed instructions on how to obtain the regression line and graph in Example 2 using a TI-83/84 Plus or a spreadsheet. Here is an outline:

TI-83/84 Plus

STAT EDIT

Values of x in L_1, and y in L_2.

Regression equation: STAT CALC option #4: `LinReg(ax+b)`

Graph: Y= VARS 5 EQ 1, then ZOOM 9

[More details on page 94.]

Spreadsheet

x-values in A2–A9, y-values in B2–B9

Graph: Highlight A2–B9. Insert a Scatter Chart.

Regression line: Add a linear trendline. [More details on page 99.]

Website

www.WanerMath.com

The following two utilities will calculate and plot regression lines (link to either from Math Tools for Chapter 1):

Simple Regression

Function Evaluator and Grapher

Solution Let's organize our work in the form of a table, where the original data are entered in the first two columns and the bottom row contains the column sums.

	x	y	xy	x^2
	0	9	0	0
	2	9	18	4
	4	10	40	16
	6	11	66	36
	8	11	88	64
	10	12	120	100
	12	13	156	144
	14	13	182	196
$\sum$ (Sum)	56	88	670	560

Because there are $n = 8$ data points, we get

$$m = \frac{n(\sum xy) - (\sum x)(\sum y)}{n(\sum x^2) - (\sum x)^2} = \frac{8(670) - (56)(88)}{8(560) - (56)^2} \approx 0.321$$

and

$$b = \frac{\sum y - m(\sum x)}{n} \approx \frac{88 - (0.321)(56)}{8} \approx 8.75.$$

So, the regression line is

$$y = 0.321x + 8.75.$$

To predict the per capita GDP in Mexico in 2016 we substitute $x = 16$ and get $y \approx 14$, or \$14,000 per capita.

Figure 30 shows the data points and the regression line (which has SSE ≈ 0.643; a lot lower than in Example 1).

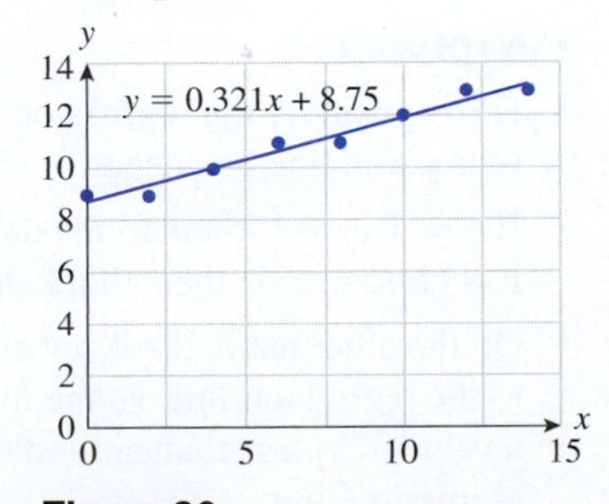

Figure 30

Coefficient of Correlation

If all the data points do not lie on one straight line, we would like to be able to measure how closely they can be approximated by a straight line. Recall that SSE measures the sum of the squares of the deviations from the regression line; therefore it constitutes a measurement of what is called "goodness of fit." (For instance, if $SSE = 0$, then all the points lie on a straight line.) However, SSE depends on the units we use to measure y, and also on the number of data points (the more data points we use, the larger SSE tends to be). Thus, while we can (and do) use SSE to compare the goodness of fit of two lines to the same data, we cannot use it to compare the goodness of fit of one line to one set of data with that of another to a different set of data.

To remove this dependency, statisticians have found a related quantity that can be used to compare the goodness of fit of lines to different sets of data. This quantity, called the **coefficient of correlation** or **correlation coefficient**, and usually denoted r, is between -1 and 1. The closer r is to -1 or 1, the better the fit. For an *exact* fit, we would have $r = -1$ (for a line with negative slope) or $r = 1$ (for a line with positive slope). For a bad fit, we would have r close to 0. Figure 31 shows several collections of data points with least-squares lines and the corresponding values of r.

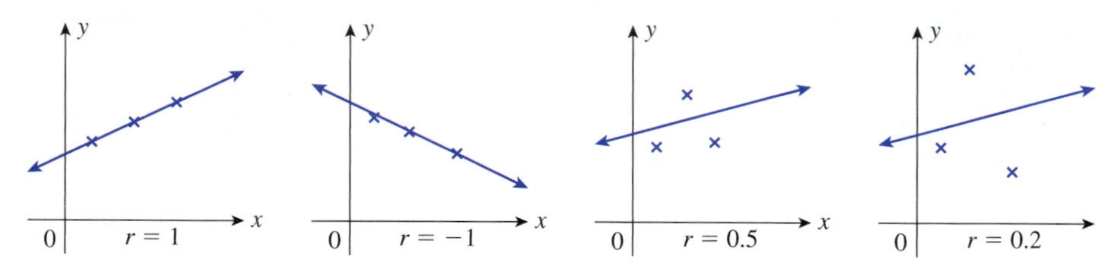

Figure 31

Correlation Coefficient

The coefficient of correlation of the n data points $(x_1, y_1), (x_2, y_2), \ldots, (x_n, y_n)$ is

$$r = \frac{n(\sum xy) - (\sum x)(\sum y)}{\sqrt{n(\sum x^2) - (\sum x)^2} \cdot \sqrt{n(\sum y^2) - (\sum y)^2}}.$$

It measures how closely the data points $(x_1, y_1), (x_2, y_2), \ldots, (x_n, y_n)$ fit the regression line. (The value r^2 is sometimes called the **coefficient of determination**.)

Interpretation

- If r is positive, the regression line has positive slope; if r is negative, the regression line has negative slope.
- If $r = 1$ or -1, then all the data points lie exactly on the regression line; if it is close to ± 1, then all the data points are close to the regression line.
- On the other hand, if r is not close to ± 1, then the data points are not close to the regression line, so the fit is not a good one. As a general rule of thumb, a value of $|r|$ less than around 0.8 indicates a poor fit of the data to the regression line.

using Technology

See the Technology Guides at the end of the chapter for detailed instructions on how to obtain the correlation coefficient in Example 3 using a TI-83/84 Plus or a spreadsheet. Here is an outline:

TI-83/84 Plus

[2ND] [CATALOG] `DiagnosticOn`
Then [STAT] CALC option #4: `LinReg(ax+b)` [More details on page 95.]

Spreadsheet

Add a trendline and select the option to "Display R-squared value on chart."
[More details and other alternatives on page 100.]

Website

www.WanerMath.com
The following two utilities will show regression lines and also r^2 (link to either from Math Tools for Chapter 1):

Simple Regression
Function Evaluator and Grapher

EXAMPLE 3 Computing the Coefficient of Correlation

Find the correlation coefficient for the data in Example 2. Is the regression line a good fit?

Solution The formula for r requires $\sum x, \sum x^2, \sum xy, \sum y$, and $\sum y^2$. We have all of these except for $\sum y^2$, which we find in a new column as shown.

	x	y	xy	x^2	y^2
	0	9	0	0	81
	2	9	18	4	81
	4	10	40	16	100
	6	11	66	36	121
	8	11	88	64	121
	10	12	120	100	144
	12	13	156	144	169
	14	13	182	196	169
$\sum$ (Sum)	56	88	670	560	986

Substituting these values into the formula, we get

$$r = \frac{n(\sum xy) - (\sum x)(\sum y)}{\sqrt{n(\sum x^2) - (\sum x)^2} \cdot \sqrt{n(\sum y^2) - (\sum y)^2}}$$

$$= \frac{8(670) - (56)(88)}{\sqrt{8(560) - 56^2} \cdot \sqrt{8(986) - 88^2}}$$

$$\approx 0.982.$$

As r is close to 1, the fit is a fairly good one; that is, the original points lie nearly along a straight line, as can be confirmed from the graph in Example 2.

1.4 EXERCISES

Access end-of-section exercises online at **www.webassign.net**

CHAPTER 1 REVIEW

KEY CONCEPTS

Website www.WanerMath.com

Go to the Website at www.WanerMath.com to find a comprehensive and interactive Web-based summary of Chapter 1.

1.1 Functions from the Numerical, Algebraic, and Graphical Viewpoints

Real-valued function f of a real-valued variable x, domain *p. 38*

Independent and dependent variables *p. 38*

Graph of the function f *p. 38*

Numerically specified function *p. 38*

Graphically specified function *p. 38*

Algebraically defined function *p. 38*

Piecewise-defined function *p. 43*

Vertical line test *p. 45*

Common types of algebraic functions and their graphs *p. 46*

1.2 Functions and Models

Mathematical model *p. 48*

Analytical model *p. 49*

Curve-fitting model *p. 49*

Cost, revenue, and profit; marginal cost, revenue, and profit; break-even point *pp. 50–51*

Demand, supply, and equilibrium price *pp. 53–54*

Selecting a model *p. 56*

Compound interest *p. 58*

Exponential growth and decay *p. 60*

Algebra of functions (sum, difference, product, quotient) *p. 62*

1.3 Linear Functions and Models

Linear function $f(x) = mx + b$ *p. 63*

Change in q: $\Delta q = q_2 - q_1$ *p. 64*

Slope of a line:

$$m = \frac{\Delta y}{\Delta x} = \frac{\text{Change in } y}{\text{Change in } x} \quad p.\ 65$$

Interpretations of m *p. 65*

Interpretation of b: y-intercept *p. 65*

Recognizing linear data *p. 66*

Computing the slope of a line *p. 67*

Slopes of horizontal and vertical lines *p. 68*

Computing the y-intercept *p. 68*

Linear modeling *p. 69*

Linear cost *p. 70*

Linear demand *p. 70*

Linear change over time; rate of change; velocity *p. 73*

General linear models *p. 73*

1.4 Linear Regression

Observed and predicted values *p. 75*

Residuals and sum-of-squares error (SSE) *p. 76*

Regression line (least-squares line, best-fit line) *p. 78*

Correlation coefficient; coefficient of determination *p. 80*

REVIEW EXERCISES

In Exercises 1–4, use the graph of the function f to find approximations of the given values.

1.

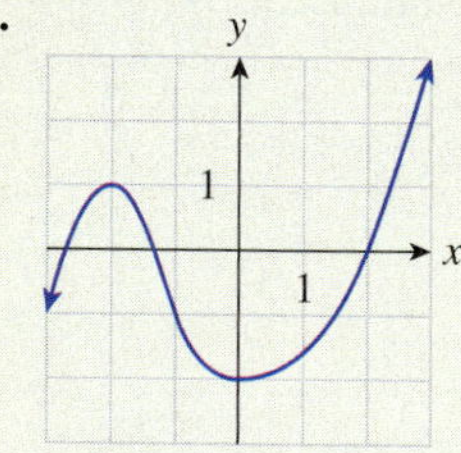

a. $f(-2)$ **b.** $f(0)$ **c.** $f(2)$ **d.** $f(2) - f(-2)$

2.

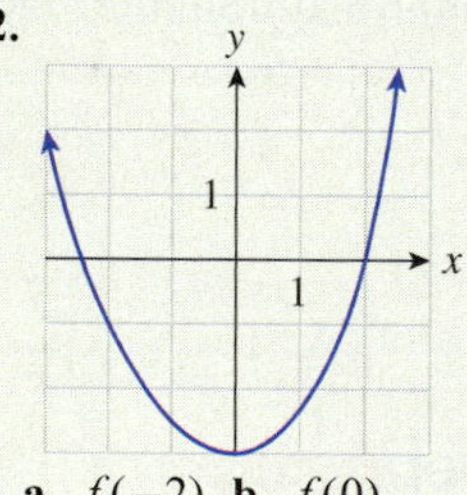

a. $f(-2)$ **b.** $f(0)$ **c.** $f(2)$ **d.** $f(2) - f(-2)$

3.

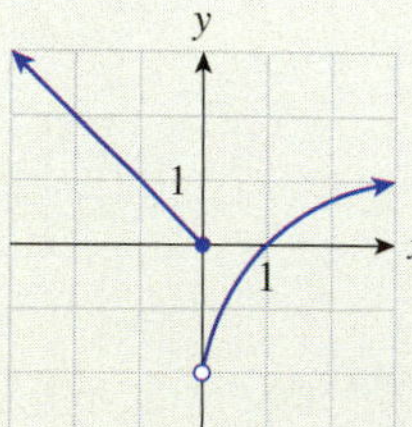

a. $f(-1)$ **b.** $f(0)$ **c.** $f(1)$ **d.** $f(1) - f(-1)$

4.

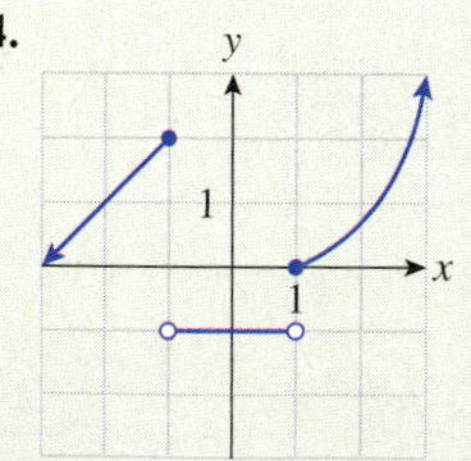

a. $f(-1)$ **b.** $f(0)$ **c.** $f(1)$ **d.** $f(1) - f(-1)$

In Exercises 5–8, graph the given function or equation.

5. $y = -2x + 5$

6. $2x - 3y = 12$

7. $y = \begin{cases} \frac{1}{2}x & \text{if } -1 \le x \le 1 \\ x - 1 & \text{if } 1 < x \le 3 \end{cases}$

8. $f(x) = 4x - x^2$ with domain $[0, 4]$

In Exercises 9–14, decide whether the specified values come from a linear, quadratic, exponential, or absolute value function.

9.

x	−2	0	1	2	4
$f(x)$	4	2	1	0	2

10.

x	−2	0	1	2	4
$g(x)$	−5	−3	−2	−1	1

11.

x	−2	0	1	2	4
$h(x)$	1.5	1	0.75	0.5	0

12.

x	−2	0	1	2	4
$k(x)$	0.25	1	2	4	16

13.

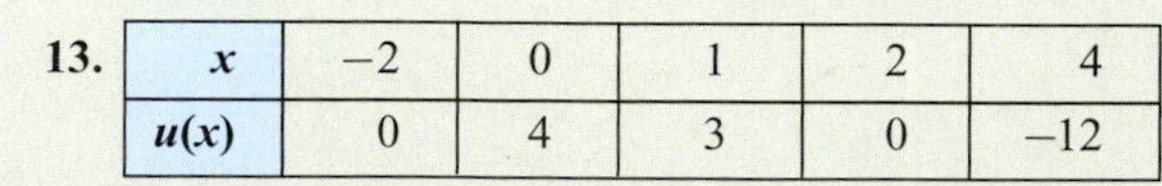

x	−2	0	1	2	4
$u(x)$	0	4	3	0	−12

14.

x	-2	0	1	2	4
$w(x)$	32	8	4	2	0.5

In Exercises 15–22, find the equation of the specified line.

15. Through (3, 2) with slope -3

16. Through $(-2, 4)$ with slope -1

17. Through $(1, -3)$ and (5, 2)

18. Through $(-1, 2)$ and (1, 0)

19. Through (1, 2) parallel to $x - 2y = 2$

20. Through $(-3, 1)$ parallel to $-2x - 4y = 5$

21. With slope 4 crossing $2x - 3y = 6$ at its x-intercept

22. With slope 1/2 crossing $3x + y = 6$ at its x-intercept

In Exercises 23 and 24, determine which of the given lines better fits the given points.

23. $(-1, 1), (1, 2), (2, 0)$; $y = -x/2 + 1$ or $y = -x/4 + 1$

24. $(-2, -1), (-1, 1), (0, 1), (1, 2), (2, 4), (3, 3)$; $y = x + 1$ or $y = x/2 + 1$

In Exercises 25 and 26, find the line that best fits the given points and compute the correlation coefficient.

25. $(-1, 1), (1, 2), (2, 0)$

26. $(-2, -1), (-1, 1), (0, 1), (1, 2), (2, 4), (3, 3)$

APPLICATIONS: OHaganBooks.com

27. ***Web Site Traffic*** John Sean O'Hagan is CEO of the online bookstore OHaganBooks.com and notices that, since the establishment of the company Web site six years ago ($t = 0$), the number of visitors to the site has grown quite dramatically, as indicated by the following table:

Year t	0	1	2	3	4	5	6
Web site Traffic $V(t)$ (visits/day)	100	300	1,000	3,300	10,500	33,600	107,400

a. Graph the function V as a function of time t. Which of the following types of function seem to fit the curve best: linear, quadratic, or exponential?

b. Compute the ratios $\frac{V(1)}{V(0)}, \frac{V(2)}{V(1)}, \dots,$ and $\frac{V(6)}{V(5)}$. What do you notice?

c. Use the result of part (b) to predict Web site traffic next year (to the nearest 100).

28. ***Publishing Costs*** Marjory Maureen Duffin is CEO of publisher Duffin House, a major supplier of paperback titles to OHaganBooks.com. She notices that publishing costs over the past five years have varied considerably as indicated by the following table, which shows the average cost to the company of publishing a paperback novel (t is time in years, and the current year is $t = 5$):

Year t	0	1	2	3	4	5
Cost $C(t)$	\$5.42	\$5.10	\$5.00	\$5.12	\$5.40	\$5.88

a. Graph the function C as a function of time t. Which of the following types of function seem to fit the curve best: linear, quadratic, or exponential?

b. Compute the differences $C(1) - C(0), C(2) - C(1), \dots,$ and $C(5) - C(4)$, rounded to one decimal place. What do you notice?

c. Use the result of part (b) to predict the cost of producing a paperback novel next year.

29. ***Web Site Stability*** John O'Hagan is considering upgrading the Web server equipment at OHaganBooks.com because of frequent crashes. The tech services manager has been monitoring the frequency of crashes as a function of Web site traffic (measured in thousands of visits per day) and has obtained the following model:

$$c(x) = \begin{cases} 0.03x + 2 & \text{if } 0 \le x \le 50 \\ 0.05x + 1 & \text{if } x > 50 \end{cases}$$

where $c(x)$ is the average number of crashes in a day in which there are x thousand visitors.

a. On average, how many times will the Web site crash on a day when there are 10,000 visits? 50,000 visits? 100,000 visits?

b. What does the coefficient 0.03 tell you about the Web site's stability?

c. Last Friday, the Web site went down 8 times. Estimate the number of visits that day.

30. ***Book Sales*** As OHaganBooks.com has grown in popularity, the sales manager has been monitoring book sales as a function of the Web site traffic (measured in thousands of visits per day) and has obtained the following model:

$$s(x) = \begin{cases} 1.55x & \text{if } 0 \le x \le 100 \\ 1.75x - 20 & \text{if } 100 < x \le 250 \end{cases}$$

where $s(x)$ is the average number of books sold in a day in which there are x thousand visitors.

a. On average, how many books per day does the model predict OHaganBooks.com will sell when it has 60,000 visits in a day? 100,000 visits in a day? 160,000 visits in a day?

b. What does the coefficient 1.75 tell you about book sales?

c. According to the model, approximately how many visitors per day will be needed in order to sell an average of 300 books per day?

31. ***New Users*** The number of registered users at OHaganBooks.com has increased substantially over the past few months. The following table shows the number of new users registering each month for the past six months:

Month t	1	2	3	4	5	6
New Users (thousands)	12.5	37.5	62.5	72.0	74.5	75.0

a. Which of the following models best approximates the data?

(A) $n(t) = \dfrac{300}{4 + 100(5^{-t})}$ **(B)** $n(t) = 13.3t + 8.0$

(C) $n(t) = -2.3t^2 + 30.0t - 3.3$

(D) $n(t) = 7(3^{0.5t})$

b. What do each of the above models predict for the number of new users in the next few months: rising, falling, or leveling off?

32. ***Purchases*** OHaganBooks.com has been promoting a number of books published at Duffin House. The following table shows the number of books purchased each month from Duffin House for the past five months:

Month t	1	2	3	4	5
Purchases (books)	1,330	520	520	1,340	2,980

a. Which of the following models best approximates the data?

(A) $n(t) = \dfrac{3,000}{1 + 12(2^{-t})}$ **(B)** $n(t) = \dfrac{2,000}{4.2 - 0.7t}$

(C) $n(t) = 300(1.6^t)$

(D) $n(t) = 100(4.1t^2 - 20.4t + 29.5)$

b. What do each of the above models predict for the number of new users in the next few months: rising, falling, leveling off, or something else?

33. ***Internet Advertising*** Several months ago. John O'Hagan investigated the effect on the popularity of OHaganBooks.com of placing banner ads at well-known Internet portals. The following model was obtained from available data:

$v(c) = -0.000005c^2 + 0.085c + 1,750$ new visits per day

where c is the monthly expenditure on banner ads.

a. John O'Hagan is considering increasing expenditure on banner ads from the current level of \$5,000 to \$6,000 per month. What will be the resulting effect on Web site popularity?

b. According to the model, would the Web site popularity continue to grow at the same rate if he continued to raise expenditure on advertising \$1,000 each month? Explain.

c. Does this model give a reasonable prediction of traffic at expenditures larger than \$8,500 per month? Why?

34. ***Production Costs*** Over at Duffin House, Marjory Duffin is trying to decide on the size of the print runs for the best-selling new fantasy novel *Larry Plotter and the Simplex Method*. The following model shows a calculation of the total cost to produce a million copies of the novel, based on an analysis of setup and storage costs:

$$c(n) = 0.0008n^2 - 72n + 2,000,000 \text{ dollars}$$

where n is the print run size (the number of books printed in each run).

a. What would be the effect on cost if the run size was increased from 20,000 to 30,000?

b. Would increasing the run size in further steps of 10,000 result in the same changes in the total cost? Explain.

c. What approximate run size would you recommend that Marjoy Duffin use for a minimum cost?

35. ***Internet Advertising*** When OHaganBooks.com actually went ahead and increased Internet advertising from \$5,000 per month to \$6,000 per month (see Exercise 33) it was noticed that the number of new visits increased from an estimated 2,050 per day to 2,100 per day. Use this information to construct a linear model giving the average number v of new visits per day as a function of the monthly advertising expenditure c.

a. What is the model?

b. Based on the model, how many new visits per day could be anticipated if OHaganBooks.com budgets \$7,000 per month for Internet advertising?

c. The goal is to eventually increase the number of new visits to 2,500 per day. Based on the model, how much should be spent on Internet advertising in order to accomplish this?

36. ***Production Costs*** When Duffin House printed a million copies of *Larry Plotter and the Simplex Method* (see Exercise 34), it used print runs of 20,000, which cost the company \$880,000. For the sequel, *Larry Plotter and the Simplex Method, Phase 2* it used print runs of 40,000 which cost the company \$550,000. Use this information to construct a linear model giving the production cost c as a function of the run size n.

a. What is the model?

b. Based on the model, what would print runs of 25,000 have cost the company?

c. Marjory Duffin has decided to budget \$418,000 for production of the next book in the *Simplex Method* series. Based on the model, how large should the print runs be to accomplish this?

37. ***Recreation*** John O'Hagan has just returned from a sales convention at Puerto Vallarta, Mexico where, in order to win a bet he made with Marjory Duffin (Duffin House was also at the convention), he went bungee jumping at a nearby mountain retreat. The bungee cord he used had the property that a person weighing 70 kg would drop a total distance of 74.5 meters, while a 90 kg person would drop 93.5 meters. Express the distance d a jumper drops as a linear function of the jumper's weight w. John OHagan dropped 90 m. What was his approximate weight?

38. ***Crickets*** The mountain retreat near Puerto Vallarta was so quiet at night that all one could hear was the chirping of the snowy tree crickets. These crickets behave in a rather interesting way: The rate at which they chirp depends linearly on the temperature. Early in the evening, John O'Hagan counted 140 chirps/minute and noticed that the temperature was 80°F. Later in the evening the temperature dropped to 75°F, and the chirping slowed down to 120 chirps/minute. Express the temperature T as a function of the rate of chirping r. The temperature that night dropped to a low of 65°F. At approximately what rate were the crickets chirping at that point?

39. ***Break-Even Analysis*** OHaganBooks.com has recently decided to start selling music albums online through a service it calls *o'Tunes*.[11] Users pay a fee to download an entire music album. Composer royalties and copyright fees cost an average of $5.50 per album, and the cost of operating and maintaining *o'Tunes* amounts to $500 per week. The company is currently charging customers $9.50 per album.

a. What are the associated (weekly) cost, revenue, and profit functions?
b. How many albums must be sold per week in order to make a profit?
c. If the charge is lowered to $8.00 per album, how many albums must be sold per week in order to make a profit?

40. ***Break-Even Analysis*** OHaganBooks.com also generates revenue through its *o'Books* e-book service. Author royalties and copyright fees cost the company an average of $4 per novel, and the monthly cost of operating and maintaining the service amounts to $900 per month. The company is currently charging readers $5.50 per novel.

a. What are the associated cost, revenue, and profit functions?
b. How many novels must be sold per month in order to break even?
c. If the charge is lowered to $5.00 per novel, how many books must be sold in order to break even?

41. ***Demand and Profit*** In order to generate a profit from its new *o'Tunes* service, OHaganBooks.com needs to know how the demand for music albums depends on the price it charges. During the first week of the service, it was charging $7 per album, and sold 500. Raising the price to $9.50 had the effect of lowering demand to 300 albums per week.

a. Use the given data to construct a linear demand equation.
b. Use the demand equation you constructed in part (a) to estimate the demand if the price was raised to $12 per album.

[11]The (highly original) name was suggested to John O'Hagan by Marjory Duffin over cocktails one evening.

c. Using the information on cost given in Exercise 39, determine which of the three prices ($7, $9.50 and $12) would result in the largest weekly profit, and the size of that profit.

42. ***Demand and Profit*** In order to generate a profit from its *o'Books* e-book service, OHaganBooks.com needs to know how the demand for novels depends on the price it charges. During the first month of the service, it was charging $10 per novel, and sold 350. Lowering the price to $5.50 per novel had the effect of increasing demand to 620 novels per month.

a. Use the given data to construct a linear demand equation.
b. Use the demand equation you constructed in part (a) to estimate the demand if the price was raised to $15 per novel.
c. Using the information on cost given in Exercise 40, determine which of the three prices ($5.50, $10 and $15) would result in the largest profit, and the size of that profit.

43. ***Demand*** OHaganBooks.com has tried selling music albums on *o'Tunes* at a variety of prices, with the following results:

Price	$8.00	$8.50	$10	$11.50
Demand (Weekly sales)	440	380	250	180

a. Use the given data to obtain a linear regression model of demand.
b. Use the demand model you constructed in part (a) to estimate the demand if the company charged $10.50 per album. (Round the answer to the nearest album.)

44. ***Demand*** OHaganBooks.com has tried selling novels through *o'Books* at a variety of prices, with the following results:

Price	$5.50	$10	$11.50	$12
Demand (Monthly sales)	620	350	350	300

a. Use the given data to obtain a linear regression model of demand.
b. Use the demand model you constructed in part (a) to estimate the demand if the company charged $8 per novel. (Round the answer to the nearest novel.)

Case Study

Modeling Spending on Internet Advertising

Jeff Titcomb/Photographer's Choice / Getty Images

You are the new director of Impact Advertising Inc.'s Internet division, which has enjoyed a steady 0.25% of the Internet advertising market. You have drawn up an ambitious proposal to expand your division in light of your anticipation that Internet advertising will continue to skyrocket. However, upper management sees things differently and, based on the following email, does not seem likely to approve the budget for your proposal.

TO: JCheddar@impact.com (J. R. Cheddar)
CC: CVODoylePres@impact.com (C. V. O'Doyle, CEO)
FROM: SGLombardoVP@impact.com (S. G. Lombardo, VP Financial Affairs)
SUBJECT: Your Expansion Proposal
DATE: May 30, 2014

Hi John:

Your proposal reflects exactly the kind of ambitious planning and optimism we like to see in our new upper management personnel. Your presentation last week was most impressive, and obviously reflected a great deal of hard work and preparation.

I am in full agreement with you that Internet advertising is on the increase. Indeed, our Market Research department informs me that, based on a regression of the most recently available data, Internet advertising revenue in the United States will continue to grow at a rate of approximately $2.7 billion per year. This translates into approximately $6.75 million in increased revenues per year for Impact, given our 0.25% market share. This rate of expansion is exactly what our planned 2015 budget anticipates. Your proposal, on the other hand, would require a budget of approximately *twice* the 2015 budget allocation, even though your proposal provides no hard evidence to justify this degree of financial backing.

At this stage, therefore, I am sorry to say that I am inclined not to approve the funding for your project, although I would be happy to discuss this further with you. I plan to present my final decision on the 2015 budget at next week's divisional meeting.

Regards, Sylvia

Refusing to admit defeat, you contact the Market Research department and request the details of their projections on Internet advertising. They fax you the following information:[12]

Year	2007	2008	2009	2010	2011	2012	2013	2014
Internet Advertising Revenue ($ Billion)	21.2	23.4	22.7	25.8	28.5	32.6	36	40.5

Regression Model: $y = 2.744x + 19.233$ (x = time in years since 2007)

Correlation Coefficient: $r = 0.970$

Now you see where the VP got that $2.7 billion figure: The slope of the regression equation is close to 2.7, indicating a rate of increase of about $2.7 billion per year. Also, the correlation coefficient is very high—an indication that the linear model fits the data well. In view of this strong evidence, it seems difficult to argue that revenues will increase by significantly more than the projected $2.7 billion per year. To get a

[12]The 2011–2014 figures are projections by eMarketer. Source: www.eMarketer.com.

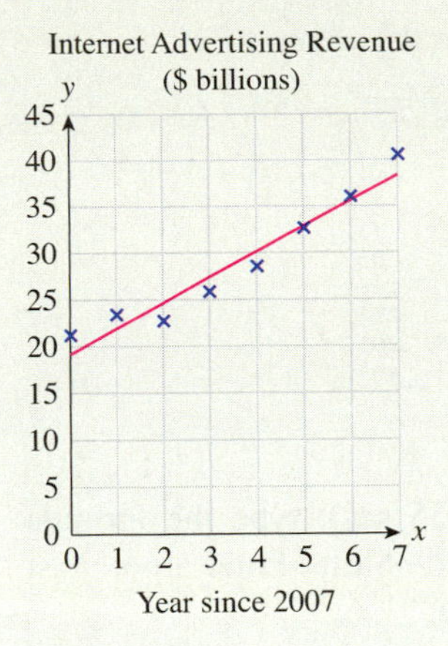

Figure 32

better picture of what's going on, you decide to graph the data together with the regression line in your spreadsheet. What you get is shown in Figure 32. You immediately notice that the data points seem to suggest a curve, and not a straight line. Then again, perhaps the suggestion of a curve is an illusion. Thus there are, you surmise, two possible interpretations of the data:

1. (Your first impression) As a function of time, Internet advertising revenue is nonlinear, and is in fact accelerating (the rate of change is increasing), so a linear model is inappropriate.
2. (Devil's advocate) Internet advertising revenue *is* a linear function of time; the fact that the points do not lie on the regression line is simply a consequence of random factors that do not reflect a long-term trend, such as world events, mergers and acquisitions, short-term fluctuations in economy or the stock market, etc.

You suspect that the VP will probably opt for the second interpretation and discount the graphical evidence of accelerating growth by claiming that it is an illusion: a "statistical fluctuation." That is, of course, a possibility, but you wonder how likely it really is.

For the sake of comparison, you decide to try a regression based on the simplest nonlinear model you can think of—a quadratic function.

$$y = ax^2 + bx + c$$

Your spreadsheet allows you to fit such a function with a click of the mouse. The result is the following.

$$y = 0.3208x^2 + 0.4982x + 21.479 \quad (x = \text{number of years since 2007})$$

$$r = 0.996$$

See Note.*

Figure 33 shows the graph of the regression function together with the original data.

Aha! The fit is visually far better, and the correlation coefficient is even higher! Further, the quadratic model predicts 2015 revenue as

$$y = 0.3208(8)^2 + 0.4982(8) + 21.479 \approx \$46.0 \text{ billion},$$

which is \$5.5 billion above the 2014 spending figure in the table above. Given Impact Advertising's 0.25% market share, this translates into an increase in revenues of \$13.75 million, which is about double the estimate predicted by the linear model!

You quickly draft an email to Lombardo, and are about to click "Send" when you decide, as a precaution, to check with a colleague who is knowledgeable in statistics. He tells you to be cautious: The value of r will always tend to increase if you pass from a linear model to a quadratic one because of the increase in "degrees of freedom."† A good way to test whether a quadratic model is more appropriate than a linear one is to compute a statistic called the "p-value" associated with the coefficient of x^2. A low value of p indicates a high degree of confidence that the coefficient of x^2 cannot be zero (see below). Notice that if the coefficient of x^2 *is* zero, then you have a linear model.

You can, your colleague explains, obtain the p-value using your spreadsheet as follows (the method we describe here works on all the popular spreadsheets, including *Excel, Google Docs,* and *Open Office Calc*).

* Note that this r is *not* the linear correlation coefficient we defined on page 80; what this r measures is how closely the *quadratic* regression model fits the data.

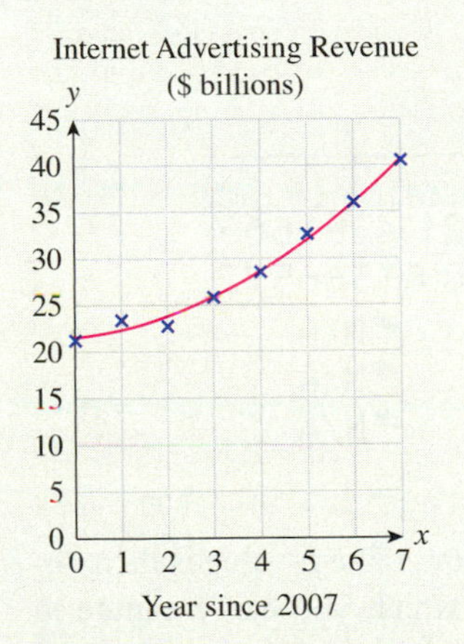

Figure 33

† The number of degrees of freedom in a regression model is 1 less than the number of coefficients. For a linear model, it is 1 (there are two coefficients: the slope m and the intercept b), and for a quadratic model it is 2. For a detailed discussion, consult a text on regression analysis.

First, set up the data in columns, with an extra column for the values of x^2:

◇	A	B	C
1	y	x	x^2
2	21.2	0	0
3	23.4	1	1
4	22.7	2	4
5	25.8	3	9
6	28.5	4	16
7	32.6	5	25
8	36	6	36
9	40.5	7	49

Then, highlight a vacant 5×3 block (the block E1:G5 say), type the formula `=LINEST(A2:A9,B2:C9,,TRUE)`, and press Cntl+Shift+Enter (not just Enter!). You will see a table of statistics like the following:

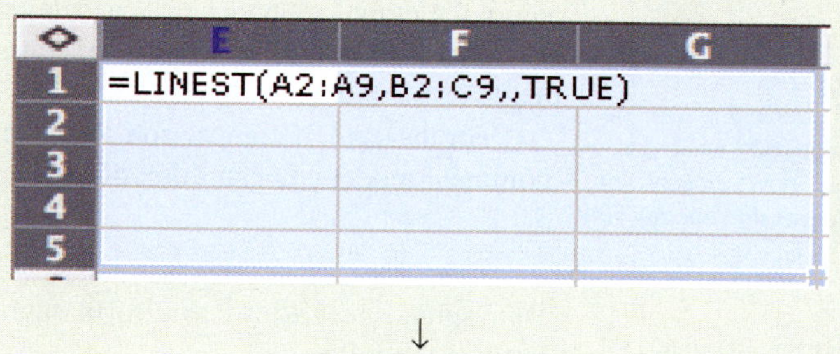

◇	E	F	G
1	=LINEST(A2:A9,B2:C9,,TRUE)		
2			
3			
4			
5			

↓

Cntl+Shift+Enter

◇	E	F	G
1	0.32083333	0.49821429	21.4791667
2	0.05808804	0.42288732	0.63366572
3	0.99157395	0.75290706	#N/A
4	294.198816	5	#N/A
5	333.544405	2.83434524	#N/A

(Notice the coefficients of the quadratic model in the first row.) The p-value is then obtained by the formula `=TDIST(ABS(E1/E2),F4,2)`, which you can compute in any vacant cell. You should get $p \approx 0.00267$.

Q: *What does p actually measure?*

A: *Roughly speaking,* $1 - p \approx 0.997733$ gives the degree of confidence you can have (99.7733%) in asserting that the coefficient of x^2 is not zero. (Technically, p is the probability—allowing for random fluctuation in the data—that, if the coefficient of x^2 were in fact zero, the ratio E1/E2 could be as large as it is.)

In short, you can go ahead and send your email with almost 100% confidence!

EXERCISES

Suppose you are given the following data for the spending on Internet advertising in a hypothetical country in which Impact Advertising also has a 0.25% share of the market.

Year	2010	2011	2012	2013	2014	2015	2016
Spending on Advertising ($ Billion)	0	0.3	1.5	2.6	3.4	4.3	5.0

1. Obtain a linear regression model and the correlation coefficient r. (Take t to be time in years since 2010.) According to the model, at what rate is spending on Internet Advertising increasing in this country? How does this translate to annual revenues for Impact Advertising?
2. Use a spreadsheet or other technology to graph the data together with the best-fit line. Does the graph suggest a quadratic model (parabola)?
3. Test your impression in the preceding exercise by using technology to fit a quadratic function and graphing the resulting curve together with the data. Does the graph suggest that the quadratic model is appropriate?
4. Perform a regression analysis using the quadratic model and find the associated p-value. What does it tell you about the appropriateness of a quadratic model?

TI-83/84 Plus Technology Guide

Section 1.1

Example 1(a) and (c) (page 40) The total number of iPods sold by Apple up to the end of year x can be approximated by $f(x) = 4x^2 + 16x + 2$ million iPods $(0 \le x \le 6)$, where $x = 0$ represents 2003. Compute $f(0)$, $f(2)$, $f(4)$, and $f(6)$, and obtain the graph of f.

Solution with Technology

You can use the Y= screen to enter an algebraically defined function.

1. Enter the function in the Y= screen, as

`Y1 = 4X^2+16X+2`

or $\quad$ `Y1 = 4X`^{2}`+16X+2`

(See Chapter 0 for a discussion of technology formulas.)

2. To evaluate $f(0)$, for example, enter $Y_1(0)$ in the Home screen to evaluate the function Y_1 at 0. Alternatively, you can use the table feature: After entering the function under Y_1, press [2ND] [TBLSET], and set `Indpnt` to `Ask`. (You do this once and for all; it will permit you to specify values for x in the table screen.) Then, press [2ND] [TABLE], and you will be able to evaluate the function at several values of x. Below (top) is a table showing the values requested:

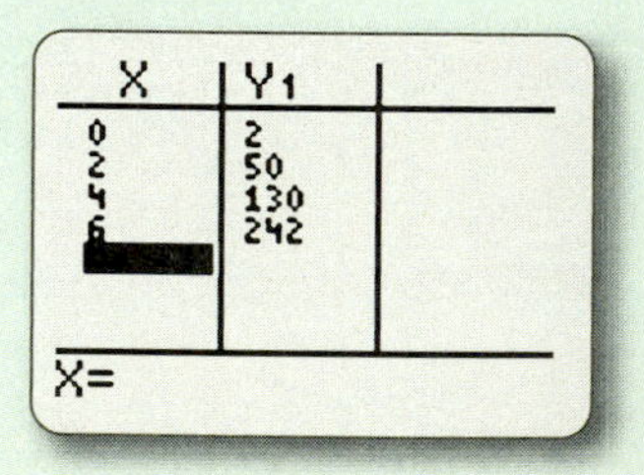

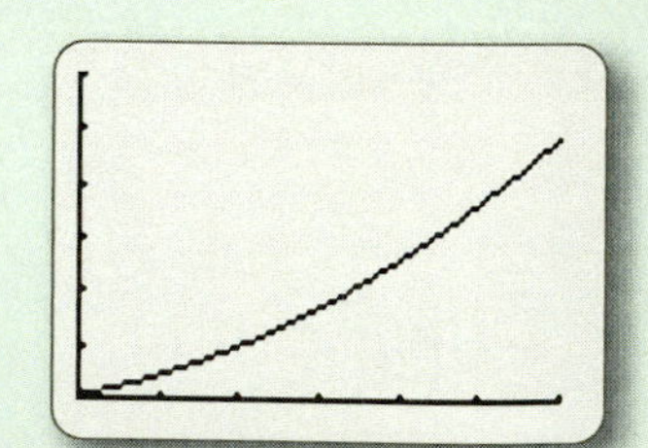

3. To obtain the graph above press [WINDOW], set Xmin = 0, Xmax = 6 (the range of x-values we are interested in), Ymin = 0, Ymax = 300 (we estimated Ymin and Ymax from the corresponding set of y-values in the table) and press [GRAPH] to obtain the curve. Alternatively, you can avoid having to estimate Ymin and Ymax by pressing ZoomFit ([ZOOM] [0]), which automatically sets Ymin and Ymax to the smallest and greatest values of y in the specified range for x.

Example 2 (page 43) The number $n(t)$ of Facebook members can be approximated by the following function of time t in years ($t = 0$ represents January 2004):

$$n(t) = \begin{cases} 4t & \text{if } 0 \le t \le 3 \\ 12 + 23(t-3)^{2.7} & \text{if } 3 < t \le 6 \end{cases} \quad \text{million members.}$$

Obtain a table showing the values $n(t)$ for $t = 0, \ldots, 6$ and also obtain the graph of n.

Solution with Technology

You can enter a piecewise-defined function using the logical inequality operators <, >, ≤, and ≥, which are found by pressing [2ND] [TEST]:

1. Enter the function n in the Y= screen as:

```
Y1=(X≤3)*(4X)+(X>3)*(12+23*abs(X-
3)^2.7)
```

When x is less than or equal to 3, the logical expression (`X≤3`) evaluates to 1 because it is true, and the expression (`X>3`) evaluates to 0 because it is false. The value of the function is therefore given by the expression (`4X`). When x is greater than 3, the expression (`X≤3`) evaluates to 0 while the expression (`X>3`) evaluates to 1, so the value of the function is given by the expression (`12+23*abs(X-3)^2.7`). (The reason we use the `abs` in the formula is to prevent an error in evaluating $(x-3)^{2.7}$ when $x < 3$; even though we don't use that formula when $x < 3$, we are in fact evaluating it and multiplying it by zero.)

2. As in Example 1, use the Table feature to compute several values of the function at once by pressing [2ND] [TABLE].

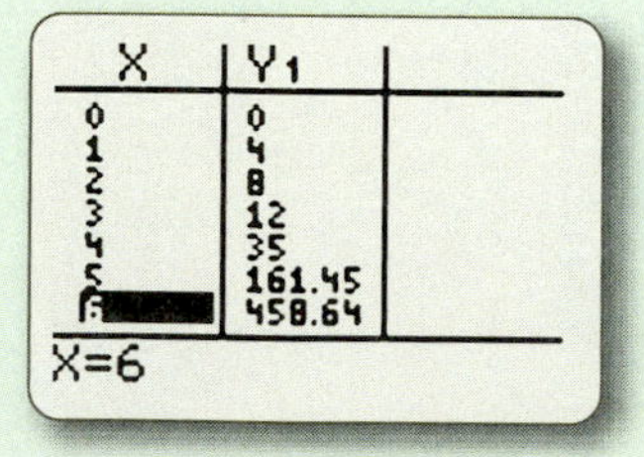

3. To obtain the graph, we proceed as in Example 1: Press [WINDOW], set Xmin = 0, Xmax = 6 (the range of x-values we are interested in), Ymin = 0, Ymax = 500 (see the y-values in the table) and press [GRAPH].

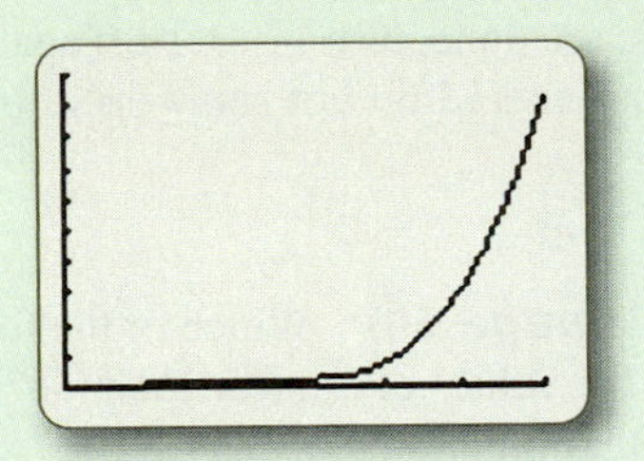

Section 1.2

Example 4(a) (page 54) The demand and supply curves for private schools in Michigan are $q = 77.8p^{-0.11}$ and $q = 30.4 + 0.006p$ thousand students, respectively $(200 \le p \le 2{,}200)$, where p is the net tuition cost in dollars. Graph the demand and supply curves on the same set of axes. Use your graph to estimate, to the nearest \$100, the tuition at which the demand equals the supply (equilibrium price). Approximately how many students will be accommodated at that price?

Solution with Technology

To obtain the graphs of demand and supply:

1. Enter `Y1=77.8*X^(-0.11)` and `Y2=30.4+0.006*X` in the "Y=" screen.
2. Press WINDOW, enter Xmin = 200, Xmax = 2200, Ymin = 0, Ymax = 50 and press GRAPH for the graph shown below:

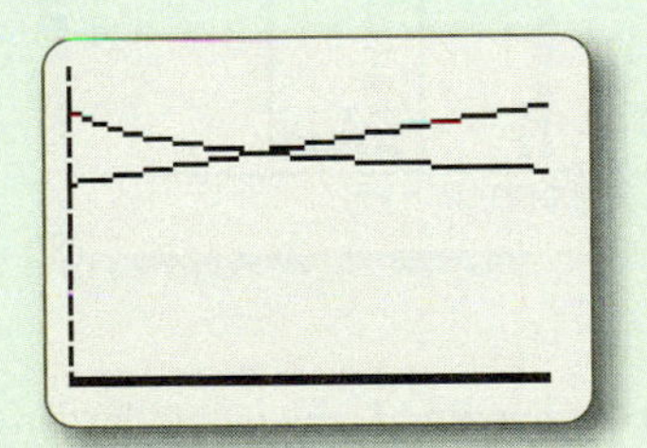

3. To estimate the equilibrium price, press TRACE and use the arrow keys to follow the curve to the approximate point of intersection (around X = 1008) as shown below.

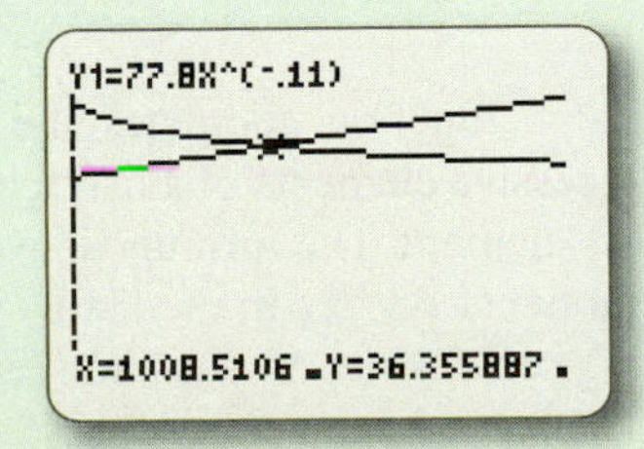

4. For a more accurate estimate, zoom in by pressing ZOOM and selecting Option 1 `ZBox`.
5. Move the curser to a point slightly above and to the left of the intersection, press ENTER, and then move the curser to a point slightly below and to the right and press ENTER again to obtain a box.

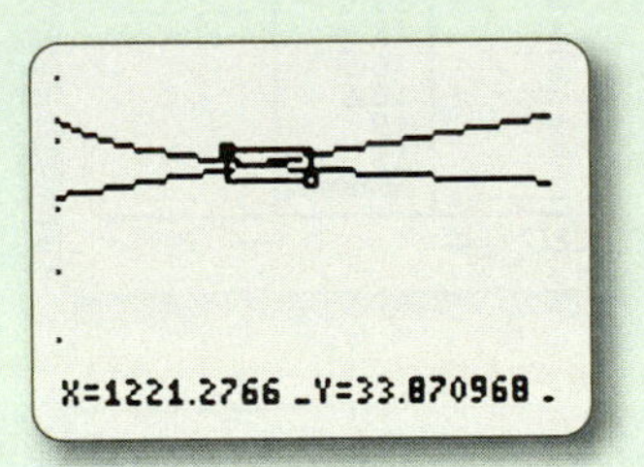

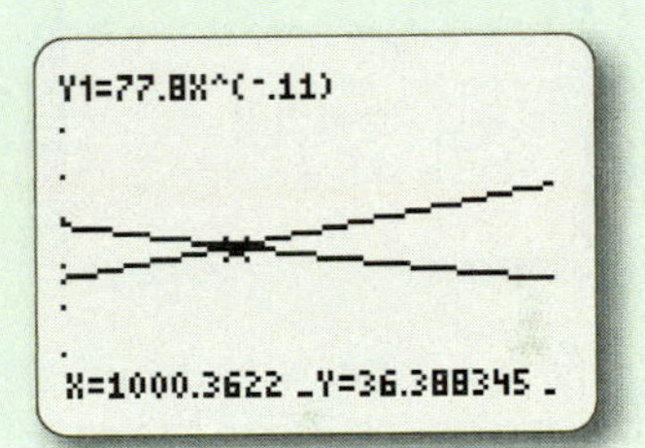

6. Now press ENTER again for a zoomed-in view of the intersection.
7. You can now use TRACE to obtain the intersection coordinates more accurately: X ≈ 1,000, representing a tuition cost of \$1,000. The associated demand is the Y-coordinate: around 36.4 thousand students.

Example 5(a) (page 56) The following table shows annual sales, in billions of dollars, by *Nike* from 2005 through 2010 ($t = 0$ represents 2005):

t	0	1	2	3	4	5
Sales (\$ billion)	13.5	15	16.5	18.5	19	19

Consider the following four models:

(1) $s(t) = 14 + 1.2t$ Linear model

(2) $s(t) = 13 + 2.2t - 0.2t^2$ Quadratic model

(3) $s(t) = 14(1.07^t)$ Exponential model

(4) $s(t) = \dfrac{19.5}{1 + 0.48(1.8^{-t})}$ Logistic model

a. Which models fit the data significantly better than the rest?

b. Of the models you selected in part (a), which gives the most reasonable prediction for 2013?

TECHNOLOGY GUIDE

Solution with Technology

1. First enter the actual revenue data in the stat list editor (STAT EDIT) with the values of t in L_1, and the values of $s(t)$ in L_2.

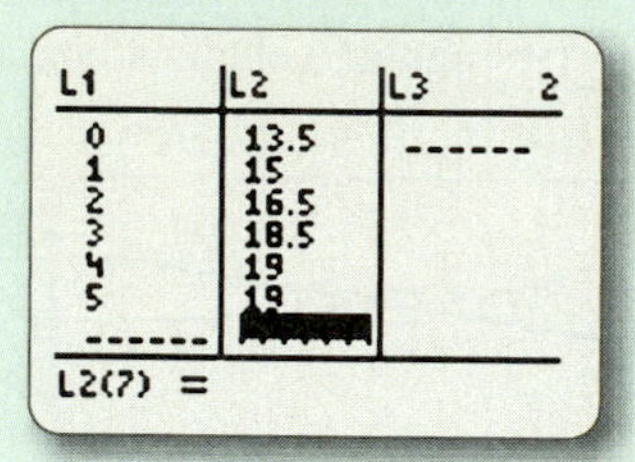

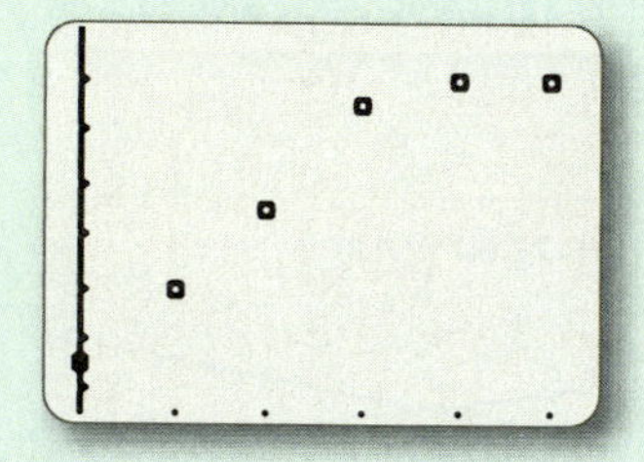

2. Now go to the Y= window and turn `Plot1` on by selecting it and pressing ENTER. (You can also turn it on in the 2ND STAT PLOT screen.) Then press ZoomStat (ZOOM 9) to obtain a plot of the points (above).
3. To see any of the four curves plotted along with the points, enter its formula in the Y= screen (for instance, Y_1=13+2.2x−0.2X^2 for the second model) and press GRAPH (figure on top below).

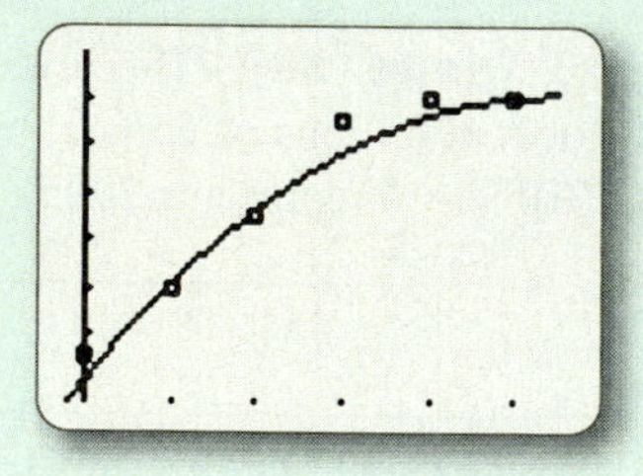

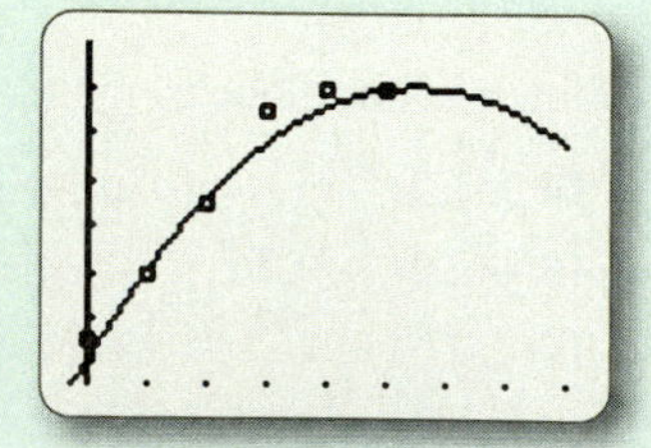

4. To see the extrapolation of the curve to 2013, just change Xmax to 8 (in the WINDOW screen) and press GRAPH again (lower figure above).
5. Now change Y_1 to see similar graphs for the remaining curves.
6. When you are done, turn `Plot1` off again so that the points you entered do not show up in other graphs.

Section 1.3

Example 1 (page 66) Which of the following two tables gives the values of a linear function? What is the formula for that function?

x	0	2	4	6	8	10	12
$f(x)$	3	−1	−3	−6	−8	−13	−15

x	0	2	4	6	8	10	12
$g(x)$	3	−1	−5	−9	−13	−17	−21

Solution with Technology

We can use the "List" feature in the TI-83/84 Plus to automatically compute the successive quotients $m = \Delta y/\Delta x$ for either f or g as follows:

1. Use the stat list editor (STAT EDIT) to enter the values of x and $f(x)$ in the first two columns, called L_1 and L_2, as shown in the screenshot below. (If there is already data in a column you want to use, you can clear it by highlighting the column heading (e.g., L_1) using the arrow key, and pressing CLEAR ENTER .)

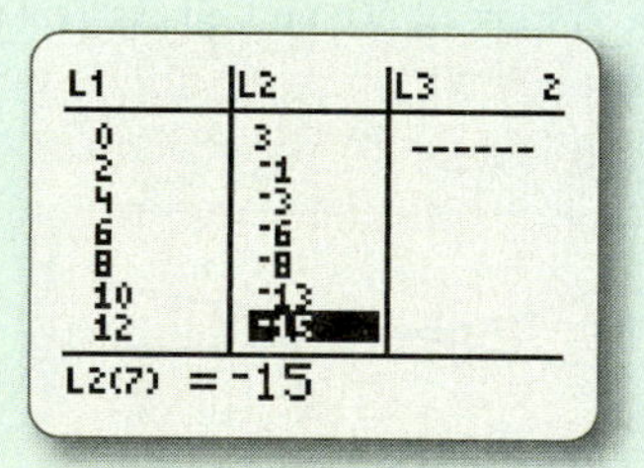

2. Highlight the heading L_3 by using the arrow keys, and enter the following formula (with the quotes, as explained below):

"ΔList(L_2)/ΔList(L_1)"   ΔList is found under 2ND LIST OPS. L_1 is 2ND 1

The "ΔList" function computes the differences between successive elements of a list, returning a list with one less element. The formula above then computes the quotients $\Delta y/\Delta x$ in the list L_3 as shown in the following screenshot. As you can see in the third column, $f(x)$ is not linear.

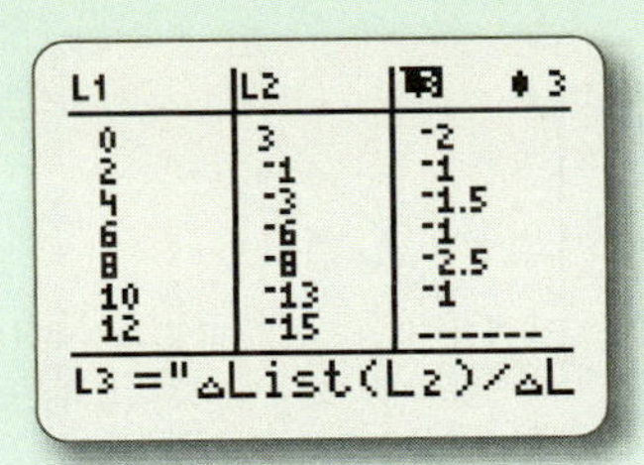

3. To redo the computation for $g(x)$, all you need to do is edit the values of L_2 in the stat list editor. By putting quotes around the formula we used for L_3, we told the calculator to remember the formula, so it automatically recalculates the values.

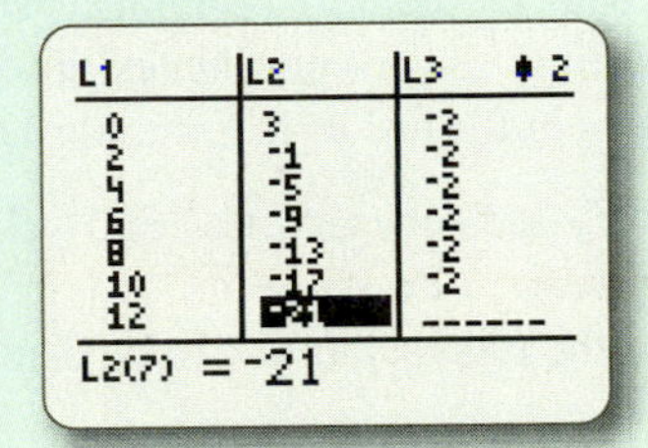

Example 2(a) (page 68) Find the equation of the line through the points (1, 2) and (3, −1).

Solution with Technology

1. Enter the coordinates of the given points in the stat list editor (STAT EDIT) with the values of x in L_1, and the values of y in L_2.

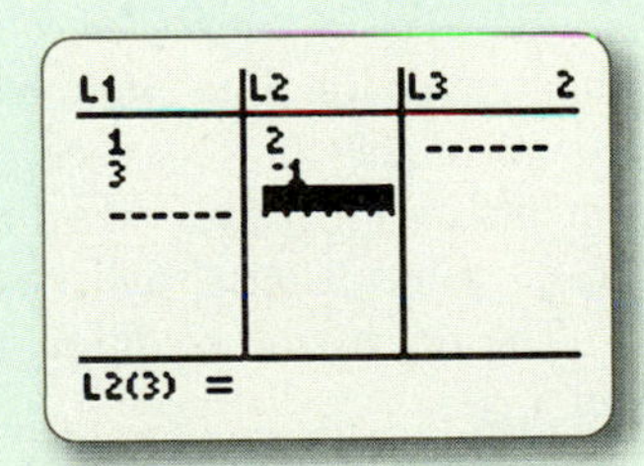

2. To compute the slope, enter the following formula in the Home screen:

   `(L2(2)-L2(1))/(L1(2)-L1(1))→M`

   L_1 and L_2 are under 2ND LIST and the arrow is STO

3. Then, to compute the y-intercept, enter

   `L2(1)-M*L1(1)`

Section 1.4

Example 1 (page 76) Using the data on the per capita GDP in Mexico given at the beginning of Section 1.4, compute SSE, the sum-of-squares error, for the linear models $y = 0.5t + 8$ and $y = 0.25t + 9$, and graph the data with the given models.

Solution with Technology

We can use the "List" feature in the TI-83/84 Plus to automate the computation of SSE.

1. Use the stat list editor (STAT EDIT) to enter the given data in the lists L_1 and L_2, as shown in the first screenshot below. (If there is already data in a column you want to use, you can clear it by highlighting the column heading (e.g., L_1) using the arrow key, and pressing CLEAR ENTER .)
2. To compute the predicted values, highlight the heading L_3 using the arrow keys, and enter the following formula for the predicted values (figure on the top below):

   `0.5*L1+8` L_1 is 2ND 1

   Pressing ENTER again will fill column 3 with the predicted values (below bottom). Note that only seven of the eight data points can be seen on the screen at one time.

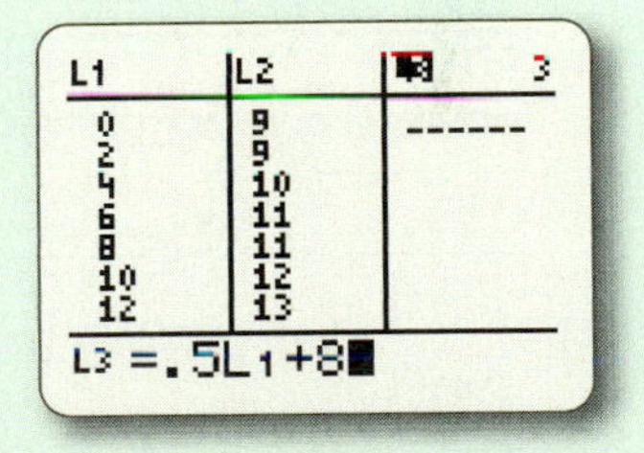

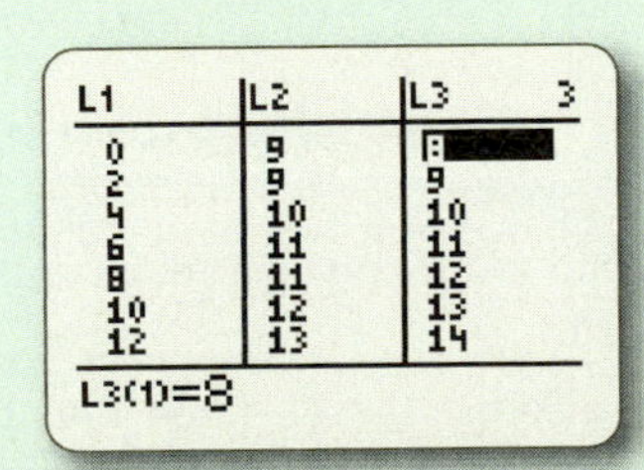

3. Highlight the heading L_4 and enter the following formula (including the quotes):

   `"(L2-L3)^2"`   Squaring the residuals

4. Pressing ENTER will fill L_4 with the squares of the residuals. (Putting quotes around the formula will allow us to easily check the second model, as we shall see.)

5. To compute SSE, the sum of the entries in L_4, go to the home screen and enter `sum(L4)` (see below; "sum" is under 2ND LIST MATH.)

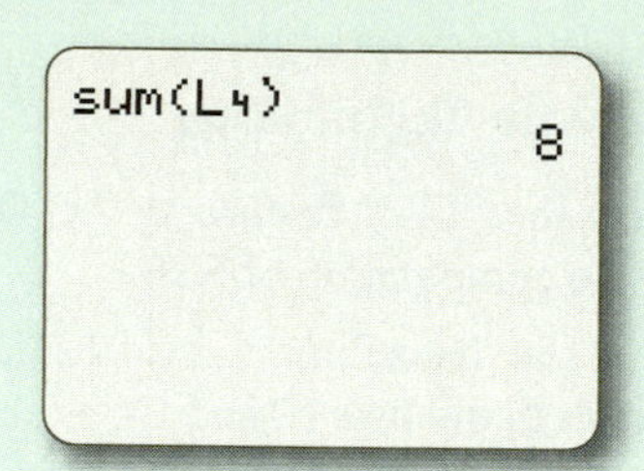

6. To check the second model, go back to the List screen, highlight the heading L_3, enter the formula for the second model, `0.25*L1+9`, and press ENTER. Because we put quotes around the formula for the residuals in L_4, the TI-83/84 Plus will remember the formula and automatically recalculate the values (below top). On the home screen we can again calculate `sum(L4)` to get SSE for the second model (below bottom).

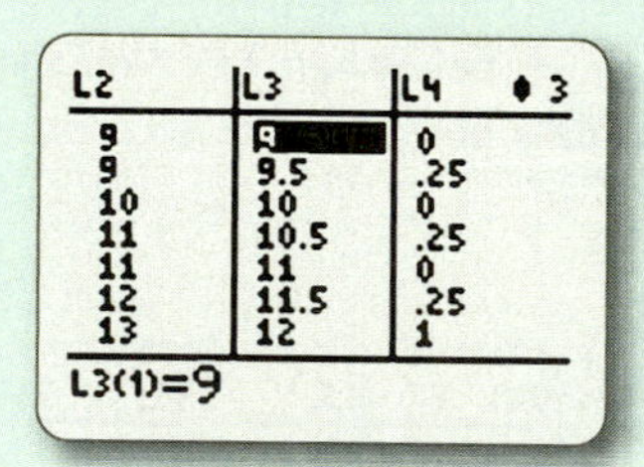

L2	L3	L4
9	9	0
9	9.5	.25
10	10	0
11	10.5	.25
11	11	0
12	11.5	.25
13	12	1

L3(1)=9

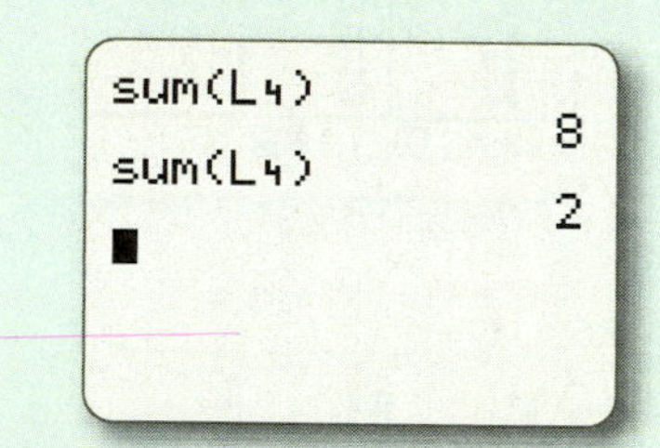

The second model gives a much smaller SSE, so is the better fit.

7. You can also use the TI-83/84 Plus to plot both the original data points and the two lines (see below). Turn `Plot1` on in the STAT PLOT window, obtained by pressing 2ND STAT PLOT. To show the lines, enter them in the "Y=" screen as usual. To obtain a convenient window showing all the points and the lines, press ZOOM and choose 9: `ZoomStat`.

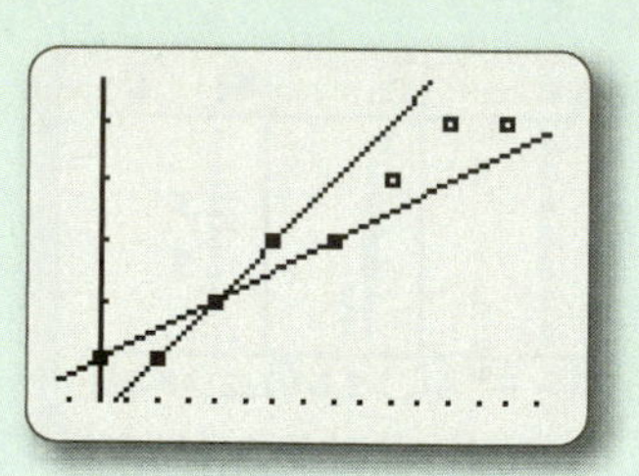

Example 2 (page 79) Use the data on the per capita GDP in Mexico to find the best-fit linear model.

Solution with Technology

1. Enter the data in the TI-83/84 Plus using the List feature, putting the x-coordinates in L_1 and the y-coordinates in L_2, just as in Example 1.

2. Press STAT, select CALC, and choose `#4: LinReg(ax+b)`. Pressing ENTER will cause the equation of the regression line to be displayed in the home screen:

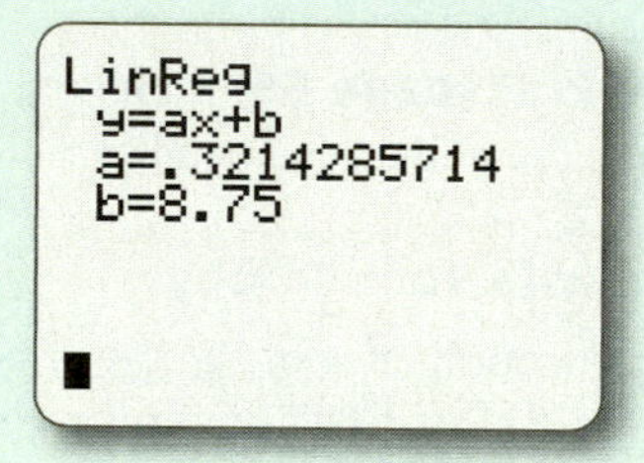

So, the regression line is $y \approx 0.321x + 8.75$.

3. To graph the regression line without having to enter it by hand in the `"Y="` screen, press Y=, clear the contents of Y_1, press VARS, choose `#5: Statistics`, select `EQ`, and then choose `#1:RegEQ`. The regression equation will then be entered under Y_1.

4. To simultaneously show the data points, press 2ND STATPLOT and turn `Plot1` on as in Example 1. To obtain a convenient window showing all the points and the line (see below), press ZOOM and choose `#9: ZoomStat`.

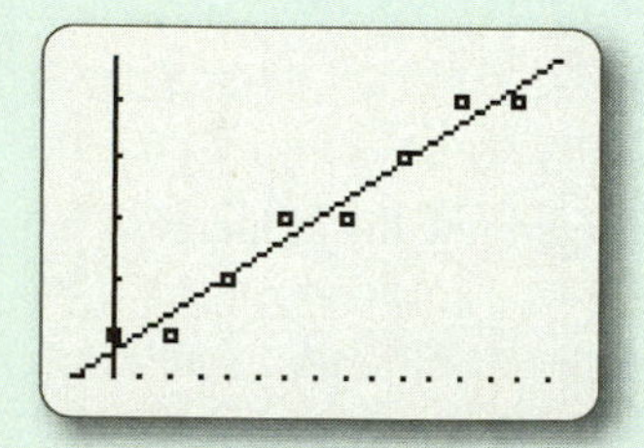

Example 3 (page 81) Find the correlation coefficient for the data in Example 2.

Solution with Technology

To find the correlation coefficient using a TI-83/84 Plus you need to tell the calculator to show you the coefficient at the same time that it shows you the regression line.

1. Press [2ND] [CATALOG] and select `DiagnosticOn` from the list. The command will be pasted to the home screen, and you should then press [ENTER] to execute the command.
2. Once you have done this, the "`LinReg(ax+b)`" command (see the discussion for Example 2) will show you not only a and b, but r and r^2 as well:

```
LinReg
 y=ax+b
 a=.3214285714
 b=8.75
 r²=.9642857143
 r=.9819805061
```


SPREADSHEET Technology Guide

Section 1.1

Example 1(a) and (c) (page 40) The total number of iPods sold by Apple up to the end of year x can be approximated by $f(x) = 4x^2 + 16x + 2$ million iPods ($0 \le x \le 6$), where $x = 0$ represents 2003. Compute $f(0)$, $f(2)$, $f(4)$, and $f(6)$, and obtain the graph of f.

Solution with Technology

To create a table of values of f using a spreadsheet:

1. Set up two columns: one for the values of x and one for the values of $f(x)$. Then enter the sequence of values 0, 2, 4, 6 in the x column as shown.

	A	B
1	x	f(x)
2	0	
3	2	
4	4	
5	6	

2. Now we enter a formula for $f(x)$ in cell B2 (below). The technology formula is `4*x^2+16*x+2`. To use this formula in a spreadsheet, we modify it slightly:

`=4*A2^2+16*A2+2` Spreadsheet version of tech formula

Notice that we have preceded the Excel formula by an equals sign (=) and replaced each occurrence of x by the name of the cell holding the value of x (cell A2 in this case).

	A	B	C
1	x	f(x)	
2	0	=4*A2^2+16*A2+2	
3	2		
4	4		
5	6		

Note Instead of typing in the name of the cell "A2" each time, you can simply click on the cell A2, and "A2" will be automatically inserted. ■

3. Now highlight cell B2 and drag the **fill handle** (the little square at the lower right-hand corner of the selection) down until you reach Row 5 as shown below on the top, to obtain the result shown on the bottom.

	A	B	C
1	x	f(x)	
2	0	=4*A2^2+16*A2+2	
3	2		
4	4		
5	6		

	A	B
1	x	f(x)
2	0	2
3	2	50
4	4	130
5	6	242

4. To graph the data, highlight A1 through B5, and insert a "Scatter chart" (the exact method of doing this depends on the specific version of the spreadsheet program). When choosing the style of the chart,

choose a style that shows points connected by lines (if possible) to obtain a graph something like the following:

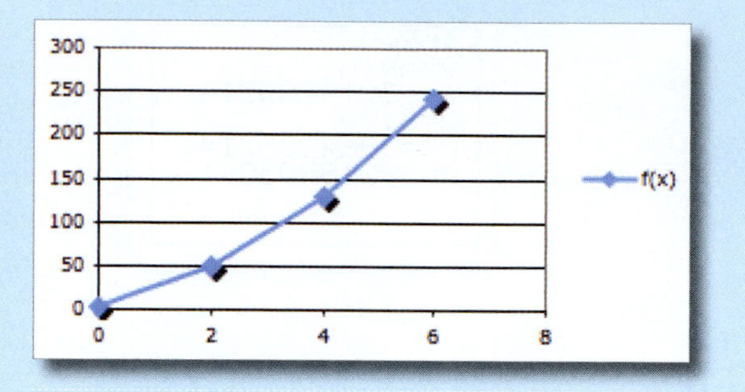

Example 2 (page 43) The number $n(t)$ of Facebook members can be approximated by the following function of time t in years ($t = 0$ represents January 2004):

$$n(t) = \begin{cases} 4t & \text{if } 0 \le t \le 3 \\ 12 + 23(t-3)^{2.7} & \text{if } 3 < t \le 6 \end{cases} \quad \text{million members.}$$

Obtain a table showing the values $n(t)$ for $t = 0, \ldots, 6$ and also obtain the graph of n.

Solution with Technology

We can generate a table of values of $n(t)$ for $t = 0, 1, \ldots, 6$ as follows:

1. Set up two columns; one for the values of t and one for the values of $n(t)$, and enter the values 0, 1, . . . , 6 in the t column as shown in the first screenshot below:
2. We must now enter the formula for n in cell B2. The following formula defines the function n in Excel:

```
=(x<=3)*(4*x)+(x>3)*(12+23*abs
 (x-3)^2.7)
```

When x is less than or equal to 3, the logical expression `(x≤3)` evaluates to 1 because it is true, and the expression `(x>3)` evaluates to 0 because it is false. The value of the function is therefore given by the expression (`4*x`). When x is greater than 3, the expression `(x≤3)` evaluates to 0 while the expression `(x>3)` evaluates to 1, so the value of the function is given by the expression `(12+23*abs(x-3)^2.7)`. (The reason we use the `abs` in the formula is to prevent an error in evaluating $(x-3)^{2.7}$ when $x < 3$; even though we don't use that formula when $x < 3$, we are in fact evaluating it and multiplying it by zero.) We therefore enter the formula

```
=(A2<=3)*(4*A2)+(A2>3)*(12+23
 *ABS(A2-3)^2.7)
```

in cell B2 and then copy down to cell B8 (below top) to obtain the result shown on the bottom:

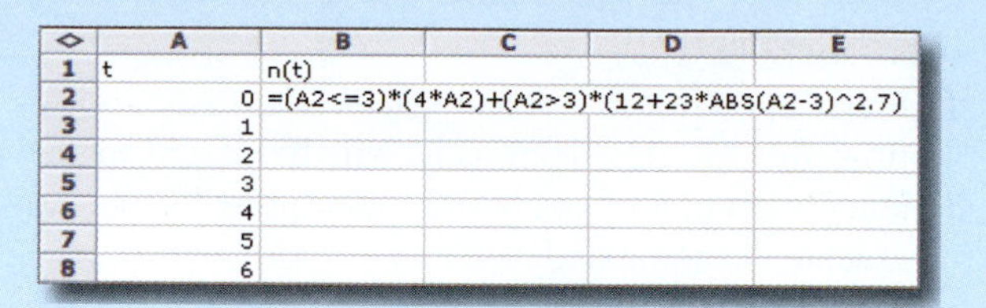

	A	B	C	D	E
1	t	n(t)			
2	0	=(A2<=3)*(4*A2)+(A2>3)*(12+23*ABS(A2-3)^2.7)			
3	1				
4	2				
5	3				
6	4				
7	5				
8	6				

	A	B
1	t	n(t)
2	0	0
3	1	4
4	2	8
5	3	12
6	4	35
7	5	161.454
8	6	458.638

3. To graph the data, highlight A1 through B8, and insert a "Scatter chart" as in Example 1 to obtain the result shown below:

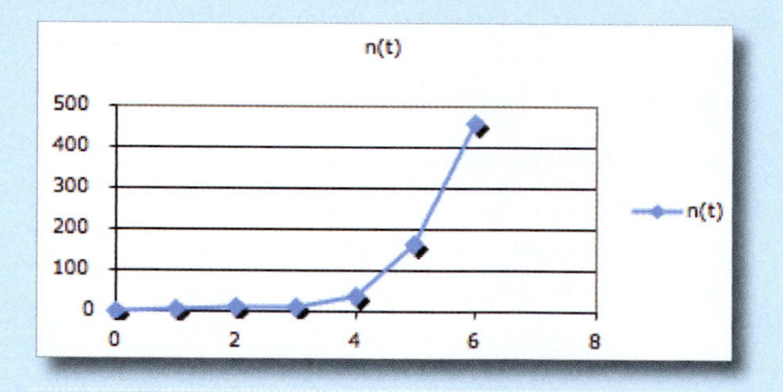

Section 1.2

Example 4(a) (page 54) The demand and supply curves for private schools in Michigan are $q = 77.8p^{-0.11}$ and $q = 30.4 + 0.006p$ thousand students, respectively ($200 \le p \le 2{,}200$), where p is the net tuition cost in dollars. Graph the demand and supply curves on the same set of axes. Use your graph to estimate, to the nearest \$100, the tuition at which the demand equals the supply (equilibrium price). Approximately how many students will be accommodated at that price?

Solution with Technology

To obtain the graphs of demand and supply:

1. Enter the headings p, Demand, and Supply in cells A1–C1 and the p-values 200, 300, . . . , 2,200 in A2–A22.

2. Next, enter the formulas for the demand and supply functions in cells B2 and C2.

Demand: `=77.8*A2^(-0.11)` in cell B2

Supply: `=30.4+0.006*A2` in cell C2

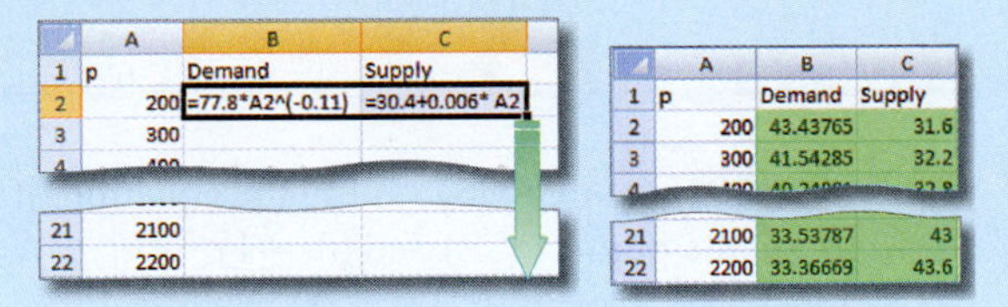

3. To graph the data, highlight A1 through C22, and insert a Scatter chart:

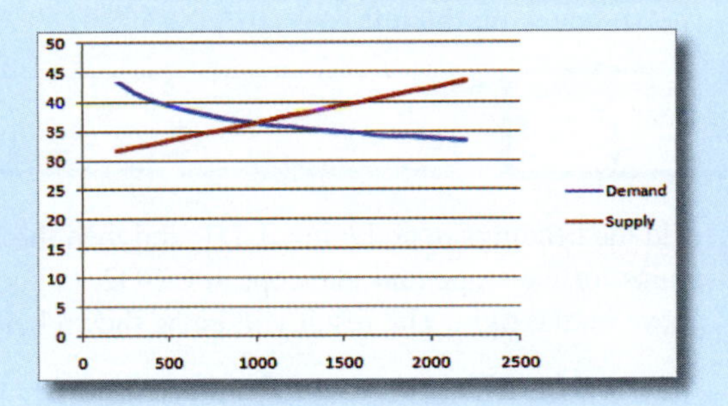

4. If you place the cursor as close as you can get to the intersection point (or just look at the table of values), you will see that the curves cross close to $p = \$1{,}000$ (to the nearest \$100).

5. To more accurately determine where the curves cross, you can narrow down the range of values shown on the x-axis by changing the p-values to 990, 991, . . . , 1010.

Example 5(a) (page 56) The following table shows annual sales, in billions of dollars, by *Nike* from 2005 through 2010 ($t = 0$ represents 2005):

t	0	1	2	3	4	5
Sales ($ billion)	13.5	15	16.5	18.5	19	19

Consider the following four models:

(1) $s(t) = 14t + 1.2t$ Linear model

(2) $s(t) = 13 + 2.2t - 0.2t^2$ Quadratic model

(3) $s(t) = 14(1.07^t)$ Exponential model

(4) $s(t) = \dfrac{19.5}{1 + 0.48(1.8^{-t})}$ Logistic model

a. Which models fit the data significantly better than the rest?

b. Of the models you selected in part (a), which gives the most reasonable prediction for 2013?

Solution with Technology

1. First create a scatter plot of the given data by tabulating the data as shown below, and selecting the Insert tab and choosing a "Scatter" chart:

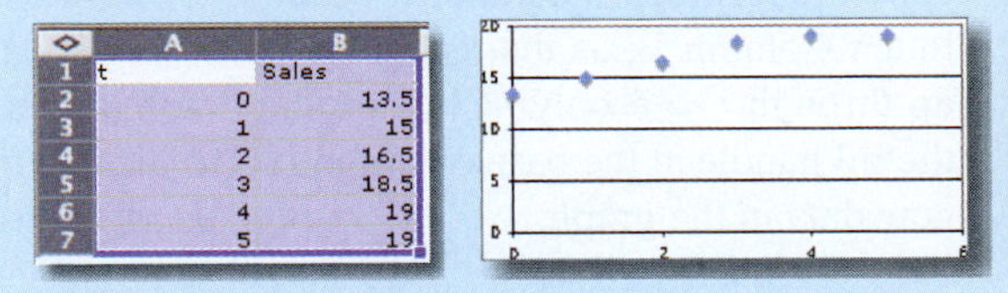

2. In column C use the formula for the model you are interested in seeing; for example, model (2): `=13+2.2*A2-0.2*A2^2`

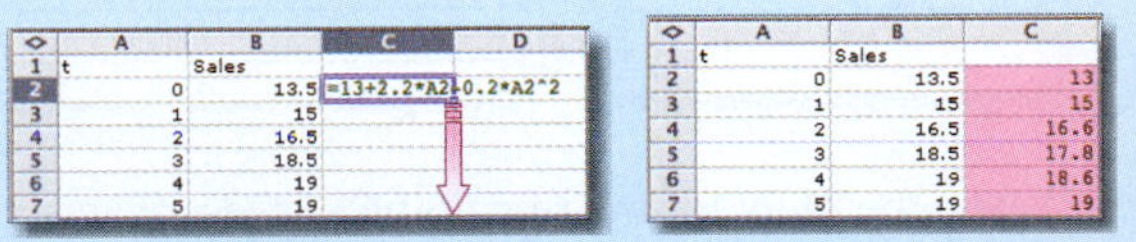

3. To adjust the graph to include the graph of the model you have added, you need to change the graph data from \$A\$2:\$B\$7 to \$A\$2:\$C\$7 so as to include column C. In Excel you can obtain this by right-clicking on the graph to select "Source Data". In OpenOffice, double-click on the graph and then right-click it to choose "Data Ranges". In Excel, you can also click once on the graph—the effect will be to outline the data you have graphed in columns A and B—and then use the fill handle at the bottom of Column B to extend the selection to Column C as shown:

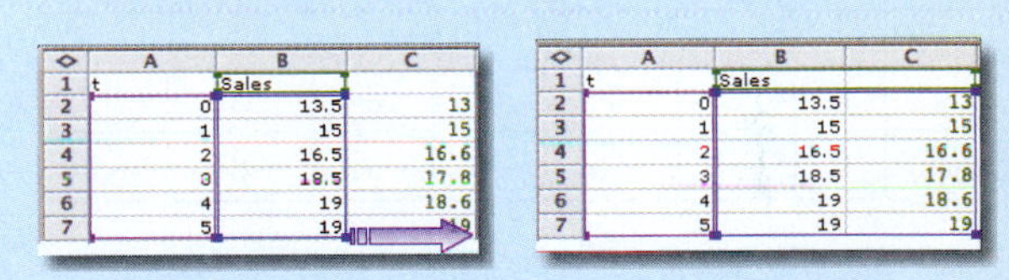

The graph will now include markers showing the values of both the actual sales and the model you inserted in Column C.

4. Right-click on any of the markers corresponding to column B in the graph (in OpenOffice you would first double-click on the graph), select "Format data series" to add lines connecting the points and remove the markers. The effect will be as shown below, with the model represented by a curve and the actual data points represented by dots:

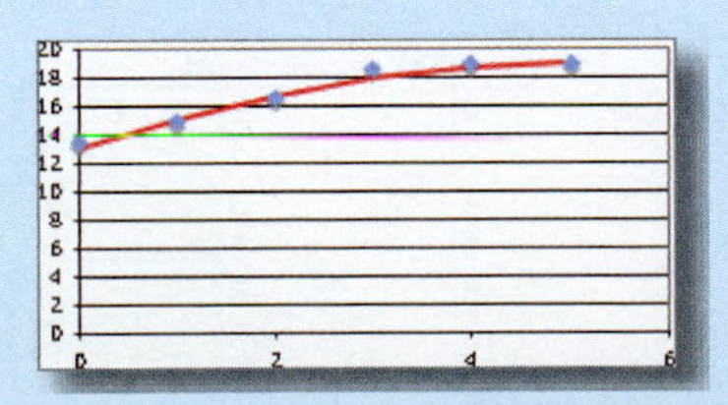

5. To see the extrapolation of the curve to 2013, add the values 6, 7, 8 to Column A. The values of $s(t)$ may automatically be computed in Column C as you type, depending on the spreadsheet. If not, you will need to copy the formula in column C down to C10. (Do not touch Column B, as that contains the observed data up through $t = 5$ only.) Click on the graph, and use the fill handle at the base of Column C to include the new data in the graph:

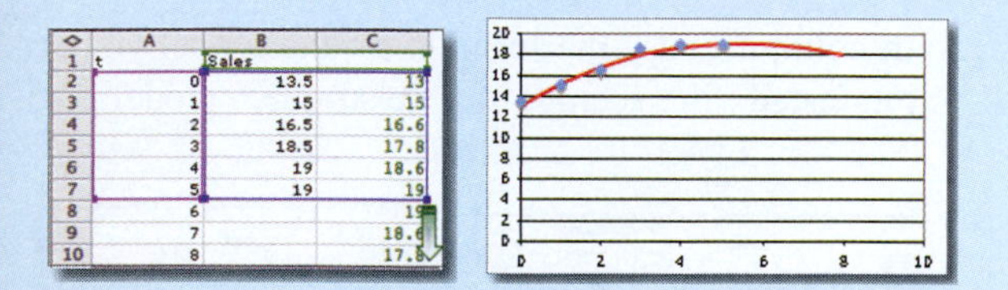

	A	B	C
1	t	Sales	
2	0	13.5	13
3	1	15	15
4	2	16.5	16.6
5	3	18.5	17.8
6	4	19	18.6
7	5	19	19
8	6		19
9	7		18.6
10	8		17.8

6. To see the plots for the remaining curves, change the formula in Column B (and don't forget to copy the new formula down to cell C10 when you do so).

Section 1.3

Example 1 (page 66) Which of the following two tables gives the values of a linear function? What is the formula for that function?

x	0	2	4	6	8	10	12
$f(x)$	3	−1	−3	−6	−8	−13	−15

x	0	2	4	6	8	10	12
$g(x)$	3	−1	−5	−9	−13	−17	−21

Solution with Technology

1. The following worksheet shows how you can compute the successive quotients $m = \Delta y/\Delta x$, and hence check whether a given set of data shows a linear relationship, in which case all the quotients will be the same. (The shading indicates that the formula is to be copied down only as far as cell C7. Why not cell C8?)

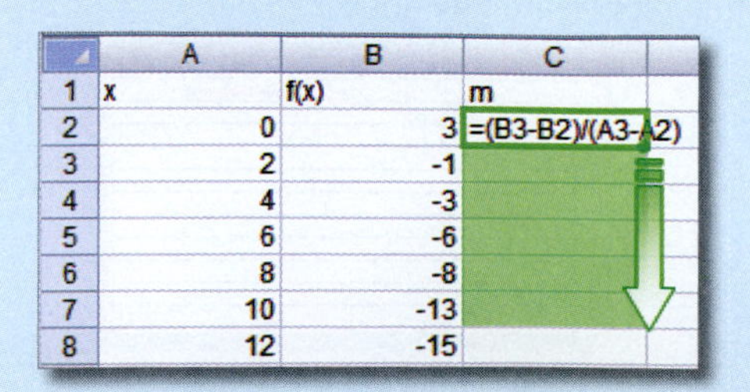

	A	B	C
1	x	f(x)	m
2	0	3	=(B3-B2)/(A3-A2)
3	2	-1	
4	4	-3	
5	6	-6	
6	8	-8	
7	10	-13	
8	12	-15	

2. Here are the results for both f and g.

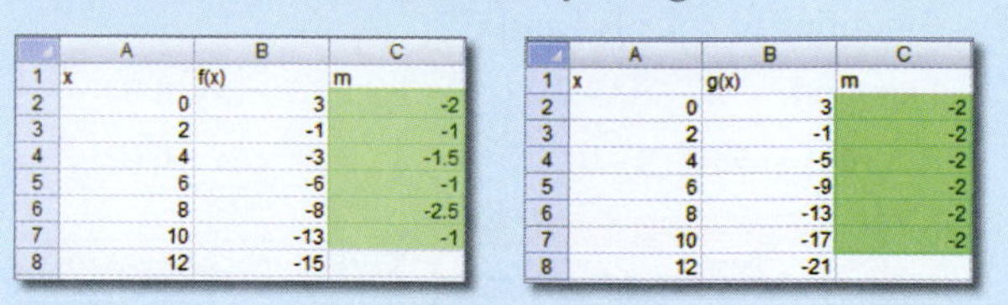

	A	B	C
1	x	f(x)	m
2	0	3	-2
3	2	-1	-1
4	4	-3	-1.5
5	6	-6	-1
6	8	-8	-2.5
7	10	-13	-1
8	12	-15	

	A	B	C
1	x	g(x)	m
2	0	3	-2
3	2	-1	-2
4	4	-5	-2
5	6	-9	-2
6	8	-13	-2
7	10	-17	-2
8	12	-21	

Example 2(a) (page 68) Find the equation of the line through the points (1, 2) and (3, −1).

Solution with Technology

1. Enter the x- and y-coordinates in columns A and B, as shown below on the left.

	A	B
1	x	y
2	1	2
3	3	-1

	A	B	C	D
1	x	y	m	b
2	1	2	=(B3-B2)/(A3-A2)	=B2-C2*A2
3	3	-1	Slope	Intercept

2. Add the headings m and b in C1-D1, and then the formulas for the slope and intercept in C2-D2, as shown above on the right. The result will be as shown below:

	A	B	C	D
1	x	y	m	b
2	1	2	-1.5	3.5
3	3	-1	Slope	Intercept

Section 1.4

Example 1 (page 76) Using the data on the per capita GDP in Mexico given at the beginning of Section 1.4, compute SSE, the sum-of-squares error, for the linear models $y = 0.5t + 8$ and $y = 0.25t + 9$, and graph the data with the given models.

Solution with Technology

1. Begin by setting up your worksheet with the observed data in two columns, t and y, and the predicted data for the first model in the third.

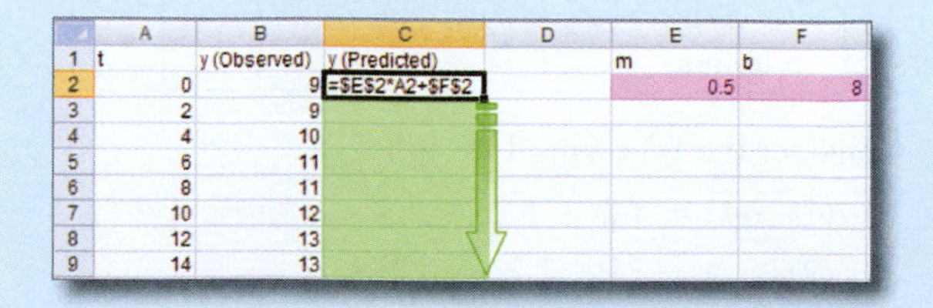

	A	B	C	D	E	F
1	t	y (Observed)	y (Predicted)		m	b
2	0	9	=$E$2*A2+$F$2		0.5	8
3	2	9				
4	4	10				
5	6	11				
6	8	11				
7	10	12				
8	12	13				
9	14	13				

2. Notice that, instead of using the numerical equation for the first model in column C, we used absolute references to the cells containing the slope m and the intercept b. This way, we can switch from one linear model to the next by changing only m and b in cells E2 and F2. (We have deliberately left column D empty in anticipation of the next step.)

3. In column D we compute the squares of the residuals using the Excel formula `=(B2-C2)^2`.

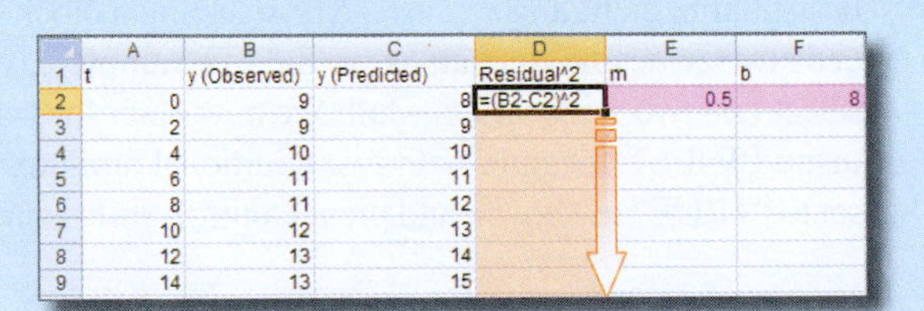

	A	B	C	D	E	F
1	t	y (Observed)	y (Predicted)	Residual^2	m	b
2	0	9	8	=(B2-C2)^2	0.5	8
3	2	9	9			
4	4	10	10			
5	6	11	11			
6	8	11	12			
7	10	12	13			
8	12	13	14			
9	14	13	15			

4. We now compute SSE in cell F4 by summing the entries in column D:

	A	B	C	D	E	F
1	t	y (Observed)	y (Predicted)	Residual^2	m	b
2	0	9	8	1	0.5	8
3	2	9	9	0		
4	4	10	10	0	SSE:	=SUM(D2:D9)
5	6	11	11	0		
6	8	11	12	1		
7	10	12	13	1		
8	12	13	14	1		
9	14	13	15	4		

5. Here is the completed spreadsheet:

	A	B	C	D	E	F
1	t	y (Observed)	y (Predicted)	Residual^2	m	b
2	0	9	8	1	0.5	8
3	2	9	9	0		
4	4	10	10	0	SSE:	8
5	6	11	11	0		
6	8	11	12	1		
7	10	12	13	1		
8	12	13	14	1		
9	14	13	15	4		

6. Changing m to 0.25 and b to 9 gives the sum of squares error for the second model, SSE $= 2$.

	A	B	C	D	E	F
1	t	y (Observed)	y (Predicted)	Residual^2	m	b
2	0	9	9	0	0.25	9
3	2	9	9.5	0.25		
4	4	10	10	0	SSE:	2
5	6	11	10.5	0.25		
6	8	11	11	0		
7	10	12	11.5	0.25		
8	12	13	12	1		
9	14	13	12.5	0.25		

7. To plot both the original data points and each of the two lines, use a scatter plot to graph the data in columns A through C in each of the last two worksheets above.

$y = 0.5t + 8$ $\qquad$ $y = 0.25t + 9$

Example 2 (page 79) Use the data on the per capita GDP in Mexico to find the best-fit linear model.

Solution with Technology

Here are two spreadsheet shortcuts for linear regression; one graphical and one based on a spreadsheet formula:

Using a Trendline

1. Start with the original data and insert a scatter plot (below left and right).

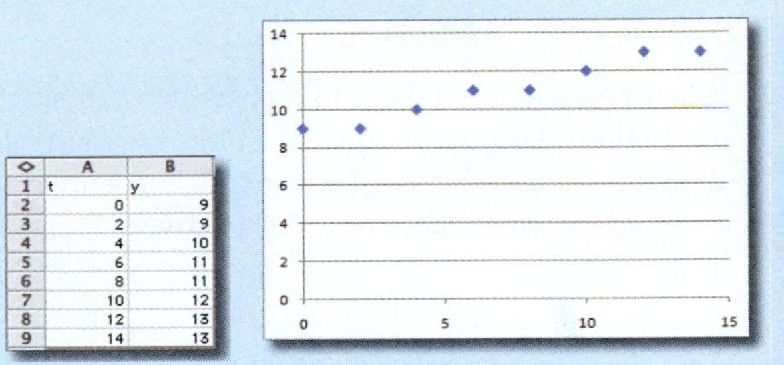

	A	B
1	t	y
2	0	9
3	2	9
4	4	10
5	6	11
6	8	11
7	10	12
8	12	13
9	14	13

2. Insert a "linear trendline", choosing the option to display the equation on the chart. The method for doing so varies from spreadsheet to spreadsheet.[13] In Excel, you can right-click on one of the points in the graph and choose "Add Trendline" (in OpenOffice you would first double-click on the graph). Then, under "Trendline Options", select "Display Equation on chart". The procedure for OpenOffice is almost identical, but you first need to double-click on the graph. The result is shown below.

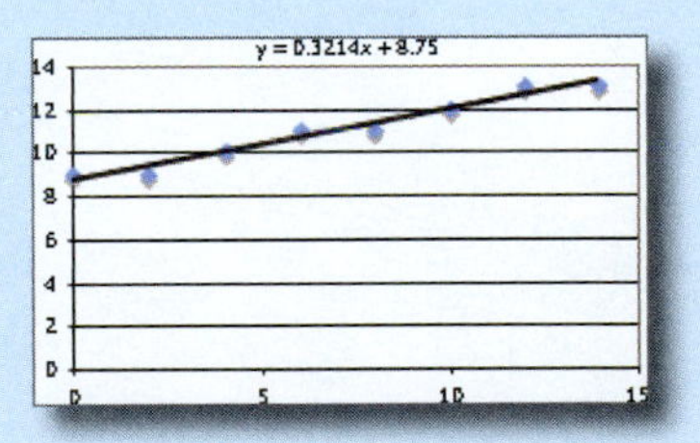

Using a Formula

1. Enter your data as above, and select a block of unused cells two wide and one tall; for example, C2:D2. Then enter the formula

```
=LINEST(B2:B9,A2:A9)
```

[13] At the time of this writing, Google Docs has no trendline feature for its spreadsheet, so you would need to use the formula method.

as shown on the left. Then press Control-Shift-Enter. The result should appear as on the right, with m and b appearing in cells C2 and D2 as shown:

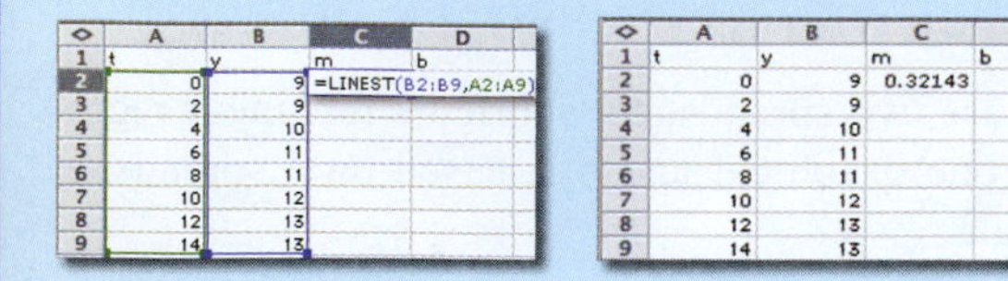

	A	B	C	D
1	t	y	m	b
2	0	9	=LINEST(B2:B9,A2:A9)	
3	2	9		
4	4	10		
5	6	11		
6	8	11		
7	10	12		
8	12	13		
9	14	13		

	A	B	C	D
1	t	y	m	b
2	0	9	0.32143	8.75
3	2	9		
4	4	10		
5	6	11		
6	8	11		
7	10	12		
8	12	13		
9	14	13		

Example 3 (page 81) Find the correlation coefficient for the data in Example 2.

Solution with Technology

1. When you add a trendline to a chart you can select the option "Display R-squared value on chart" to show the value of r^2 on the chart (it is common to examine r^2, which takes on values between 0 and 1, instead of r).
2. Alternatively, the LINEST function we used above in 2 can be used to display quite a few statistics about a best-fit line, including r^2. Instead of selecting a block of cells two wide and one tall as we did in Example 2, we select one two wide and *five* tall. We now enter the requisite LINEST formula with two additional arguments set to "TRUE" as shown, and press Control-Shift-Enter:

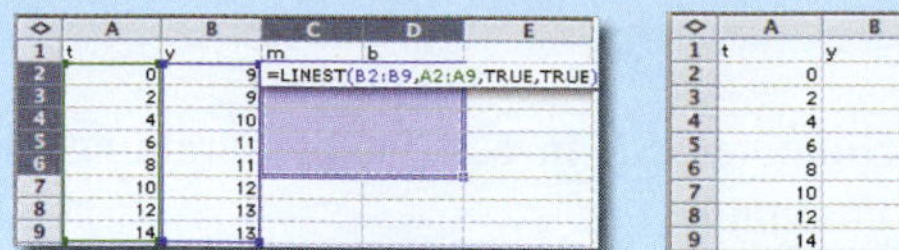

	A	B	C	D	E
1	t	y	m	b	
2	0	9	=LINEST(B2:B9,A2:A9,TRUE,TRUE)		
3	2	9			
4	4	10			
5	6	11			
6	8	11			
7	10	12			
8	12	13			
9	14	13			

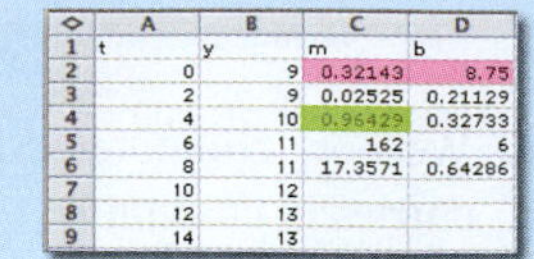

	A	B	C	D
1	t	y	m	b
2	0	9	0.32143	8.75
3	2	9	0.02525	0.21129
4	4	10	0.96429	0.32733
5	6	11	162	6
6	8	11	17.3571	0.64286
7	10	12		
8	12	13		
9	14	13		

The values of m and b appear in cells C2 and D2 as before, and the value of r^2 in cell C4. (Among the other numbers shown is SSE in cell D6. For the meanings of the remaining numbers shown, do a Web search for "LINEST"; you will see numerous articles, including many that explain all the terms. A good course in statistics wouldn't hurt, either.)

2

The Mathematics of Finance

 **Website**

www.WanerMath.com

At the Website you will find:

- A detailed chapter summary
- A true/false quiz
- A time value of money utility
- Excel tutorials for every section

Case Study Adjustable Rate and Subprime Mortgages

Mr. and Mrs. Wong have an appointment tomorrow with you, their investment counselor, to discuss their plan to purchase a $400,000 house in Orlando, Florida. Their combined annual income is $80,000 per year, which they estimate will increase by 4% annually over the foreseeable future, and they are considering three different specialty 30-year mortgages:

Hybrid: The interest is fixed at a low introductory rate of 4% for 5 years.

Interest-Only: During the first 5 years, the rate is set at 4.2% and no principal is paid.

Negative Amortization: During the first 5 years, the rate is set at 4.7% based on a principal of 60% of the purchase price of the home.

How would you advise them?

Andy Dean Photography/Shutterstock

Introduction

A knowledge of the mathematics of investments and loans is important not only for business majors but also for everyone who deals with money, which is all of us. This chapter is largely about *interest*: interest paid by an investment, interest paid on a loan, and variations on these.

We focus on three forms of investment: investments that pay simple interest, investments in which interest is compounded, and annuities. An investment that pays *simple interest* periodically gives interest directly to the investor, perhaps in the form of a monthly check. If instead, the interest is reinvested, the interest is *compounded*, and the value of the account grows as the interest is added. An *annuity* is an investment earning compound interest into which periodic payments are made or from which periodic withdrawals are made; in the case of periodic payments, such an investment is more commonly called a *sinking fund*. From the point of view of the lender, a loan is a kind of annuity.

We also look at bonds, the primary financial instrument used by companies and governments to raise money. Although bonds nominally pay simple interest, determining their worth, particularly in the secondary market, requires an annuity calculation.

2.1 Simple Interest

You deposit \$1,000, called the **principal** or **present value**, into a savings account. The bank pays you 5% interest, in the form of a check, each year. How much interest will you earn each year? Because the bank pays you 5% interest each year, your annual (or yearly) interest will be 5% of \$1,000, or $1{,}000 \times 0.05 = \$50$.

Generalizing this calculation, call the present value PV and the interest rate (expressed as a decimal) r. Then INT, the annual interest paid to you, is given by*

$$INT = PVr.$$

If the investment is made for a period of t years, then the total interest accumulated is t times this amount, which gives us the following:

* Multiletter variables like PV and INT used here may be unusual in a math textbook but are almost universally used in finance textbooks, calculators (such as the TI-83/84 Plus), and such places as study guides for the finance portion of the Society of Actuaries exams. Just watch out for expressions like PVr, which is the product of two things, PV and r, not three.

Simple Interest

The **simple interest** on an investment (or loan) of PV dollars at an annual interest rate of r for a period of t years is

$$INT = PVrt.$$

Quick Example

The simple interest over a period of 4 years on a \$5,000 investment earning 8% per year is

$$\begin{aligned} INT &= PVrt \\ &= (5{,}000)(0.08)(4) = \$1{,}600. \end{aligned}$$

Given your \$1,000 investment at 5% simple interest, how much money will you have after 2 years? To find the answer, we need to add the accumulated interest to the principal to get the **future value** (FV) of your deposit.

$$FV = PV + INT = \$1{,}000 + (1{,}000)(0.05)(2) = \$1{,}100$$

In general, we can compute the future value as follows:

$$FV = PV + INT = PV + PVrt = PV(1 + rt).$$

Future Value for Simple Interest

The **future value** of an investment of PV dollars at an annual simple interest rate of r for a period of t years is

$$FV = PV(1 + rt).$$

Quick Examples

1. The value, at the end of 4 years, of a \$5,000 investment earning 8% simple interest per year is

$$FV = PV(1 + rt)$$
$$= 5{,}000[1 + (0.08)(4)] = \$6{,}600.$$

2. Writing the future value in Quick Example 1 as a function of time, we get

$$FV = 5{,}000(1 + 0.08t)$$
$$= 5{,}000 + 400t,$$

which is a linear function of time t. The intercept is $PV = \$5{,}000$, and the slope is the annual interest, \$400 per year.

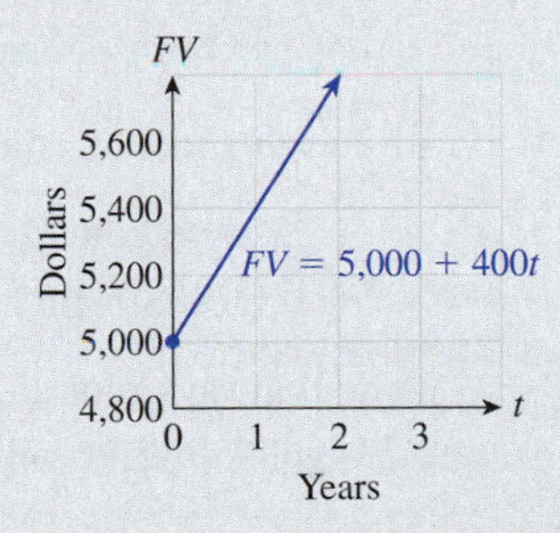

In general: *Simple interest growth is a linear function of time, with intercept given by the present value and slope given by annual interest.*

EXAMPLE 1 Savings Accounts

In November 2011, the Bank of Montreal was paying 1.30% interest on savings accounts.[1] If the interest is paid as simple interest, find the future value of a \$2,000 deposit in 6 years. What is the total interest paid over the period?

[1] Source: Canoe Money (http://money.canoe.ca/rates/savings.html)

Solution We use the future value formula:

$$FV = PV(1 + rt)$$
$$= 2{,}000[1 + (0.013)(6)] = 2{,}000[1.078] = \$2{,}156.$$

The total interest paid is given by the simple interest formula:

$$INT = PVrt$$
$$= (2{,}000)(0.013)(6) = \$156.$$

Note To find the interest paid, we could also have computed

$$INT = FV - PV = 2{,}156 - 2{,}000 = \$156. \blacksquare$$

Before we go on... In the preceding example, we could look at the future value as a function of time:

$$FV = 2{,}000(1 + 0.013t) = 2{,}000 + 26t.$$

Thus, the future value is growing linearly at a rate of $26 per year. ■

EXAMPLE 2 Bridge Loans

When "trading up," homeowners sometimes have to buy a new house before they sell their old house. One way to cover the costs of the new house until they get the proceeds from selling the old house is to take out a short-term *bridge loan*. Suppose a bank charges 12% simple annual interest on such a loan. How much will be owed at the maturation (the end) of a 90-day bridge loan of $90,000?

Solution We use the future value formula

$$FV = PV(1 + rt)$$

with $t = 90/365$, the fraction of a year represented by 90 days:

$$FV = 90{,}000[1 + (0.12)(90/365)]$$
$$= \$92{,}663.01.$$

(We will always round our answers to the nearest cent after calculation. Be careful not to round intermediate results.)

Before we go on... Many banks use 360 days for this calculation rather than 365, which makes a "year" for the purposes of the loan slightly shorter than a calendar year. The effect is to increase the amount owed:

$$FV = 90{,}000[1 + (0.12)(90/360)] = \$92{,}700$$ ■

One of the primary ways companies and governments raise money is by selling **bonds**. At its most straightforward, a corporate bond promises to pay simple interest, usually twice a year, for a length of time until it **matures**, at which point it returns the original investment to the investor (U.S. Treasury notes and bonds are similar). Things get more complicated when the selling price is negotiable, as we will see later in this chapter.

EXAMPLE 3 Corporate Bonds

The Megabucks Corporation is issuing 10-year bonds paying an annual rate of 6.5%. If you buy $10,000 worth of bonds, how much interest will you earn every 6 months, and how much interest will you earn over the life of the bonds?

Solution Using the simple interest formula, every 6 months you will receive

$$\begin{aligned} INT &= PVrt \\ &= (10{,}000)(0.065)\left(\frac{1}{2}\right) = \$325. \end{aligned}$$

Over the 10-year life of the bonds, you will earn

$$\begin{aligned} INT &= PVrt \\ &= (10{,}000)(0.065)(10) = \$6{,}500 \end{aligned}$$

in interest. So, at the end of 10 years, when your original investment is returned to you, your $10,000 will have turned into $16,500.

We often want to turn an interest calculation around: Rather than starting with the present value and finding the future value, there are times when we know the future value and need to determine the present value. Solving the future value formula for PV gives us the following.

Present Value for Simple Interest

The present value of an investment at an annual simple interest rate of r for a period of t years, with future value FV, is

$$PV = \frac{FV}{1 + rt}.$$

Quick Example

If an investment earns 5% simple interest and will be worth $1,000 in 4 years, then its present value (its initial value) is

$$\begin{aligned} PV &= \frac{FV}{1 + rt} \\ &= \frac{1{,}000}{1 + (0.05)(4)} = \$833.33. \end{aligned}$$

Here is a typical example. U.S. Treasury bills (T-bills) are short-term investments (up to 1 year) that pay you a set amount after a period of time; what you pay to buy a T-bill depends on the interest rate.

EXAMPLE 4 Treasury Bills

A U.S. Treasury bill paying $10,000 after 6 months earns 3.67% simple annual interest. How much did it cost to buy?

Solution The future value of the T-bill is $10,000; the price we paid is its present value. We know that

$$FV = \$10,000$$
$$r = 0.0367$$

and

$$t = 0.5.$$

Substituting into the present value formula, we have

$$PV = \frac{10,000}{1 + (0.0367)(0.5)} = 9,819.81$$

so we paid $9,819.81 for the T-bill.

➡ **Before we go on...** The simplest way to find the interest earned on the T-bill is by subtraction:

$$INT = FV - PV = 10,000 - 9,819.81 = \$180.19.$$

So, after 6 months we received back $10,000, which is our original investment plus $180.19 in interest. ■

Here is some additional terminology on Treasury bills:

Treasury Bills (T-Bills): Maturity Value, Discount Rate, and Yield

The **maturity value** of a T-bill is the amount of money it will pay at the end of its life, that is, upon **maturity**.

Quick Example

A 1-year $10,000 T-bill has a maturity value of $10,000, and so will pay you $10,000 after one year.

The cost of a T-bill is generally less than its maturity value.* In other words, a T-bill will generally sell at a *discount*, and the **discount rate** is the *annualized* percentage of this discount; that is, the percentage is adjusted to give an annual percentage. (See Quick Examples 2 and 3.)

* An exception occurred during the financial meltdown of 2008, when T-bills were heavily in demand as "safe haven" investments and were sometimes selling at—or even above—their maturity values.

Quick Examples

1. A 1-year $10,000 T-bill with a discount rate of 5% will sell for 5% less than its maturity value of $10,000, that is, for $9,500.
2. A 6-month $10,000 T-bill with a discount rate of 5% will sell at an actual discount of half of that—2.5% less than its maturity value, or $9,750—because 6 months is half of a year.
3. A 3-month $10,000 T-bill with a discount rate of 5% will sell at an actual discount of a fourth of that: 1.25% less than its maturity value, or $9,875.

The annual **yield** of a T-bill is the simple annual interest rate an investor earns when the T-bill matures, as calculated in the next example.

EXAMPLE 5 Treasury Bills

A T-bill paying \$10,000 after 6 months sells at a discount rate of 3.6%. What does it sell for? What is the annual yield?

Solution The (annualized) discount rate is 3.6%; so, for a 6-month bill, the actual discount will be half of that: $3.6\%/2 = 1.8\%$ below its maturity value. This makes the selling price

$$10{,}000 - (0.018)(10{,}000) = \$9{,}820. \qquad \text{Maturity value} - \text{Discount}$$

To find the annual yield, note that the present value of the investment is the price the investor pays, \$9,820, and the future value is its maturity value, \$10,000 six months later. So,

$$PV = \$9{,}820 \qquad FV = \$10{,}000 \qquad t = 0.5$$

and we wish to find the annual interest rate r. Substituting in the future value formula, we get

$$FV = PV(1 + rt)$$
$$10{,}000 = 9{,}820(1 + 0.5r)$$

so

$$1 + 0.5r = 10{,}000/9{,}820$$

and

$$r = (10{,}000/9{,}820 - 1)/0.5 \approx 0.0367.$$

Thus, the T-bill is paying 3.67% simple annual interest, so we say that its annual yield is 3.67%.

Before we go on... The T-bill in Example 5 is the same one as in Example 4 (with a bit of rounding). The yield and the discount rate are two different ways of telling what the investment pays. One of the Communication and Reasoning Exercises for this section asks you to find a formula for the yield in terms of the discount rate. ■

Fees on loans can also be thought of as a form of interest.

EXAMPLE 6 Tax Refunds

You are expecting a tax refund of \$800. Because it may take up to 6 weeks to get the refund, your tax preparation firm offers, for a fee of \$40, to give you an "interest-free" loan of \$800 to be paid back with the refund check. If we think of the fee as interest, what simple annual interest rate is the firm actually charging?

Solution If we view the \$40 as interest, then the future value of the loan (the value of the loan to the firm, or the total you will pay the firm) is \$840. Thus, we have

$$FV = 840$$
$$PV = 800$$
$$t = 6/52 \qquad \text{Using 52 weeks in a year}$$

and we wish to find r. Substituting, we get

$$FV = PV(1 + rt)$$
$$840 = 800(1 + 6r/52) = 800 + \frac{4{,}800r}{52}$$

so

$$\frac{4{,}800r}{52} = 840 - 800 = 40$$
$$r = \frac{40 \times 52}{4{,}800} \approx 0.43.$$

In other words, the firm is charging you 43% annual interest! Save your money and wait 6 weeks for your refund.

2.1 EXERCISES

Access end-of-section exercises online at **www.webassign.net** 

2.2 Compound Interest

You deposit \$1,000 into a savings account. The bank pays you 5% interest, which it deposits into your account, or **reinvests**, at the end of each year. At the end of 5 years, how much money will you have accumulated? Let us compute the amount you have at the end of each year. At the end of the first year, the bank will pay you simple interest of 5% on your \$1,000, which gives you

$$PV(1 + rt) = 1{,}000(1 + 0.05)$$
$$= \$1{,}050.$$

At the end of the second year, the bank will pay you another 5% interest, but this time computed on the total in your account, which is \$1,050. Thus, you will have a total of

$$1{,}050(1 + 0.05) = \$1{,}102.50.$$

If you were being paid simple interest on your original \$1,000, you would have only \$1,100 at the end of the second year. The extra \$2.50 is the interest earned on the \$50 interest added to your account at the end of the first year. Having interest earn interest is called **compounding** the interest. We could continue like this until the end of the fifth year, but notice what we are doing: Each year we are multiplying by $1 + 0.05$. So, at the end of 5 years, you will have

$$1{,}000(1 + 0.05)^5 \approx \$1{,}276.28.$$

It is interesting to compare this to the amount you would have if the bank paid you simple interest:

$$1{,}000(1 + 0.05 \times 5) = \$1{,}250.00.$$

The extra \$26.28 is again the effect of compounding the interest.

* This is approximate. They will actually give you 31/365 of the 5% at the end of January and so on, but it's simpler and reasonably accurate to call it 1/12.

Banks often pay interest more often than once a year. Paying interest quarterly (four times per year) or monthly is common. If your bank pays interest monthly, how much will your \$1,000 deposit be worth after 5 years? The bank will not pay you 5% interest every month, but will give you 1/12 of that,* or 5/12% interest each month. Thus, instead of multiplying by $1 + 0.05$ every year, we should multiply by $1 + 0.05/12$ each month. Because there are $5 \times 12 = 60$ months in 5 years, the total amount you will have at the end of 5 years is

$$1{,}000\left(1 + \frac{0.05}{12}\right)^{60} \approx \$1{,}283.36.$$

Compare this to the \$1,276.28 you would get if the bank paid the interest every year. You earn an extra \$7.08 if the interest is paid monthly because interest gets into your account and starts earning interest earlier. The amount of time between interest payments is called the **compounding period**.

The following table summarizes the results above.

Time in Years	*Amount with Simple Interest*	*Amount with Annual Compounding*	*Amount with Monthly Compounding*
0	\$1,000	\$1,000	\$1,000
1	$1{,}000(1 + 0.05)$ $= \$1{,}050$	$1{,}000(1 + 0.05)$ $= \$1{,}050$	$1{,}000(1 + 0.05/12)^{12}$ $= \$1{,}051.16$
2	$1{,}000(1 + 0.05 \times 2)$ $= \$1{,}100$	$1{,}000(1 + 0.05)^2$ $= \$1{,}102.50$	$1{,}000(1 + 0.05/12)^{24}$ $= \$1{,}104.94$
5	$1{,}000(1 + 0.05 \times 5)$ $= \$1{,}250$	$1{,}000(1 + 0.05)^5$ $= \$1{,}276.28$	$1{,}000(1 + 0.05/12)^{60}$ $= \$1{,}283.36$

The preceding calculations generalize easily to give the general formula for future value when interest is compounded.

Future Value for Compound Interest

The future value of an investment of PV dollars earning interest at an annual rate of r compounded (reinvested) m times per year for a period of t years is

$$FV = PV\left(1 + \frac{r}{m}\right)^{mt}$$

or

$$FV = PV(1 + i)^n,$$

where $i = r/m$ is the interest paid each compounding period and $n = mt$ is the total number of compounding periods.

Quick Examples

1. To find the future value after 5 years of a \$10,000 investment earning 6% interest, with interest reinvested every month, we set $PV = 10{,}000$, $r = 0.06$, $m = 12$, and $t = 5$. Thus,

$$FV = PV\left(1 + \frac{r}{m}\right)^{mt} = 10{,}000\left(1 + \frac{0.06}{12}\right)^{60} \approx \$13{,}488.50.$$

2. Writing the future value in Quick Example 1 as a function of time, we get

$$FV = 10{,}000\left(1 + \frac{0.06}{12}\right)^{12t}$$
$$= 10{,}000(1.005)^{12t},$$

which is an **exponential** function of time t.

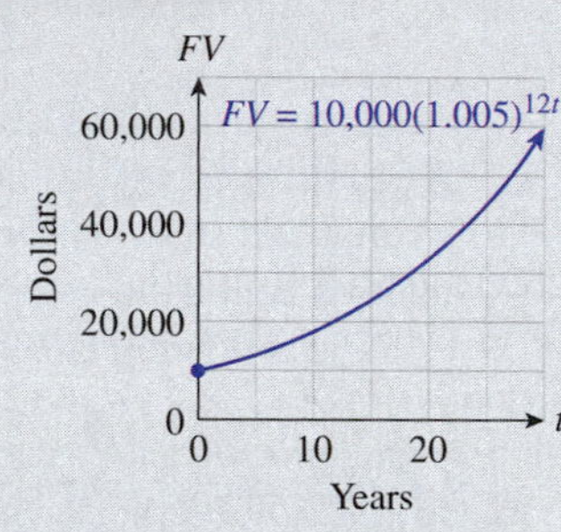

In general: *Compound interest growth is an exponential function of time.*

using Technology

All three technologies discussed in this book have built-in mathematics of finance capabilities. See the Technology Guides at the end of the chapter for details on using a TI-83/84 Plus or a spreadsheet to do the calculations in Example 1. [Details: TI-83/84 Plus: page 134, Spreadsheet: page 138]

Website
www.WanerMath.com
→ On Line Utilities
→ Time Value of Money Utility

This utility is similar to the TVM Solver on the TI-83/84 Plus. To compute the future value, enter the values shown, and press "Compute" next to FV.

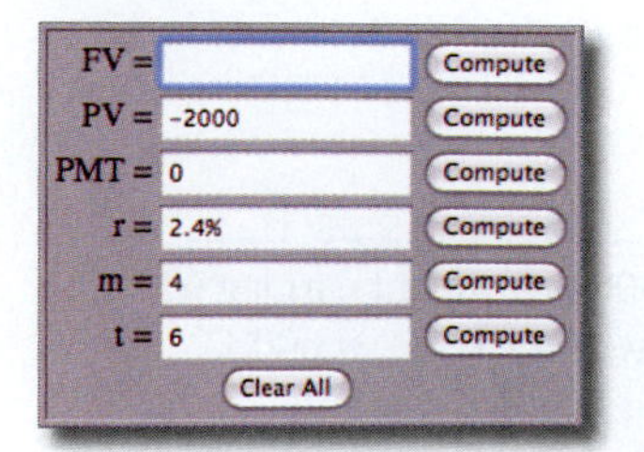

(For an explanation of the terms, see the Technology Guide for the TI-83/84 Plus.)

EXAMPLE 1 Savings Accounts

In November 2011, the Bank of Montreal was paying 1.30% interest on savings accounts.[2] If the interest is compounded quarterly, find the future value of a $2,000 deposit in 6 years. What is the total interest paid over the period?

Solution We use the future value formula with $m = 4$:

$$FV = PV\left(1 + \frac{r}{m}\right)^{mt}$$
$$= 2{,}000\left(1 + \frac{0.0130}{4}\right)^{4\times 6} \approx \$2{,}161.97.$$

The total interest paid is

$$INT = FV - PV = 2{,}161.97 - 2{,}000 = \$161.97.$$

Example 1 illustrates the concept of the **time value of money**: A given amount of money received now will usually be worth a different amount to us than the same amount received some time in the future. In the example above, we can say that $2,000 received now is worth the same as $2,161.97 received 6 years from now, because if we receive $2,000 now, we can turn it into $2,161.97 by the end of 6 years.

We often want to know, for some amount of money in the future, what is the equivalent value at present. As we did for simple interest, we can solve the future value formula for the present value and obtain the following formula.

[2]Source: Canoe Money (http://money.canoe.ca/rates/savings.html)

Present Value for Compound Interest

The present value of an investment earning interest at an annual rate of r compounded m times per year for a period of t years, with future value FV, is

$$PV = \frac{FV}{\left(1 + \frac{r}{m}\right)^{mt}}$$

or

$$PV = \frac{FV}{(1+i)^n} = FV(1+i)^{-n},$$

where $i = r/m$ is the interest paid each compounding period and $n = mt$ is the total number of compounding periods.

Quick Example

To find the amount we need to invest in an investment earning 12% per year, compounded annually, so that we will have \$1 million in 20 years, use $FV = \$1{,}000{,}000$, $r = 0.12$, $m = 1$, and $t = 20$:

$$PV = \frac{FV}{\left(1 + \frac{r}{m}\right)^{mt}} = \frac{1{,}000{,}000}{(1+0.12)^{20}} \approx \$103{,}666.77.$$

Put another way, \$1,000,000 20 years from now is worth only \$103,666.77 to us now, if we have a 12% investment available.

In the preceding section, we mentioned that a bond pays interest until it reaches maturity, at which point it pays you back an amount called its **maturity value** or **par value**. The two parts, the interest and the maturity value, can be separated and sold and traded by themselves. A **zero coupon bond** is a form of corporate bond that pays no interest during its life but, like U.S. Treasury bills, promises to pay you the maturity value when it reaches maturity. Zero coupon bonds are often created by removing or *stripping* the interest coupons from an ordinary bond, so are also known as **strips**. Zero coupon bonds sell for less than their maturity value, and the return on the investment is the difference between what the investor pays and the maturity value. Although no interest is actually paid, we measure the return on investment by thinking of the interest rate that would make the selling price (the present value) grow to become the maturity value (the future value).*

* The IRS refers to this kind of interest as **original issue discount (OID)** and taxes it as if it were interest actually paid to you each year.

EXAMPLE 2 Zero Coupon Bonds

Megabucks Corporation is issuing 10-year zero coupon bonds. How much would you pay for bonds with a maturity value of \$10,000 if you wish to get a return of 6.5% compounded annually?†

† The return investors look for depends on a number of factors, including risk (the chance that the company will go bankrupt and you will lose your investment); the higher the risk, the higher the return. U.S. Treasuries are considered risk free because the federal government has never defaulted on its debts. On the other hand, so-called junk bonds are high-risk investments (below investment grade) and have correspondingly high yields.

Solution As we said earlier, we think of a zero coupon bond as if it were an account earning (compound) interest. We are asked to calculate the amount you will pay for the bond—the present value PV. We have

$$\begin{aligned} FV &= \$10{,}000 \\ r &= 0.065 \\ t &= 10 \\ m &= 1. \end{aligned}$$

See the Technology Guides at the end of the chapter for details on using TVM Solver on the TI-83/84 Plus or the built-in finance functions in spreadsheets to do the calculations in Example 2. [Details: TI-83/84 Plus: page 134, Spreadsheet: page 139]

Website
www.WanerMath.com
→ On Line Utilities
→ Time Value of Money Utility
This utility is similar to the TVM Solver on the TI-83/84 Plus. To compute the present value, enter the values shown, and press "Compute" next to PV.

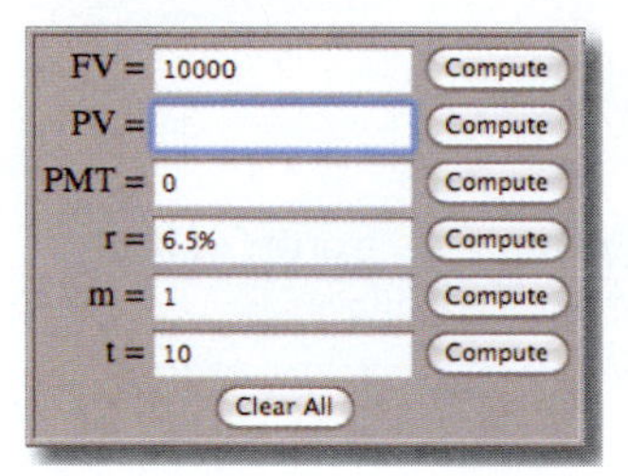

We can now use the present value formula:

$$PV = \frac{FV}{\left(1 + \frac{r}{m}\right)^{mt}}$$

$$PV = \frac{10,000}{\left(1 + \frac{0.065}{1}\right)^{10\times 1}} \approx \$5,327.26.$$

Thus, you should pay $5,327.26 to get a return of 6.5% annually.

Before we go on... Particularly in financial applications, you will hear the word **"discounted"** in place of "compounded" when discussing present value. Thus, the result of Example 2 might be phrased, "The present value of $10,000 to be received 10 years from now, with an interest rate of 6.5% discounted annually, is $5,327.26." ■

Time value of money calculations are often done to take into account inflation, which behaves like compound interest. Suppose, for example, that inflation is running at 5% per year. Then prices will increase by 5% each year, so if PV represents the price now, the price one year from now will be 5% higher, or $PV(1 + 0.05)$. The price a year from then will be 5% higher still, or $PV(1 + 0.05)^2$. Thus, the effects of inflation are compounded just as reinvested interest is.

EXAMPLE 3 Inflation

Inflation in East Avalon is 5% per year. TruVision television sets cost $200 today. How much will a comparable set cost 2 years from now?

Solution To find the price of a television set 2 years from now, we compute the future value of $200 at an inflation rate of 5% compounded yearly:

$$FV = 200(1 + 0.05)^2 = \$220.50.$$

EXAMPLE 4 Constant Dollars

Inflation in North Avalon is 6% per year. Which is really more expensive, a car costing $20,000 today or one costing $22,000 in 3 years?

Solution We cannot compare the two costs directly because inflation makes $1 today worth more (it buys more) than a dollar 3 years from now. We need the two prices expressed in comparable terms, so we convert to **constant dollars**. We take the car costing $22,000 three years from now and ask what it would cost in today's dollars. In other words, we convert the future value of $22,000 to its present value:

$$\begin{aligned} PV &= FV(1 + i)^{-n} \\ &= 22,000(1 + 0.06)^{-3} \\ &\approx \$18,471.62. \end{aligned}$$

Thus, the car costing $22,000 in 3 years actually costs less, after adjusting for inflation, than the one costing $20,000 now.

➡ **Before we go on...** In the presence of inflation, the only way to compare prices at different times is to convert all prices to constant dollars. We pick some fixed time and compute future or present values as appropriate to determine what things would have cost at that time. ■

There are some other interesting calculations related to compound interest, besides present and future values.

EXAMPLE 5 Effective Interest Rate

You have just won $1 million in the lottery and are deciding what to do with it during the next year before you move to the South Pacific. Bank Ten offers 10% interest, compounded annually, while Bank Nine offers 9.8% compounded monthly. In which should you deposit your money?

Solution Let's calculate the future value of your $1 million after one year in each of the banks:

$$\text{Bank Ten:} \quad FV = 1(1 + 0.10)^1 = \$1.1 \text{ million}$$

$$\text{Bank Nine:} \quad FV = 1\left(1 + \frac{0.098}{12}\right)^{12} = \$1.1025 \text{ million.}$$

Bank Nine turns out to be better: It will pay you a total of $102,500 in interest over the year, whereas Bank Ten will pay only $100,000 in interest.

Another way of looking at the calculation in Example 5 is that Bank Nine gave you a total of 10.25% interest on your investment over the year. We call 10.25% the **effective interest rate** of the investment (also referred to as the **annual percentage yield**, or **APY** in the banking industry); the stated 9.8% is called the **nominal** interest rate. In general, to best compare two different investments, it is wisest to compare their *effective*—rather than nominal—interest rates.

Notice that we got 10.25% by computing

$$\left(1 + \frac{0.098}{12}\right)^{12} = 1.1025$$

and then subtracting 1 to get 0.1025, or 10.25%. Generalizing, we get the following formula.

Effective Interest Rate

The effective interest rate r_{eff} of an investment paying a nominal interest rate of r_{nom} compounded m times per year is

$$r_{\text{eff}} = \left(1 + \frac{r_{\text{nom}}}{m}\right)^m - 1.$$

To compare rates of investments with different compounding periods, always compare the effective interest rates rather than the nominal rates.

Quick Example

To calculate the effective interest rate of an investment that pays 8% per year, with interest reinvested monthly, set $r_{\text{nom}} = 0.08$ and $m = 12$, to obtain

$$r_{\text{eff}} = \left(1 + \frac{0.08}{12}\right)^{12} - 1 \approx 0.0830, \text{ or } 8.30\%.$$

using Technology

See the Technology Guides at the end of the chapter for details on using TVM Solver on the TI-83/84 Plus or the built-in finance functions in spreadsheets to do the calculations in Example 6. [Details: TI-83/84 Plus: page 135, Spreadsheet: page 139]

Website

www.WanerMath.com

→ On Line Utilities

→ Time Value of Money Utility

This utility is similar to the TVM Solver on the TI-83/84 Plus. To compute the time needed, enter the values shown, and press "Compute" next to t.

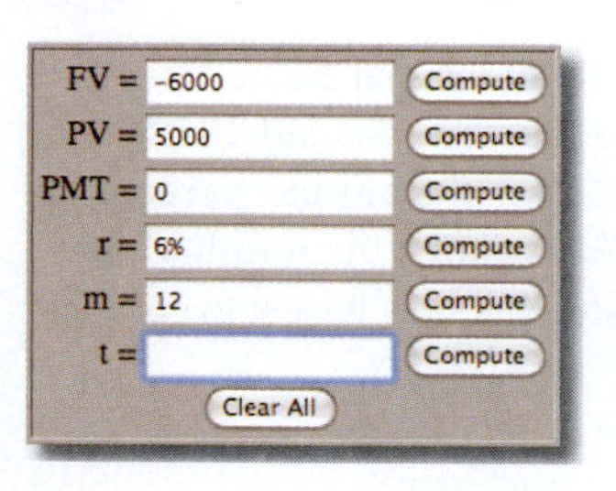

EXAMPLE 6 How Long to Invest

You have $5,000 to invest at 6% interest compounded monthly. How long will it take for your investment to grow to $6,000?

Solution If we use the future value formula, we already have the values

$$\begin{aligned} FV &= 6{,}000 \\ PV &= 5{,}000 \\ r &= 0.06 \\ m &= 12. \end{aligned}$$

Substituting, we get

$$6{,}000 = 5{,}000\left(1 + \frac{0.06}{12}\right)^{12t}.$$

If you are familiar with logarithms, you can solve explicitly for t as follows:

$$\left(1 + \frac{0.06}{12}\right)^{12t} = \frac{6{,}000}{5{,}000} = 1.2$$

$$\log\left(1 + \frac{0.06}{12}\right)^{12t} = \log 1.2$$

$$12t \log\left(1 + \frac{0.06}{12}\right) = \log 1.2$$

$$t = \frac{\log 1.2}{12 \log\left(1 + \frac{0.06}{12}\right)} \approx 3.046 \approx 3 \text{ years.}$$

Another approach is to use a bit of trial and error to find the answer. Let's see what the future value is after 2, 3, and 4 years:

$$5{,}000\left(1 + \frac{0.06}{12}\right)^{12\times 2} = 5{,}635.80 \quad \text{Future value after 2 years}$$

$$5{,}000\left(1 + \frac{0.06}{12}\right)^{12\times 3} = 5{,}983.40 \quad \text{Future value after 3 years}$$

$$5{,}000\left(1 + \frac{0.06}{12}\right)^{12\times 4} = 6{,}352.45. \quad \text{Future value after 4 years}$$

From these calculations, it looks as if the answer should be a bit more than 3 years, but is certainly between 3 and 4 years. We could try 3.5 years next and then narrow it down systematically until we have a pretty good approximation of the correct answer.

Graphing calculators and spreadsheets give us alternative methods of solution.

FAQs

Recognizing When to Use Compound Interest and the Meaning of Present Value

Q: *How do I distinguish a problem that calls for compound interest from one that calls for simple interest?*

A: Study the scenario to ascertain whether the interest is being withdrawn as it is earned or reinvested (deposited back into the account). If the interest is being withdrawn, the problem is calling for simple interest because the interest is not itself earning interest. If it is being reinvested, the problem is calling for compound interest.

Q: *How do I distinguish present value from future value in a problem?*

A: The present value always refers to the value of an investment before any interest is included (or, in the case of a depreciating investment, before any depreciation takes place). As an example, the future value of a bond is its maturity value. The value of $1 today in constant 2010 dollars is its present value (even though 2010 is in the past).

2.2 EXERCISES

Access end-of-section exercises online at **www.webassign.net**

2.3 Annuities, Loans, and Bonds

A typical defined-contribution pension fund works as follows*: Every month while you work, you and your employer deposit a certain amount of money in an account. This money earns (compound) interest from the time it is deposited. When you retire, the account continues to earn interest, but you may then start withdrawing money at a rate calculated to reduce the account to zero after some number of years. This account is an example of an **annuity**, an account earning interest into which you make periodic deposits or from which you make periodic withdrawals. In common usage, the term "annuity" is used for an account from which you make withdrawals. There are various terms used for accounts into which you make payments, based on their purpose. Examples include **savings account**, **pension fund**, and **sinking fund**. A sinking fund is generally used by businesses or governments to accumulate money to pay off an anticipated debt, but we'll use the term to refer to any account into which you make periodic payments.

* Defined-contribution pension plans have largely replaced the defined-benefit pensions that were once the norm in private industry. In a defined-benefit plan, the size of your pension is guaranteed; it is typically a percentage of your final working salary. In a defined-contribution plan, the size of your pension depends on how well your investments do.

Sinking Funds

Suppose you make a payment of \$100 at the end of every month into an account earning 3.6% interest per year, compounded monthly. This means that your investment is earning $3.6\%/12 = 0.3\%$ per month. We write $i = 0.036/12 = 0.003$. What will be the value of the investment at the end of 2 years (24 months)?

Think of the deposits separately. Each earns interest from the time it is deposited, and the total accumulated after 2 years is the sum of these deposits and the interest they earn. In other words, the accumulated value is the sum of the future values of the deposits, taking into account how long each deposit sits in the account. Figure 1 shows a timeline with the deposits and the contribution of each to the final value. For example, the very last deposit (at the end of month 24) has no time to earn interest, so it contributes only \$100. The very first deposit, which earns interest for 23 months, by the future value formula for compound interest contributes $\$100(1+0.003)^{23}$ to the total. Adding together all of the future values gives us the total future value:

$$FV = 100 + 100(1+0.003) + 100(1+0.003)^2 + \cdots + 100(1+0.003)^{23}$$
$$= 100[1 + (1+0.003) + (1+0.003)^2 + \cdots + (1+0.003)^{23}]$$

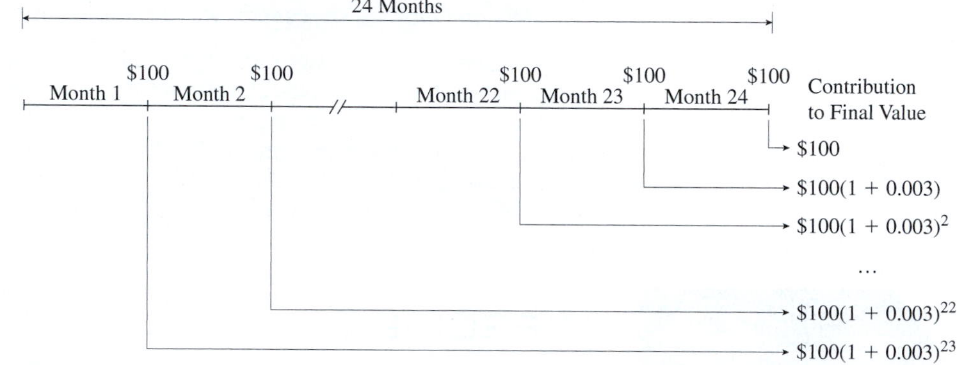

Figure 1

Fortunately, this sort of sum is well-known (to mathematicians, anyway*) and there is a convenient formula for its value†:

$$1 + x + x^2 + \cdots + x^{n-1} = \frac{x^n - 1}{x - 1}.$$

In our case, with $x = 1 + 0.003$, this formula allows us to calculate the future value:

$$FV = 100\frac{(1+0.003)^{24} - 1}{(1+0.003) - 1} = 100\frac{(1.003)^{24} - 1}{0.003} \approx \$2{,}484.65.$$

It is now easy to generalize this calculation.

* It is called a **geometric series**.

† The quickest way to convince yourself that this formula is correct is to multiply out $(x-1)(1+x+x^2+\cdots+x^{n-1})$ and see that you get $x^n - 1$. You should also try substituting some numbers. For example, $1 + 3 + 3^2 = 13 = (3^3 - 1)/(3 - 1)$.

Future Value of a Sinking Fund

A **sinking fund** is an account earning compound interest into which you make periodic deposits. Suppose that the account has an annual rate of r compounded m times per year, so that $i = r/m$ is the interest rate per compounding period. If you make a payment of PMT at the end of each period, then the future value after t years, or $n = mt$ periods, will be

$$FV = PMT\frac{(1+i)^n - 1}{i}.$$

Quick Example

At the end of each month you deposit \$50 into an account earning 2% annual interest compounded monthly. To find the future value after 5 years, we use $i = 0.02/12$ and $n = 12 \times 5 = 60$ compounding periods, so

$$FV = 50\frac{(1 + 0.02/12)^{60} - 1}{0.02/12} = \$3{,}152.37.$$

using Technology

To automate the computations in Example 1 using a graphing calculator or a spreadsheet, see the Technology Guides at the end of the chapter. Outline:

TI-83/84 Plus

[APPS] `1:Finance`, then `1:TVM Solver`
N = 120, I% = 5, PV = −5000, PMT = −100, P/Y = 12, C/Y = 12
With cursor on FV line, [ALPHA] [SOLVE] [More details on page 135.]

Spreadsheet

`=FV(5/12,10*12,-100,-5000)`
[More details on page 139.]

Website

www.WanerMath.com
→ On Line Utilities
→ Time Value of Money Utility

Enter the values shown, and press "Compute" next to FV.

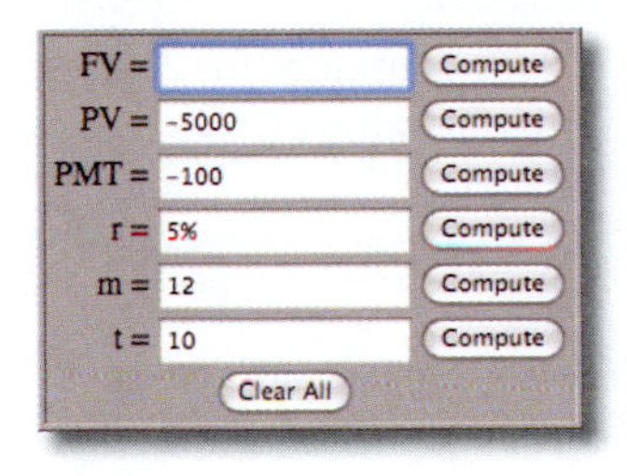

EXAMPLE 1 Retirement Account

Your retirement account has \$5,000 in it and earns 5% interest per year compounded monthly. Every month for the next 10 years you will deposit \$100 into the account. How much money will there be in the account at the end of those 10 years?

Solution This is a sinking fund with $PMT = \$100$, $r = 0.05$, $m = 12$, so $i = 0.05/12$, and $n = 12 \times 10 = 120$. Ignoring for the moment the \$5,000 already in the account, your payments have the following future value:

$$\begin{aligned} FV &= PMT\frac{(1+i)^n - 1}{i} \\ &= 100\frac{(1+0.05/12)^{120} - 1}{0.05/12} \\ &\approx \$15{,}528.23. \end{aligned}$$

What about the \$5,000 that was already in the account? That sits there and earns interest, so we need to find its future value as well, using the compound interest formula:

$$\begin{aligned} FV &= PV(1+i)^n \\ &= 5{,}000(1+0.05/12)^{120} \\ &= \$8{,}235.05. \end{aligned}$$

Hence, the total amount in the account at the end of 10 years will be

$$\$15{,}528.23 + 8{,}235.05 = \$23{,}763.28.$$

Sometimes we know what we want the future value to be and need to determine the payments necessary to achieve that goal. We can simply solve the future value formula for the payment.

Payment Formula for a Sinking Fund

Suppose that an account has an annual rate of r compounded m times per year, so that $i = r/m$ is the interest rate per compounding period. If you want to accumulate a total of FV in the account after t years, or $n = mt$ periods, by making payments of PMT at the end of each period, then each payment must be

$$PMT = FV\frac{i}{(1+i)^n - 1}.$$

using Technology

To automate the computations in Example 2 using a graphing calculator or a spreadsheet, see the Technology Guides at the end of the chapter. Outline:

TI-83/84 Plus

[APPS] `1:Finance`, then `1:TVM Solver`
N = 68, I% = 4, PV = 0, FV = 100000, P/Y = 4, C/Y = 4
With cursor on PMT line, [ALPHA] [SOLVE]
[More details on page 135.]

EXAMPLE 2 Education Fund

Tony and Maria have just had a son, José Phillipe. They establish an account to accumulate money for his college education, in which they would like to have \$100,000 after 17 years. If the account pays 4% interest per year compounded quarterly, and they make deposits at the end of every quarter, how large must each deposit be for them to reach their goal?

Spreadsheet
```
=PMT(4/4,17*4,0,100000)
```
[More details on page 140.]

Website
www.WanerMath.com
→ On Line Utilities
→ Time Value of Money Utility
Enter the values shown, and press "Compute" next to PMT.

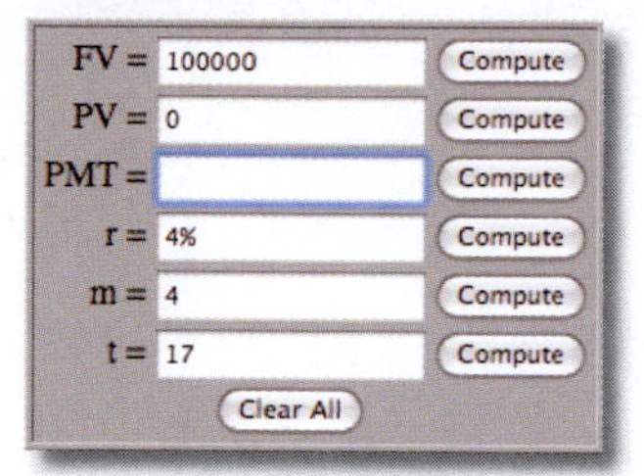

Solution This is a sinking fund with $FV = \$100{,}000$, $m = 4$, $n = 4 \times 17 = 68$, and $r = 0.04$, so $i = 0.04/4 = 0.01$. From the payment formula, we get

$$PMT = 100{,}000\frac{0.01}{(1+0.01)^{68}-1} \approx \$1{,}033.89.$$

So, Tony and Maria must deposit $1,033.89 every quarter in order to meet their goal.

Annuities

Suppose we deposit an amount PV now in an account earning 3.6% interest per year, compounded monthly. Starting 1 month from now, the bank will send us monthly payments of $100. What must PV be so that the account will be drawn down to $0 in exactly 2 years?

As before, we write $i = r/m = 0.036/12 = 0.003$, and we have $PMT = 100$. The first payment of $100 will be made 1 month from now, so its present value is

$$\frac{PMT}{(1+i)^n} = \frac{100}{1+0.003} = 100(1+0.003)^{-1} \approx \$99.70.$$

In other words, that much of the original PV goes toward funding the first payment. The second payment, 2 months from now, has a present value of

$$\frac{PMT}{(1+i)^n} = \frac{100}{(1+0.003)^2} = 100(1+0.003)^{-2} \approx \$99.40.$$

That much of the original PV funds the second payment. This continues for 2 years, at which point we receive the last payment, which has a present value of

$$\frac{PMT}{(1+i)^n} = \frac{100}{(1+0.003)^{24}} = 100(1+0.003)^{-24} \approx \$93.06$$

and that exhausts the account. Figure 2 shows a timeline with the payments and the present value of each.

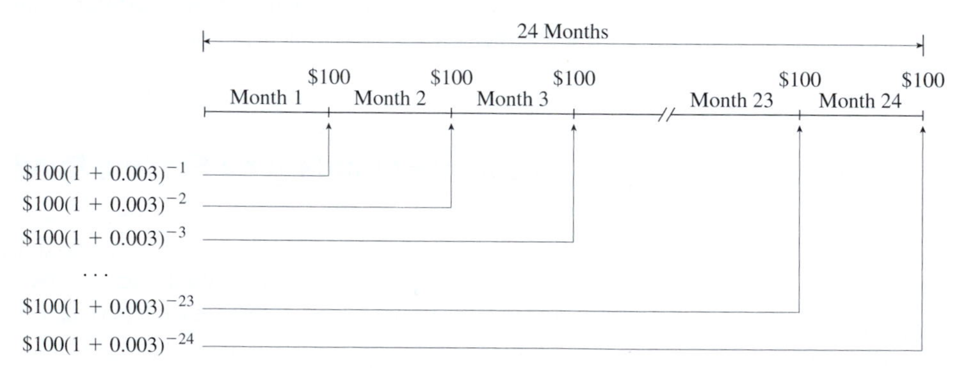

Figure 2

Because PV must be the sum of these present values, we get

$$\begin{aligned} PV &= 100(1+0.003)^{-1} + 100(1+0.003)^{-2} + \cdots + 100(1+0.003)^{-24} \\ &= 100[(1+0.003)^{-1} + (1+0.003)^{-2} + \cdots + (1+0.003)^{-24}]. \end{aligned}$$

We can again find a simpler formula for this sum:

$$\begin{aligned} x^{-1} + x^{-2} + \cdots + x^{-n} &= \frac{1}{x^n}(x^{n-1} + x^{n-2} + \cdots + 1) \\ &= \frac{1}{x^n} \cdot \frac{x^n - 1}{x - 1} = \frac{1 - x^{-n}}{x - 1}. \end{aligned}$$

So, in our case,

$$PV = 100\frac{1-(1+0.003)^{-24}}{(1+0.003)-1}$$

or

$$PV = 100\frac{1-(1.003)^{-24}}{0.003} \approx \$2{,}312.29.$$

If we deposit \$2,312.29 initially and the bank sends us \$100 per month for 2 years, our account will be exhausted at the end of that time.

Generalizing, we get the following formula:

Present Value of an Annuity

An **annuity** is an account earning compound interest from which periodic withdrawals are made. Suppose that the account has an annual rate of r compounded m times per year, so that $i = r/m$ is the interest rate per compounding period. Suppose also that the account starts with a balance of PV. If you receive a payment of PMT at the end of each compounding period, and the account is down to \$0 after t years, or $n = mt$ periods, then

$$PV = PMT\frac{1-(1+i)^{-n}}{i}.$$

Quick Example

At the end of each month you want to withdraw \$50 from an account earning 2% annual interest compounded monthly. If you want the account to last for 5 years (60 compounding periods), it must have the following amount to begin with:

$$PV = 50\frac{1-(1+0.02/12)^{-60}}{0.02/12} = \$2{,}852.62.$$

Note If you make your withdrawals at the end of each compounding period, as we've discussed so far, you have an **ordinary annuity**. If, instead, you make withdrawals at the beginning of each compounding period, you have an **annuity due**. Because each payment occurs one period earlier, there is one less period in which to earn interest, hence the present value must be larger by a factor of $(1 + i)$ to fund each payment. So, the present value formula for an annuity due is

$$PV = PMT(1+i)\frac{1-(1+i)^{-n}}{i}.$$

In this book, we will concentrate on ordinary annuities. ■

EXAMPLE 3 Trust Fund

You wish to establish a trust fund from which your niece can withdraw \$2,000 every 6 months for 15 years, at the end of which time she will receive the remaining money in the trust, which you would like to be \$10,000. The trust will be invested at 7% per year compounded every 6 months. How large should the trust be?

using Technology

To automate the computations in Example 3 using a graphing calculator or a spreadsheet, see the Technology Guides at the end of the chapter. Outline:

TI-83/84 Plus

APPS 1:Finance, then 1:TVM Solver
N = 30, I% = 7, PMT = 2000, FV = 10000, P/Y = 2, C/Y = 2
With cursor on PV line.
ALPHA SOLVE
[More details on page 136.]

Spreadsheet

=PV(7/2,15*2,2000,10000)
[More details on page 140.]

Website

www.WanerMath.com
→ On Line Utilities
→ Time Value of Money Utility

Enter the values shown, and press "Compute" next to PV.

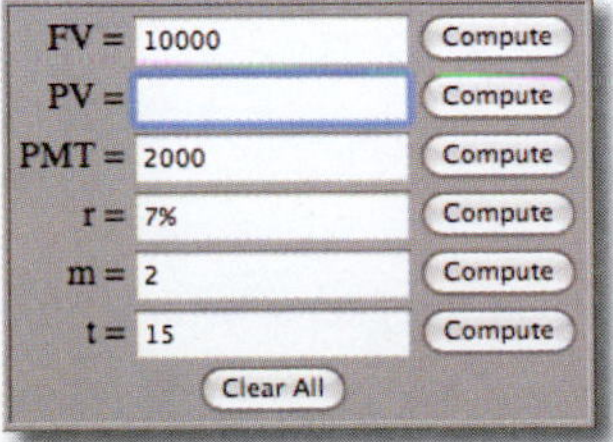

Solution We view this account as having two parts, one funding the semiannual payments and the other funding the \$10,000 lump sum at the end. The amount of money necessary to fund the semiannual payments is the present value of an annuity, with $PMT = 2{,}000$, $r = 0.07$, and $m = 2$, so $i = 0.07/2 = 0.035$, and $n = 2 \times 15 = 30$. Substituting gives

$$\begin{aligned} PV &= 2{,}000\frac{1-(1+0.035)^{-30}}{0.035} \\ &= \$36{,}784.09. \end{aligned}$$

To fund the lump sum of \$10,000 after 15 years, we need the present value of \$10,000 under compound interest:

$$\begin{aligned} PV &= 10{,}000(1+0.035)^{-30} \\ &= \$3{,}562.78. \end{aligned}$$

Thus the trust should start with \$36,784.09 + 3,562.78 = \$40,346.87.

Sometimes we know how much money we begin with and for how long we want to make withdrawals. We then want to determine the amount of money we can withdraw each period. For this, we simply solve the present value formula for the payment.

using Technology

To automate the computations in Example 4 using a graphing calculator or a spreadsheet, see the Technology Guides at the end of the chapter. Outline:

TI-83/84 Plus

`APPS` `1:Finance`, then `1:TVM Solver`

N = 16, I% = 4, PV = −100000, FV = 0, P/Y = 4, C/Y = 4

With cursor on PMT line, `ALPHA` `SOLVE`

[More details on page 136.]

Spreadsheet

`=PMT(4/4,4*4,-100000,0)`

[More details on page 140.]

Website

www.WanerMath.com

→ On Line Utilities

→ Time Value of Money Utility

Enter the values shown, and press "Compute" next to PMT.

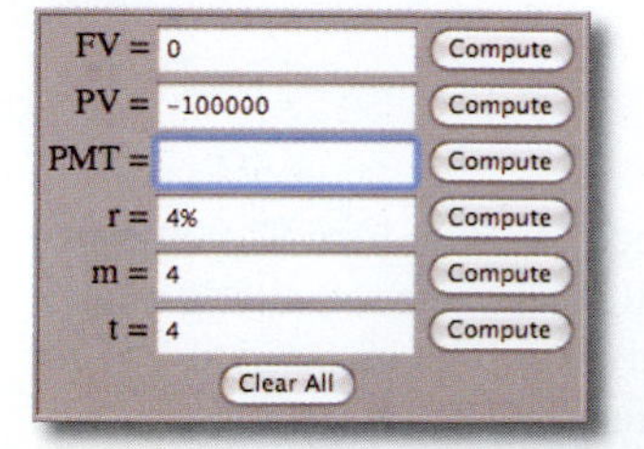

Payment Formula for an Ordinary Annuity

Suppose that an account has an annual rate of r compounded m times per year, so that $i = r/m$ is the interest rate per compounding period. Suppose also that the account starts with a balance of PV. If you want to receive a payment of PMT at the end of each compounding period, and the account is down to \$0 after t years, or $n = mt$ periods, then

$$PMT = PV\frac{i}{1-(1+i)^{-n}}.$$

EXAMPLE 4 Education Fund

Tony and Maria (see Example 2), having accumulated \$100,000 for José Phillipe's college education, would now like to make quarterly withdrawals over the next 4 years. How much money can they withdraw each quarter in order to draw down the account to zero at the end of the 4 years? (Recall that the account pays 4% interest compounded quarterly.)

Solution Now Tony and Maria's account is acting as an annuity with a present value of \$100,000. So, $PV = \$100{,}000$, $r = 0.04$, and $m = 4$, giving $i = 0.04/4 = 0.01$, and $n = 4 \times 4 = 16$. We use the payment formula to get

$$PMT = 100{,}000\frac{0.01}{1-(1+0.01)^{-16}} \approx \$6{,}794.46.$$

So, they can withdraw \$6,794.46 each quarter for 4 years, at the end of which time their account balance will be 0.

EXAMPLE 5 Saving for Retirement

Jane Q. Employee has just started her new job with *Big Conglomerate, Inc.*, and is already looking forward to retirement. BCI offers her as a pension plan an annuity that is guaranteed to earn 6% annual interest compounded monthly. She plans to work for 40 years before retiring and would then like to be able to draw an income of \$7,000 per month for 20 years. How much do she and BCI together have to deposit per month into the fund to accomplish this?

Solution Here we have the situation we described at the beginning of the section: a sinking fund accumulating money to be used later as an annuity. We know the desired payment out of the annuity, so we work backward. The first thing we need to do is calculate the present value of the annuity required to make the pension payments. We use the annuity present value formula with $PMT = 7{,}000$, $i = r/m = 0.06/12 = 0.005$ and $n = 12 \times 20 = 240$.

$$\begin{aligned} PV &= PMT\frac{1-(1+i)^{-n}}{i} \\ &= 7{,}000\frac{1-(1+0.005)^{-240}}{0.005} \approx \$977{,}065.40 \end{aligned}$$

This is the total that must be accumulated in the sinking fund during the 40 years she plans to work. In other words, this is the *future* value, FV, of the sinking fund. (Thus, the present value in the first step of our calculation is the future value in the second step.) To determine the payments necessary to accumulate this amount, we use the sinking fund payment formula with $FV = 977{,}065.40$, $i = 0.005$, and $n = 12 \times 40 = 480$.

$$\begin{aligned} PMT &= FV\frac{i}{(1+i)^n - 1} \\ &= 977{,}065.40\frac{0.005}{1.005^{480} - 1} \\ &\approx \$490.62 \end{aligned}$$

So, if she and BCI collectively deposit \$490.62 per month into her retirement fund, she can retire with the income she desires.

Installment Loans

In a typical installment loan, such as a car loan or a home mortgage, we borrow an amount of money and then pay it back with interest by making fixed payments (usually every month) over some number of years. From the point of view of the lender, this is an annuity. Thus, loan calculations are identical to annuity calculations.

EXAMPLE 6 Home Mortgages

Marc and Mira are buying a house, and have taken out a 30-year, \$90,000 mortgage at 8% interest per year. What will their monthly payments be?

Solution From the bank's point of view, a mortgage is an annuity. In this case, the present value is $PV = \$90{,}000$, $r = 0.08$, $m = 12$, and $n = 12 \times 30 = 360$. To find the payments, we use the payment formula:

$$PMT = 90{,}000\frac{0.08/12}{1-(1+0.08/12)^{-360}} \approx \$660.39.$$

The word "mortgage" comes from the French for "dead pledge." The process of paying off a loan is called **amortizing** the loan, meaning to kill the debt owed.

EXAMPLE 7 Amortization Schedule

Continuing Example 6: Mortgage interest is tax deductible, so it is important to know how much of a year's mortgage payments represents interest. How much interest will Marc and Mira pay in the first year of their mortgage?

Solution Let us calculate how much of each month's payment is interest and how much goes to reducing the outstanding principal. At the end of the first month Marc and Mira must pay 1 month's interest on \$90,000, which is

$$\$90{,}000 \times \frac{0.08}{12} = \$600.$$

The remainder of their first monthly payment, \$660.39 − 600 = \$60.39, goes to reducing the principal. Thus, in the second month the outstanding principal is \$90,000 − 60.39 = \$89,939.61, and part of their second monthly payment will be for the interest on this amount, which is

$$\$89{,}939.61 \times \frac{0.08}{12} \approx \$599.60.$$

The remaining \$660.39 − \$599.60 = \$60.79 goes to further reduce the principal. If we continue this calculation for the 12 months of the first year, we get the beginning of the mortgage's **amortization schedule**.

using Technology

To automate the construction of the amortization schedule in Example 7 using a graphing calculator or a spreadsheet, see the Technology Guides at the end of the chapter. [Details: TI-83/84 Plus: page 137, Spreadsheet: page 141]

Month	*Interest Payment*	*Payment on Principal*	*Outstanding Principal*
0			\$90,000.00
1	\$600.00	\$60.39	89,939.61
2	599.60	60.79	89,878.82
3	599.19	61.20	89,817.62
4	598.78	61.61	89,756.01
5	598.37	62.02	89,693.99
6	597.96	62.43	89,631.56
7	597.54	62.85	89,568.71
8	597.12	63.27	89,505.44
9	596.70	63.69	89,441.75
10	596.28	64.11	89,377.64
11	595.85	64.54	89,313.10
12	595.42	64.97	89,248.13
Total	\$7,172.81	\$751.87	

As we can see from the totals at the bottom of the columns, Marc and Mira will pay a total of \$7,172.81 in interest in the first year.

Bonds

Suppose that a corporation offers a 10-year bond paying 6.5% with payments every 6 months. As we saw in Example 3 of Section 2.1, this means that if we pay \$10,000 for bonds with a maturity value of \$10,000, we will receive $6.5/2 = 3.25\%$ of \$10,000, or \$325, every 6 months for 10 years, at the end of which time the corporation will give us the original \$10,000 back. But bonds are rarely sold at their maturity value. Rather, they are auctioned off and sold at a price the bond market determines they are worth.

For example, suppose that bond traders are looking for an investment that has a **rate of return** or **yield** of 7% rather than the stated 6.5% (sometimes called the **coupon interest rate** to distinguish it from the rate of return). How much would they be willing to pay for the bonds above with a maturity value of \$10,000? Think of the bonds as an investment that will pay the owner \$325 every 6 months for 10 years, and will pay an additional \$10,000 on maturity at the end of the 10 years. We can treat the \$325 payments as if they come from an annuity and determine how much an investor would pay for such an annuity if it earned 7% compounded semiannually. Separately, we determine the present value of an investment worth \$10,000 ten years from now, if it earned 7% compounded semiannually. For the first calculation, we use the annuity present value formula, with $i = 0.07/2$ and $n = 2 \times 10 = 20$.

$$\begin{aligned} PV &= PMT\frac{1-(1+i)^{-n}}{i} \\ &= 325\frac{1-(1+0.07/2)^{-20}}{0.07/2} \\ &= \$4{,}619.03 \end{aligned}$$

For the second calculation, we use the present value formula for compound interest:

$$\begin{aligned} PV &= 10{,}000(1+0.07/2)^{-20} \\ &= \$5{,}025.66. \end{aligned}$$

Thus, an investor looking for a 7% return will be willing to pay \$4,619.03 for the semiannual payments of \$325 and \$5,025.66 for the \$10,000 payment at the end of 10 years, for a total of $\$4{,}619.03 + 5{,}025.66 = \$9{,}644.69$ for the \$10,000 bond.

EXAMPLE 8 Bonds

Suppose that bond traders are looking for only a 6% yield on their investment. How much would they pay per \$10,000 for the 10-year bonds above, which have a coupon interest rate of 6.5% and pay interest every 6 months?

Solution We redo the calculation with $r = 0.06$. For the annuity calculation we now get

$$PV = 325\frac{1-(1+0.06/2)^{-20}}{0.06/2} = \$4{,}835.18.$$

For the compound interest calculation we get

$$PV = 10{,}000(1+0.06/2)^{-20} = \$5{,}536.76.$$

Thus, traders would be willing to pay a total of $\$4{,}835.18 + 5{,}536.76 = \$10{,}371.94$ for bonds with a maturity value of \$10,000.

➡ **Before we go on...** Notice how the selling price of the bonds behaves as the desired yield changes. As desired yield goes up, the price of the bonds goes down, and as desired yield goes down, the price of the bonds goes up. When the desired yield equals the coupon interest rate, the selling price will equal the maturity value. Therefore, when the yield is higher than the coupon interest rate, the price of the bond will be below its maturity value, and when the yield is lower than the coupon interest rate, the price will be above the maturity value.

As we've mentioned before, the desired yield depends on many factors, but it generally moves up and down with prevailing interest rates. And interest rates have historically gone up and down cyclically. The effect on the value of bonds can be quite dramatic (see Exercises 65 and 66 in the preceding section). Because bonds can be sold again once bought, someone who buys bonds while interest rates are high and then resells them when interest rates decline can make a healthy profit. ■

using Technology

To automate the computations in Example 9 using a graphing calculator or a spreadsheet, see the Technology Guides at the end of the chapter. Outline:

TI-83/84 Plus

`APPS` `1:Finance`, then `1:TVM Solver`
N = 40, PV = −9800, PMT = 250, FV = 10000, P/Y = 2, C/Y = 2
With cursor on I% line, `ALPHA` `SOLVE`
[More details on page 138.]

Spreadsheet

`=RATE(20*2,250,-9800,10000)*2`
[More details on page 142.]

 **Website**

www.WanerMath.com
→ On Line Utilities
→ Time Value of Money Utility

Enter the values shown, and press "Compute" next to r.

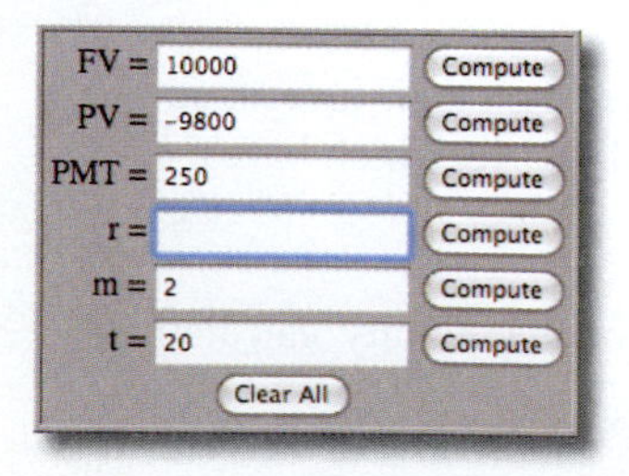

EXAMPLE 9 Rate of Return on a Bond

Suppose that a 5%, 20-year bond sells for \$9,800 per \$10,000 maturity value. What rate of return will investors get?

Solution Assuming the usual semiannual payments, we know the following about the annuity calculation:

$$PMT = 0.05 \times 10{,}000/2 = 250$$
$$n = 20 \times 2 = 40.$$

What we do not know is r or i, the annual or semiannual rate of return, respectively. So, we write

$$PV = 250\frac{1-(1+i)^{-40}}{i}.$$

For the compound interest calculation, we know $FV = 10{,}000$ and $n = 40$ again, so we write

$$PV = 10{,}000(1+i)^{-40}.$$

Adding these together should give the selling price of \$9,800:

$$250\frac{1-(1+i)^{-40}}{i} + 10{,}000(1+i)^{-40} = 9{,}800.$$

This equation cannot be solved for i directly. The best we can do by hand is a sort of trial-and-error approach, substituting a few values for i in the left-hand side of the above equation to get an estimate; we use the fact that, because the selling price is below the maturity value, r must be larger than 0.05, so i must be more than 0.025:

i	0.025	0.03	0.035
$250\frac{1-(1+i)^{-40}}{i} + 10{,}000(1+i)^{-40}$	10,000	8,844	7,864

Since we want the value to be 9,800, we see that the correct answer is somewhere between $i = 0.025$ and $i = 0.03$. Let us try the value midway between 0.025 and 0.03; namely $i = 0.0275$.

i	0.025	0.0275	0.03
$250\frac{1-(1+i)^{-40}}{i} + 10{,}000(1+i)^{-40}$	10,000	9,398	8,844

Now we know that the correct value of i is somewhere between 0.025 and 0.0275, so we can choose for our next estimate of i the number midway between them: 0.02625. We could continue in this fashion to obtain i as accurately as we like. In fact, $i \approx 0.02581$, corresponding to an annual rate of return of approximately 5.162%.

FAQs

Which Formula to Use

Q: *We have retirement accounts, trust funds, loans, bonds, and so on. Some are sinking funds, others are annuities. How do we distinguish among them, so we can tell which formula to use?*

A: In general, remember that a sinking fund is an interest-bearing fund into which payments are made, while an annuity is an interest-bearing fund from which money is withdrawn. Here is a list of some of the accounts we have discussed in this section:

- *Retirement Accounts* A retirement account is a sinking fund while payments are being made into the account (prior to retirement) and an annuity while a pension is being withdrawn (after retirement).
- *Education Funds* These are similar to retirement accounts.
- *Trust Funds* A trust fund is an annuity if periodic withdrawals are made.
- *Installment Loans* We think of an installment loan as an investment a bank makes in the lender. In this way, the lender's payments can be viewed as the bank's withdrawals, and so a loan is an annuity.
- *Bonds* A bond pays regular fixed amounts until it matures, at which time it pays its maturity value. We think of the bond as an annuity coupled with a compound interest investment funding the payment of the maturity value. We can then determine its present value based on the current market interest rate.

From a mathematical point of view, sinking funds and annuities are really the same thing. See the Communication and Reasoning Exercises for this section for more about this.

2.3 EXERCISES

Access end-of-section exercises online at **www.webassign.net** ENHANCED WebAssign

CHAPTER 2 REVIEW

KEY CONCEPTS

 Website www.WanerMath.com

Go to the Website at www.WanerMath.com to find a comprehensive and interactive Web-based summary of Chapter 2.

2.1 Simple Interest

Simple interest *p. 102*
Future value *p. 103*
Bond, maturity *p. 104*
Present value *p. 105*
Treasury bills (T-bills); maturity value, discount rate, yield *p. 106*

2.2 Compound Interest

Compound interest *p. 108*
Future value for compound interest *p. 109*
Present value for compound interest *p. 111*
Zero coupon bond or strip *p. 111*
Inflation, constant dollars *p. 112*
Effective interest rate, annual percentage yield (APY) *p. 113*

2.3 Annuities, Loans, and Bonds

Annuity, sinking fund *p. 115*
Future value of a sinking fund *p. 116*
Payment formula for a sinking fund *p. 117*
Present value of an annuity *p. 119*
Ordinary annuity, annuity due *p. 119*
Payment formula for an annuity *p. 120*
Installment loan *p. 121*
Amortization schedule *p. 122*
Bond *p. 123*

REVIEW EXERCISES

In each of Exercises 1–6, find the future value of the investment.

1. $6,000 for 5 years at 4.75% simple annual interest

2. $10,000 for 2.5 years at 5.25% simple annual interest

3. $6,000 for 5 years at 4.75% compounded monthly

4. $10,000 for 2.5 years at 5.25% compounded semiannually

5. $100 deposited at the end of each month for 5 years, at 4.75% interest compounded monthly

6. $2,000 deposited at the end of each half-year for 2.5 years, at 5.25% interest compounded semiannually

In each of Exercises 7–12, find the present value of the investment.

7. Worth $6,000 after 5 years at 4.75% simple annual interest

8. Worth $10,000 after 2.5 years at 5.25% simple annual interest

9. Worth $6,000 after 5 years at 4.75% compounded monthly

10. Worth $10,000 after 2.5 years at 5.25% compounded semiannually

11. Funding $100 withdrawals at the end of each month for 5 years, at 4.75% interest compounded monthly

12. Funding $2,000 withdrawals at the end of each half-year for 2.5 years, at 5.25% interest compounded semiannually

In each of Exercises 13–18, find the amounts indicated.

13. The monthly deposits necessary to accumulate $12,000 after 5 years in an account earning 4.75% compounded monthly

14. The semiannual deposits necessary to accumulate $20,000 after 2.5 years in an account earning 5.25% compounded semiannually

15. The monthly withdrawals possible over 5 years from an account earning 4.75% compounded monthly and starting with $6,000

16. The semiannual withdrawals possible over 2.5 years from an account earning 5.25% compounded semiannually and starting with $10,000

17. The monthly payments necessary on a 5-year loan of $10,000 at 4.75%

18. The semiannual payments necessary on a 2.5-year loan of $15,000 at 5.25%

19. How much would you pay for a $10,000, 5-year, 6% bond if you want a return of 7%? (Assume that the bond pays interest every 6 months.)

20. How much would you pay for a $10,000, 5-year, 6% bond if you want a return of 5%? (Assume that the bond pays interest every 6 months.)

21. T A $10,000, 7-year, 5% bond sells for $9,800. What return does it give you? (Assume that the bond pays interest every 6 months.)

22. T A $10,000, 7-year, 5% bond sells for $10,200. What return does it give you? (Assume that the bond pays interest every 6 months.)

In each of Exercises 23–28, find the time requested, to the nearest 0.1 year.

23. The time it would take $6,000 to grow to $10,000 at 4.75% simple annual interest

24. The time it would take $10,000 to grow to $15,000 at 5.25% simple annual interest

25. T The time it would take $6,000 to grow to $10,000 at 4.75% interest compounded monthly

26. T The time it would take $10,000 to grow to $15,000 at 5.25% interest compounded semiannually

27. T The time it would take to accumulate $10,000 by depositing $100 at the end of each month in an account earning 4.75% interest compounded monthly

28. T The time it would take to accumulate $15,000 by depositing $2,000 at the end of each half-year in an account earning 5.25% compounded semiannually

APPLICATIONS

Stock Investments *Exercises 29–34 are based on the following table, which shows some values of ABCromD (ABCD) stock:*

Dec. 2002	Aug. 2004	Mar. 2005	May 2005	Aug. 2005	Dec. 2005
3.28	16.31	33.95	21.00	30.47	7.44
Jan. 2007	**Mar. 2008**	**Oct. 2008**	**Nov. 2009**	**Feb. 2010**	**Aug. 2010**
12.36	7.07	11.44	33.53	44.86	45.74

29. Marjory Duffin purchased ABCD stock in December 2002 and sold it in August 2010. Calculate her annual return on a simple interest basis to the nearest 0.01%.

30. John O'Hagan purchased ABCD stock in March 2005 and sold it in January 2007. Calculate his annual return on a simple interest basis to the nearest 0.01%.

31. Suppose Marjory Duffin had bought ABCD stock in January 2007. If she had later sold at one of the dates in the table, which of those dates would have given her the largest annual return on a simple interest basis, and what would that return have been?

32. Suppose John O'Hagan had purchased ABCD stock in August 2004. If he had later sold at one of the dates in the table, which of those dates would have given him the largest annual loss on a simple interest basis, and what would that loss have been?

33. Did ABCD stock undergo simple interest increase in the period December 2002 through March 2005? (Give a reason for your answer.)

34. If ABCD stock underwent simple interest increase from February 2010 through August 2010 and into 2011, what would the price have been in December 2011?

35. ***Revenue*** Total Online revenues at OHaganBooks.com during 1999, its first year of operation, amounted to \$150,000. After December, 1999, revenues increased by a steady 20% each year. Track OHaganBooks.com's revenues for the subsequent 5 years, assuming that this rate of growth continued. During which year did the revenue surpass \$300,000?

36. ***Net Income*** Unfortunately, the picture for net income was not so bright: The company lost \$20,000 in the fourth quarter of 1999. However, the quarterly loss decreased at an annual rate of 15%. How much did the company lose during the third quarter of 2001?

37. ***Stocks*** In order to finance anticipated expansion, CEO John O'Hagan is considering making a public offering of OHaganBooks.com shares at \$3.00 a share. O'Hagan is not sure how many shares to offer, but would like the shares to reach a total market value of at least \$500,000 6 months after the initial offering. He estimates that the value of the stock will double in the first day of trading, and then appreciate at around 8% per month for the first 6 months. How many shares of stock should the company offer?

38. ***Stocks*** Unfortunately, renewed panic about the Monaco debt crisis cause the U.S. stock market to plunge on the very first day of OHaganBooks.com's initial public offering (IPO) of 600,000 shares at \$3.00 per share, and the shares ended the trading day 60% lower. Subsequently, as the Monaco debt crisis worsened, the stock depreciated by 10% per week during the subsequent 5 weeks. What was the total market value of the stocks at the end of 5 weeks?

39. ***Loans*** OHaganBooks.com is seeking a \$250,000 loan to finance its continuing losses. One of the best deals available is offered by Industrial Bank, which offers a 10-year 9.5% loan. What would the monthly payments be for this loan?

40. ***Loans*** (See Exercise 39.) Expansion Loans offers an 8-year 6.5% loan. What would the monthly payments be for the \$250,000 loan from Expansion?

41. ***Loans*** (See Exercise 39.) OHaganBooks.com can afford to pay only \$3,000 per month to service its debt. What, to the nearest dollar, is the largest amount the company can borrow from Industrial Bank?

42. ***Loans*** (See Exercise 40.) OHaganBooks.com can afford to pay only \$3,000 per month to service its debt. What, to the nearest dollar, is the largest amount the company can borrow from Expansion Loans?

43. T ***Loans*** (See Exercise 39.) What interest rate would Industrial Bank have to offer in order to meet the company's loan requirements at a price (no more than \$3,000 per month) it can afford?

44. T ***Loans*** (See Exercise 40.) What interest rate would Expansion Loans have to offer in order to meet the company's loan requirements at a price (no more than \$3,000 per month) it can afford?

Retirement Planning *OHaganBooks.com has just introduced a retirement package for the employees. Under the annuity plan operated by Sleepy Hollow, the monthly contribution by the company on behalf of each employee is \$800. Each employee can then supplement that amount through payroll deductions. The current rate of return of Sleepy Hollow's retirement fund is 7.3%. Use this information in Exercises 45–52.*

45. Jane Callahan, the Web site developer at OHaganBooks.com, plans to retire in 10 years. She contributes \$1,000 per month to the plan (in addition to the company contribution of \$800). Currently, there is \$50,000 in her retirement annuity. How much (to the nearest dollar) will it be worth when she retires?

46. Percy Egan, the assistant Web site developer at OHaganBooks .com, plans to retire in 8 years. He contributes \$950 per month to the plan (in addition to the company contribution of \$800). Currently, there is \$60,000 in his retirement annuity. How much (to the nearest dollar) will it be worth when he retires?

47. When she retires, how much of Jane Callahan's retirement fund will have resulted from the company contribution? (See Exercise 45. The company did not contribute toward the \$50,000 Callahan now has.)

48. When he retires, how much of Percy Egan's retirement fund will have resulted from the company contribution? (See Exercise 46. The company began contributing $800 per month to his retirement fund when he was hired 5 years ago. Assume the rate of return from Sleepy Hollow has been unchanged.)

49. (See Exercise 45.) Jane Callahan actually wants to retire with $500,000. How much should she contribute each month to the annuity?

50. (See Exercise 46.) Percy Egan actually wants to retire with $600,000. How much should he contribute each month to the annuity?

51. (See Exercise 45.) On second thought, Callahan wants to be in a position to draw at least $5,000 per month for 30 years after her retirement. She feels she can invest the proceeds of her retirement annuity at 8.7% per year in perpetuity. Given the information in Exercise 45, how much will she need to contribute to the plan starting now?

52. (See Exercise 46.) On second thought, Egan wants to be in a position to draw at least $6,000 per month for 25 years after his retirement. He feels he can invest the proceeds of his retirement annuity at 7.8% per year in perpetuity. Given the information in Exercise 46, how much will he need to contribute to the plan starting now?

Actually, Jane Callahan is quite pleased with herself; 1 year ago she purchased a $50,000 government bond paying 7.2% per year (with interest paid every 6 months) and maturing in 10 years, and interest rates have come down since then.

53. The current interest rate on 10-year government bonds is 6.3%. If she were to auction the bond at the current interest rate, how much would she get?

54. If she holds on to the bond for 6 more months and the interest rate drops to 6%, how much will the bond be worth then?

55. T Jane suspects that interest rates will come down further during the next 6 months. If she hopes to auction the bond for $54,000 in 6 months' time, what will the interest rate need to be at that time?

56. T If, in 6 months' time, the bond is auctioned for only $52,000, what will the interest rate be at that time?

Case Study

Adjustable Rate and Subprime Mortgages

Andy Dean Photography/Shutterstock

The term **subprime mortgage** refers to mortgages given to home buyers with a heightened perceived risk of default, as when, for instance, the price of the home being purchased is higher than the borrower can reasonably afford. Such loans are typically **adjustable rate** loans, meaning that the lending rate varies through the duration of the loan.* Subprime adjustable rate loans typically start at artificially low "teaser rates" that the borrower can afford, but then increase significantly over the life of the mortgage. The U.S. real estate bubble of 2000–2005 led to a frenzy of subprime lending, the rationale being that a borrower having trouble meeting mortgage payments could either sell the property at a profit or re-finance the loan, or the lending institution could earn a hefty profit by repossessing the property in the event of foreclosure.

Mr. and Mrs. Wong have an appointment tomorrow with you, their investment counselor, to discuss their plan to purchase a $400,000 house in Orlando, Florida. They have saved $20,000 for a down payment, so want to take out a $380,000 mortgage. Their combined annual income is $80,000 per year, which they estimate will increase by 4% annually over the foreseeable future, and they are considering three different specialty 30-year mortgages:

Hybrid: The interest is fixed at a low introductory rate of 4% for 5 years, and then adjusts annually to 5% over the U.S. federal funds rate.†

Interest-Only: During the first 5 years, the rate is set at 4.2% and no principal is paid. After that time, the mortgage adjusts annually to 5% over the U.S. federal funds rate.

Negative Amortization: During the first 5 years, the rate is set at 4.7% based on a principal of 60% of the purchase price of the home, with the result that the balance on the principal actually grows during this period. After that time, the mortgage adjusts annually to 5% over the U.S. federal funds rate.

* In an adjustable rate mortgage, the payments are recalculated each time the interest rate changes, based on the assumption that the new interest rate will be unchanged for the remaining life of the loan. We say that the loan is **re-amortized** at the new rate.

† The U.S. federal funds rate is the rate banks charge each other for loans and is often used to set rates for other loans. Manipulating this rate is one way the U.S. Federal Reserve regulates the money supply.

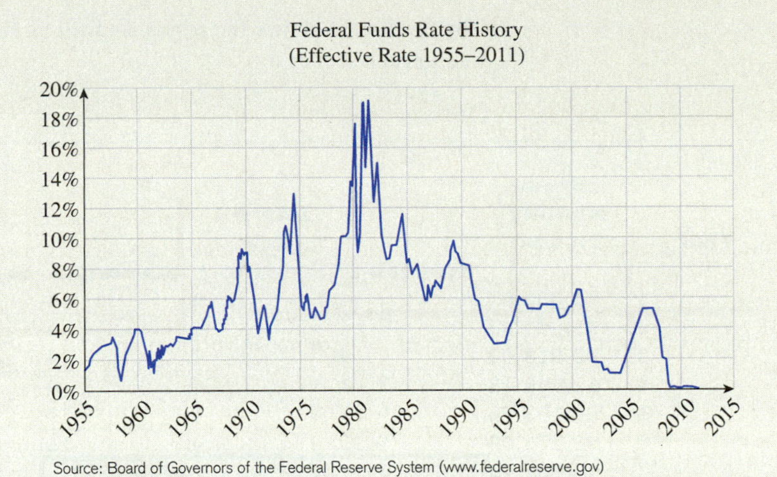

Figure 3

You decide that you should create an Excel worksheet that will compute the monthly payments for the three types of loan. Of course, you have no way of predicting what the U.S. federal funds rate will be in over the next 30 years (see Figure 3 for historical values), so you decide to include three scenarios for the federal funds rate in each case:

Scenario 1: Federal funds rate is 4.25% in year 6 and then increases by 0.25% per year.

Scenario 2: Federal funds rate is steady at 4% during the term of the loan.

Scenario 3: Federal funds rate is 15% in year 6 and then decreases by 0.25% per year.

Each worksheet will show month-by-month payments for the specific type of loan. Typically, to be affordable, payments should not exceed 28% of gross monthly income, so you will tabulate that quantity as well.

Hybrid Loan: You begin to create your worksheet by estimating 28% of the Wongs' monthly income, assuming a 4% increase each year (the income is computed using the compound interest formula for annual compounding):

	A	B	C	D	E	F	G	H
1	Year	28% of Monthly Income		Interest Rate			Monthly Payment	
2			Scenario 1	Scenario 2	Scenario 3	Scenario 1	Scenario 2	Scenario 3
3	1	=0.28*80000*(1.04)^(A3-1)/12						
4	2							
5	3							
30	28							
31	29							
32	30							

The next sheet shows the result, as well as the formulas for computing the interest rate in each scenario.

	A	B	C	D	E	F	G	H
1	Year	28% of Monthly Income		Interest Rate			Monthly Payment	
2			Scenario 1	Scenario 2	Scenario 3	Scenario 1	Scenario 2	Scenario 3
3	1	$1,866.67	4	4	4			
4	2	$1,941.33	4	4	4			
5	3	$2,018.99	4	4	4			
6	4	$2,099.75	4	4	4			
7	5	$2,183.74	4	4	4			
8	6	$2,271.09	9.25	9	15			
9	7	$2,361.93	=C8+0.25	=D8	=E8-0.25			
10	8	$2,456.41						
30	28	$5,382.29						
31	29	$5,597.58						
32	30	$5,821.48						

To compute the monthly payment, you decide to use the built-in function PMT, which has the format

$$PMT(i, n, PV, [FV], [type]),$$

where i = interest per period, n = total number of periods of the loan, PV = present value, FV = future value (optional); the *type*, also optional, is 0 or omitted if payments are at the end of each period, and 1 if at the start of each period. The present value will be the outstanding principal owed on the home each time the rate is changed and so that too will need to be known. During the first 5 years we can use as the present value the original cost of the home, but each year thereafter, the loan is re-amortized at the new interest rate, and so the outstanding principal will need to be computed. Although Excel has a built-in function that calculates payment on the principal, it returns only the payment for a single period (month), so without creating a month-by-month amortization table it would be difficult to use this function to track the outstanding principal. On the other hand, the total outstanding principal at any point in time can be computed using the future value formula FV. You decide to add three more columns to your Excel worksheet to show the principal outstanding at the start of each year. Here is the spreadsheet with the formulas for the payments and outstanding principal for the first 5 years.

	A	B	C	D	E	F	G	H	I	J	K
1	Year	28% of Monthly Income		Interest Rate			Monthly Payment			Balance on Principal	
2			Scenario 1	Scenario 2	Scenario 3	Scenario 1	Scenario 2	Scenario 3	Scenario 1	Scenario 2	Scenario 3
3	1	$1,866.67	4	4	4	=-PMT(C3/1200,360,I$3)			$380,000.00	$380,000.00	$380,000.00
4	2	$1,941.33	4	4	4				=FV(C3/1200,12*$A3,F3,-I$3)		
5	3	$2,018.99	4	4	4						
6	4	$2,099.75	4	4	4						
7	5	$2,183.74	4	4	4						
8	6	$2,271.09	9.25	9	15						
9	7	$2,361.93	9.5	9	14.75						
10	8	$2,456.41	9.75	9	14.5						

The two formulas will each be copied across to the adjacent two cells for the other scenarios. A few things to notice: The negative sign before *PMT* converts the negative quantity returned by *PMT* to a positive amount. The dollar sign in I$3 in the *PMT* formula fixes the present value for each year at the original cost of the home for the first 5 years, during which payments are computed as for a fixed rate loan. In the formula for the balance on principal at the start of each year, the number of periods is the total number of months up through the preceding year, and the present value is the same initial price of the home each year during the 5-year fixed rate period.

The next sheet shows the calculated results for the fixed-rate period, and the new formulas to be added for the adjustable rate period starting with the sixth year.

	C	D	E	F	G	H	I	J	K
1		Interest Rate			Monthly Payment			Balance on Principal	
2	Scenario 1	Scenario 2	Scenario 3	Scenario 1	Scenario 2	Scenario 3	Scenario 1	Scenario 2	Scenario 3
3	4	4	4	$1,814.18	$1,814.18	$1,814.18	$380,000.00	$380,000.00	$380,000.00
4	4	4	4	$1,814.18	$1,814.18	$1,814.18	$373,308.06	$373,308.06	$373,308.06
5	4	4	4	$1,814.18	$1,814.18	$1,814.18	$366,343.48	$366,343.48	$366,343.48
6	4	4	4	$1,814.18	$1,814.18	$1,814.18	$359,095.16	$359,095.16	$359,095.16
7	4	4	4	$1,814.18	$1,814.18	$1,814.18	$351,551.52	$351,551.52	$351,551.52
8	9.25	9	15	=-PMT(C8/1200,360-12*$A7,I8)			=FV(C7/1200,12,F7,-I7)		
9	9.5	9	14.75						
10	9.75	9	14.5						

Notice the changes: The loan is re-amortized each year starting with year 6, and the payment calculation needs to take into account the reduced, remaining lifetime of the loan each time. You now copy these formulas across for the remaining two scenarios, and then copy all six formulas down the remaining rows to complete the calculation. Following is a portion of the complete worksheet showing, in red, those years during which the monthly payment will exceed 28% of the gross monthly income.

	A	B	C	D	E	F	G	H	I	J	K
1	Year	28% of Monthly Income		Interest Rate			Monthly Payment			Balance on Principal	
2			Scenario 1	Scenario 2	Scenario 3	Scenario 1	Scenario 2	Scenario 3	Scenario 1	Scenario 2	Scenario 3
3	1	$1,866.67	4	4	4	$1,814.18	$1,814.18	$1,814.18	$380,000.00	$380,000.00	$380,000.00
4	2	$1,941.33	4	4	4	$1,814.18	$1,814.18	$1,814.18	$373,308.06	$373,308.06	$373,308.06
5	3	$2,018.99	4	4	4	$1,814.18	$1,814.18	$1,814.18	$366,343.48	$366,343.48	$366,343.48
6	4	$2,099.75	4	4	4	$1,814.18	$1,814.18	$1,814.18	$359,095.16	$359,095.16	$359,095.16
7	5	$2,183.74	4	4	4	$1,814.18	$1,814.18	$1,814.18	$351,551.52	$351,551.52	$351,551.52
8	6	$2,271.09	9.25	9	15	$2,943.39	$2,884.32	$4,402.22	$343,700.55	$343,700.55	$343,700.55
9	7	$2,361.93	9.5	9	14.75	$3,001.60	$2,884.32	$4,336.46	$340,018.68	$339,866.12	$342,337.80
10	8	$2,456.41	9.75	9	14.5	$3,058.89	$2,884.32	$4,271.84	$336,135.05	$335,671.99	$340,686.40
11	9	$2,554.66	10	9	14.25	$3,115.17	$2,884.32	$4,208.48	$332,020.97	$331,084.43	$338,694.96
12	10	$2,656.85	10.25	9	14	$3,170.39	$2,884.32	$4,146.53	$327,643.98	$326,066.52	$336,305.10
13	11	$2,763.12	10.5	9	13.75	$3,224.44	$2,884.32	$4,086.13	$322,967.19	$320,577.89	$333,450.89
14	12	$2,873.65	10.75	9	13.5	$3,277.24	$2,884.32	$4,027.41	$317,948.51	$314,574.40	$330,058.36
15	13	$2,988.59	11	9	13.25	$3,328.70	$2,884.32	$3,970.53	$312,539.70	$308,007.73	$326,045.01
16	14	$3,108.14	11.25	9	13	$3,378.71	$2,884.32	$3,915.64	$306,685.30	$300,825.07	$321,319.42
17	15	$3,232.46	11.5	9	12.75	$3,427.16	$2,884.32	$3,862.09	$300,321.34	$292,968.62	$315,780.91
18	16	$3,361.76	11.75	9	12.5	$3,473.93	$2,884.32	$3,812.43	$293,373.72	$284,375.19	$309,319.29
19	17	$3,496.23	12	9	12.25	$3,518.89	$2,884.32	$3,764.40	$285,756.40	$274,975.63	$301,814.76
20	18	$3,636.08	12.25	9	12	$3,561.90	$2,884.32	$3,718.94	$277,369.15	$264,694.33	$293,137.86
21	19	$3,781.52	12.5	9	11.75	$3,602.81	$2,884.32	$3,676.20	$268,094.83	$253,448.57	$283,149.60
22	20	$3,932.79	12.75	9	11.5	$3,641.47	$2,884.32	$3,636.32	$257,796.15	$241,147.88	$271,701.73

In the third scenario, the Wongs' payments would more than double at the start of the sixth year, and remain above what they can reasonably afford for 13 more years. Even if the Fed rate were to remain at the low rate of 4%, the monthly payments would still jump to above what the Wongs can afford at the start of the sixth year.

Interest-Only Loan: Here, the only change in the worksheet constructed previously is the computation of the payments for the first 5 years; because the loan is interest-only during this period, the monthly payment is computed as simple interest at 4.2% on the $380,000 loan for a 30-year period:

$$INT = PVr = 380{,}000 \times .042/12 = \$1{,}330.00$$

The formula you could use in the spreadsheet in cell F3 is `=C3/1200*I$3`, and then copy this across and down the entire block of payments for the first 5 years. The rest of the spreadsheet (including the balance on principal) will adjust itself accordingly with the formulas you had for the hybrid loan. Below is a portion of the result, with a lot more red than in the hybrid loan case!

	A	B	C	D	E	F	G	H	I	J	K
1	Year	28% of Monthly Income		Interest Rate			Monthly Payment			Balance on Principal	
2			Scenario 1	Scenario 2	Scenario 3	Scenario 1	Scenario 2	Scenario 3	Scenario 1	Scenario 2	Scenario 3
3	1	$1,866.67	4.2	4.2	4.2	$1,330.00	$1,330.00	$1,330.00	$380,000.00	$380,000.00	$380,000.00
4	2	$1,941.33	4.2	4.2	4.2	$1,330.00	$1,330.00	$1,330.00	$380,000.00	$380,000.00	$380,000.00
5	3	$2,018.99	4.2	4.2	4.2	$1,330.00	$1,330.00	$1,330.00	$380,000.00	$380,000.00	$380,000.00
6	4	$2,099.75	4.2	4.2	4.2	$1,330.00	$1,330.00	$1,330.00	$380,000.00	$380,000.00	$380,000.00
7	5	$2,183.74	4.2	4.2	4.2	$1,330.00	$1,330.00	$1,330.00	$380,000.00	$380,000.00	$380,000.00
8	6	$2,271.09	9.25	9	15	$3,254.25	$3,188.95	$4,867.16	$380,000.00	$380,000.00	$380,000.00
9	7	$2,361.93	9.5	9	14.75	$3,318.61	$3,188.95	$4,794.45	$375,929.28	$375,760.60	$378,493.33
10	8	$2,456.41	9.75	9	14.5	$3,381.95	$3,188.95	$4,723.00	$371,635.48	$371,123.52	$376,667.51
11	9	$2,554.66	10	9	14.25	$3,444.18	$3,188.95	$4,652.96	$367,086.90	$366,051.44	$374,465.75
12	10	$2,656.85	10.25	9	14	$3,505.22	$3,188.95	$4,584.46	$362,247.64	$360,503.57	$371,823.49
13	11	$2,763.12	10.5	9	13.75	$3,564.98	$3,188.95	$4,517.68	$357,076.92	$354,435.28	$368,667.84
14	12	$2,873.65	10.75	9	13.5	$3,623.37	$3,188.95	$4,452.76	$351,528.19	$347,797.73	$364,917.01
15	13	$2,988.59	11	9	13.25	$3,680.26	$3,188.95	$4,389.88	$345,548.14	$340,537.54	$360,479.80
16	14	$3,108.14	11.25	9	13	$3,735.55	$3,188.95	$4,329.19	$339,075.44	$332,596.29	$355,255.12
17	15	$3,232.46	11.5	9	12.75	$3,789.12	$3,188.95	$4,270.87	$332,039.35	$323,910.09	$349,131.66
18	16	$3,361.76	11.75	9	12.5	$3,840.82	$3,188.95	$4,215.07	$324,357.97	$314,409.07	$341,987.61
19	17	$3,496.23	12	9	12.25	$3,890.53	$3,188.95	$4,161.97	$315,936.16	$304,016.79	$333,690.51
20	18	$3,636.08	12.25	9	12	$3,938.08	$3,188.95	$4,111.71	$306,663.10	$292,649.65	$324,097.21
21	19	$3,781.52	12.5	9	11.75	$3,983.32	$3,188.95	$4,064.46	$296,409.28	$280,216.18	$313,054.05
22	20	$3,932.79	12.75	9	11.5	$4,026.06	$3,188.95	$4,020.37	$285,022.93	$266,616.37	$300,397.13
23	21	$4,090.10	13	9	11.25	$4,066.11	$3,188.95	$3,979.57	$272,325.66	$251,740.81	$285,952.80
24	22	$4,253.70	13.25	9	11	$4,103.29	$3,188.95	$3,942.23	$258,107.21	$235,469.81	$269,538.33

In all three scenarios, this type of mortgage is worse for the Wongs than the hybrid loan; in particular, their payments in Scenario 3 would jump to more than double what they can afford at the start of the sixth year.

Negative Amortization Loan: Again, the only change in the worksheet is the computation of the payments for the first 5 years. This time, the loan amortizes negatively during the initial 5-year period, so the payment formula in this period is adjusted to reflect this

```
=-PMT(C3/1200,360,I$3*0.6).
```

	A	B	C	D	E	F	G	H	I	J	K
1	Year	28% of Monthly Income		Interest Rate			Monthly Payment			Balance on Principal	
2			Scenario 1	Scenario 2	Scenario 3	Scenario 1	Scenario 2	Scenario 3	Scenario 1	Scenario 2	Scenario 3
3	1	$1,866.67	4.7	4.7	4.7	$1,182.49	$1,182.49	$1,182.49	$380,000.00	$380,000.00	$380,000.00
4	2	$1,941.33	4.7	4.7	4.7	$1,182.49	$1,182.49	$1,182.49	$383,750.17	$383,750.17	$383,750.17
5	3	$2,018.99	4.7	4.7	4.7	$1,182.49	$1,182.49	$1,182.49	$387,680.45	$387,680.45	$387,680.45
6	4	$2,099.75	4.7	4.7	4.7	$1,182.49	$1,182.49	$1,182.49	$391,799.48	$391,799.48	$391,799.48
7	5	$2,183.74	4.7	4.7	4.7	$1,182.49	$1,182.49	$1,182.49	$396,116.32	$396,116.32	$396,116.32
8	6	$2,271.09	9.25	9	15	$3,431.01	$3,362.16	$5,131.53	$400,640.49	$400,640.49	$400,640.49
9	7	$2,361.93	9.5	9	14.75	$3,498.87	$3,362.16	$5,054.87	$396,348.66	$396,170.82	$399,051.98
10	8	$2,456.41	9.75	9	14.5	$3,565.64	$3,362.16	$4,979.54	$391,821.64	$391,281.87	$397,127.00
11	9	$2,554.66	10	9	14.25	$3,631.26	$3,362.16	$4,905.69	$387,025.99	$385,934.29	$394,805.64
12	10	$2,656.85	10.25	9	14	$3,695.62	$3,362.16	$4,833.48	$381,923.88	$380,085.08	$392,019.86
13	11	$2,763.12	10.5	9	13.75	$3,758.62	$3,362.16	$4,763.06	$376,472.30	$373,687.17	$388,692.80
14	12	$2,873.65	10.75	9	13.5	$3,820.18	$3,362.16	$4,694.62	$370,622.18	$366,689.09	$384,738.24
15	13	$2,988.59	11	9	13.25	$3,880.16	$3,362.16	$4,628.32	$364,317.31	$359,034.55	$380,060.01
16	14	$3,108.14	11.25	9	13	$3,938.46	$3,362.16	$4,564.34	$357,493.03	$350,661.95	$374,551.54
17	15	$3,232.46	11.5	9	12.75	$3,994.93	$3,362.16	$4,502.85	$350,074.77	$341,503.95	$368,095.48
18	16	$3,361.76	11.75	9	12.5	$4,049.45	$3,362.16	$4,444.02	$341,976.15	$331,486.86	$360,563.39
19	17	$3,496.23	12	9	12.25	$4,101.85	$3,362.16	$4,388.03	$333,096.90	$320,530.10	$351,815.60
20	18	$3,636.08	12.25	9	12	$4,151.99	$3,362.16	$4,335.05	$323,320.15	$308,545.52	$341,701.22
21	19	$3,781.52	12.5	9	11.75	$4,199.68	$3,362.16	$4,285.23	$312,509.37	$295,436.71	$330,058.23
22	20	$3,932.79	12.75	9	11.5	$4,244.74	$3,362.16	$4,238.74	$300,504.55	$281,098.20	$316,713.83
23	21	$4,090.10	13	9	11.25	$4,286.97	$3,362.16	$4,195.73	$287,117.60	$265,414.64	$301,484.92
24	22	$4,253.70	13.25	9	11	$4,326.17	$3,362.16	$4,156.36	$272,126.84	$248,259.85	$284,178.86
25	23	$4,423.85	13.5	9	10.75	$4,362.10	$3,362.16	$4,120.77	$255,270.30	$229,495.82	$264,594.34
26	24	$4,600.80	13.75	9	10.5	$4,394.53	$3,362.16	$4,089.09	$236,237.47	$208,971.60	$242,522.55

Clearly the Wongs should steer clear of this type of loan in order to be able to continue to afford making payments!

In short, it seems unlikely that the Wongs will be able to afford payments on any of the three mortgages in question, and you decide to advise them to either seek a less expensive home or wait until their income has appreciated to enable them to afford a home of this price.

EXERCISES

1. **T** In the case of a hybrid loan, what would the federal funds rate have to be in Scenario 2 to ensure that the Wongs can afford to make all payments?
2. **T** Repeat the preceding exercise in the case of an interest-only loan.
3. **T** Repeat the preceding exercise in the case of a negative-amortization loan.
4. **T** What home price, to the nearest $5,000, could the Wongs afford if they took out a hybrid loan, regardless of scenario? HINT [Adjust the original value of the loan on your spreadsheet to obtain the desired result.]
5. **T** What home price, to the nearest $5,000, could the Wongs afford if they took out a negative-amortization loan, regardless of scenario? HINT [Adjust the original value of the loan on your spreadsheet to obtain the desired result.]
6. **T** How long would the Wongs need to wait before they could afford to purchase a $400,000 home, assuming that their income continues to increase as above, they still have $20,000 for a down payment, and the mortgage offers remain the same?

TI-83/84 Plus Technology Guide

Section 2.2

Example 1 (page 110) In November 2011, the Bank of Montreal was paying 1.30% interest on savings accounts. If the interest is compounded quarterly, find the future value of a $2,000 deposit in 6 years. What is the total interest paid over the period?

Solution with Technology

We could calculate the future value using the TI-83/84 Plus by entering

```
2000(1+0.0130/4)^(4*6)
```

However, the TI-83/84 Plus has this and other useful calculations built into its TVM (Time Value of Money) Solver.

1. Press [APPS] then choose item `1:Finance...` and then choose item `1:TVM Solver...`. This brings up the TVM Solver window.

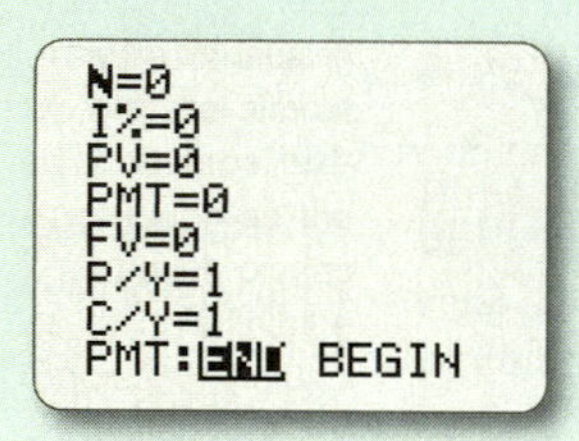

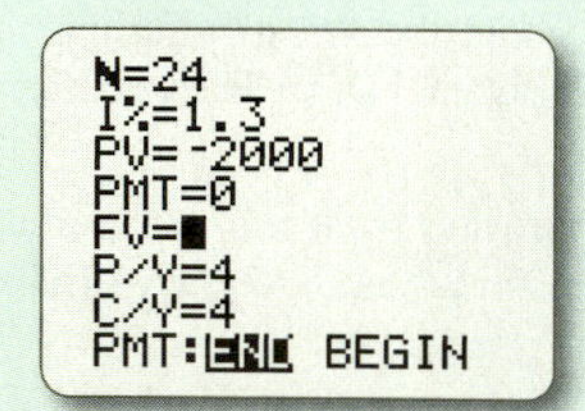

The second screen shows the values you should enter for this example. The various variables are:

`N`	Number of compounding periods
`I%`	Annual interest rate, as percent, not decimal
`PV`	Negative of present value
`PMT`	Payment per period (0 in this section)
`FV`	Future value
`P/Y`	Payments per year
`C/Y`	Compounding periods per year
`PMT:`	Not used in this section

Several things to notice:

- *I%* is the *annual* interest rate, corresponding to r, not i, in the compound interest formula.
- The present value, *PV*, is entered as a negative number. In general, when using the TVM Solver, any amount of money you give to someone else (such as the $2,000 you deposit in the bank) will be a negative number, whereas any amount of money someone gives to you (such as the future value of your deposit, which the bank will give back to you) will be a positive number.
- *PMT* is not used in this example (it will be used in the next section) and should be 0.
- *FV* is the future value, which we shall compute in a moment; it doesn't matter what you enter now.
- *P/Y* and *C/Y* stand for payments per year and compounding periods per year, respectively: They should both be set to the number of compounding periods per year for compound interest problems (setting *P/Y* automatically sets *C/Y* to the same value).
- *PMT*: *END* or *BEGIN* is not used in this example and it doesn't matter which you select.

2. To compute the future value, use the up or down arrow to put the cursor on the *FV* line, then press [ALPHA] [SOLVE].

Example 2 (page 111) Megabucks Corporation is issuing 10-year zero coupon bonds. How much would you pay for bonds with a maturity value of $10,000 if you wish to get a return of 6.5% compounded annually?

Solution with Technology

To compute the present value using a TI-83/84 Plus:

1. Enter the numbers shown below (top) in the TVM Solver window.
2. Put the cursor on the PV line, and press ALPHA SOLVE.

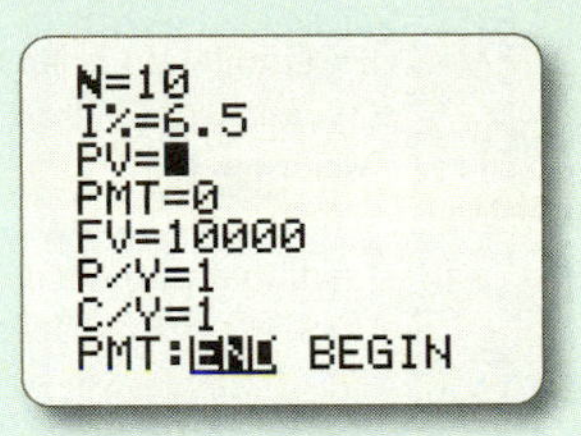


```
N=10
I%=6.5
▪PV=-5327.260355
PMT=0
FV=10000
P/Y=1
C/Y=1
PMT:END BEGIN
```

Why is the present value given as negative?

Example 6 (page 114) You have $5,000 to invest at 6% interest compounded monthly. How long will it take for your investment to grow to $6,000?

Solution with Technology

1. Enter the numbers shown below (top) in the TVM Solver window.
2. Put the cursor on the N line, and press ALPHA SOLVE.

```
N=
I%=6
PV=-5000
PMT=0
FV=6000
P/Y=12
C/Y=12
PMT:END BEGIN
```

```
▪N=36.55539636
I%=6
PV=-5000
PMT=0
FV=6000
P/Y=12
C/Y=12
PMT:END BEGIN
```

Recall that I% is the annual interest rate, corresponding to r in the formula, but N is the number of compounding periods, so number of months in this example. Thus, you will need to invest your money for about 36.5 months, or just over 3 years, before it grows to $6,000.

Section 2.3

Example 1 (page 117) Your retirement account has $5,000 in it and earns 5% interest per year compounded monthly. Every month for the next 10 years, you will deposit $100 into the account. How much money will there be in the account at the end of those 10 years?

Solution with Technology

We can use the TVM Solver in the TI-83/84 Plus to calculate future values like these:

1. The TVM Solver allows you to put the $5,000 already in the account as the present value of the account. Following the TI-83/84 Plus's usual convention, set PV to the *negative* of the present value because this is money you paid into the account.
2. Likewise, set PMT to −100 since you are paying $100 each month.
3. Set the number of payment and compounding periods to 12 per year.
4. Set the payments to be made at the end of each period.
5. With the cursor on the FV line, press ALPHA SOLVE to find the future value.

```
N=120
I%=5
PV=-5000
PMT=-100
FV=
P/Y=12
C/Y=12
PMT:END BEGIN
```

```
N=120
I%=5
PV=-5000
PMT=-100
▪FV=23763.27543
P/Y=12
C/Y=12
PMT:END BEGIN
```

Example 2 (page 117) Tony and Maria have just had a son, José Phillipe. They establish an account to accumulate money for his college education. They would like to have $100,000 in this account after 17 years. If the account pays 4% interest per year compounded quarterly, and they make deposits at the end of every quarter, how large must each deposit be for them to reach their goal?

TECHNOLOGY GUIDE

Solution with Technology

1. In the TVM Solver in the TI-83/84 Plus, enter the values shown below.
2. Solve for *PMT*.

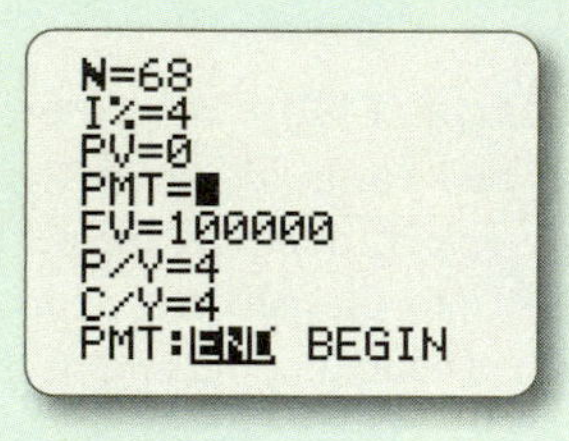

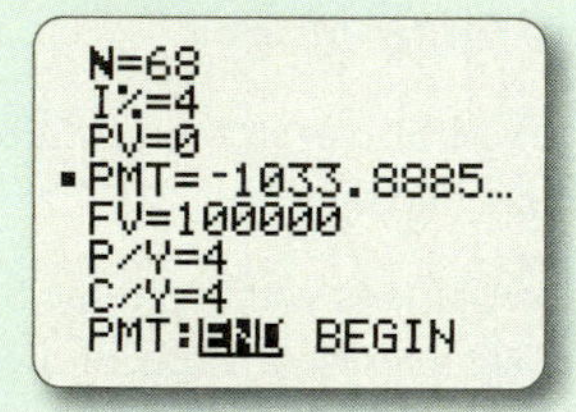

Why is *PMT* negative?

Example 3 (page 119) You wish to establish a trust fund from which your niece can withdraw $2,000 every 6 months for 15 years, at which time she will receive the remaining money in the trust, which you would like to be $10,000. The trust will be invested at 7% per year compounded every 6 months. How large should the trust be?

Solution with Technology

1. In the TVM Solver in the TI-83/84 Plus, enter the values shown below.
2. Solve for *PV*.

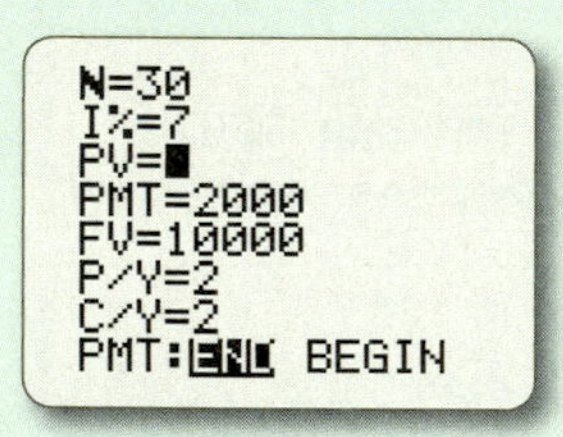

The payment and future value are positive because you (or your niece) will be receiving these amounts from the investment.

Note We have assumed that your niece receives the withdrawals at the end of each compounding period, so that the trust fund is an ordinary annuity. If, instead, she receives the payments at the beginning of each compounding period, it is an annuity due. You switch between the two types of annuity by changing PMT: END at the bottom to PMT: BEGIN. ■

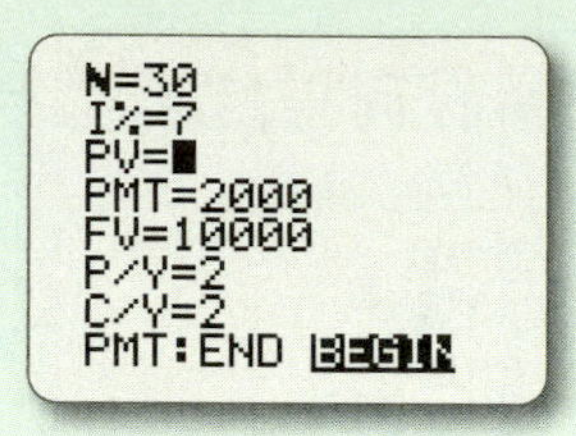

As mentioned in the text, the present value must be higher to fund payments at the beginning of each period, because the money in the account has less time to earn interest.

Example 4 (page 120) Tony and Maria (see Example 2), having accumulated $100,000 for José Phillipe's college education, would now like to make quarterly withdrawals over the next 4 years. How much money can they withdraw each quarter in order to draw down the account to zero at the end of the 4 years? (Recall that the account pays 4% interest compounded quarterly.)

Solution with Technology

1. Enter the values shown below in the TI-83/84 Plus TVM Solver.
2. Solve for *PMT*.

The present value is negative because Tony and Maria do not possess it; the bank does.

Example 7 (page 122) Marc and Mira are buying a house and have taken out a 30-year, \$90,000 mortgage at 8% interest per year. Mortgage interest is tax deductible, so it is important to know how much of a year's mortgage payments represents interest. How much interest will Marc and Mira pay in the first year of their mortgage?

Solution with Technology

The TI-83/84 Plus has built-in functions to compute the values in an amortization schedule.

1. First, use the TVM Solver to find the monthly payment.

```
N=360
I%=8
PV=90000
▪PMT=-660.38811…
FV=0
P/Y=12
C/Y=12
PMT:END BEGIN
```

Three functions correspond to the last three columns of the amortization schedule given in the text: `ΣInt`, `ΣPrn`, and `bal`, (found in the `Finance` menu accessed through APPS). They all require that the values of I%, PV, and PMT be entered or calculated ahead of time; calculating the payment in the TVM Solver in Step 1 accomplishes this.

2. Use `ΣInt(m,n,2)` to compute the sum of the interest payments from payment m through payment n. For example,

   `ΣInt(1,12,2)`

   will return $-7{,}172.81$, the total paid in interest in the first year, which answers the question asked in this example. (The last argument, 2, tells the calculator to round all intermediate calculations to two decimal places—that is, the nearest cent—as would the mortgage lender.)

3. Use `ΣPrn(m,n,2)` to compute the sum of the payments on principal from payment m through payment n. For example,

   `ΣPrn(1,12,2)`

   will return -751.87, the total paid on the principal in the first year.

4. Finally, `bal(n,2)` finds the balance of the principal outstanding after n payments. For example,

   `bal(12,2)`

   will return the value 89,248.13, the balance remaining at the end of one year.

5. To construct an amortization schedule as in the text, make sure that `FUNC` is selected in the MODE window; then enter the functions in the Y= window as shown below.

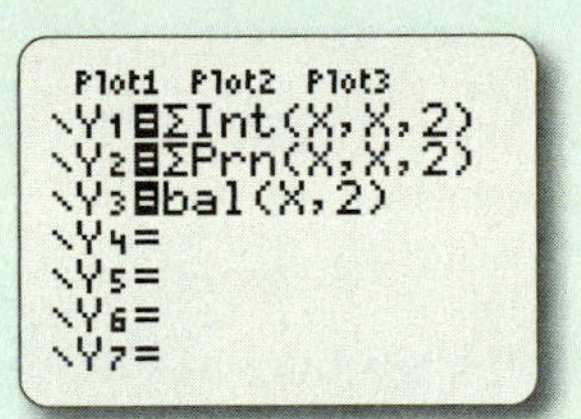

6. Press 2ND TBLSET and enter the values shown here.

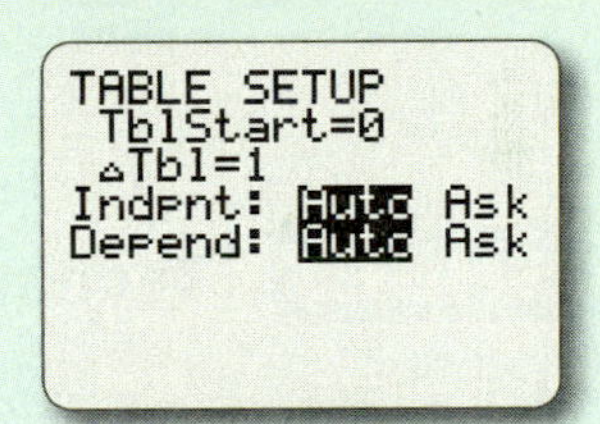

7. Press 2ND TABLE, to get the table shown here.

X	Y1	Y2
0	ERROR	ERROR
1	-600	-60.39
2	-599.6	-60.79
3	-599.2	-61.2
4	-598.8	-61.61
5	-598.4	-62.02
6	-598	-62.43

Y1=-600

X	Y2	Y3
0	ERROR	90000
1	-60.39	89940
2	-60.79	89879
3	-61.2	89818
4	-61.61	89756
5	-62.02	89694
6	-62.43	89632

Y3=89939.61

The column labeled `X` gives the month, the column labeled Y_1 gives the interest payment for each month, the column labeled Y_2 gives the payment on principal for each month, and the column labeled Y_3 (use the right arrow button to make it visible) gives the outstanding principal.

8. To see later months, use the down arrow. As you can see, some of the values will be rounded in the table, but by selecting a value (as the outstanding principal at the end of the first month is selected in the second screen) you can see its exact value at the bottom of the screen.

Example 9 (page 124) Suppose that a 5%, 20-year bond sells for $9,800 per $10,000 maturity value. What rate of return will investors get?

Solution with Technology

We can use the TVM Solver in the TI-83/84 Plus to find the interest rate just as we use it to find any other one of the variables.

1. Enter the values shown in the TVM Solver window.

2. Solve for *I%*. (Recall that *I%* is the annual interest rate, corresponding to *r* in the formula.)

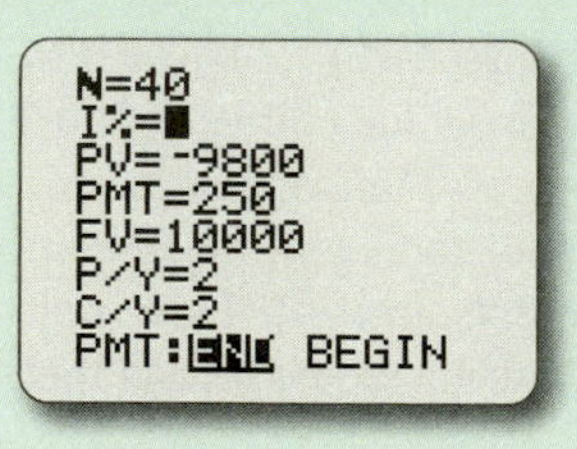


```
N=40
▪I%=5.161519731
PV=-9800
PMT=250
FV=10000
P/Y=2
C/Y=2
PMT:END BEGIN
```

Thus, at $9,800 per $10,000 maturity value, these bonds yield 5.162% compounded semiannually.

SPREADSHEET Technology Guide

Section 2.2

Example 1 (page 110) In November 2011, the Bank of Montreal was paying 1.30% interest on savings accounts. If the interest is compounded quarterly, find the future value of a $2,000 deposit in 6 years. What is the total interest paid over the period?

Solution with Technology

You can either compute compound interest directly or use financial functions built into your spreadsheet. The following worksheet has more than we need for this example, but will be useful for other examples in this and the next section.

	A	B	C	D
1		Entered	Calculated	
2	Rate	1.30%		
3	Years	6		
4	Payment	$0.00		
5	Present Value	-$2,000.00		
6	Future Value		=FV(B2/B7,B3*B7,B4,B5)	
7	Periods per year	4		

For this example the payment amount in B4 should be 0 (we shall use it in the next section).

1. Enter the other numbers as shown. As with other technologies, like the TVM Solver in the TI-83/84 Plus calculator, money that you pay to others (such as the $2,000 you deposit in the bank) should be entered as negative, whereas money that is paid to you is positive.

2. The formula entered in C6 uses the built-in FV function to calculate the future value based on the entries in column B. This formula has the following format:

=FV(*i*,*n*,*PMT*,*PV*)

i = interest per period — We use B2/B7 for the interest.

n = number of periods — We use B3*B7 for the number of periods.

PMT = payment per period — The payment is 0 (cell B4).

PV = present value — The present value is in cell B5.

Instead of using the built-in FV function, we could use

```
=-B5*(1+B2/B7)^(B3*B7)
```

based on the future value formula for compound interest. After calculation the result will appear in cell C6.

	A	B	C	D
1		Entered	Calculated	
2	Rate	1.30%		
3	Years	6		
4	Payment	$0.00		
5	Present Value	-$2,000.00		
6	Future Value		$2,161.97	
7	Periods per year	4		

Note that we have formatted the cells B4:C6 as currency with two decimal places. If you change the values in column B, the future value in column C will be automatically recalculated.

Example 2 (page 111) Megabucks Corporation is issuing 10-year zero coupon bonds. How much would you pay for bonds with a maturity value of $10,000 if you wish to get a return of 6.5% compounded annually?

Solution with Technology

You can compute present value in your spreadsheet using the PV worksheet function. The following worksheet is similar to the one in the preceding example, except that we have entered a formula for computing the present value from the entered values.

	A	B	C	D
1		Entered	Calculated	
2	Rate	6.50%		
3	Years	10		
4	Payment	$0.00		
5	Present Value		=PV(B2/B7,B3*B7,B4,B6)	
6	Future Value	$10,000.00		
7	Periods per year	1		

The next worksheet shows the calculated value.

	A	B	C	D
1		Entered	Calculated	
2	Rate	6.50%		
3	Years	10		
4	Payment	$0.00		
5	Present Value		-$5,327.26	
6	Future Value	$10,000.00		
7	Periods per year	1		

Why is the present value negative?

Example 6 (page 114) You have $5,000 to invest at 6% interest compounded monthly. How long will it take for your investment to grow to $6,000?

Solution with Technology

You can compute the requisite length of an investment in your spreadsheet using the NPER worksheet function. The following worksheets show the calculation.

	A	B	C	D
1		Entered	Calculated	
2	Rate	6.00%		
3	Years		=NPER(B2/B7,B4,B5,B6)/B7	
4	Payment	$0.00		
5	Present Value	-$5,000.00		
6	Future Value	$6,000.00		
7	Periods per year	12		

	A	B	C	D
1		Entered	Calculated	
2	Rate	6.00%		
3	Years		3.04628303	
4	Payment	$0.00		
5	Present Value	-$5,000.00		
6	Future Value	$6,000.00		
7	Periods per year	12		

The NPER function computes the number of compounding periods, months in this case, so we divide by B7, the number of periods per year, to calculate the number of years, which appears as 3.046. So, you need to invest your money for just over 3 years for it to grow to $6,000.

Section 2.3

Example 1 (page 117) Your retirement account has $5,000 in it and earns 5% interest per year compounded monthly. Every month for the next 10 years you will deposit $100 into the account. How much money will there be in the account at the end of those 10 years?

Solution with Technology

We can use exactly the same worksheet that we used in Example 1 in the preceding section. In fact, we included the "Payment" row in that worksheet just for this purpose.

	A	B	C	D
1		Entered	Calculated	
2	Rate	5.00%		
3	Years	10		
4	Payment	-$100.00		
5	Present Value	-$5,000.00		
6	Future Value		=FV(B2/B7,B3*B7,B4,B5)	
7	Periods per year	12		

	A	B	C	D
1		Entered	Calculated	
2	Rate	5.00%		
3	Years	10		
4	Payment	-$100.00		
5	Present Value	-$5,000.00		
6	Future Value		$23,763.28	
7	Periods per year	12		

TECHNOLOGY GUIDE

Note that the FV function allows us to enter, as the last argument, the amount of money already in the account. Following the usual convention, we enter the present value and the payment as *negative*, because these are amounts you pay into the account.

Example 2 (page 117) Tony and Maria have just had a son, José Phillipe. They establish an account to accumulate money for his college education, in which they would like to have $100,000 after 17 years. If the account pays 4% interest per year compounded quarterly, and they make deposits at the end of every quarter, how large must each deposit be for them to reach their goal?

Solution with Technology

Use the following worksheet, in which the PMT worksheet function is used to calculate the required payments.

	A	B	C	D
1		Entered	Calculated	
2	Rate	4.00%		
3	Years	17		
4	Payment		=PMT(B2/B7,B3*B7,B5,B6)	
5	Present Value	$0.00		
6	Future Value	$100,000.00		
7	Periods per year	4		

	A	B	C	D
1		Entered	Calculated	
2	Rate	4.00%		
3	Years	17		
4	Payment		-$1,033.89	
5	Present Value	$0.00		
6	Future Value	$100,000.00		
7	Periods per year	4		

Why is the payment negative?

Example 3 (page 119) You wish to establish a trust fund from which your niece can withdraw $2,000 every 6 months for 15 years, at which time she will receive the remaining money in the trust, which you would like to be $10,000. The trust will be invested at 7% per year compounded every 6 months. How large should the trust be?

Solution with Technology

You can use the same worksheet as in Example 2 in Section 2.2.

	A	B	C	D
1		Entered	Calculated	
2	Rate	7.00%		
3	Years	15		
4	Payment	$2,000.00		
5	Present Value		=PV(B2/B7,B3*B7,B4,B6)	
6	Future Value	$10,000.00		
7	Periods per year	2		

	A	B	C	D
1		Entered	Calculated	
2	Rate	7.00%		
3	Years	15		
4	Payment	$2,000.00		
5	Present Value		-$40,346.87	
6	Future Value	$10,000.00		
7	Periods per year	2		

The payment and future value are positive because you (or your niece) will be receiving these amounts from the investment.

Note We have assumed that your niece receives the withdrawals at the end of each compounding period, so that the trust fund is an ordinary annuity. If, instead, she receives the payments at the beginning of each compounding period, it is an annuity due. You switch to an annuity due by adding an optional last argument of 1 to the PV function (and similarly for the other finance functions in spreadsheets).

	A	B	C	D
1		Entered	Calculated	
2	Rate	7.00%		
3	Years	15		
4	Payment	$2,000.00		
5	Present Value		=PV(B2/B7,B3*B7,B4,B6,1)	
6	Future Value	$10,000.00		
7	Periods per year	2		

	A	B	C	D
1		Entered	Calculated	
2	Rate	7.00%		
3	Years	15		
4	Payment	$2,000.00		
5	Present Value		-$41,634.32	
6	Future Value	$10,000.00		
7	Periods per year	2		

As mentioned in the text, the present value must be higher to fund payments at the beginning of each period, because the money in the account has less time to earn interest. ■

Example 4 (page 120) Tony and Maria (see Example 2), having accumulated $100,000 for José Phillipe's college education, would now like to make quarterly withdrawals over the next 4 years. How much money can they withdraw each quarter in order to draw down the account to zero at the end of the four years? (Recall that the account pays 4% interest compounded quarterly.)

Solution with Technology

You can use the same worksheet as in Example 2.

	A	B	C	D
1		Entered	Calculated	
2	Rate	4.00%		
3	Years	4		
4	Payment		=PMT(B2/B7,B3*B7,B5,B6)	
5	Present Value	-$100,000.00		
6	Future Value	$0.00		
7	Periods per year	4		

	A	B	C	D
1		Entered	Calculated	
2	Rate	4.00%		
3	Years	4		
4	Payment		$6,794.46	
5	Present Value	-$100,000.00		
6	Future Value	$0.00		
7	Periods per year	4		

The present value is negative since Tony and Maria do not possess it; the bank does.

Example 7 (page 122) Marc and Mira are buying a house and have taken out a 30-year, $90,000 mortgage at 8% interest per year. Mortgage interest is tax deductible, so it is important to know how much of a year's mortgage payments represents interest. How much interest will Marc and Mira pay in the first year of their mortgage?

Solution with Technology

We construct an amortization schedule with which we can answer the question.

1. Begin with the worksheet below.

	A	B	C	D	E	F	G	H
1	Month	Interest Payment	Payment on Principal	Outstanding Principal				
2	0			$90,000.00	Rate	8%		
3					Years	30		
4					Payment	=DOLLAR(-PMT(F2/12,F3*12,D2))		

	A	B	C	D	E	F
1	Month	Interest Payment	Payment on Principal	Outstanding Principal		
2	0			$90,000.00	Rate	8%
3					Years	30
4					Payment	$660.39

Note the formula for the monthly payment:

```
=DOLLAR(-PMT(F2/12,F3*12,D2))
```

The function DOLLAR rounds the payment to the nearest cent, as the bank would.

2. Calculate the interest owed at the end of the first month using the formula

```
=DOLLAR(D2*F$2/12)
```

in cell B3.

3. The payment on the principal is the remaining part of the payment, so enter

```
=F$4-B3
```

in cell C3.

4. Calculate the outstanding principal by subtracting the payment on the principal from the previous outstanding principal, by entering

```
=D2-C3
```

in cell D3.

5. Copy the formulas in cells B3, C3, and D3 into the cells below them to continue the table.

	A	B	C	D	E	F
1	Month	Interest Payment	Payment on Principal	Outstanding Principal		
2	0			$90,000.00	Rate	8%
3	1	$600.00	$60.39	$89,939.61	Years	30
4	2				Payment	$660.39
5	3					
6	4					
7	5					
8	6					
9	7					
10	8					
11	9					
12	10					
13	11					
14	12					

The result should be something like the following:

	A	B	C	D	E	F
1	Month	Interest Payment	Payment on Principal	Outstanding Principal		
2	0			$90,000.00	Rate	8%
3	1	$600.00	$60.39	$89,939.61	Years	30
4	2	$599.60	$60.79	$89,878.82	Payment	$660.39
5	3	$599.19	$61.20	$89,817.62		
6	4	$598.78	$61.61	$89,756.01		
7	5	$598.37	$62.02	$89,693.99		
8	6	$597.96	$62.43	$89,631.56		
9	7	$597.54	$62.85	$89,568.71		
10	8	$597.12	$63.27	$89,505.44		
11	9	$596.70	$63.69	$89,441.75		
12	10	$596.28	$64.11	$89,377.64		
13	11	$595.85	$64.54	$89,313.10		
14	12	$595.42	$64.97	$89,248.13		

6. Adding the calculated interest payments gives us the total interest paid in the first year: $7,172.81.

Note Spreadsheets have built-in functions that compute the interest payment (IPMT) or the payment on the principle (PPMT) in a given period. We could also have used the built-in future value function (FV) to calculate the outstanding principal each month. The main problem with using these functions is that, in a sense, they are too accurate. They do not take into account the fact that payments and interest are rounded to the nearest cent. Over time, this rounding causes the actual value of the outstanding principal to differ from what the FV function would tell us. In fact, because the actual payment is rounded slightly upward (to $660.39 from 660.38811...), the principal is reduced slightly faster than necessary and a last payment of $660.39 would be $2.95 larger than needed to clear out the debt. The lender would reduce the last payment by $2.95 for this reason; Marc and Mira will pay only $657.44 for their final payment. This is common: The last payment on an installment loan is usually slightly larger or smaller than the others, to compensate for the rounding of the monthly payment amount. ■

Example 9 (page 124) Suppose that a 5%, 20-year bond sells for $9,800 per $10,000 maturity value. What rate of return will investors get?

Solution with Technology

Use the following worksheet, in which the RATE worksheet function is used to calculate the interest rate.

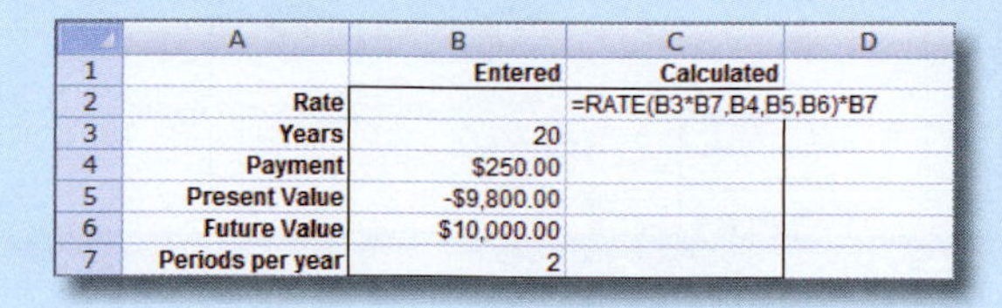

	A	B	C	D
1		Entered	Calculated	
2	Rate		=RATE(B3*B7,B4,B5,B6)*B7	
3	Years	20		
4	Payment	$250.00		
5	Present Value	-$9,800.00		
6	Future Value	$10,000.00		
7	Periods per year	2		

	A	B	C	D
1		Entered	Calculated	
2	Rate		5.162%	
3	Years	20		
4	Payment	$250.00		
5	Present Value	-$9,800.00		
6	Future Value	$10,000.00		
7	Periods per year	2		

3

Systems of Linear Equations and Matrices

 **Website**

www.WanerMath.com

At the Website you will find:

- Section-by-section tutorials, including game tutorials with randomized quizzes
- A detailed chapter summary
- A true/false quiz
- Additional review exercises
- Graphers, Excel tutorials, TI-83/84 Plus programs
- A Web page that pivots and does row operations
- An Excel worksheet that pivots and does row operations

Case Study Hybrid Cars—Optimizing the Degree of Hybridization

You are involved in new model development at a major automobile company. The company is planning to introduce two new plug-in hybrid electric vehicles: the subcompact "Green Town Hopper" and the midsize "Electra Supreme," and your department must decide on the degree of hybridization (DOH) for each of these models that will result in the largest reduction in gasoline consumption. The data you have available show the gasoline saving for only three values of the DOH. **How do you estimate the optimal value?**

Oleksiy Maksymenko/Alamy

Introduction

In Chapter 1 we studied single functions and equations. In this chapter we seek solutions to **systems** of two or more equations. For example, suppose we need to *find two numbers whose sum is* 3 *and whose difference is* 1. In other words, we need to find two numbers x and y such that $x + y = 3$ and $x - y = 1$. The only solution turns out to be $x = 2$ and $y = 1$, a solution you might easily guess. But, how do we know that this is the only solution, and how do we find solutions systematically? When we restrict ourselves to systems of *linear* equations, there is a very elegant method for determining the number of solutions and finding them all. Moreover, as we will see, many real-world applications give rise to just such systems of linear equations.

We begin in Section 3.1 with systems of two linear equations in two unknowns and some of their applications. In Section 3.2 we study a powerful matrix method, called *row reduction*, for solving systems of linear equations in any number of unknowns. In Section 3.3 we look at more applications.

Computers have been used for many years to solve the large systems of equations that arise in the real world. You probably already have access to devices that will do the row operations used in row reduction. Many graphing calculators can do them, as can spreadsheets and various special-purpose applications, including utilities available at the Website. Using such a device or program makes the calculations quicker and helps avoid arithmetic mistakes. Then there are programs (and calculators) into which you simply feed the system of equations and out pop the solutions. We can think of what we do in this chapter as looking inside the "black box" of such a program. More important, we talk about how, starting from a real-world problem, to get the system of equations to solve in the first place. No computer will do this conversion for us yet.

3.1 Systems of Two Equations in Two Unknowns

Suppose you have \$3 in your pocket to spend on snacks and a drink. If x represents the amount you'll spend on snacks and y represents the amount you'll spend on a drink, you can say that $x + y = 3$. On the other hand, if for some reason you want to spend \$1 more on snacks than on your drink, you can also say that $x - y = 1$. These are simple examples of **linear equations in two unknowns**.

Linear Equations in Two Unknowns

A **linear equation in two unknowns** is an equation that can be written in the form

$$ax + by = c$$

with a, b, and c being real numbers. The number a is called the **coefficient of x** and b is called the **coefficient of y**. A **solution** of an equation consists of a pair of numbers: a value for x and a value for y that satisfy the equation.

Quick Example

In the linear equation $3x - y = 15$, the coefficients are $a = 3$ and $b = -1$. The point $(x, y) = (5, 0)$ is a solution, because $3(5) - (0) = 15$.

In fact, a single linear equation such as $3x - y = 15$ has infinitely many solutions: We could solve for $y = 3x - 15$ and then, for every value of x we choose, we can get the corresponding value of y, giving a solution (x, y). As we saw in Chapter 1, these solutions are the points on a straight line, the *graph* of the equation.

In this section we are concerned with pairs (x, y) that are solutions of *two* linear equations at the same time. For example, $(2, 1)$ is a solution of both of the equations $x + y = 3$ and $x - y = 1$, because substituting $x = 2$ and $y = 1$ into these equations gives $2 + 1 = 3$ (true) and $2 - 1 = 1$ (also true), respectively. So, in the simple example we began with, you could spend \$2 on snacks and \$1 on a drink.

In the following set of examples, you will see how to graphically and algebraically solve a system of two linear equations in two unknowns. Then we'll return to some more interesting applications.

EXAMPLE 1 Two Ways of Solving a System: Graphically and Algebraically

Find all solutions (x, y) of the following system of two equations:

$$x + y = 3$$
$$x - y = 1.$$

Solution We will see how to find the solution(s) in two ways: graphically and algebraically. Remember that a solution is a pair (x, y) that simultaneously satisfies *both* equations.

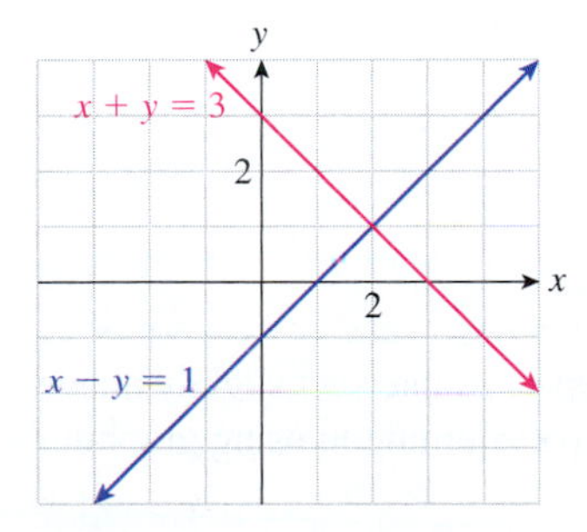

Figure 1

Method 1: Graphical We already know that the solutions of a single linear equation are the points on its graph, which is a straight line. For a point to represent a solution of two linear equations, it must lie simultaneously on both of the corresponding lines. In other words, it must be a point where the two lines cross, or intersect. A look at Figure 1 should convince us that the lines cross only at the point $(2, 1)$, so this is the only possible solution.

Method 2: Algebraic In the algebraic approach, we try to combine the equations in such a way as to eliminate one variable. In this case, notice that if we add the left-hand sides of the equations, the terms with y are eliminated. So, we add the first equation to the second (that is, add the left-hand sides and add the right-hand sides*):

$$\begin{aligned} x + y &= 3 \\ x - y &= 1 \\ \hline 2x + 0 &= 4 \\ 2x \quad &= 4 \\ x \quad &= 2. \end{aligned}$$

* We can add these equations because when we add equal amounts to both sides of an equation, the results are equal. That is, if $A = B$ and $C = D$, then $A + C = B + D$.

Now that we know that x has to be 2, we can substitute back into either equation to find y. Choosing the first equation (it doesn't matter which we choose), we have

$$\begin{aligned} 2 + y &= 3 \\ y &= 3 - 2 \\ &= 1. \end{aligned}$$

We have found that the only possible solution is $x = 2$ and $y = 1$, or

$$(x, y) = (2, 1).$$

 **using Technology**

See the Technology Guides at the end of the chapter for details on the graphical solution of Example 1 using a TI-83/84 Plus or a spreadsheet. Here is an outline:

TI-83/84 Plus

`Y1=-X+3 Y2=X-1`

Graph: [WINDOW]; Xmin = −4, Xmax = 4; [ZOOM] [0]

Trace to estimate the point of intersection.

[More details on page 181.]

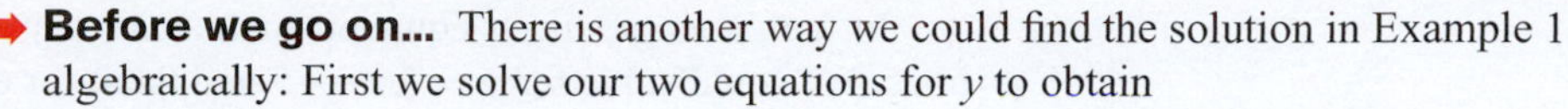

Spreadsheet
Headings x, $y1$, and $y2$ in A1–C1, x-values -4, 4 in A2, A3
`=-A2+3` and `=A2-1` in B1 and C1; copy down to B2–C2
Graph the data in columns A–C with a line-segment scatter plot. To see a closer view change the values in A2–A3.
[More details on page 183.]

Website
www.WanerMath.com
→ On Line Utilities
→ Function Evaluator and Grapher
Enter `-x+3` for y_1 and `x-1` for y_2. Set Xmin $= -4$, Xmax $= 4$ and press "Plot Graphs."
Click on the graph and use the trace arrows below to estimate the point of intersection.

➡ **Before we go on...** There is another way we could find the solution in Example 1 algebraically: First we solve our two equations for y to obtain

$$y = -x + 3$$
$$y = x - 1$$

and then we equate the right-hand sides to get

$$-x + 3 = x - 1,$$

which we solve to find $x = 2$. The value for y is then found by substituting $x = 2$ in either equation.

Q: *So why don't we solve all systems of equations this way?*

A: The elimination method extends more easily to systems with more equations and unknowns. It is the basis for the matrix method of solving systems—a method we discuss in Section 3.2. So, we shall use it exclusively for the rest of this section. ■

Example 2 illustrates the drawbacks of the graphical method.

EXAMPLE 2 Solving a System: Algebraically vs. Graphically

Solve the system

$$3x + 5y = 0$$
$$2x + 7y = 1.$$

Solution

Method 1: Graphical First, solve for y, obtaining $y = -\frac{3}{5}x$ and $y = -\frac{2}{7}x + \frac{1}{7}$. Graphing these equations, we get Figure 2. The lines appear to intersect slightly above and to the left of the origin. Redrawing with a finer scale (or zooming in using graphing technology), we can get the graph in Figure 3.

If we look carefully at Figure 3, we see that the graphs intersect near $(-0.45, 0.27)$. Is the point of intersection *exactly* $(-0.45, 0.27)$? (Substitute these values into the equations to find out.) In fact, it is impossible to find the exact solution of this system graphically, but we now have a ballpark answer that we can use to help check the following algebraic solution.

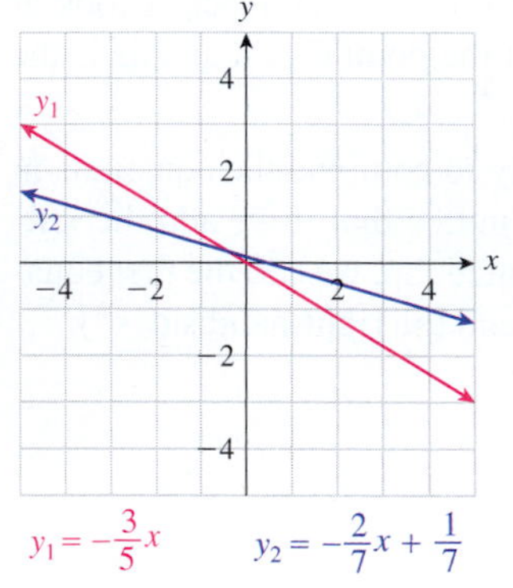

Figure 2

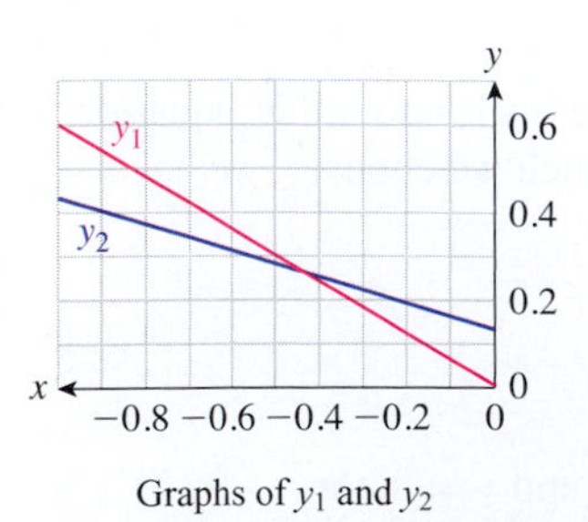

Figure 3

Method 2: Algebraic We first see that adding the equations is not going to eliminate either x or y. Notice, however, that if we multiply (both sides of) the first equation by 2 and the second by -3, the coefficients of x will become 6 and -6. *Then* if we add them, x will be eliminated. So we proceed as follows:

$$2(3x + 5y) = 2(0)$$
$$-3(2x + 7y) = -3(1)$$

gives

$$6x + 10y = 0$$
$$-6x - 21y = -3.$$

Adding these equations, we get

$$-11y = -3$$

so that

$$y = \frac{3}{11} = 0.\overline{27}.$$

Substituting $y = \frac{3}{11}$ in the first equation gives

$$3x + 5\left(\frac{3}{11}\right) = 0$$

$$3x = -\frac{15}{11}$$

$$x = -\frac{5}{11} = -0.\overline{45}.$$

The solution is $(x, y) = \left(-\frac{5}{11}, \frac{3}{11}\right) = (-0.\overline{45}, 0.\overline{27})$.

Notice that the algebraic method gives us the exact solution that we could not find with the graphical method. Still, we can check that the graph and our algebraic solution agree to the accuracy with which we can read the graph. To be absolutely sure that our answer is correct, we should check it:

$$3\left(-\frac{5}{11}\right) + 5\left(\frac{3}{11}\right) = -\frac{15}{11} + \frac{15}{11} = 0 \quad ✔$$

$$2\left(-\frac{5}{11}\right) + 7\left(\frac{3}{11}\right) = -\frac{10}{11} + \frac{21}{11} = 1. \quad ✔$$

Get in the habit of checking your answers.

Before we go on...

Q: *In solving the system in Example 2 algebraically, we multiplied (both sides of) the equations by numbers. How does that affect their graphs?*

A: Multiplying both sides of an equation by a nonzero number has no effect on its solutions, so the graph (which represents the set of all solutions) is unchanged. ■

Before doing some more examples, let's summarize what we've said about solving systems of equations.

Graphical Method for Solving a System of Two Equations in Two Unknowns

Graph both equations on the same graph. (For example, solve each for y to find the slope and y-intercept.) A point of intersection gives the solution to the system. To find the point, you may need to adjust the range of x-values you use. To find the point accurately you may need to use a smaller range (or zoom in if using technology).

Algebraic Method for Solving a System of Two Equations in Two Unknowns

Multiply each equation by a nonzero number so that the coefficients of x are the same in absolute value but opposite in sign. Add the two equations to eliminate x; this gives an equation in y that we can solve to find its value. Substitute this value of y into one of the original equations to find the value of x. (Note that we could eliminate y first instead of x if it's more convenient.)

Sometimes, something appears to go wrong with these methods. The following examples show what can happen.

EXAMPLE 3 Inconsistent System

Solve the system

$$x - 3y = 5$$
$$-2x + 6y = 8.$$

Solution To eliminate x, we multiply the first equation by 2 and then add:

$$2x - 6y = 10$$
$$-2x + 6y = 8.$$

Adding gives

$$0 = 18.$$

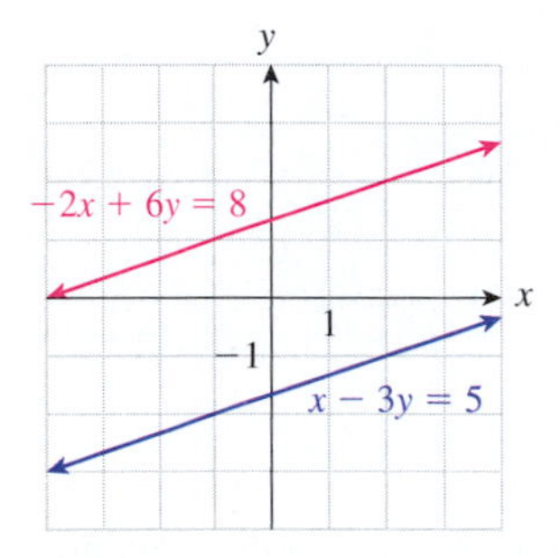

Figure 4

But this is absurd! This calculation shows that if we had two numbers x and y that satisfied both equations, it would be true that $0 = 18$. As 0 is *not* equal to 18, there can be no such numbers x and y. In other words, *the system has no solutions*, and is called an **inconsistent system**.

In slope-intercept form these lines are $y = \frac{1}{3}x - \frac{5}{3}$ and $y = \frac{1}{3}x + \frac{4}{3}$. Notice that they have the same slope but different y-intercepts. This means that they are parallel, but different lines. Plotting them confirms this fact (Figure 4).

Because they are parallel, they do not intersect. A solution must be a point of intersection, so we again conclude that there is no solution.

EXAMPLE 4 Redundant System

Solve the system

$$x + \ \ y = 2$$
$$2x + 2y = 4.$$

Solution Multiplying the first equation by -2 gives

$$-2x - 2y = -4$$
$$2x + 2y = 4.$$

Adding gives the not-very-enlightening result

$$0 = 0.$$

Now what has happened? Looking back at the original system, we note that the second equation is really the first equation in disguise. (It is the first equation multiplied by 2.) Put another way, if we solve both equations for y, we find that, in slope-intercept form, both equations become the same:

$$y = -x + 2.$$

The second equation gives us the same information as the first, so we say that this is a **redundant**, or **dependent system**. In other words, we really have only one equation in two unknowns. From Chapter 1, we know that a single linear equation in two unknowns has infinitely many solutions, one for each value of x. (Recall that to get the corresponding solution for y, we solve the equation for y and substitute the x-value.) The entire set of solutions can be summarized as follows:

x is arbitrary

$y = 2 - x$. Solve the first equation for y.

This set of solutions is called the **general solution** because it includes all possible solutions. When we write the general solution this way, we say that we have a **parameterized solution** and that x is the **parameter**.

We can also write the general solution as

$$(x, 2 - x) \quad x \text{ arbitrary}.$$

Different choices of the parameter x lead to different **particular solutions**. For instance, choosing $x = 3$ gives the particular solution

$$(x, y) = (3, -1).$$

Because there are infinitely many values of x from which to choose, there are infinitely many solutions.

What does this system of equations look like graphically? The two equations are really the same, so their graphs are identical, each being the line with x-intercept 2 and y-intercept 2. The "two" lines intersect at every point, so there is a solution for each point on the common line. In other words, we have a "whole line of solutions" (Figure 5).

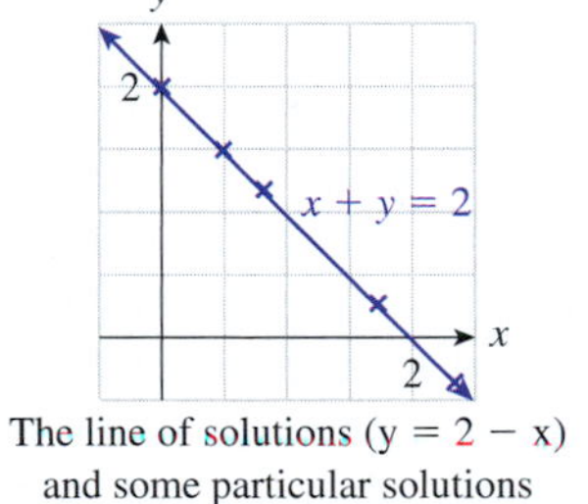

The line of solutions (y = 2 − x) and some particular solutions

Figure 5

We could also have solved the first equation for x instead and used y as a parameter, obtaining another form of the general solution:

$$(2 - y, y) \quad y \text{ arbitrary}. \qquad \text{Alternate form of the general solution}$$

We summarize the three possible outcomes we have encountered.

Possible Outcomes for a System of Two Linear Equations in Two Unknowns

1. **A single (or *unique*) solution** This happens when the lines corresponding to the two equations are distinct and not parallel so that they intersect at a single point. (See Example 1.)
2. **No solution** This happens when the two lines are parallel. We say that the system is **inconsistent**. (See Example 3.)
3. **An infinite number of solutions** This occurs when the two equations represent the same straight line, and we say that such a system is **redundant**, or **dependent**. In this case, we can represent the solutions by choosing one variable arbitrarily and solving for the other. (See Example 4.)

In cases 1 and 3, we say that the system of equations is **consistent** because it has at least one solution.

You should think about straight lines and convince yourself that these are the only three possibilities.

APPLICATIONS

EXAMPLE 5 Blending

Acme Baby Foods mixes two strengths of apple juice. One quart of Beginner's juice is made from 30 fluid ounces of water and 2 fluid ounces of apple juice concentrate. One quart of Advanced juice is made from 20 fluid ounces of water and 12 fluid ounces of concentrate. Every day Acme has available 30,000 fluid ounces of water and 3,600 fluid ounces of concentrate. If the company wants to use all the water and concentrate, how many quarts of each type of juice should it mix?

Solution In all applications we follow the same general strategy.

1. ***Identify and label the unknowns.*** What are we asked to find? To answer this question, it is common to respond by saying, "The unknowns are Beginner's juice and Advanced juice." Quite frankly, this is a baffling statement. Just what is unknown about juice? We need to be more precise:

   The unknowns are (1) *the* ***number of quarts*** *of Beginner's juice and* (2) *the* ***number of quarts*** *of Advanced juice made each day.*

   So, we label the unknowns as follows: Let

   $x =$ number of quarts of Beginner's juice made each day

   $y =$ number of quarts of Advanced juice made each day.

2. ***Use the information given to set up equations in the unknowns.*** This step is trickier, and the strategy varies from problem to problem. Here, the amount of juice the company can make is constrained by the fact that they have limited amounts of water and concentrate. This example shows a kind of application we will often see, and it is helpful in these problems to use a table to record the amounts of the resources used.

	Beginner's (x)	*Advanced (y)*	*Available*
Water (fl oz)	30	20	30,000
Concentrate (fl oz)	2	12	3,600

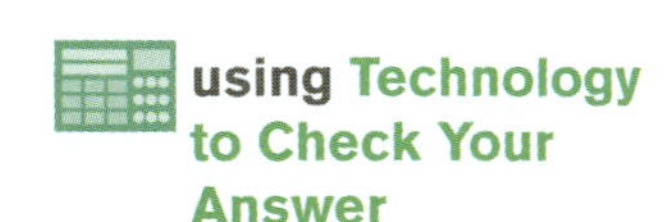

using Technology to Check Your Answer

Website
www.WanerMath.com
→ Math Tools for Chapter 3
→ Pivot and Gauss-Jordan Tool

To check Example 5 enter the coefficients of x and y and the right-hand sides of the two equations as shown (no commas in numbers!):

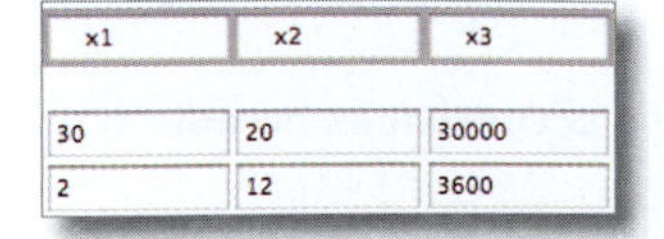

x1	x2	x3
30	20	30000
2	12	3600

Then press "Reduce Completely". The answer will appear in the last column. (In the next section we will see how this works.)

We can now set up an equation for each of the items listed in the left column of the table.

Water: We read across the first row. If Acme mixes x quarts of Beginner's juice, each quart using 30 fluid ounces of water, and y quarts of Advanced juice, each using 20 fluid ounces of water, it will use a total of $30x + 20y$ fluid ounces of water. But we are told that the total has to be 30,000 fluid ounces. Thus, $30x + 20y = 30{,}000$. This is our first equation.

Concentrate: We read across the second row. If Acme mixes x quarts of Beginner's juice, each using 2 fluid ounces of concentrate, and y quarts of Advanced juice, each using 12 fluid ounces of concentrate, it will use a total of $2x + 12y$ fluid ounces of concentrate. But we are told that the total has to be 3,600 fluid ounces. Thus, $2x + 12y = 3{,}600$.

Now we have two equations:

$$30x + 20y = 30{,}000$$
$$2x + 12y = 3{,}600.$$

To make the numbers easier to work with, let's divide (both sides of) the first equation by 10 and the second by 2:

$$3x + 2y = 3{,}000$$
$$x + 6y = 1{,}800.$$

We can now eliminate x by multiplying the second equation by -3 and adding:

$$\begin{aligned} 3x + 2y &= 3{,}000 \\ -3x - 18y &= -5{,}400 \\ \hline -16y &= -2{,}400. \end{aligned}$$

So, $y = 2{,}400/16 = 150$. Substituting this into the equation $x + 6y = 1{,}800$ gives $x + 900 = 1{,}800$, and so $x = 900$. The solution is $(x, y) = (900, 150)$. In other words, the company should mix 900 quarts of Beginner's juice and 150 quarts of Advanced juice.

EXAMPLE 6 Blending

SuperStock

A medieval alchemist's love potion calls for a number of eyes of newt and toes of frog, the total being 20, but with twice as many newt eyes as frog toes. How many of each is required?

Solution As in the preceding example, the first step is to identify and label the unknowns. Let

$$x = \text{number of newt eyes}$$
$$y = \text{number of frog toes.}$$

As for the second step—setting up the equations—a table is less appropriate here than in the preceding example. Instead, we translate each phrase of the problem into an equation. The first sentence tells us that the total number of eyes and toes is 20. Thus,

$$x + y = 20.$$

The end of the first sentence gives us more information, but the phrase "twice as many newt eyes as frog toes" is a little tricky: does it mean that $2x = y$ or that $x = 2y$? We can decide which by rewording the statement using the phrases "the *number of* newt eyes," which is x, and "the *number of* frog toes," which is y. Rephrased, the statement reads:

The ***number of*** *newt eyes is twice the* ***number of*** *frog toes.*

(Notice how the word "twice" is forced into a different place.) With this rephrasing, we can translate directly into algebra:

$$x = 2y.$$

In standard form ($ax + by = c$), this equation reads

$$x - 2y = 0.$$

Thus, we have the two equations:

$$\begin{aligned} x + y &= 20 \\ x - 2y &= 0. \end{aligned}$$

To eliminate x, we multiply the second equation by -1 and then add:

$$\begin{aligned} x + y &= 20 \\ -x + 2y &= 0. \end{aligned}$$

We'll leave it to you to finish solving the system and find that $x = 13\frac{1}{3}$ and $y = 6\frac{2}{3}$.

So, the recipe calls for exactly $13\frac{1}{3}$ eyes of newt and $6\frac{2}{3}$ toes of frog. The alchemist needs a very sharp scalpel and a very accurate balance (not to mention a very strong stomach).

We saw in Chapter 1 that the *equilibrium price* of an item (the price at which supply equals demand) and the *break-even point* (the number of items that must be sold to break even) can both be described as the intersection points of two graphs. If the graphs are straight lines, what we need to do to find the intersection is solve a system of two linear equations in two unknowns, as illustrated in the following problem.

EXAMPLE 7 Equilibrium Price

The demand for refrigerators in West Podunk is given by

$$q = -\frac{p}{10} + 100,$$

where q is the number of refrigerators that the citizens will buy each year if the refrigerators are priced at p dollars each. The supply is

$$q = \frac{p}{20} + 25,$$

where now q is the number of refrigerators the manufacturers will be willing to ship into town each year if they are priced at p dollars each. Find the equilibrium price and the number of refrigerators that will be sold at that price.

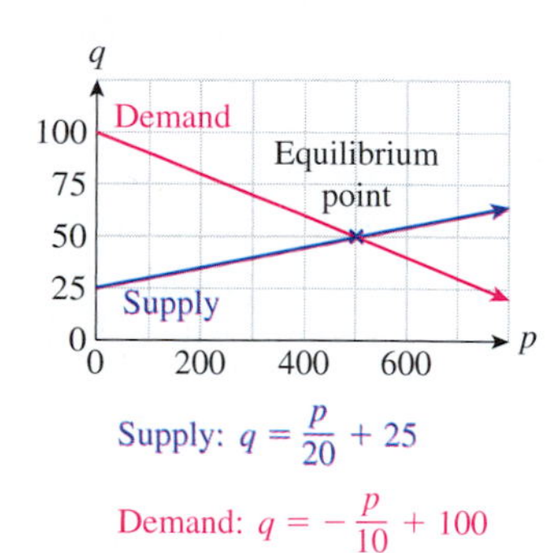

Figure 6

Solution Figure 6 shows the demand and supply curves. The equilibrium price occurs at the point where these two lines cross, which is where demand equals supply. The graph suggests that the equilibrium price is \$500, and zooming in confirms this.

To solve this system algebraically, first write both equations in standard form:

$$\frac{p}{10} + q = 100$$
$$-\frac{p}{20} + q = 25.$$

We can clear fractions and also prepare to eliminate p if we multiply the first equation by 10 and the second by 20:

$$p + 10q = 1{,}000$$
$$-p + 20q = 500,$$

and so:

$$30q = 1{,}500$$
$$q = 50.$$

Substituting this value of q into either equation gives us $p = 500$. Thus, the equilibrium price is \$500, and 50 refrigerators will be sold at this price.

We could also have solved this system of equations by setting the two expressions for q (the supply and the demand) equal to each other:

$$-p/10 + 100 = p/20 + 25$$

and then solving for p. (See the **Before we go on** discussion at the end of Example 1.)

FAQs

Setting Up the Equations

Q: *Looking through these examples, I notice that in some, we can tabulate the information given and read off the equations (as in Example 5), whereas in others (like Example 6), we have to reword each sentence to turn it into an equation. How do I know what approach to use?*

A: There is no hard-and-fast rule, and indeed some applications might call for a bit of each approach. However, it is generally not hard to see when it would be useful to tabulate values: Lists of the numbers of ingredients or components generally lend themselves to tabulation, whereas phrases like "twice as many of these as those" generally require direct translation into equations (after rewording if necessary).

3.1 EXERCISES

Access end-of-section exercises online at **www.webassign.net**

3.2 Using Matrices to Solve Systems of Equations

In this section we describe a systematic method for solving systems of equations that makes solving large systems of equations in any number of unknowns straightforward. Although this method may seem a little cumbersome at first, it will prove *immensely* useful in this and the next several chapters. First, some terminology:

Linear Equation

A linear equation in the n variables $x_1, x_2, \ldots, x_n$ has the form

$$a_1x_1 + \cdots + a_nx_n = b. \qquad (a_1, a_2, \ldots, a_n, b \text{ constants})$$

The numbers $a_1, a_2, \ldots, a_n$ are called the **coefficients**, and the number b is called the **constant term**, or **right-hand side**.

Quick Examples

1. $3x - 5y = 0$ — Linear equation in x and y. Coefficients: 3, −5 Constant term: 0

2. $x + 2y - z = 6$ — Linear equation in x, y, z. Coefficients: 1, 2, −1 Constant term: 6

3. $30x_1 + 18x_2 + x_3 + x_4 = 19$ — Linear equation in x_1, x_2, x_3, x_4. Coefficients: 30, 18, 1, 1 Constant term: 19

Note When the number of variables is small, we will almost always use $x, y, z, \ldots$ (as in Quick Examples 1 and 2) rather than $x_1, x_2, x_3, \ldots$ as the names of the variables. ■

Notice that a linear equation in any number of unknowns (for example, $2x - y = 3$) is entirely determined by its coefficients and its constant term. In other words, if we were simply given the row of numbers

$$[2 \quad -1 \quad 3]$$

we could easily reconstruct the original linear equation by multiplying the first number by x, the second by y, and inserting a plus sign and an equals sign, as follows:

$$2 \cdot x + (-1) \cdot y = 3$$

or $\quad 2x - y = 3.$

Similarly, the equation

$$-4x + 2y = 0$$

is represented by the row

$$[-4 \quad 2 \quad 0].$$

and the equation

$$-3y = \frac{1}{4}$$

is represented by

$$\begin{bmatrix} 0 & -3 & \frac{1}{4} \end{bmatrix}.$$

As the last example shows, the first number is always the coefficient of x and the second is the coefficient of y. If an x or a y is missing, we write a zero for its coefficient. We shall call such a row the **coefficient row** of an equation.

If we have a system of equations, for example the system

$$\begin{aligned} 2x - y &= 3 \\ -x + 2y &= -4, \end{aligned}$$

we can put the coefficient rows together like this:

$$\begin{bmatrix} 2 & -1 & 3 \\ -1 & 2 & -4 \end{bmatrix}.$$

We call this the **augmented matrix** of the system of equations. The term "augmented" means that we have included the right-hand sides 3 and -4. We will often drop the word "augmented" and simply refer to the matrix of the system. A **matrix** (plural: **matrices**) is nothing more than a rectangular array of numbers as above.

Matrix, Augmented Matrix

A **matrix** is a rectangular array of numbers. The **augmented matrix** of a system of linear equations is the matrix whose rows are the coefficient rows of the equations.

Quick Example

The augmented matrix of the system

$$\begin{aligned} x + y &= 3 \\ x - y &= 1 \end{aligned}$$

is $\begin{bmatrix} 1 & 1 & 3 \\ 1 & -1 & 1 \end{bmatrix}.$

We'll be studying matrices in more detail in Chapter 4.

Q: *What good are coefficient rows and matrices?*

A: Think about what we do when we multiply both sides of an equation by a number. For example, consider multiplying both sides of the equation $2x - y = 3$ by -2 to get $-4x + 2y = -6$. All we are really doing is multiplying the coefficients and the right-hand side by -2. This corresponds to *multiplying the row* $[2\ \ -1\ \ 3]$ *by* -2, that is, multiplying every number in the row by -2. We shall see that any manipulation we want to do with equations can be done instead with rows, and this fact leads to a method of solving equations that is systematic and generalizes easily to larger systems.

Here is the same operation both in the language of equations and the language of rows. (We refer to the equation here as *Equation* 1, or simply E_1 for short, and to the row as *Row* 1, or R_1.)

	Equation	*Row*	
	E_1: $2x - y = 3$	$[2\ \ -1\ \ 3]$	R_1
Multiply by -2:	$(-2)E_1$: $-4x + 2y = -6$	$[-4\ \ 2\ \ -6]$	$(-2)R_1$

Multiplying both sides of an equation by the number a corresponds to multiplying the coefficient row by a.

Now look at what we do when we add two equations:

	Equation	*Row*	
	E_1: $2x - y = 3$	$[2\ \ -1\ \ 3]$	R_1
	E_2: $-x + 2y = -4$	$[-1\ \ 2\ \ -4]$	R_2
Add:	$E_1 + E_2$: $x + y = -1$	$[1\ \ 1\ \ -1]$	$R_1 + R_2$

All we are really doing is *adding the corresponding entries in the rows*, or *adding the rows*. In other words,

Adding two equations corresponds to adding their coefficient rows.

In short, the manipulations of equations that we saw in the preceding section can be done more easily with rows in a matrix because we don't have to carry x, y, and other unnecessary notation along with us; x and y can always be inserted at the end if desired.

The manipulations we are talking about are known as **row operations**. In particular, we use three **elementary row operations**.

* We are using the term *elementary row operations* a little more freely than most books. Some mathematicians insist that $a = 1$ in an operation of Type 2, but the less restrictive version is very useful.

† Multiplying an equation or row by zero gives us the not very surprising result $0 = 0$. In fact, we lose any information that the equation provided, which usually means that the resulting system has more solutions than the original system.

Elementary Row Operations*

Type 1: Replacing R_i by aR_i (where $a \neq 0$)†
In words: multiplying or dividing a row by a nonzero number.

Type 2: Replacing R_i by $aR_i \pm bR_j$ (where $a \neq 0$)
Multiplying a row by a nonzero number and adding or subtracting a multiple of another row.

using Technology

See the Technology Guides at the end of the chapter to see how to do row operations using a TI-83/84 Plus or a spreadsheet.

Website

www.WanerMath.com

→ Math Tools for Chapter 3

→ Pivot and Gauss-Jordan Tool

Enter the matrix in columns $x1, x2, x3, \ldots$. To do a row operation, type the instruction(s) next to the row(s) you are changing as shown, and press "Do Row Ops" *once*.

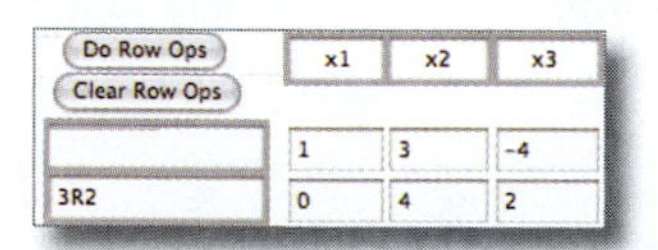

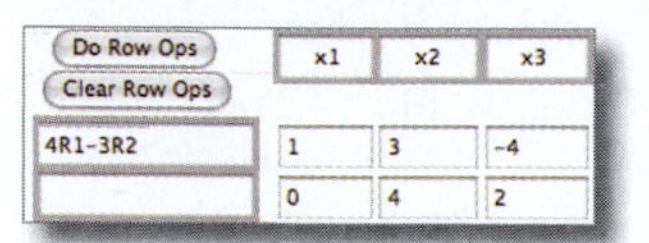

Type 3: Switching the order of the rows

This corresponds to switching the order in which we write the equations; occasionally this will be convenient.

For Types 1 and 2, we write the instruction for the row operation *next to the row we wish to replace.* (See the Quick Examples below.)

Quick Examples

Type 1: $\begin{bmatrix} 1 & 3 & -4 \\ 0 & 4 & 2 \end{bmatrix} \begin{matrix} \\ 3R_2 \end{matrix} \rightarrow \begin{bmatrix} 1 & 3 & -4 \\ 0 & 12 & 6 \end{bmatrix}$ Replace R_2 by $3R_2$.

Type 2: $\begin{bmatrix} 1 & 3 & -4 \\ 0 & 4 & 2 \end{bmatrix} \begin{matrix} 4R_1 - 3R_2 \\ \\ \end{matrix} \rightarrow \begin{bmatrix} 4 & 0 & -22 \\ 0 & 4 & 2 \end{bmatrix}$ Replace R_1 by $4R_1 - 3R_2$.

Type 3: $\begin{bmatrix} 1 & 3 & -4 \\ 0 & 4 & 2 \\ 1 & 2 & 3 \end{bmatrix} R_1 \leftrightarrow R_2 \rightarrow \begin{bmatrix} 0 & 4 & 2 \\ 1 & 3 & -4 \\ 1 & 2 & 3 \end{bmatrix}$ Switch R_1 and R_2.

One very important fact about the elementary row operations is that they do not change the solutions of the corresponding system of equations. In other words, the new system of equations that we get by applying any one of these operations will have exactly the same solutions as the original system: It is easy to see that numbers that make the original equations true will also make the new equations true, because each of the elementary row operations corresponds to a valid operation on the original equations. That any solution of the new system is a solution of the old system follows from the fact that these row operations are *invertible:* The effects of a row operation can be reversed by applying another row operation, called its **inverse**. Here are some examples of this invertibility. (Try them out in the above Quick Examples.)

Operation	*Inverse Operation*
Replace R_2 by $3R_2$.	Replace R_2 by $\frac{1}{3}R_2$.
Replace R_1 by $4R_1 - 3R_2$.	Replace R_1 by $\frac{1}{4}R_1 + \frac{3}{4}R_2$.
Switch R_1 and R_2.	Switch R_1 and R_2.

Our objective, then, is to use row operations to change the system we are given into one with exactly the same set of solutions in which it is easy to see what the solutions are.

Solving Systems of Equations by Using Row Operations

Now we put rows to work for us in solving systems of equations. Let's start with a complicated-looking system of equations:

$$-\frac{2x}{3} + \frac{y}{2} = -3$$
$$\frac{x}{4} - y = \frac{11}{4}.$$

We begin by writing the matrix of the system:

$$\begin{bmatrix} -\frac{2}{3} & \frac{1}{2} & -3 \\ \frac{1}{4} & -1 & \frac{11}{4} \end{bmatrix}.$$

Now what do we do with this matrix?

Step 1 ***Clear the fractions and/or decimals (if any) using operations of Type* 1.** To clear the fractions, we multiply the first row by 6 and the second row by 4. We record the operations by writing the symbolic form of an operation next to the row it will change, as follows.

$$\begin{bmatrix} -\frac{2}{3} & \frac{1}{2} & -3 \\ \frac{1}{4} & -1 & \frac{11}{4} \end{bmatrix} \begin{matrix} 6R_1 \\ 4R_2 \end{matrix}$$

By this we mean that we will replace the first row by $6R_1$ and the second by $4R_2$. Doing these operations gives

$$\begin{bmatrix} -4 & 3 & -18 \\ 1 & -4 & 11 \end{bmatrix}.$$

Step 2 ***Designate the first nonzero entry in the first row as the* pivot.** In this case we designate the entry -4 in the first row as the "pivot" by putting a box around it:

$$\begin{bmatrix} \boxed{-4} & 3 & -18 \\ 1 & -4 & 11 \end{bmatrix}. \quad \leftarrow \text{Pivot row}$$

↑
Pivot column

Q: *What is a "pivot"?*

A: A **pivot** is an entry in a matrix that is used to "clear a column." (See Step 3.) In this procedure, we will always select the first nonzero entry of a row as our pivot. In Chapter 5, when we study the simplex method, we will select our pivots differently.

Step 3 ***Use the pivot to clear its column using operations of Type* 2.** By **clearing a column**, we mean changing the matrix so that the pivot is the only nonzero number in its column. The procedure of clearing a column using a designated pivot is also called **pivoting**.

$$\begin{bmatrix} \boxed{-4} & 3 & -18 \\ 0 & \# & \# \end{bmatrix} \quad \leftarrow \text{Desired row 2 (the "\#"s stand for as yet unknown numbers)}$$

↑
Cleared pivot column

We want to replace R_2 by a row of the form $aR_2 \pm bR_1$ to get a zero in column 1. Moreover—and this will be important when we discuss the simplex method in Chapter 5—*we are going to choose positive values for both a and b.** We need to choose a and b so that we get the desired cancellation. We can do this quite mechanically as follows:

* Thus, the only place a negative sign may appear is between aR_2 and bR_1 as indicated in the formula $aR_2 \pm bR_1$.

a. Write the name of the row you need to change on the left and that of the pivot row on the right.

$$\underset{\substack{\uparrow \\ \text{Row to change}}}{\boldsymbol{R_2}} \qquad \underset{\substack{\uparrow \\ \text{Pivot row}}}{\boldsymbol{R_1}}$$

b. Focus on the pivot column, $\begin{bmatrix} -4 \\ 1 \end{bmatrix}$. Multiply each row by the *absolute value* of the entry currently in the other. (We are not permitting a or b to be negative.)

$$\underset{\substack{\uparrow \\ \text{From Row 1}}}{4R_2} \qquad \underset{\substack{\uparrow \\ \text{From Row 2}}}{1R_1}$$

The effect is to make the two entries in the pivot column numerically the same. Sometimes, you can accomplish this by using smaller values of a and b.

c. If the entries in the pivot column have opposite signs, insert a plus (+). If they have the same sign, insert a minus (−). Here, we get the instruction

$$4R_2 + 1R_1,$$

or simply $4R_2 + R_1$.

d. Write the operation next to the row you want to change, and then replace that row using the operation:

$$\begin{bmatrix} \boxed{-4} & 3 & -18 \\ 1 & -4 & 11 \end{bmatrix} \begin{matrix} \\ 4R_2 + 1R_1 \end{matrix} \to \begin{bmatrix} -4 & 3 & -18 \\ 0 & -13 & 26 \end{bmatrix}.$$

We have cleared the pivot column and completed Step 3.

Note In general, the row operation you use should always have the following form*:

$$\underset{\substack{\uparrow \\ \text{Row to change}}}{aR_c} \quad \pm \quad \underset{\substack{\uparrow \\ \text{Pivot row}}}{bR_p}$$

with a and b both positive. ■

* We are deviating somewhat from the traditional procedure here. It is traditionally recommended first to divide the pivot row by the pivot, turning the pivot into a 1. This allows us to use $a = 1$, but usually results in fractions. The procedure we use here is easier for hand calculations and, we feel, mathematically more elegant, because it eliminates the need to introduce fractions. See the end of this section for an example done using the traditional procedure.

The next step is one that can be performed at any time.

Simplification Step (Optional) ***If, at any stage of the process, all the numbers in a row are multiples of an integer, divide by that integer***—a Type 1 operation.

This is an optional but extremely helpful step: It makes the numbers smaller and easier to work with. In our case, the entries in R_2 are divisible by 13, so we divide that row by 13. (Alternatively, we could divide by −13. Try it.)

$$\begin{bmatrix} -4 & 3 & -18 \\ 0 & -13 & 26 \end{bmatrix} \begin{matrix} \\ \frac{1}{13}R_2 \end{matrix} \to \begin{bmatrix} -4 & 3 & -18 \\ 0 & -1 & 2 \end{bmatrix}$$

Step 4 ***Select the first nonzero number in the second row as the pivot, and clear its column.*** Here we have combined two steps in one: selecting the new pivot and clearing the column (pivoting). The pivot is shown below, as well as the desired result when the column has been cleared:

$$\begin{bmatrix} -4 & 3 & -18 \\ 0 & \boxed{-1} & 2 \end{bmatrix} \to \begin{bmatrix} \# & 0 & \# \\ 0 & -1 & 2 \end{bmatrix}. \leftarrow \text{desired row}$$

(↑ Pivot column; ↑ Cleared pivot column)

We now wish to get a 0 in place of the 3 in the pivot column. Let's run once again through the mechanical steps to get the row operation that accomplishes this.

a. Write the name of the row you need to change on the left and that of the pivot row on the right:

$$\underset{\substack{\uparrow \\ \text{Row to change}}}{\mathbf{R_1}} \qquad \underset{\substack{\uparrow \\ \text{Pivot row}}}{\mathbf{R_2}}$$

b. Focus on the pivot column, $\begin{bmatrix} 3 \\ -1 \end{bmatrix}$. Multiply each row by the absolute value of the entry currently in the other:

$$\underset{\text{From Row 2}}{\overset{1R_1}{\uparrow}} \qquad\qquad \underset{\text{From Row 1}}{\overset{3R_2}{\uparrow}}$$

c. If the entries in the pivot column have opposite signs, insert a plus (+). If they have the same sign, insert a minus (−). Here, we get the instruction

$$1R_1 + 3R_2.$$

d. Write the operation next to the row you want to change and then replace that row using the operation.

$$\begin{bmatrix} -4 & 3 & -18 \\ 0 & \boxed{-1} & 2 \end{bmatrix} \begin{matrix} R_1 + 3R_2 \\ \end{matrix} \rightarrow \begin{bmatrix} -4 & 0 & -12 \\ 0 & -1 & 2 \end{bmatrix}$$

Now we are essentially done, except for one last step.

Final Step ***Using operations of Type* 1*, turn each pivot (the first nonzero entry in each row) into a* 1.** We can accomplish this by dividing the first row by −4 and multiplying the second row by −1:

$$\begin{bmatrix} -4 & 0 & -12 \\ 0 & -1 & 2 \end{bmatrix} \begin{matrix} -\frac{1}{4}R_1 \\ -R_2 \end{matrix} \rightarrow \begin{bmatrix} 1 & 0 & 3 \\ 0 & 1 & -2 \end{bmatrix}.$$

The matrix now has the following nice form:

$$\begin{bmatrix} \boxed{1} & 0 & \# \\ 0 & \boxed{1} & \# \end{bmatrix}.$$

(This is the form we will always obtain with two equations in two unknowns when there is a unique solution.) This form is nice because, when we translate back into equations, we get

$$1x + 0y = 3$$
$$0x + 1y = -2.$$

In other words,

$$x = 3 \text{ and } y = -2$$

and so we have found the solution, which we can also write as $(x, y) = (3, -2)$.

The procedure we've just demonstrated is called **Gauss-Jordan* reduction** or **row reduction.** It may seem too complicated a way to solve a system of two equations in two unknowns, and it is. However, for systems with more equations and more unknowns, it is very efficient.

In Example 1 below we use row reduction to solve a system of linear equations in *three* unknowns: x, y, and z. Just as for a system in two unknowns, a **solution** of a system in any number of unknowns consists of values for each of the variables that, when substituted, satisfy all of the equations in the system. Again, just as for a system in two unknowns, any system of linear equations in any number of unknowns has either no solution, exactly one solution, or infinitely many solutions. There are no other possibilities.

* Gauss-Jordan reduction is named after Carl Friedrich Gauss (1777–1855) and Wilhelm Jordan (1842–1899). Gauss was one of the great mathematicians, making fundamental contributions to number theory, analysis, probability, and statistics, as well as many fields of science. Gauss also made contributions to a method of solving systems of equations that has become known as "Gaussian elimination," even though this method had been described by Isaac Newton in 1707 and, independently, had been known to the Chinese more than 2,000 years ago. (See *Mathematicians of Gaussian Elimination,* Notices of the American Mathematical Society, June/July 2011, for a history of Gaussian elimination.) The method we are showing you here, Gauss-Jordan reduction, is Jordan's variation of Gaussian elimination, first published in 1887.

Solving a system in three unknowns graphically would require the graphing of planes (flat surfaces) in three dimensions. (The graph of a linear equation in three unknowns is a flat surface.) The use of row reduction makes three-dimensional graphing unnecessary.

EXAMPLE 1 Solving a System by Gauss-Jordan Reduction

Solve the system

$$\begin{aligned} x - y + 5z &= -6 \\ 3x + 3y - z &= 10 \\ x + 3y + 2z &= 5. \end{aligned}$$

Solution The augmented matrix for this system is

$$\left[\begin{array}{ccc|c} 1 & -1 & 5 & -6 \\ 3 & 3 & -1 & 10 \\ 1 & 3 & 2 & 5 \end{array}\right].$$

Note that the columns correspond to x, y, z, and the right-hand side, respectively. We begin by selecting the pivot in the first row and clearing its column. Remember that clearing the column means that we turn *all* other numbers in the column into zeros. Thus, to clear the column of the first pivot, we need to change two rows, setting up the row operations in exactly the same way as above.

$$\left[\begin{array}{ccc|c} \boxed{1} & -1 & 5 & -6 \\ 3 & 3 & -1 & 10 \\ 1 & 3 & 2 & 5 \end{array}\right] \begin{array}{l} \\ R_2 - 3R_1 \\ R_3 - R_1 \end{array} \rightarrow \left[\begin{array}{ccc|c} 1 & -1 & 5 & -6 \\ 0 & 6 & -16 & 28 \\ 0 & 4 & -3 & 11 \end{array}\right]$$

Notice that both row operations have the required form

$$aR_c \pm bR_1$$

$\uparrow$ Row to change $\quad$ $\uparrow$ Pivot row

with a and b both positive.

Now we use the optional simplification step to simplify R_2:

$$\left[\begin{array}{ccc|c} 1 & -1 & 5 & -6 \\ 0 & 6 & -16 & 28 \\ 0 & 4 & -3 & 11 \end{array}\right] \begin{array}{l} \\ \frac{1}{2}R_2 \\ \\ \end{array} \rightarrow \left[\begin{array}{ccc|c} 1 & -1 & 5 & -6 \\ 0 & 3 & -8 & 14 \\ 0 & 4 & -3 & 11 \end{array}\right].$$

Next, we select the pivot in the second row and clear its column:

$$\left[\begin{array}{ccc|c} 1 & -1 & 5 & -6 \\ 0 & \boxed{3} & -8 & 14 \\ 0 & 4 & -3 & 11 \end{array}\right] \begin{array}{l} 3R_1 + R_2 \\ \\ 3R_3 - 4R_2 \end{array} \rightarrow \left[\begin{array}{ccc|c} 3 & 0 & 7 & -4 \\ 0 & 3 & -8 & 14 \\ 0 & 0 & 23 & -23 \end{array}\right].$$

R_1 and R_3 are to be changed.
R_2 is the pivot row.

We simplify R_3.

$$\left[\begin{array}{ccc|c} 3 & 0 & 7 & -4 \\ 0 & 3 & -8 & 14 \\ 0 & 0 & 23 & -23 \end{array}\right] \begin{array}{l} \\ \\ \frac{1}{23}R_3 \end{array} \rightarrow \left[\begin{array}{ccc|c} 3 & 0 & 7 & -4 \\ 0 & 3 & -8 & 14 \\ 0 & 0 & 1 & -1 \end{array}\right]$$

 **using Technology**

See the Technology Guides at the end of the chapter to see how to use a TI-83/84 Plus or a spreadsheet to solve this system of equations.

 **Website**

www.WanerMath.com

Everything for Finite Math

→ Math Tools for Chapter 3

→ Pivot and Gauss-Jordan Tool

Enter the augmented matrix in columns $x1$, $x2$, $x3$, ... At each step, type in the row operations exactly as written above next to the rows to which they apply. For example, for the first step type:
`R2-3R1` next to Row 2, and
`R3-R1` next to Row 3.
Press "Do Row Ops" once, and then "Clear Row Ops" to prepare for the next step.
For the last step, type
`(1/3)R1` next to Row 1, and
`(1/3)R2` next to Row 2.
Press "Do Row Ops" once.
The utility also does other things, such as automatic pivoting and complete reduction in one step. Use these features to check your work.

Now we select the pivot in the third row and clear its column:

$$\left[\begin{array}{ccc|c} 3 & 0 & 7 & -4 \\ 0 & 3 & -8 & 14 \\ 0 & 0 & \boxed{1} & -1 \end{array}\right] \begin{array}{l} R_1 - 7R_3 \\ R_2 + 8R_3 \\ \end{array} \rightarrow \left[\begin{array}{ccc|c} 3 & 0 & 0 & 3 \\ 0 & 3 & 0 & 6 \\ 0 & 0 & 1 & -1 \end{array}\right].$$

R_1 and R_2 are to be changed.
R_3 is the pivot row.

Finally, we turn all the pivots into 1s:

$$\left[\begin{array}{ccc|c} 3 & 0 & 0 & 3 \\ 0 & 3 & 0 & 6 \\ 0 & 0 & 1 & -1 \end{array}\right] \begin{array}{l} \frac{1}{3}R_1 \\ \frac{1}{3}R_2 \\ \end{array} \rightarrow \left[\begin{array}{ccc|c} 1 & 0 & 0 & 1 \\ 0 & 1 & 0 & 2 \\ 0 & 0 & 1 & -1 \end{array}\right].$$

The matrix is now reduced to a simple form, so we translate back into equations to obtain the solution:

$$x = 1,\ y = 2,\ z = -1, \text{ or } (x, y, z) = (1, 2, -1).$$

Notice the form of the very last matrix in the example:

$$\left[\begin{array}{ccc|c} 1 & 0 & 0 & \# \\ 0 & 1 & 0 & \# \\ 0 & 0 & 1 & \# \end{array}\right].$$

The 1s are on the **(main) diagonal** of the matrix; the goal in Gauss-Jordan reduction is to reduce our matrix to this form. If we can do so, then we can easily read off the solution, as we saw in Example 1. However, as we will see in several examples in this section, it is not always possible to achieve this ideal state. After Example 6, we will give a form that is always possible to achieve.

EXAMPLE 2 Solving a System by Gauss-Jordan Reduction

Solve the system:

$$\begin{aligned} 2x + y + 3z &= 1 \\ 4x + 2y + 4z &= 4 \\ x + 2y + z &= 4. \end{aligned}$$

Solution

$$\left[\begin{array}{ccc|c} \boxed{2} & 1 & 3 & 1 \\ 4 & 2 & 4 & 4 \\ 1 & 2 & 1 & 4 \end{array}\right] \begin{array}{l} \\ R_2 - 2R_1 \\ 2R_3 - R_1 \end{array} \rightarrow \left[\begin{array}{ccc|c} 2 & 1 & 3 & 1 \\ 0 & 0 & -2 & 2 \\ 0 & 3 & -1 & 7 \end{array}\right]$$

Now we have a slight problem: The number in the position where we would like to have a pivot—the second column of the second row—is a zero and thus cannot be a pivot. There are two ways out of this problem. One is to move on to the third column and pivot on the −2. Another is to switch the order of the second and third rows so that we can use the 3 as a pivot. We will do the latter.

$$\begin{bmatrix} 2 & 1 & 3 & 1 \\ 0 & 0 & -2 & 2 \\ 0 & 3 & -1 & 7 \end{bmatrix} \begin{matrix} \\ \\ R_2 \leftrightarrow R_3 \end{matrix} \rightarrow \begin{bmatrix} 2 & 1 & 3 & 1 \\ 0 & \boxed{3} & -1 & 7 \\ 0 & 0 & -2 & 2 \end{bmatrix} \begin{matrix} 3R_1 - R_2 \\ \\ \\ \end{matrix}$$

$$\rightarrow \begin{bmatrix} 6 & 0 & 10 & -4 \\ 0 & 3 & -1 & 7 \\ 0 & 0 & -2 & 2 \end{bmatrix} \begin{matrix} \\ \\ -\frac{1}{2}R_3 \end{matrix} \rightarrow \begin{bmatrix} 6 & 0 & 10 & -4 \\ 0 & 3 & -1 & 7 \\ 0 & 0 & \boxed{1} & -1 \end{bmatrix} \begin{matrix} R_1 - 10R_3 \\ R_2 + R_3 \\ \\ \end{matrix}$$

$$\rightarrow \begin{bmatrix} 6 & 0 & 0 & 6 \\ 0 & 3 & 0 & 6 \\ 0 & 0 & 1 & -1 \end{bmatrix} \begin{matrix} \frac{1}{6}R_1 \\ \frac{1}{3}R_2 \\ \\ \end{matrix} \rightarrow \begin{bmatrix} 1 & 0 & 0 & 1 \\ 0 & 1 & 0 & 2 \\ 0 & 0 & 1 & -1 \end{bmatrix}$$

Thus, the solution is $(x, y, z) = (1, 2, -1)$, as you can check in the original system.

EXAMPLE 3 Inconsistent System

Solve the system:

$$\begin{aligned} x + y + z &= 1 \\ 2x - y + z &= 0 \\ 4x + y + 3z &= 3. \end{aligned}$$

Solution

$$\begin{bmatrix} \boxed{1} & 1 & 1 & 1 \\ 2 & -1 & 1 & 0 \\ 4 & 1 & 3 & 3 \end{bmatrix} \begin{matrix} \\ R_2 - 2R_1 \\ R_3 - 4R_1 \end{matrix} \rightarrow \begin{bmatrix} 1 & 1 & 1 & 1 \\ 0 & \boxed{-3} & -1 & -2 \\ 0 & -3 & -1 & -1 \end{bmatrix} \begin{matrix} 3R_1 + R_2 \\ \\ R_3 - R_2 \end{matrix}$$

$$\rightarrow \begin{bmatrix} 3 & 0 & 2 & 1 \\ 0 & -3 & -1 & -2 \\ 0 & 0 & 0 & 1 \end{bmatrix}$$

Stop. That last row translates into $0 = 1$, which is nonsense, and so, as in Example 3 in Section 3.1, we can say that this system has no solution. We also say, as we did for systems with only two unknowns, that a system with no solution is **inconsistent**. A system with at least one solution is **consistent**.

Before we go on...

Q: *How, exactly, does the nonsensical equation $0 = 1$ tell us that there is no solution of the system in Example 3?*

A: Here is an argument similar to that in Example 3 in Section 3.1: If there *were* three numbers x, y, and z satisfying the original system of equations, then manipulating the equations according to the instructions in the row operations above leads us to conclude that $0 = 1$. Because 0 is *not* equal to 1, there can be no such numbers x, y, and z. ■

EXAMPLE 4 Infinitely Many Solutions

Solve the system:

$$\begin{aligned} x + y + z &= 1 \\ \frac{1}{4}x - \frac{1}{2}y + \frac{3}{4}z &= 0 \\ x + 7y - 3z &= 3. \end{aligned}$$

Solution

$$\begin{bmatrix} 1 & 1 & 1 & 1 \\ \frac{1}{4} & -\frac{1}{2} & \frac{3}{4} & 0 \\ 1 & 7 & -3 & 3 \end{bmatrix} \begin{matrix} \\ 4R_2 \\ \\ \end{matrix} \rightarrow \begin{bmatrix} \boxed{1} & 1 & 1 & 1 \\ 1 & -2 & 3 & 0 \\ 1 & 7 & -3 & 3 \end{bmatrix} \begin{matrix} \\ R_2 - R_1 \\ R_3 - R_1 \end{matrix}$$

$$\rightarrow \begin{bmatrix} 1 & 1 & 1 & 1 \\ 0 & -3 & 2 & -1 \\ 0 & 6 & -4 & 2 \end{bmatrix} \begin{matrix} \\ \\ \frac{1}{2}R_3 \end{matrix} \rightarrow \begin{bmatrix} 1 & 1 & 1 & 1 \\ 0 & \boxed{-3} & 2 & -1 \\ 0 & 3 & -2 & 1 \end{bmatrix} \begin{matrix} 3R_1 + R_2 \\ \\ R_3 + R_2 \end{matrix}$$

$$\rightarrow \begin{bmatrix} 3 & 0 & 5 & 2 \\ 0 & -3 & 2 & -1 \\ 0 & 0 & 0 & 0 \end{bmatrix}$$

There are no nonzero entries in the third row, so there can be no pivot in the third row. We skip to the final step and turn the pivots we did find into 1s.

$$\begin{bmatrix} 3 & 0 & 5 & 2 \\ 0 & -3 & 2 & -1 \\ 0 & 0 & 0 & 0 \end{bmatrix} \begin{matrix} \frac{1}{3}R_1 \\ -\frac{1}{3}R_2 \\ \\ \end{matrix} \rightarrow \begin{bmatrix} 1 & 0 & \frac{5}{3} & \frac{2}{3} \\ 0 & 1 & -\frac{2}{3} & \frac{1}{3} \\ 0 & 0 & 0 & 0 \end{bmatrix}$$

Now we translate back into equations and obtain:

$$\begin{aligned} x \quad + \tfrac{5}{3}z &= \tfrac{2}{3} \\ y - \tfrac{2}{3}z &= \tfrac{1}{3} \\ 0 &= 0. \end{aligned}$$

But how does this help us find a solution? The last equation doesn't tell us anything useful, so we ignore it. The thing to notice about the other equations is that we can easily solve the first equation for x and the second for y, obtaining

$$\begin{aligned} x &= \tfrac{2}{3} - \tfrac{5}{3}z \\ y &= \tfrac{1}{3} + \tfrac{2}{3}z. \end{aligned}$$

This is the solution! We can choose z to be any number and get corresponding values for x and y from the formulas above. This gives us infinitely many different solutions. Thus, the general solution (see Example 4 in Section 3.1) is

$$\begin{aligned} x &= \tfrac{2}{3} - \tfrac{5}{3}z \\ y &= \tfrac{1}{3} + \tfrac{2}{3}z \\ z &\text{ is arbitrary.} \end{aligned}$$

General solution

We can also write the general solution as

$$\left(\tfrac{2}{3} - \tfrac{5}{3}z,\ \tfrac{1}{3} + \tfrac{2}{3}z,\ z\right) \quad z \text{ arbitrary.}$$

General solution

This general solution has z as the parameter. Specific choices of values for the parameter z give particular solutions. For example, the choice $z = 6$ gives the particular solution

$$\begin{aligned} x &= \tfrac{2}{3} - \tfrac{5}{3}(6) = -\tfrac{28}{3} \\ y &= \tfrac{1}{3} + \tfrac{2}{3}(6) = \tfrac{13}{3} \\ z &= 6, \end{aligned}$$

Particular solution

while the choice $z = 0$ gives the particular solution $(x, y, z) = \left(\tfrac{2}{3}, \tfrac{1}{3}, 0\right)$.

Note that, unlike the system given in the preceding example, the system given in this example does have solutions, and is thus *consistent*.

Before we go on... Why were there infinitely many solutions to Example 4? The reason is that the third equation was really a combination of the first and second equations to begin with, so we effectively had only two equations in three unknowns.* Choosing a specific value for z (say, $z = 6$) has the effect of supplying the "missing" equation. ■

* In fact, you can check that the third equation, E_3, is equal to $3E_1 - 8E_2$. Thus, the third equation could have been left out because it conveys no more information than the first two. The process of row reduction always eliminates such a redundancy by creating a row of zeros.

Q: *How do we know when there are infinitely many solutions?*

A: When there are solutions (we have a consistent system, unlike the one in Example 3), and when the matrix we arrive at by row reduction has fewer pivots than there are unknowns. In Example 4 we had three unknowns but only two pivots.

Q: *How do we know which variables to use as parameters in a parameterized solution?*

A: The variables to use as parameters are those in the columns without pivots. In Example 4 there were pivots in the x and y columns, but no pivot in the z column, and it was z that we used as a parameter.

EXAMPLE 5 Four Unknowns

Solve the system:

$$\begin{aligned} x + 3y + 2z - w &= 6 \\ 2x + 6y + 6z + 3w &= 16 \\ x + 3y - 2z - 11w &= -2 \\ 2x + 6y + 8z + 8w &= 20. \end{aligned}$$

Solution

$$\begin{bmatrix} \boxed{1} & 3 & 2 & -1 & 6 \\ 2 & 6 & 6 & 3 & 16 \\ 1 & 3 & -2 & -11 & -2 \\ 2 & 6 & 8 & 8 & 20 \end{bmatrix} \begin{matrix} \\ R_2 - 2R_1 \\ R_3 - R_1 \\ R_4 - 2R_1 \end{matrix} \rightarrow \begin{bmatrix} 1 & 3 & 2 & -1 & 6 \\ 0 & 0 & 2 & 5 & 4 \\ 0 & 0 & -4 & -10 & -8 \\ 0 & 0 & 4 & 10 & 8 \end{bmatrix}$$

There is no pivot available in the second column, so we move on to the third column.

$$\begin{bmatrix} 1 & 3 & 2 & -1 & 6 \\ 0 & 0 & \boxed{2} & 5 & 4 \\ 0 & 0 & -4 & -10 & -8 \\ 0 & 0 & 4 & 10 & 8 \end{bmatrix} \begin{matrix} R_1 - R_2 \\ \\ R_3 + 2R_2 \\ R_4 - 2R_2 \end{matrix} \rightarrow \begin{bmatrix} 1 & 3 & 0 & -6 & 2 \\ 0 & 0 & 2 & 5 & 4 \\ 0 & 0 & 0 & 0 & 0 \\ 0 & 0 & 0 & 0 & 0 \end{bmatrix} \begin{matrix} \\ \frac{1}{2}R_2 \\ \\ \\ \end{matrix}$$

$$\rightarrow \begin{bmatrix} 1 & 3 & 0 & -6 & 2 \\ 0 & 0 & 1 & \frac{5}{2} & 2 \\ 0 & 0 & 0 & 0 & 0 \\ 0 & 0 & 0 & 0 & 0 \end{bmatrix}$$

Translating back into equations, we get:

$$\begin{aligned} x + 3y \quad - 6w &= 2 \\ z + \tfrac{5}{2}w &= 2. \end{aligned}$$

(We have not written down the equations corresponding to the last two rows, each of which is $0 = 0$.) There are no pivots in the y or w columns, so we use these two variables as parameters. We bring them over to the right-hand sides of the equations above and write the general solution as

$$x = 2 - 3y + 6w$$
y is arbitrary
$$z = 2 - \tfrac{5}{2}w$$
w is arbitrary

or

$$(x, y, z, w) = (2 - 3y + 6w, y, 2 - 5w/2, w) \quad y, w \text{ arbitrary.}$$

Before we go on... In Examples 4 and 5, you might have noticed an interesting phenomenon: If at any time in the process, two rows are equal or one is a multiple of the other, then one of those rows (eventually) becomes all zero. ■

Up to this point, we have always been given as many equations as there are unknowns. However, we shall see in the next section that some applications lead to systems where the number of equations is not the same as the number of unknowns. As the following example illustrates, such systems can be handled the same way as any other.

EXAMPLE 6 Number of Equations ≠ Number of Unknowns

Solve the system:

$$\begin{aligned} x + \quad y &= 1 \\ 13x - 26y &= -11 \\ 26x - 13y &= 2. \end{aligned}$$

Solution We proceed exactly as before and ignore the fact that there is one more equation than unknown.

$$\begin{bmatrix} \boxed{1} & 1 & 1 \\ 13 & -26 & -11 \\ 26 & -13 & 2 \end{bmatrix} \begin{matrix} \\ R_2 - 13R_1 \\ R_3 - 26R_1 \end{matrix} \rightarrow \begin{bmatrix} 1 & 1 & 1 \\ 0 & -39 & -24 \\ 0 & -39 & -24 \end{bmatrix} \begin{matrix} \\ \frac{1}{3}R_2 \\ \frac{1}{3}R_3 \end{matrix}$$

$$\rightarrow \begin{bmatrix} 1 & 1 & 1 \\ 0 & \boxed{-13} & -8 \\ 0 & -13 & -8 \end{bmatrix} \begin{matrix} 13R_1 + R_2 \\ \\ R_3 - R_2 \end{matrix} \rightarrow \begin{bmatrix} 13 & 0 & 5 \\ 0 & -13 & -8 \\ 0 & 0 & 0 \end{bmatrix} \begin{matrix} \frac{1}{13}R_1 \\ -\frac{1}{13}R_2 \\ \\ \end{matrix}$$

$$\rightarrow \begin{bmatrix} 1 & 0 & \frac{5}{13} \\ 0 & 1 & \frac{8}{13} \\ 0 & 0 & 0 \end{bmatrix}$$

Thus, the solution is $(x, y) = \left(\frac{5}{13}, \frac{8}{13}\right)$.

If, instead of a row of zeros, we had obtained, say, [0 0 6] in the last row, we would immediately have concluded that the system was inconsistent.

The fact that we wound up with a row of zeros indicates that one of the equations was actually a combination of the other two; you can check that the third equation can be obtained by multiplying the first equation by 13 and adding the result to the second. Because the third equation therefore tells us nothing that we don't already know from the first two, we call the system of equations **redundant**, or **dependent.** (Compare Example 4 in Section 3.1.)

➡ **Before we go on...** Example 5 above is another example of a redundant system; we could have started with the following smaller system of two equations in four unknowns

$$\begin{aligned} x + 3y + 2z - w &= 6 \\ 2x + 6y + 6z + 3w &= 16 \end{aligned}$$

and obtained the same general solution as we did with the larger system. Verify this by solving the smaller system. ■

The preceding examples illustrated that we cannot always reduce a matrix to the form shown before Example 2, with pivots going all the way down the diagonal. What we *can* always do is reduce a matrix to the following form:

Reduced Row Echelon Form

A matrix is said to be in **reduced row echelon form** or to be **row-reduced** if it satisfies the following properties.

P1. The first nonzero entry in each row (called the **leading entry** of that row) is a 1.

P2. The columns of the leading entries are **clear** (i.e., they contain zeros in all positions other than that of the leading entry).

P3. The leading entry in each row is to the right of the leading entry in the row above, and any rows of zeros are at the bottom.

Quick Examples

$\begin{bmatrix} 1 & 0 & 0 & 2 \\ 0 & 1 & 0 & 4 \\ 0 & 0 & 1 & -3 \end{bmatrix}$, $\begin{bmatrix} 0 & 1 & -3 \\ 0 & 0 & 0 \end{bmatrix}$, and $\begin{bmatrix} 1 & 0 & 0 & -2 \\ 0 & 0 & 1 & 4 \\ 0 & 0 & 0 & 0 \end{bmatrix}$ are row-reduced.

$\begin{bmatrix} 1 & 1 & 0 & 2 \\ 0 & 1 & 0 & 4 \\ 0 & 0 & 1 & -3 \end{bmatrix}$, $\begin{bmatrix} 0 & 1 & -3 \\ 0 & 0 & 1 \end{bmatrix}$, and $\begin{bmatrix} 0 & 0 & 1 & 4 \\ 1 & 0 & 0 & -2 \\ 0 & 0 & 0 & 0 \end{bmatrix}$ are not row-reduced.

You should check in the examples we did that the final matrices were all in reduced row echelon form.

It is an interesting and useful fact, though not easy to prove, that any two people who start with the same matrix and row-reduce it will reach exactly the same row-reduced matrix, even if they use different row operations.

The Traditional Gauss-Jordan Method (Optional)

In the version of the Gauss-Jordan method we have presented, we eliminated fractions and decimals in the first step and then worked with integer matrices, partly to make hand computation easier and partly for mathematical elegance. However, complicated fractions and decimals present no difficulty when we use technology. The following example illustrates the more traditional approach to Gauss-Jordan reduction used in many of the computer programs that solve the huge systems of equations that arise in practice.*

* Actually, for reasons of efficiency and accuracy, the methods used in commercial programs are closer to the method presented above. To learn more, consult a text on numerical methods.

EXAMPLE 7 Solving a System with the Traditional Gauss-Jordan Method

Solve the following system using the traditional Gauss-Jordan method:

$$\begin{aligned} 2x + y + 3z &= 5 \\ 3x + 2y + 4z &= 7 \\ 2x + y + 5z &= 10. \end{aligned}$$

Solution We make two changes in our method. First, there is no need to get rid of decimals (because computers and calculators can handle decimals as easily as they can integers). Second, after selecting a pivot, *divide the pivot row by the pivot value, turning the pivot into a* 1. It is easier to determine the row operations that will clear the pivot column if the pivot is a 1.

If we use technology to solve this system of equations, the sequence of matrices might look like this:

$$\begin{bmatrix} \boxed{2} & 1 & 3 & 5 \\ 3 & 2 & 4 & 7 \\ 2 & 1 & 5 & 10 \end{bmatrix} \begin{matrix} \frac{1}{2}R_1 \\ \\ \\ \end{matrix} \to \begin{bmatrix} \boxed{1} & 0.5 & 1.5 & 2.5 \\ 3 & 2 & 4 & 7 \\ 2 & 1 & 5 & 10 \end{bmatrix} \begin{matrix} \\ R_2 - 3R_1 \\ R_3 - 2R_1 \end{matrix}$$

$$\to \begin{bmatrix} 1 & 0.5 & 1.5 & 2.5 \\ 0 & \boxed{0.5} & -0.5 & -0.5 \\ 0 & 0 & 2 & 5 \end{bmatrix} 2R_2 \to \begin{bmatrix} 1 & 0.5 & 1.5 & 2.5 \\ 0 & \boxed{1} & -1 & -1 \\ 0 & 0 & 2 & 5 \end{bmatrix} \begin{matrix} R_1 - 0.5R_2 \\ \\ \\ \end{matrix}$$

$$\to \begin{bmatrix} 1 & 0 & 2 & 3 \\ 0 & 1 & -1 & -1 \\ 0 & 0 & \boxed{2} & 5 \end{bmatrix} \begin{matrix} \\ \\ \frac{1}{2}R_3 \end{matrix} \to \begin{bmatrix} 1 & 0 & 2 & 3 \\ 0 & 1 & -1 & -1 \\ 0 & 0 & \boxed{1} & 2.5 \end{bmatrix} \begin{matrix} R_1 - 2R_3 \\ R_2 + R_3 \\ \\ \end{matrix}$$

$$\to \begin{bmatrix} 1 & 0 & 0 & -2 \\ 0 & 1 & 0 & 1.5 \\ 0 & 0 & 1 & 2.5 \end{bmatrix}.$$

The solution is $(x, y, z) = (-2, 1.5, 2.5)$.

using Technology

Website
www.WanerMath.com
Follow
→ Everything for Finite Math
→ Math Tools for Chapter 3
to find the following resources:

- An online Web page that pivots and does row operations automatically
- A TI-83/84 Plus program that pivots and does other row operations
- An Excel worksheet that pivots and does row operations automatically

Q: *The solution to Example 7 looked quite easy. Why didn't we use the traditional method from the start like the other textbooks?*

A: It looked easy because we deliberately chose an example that leads to simple decimals. In all but the most contrived examples, the decimals or fractions involved get very complicated very quickly.

FAQs

Getting Unstuck, Going Round in Circles, and Knowing When to Stop

Q: *Help! I have been doing row operations on this matrix for half an hour. I have filled two pages, and I am getting nowhere. What do I do?*

A: Here is a way of keeping track of where you are *at any stage of the process* and also deciding what to do next.

Starting at the top row of your current matrix:

1. Scan along the row until you get to the leading entry: the first nonzero entry. If there is none—that is, the row is all zero—go to the next row.
2. Having located the leading entry, scan up and down its *column*. If its column is not clear (that is, it contains other nonzero entries), use your leading entry as a pivot to clear its column as in the examples in this section.
3. Now go to the next row and start again at Step 1.

When you have scanned all the rows and find that all the columns of the leading entries are clear, it means you are done (except possibly for reordering the rows so that the leading entries go from left to right as you read down the matrix, and zero rows are at the bottom).

Q: *No good. I have been following these instructions, but every time I try to clear a column, I* unclear *a column I had already cleared. What is going on?*

A: Are you using *leading entries* as pivots? Also, are you *using the pivot* to clear its column? That is, are your row operations all of the following form?

$$aR_c \pm bR_p$$

↑ Row to change ↑ Pivot row

The instruction next to the row you are changing should involve only that row and the pivot row, even though you might be tempted to use some other row instead.

Q: *Must I continue until I get a matrix that has 1s down the leading diagonal and 0s above and below?*

A: Not necessarily. You are completely done when your matrix is row-reduced: Each leading entry is a 1, the column of each leading entry is clear, and the leading entries go from left to right. You are done *pivoting* when the column of each leading entry is clear. After that, all that remains is to turn each pivot into a 1 (the "Final Step") and, if necessary, rearrange the rows.

3.2 EXERCISES

Access end-of-section exercises online at **www.webassign.net**

3.3 Applications of Systems of Linear Equations

In the examples and the exercises of this section, we consider scenarios that lead to systems of linear equations in three or more unknowns. Some of these applications will strike you as a little idealized or even contrived compared with the kinds of problems you might encounter in the real world.* One reason is that we will not have tools to handle more realistic versions of these applications until we have studied linear programming in Chapter 5.

* See the discussion at the end of the first example below.

In each example that follows, we set up the problem as a linear system and then give the solution; the emphasis in this section is on modeling a scenario by a system of linear equations and then interpreting the solution that results, rather than on obtaining the solution. For practice, you should do the row reduction necessary to get the solution.

EXAMPLE 1 Resource Allocation

The Arctic Juice Company makes three juice blends: PineOrange, using 2 quarts of pineapple juice and 2 quarts of orange juice per gallon; PineKiwi, using 3 quarts of pineapple juice and 1 quart of kiwi juice per gallon; and OrangeKiwi, using 3 quarts of orange juice and 1 quart of kiwi juice per gallon. Each day the company has 800 quarts of pineapple juice, 650 quarts of orange juice, and 350 quarts of kiwi juice available. How many gallons of each blend should it make each day if it wants to use up all of the supplies?

Solution We take the same steps to understand the problem that we took in Section 3.1. The first step is to identify and label the unknowns. Looking at the question asked in the last sentence, we see that we should label the unknowns like this:

x = number of gallons of PineOrange made each day
y = number of gallons of PineKiwi made each day
z = number of gallons of OrangeKiwi made each day.

Next, we can organize the information we are given in a table:

	PineOrange (x)	*PineKiwi (y)*	*OrangeKiwi (z)*	*Total Available*
Pineapple Juice (qt)	2	3	0	800
Orange Juice (qt)	2	0	3	650
Kiwi Juice (qt)	0	1	1	350

Notice how we have arranged the table; we have placed headings corresponding to the unknowns along the top, rather than down the side, and we have added a heading for the available totals. This gives us a table that is essentially the matrix of the system of linear equations we are looking for. (However, read the caution in the **Before we go on** section.)

using Technology

Website
www.WanerMath.com
Everything for Finite Math
→ Math Tools for Chapter 3
→ Pivot and Gauss-Jordan Tool

Once you have set up the system of equations, you can obtain the solution in a single step using the Pivot and Gauss-Jordan Tool at the Website:

Enter the augmented matrix of the system as shown, and press "Reduce Completely". You can then use the reduced matrix to write down the unique solution or general solution as discussed in Section 3.2.

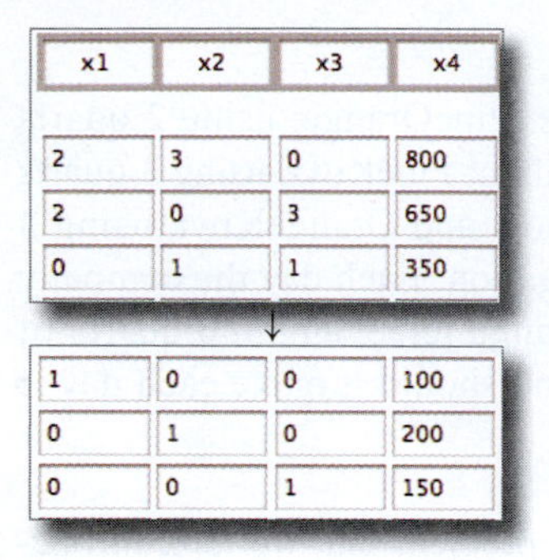

Now we read across each row of the table. The fact that we want to use exactly the amount of each juice that is available leads to the following three equations:

$$\begin{aligned} 2x + 3y &= 800 \\ 2x + 3z &= 650 \\ y + z &= 350. \end{aligned}$$

The solution of this system is $(x, y, z) = (100, 200, 150)$, so Arctic Juice should make 100 gallons of PineOrange, 200 gallons of PineKiwi, and 150 gallons of OrangeKiwi each day.

Before we go on...

Caution

We do not recommend relying on the coincidence that the table we created to organize the information in Example 1 happened to be the matrix of the system; it is too easy to set up the table "sideways" and get the wrong matrix. You should always write down the system of equations *and be sure you understand each equation*. For example, the equation $2x + 3y = 800$ in Example 1 indicates that the number of quarts of pineapple juice that will be used $(2x + 3y)$ is equal to the amount available (800 quarts). By thinking of the reason for each equation, you can check that you have the correct system. If you have the wrong system of equations to begin with, solving it won't help you.

Q: *Just how realistic is the scenario in Example 1?*

A: This is a very unrealistic scenario, for several reasons:

1. Isn't it odd that we happened to end up with exactly the same number of equations as unknowns? Real scenarios are rarely so considerate. If there had been four equations, there would in all likelihood have been no solution at all. However, we need to understand these idealized problems before we can tackle the real world.
2. Even if a real-world scenario does give the same number of equations as unknowns, there is still no guarantee that there will be a unique solution consisting of positive values. What, for instance, would we have done in this example if x had turned out to be negative?
3. The requirement that we use exactly all the ingredients would be an unreasonable constraint in real life. When we discuss linear programming, we will be able to substitute the more reasonable constraint that you use no more than is available, and we will add the more reasonable objective that you maximize profit. ■

EXAMPLE 2 Aircraft Purchases: Airbus and Boeing

A new airline has recently purchased a fleet of Airbus A330-300s, Boeing 767-200ERs, and Boeing Dreamliner 787-9s to meet an estimated demand for 4,800 seats. The A330-300s seat 320 passengers and cost \$200 million each, the 767-200ERs each seat 250 passengers and cost \$125 million each, while the Dreamliner 787-9s seat 275 passengers and cost \$200 million each.[1] The total cost of the fleet, which had twice as many Dreamliners as 767s, was \$3,100 million. How many of each type of aircraft did the company purchase?

[1] The prices are approximate 2008 prices. Prices and seating capacities found at the companies' Web sites and in Wikipedia.

Solution We label the unknowns as follows:

$x =$ number of Airbus A330-300s

$y =$ number of Boeing 767-200ERs

$z =$ number of Boeing Dreamliner 787-9s.

We must now set up the equations. We can organize some (but not all) of the given information in a table:

	A330-300	*767-200ER*	*787-9 Dreamliner*	*Total*
Capacity	320	250	275	4,800
Cost ($ million)	200	125	200	3,100

Reading across, we get the equations expressing the facts that the airline needed to seat 4,800 passengers and that it spent $3,100 million:

$$320x + 250y + 275z = 4{,}800$$
$$200x + 125y + 200z = 3{,}100.$$

There is an additional piece of information we have not yet used: the airline bought twice as many Dreamliners as 767s. As we said in Section 3.1, it is easiest to translate a statement like this into an equation if we first reword it using the phrase "the number of." Thus, we say: "The number of Dreamliners ordered was twice the number of 767s ordered," or

$$z = 2y$$
$$2y - z = 0.$$

We now have a system of three equations in three unknowns:

$$\begin{aligned} 320x + 250y + 275z &= 4{,}800 \\ 200x + 125y + 200z &= 3{,}100 \\ 2y - \quad z &= 0. \end{aligned}$$

Solving the system, we get the solution $(x, y, z) = (5, 4, 8)$. Thus, the airline ordered five A330-300s, four 767-200ERs, and eight Dreamliner 787-9s.

EXAMPLE 3 Traffic Flow

Traffic through downtown Urbanville flows through the one-way system shown in Figure 7.

Traffic counting devices installed in the road (shown as boxes) count 200 cars entering town from the west each hour, 150 leaving town on the north each hour, and 50 leaving town on the south each hour.

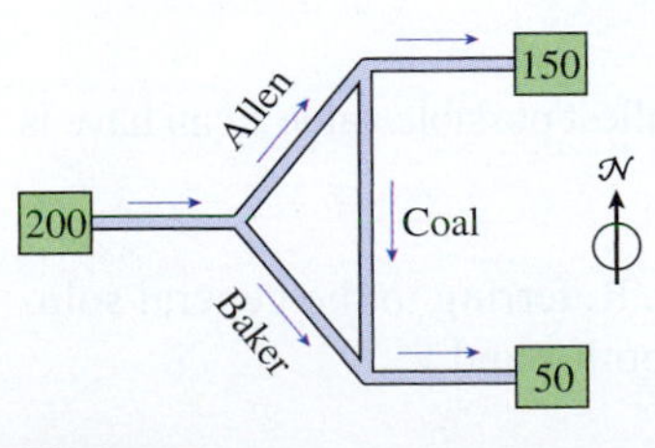

Figure 7

a. From this information, is it possible to determine how many cars drive along Allen, Baker, and Coal streets every hour?

b. What is the maximum possible traffic flow along Baker Street?

c. What is the minimum possible traffic along Allen Street?

d. What is the maximum possible traffic flow along Coal Street?

Solution

a. Our unknowns are:

x = number of cars per hour on Allen Street

y = number of cars per hour on Baker Street

z = number of cars per hour on Coal Street.

Assuming that, at each intersection, cars do not fall into a pit or materialize out of thin air, the number of cars entering each intersection has to equal the number exiting. For example, at the intersection of Allen and Baker Streets there are 200 cars entering and $x + y$ cars exiting:

$$\text{Traffic in} = \text{Traffic out}$$
$$200 = x + y.$$

At the intersection of Allen and Coal Streets, we get:

$$\text{Traffic in} = \text{Traffic out}$$
$$x = z + 150$$

and at the intersection of Baker and Coal Streets, we get:

$$\text{Traffic in} = \text{Traffic out}$$
$$y + z = 50.$$

We now have the following system of equations:

$$\begin{aligned} x + y \quad\quad &= 200 \\ x \quad\quad - z &= 150 \\ y + z &= 50. \end{aligned}$$

If we solve this system using the methods of the preceding section, we find that it has infinitely many solutions. The general solution is:

$x = z + 150$

$y = -z + 50$

z is arbitrary.

Because we do not have a unique solution, it is *not* possible to determine how many cars drive along Allen, Baker and Coal Streets every hour.

b. The traffic flow along Baker Street is measured by y. From the general solution,

$$y = -z + 50$$

where z is arbitrary. How arbitrary is z? It makes no sense for any of the variables x, y, or z to be negative in this scenario, so $z \geq 0$. Therefore, the largest possible value y can have is

$$y = -0 + 50 = 50 \text{ cars per hour.}$$

c. The traffic flow along Allen Street is measured by x. From the general solution,

$$x = z + 150,$$

where $z \geq 0$, as we saw in part (b). Therefore, the smallest possible value x can have is

$$x = 0 + 150 = 150 \text{ cars per hour.}$$

d. The traffic flow along Coal Street is measured by z. Referring to the general solution, we see that z shows up in the expressions for both x and y:

$x = z + 150$

$y = -z + 50.$

In the first of these equations, there is nothing preventing z from being as big as we like; the larger we make z, the larger x becomes. However, the second equation places a limit on how large z can be: If $z > 50$, then y is negative, which is impossible. Therefore, the largest value z can take is 50 cars per hour.

From the discussion above, we see that z is not completely arbitrary: We must have $z \geq 0$ and $z \leq 50$. Thus, z has to satisfy $0 \leq z \leq 50$ for us to get a realistic answer.

➡ **Before we go on...** Here are some questions to think about in Example 3: If you wanted to nail down x, y, and z to see where the cars are really going, how would you do it with only one more traffic counter? Would it make sense for z to be fractional? What if you interpreted x, y, and z as *average* numbers of cars per hour over a long period of time?

Traffic flow is only one kind of flow in which we might be interested. Water and electricity flows are others. In each case, to analyze the flow, we use the fact that the amount entering an intersection must equal the amount leaving it. ■

EXAMPLE 4 Transportation

A car rental company has four locations in the city: Southwest, Northeast, Southeast, and Northwest. The Northwest location has 20 more cars than it needs, and the Northeast location has 15 more cars than it needs. The Southwest location needs 10 more cars than it has, and the Southeast location needs 25 more cars than it has. It costs \$10 (in salary and gas) to have an employee drive a car from Northwest to Southwest. It costs \$20 to drive a car from Northwest to Southeast. It costs \$5 to drive a car from Northeast to Southwest, and it costs \$10 to drive a car from Northeast to Southeast. If the company will spend a total of \$475 rearranging its cars, how many cars will it drive from each of Northwest and Northeast to each of Southwest and Southeast?

Solution Figure 8 shows a diagram of this situation. Each arrow represents a route along which the rental company can drive cars. At each location is written the number of extra cars the location has or the number it needs. Along each route is written the cost of driving a car along that route.

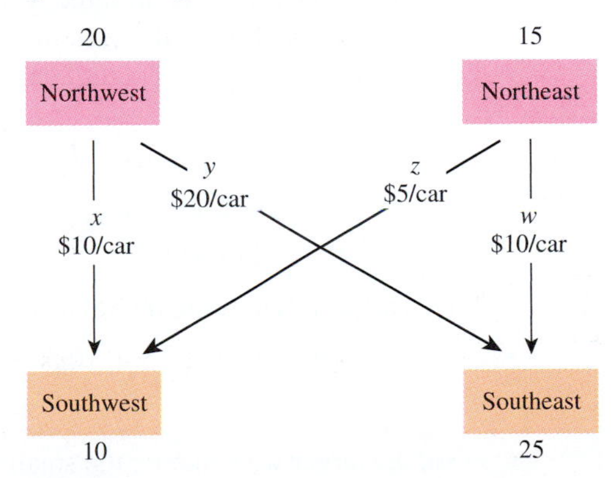

Figure 8

The unknowns are the number of cars the company will drive along each route, so we have the following four unknowns, as indicated in the figure:

x = number of cars driven from Northwest to Southwest
y = number of cars driven from Northwest to Southeast
z = number of cars driven from Northeast to Southwest
w = number of cars driven from Northeast to Southeast.

Consider the Northwest location. It has 20 more cars than it needs, so the total number of cars being driven out of Northwest should be 20. This gives us the equation

$$x + y = 20.$$

Similarly, the total number of cars being driven out of Northeast should be 15, so

$$z + w = 15.$$

Considering the number of cars needed at the Southwest and Southeast locations, we get the following two equations as well:

$$x + z = 10$$
$$y + w = 25.$$

There is one more equation that we should write down, the equation that says that the company will spend \$475:

$$10x + 20y + 5z + 10w = 475.$$

Thus, we have the following system of five equations in four unknowns:

$$\begin{aligned} x + \quad y \qquad\qquad\quad &= 20 \\ z + \quad w &= 15 \\ x \qquad + z \qquad\quad &= 10 \\ y \qquad + \quad w &= 25 \\ 10x + 20y + 5z + 10w &= 475 \end{aligned}$$

Solving this system, we find that $(x, y, z, w) = (5, 15, 5, 10)$. In words, the company will drive 5 cars from Northwest to Southwest, 15 from Northwest to Southeast, 5 from Northeast to Southwest, and 10 from Northeast to Southeast.

Before we go on... A very reasonable question to ask in Example 4 is, Can the company rearrange its cars for less than \$475? Even better, what is the least possible cost? In general, a question asking for the optimal cost may require the techniques of linear programming, which we will discuss in Chapter 5. However, in this case we can approach the problem directly. If we remove the equation that says that the total cost is \$475 and solve the system consisting of the other four equations, we find that there are infinitely many solutions and that the general solution may be written as

$$x = w - 5$$
$$y = 25 - w$$
$$z = 15 - w$$
w is arbitrary.

This allows us to write the total cost as a function of w:

$$\begin{aligned} \text{Cost} = 10x + 20y + 5z + 10w &= 10(w - 5) + 20(25 - w) + 5(15 - w) + 10w \\ &= 525 - 5w. \end{aligned}$$

So, the larger we make w, the smaller the total cost will be. The largest we can make w is 15 (why?), and if we do so we get $(x, y, z, w) = (10, 10, 0, 15)$ and a total cost of \$450. ■

3.3 EXERCISES

Access end-of-section exercises online at **www.webassign.net**

CHAPTER 3 REVIEW

KEY CONCEPTS

Website www.WanerMath.com
Go to the Website at www.WanerMath.com to find a comprehensive and interactive Web-based summary of Chapter 3.

3.1 Systems of Two Equations in Two Unknowns

Linear equation in two unknowns *p. 144*
Coefficient *p. 144*
Solution of an equation in two unknowns *p. 144*
Graphical method for solving a system of two linear equations *p. 147*
Algebraic method for solving a system of two linear equations *p. 147*
Redundant or dependent system *p. 148*
Possible outcomes for a system of two linear equations *p. 149*
Consistent system *p. 149*

3.2 Using Matrices to Solve Systems of Equations

Linear equation (in any number of unknowns) *p. 153*
Matrix *p. 154*
Augmented matrix of a system of linear equations *p. 154*
Elementary row operations *p. 155*
Pivot *p. 157*
Clearing a column; pivoting *p. 157*
Gauss-Jordan or row reduction *p. 159*
Reduced row echelon form *p. 166*

3.3 Applications of Systems of Linear Equations

Resource allocation *p. 169*
(Traffic) flow *p. 171*
Transportation *p. 173*

REVIEW EXERCISES

In each of Exercises 1–6, graph the equations and determine how many solutions the system has, if any.

1. $x + 2y = 4$
$2x - y = 1$

2. $0.2x - 0.1y = 0.3$
$0.2x + 0.2y = 0.4$

3. $\frac{1}{2}x - \frac{3}{4}y = 0$
$6x - 9y = 0$

4. $2x + 3y = 2$
$-x - 3y/2 = 1/2$

5. $x + y = 1$
$2x + y = 0.3$
$3x + 2y = \frac{13}{10}$

6. $3x + 0.5y = 0.1$
$6x + y = 0.2$
$\frac{3x}{10} - 0.05y = 0.01$

Solve each of the systems of linear equations in Exercises 7–18.

7. $x + 2y = 4$
$2x - y = 1$

8. $0.2x - 0.1y = 0.3$
$0.2x + 0.2y = 0.4$

9. $\frac{1}{2}x - \frac{3}{4}y = 0$
$6x - 9y = 0$

10. $2x + 3y = 2$
$-x - 3y/2 = 1/2$

11. $x + y = 1$
$2x + y = 0.3$
$3x + 2y = \frac{13}{10}$

12. $3x + 0.5y = 0.1$
$6x + y = 0.2$
$\frac{3x}{10} - 0.05y = 0.01$

13. $x + 2y = -3$
$x - z = 0$
$x + 3y - 2z = -2$

14. $x - y + z = 2$
$7x + y - z = 6$
$x - \frac{1}{2}y + \frac{1}{3}z = 1$
$x + y + z = 6$

15. $x - \frac{1}{2}y + z = 0$
$\frac{1}{2}x - \frac{1}{2}z = -1$
$\frac{3}{2}x - \frac{1}{2}y + \frac{1}{2}z = -1$

16. $x + y - 2z = -1$
$-2x - 2y + 4z = 2$
$0.75x + 0.75y - 1.5z = -0.75$

17. $x = \frac{1}{2}y$
$\frac{1}{2}x = -\frac{1}{2}z + 2$
$z = -3x + y$

18. $x - y + z = 1$
$y - z + w = 1$
$x + z - w = 1$
$2x + z = 3$

Exercises 19–22 are based on the following equation relating the Fahrenheit and Celsius (or centigrade) temperature scales:

$$5F - 9C = 160,$$

where F is the Fahrenheit temperature of an object and C is its Celsius temperature.

19. What temperature should an object be if its Fahrenheit and Celsius temperatures are the same?

20. What temperature should an object be if its Celsius temperature is half its Fahrenheit temperature?

21. Is it possible for the Fahrenheit temperature of an object to be 1.8 times its Celsius temperature? Explain.

22. Is it possible for the Fahrenheit temperature of an object to be 30° more than 1.8 times its Celsius temperature? Explain.

In Exercises 23–28, let x, y, z, and w represent the population in millions of four cities A, B, C, and D, respectively. Express the given statement as an equation in x, y, z, and w. If the equation is linear, say so and express it in the standard form $ax + by + cz + dw = k$.

23. The total population of the four cities is 10 million people.

24. City A has three times as many people as cities B and C combined.
25. City D is actually a ghost town; there are no people living in it.
26. The population of City A is the sum of the squares of the populations of the other three cities.
27. City C has 30% more people than City B.
28. City C has 30% fewer people than City B.

APPLICATIONS: OHaganBooks.com

Purchasing *You are the buyer for OHaganBooks.com and are considering increasing stocks of romance and horror novels at the new OHaganBooks.com warehouse in Texas. You have offers from two publishers: Duffin House and Higgins Press. Duffin offers a package of 5 horror novels and 5 romance novels for \$50, and Higgins offers a package of 5 horror and 11 romance novels for \$15. Exercises 29–32 deal with your purchasing options given these offers.*

29. How many packages should you purchase from each publisher to get exactly 4,500 horror novels and 6,600 romance novels?
30. You want to spend a total of \$50,000 on books and have promised to buy twice as many packages from Duffin as from Higgins. How many packages should you purchase from each publisher?
31. The accountant tells you that the company can actually afford to spend a total of \$90,000 on romance and horror books. She also reminds you that you had signed an agreement to spend twice as much money for books from Duffin as from Higgins. How many packages should you purchase from each publisher?
32. Upon revising her records, the accountant now tells you that the company can afford to spend a total of only \$60,000 on romance and horror books, and that it is company policy to spend the same amount of money at both publishers. How many packages should you purchase from each publisher?
33. ***Equilibrium*** The demand per year for *Finite Math the OHagan Way* is given by $q = -1{,}000p + 140{,}000$, where p is the price per book in dollars. The supply is given by $q = 2{,}000p + 20{,}000$. Find the price at which supply and demand balance.
34. ***Equilibrium*** OHaganbooks.com CEO John O'Hagan announces to a stunned audience at the annual board meeting that he is considering expanding into the jumbo jet airline manufacturing business. The demand per year for jumbo jets is given by $q = -2p + 18$ where p is the price per jet in millions of dollars. The supply is given by $q = 3p + 3$. Find the price the envisioned O'Hagan jumbo jet division should charge to balance supply and demand.
35. ***Feeding Schedules*** Billy-Sean O'Hagan is John O'Hagan's son and a freshman in college. Billy's 36-gallon tropical fish tank contains three types of carnivorous creatures: baby sharks, piranhas and squids, and he feeds them three types of delicacies: goldfish, angelfish and butterfly fish. Each baby shark can consume 1 goldfish, 2 angelfish, and 2 butterfly fish per day; each piranha can consume 1 goldfish and 3 butterfly fish per day (the piranhas are rather large as a result of their diet); while each squid can consume 1 goldfish and 1 angelfish per day. After a trip to the local pet store, he was able to feed his creatures to capacity, and noticed that 21 goldfish, 21 angelfish, and 35 butterfly fish were eaten. How many of each type of creature does he have?
36. ***Resource Allocation*** Duffin House is planning its annual Song Festival, when it will serve three kinds of delicacies: granola treats, nutty granola treats, and nuttiest granola treats. The following table shows the ingredients required (in ounces) for a single serving of each delicacy, as well as the total amount of each ingredient available.

	Granola	*Nutty Granola*	*Nuttiest Granola*	*Total Available*
Toasted Oats	1	1	5	1,500
Almonds	4	8	8	10,000
Raisins	2	4	8	4,000

The Song Festival planners at Duffin House would like to use up all the ingredients. Is this possible? If so, how many servings of each kind of delicacy can they make?

37. ***Web Site Traffic*** OHaganBooks.com has two principal competitors: JungleBooks.com and FarmerBooks.com. Combined Web site traffic at the three sites is estimated at 10,000 hits per day. Only 10% of the hits at OHaganBooks.com result in orders, whereas JungleBooks.com and FarmerBooks.com report that 20% of the hits at their sites result in book orders. Together, the three sites process 1,500 book orders per day. FarmerBooks.com appears to be the most successful of the three, and gets as many book orders as the other two combined. What is the traffic (in hits per day) at each of the sites?
38. ***Sales*** As the buyer at OHaganBooks.com you are planning to increase stocks of books about music, and have been monitoring worldwide sales. Last year, worldwide sales of books about rock, rap, and classical music amounted to \$5.8 billion. Books on rock music brought in twice as much revenue as books on rap music and they brought in 900% the revenue of books on classical music. How much revenue was earned in each of the three categories of books?
39. ***Investing in Stocks*** Billy-Sean O'Hagan is the treasurer at his college fraternity, which recently earned \$12,400 in its annual carwash fundraiser. Billy-Sean decided to invest all the proceeds in the purchase of three computer stocks: HAL, POM, and WELL.

	Price per Share	*Dividend Yield*
HAL	\$100	0.5%
POM	\$20	1.50%
WELL	\$25	0%

If the investment was expected to earn $56 in annual dividends, and he purchased a total of 200 shares, how many shares of each stock did he purchase?

40. ***Initial Public Offerings (IPOs)*** Duffin House, Higgins Press, and Sickle Publications all went public on the same day recently. John O'Hagan had the opportunity to participate in all three initial public offerings (partly because he and Marjory Duffin are good friends). He made a considerable profit when he sold all of the stock two days later on the open market. The following table shows the purchase price and percentage yield on the investment in each company.

	Purchase Price per Share	*Yield*
Duffin House (DHS)	$8	20%
Higgins Press (HPR)	$10	15%
Sickle Publications (SPUB)	$15	15%

He invested $20,000 in a total of 2,000 shares, and made a $3,400 profit from the transactions. How many shares in each company did he purchase?

41. ***Degree Requirements*** During his lunch break, John O'Hagan decides to devote some time to assisting his son Billy-Sean, who is having a terrible time coming up with a college course schedule. One reason for this is the very complicated Bulletin of Suburban State University. It reads as follows:

> *All candidates for the degree of Bachelor of Science at SSU must take a total of 124 credits from the Sciences, Fine Arts, Liberal Arts, and Mathematics,*[2] *including an equal number of Science and Fine Arts credits, and twice as many Mathematics credits as Science credits and Fine Arts credits combined, but with Liberal Arts credits exceeding Mathematics credits by exactly one-third of the number of Fine Arts credits.*

What are all the possible degree programs for Billy-Sean?

42. ***Degree Requirements*** Having finally decided on his degree program, Billy-Sean learns that the Suburban State University Senate (under pressure from the English Department) has revised the Bulletin to include a "Verbal Expression" component in place of the Fine Arts requirement in all programs (including the sciences):

> *All candidates for the degree of Bachelor of Science at SSU must take a total of 120 credits from the Liberal Arts, Sciences, Verbal Expression, and Mathematics, including an equal number of Science and Liberal Arts credits, and twice as many Verbal Expression credits as Science credits and Liberal Arts credits combined, but with Liberal Arts credits exceeding Mathematics credits by one quarter of the number of Verbal Expression Credits.*

What are now the possible degree programs for Billy-Sean?

[2]Strictly speaking, mathematics is not a science; it is the Queen of the Sciences, although we like to think of it as the Mother of all Sciences.

43. ***Network Traffic*** All book orders received at the Order Department at OHaganBooks.com are transmitted through a small computer network to the Shipping Department. The following diagram shows the network (which uses two intermediate computers as routers), together with some of the average daily traffic measured in book orders.

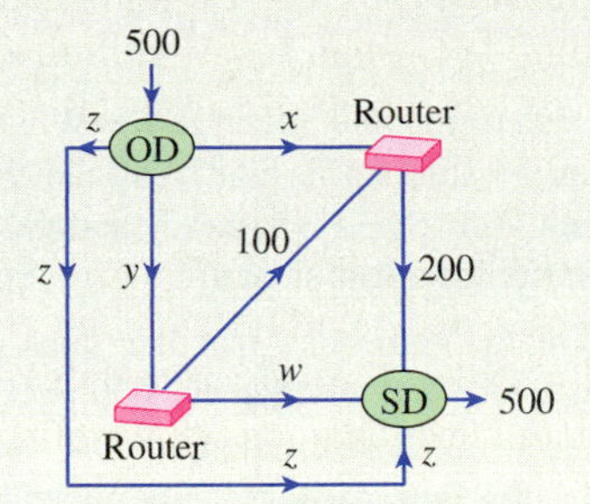

a. Set up a system of linear equations in which the unknowns give the average traffic along the paths labeled x, y, z, w, and find the general solution.

b. What is the minimum volume of traffic along y?

c. What is the maximum volume of traffic along w?

d. If there is no traffic along z, find the volume of traffic along all the paths.

e. If there is the same volume of traffic along y and z, what is the volume of traffic along w?

44. ***Business Retreats*** Marjory Duffin is planning a joint business retreat for Duffin House and OHaganBooks.com at Laguna Surf City, but is concerned about traffic conditions (she feels that too many cars tend to spoil the ambiance of a seaside retreat). She managed to obtain the following map from the Laguna Surf City Engineering Department (all the streets are one-way as indicated). The counters show traffic every five minutes.

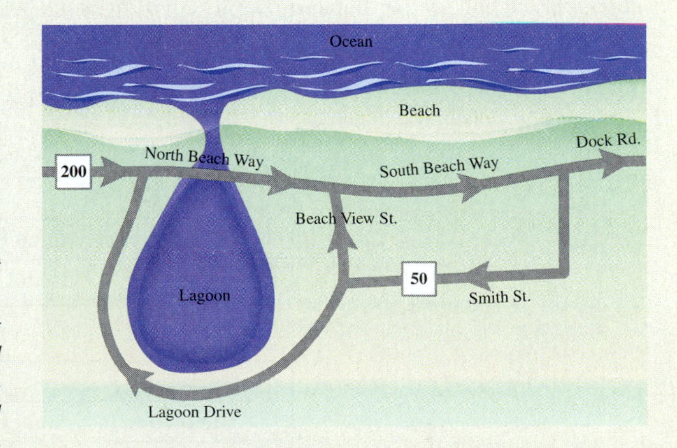

a. Set up and solve the associated system of linear equations. *Be sure to give the general solution.* (Take x = traffic along North Beach Way, y = traffic along South Beach Way, z = traffic along Beach View

St., u = traffic along Lagoon Drive and v = traffic along Dock Road.)

b. Assuming that all roads are one-way in the directions shown, what, if any, is the maximum possible traffic along Lagoon Drive?

c. The Laguna Surf City Traffic Department is considering opening up Beach View Street to two-way traffic, but an environmentalist group is concerned that this will result in increased traffic on Lagoon Drive. What, if any, is the maximum possible traffic along Lagoon Drive assuming that Beach View Street is two-way and the traffic counter readings are as shown?

45. ***Shipping*** On the same day that the sales department at Duffin House received an order for 600 packages from the OHaganBooks.com Texas headquarters, it received an additional order for 200 packages from FantasyBooks.com, based in California. Duffin House has warehouses in New York and Illinois. The Illinois warehouse is closing down and must clear all 300 packages it has in stock. Shipping costs per package of books are as follows:

New York to Texas: $20 New York to California: $50
Illinois to Texas: $30 Illinois to California: $40

Is it possible to fill both orders and clear the Illinois warehouse at a cost of $22,000? If so, how many packages should be sent from each warehouse to each online bookstore?

46. ***Transportation Scheduling*** Duffin House is about to start a promotional blitz for its new book, *Physics for the Liberal Arts.* The company has 20 salespeople stationed in Chicago and 10 in Denver, and would like to fly 15 to sales fairs at each of Los Angeles and New York. A round-trip plane flight from Chicago to LA costs $200; from Chicago to NY costs $150; from Denver to LA costs $400; and from Denver to NY costs $200. For tax reasons, Duffin House needs to budget exactly $6,500 for the total cost of the plane flights. How many salespeople should the company fly from each of Chicago and Denver to each of LA and NY?

Case Study: Hybrid Cars—Optimizing the Degree of Hybridization

Oleksiy Maksymenko/Alamy

You are involved in new model development at a major automobile company. The company is planning to introduce two new plug-in hybrid electric vehicles: the subcompact "Green Town Hopper" and the midsize "Electra Supreme," and your department must decide on the degree of hybridization (DOH) for each of these models that will result in the largest reduction in gasoline consumption. (The DOH of a vehicle is defined as the ratio of electric motor power to the total power, and typically ranges from 10% to 50%. For example, a model with a 20% DOH has an electric motor that delivers 20% of the total power of the vehicle.)

The tables below show the benefit for each of the two models, measured as the estimated reduction in annual gasoline consumption, as well as an estimate of retail cost increment, for various DOH percentages[3]: (The retail cost increment estimate is given by the formula $5{,}000 + 50(DOH - 10)$ for the Green Town Hopper and $7{,}000 + 50(DOH - 10)$ for the Electra Supreme.)

Green Town Hopper

DOH (%)	10	20	50
Reduction in Annual Consumption (gals)	180	230	200
Retail Cost Increment ($)	5,000	5,500	7,000

Electra Supreme

DOH (%)	10	20	50
Reduction in Annual Consumption (gals)	220	270	260
Retail Cost Increment ($)	7,000	7,500	9,000

[3]The figures are approximate and based on data for two actual vehicles as presented in a 2006 paper entitled *Cost-Benefit Analysis of Plug-In Hybrid Electric Vehicle Technology* by A. Simpson. Source: National Renewable Energy Laboratory, U.S. Department of Energy (www.nrel.gov).

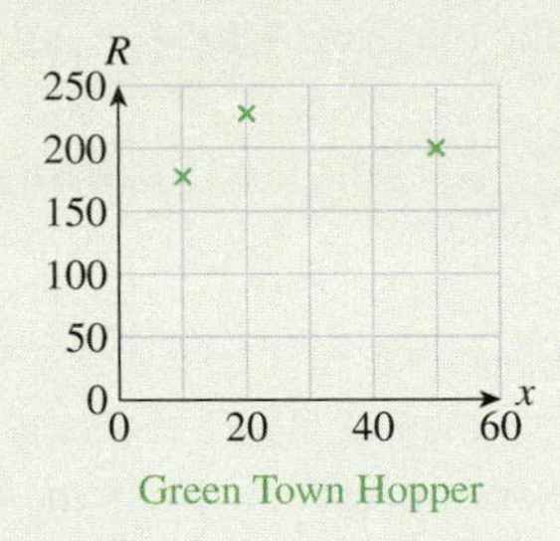

Green Town Hopper

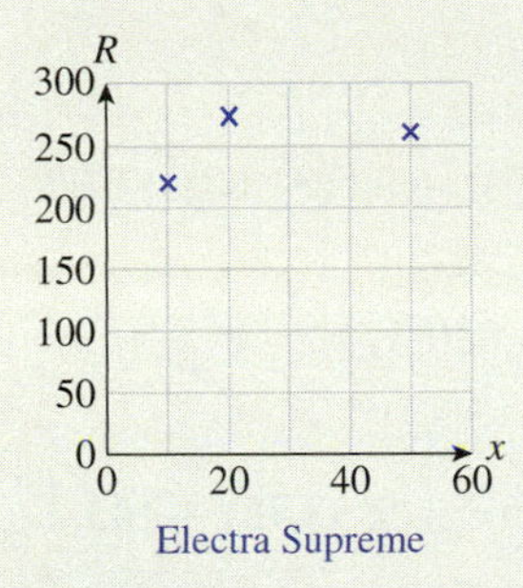

Electra Supreme

Figure 9

Notice that increasing the DOH toward 50% results in a decreased benefit. This is due in part to the need to increase the weight of the batteries while keeping the vehicle performance at a desirable level, thus necessitating a more powerful gasoline engine. The *optimum* DOH is the percentage that gives the largest reduction in gasoline consumption, and this is what you need to determine. Since the optimum DOH may not be 20%, you would like to create a mathematical model to compute the reduction R in gas consumption as a function of the DOH x. Your first inclination is to try linear equations—that is, an equation of the form

$$R = ax + b \quad (a \text{ and } b \text{ constants}),$$

but you quickly discover that the data simply won't fit, no matter what the choice of the constants. The reason for this can be seen graphically by plotting R versus x (Figure 9). In neither case do the three points lie on a straight line. In fact, the data are not even *close* to being linear. Thus, you will need curves to model these data. After giving the matter further thought, you remember something your mathematics instructor once said: The simplest curve passing through any three points not all on the same line is a parabola. Since you are looking for a simple model of the data, you decide to try a parabola. A general parabola has the equation

$$R = ax^2 + bx + c,$$

where a, b, and c are constants. The problem now is: What are a, b, and c? You decide to try substituting the values of R and x for the Green Town Hopper into the general equation, and you get the following:

$$\begin{aligned} x = 10, R = 180 &\quad \text{gives} \quad 180 = 100a + 10b + c \\ x = 20, R = 230 &\quad \text{gives} \quad 230 = 400a + 20b + c \\ x = 50, R = 200 &\quad \text{gives} \quad 200 = 2{,}500a + 50b + c. \end{aligned}$$

Now you notice that you have three linear equations in three unknowns! You solve the system:

$$a = -0.15, b = 9.5, c = 100.$$

Thus your reduction equation for the Green Town Hopper becomes

$$R = -0.15x^2 + 9.5x + 100.$$

For the Electra Supreme, you get

$$\begin{aligned} x = 10, R = 220: &\quad 220 = 100a + 10b + c \\ x = 20, R = 270: &\quad 270 = 400a + 20b + c \\ x = 50, R = 260: &\quad 260 = 2{,}500a + 50b + c. \end{aligned}$$

$$a = -0.1\bar{3}, b = 9, c = 143.\bar{3}$$

and so

$$R = -0.1\bar{3}x^2 + 9x + 143.\bar{3}.$$

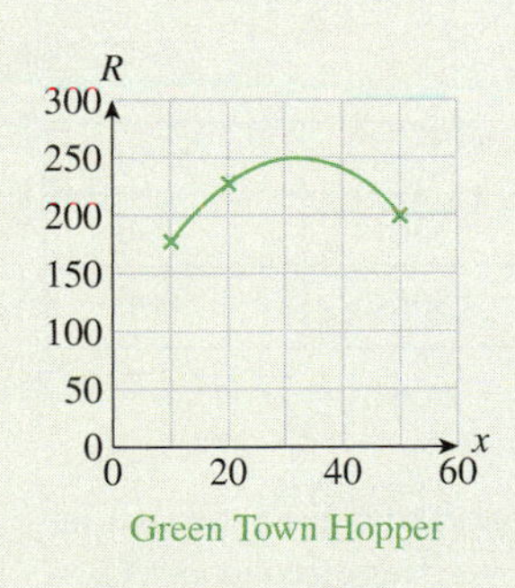

Green Town Hopper

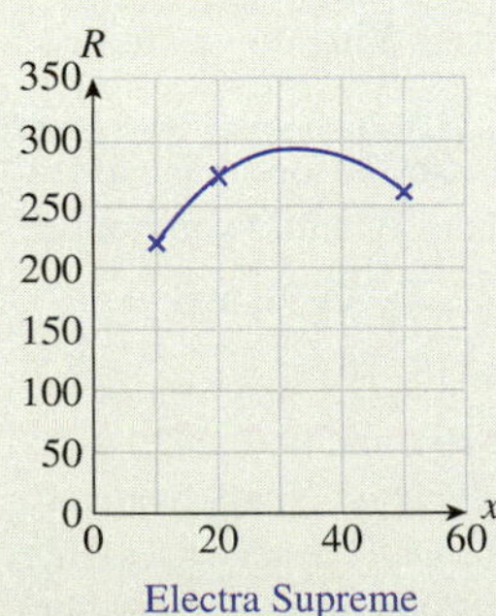

Electra Supreme

Figure 10

Figure 10 shows the parabolas superimposed on the data points. You can now estimate a value for the optimal DOH as the value of x that gives the largest benefit R.

Recalling that the x-coordinate of the vertex of the parabola $y = ax^2 + bx + c$ is $x = -\frac{b}{2a}$, you obtain the following estimates:

$$\text{Green Town Hopper: Optimal DOH} = -\frac{9.5}{2(-0.15)} \approx 31.67\%$$

$$\text{Electra Supreme: Optimal DOH} = -\frac{9}{2(-0.1\bar{3})} \approx 33.75\%.$$

You can now use the formulas given earlier to estimate the resulting reductions in gasoline consumption and increases in cost:

Green Town Hopper:

$$\text{Reduction in gasoline consumption} = R \approx -0.15(31.67)^2 + 9.5(31.67) + 100$$
$$\approx 250.4 \text{ gals/year}$$

$$\text{Retail Cost Increment} \approx 5{,}000 + 50(31.67 - 10) \approx \$6{,}080$$

Electra Supreme:

$$\text{Reduction in gasoline consumption} = R = -0.1\bar{3}(33.75)^2 + 9(33.75) + 143.\bar{3}$$
$$\approx 295.2 \text{ gals/year}$$

$$\text{Retail Cost Increment} \approx 7{,}000 + 50(33.75 - 10) \approx \$8{,}190.$$

You thus submit the following estimates: The optimal degree of hybridization for the Green Town Hopper is about 31.67% and will result in a reduction in gasoline consumption of 250.4 gals/year and a retail cost increment of around $6,080. The optimal degree of hybridization for the Electra Supreme is about 33.75% and will result in a reduction in gasoline consumption of 295.2 gals/year and a retail cost increment of around $8,190.

EXERCISES

1. Repeat the computations above for the "Earth Suburban" using the following data:

Earth Suburban

DOH (%)	10	20	50
Reduction in Annual Consumption (gals)	240	330	300
Retail Cost Increment ($)	9,000	9,500	11,000

$$\text{Retail cost increment} = 9{,}000 + 50(DOH - 10)$$

2. T Repeat the analysis for the Green Town Hopper, but this time take x to be the cost increment, in thousands of dollars. (The curve of benefit versus cost is referred to as a *cost-benefit* curve.) What value of DOH corresponds to the optimal cost? What do you notice? Comment on the answer.
3. T Repeat the analysis for the Electra Supreme, but this time use the optimal values, DOH = 33.8, Reduction = 295.2 gals/year in place of the 20% data. What do you notice?
4. Find the equation of the parabola that passes through the points (1, 2), (2, 9), and (3, 19).
5. Is there a parabola that passes through the points (1, 2), (2, 9), and (3, 16)?
6. Is there a parabola that passes though the points (1, 2), (2, 9), (3, 19), and (−1, 2)?
7. T You submit your recommendations to your manager, and she tells you, "Thank you very much, but we have additional data for the Green Town Hopper: A 30% DOH results in a saving of 255 gals/year. Please resubmit your recommendations taking this into account by tomorrow." HINT [You now have four data points on each graph, so try a general cubic instead: $R = ax^3 + bx^2 + cx + d$. Use a graph to estimate the optimal DOH.]

TI-83/84 Plus Technology Guide

Section 3.1

Example 1 (page 145) Find all solutions (x, y) of the following system of two equations:

$$x + y = 3$$
$$x - y = 1.$$

Solution with Technology

You can use a graphing calculator to draw the graphs of the two equations on the same set of axes and to check the solution. First, solve the equations for y, obtaining $y = -x + 3$ and $y = x - 1$. On the TI-83/84 Plus:

1. Set

   Y1=-X+3
   Y2=X-1

2. Decide on the range of x-values you want to use. As in Figure 1, let us choose the range $[-4, 4]$.[4]
3. In the WINDOW menu set Xmin $= -4$ and Xmax $= 4$.
4. Press ZOOM and select Zoomfit to set the y range.

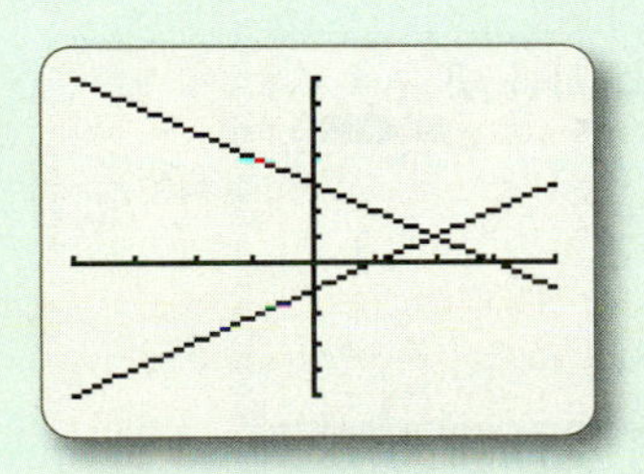

You can now zoom in for a more accurate view by choosing a smaller x-range that includes the point of intersection, like $[1.5, 2.5]$, and using Zoomfit again. You can also use the trace feature to see the coordinates of points near the point of intersection.

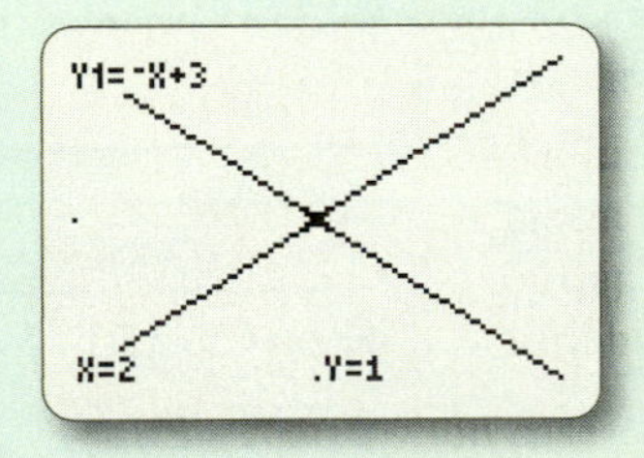

To check that (2, 1) is the correct solution, use the table feature to compare the two values of y corresponding to $x = 2$:

1. Press 2ND TABLE.
2. Set X $= 2$, and compare the corresponding values of Y1 and Y2; they should each be 1.

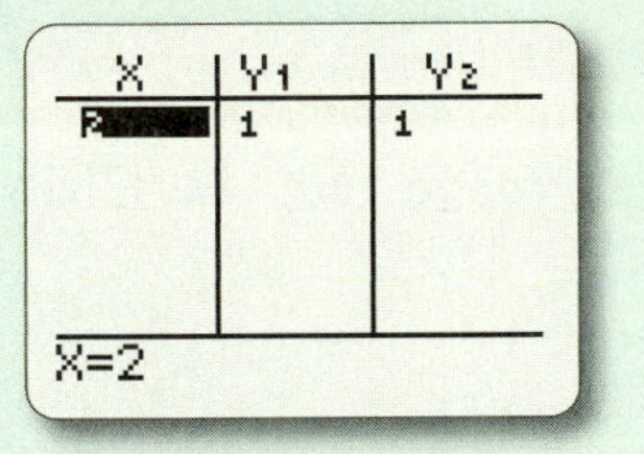

Q: *How accurate is the answer shown using the trace feature?*

A: That depends. We can increase the accuracy up to a point by zooming in on the point of intersection of the two graphs. But there is a limit to this: Most graphing calculators are capable of giving an answer correct to about 13 decimal places. This means, for instance, that, in the eyes of the TI-83/84 Plus, 2.000 000 000 000 1 is exactly the same as 2 (subtracting them yields 0). It follows that if you attempt to use a window so narrow that you need approximately 13 significant digits to distinguish the left and right edges, you will run into accuracy problems.

Section 3.2

Row Operations with a TI-83/84 Plus Start by entering the matrix into [A] using MATRIX EDIT. You can then do row operations on [A] using the following instructions (*row, *row+, and rowSwap are found in the MATRIX MATH menu).

[4] How did we come up with this interval? Trial and error. You might need to try several intervals before finding one that gives a graph showing the point of intersection clearly.

Row Operation	*TI-83/84 Plus Instruction* (Matrix name is [A])
$R_i \to kR_i$	`*row(k,[A],i)→[A]`
Example: $R_2 \to 3R_2$	`*row(3,[A],2)→[A]`
$R_i \to R_i + kR_j$	`*row+(k,[A],j,i)→[A]`
Examples: $R_1 \to R_1 - 3R_2$	`*row+(-3,[A],2,1)→[A]`
$R_1 \to 4R_1 - 3R_2$	`*row(4,[A],1)→[A]` `*row+(-3,[A],2,1)→[A]`
Swap R_i and R_j	`rowSwap([A],i,j)→[A]`
Example: Swap R_1 and R_2	`rowSwap([A],1,2)→[A]`

Example 1 (page 160) Solve the system:

$$\begin{aligned} x - y + 5z &= -6 \\ 3x + 3y - z &= 10 \\ x + 3y + 2z &= 5. \end{aligned}$$

Solution with Technology

1. Begin by entering the matrix into [A] using MATRX EDIT. Only three columns can be seen at a time; you can see the rest of the matrix by scrolling left or right.

```
MATRIX[A] 3 ×4
[1    -1    5   ]
[3    3     -1  ]
[1    3     2   ]

1,1=1
```

```
MATRIX[A] 3 ×4
 -1    5    -6 ]
 3     -1   10 ]
 3     2    5  ]

1,4=-6
```

2. Now perform the operations given in Example 1:

```
*row+(-3,[A],1,2
)→[A]
 [[1 -1 5   -6]
  [0 6  -16 28]
  [1 3  2   5 ]]
```

```
*row+(-1,[A],1,3
)→[A]
 [[1 -1 5   -6]
  [0 6  -16 28]
  [0 4  -3  11]]
```

```
*row(1/2,[A],2)→
[A]
 [[1 -1 5  -6]
  [0 3  -8 14]
  [0 4  -3 11]]
```

```
*row(3,[A],1)→[A
]
 [[3 -3 15 -18]
  [0 3  -8 14 ]
  [0 4  -3 11 ]]
```

```
*row+(1,[A],2,1)
→[A]
 [[3 0 7  -4]
  [0 3 -8 14]
  [0 4 -3 11]]
```

```
*row(3,[A],3)→[A
]
 [[3 0  7  -4]
  [0 3  -8 14]
  [0 12 -9 33]]
```

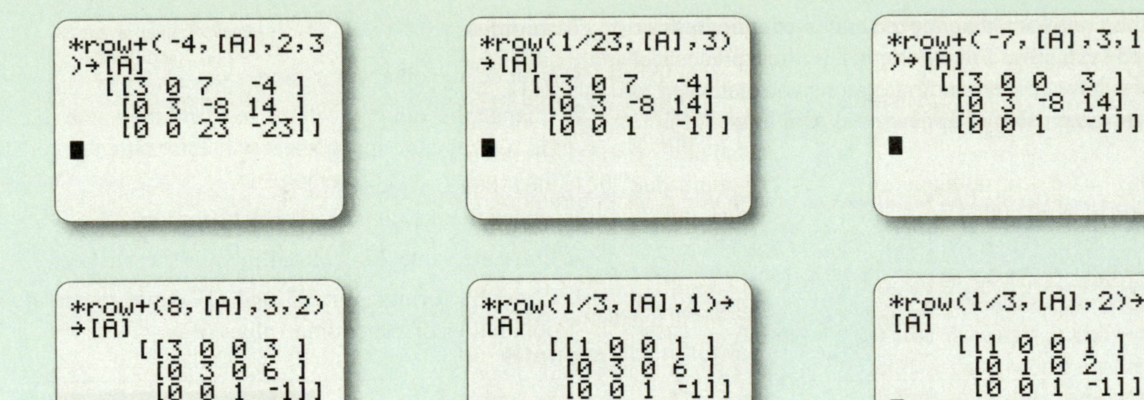

As in Example 1, we can now read the solution from the right-hand column: $x = 1, y = 2, z = -1$.

Note The TI-83/84 Plus has a function, `rref`, that gives the reduced row echelon form of a matrix in one step. (See the text for the definition of reduced row echelon form.) Internally it uses a variation of Gauss-Jordan reduction to do this. ■

SPREADSHEET Technology Guide

Section 3.1

Example 1 (page 145) Find all solutions (x, y) of the following system of two equations:

$$x + y = 3$$
$$x - y = 1.$$

Solution with Technology

You can use a spreadsheet to draw the graphs of the two equations on the same set of axes, and to check the solution.

1. Solve the equations for y, obtaining $y = -x + 3$ and $y = x - 1$.
2. To graph these lines we can use the following simple worksheet:

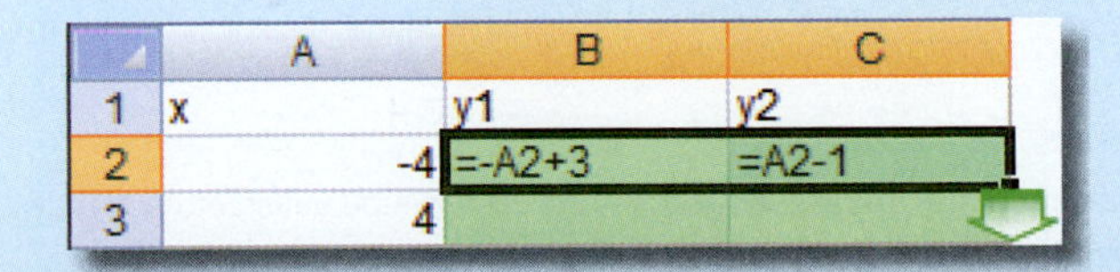

The two values of x give the x-coordinates of the two points we will use as endpoints of the lines. (We have—somewhat arbitrarily—chosen the range $[-4, 4]$ for x.) The formula for the first line, $y = -x + 3$, is in cell B2, and the formula for the second line, $y = x - 1$, is in C2.

3. Copy these two cells as shown to yield the following result:

	A	B	C
1	x	y1	y2
2	-4	7	-5
3	4	-1	3

4. For the graph, select all nine cells and create a scatter graph with line segments joining the data points. Instruct the spreadsheet to insert a chart and select the "scatter" option. In the same dialogue box, select the option that shows points connected by lines. If you are using Excel, press "Next" to bring up a new dialogue box called "Data Type," where you should make sure that the "Series in Columns" option is selected,

telling the program that the x- and y-coordinates are arranged vertically, down columns. (Other spreadsheet programs have corresponding options you may need to set.) Your graph should appear as shown below.

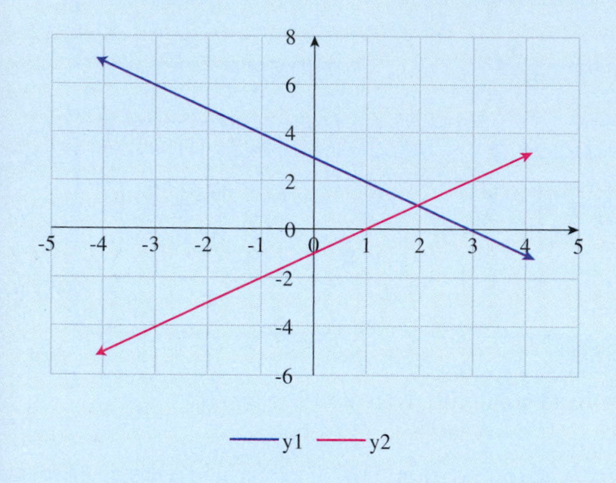

To zoom in:

1. First decide on a new x-range, say, [1, 3].
2. Change the value in cell A2 to 1 and the value in cell A3 to 3, and the spreadsheet will automatically update the y-values and the graph.*

To check that (2, 1) is the correct solution:

1. Enter the value 2 in cell A4 in your spreadsheet.
2. Copy the formulas in B3 and C3 down to row 4 to obtain the corresponding values of y.

	A	B	C
1	x	y1	y2
2	-4	7	-5
3	4	-1	3
4	2	1	1

Because the values of y agree, we have verified that (2, 1) is a solution of both equations.

Section 3.2

Row Operations with a Spreadsheet To a spreadsheet, a block of data with one or more rows or columns is an **array**, and a spreadsheet has built in the capability to handle arrays in much the same way as it handles single cells. Consider the following example:

$$\begin{bmatrix} 1 & 3 & -4 \\ 0 & 4 & 2 \end{bmatrix} 3R_2 \rightarrow \begin{bmatrix} 1 & 3 & -4 \\ 0 & 12 & 6 \end{bmatrix}.$$

Replace R_2 by $3R_2$.

1. Enter the original matrix in a convenient location, say the cells A1 through C2.
2. To get the first row of the new matrix, which is simply a copy of the first row of the old matrix, decide where you want to place the new matrix, and highlight *the whole row of the new matrix,* say A4:C4.
3. Enter the formula `=A1:C1` (the easiest way to do this is to type "=" and then use the mouse to select cells A1 through C1, i.e., the first row of the old matrix).

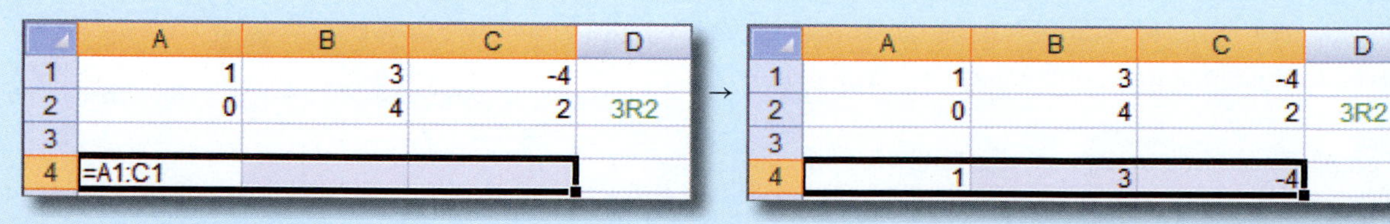

	A	B	C	D
1	1	3	-4	
2	0	4	2	3R2
3				
4	=A1:C1			

→

	A	B	C	D
1	1	3	-4	
2	0	4	2	3R2
3				
4	1	3	-4	

Enter formula for new first row.

Press Control-Shift-Enter.

* **NOTE** Your spreadsheet will set the y-range of the graph automatically, and the range it chooses may not always be satisfactory. It may sometimes be necessary to "narrow down" the range of one or both axes (in Excel, double-click on the axis in question).

4. Press **Control-Shift-Enter** (instead of "Enter" alone) and the whole row will be copied.[5] Pressing Control-Shift-Enter tells the spreadsheet that your formula is an *array formula,* one that returns an array rather than a single number. Once entered, the spreadsheet will show an array formula enclosed in "curly braces." (Note that you must also use Control-Shift-Enter to delete any array you create: Select the block you wish to delete and press Delete followed by Control-Shift-Enter.)
5. Similarly, to get the second row, select cells A5 through C5, where the new second row will go, enter the formula `=3*A2:C2`, and press Control-Shift-Enter. (Again, the easiest way to enter the formula is to type "`=3*`" and then select the cells A2 through C2 using the mouse.)

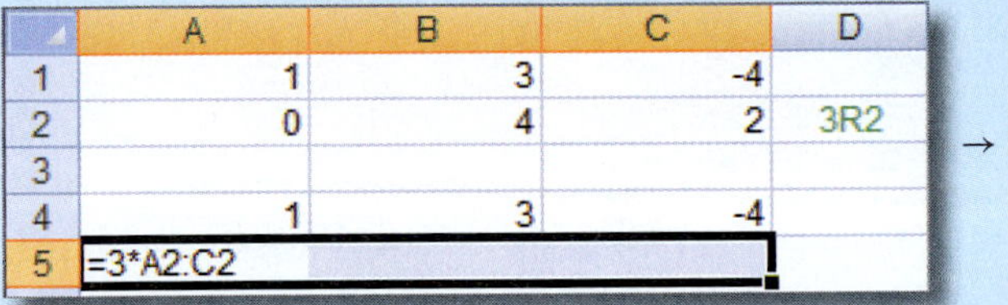

	A	B	C	D
1	1	3	-4	
2	0	4	2	3R2
3				
4	1	3	-4	
5	=3*A2:C2			

Enter formula for new second row.

→

	A	B	C	D
1	1	3	-4	
2	0	4	2	3R2
3				
4	1	3	-4	
5	0	12	6	

Press Control-Shift-Enter.

We can perform the following operation in a similar way:

$$\begin{bmatrix} 1 & 3 & -4 \\ 0 & 4 & 2 \end{bmatrix} \begin{matrix} 4R_1 - 3R_2 \\ \end{matrix} \rightarrow \begin{bmatrix} 4 & 0 & -22 \\ 0 & 4 & 2 \end{bmatrix}.$$

Replace R_1 by $4R_1 - 3R_2$.

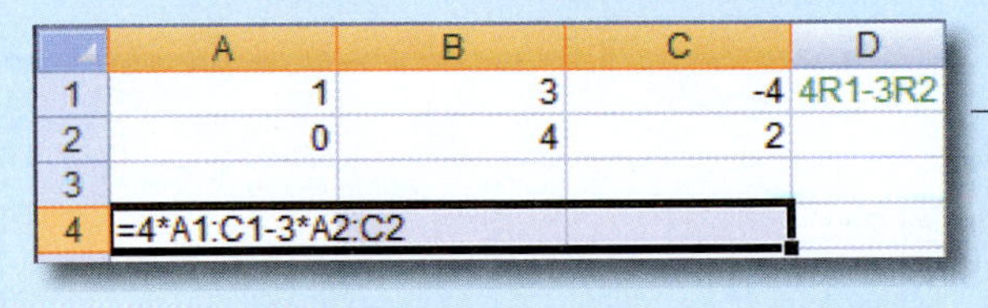

	A	B	C	D
1	1	3	-4	4R1-3R2
2	0	4	2	
3				
4	=4*A1:C1-3*A2:C2			

→

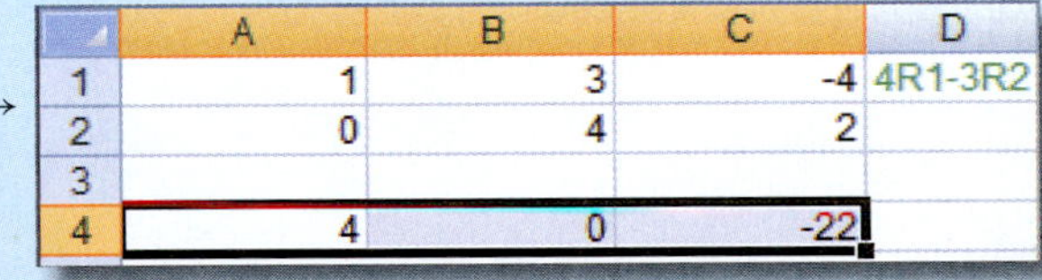

	A	B	C	D
1	1	3	-4	4R1-3R2
2	0	4	2	
3				
4	4	0	-22	

(To easily enter the formula for $4R_1 - 3R_2$, type "`=4*`", select the first row `A1:C1` using the mouse, type "`-3*`", and then select the second row `A2:C2` using the mouse.)

Example 1 (page 160) Solve the system:

$$\begin{aligned} x - y + 5z &= -6 \\ 3x + 3y - z &= 10 \\ x + 3y + 2z &= 5. \end{aligned}$$

[5]Note that on a Mac, Command-Enter and Command-Shift-Enter have the same effect as Control-Shift-Enter.

TECHNOLOGY GUIDE

Solution with Technology

Here is the complete row reduction as it would appear in a spreadsheet.

	A	B	C	D	E
1	1	-1	5	-6	
2	3	3	-1	10	R2 - 3R1
3	1	3	2	5	R3 - R1
4					
5	=A1:D1				
6	=A2:D2-3*A1:D1				
7	=A3:D3-A1:D1				

	A	B	C	D	E
1	1	-1	5	-6	
2	3	3	-1	10	R2 - 3R1
3	1	3	2	5	R3 - R1
4					
5	1	-1	5	-6	
6	0	6	-16	28	(1/2)R2
7	0	4	-3	11	
8					
9	=A5:D5				
10	=(1/2)*A6:D6				
11	=A7:D7				

8					
9	1	-1	5	-6	3R1 + R2
10	0	3	-8	14	
11	0	4	-3	11	3R3 - 4R2
12					
13	=3*A9:D9+A10:D10				
14	=A10*D10				
15	=3*A11:D11-4*A10:D10				

12					
13	3	0	7	-4	
14	0	3	-8	14	
15	0	0	23	-23	(1/23)R3
16					
17	=A13:D13				
18	=A14:D14				
19	=(1/23)*A15:D15				

16					
17	3	0	7	-4	R1 - 7R3
18	0	3	-8	14	R2 + 8R3
19	0	0	1	-1	
20					
21	=A17:D17-7*A19:D19				
22	=A18:D18+8*A19:D19				
23	=A19:D19				

20					
21	3	0	0	3	(1/3)R1
22	0	3	0	6	(1/3)R2
23	0	0	1	-1	
24					
25	=(1/3)*A21:D21				
26	=(1/3)*A22:D22				
27	=A23:D23				

24					
25	1	0	0	1	
26	0	1	0	2	
27	0	0	1	-1	

As in Example 1, we can now read the solution from the right hand column: $x = 1, y = 2, z = -1$. What do you notice if you change the entries in cells D1, D2, and D3?

4

Matrix Algebra and Applications

Website

www.WanerMath.com

At the Website you will find:

- Section-by-section tutorials
- A detailed chapter summary
- A true/false quiz
- Additional review exercises
- A matrix algebra utility
- A game theory utility
- The following optional extra sections:

  Determinants

  Using Determinants to Solve Systems: Cramer's Rule

Case Study Predicting Market Share

You are the sales director at *Selular*, a cellphone provider, and things are not looking good for your company: The recently launched *iClone* competitor is beginning to chip away at *Selular's* market share. Particularly disturbing are rumors of fierce brand loyalty by *iClone* customers, with several bloggers suggesting that *iClone* retains close to 100% of their customers. Worse, you will shortly be presenting a sales report to the board of directors, and the CEO has "suggested" that your report include 2-, 5-, and 10-year projections of *Selular's* market share given the recent impact on the market by *iClone*, and also a "worst-case" scenario projecting what would happen if *iClone* customers are so loyal that none of them ever switch services. You have results from two market surveys, taken one quarter apart, of the major cellphone providers, which shows the current market shares and the percentages of subscribers who switched from one service to another during the quarter. **How should you respond?**

GOH CHAI HIN/AFP/Newscom

Introduction

We used matrices in Chapter 3 simply to organize our work. It is time we examined them as interesting objects in their own right. There is much that we can do with matrices besides row operations: We can add, subtract, multiply, and even, in a sense, "divide" matrices. We use these operations to study game theory and input-output models in this chapter, and Markov chains in a later chapter.

Many calculators, spreadsheets, and other computer programs can do these matrix operations, which is a big help in doing calculations. However, we need to know how these operations are defined to see why they are useful and to understand which to use in any particular application.

4.1 Matrix Addition and Scalar Multiplication

Let's start by formally defining what a matrix is and introducing some basic terms.

Matrix, Dimension, and Entries

An $m \times n$ **matrix** A is a rectangular array of real numbers with m rows and n columns. We refer to m and n as the **dimensions** of the matrix. The numbers that appear in the matrix are called its **entries**. We customarily use capital letters $A, B, C, \ldots$ for the names of matrices.

Quick Examples

1. $A = \begin{bmatrix} 2 & 0 & 1 \\ 33 & -22 & 0 \end{bmatrix}$ is a 2×3 matrix because it has two rows and three columns.

2. $B = \begin{bmatrix} 2 & 3 \\ 10 & 44 \\ -1 & 3 \\ 8 & 3 \end{bmatrix}$ is a 4×2 matrix because it has four rows and two columns.*

The entries of A are 2, 0, 1, 33, −22, and 0. The entries of B are the numbers 2, 3, 10, 44, −1, 3, 8, and 3.

* Remember that the number of rows is given first and the number of columns second. An easy way to remember this is to think of the acronym "RC" for "Row then Column."

Referring to the Entries of a Matrix

There is a systematic way of referring to particular entries in a matrix. If i and j are numbers, then the entry in the ith row and jth column of the matrix A is called the ***ij*th entry** of A. We usually write this entry as a_{ij} or A_{ij}. (If the matrix was called B, we would write its ijth entry as b_{ij} or B_{ij}.) Notice that this follows the "RC" convention: The row number is specified first and the column number second.

Quick Example

With $A = \begin{bmatrix} 2 & 0 & 1 \\ 33 & -22 & 0 \end{bmatrix}$,

$a_{13} = 1$ First row, third column

$a_{21} = 33.$ Second row, first column

According to the labeling convention, the entries of the matrix A above are

$$A = \begin{bmatrix} a_{11} & a_{12} & a_{13} \\ a_{21} & a_{22} & a_{23} \end{bmatrix}.$$

In general, the $m \times n$ matrix A has its entries labeled as follows:

$$A = \begin{bmatrix} a_{11} & a_{12} & a_{13} & \dots & a_{1n} \\ a_{21} & a_{22} & a_{23} & \dots & a_{2n} \\ \vdots & \vdots & \vdots & \ddots & \vdots \\ a_{m1} & a_{m2} & a_{m3} & \dots & a_{mn} \end{bmatrix}.$$

We say that two matrices A and B are **equal** if they have the same dimensions and the corresponding entries are equal. Note that a 3×4 matrix can never equal a 3×5 matrix because they do not have the same dimensions.

EXAMPLE 1 Matrix Equality

Let $A = \begin{bmatrix} 7 & 9 & x \\ 0 & -1 & y+1 \end{bmatrix}$ and $B = \begin{bmatrix} 7 & 9 & 0 \\ 0 & -1 & 11 \end{bmatrix}$. Find the values of x and y such that $A = B$.

Solution For the two matrices to be equal, we must have corresponding entries equal, so

$x = 0$ $\quad a_{13} = b_{13}$

$y + 1 = 11$ or $y = 10.$ $\quad a_{23} = b_{23}$

Before we go on... Note in Example 1 that the matrix equation

$$\begin{bmatrix} 7 & 9 & x \\ 0 & -1 & y+1 \end{bmatrix} = \begin{bmatrix} 7 & 9 & 0 \\ 0 & -1 & 11 \end{bmatrix}$$

is really six equations in one: $7 = 7$, $9 = 9$, $x = 0$, $0 = 0$, $-1 = -1$, and $y + 1 = 11$. We used only the two that were interesting. ■

Row Matrix, Column Matrix, and Square Matrix

A matrix with a single row is called a **row matrix**, or **row vector**. A matrix with a single column is called a **column matrix** or **column vector**. A matrix with the same number of rows as columns is called a **square matrix**.

Quick Examples

The 1×5 matrix $C = [3 \quad -4 \quad 0 \quad 1 \quad -11]$ is a row matrix.

The 4×1 matrix $D = \begin{bmatrix} 2 \\ 10 \\ -1 \\ 8 \end{bmatrix}$ is a column matrix.

The 3×3 matrix $E = \begin{bmatrix} 1 & -2 & 0 \\ 0 & 1 & 4 \\ -4 & 32 & 1 \end{bmatrix}$ is a square matrix.

Matrix Addition and Subtraction

The first matrix operations we discuss are matrix addition and subtraction. The rules for these operations are simple.

Matrix Addition and Subtraction

Two matrices can be added (or subtracted) if and only if they have the same dimensions. To add (or subtract) two matrices of the same dimensions, we add (or subtract) the corresponding entries. More formally, if A and B are $m \times n$ matrices, then $A + B$ and $A - B$ are the $m \times n$ matrices whose entries are given by:

$(A + B)_{ij} = A_{ij} + B_{ij}$ ijth entry of the sum = sum of the ijth entries

$(A - B)_{ij} = A_{ij} - B_{ij}$. ijth entry of the difference = difference of the ijth entries

Visualizing Matrix Addition

$$\begin{bmatrix} 2 & -3 \\ 1 & 0 \end{bmatrix} + \begin{bmatrix} 1 & 1 \\ -2 & 1 \end{bmatrix} = \begin{bmatrix} 3 & -2 \\ -1 & 1 \end{bmatrix}$$

Quick Examples

1. $\begin{bmatrix} 2 & -3 \\ 1 & 0 \\ -1 & 3 \end{bmatrix} + \begin{bmatrix} 9 & -5 \\ 0 & 13 \\ -1 & 3 \end{bmatrix} = \begin{bmatrix} 11 & -8 \\ 1 & 13 \\ -2 & 6 \end{bmatrix}$ Corresponding entries added

2. $\begin{bmatrix} 2 & -3 \\ 1 & 0 \\ -1 & 3 \end{bmatrix} - \begin{bmatrix} 9 & -5 \\ 0 & 13 \\ -1 & 3 \end{bmatrix} = \begin{bmatrix} -7 & 2 \\ 1 & -13 \\ 0 & 0 \end{bmatrix}$ Corresponding entries subtracted

using Technology

Technology can be used to enter, add, and subtract matrices. Here is an outline (see the Technology Guides at the end of the chapter for additional details on using a TI-83/84 Plus or a spreadsheet):

TI-83/84 Plus

Entering a matrix: MATRX; EDIT
Select a name, ENTER; type in the entries.
Adding two matrices:
Home Screen: [A]+[B]
(Use MATRX; NAMES to enter them on the home screen.)
[More details on page 241.]

Spreadsheet

Entering a matrix: Type entries in a convenient block of cells.
Adding two matrices: Highlight block where you want the answer to appear.
Type "="; highlight first matrix; type "+"; highlight second matrix; press Control+Shift+Enter
[More details on page 244.]

Website

www.WanerMath.com
Student Home
→ On Line Utilities
→ Matrix Algebra Tool

Enter matrices as shown (use a single letter for the name).

```
Enter your matrices here.
J = [20, 15
10, 12
8, 4]

F = [23, 12,
8, 12
4, 5]
```

To compute their difference, type F-J in the formula box and press "Compute". (You can enter multiple formulas separated by commas in the formula box. For instance, F+J, F-J will compute both the sum and the difference. *Note:* The utility is case sensitive, so be consistent.)

EXAMPLE 2 Sales

The A-Plus auto parts store chain has two outlets, one in Vancouver and one in Quebec. Among other things, it sells wiper blades, windshield cleaning fluid, and floor mats. The monthly sales of these items at the two stores for two months are given in the following tables:

January Sales

	Vancouver	Quebec
Wiper Blades	20	15
Cleaning Fluid (bottles)	10	12
Floor Mats	8	4

February Sales

	Vancouver	Quebec
Wiper Blades	23	12
Cleaning Fluid (bottles)	8	12
Floor Mats	4	5

Use matrix arithmetic to calculate the change in sales of each product in each store from January to February.

Solution The tables suggest two matrices:

$$J = \begin{bmatrix} 20 & 15 \\ 10 & 12 \\ 8 & 4 \end{bmatrix} \quad \text{and} \quad F = \begin{bmatrix} 23 & 12 \\ 8 & 12 \\ 4 & 5 \end{bmatrix}.$$

To compute the change in sales of each product for both stores, we want to subtract corresponding entries in these two matrices. In other words, we want to compute the difference of the two matrices:

$$F - J = \begin{bmatrix} 23 & 12 \\ 8 & 12 \\ 4 & 5 \end{bmatrix} - \begin{bmatrix} 20 & 15 \\ 10 & 12 \\ 8 & 4 \end{bmatrix} = \begin{bmatrix} 3 & -3 \\ -2 & 0 \\ -4 & 1 \end{bmatrix}.$$

Thus, the change in sales of each product is the following:

	Vancouver	Quebec
Wiper Blades	3	−3
Cleaning Fluid (bottles)	−2	0
Floor Mats	−4	1

Scalar Multiplication

A matrix A can be added to itself because the expression $A + A$ is the sum of two matrices that have the same dimensions. When we compute $A + A$, we end up doubling every entry in A. So we can think of the expression $2A$ as telling us to *multiply every element in A by* 2.

In general, to multiply a matrix by a number, multiply every entry in the matrix by that number. For example,

$$6\begin{bmatrix} \frac{5}{2} & -3 \\ 1 & 0 \\ -1 & \frac{5}{6} \end{bmatrix} = \begin{bmatrix} 15 & -18 \\ 6 & 0 \\ -6 & 5 \end{bmatrix}.$$

It is traditional when talking about matrices to call individual numbers **scalars**. For this reason, we call the operation of multiplying a matrix by a number **scalar multiplication**.

using Technology

Technology can be used to compute scalar multiples:

TI-83/84 Plus

Enter the matrix [A] using MATRX; EDIT

Home Screen: 0.65[A]

Spreadsheets

Enter the matrix A in a convenient 3×2 block of cells.

Highlight block where you want the answer to appear.

Type =0.65*; highlight matrix A; press Control+Shift+Enter.

Website

www.WanerMath.com

Student Home

→ On Line Utilities

→ Matrix Algebra Tool

Enter the matrix A as shown.

```
Enter your matrices here.
A = [140, 105
30, 36
96, 48]
```

Type 0.65*A in the formula box and press "Compute".

EXAMPLE 3 Sales

The revenue generated by sales in the Vancouver and Quebec branches of the A-Plus auto parts store (see Example 2) was as follows:

January Sales in Canadian Dollars

	Vancouver	Quebec
Wiper Blades	140.00	105.00
Cleaning Fluid	30.00	36.00
Floor Mats	96.00	48.00

If the Canadian dollar was worth \$0.65 U.S. at the time, compute the revenue in U.S. dollars.

Solution We need to multiply each revenue figure by 0.65. Let A be the matrix of revenue figures in Canadian dollars:

$$A = \begin{bmatrix} 140.00 & 105.00 \\ 30.00 & 36.00 \\ 96.00 & 48.00 \end{bmatrix}.$$

The revenue figures in U.S. dollars are then given by the scalar multiple

$$0.65A = 0.65 \begin{bmatrix} 140.00 & 105.00 \\ 30.00 & 36.00 \\ 96.00 & 48.00 \end{bmatrix} = \begin{bmatrix} 91.00 & 68.25 \\ 19.50 & 23.40 \\ 62.40 & 31.20 \end{bmatrix}.$$

In other words, in U.S. dollars, \$91 worth of wiper blades was sold in Vancouver, \$68.25 worth of wiper blades was sold in Quebec, and so on.

Formally, scalar multiplication is defined as follows:

Scalar Multiplication

If A is an $m \times n$ matrix and c is a real number, then cA is the $m \times n$ matrix obtained by multiplying all the entries of A by c. (We usually use lowercase letters $c, d, e, \ldots$ to denote scalars.) Thus, the ijth entry of cA is given by

$$(cA)_{ij} = c(A_{ij}).$$

In words, this rule is: To get the ijth entry of cA, multiply the ijth entry of A by c.

EXAMPLE 4 Combining Operations

Let $A = \begin{bmatrix} 2 & -1 & 0 \\ 3 & 5 & -3 \end{bmatrix}$, $B = \begin{bmatrix} 1 & 3 & -1 \\ 5 & -6 & 0 \end{bmatrix}$, and $C = \begin{bmatrix} x & y & w \\ z & t+1 & 3 \end{bmatrix}$.

Evaluate the following: $4A$, xB, and $A + 3C$.

Solution First, we find $4A$ by multiplying each entry of A by 4:

$$4A = 4\begin{bmatrix} 2 & -1 & 0 \\ 3 & 5 & -3 \end{bmatrix} = \begin{bmatrix} 8 & -4 & 0 \\ 12 & 20 & -12 \end{bmatrix}.$$

Similarly, we find xB by multiplying each entry of B by x:

$$xB = x\begin{bmatrix} 1 & 3 & -1 \\ 5 & -6 & 0 \end{bmatrix} = \begin{bmatrix} x & 3x & -x \\ 5x & -6x & 0 \end{bmatrix}.$$

We get $A + 3C$ in two steps as follows:

$$\begin{aligned} A + 3C &= \begin{bmatrix} 2 & -1 & 0 \\ 3 & 5 & -3 \end{bmatrix} + 3\begin{bmatrix} x & y & w \\ z & t+1 & 3 \end{bmatrix} \\ &= \begin{bmatrix} 2 & -1 & 0 \\ 3 & 5 & -3 \end{bmatrix} + \begin{bmatrix} 3x & 3y & 3w \\ 3z & 3t+3 & 9 \end{bmatrix} \\ &= \begin{bmatrix} 2+3x & -1+3y & 3w \\ 3+3z & 3t+8 & 6 \end{bmatrix}. \end{aligned}$$

Addition and scalar multiplication of matrices have nice properties, reminiscent of the properties of addition and multiplication of real numbers. Before we state them, we need to introduce some more notation.

If A is any matrix, then $-A$ is the matrix $(-1)A$. In other words, $-A$ is A multiplied by the scalar -1. This amounts to changing the signs of all the entries in A. For example,

$$-\begin{bmatrix} 4 & -2 & 0 \\ 6 & 10 & -6 \end{bmatrix} = \begin{bmatrix} -4 & 2 & 0 \\ -6 & -10 & 6 \end{bmatrix}.$$

For any two matrices A and B, $A - B$ is the same as $A + (-B)$. (Why?)

Also, a **zero matrix** is a matrix all of whose entries are zero. Thus, for example, the 2×3 zero matrix is

$$O = \begin{bmatrix} 0 & 0 & 0 \\ 0 & 0 & 0 \end{bmatrix}.$$

Now we state the most important properties of the operations that we have been talking about:

Properties of Matrix Addition and Scalar Multiplication

If A, B, and C are any $m \times n$ matrices and if O is the zero $m \times n$ matrix, then the following hold:

$A + (B + C) = (A + B) + C$	*Associative law*
$A + B = B + A$	*Commutative law*
$A + O = O + A = A$	*Additive identity law*
$A + (-A) = O = (-A) + A$	*Additive inverse law*
$c(A + B) = cA + cB$	*Distributive law*
$(c + d)A = cA + dA$	*Distributive law*
$1A = A$	*Scalar unit*
$0A = O$	*Scalar zero*

These properties would be obvious if we were talking about addition and multiplication of *numbers*, but here we are talking about addition and multiplication of *matrices*. We are using "+" to mean something new: matrix addition. There is no reason why matrix addition has to obey *all* the properties of addition of numbers. It happens that it does obey many of them, which is why it is convenient to call it *addition* in the first place. This means that we can manipulate equations involving matrices in much the same way that we manipulate equations involving numbers. One word of caution: We haven't yet discussed how to multiply matrices, and it probably isn't what you think. It will turn out that multiplication of matrices does *not* obey all the same properties as multiplication of numbers.

Transposition

We mention one more operation on matrices:

Transposition

If A is an $m \times n$ matrix, then its **transpose** is the $n \times m$ matrix obtained by writing its rows as columns, so that the ith row of the original matrix becomes the ith column of the transpose. We denote the transpose of the matrix A by A^T.

Visualizing Transposition

$$\begin{bmatrix} 2 & -3 \\ 1 & 0 \\ 5 & 1 \end{bmatrix} \qquad \begin{bmatrix} 2 & 1 & 5 \\ -3 & 0 & 1 \end{bmatrix}$$

Quick Examples

1. Let $A = \begin{bmatrix} 2 & 0 & 1 & 0 \\ 33 & -22 & 0 & 5 \\ 1 & -1 & 2 & -2 \end{bmatrix}$. Then $A^T = \begin{bmatrix} 2 & 33 & 1 \\ 0 & -22 & -1 \\ 1 & 0 & 2 \\ 0 & 5 & -2 \end{bmatrix}$.

3 × 4 matrix   4 × 3 matrix

2. $[-1 \quad 1 \quad 2]^T = \begin{bmatrix} -1 \\ 1 \\ 2 \end{bmatrix}$.

1 × 3 matrix   3 × 1 matrix

using Technology

Technology can be used to compute transposes like Quick Example 1 on the right:

TI-83/84 Plus

Enter the matrix `[A]` using `MATRX`; EDIT

Home Screen: `[A]`^{T}

(The symbol ^{T} is found in `MATRX`; MATH.)

Spreadsheet

Enter the matrix A in a convenient 3 × 4 block of cells (e.g., A1–D3).

Highlight a 4 × 3 block for the answer (e.g., A5–C8).

Type `=TRANSPOSE(A1:D3)`

Press Control+Shift+Enter

Website

www.WanerMath.com

Student Home

→ On Line Utilities

→ Matrix Algebra Tool

Enter the matrix A as shown.

```
Enter your matrices here.
A = [2, 0, 1, 0
33, -22, 0, 5
1, -1, 2, -2]
```

Type `A^T` in the formula box and press "Compute".

Properties of Transposition

If A and B are $m \times n$ matrices, then the following hold:

$$(A + B)^T = A^T + B^T$$
$$(cA)^T = c(A^T)$$
$$(A^T)^T = A.$$

To see why the laws of transposition are true, let us consider the first one: $(A + B)^T = A^T + B^T$. The left-hand side is the transpose of $A + B$, and so is obtained by first adding A and B, and then writing the rows as columns. This is the same as first writing the rows of A and B individually as columns before adding, which gives the right-hand side. Similar arguments can be used to establish the other laws of transposition.

4.1 EXERCISES

Access end-of-section exercises online at **www.webassign.net**

4.2 Matrix Multiplication

Suppose we download 3 movies at \$10 each and 5 Chopin albums at \$8 each. We calculate our total cost by computing the products' price × quantity and adding:

$$\text{Cost} = 10 \times 3 + 8 \times 5 = \$70.$$

Let us instead put the prices in a row vector

$$P = [10 \quad 8] \qquad \text{The price matrix}$$

and the quantities purchased in a column vector,

$$Q = \begin{bmatrix} 3 \\ 5 \end{bmatrix}. \qquad \text{The quantity matrix}$$

Q: *Why a row and a column instead of, say, two rows?*

A: It's rather a long story, but mathematicians found that it works best this way . . .

Because P represents the prices of the items we are purchasing and Q represents the quantities, it would be useful if the product PQ represented the total cost, a *single number* (which we can think of as a 1×1 matrix). For this to work, PQ should be calculated the same way we calculated the total cost:

$$PQ = [10 \quad 8] \begin{bmatrix} 3 \\ 5 \end{bmatrix} = [10 \times 3 + 8 \times 5] = [70].$$

Notice that we obtain the answer by multiplying each entry in P (going from left to right) by the corresponding entry in Q (going from top to bottom) and then adding the results.

The Product *Row* × *Column*

The **product** AB of a row matrix A and a column matrix B is a 1×1 matrix. The length of the row in A must match the length of the column in B for the product to be defined. To find the product, multiply each entry in A (going from left to right) by the corresponding entry in B (going from top to bottom) and then add the results.

Visualizing Matrix Multiplication

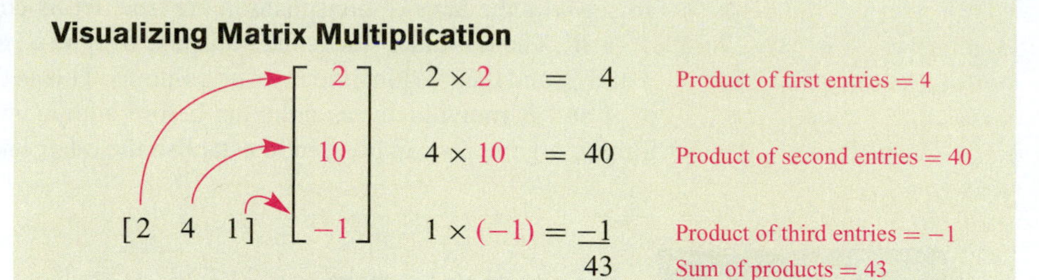

Quick Examples

1. $[2 \quad 1]\begin{bmatrix} -3 \\ 1 \end{bmatrix} = [2 \times (-3) + 1 \times 1] = [-6 + 1] = [-5]$

2. $[2 \quad 4 \quad 1]\begin{bmatrix} 2 \\ 10 \\ -1 \end{bmatrix} = [2 \times 2 + 4 \times 10 + 1 \times (-1)] = [4 + 40 + (-1)]$

$= [43]$

Notes

1. In the discussion so far, *the row is on the left and the column is on the right* (RC again). (Later we will consider products where the column matrix is on the left and the row matrix is on the right.)
2. The row size has to match the column size. This means that, if we have a 1×3 row on the left, then the column on the right must be 3×1 in order for the product to make sense. For example, the product

   $$[a \quad b \quad c]\begin{bmatrix} x \\ y \end{bmatrix}$$

   is not defined. ■

EXAMPLE 1 Revenue

The A-Plus auto parts store mentioned in examples in the previous section had the following sales in its Vancouver store:

	Vancouver
Wiper Blades	20
Cleaning Fluid (bottles)	10
Floor Mats	8

The store sells wiper blades for \$7.00 each, cleaning fluid for \$3.00 per bottle, and floor mats for \$12.00 each. Use matrix multiplication to find the total revenue generated by sales of these items.

Solution We need to multiply each sales figure by the corresponding price and then add the resulting revenue figures. We represent the sales by a column vector, as suggested by the table:

$$Q = \begin{bmatrix} 20 \\ 10 \\ 8 \end{bmatrix}.$$

We put the selling prices in a row vector:

$$P = [7.00 \quad 3.00 \quad 12.00].$$

We can now compute the total revenue as the product

$$R = PQ = [7.00 \quad 3.00 \quad 12.00] \begin{bmatrix} 20 \\ 10 \\ 8 \end{bmatrix}$$

$$= [140.00 + 30.00 + 96.00] = [266.00].$$

So, the sale of these items generated a total revenue of $266.00.

Note We could also have written the quantity sold as a row vector (which would be Q^T) and the prices as a column vector (which would be P^T) and then multiplied them in the opposite order ($Q^T P^T$). Try this. ■

EXAMPLE 2 Relationship with Linear Equations

a. Represent the matrix equation

$$[2 \quad -4 \quad 1] \begin{bmatrix} x \\ y \\ z \end{bmatrix} = [5]$$

as an ordinary equation.

b. Represent the linear equation $3x + y - z + 2w = 8$ as a matrix equation.

Solution

a. If we perform the multiplication on the left, we get the 1×1 matrix $[2x - 4y + z]$. Thus the equation may be rewritten as

$$[2x - 4y + z] = [5]. \qquad \text{1} \times \text{1 matrix on the left} = \text{1} \times \text{1 matrix on the right}$$

Saying that these two 1×1 matrices are equal means that their entries are equal, so we get the equation

$$2x - 4y + z = 5.$$

b. This is the reverse of part (a):

$$[3 \quad 1 \quad -1 \quad 2] \begin{bmatrix} x \\ y \\ z \\ w \end{bmatrix} = [8].$$

➡ **Before we go on...** The row matrix $[3 \quad 1 \quad -1 \quad 2]$ in Example 2 is the row of **coefficients** of the original equation. (See Section 3.1.) ■

Now to the general case of matrix multiplication:

The Product of Two Matrices: General Case

In general, for matrices A and B, we can take the product AB only if the number of columns of A equals the number of rows of B (so that we can multiply the rows of A by the columns of B as above). The product AB is then obtained by taking its ijth entry to be

ijth entry of AB = Row i of A × Column j of B. As defined above

Quick Examples

(R stands for row; C stands for column.)

1. $$R_1 \rightarrow [2 \quad 0 \quad -1 \quad 3] \begin{bmatrix} 1 & 1 & -8 \\ 1 & -6 & 0 \\ 0 & 5 & 2 \\ -3 & 8 & 1 \end{bmatrix} = [R_1 \times C_1 \quad R_1 \times C_2 \quad R_1 \times C_3] = [-7 \quad 21 \quad -15]$$
(with C_1, C_2, C_3 marking the columns)

2. $$\begin{matrix} R_1 \rightarrow \\ R_2 \rightarrow \end{matrix} \begin{bmatrix} 1 & -1 \\ 0 & 2 \end{bmatrix} \begin{bmatrix} 3 & 0 \\ 5 & -1 \end{bmatrix} = \begin{bmatrix} R_1 \times C_1 & R_1 \times C_2 \\ R_2 \times C_1 & R_2 \times C_2 \end{bmatrix} = \begin{bmatrix} -2 & 1 \\ 10 & -2 \end{bmatrix}$$
(with C_1, C_2 marking the columns)

In matrix multiplication we always take

Rows on the left × Columns on the right.

Look at the dimensions in the two Quick Examples above.

Match ↓ ↓ $(1 \times 4)(4 \times 3) \rightarrow 1 \times 3$ Match ↓ ↓ $(2 \times 2)(2 \times 2) \rightarrow 2 \times 2$

The fact that the number of columns in the left-hand matrix equals the number of rows in the right-hand matrix amounts to saying that the middle two numbers must match as above. If we "cancel" the middle matching numbers, we are left with the dimensions of the product.

Before continuing with examples, we state the rule for matrix multiplication formally.

Multiplication of Matrices: Formal Definition

If A is an $m \times n$ matrix and B is an $n \times k$ matrix, then the product AB is the $m \times k$ matrix whose ijth entry is the product

Row i of A × Column j of B

$$(AB)_{ij} = [a_{i1} \quad a_{i2} \quad a_{i3} \ldots a_{in}] \begin{bmatrix} b_{1j} \\ b_{2j} \\ b_{3j} \\ \vdots \\ b_{nj} \end{bmatrix} = a_{i1}b_{1j} + a_{i2}b_{2j} + a_{i3}b_{3j} + \cdots + a_{in}b_{nj}.$$

using Technology

Technology can be used to multiply the matrices in Example 3. Here is an outline for part (a) (see the Technology Guides at the end of the chapter for additional details on using a TI-83/84 Plus or a spreadsheet):

TI-83/84 Plus

Enter the matrices A and B using MATRX; EDIT

Home Screen: `[A]*[B]`

[More details on page 241.]

Spreadsheet

Enter the matrices in convenient blocks of cells (e.g., A1–D2 and F1–H4).

Highlight a 2×3 block for the answer.

Type `=MMULT(A1:D2,F1:H4)`

Press Control+Shift+Enter

[More details on page 245.]

Website

www.WanerMath.com

Student Home

→ On Line Utilities

→ Matrix Algebra Tool

Enter matrices as shown:

```
Enter your matrices here.
A = [2, 0, -1, 3
1, -1, 2, -2]

B = [1, 1, -8
1, 0, 0
0, 5, 2
-2, 8, -1]
```

Type `A*B` in the formula box and press "Compute". If you try to multiply two matrices whose product is not defined, you will get an error alert box telling you that.

EXAMPLE 3 Matrix Product

Calculate:

a. $\begin{bmatrix} 2 & 0 & -1 & 3 \\ 1 & -1 & 2 & -2 \end{bmatrix} \begin{bmatrix} 1 & 1 & -8 \\ 1 & 0 & 0 \\ 0 & 5 & 2 \\ -2 & 8 & -1 \end{bmatrix}$ **b.** $\begin{bmatrix} -3 \\ 1 \end{bmatrix} [2 \quad 1]$

Solution

a. Before we start the calculation, we check that the dimensions of the matrices match up.

$$\overset{2 \times 4}{\begin{bmatrix} 2 & 0 & -1 & 3 \\ 1 & -1 & 2 & -2 \end{bmatrix}} \overset{4 \times 3}{\begin{bmatrix} 1 & 1 & -8 \\ 1 & 0 & 0 \\ 0 & 5 & 2 \\ -2 & 8 & -1 \end{bmatrix}}$$

(Match: the 4s)

The product of the two matrices is defined, and the product will be a 2×3 matrix (we remove the matching 4s: $(2 \times 4)(4 \times 3) \rightarrow 2 \times 3$). To calculate the product, we follow the previous prescription:

$$\begin{matrix} R_1 \rightarrow \\ R_2 \rightarrow \end{matrix} \begin{bmatrix} 2 & 0 & -1 & 3 \\ 1 & -1 & 2 & -2 \end{bmatrix} \overset{C_1 \; C_2 \; C_3}{\begin{bmatrix} 1 & 1 & -8 \\ 1 & 0 & 0 \\ 0 & 5 & 2 \\ -2 & 8 & -1 \end{bmatrix}} = \begin{bmatrix} R_1 \times C_1 & R_1 \times C_2 & R_1 \times C_3 \\ R_2 \times C_1 & R_2 \times C_2 & R_2 \times C_3 \end{bmatrix}$$

$$= \begin{bmatrix} -4 & 21 & -21 \\ 4 & -5 & -2 \end{bmatrix}.$$

b. The dimensions of the two matrices given are 2×1 and 1×2. Because the 1s match, the product is defined, and the result will be a 2×2 matrix.

$$\begin{matrix} R_1 \rightarrow \\ R_2 \rightarrow \end{matrix} \begin{bmatrix} -3 \\ 1 \end{bmatrix} \overset{C_1 \; C_2}{[2 \quad 1]} = \begin{bmatrix} R_1 \times C_1 & R_1 \times C_2 \\ R_2 \times C_1 & R_2 \times C_2 \end{bmatrix} = \begin{bmatrix} -6 & -3 \\ 2 & 1 \end{bmatrix}.$$

Note In part (a) we *cannot* multiply the matrices in the opposite order—the dimensions do not match. We say simply that the product in the opposite order is **not defined**. In part (b) we *can* multiply the matrices in the opposite order, but we would get a 1×1 matrix if we did so. Thus, order is important when multiplying matrices. In general, if AB is defined, then BA need not even be defined. If BA is also defined, it may not have the same dimensions as AB. And even if AB and BA have the same dimensions, they may have different entries. (See the next example.) ■

EXAMPLE 4 *AB* versus *BA*

Let $A = \begin{bmatrix} 1 & -1 \\ 0 & 2 \end{bmatrix}$ and $B = \begin{bmatrix} 3 & 0 \\ 5 & -1 \end{bmatrix}$. Find AB and BA.

Solution Note first that A and B are both 2×2 matrices, so the products AB and BA are both defined and are both 2×2 matrices—unlike the case in Example 3(b). We first calculate AB:

$$AB = \begin{bmatrix} 1 & -1 \\ 0 & 2 \end{bmatrix}\begin{bmatrix} 3 & 0 \\ 5 & -1 \end{bmatrix} = \begin{bmatrix} -2 & 1 \\ 10 & -2 \end{bmatrix}.$$

Now let's calculate BA:

$$BA = \begin{bmatrix} 3 & 0 \\ 5 & -1 \end{bmatrix}\begin{bmatrix} 1 & -1 \\ 0 & 2 \end{bmatrix} = \begin{bmatrix} 3 & -3 \\ 5 & -7 \end{bmatrix}.$$

Notice that BA has no resemblance to AB! Thus, we have discovered that, even for square matrices:

Matrix multiplication is not commutative.

In other words, $AB \neq BA$ in general, even when AB and BA both exist and have the same dimensions. (There are instances when $AB = BA$ for particular matrices A and B, but this is an exception, not the rule.)

EXAMPLE 5 Revenue

January sales at the A-Plus auto parts stores in Vancouver and Quebec are given in the following table.

	Vancouver	Quebec
Wiper Blades	20	15
Cleaning Fluid (bottles)	10	12
Floor Mats	8	4

The usual selling prices for these items are \$7.00 each for wiper blades, \$3.00 per bottle for cleaning fluid, and \$12.00 each for floor mats. The discount prices for A-Plus Club members are \$6.00 each for wiper blades, \$2.00 per bottle for cleaning fluid, and \$10.00 each for floor mats. Use matrix multiplication to compute the total revenue at each store, assuming first that all items were sold at the usual prices, and then that they were all sold at the discount prices.

Solution We can do all of the requested calculations at once with a single matrix multiplication. Consider the following two labeled matrices.

$$Q = \begin{array}{c} \\ \textbf{Wb} \\ \textbf{Cf} \\ \textbf{Fm} \end{array} \begin{array}{c} \begin{array}{cc} \textbf{V} & \textbf{Q} \end{array} \\ \begin{bmatrix} 20 & 15 \\ 10 & 12 \\ 8 & 4 \end{bmatrix} \end{array}$$

$$P = \begin{array}{c} \\ \textbf{Usual} \\ \textbf{Discount} \end{array} \begin{array}{c} \begin{array}{ccc} \textbf{Wb} & \textbf{Cf} & \textbf{Fm} \end{array} \\ \begin{bmatrix} 7.00 & 3.00 & 12.00 \\ 6.00 & 2.00 & 10.00 \end{bmatrix} \end{array}$$

The first matrix records the quantities sold, while the second records the sales prices under the two assumptions. To compute the revenue at both stores under the two different assumptions, we calculate $R = PQ$.

$$R = PQ = \begin{bmatrix} 7.00 & 3.00 & 12.00 \\ 6.00 & 2.00 & 10.00 \end{bmatrix} \begin{bmatrix} 20 & 15 \\ 10 & 12 \\ 8 & 4 \end{bmatrix}$$

$$= \begin{bmatrix} 266.00 & 189.00 \\ 220.00 & 154.00 \end{bmatrix}$$

We can label this matrix as follows:

$$R = \begin{array}{r} \\ \textbf{Usual} \\ \textbf{Discount} \end{array} \begin{array}{c} \quad\textbf{V} \qquad\quad \textbf{Q} \\ \begin{bmatrix} 266.00 & 189.00 \\ 220.00 & 154.00 \end{bmatrix} \end{array}.$$

In other words, if the items were sold at the usual price, then Vancouver had a revenue of \$266 while Quebec had a revenue of \$189, and so on.

Before we go on... In Example 5 we were able to calculate PQ because the dimensions matched correctly: $(2 \times 3)(3 \times 2) \rightarrow 2 \times 2$. We could also have multiplied them in the opposite order and gotten a 3×3 matrix. Would the product QP be meaningful? In an application like this, not only do the dimensions have to match, but also the *labels* have to match for the result to be meaningful. The labels on the three columns of P are the parts that were sold, and these are also the labels on the three rows of Q. Therefore, we can "cancel labels" at the same time that we cancel the dimensions in the product. However, the labels on the two columns of Q do not match the labels on the two rows of P, and there is no useful interpretation of the product QP in this situation. ■

There are very special square matrices of every size: $1 \times 1, 2 \times 2, 3 \times 3$, and so on, called the **identity** matrices.

Identity Matrix

The $n \times n$ identity matrix I is the matrix with 1s down the **main diagonal** (the diagonal starting at the top left) and 0s everywhere else. In symbols,

$$I_{ii} = 1, \quad \text{and}$$
$$I_{ij} = 0 \quad \text{if } i \neq j.$$

Quick Examples

1. 1×1 identity matrix $\quad I = [1]$

2. 2×2 identity matrix $\quad I = \begin{bmatrix} 1 & 0 \\ 0 & 1 \end{bmatrix}$

3. 3×3 identity matrix $\quad I = \begin{bmatrix} 1 & 0 & 0 \\ 0 & 1 & 0 \\ 0 & 0 & 1 \end{bmatrix}$

4. 4×4 identity matrix $\quad I = \begin{bmatrix} 1 & 0 & 0 & 0 \\ 0 & 1 & 0 & 0 \\ 0 & 0 & 1 & 0 \\ 0 & 0 & 0 & 1 \end{bmatrix}$

Note Identity matrices are always square matrices, meaning that they have the same number of rows as columns. There is no such thing, for example, as the "2×4 identity matrix."

The next example shows why I is interesting. ■

using Technology

Technology can be used to obtain an identity matrix as in Example 6. Here is an outline (see the Technology Guides at the end of the chapter for additional details on using a TI-83/84 Plus or a spreadsheet):

TI-83/84 Plus

3×3 identity matrix: `identity(3)`
Obtained from MATRX; MATH
[More details on page 242.]

Spreadsheet

Click on a cell for the top left corner (e.g., B1).
Type `=IF(ROW(B1)-ROW($B$1)=COLUMN(B1)-COLUMN($B$1),1,0)`
Copy across and down.
[More details on page 245.]

Website

www.WanerMath.com
Student Home
→ On Line Utilities
→ Matrix Algebra Tool

Just use `I` in the formula box to refer to the identity matrix of any dimension. The program will choose the correct dimension in the context of the formula. For example, if A is a 3×3 matrix, then the expression `I-A` uses the 3×3 identity matrix for I.

EXAMPLE 6 Identity Matrix

Evaluate the products AI and IA, where $A = \begin{bmatrix} a & b & c \\ d & e & f \\ g & h & i \end{bmatrix}$ and I is the 3×3 identity matrix.

Solution

First notice that A is arbitrary; it could be any 3×3 matrix.

$$AI = \begin{bmatrix} a & b & c \\ d & e & f \\ g & h & i \end{bmatrix} \begin{bmatrix} 1 & 0 & 0 \\ 0 & 1 & 0 \\ 0 & 0 & 1 \end{bmatrix} = \begin{bmatrix} a & b & c \\ d & e & f \\ g & h & i \end{bmatrix}$$

and

$$IA = \begin{bmatrix} 1 & 0 & 0 \\ 0 & 1 & 0 \\ 0 & 0 & 1 \end{bmatrix} \begin{bmatrix} a & b & c \\ d & e & f \\ g & h & i \end{bmatrix} = \begin{bmatrix} a & b & c \\ d & e & f \\ g & h & i \end{bmatrix}.$$

In both cases, the answer is the matrix A we started with. In symbols,

$$AI = A$$

and

$$IA = A$$

no matter which 3×3 matrix A you start with. Now this should remind you of a familiar fact from arithmetic:

$$a \cdot 1 = a$$

and

$$1 \cdot a = a.$$

That is why we call the matrix I the 3×3 *identity* matrix, because it appears to play the same role for 3×3 matrices that the identity 1 does for numbers.

➡ **Before we go on...** Try a similar calculation using 2×2 matrices: Let $A = \begin{bmatrix} a & b \\ c & d \end{bmatrix}$, let I be the 2×2 identity matrix, and check that $AI = IA = A$. In fact, the equation

$$AI = IA = A$$

works for square matrices of every dimension. It is also interesting to notice that $AI = A$ if I is the 2×2 identity matrix and A is any 3×2 matrix (try one). In fact, if I is any identity matrix, then $AI = A$ whenever the product is defined, and $IA = A$ whenever this product is defined. ■

We can now add to the list of properties we gave for matrix arithmetic at the end of Section 4.1 by writing down properties of matrix multiplication. In stating these properties, we shall assume that all matrix products we write are defined—that is, that the matrices have correctly matching dimensions. The first eight properties are the ones we've already seen; the rest are new.

Properties of Matrix Addition and Multiplication

If A, B, and C are matrices, if O is a zero matrix, and if I is an identity matrix, then the following hold:

$A+(B+C)=(A+B)+C$	*Additive associative law*
$A+B=B+A$	*Additive commutative law*
$A+O=O+A=A$	*Additive identity law*
$A+(-A)=O=(-A)+A$	*Additive inverse law*
$c(A+B)=cA+cB$	*Distributive law*
$(c+d)A=cA+dA$	*Distributive law*
$1A=A$	*Scalar unit*
$0A=O$	*Scalar zero*
$A(BC)=(AB)C$	*Multiplicative associative law*
$c(AB)=(cA)B$	*Multiplicative associative law*
$c(dA)=(cd)A$	*Multiplicative associative law*
$AI=IA=A$	*Multiplicative identity law*
$A(B+C)=AB+AC$	*Distributive law*
$(A+B)C=AC+BC$	*Distributive law*
$OA=AO=O$	*Multiplication by zero matrix*

Note that we have not included a multiplicative commutative law for matrices, because the equation $AB = BA$ does not hold in general. In other words, matrix multiplication is *not* exactly like multiplication of numbers. (You have to be a little careful because it is easy to apply the commutative law without realizing it.)

We should also say a bit more about transposition. Transposition and multiplication have an interesting relationship. We write down the properties of transposition again, adding one new one.

Properties of Transposition

$$(A+B)^T = A^T + B^T$$
$$(cA)^T = c(A^T)$$
$$(AB)^T = B^T A^T$$

Notice the change in order in the last one. The order is crucial.

Quick Examples

1. $\left(\begin{bmatrix} 1 & -1 \\ 0 & 2 \end{bmatrix}\begin{bmatrix} 3 & 0 \\ 5 & -1 \end{bmatrix}\right)^T = \begin{bmatrix} -2 & 1 \\ 10 & -2 \end{bmatrix}^T = \begin{bmatrix} -2 & 10 \\ 1 & -2 \end{bmatrix}$ $\quad (AB)^T$

2. $\begin{bmatrix} 3 & 0 \\ 5 & -1 \end{bmatrix}^T \begin{bmatrix} 1 & -1 \\ 0 & 2 \end{bmatrix}^T = \begin{bmatrix} 3 & 5 \\ 0 & -1 \end{bmatrix}\begin{bmatrix} 1 & 0 \\ -1 & 2 \end{bmatrix} = \begin{bmatrix} -2 & 10 \\ 1 & -2 \end{bmatrix}$ $\quad B^T A^T$

3. $\begin{bmatrix} 1 & -1 \\ 0 & 2 \end{bmatrix}^T \begin{bmatrix} 3 & 0 \\ 5 & -1 \end{bmatrix}^T = \begin{bmatrix} 1 & 0 \\ -1 & 2 \end{bmatrix}\begin{bmatrix} 3 & 5 \\ 0 & -1 \end{bmatrix} = \begin{bmatrix} 3 & 5 \\ -3 & -7 \end{bmatrix}$ $\quad A^T B^T$

These properties give you a glimpse of the field of mathematics known as **abstract algebra**. Algebraists study operations like these that resemble the operations on numbers but differ in some way, such as the lack of commutativity for multiplication seen here.

We end this section with more on the relationship between linear equations and matrix equations, which is one of the important applications of matrix multiplication.

EXAMPLE 7 Matrix Form of a System of Linear Equations

a. If

$$A = \begin{bmatrix} 1 & -2 & 3 \\ 2 & 0 & -1 \\ -3 & 1 & 1 \end{bmatrix}, \quad X = \begin{bmatrix} x \\ y \\ z \end{bmatrix}, \quad \text{and } B = \begin{bmatrix} 3 \\ -1 \\ 0 \end{bmatrix},$$

rewrite the matrix equation $AX = B$ as a system of linear equations.

b. Express the following system of equations as a matrix equation of the form $AX = B$:

$$\begin{aligned} 2x + y &= 3 \\ 4x - y &= -1. \end{aligned}$$

Solution

a. The matrix equation $AX = B$ is

$$\begin{bmatrix} 1 & -2 & 3 \\ 2 & 0 & -1 \\ -3 & 1 & 1 \end{bmatrix} \begin{bmatrix} x \\ y \\ z \end{bmatrix} = \begin{bmatrix} 3 \\ -1 \\ 0 \end{bmatrix}.$$

As in Example 2(a), we first evaluate the left-hand side and then set it equal to the right-hand side.

$$\begin{bmatrix} 1 & -2 & 3 \\ 2 & 0 & -1 \\ -3 & 1 & 1 \end{bmatrix} \begin{bmatrix} x \\ y \\ z \end{bmatrix} = \begin{bmatrix} x - 2y + 3z \\ 2x - z \\ -3x + y + z \end{bmatrix}$$

$$\begin{bmatrix} x - 2y + 3z \\ 2x - z \\ -3x + y + z \end{bmatrix} = \begin{bmatrix} 3 \\ -1 \\ 0 \end{bmatrix}$$

Because these two matrices are equal, their corresponding entries must be equal.

$$\begin{aligned} x - 2y + 3z &= 3 \\ 2x \qquad - z &= -1 \\ -3x + y + z &= 0 \end{aligned}$$

In other words, the matrix equation $AX = B$ is equivalent to this system of linear equations. Notice that the coefficients of the left-hand sides of these equations are the entries of the matrix A. We call A the **coefficient matrix** of the system of equations. The entries of X are the unknowns and the entries of B are the right-hand sides 3, -1, and 0.

b. As we saw in part (a), the coefficient matrix A has entries equal to the coefficients of the left-hand sides of the equations. Thus,

$$A = \begin{bmatrix} 2 & 1 \\ 4 & -1 \end{bmatrix}.$$

X is the column matrix consisting of the unknowns, while B is the column matrix consisting of the right-hand sides of the equations, so

$$X = \begin{bmatrix} x \\ y \end{bmatrix} \quad \text{and} \quad B = \begin{bmatrix} 3 \\ -1 \end{bmatrix}.$$

The system of equations can be rewritten as the matrix equation $AX = B$ with this A, X, and B.

This translation of systems of linear equations into matrix equations is really the first step in the method of solving linear equations discussed in Chapter 3. There we worked with the **augmented matrix** of the system, which is simply A with B adjoined as an extra column.

Q: *When we write a system of equations as $AX = B$, couldn't we solve for the unknown X by dividing both sides by A?*

A: If we interpret division as multiplication by the inverse (for example, $2 \div 3 = 2 \times 3^{-1}$), we shall see in the next section that *certain* systems of the form $AX = B$ can be solved in this way, by multiplying both sides by A^{-1}. We first need to discuss what we mean by A^{-1} and how to calculate it.

4.2 EXERCISES

Access end-of-section exercises online at **www.webassign.net** ENHANCED WebAssign

4.3 Matrix Inversion

Now that we've discussed matrix addition, subtraction, and multiplication, you may well be wondering about matrix *division*. In the realm of real numbers, division can be thought of as a form of multiplication: Dividing 3 by 7 is the same as multiplying 3 by $1/7$, the inverse of 7. In symbols, $3 \div 7 = 3 \times (1/7)$, or 3×7^{-1}. In order to imitate division of real numbers in the realm of matrices, we need to discuss the multiplicative **inverse,** A^{-1}, of a matrix A.

Note Because multiplication of real numbers is commutative, we can write, for example, $\frac{3}{7}$ as either 3×7^{-1} or $7^{-1} \times 3$. In the realm of matrices, multiplication is not commutative, so from now on we shall *never* talk about "division" of matrices (by "$\frac{B}{A}$" should we mean $A^{-1}B$ or BA^{-1}?). ■

Before we try to find the inverse of a matrix, we must first know exactly what we *mean* by the inverse. Recall that the inverse of a number a is the number, often written a^{-1}, with the property that $a^{-1} \cdot a = a \cdot a^{-1} = 1$. For example, the inverse of 76 is the number $76^{-1} = 1/76$, because $(1/76) \cdot 76 = 76 \cdot (1/76) = 1$. This is the number calculated by the x^{-1} button found on most calculators. Not all numbers have an inverse. For example—and this is the only example—the number 0 has no inverse, because you cannot get 1 by multiplying 0 by anything.

The inverse of a matrix is defined similarly. To make life easier, we shall restrict attention to **square** matrices, matrices that have the same number of rows as columns.*

* Nonsquare matrices *cannot* have inverses in the sense that we are talking about here. This is not a trivial fact to prove.

Inverse of a Matrix

The **inverse** of an $n \times n$ matrix A is that $n \times n$ matrix A^{-1} which, when multiplied by A on either side, yields the $n \times n$ identity matrix I. Thus,

$$AA^{-1} = A^{-1}A = I.$$

If A has an inverse, it is said to be **invertible**. Otherwise, it is said to be **singular**.

Quick Examples

1. The inverse of the 1×1 matrix $[3]$ is $[1/3]$, because $[3][1/3] = [1] = [1/3][3]$.
2. The inverse of the $n \times n$ identity matrix I is I itself, because $I \times I = I$. Thus, $I^{-1} = I$.
3. The inverse of the 2×2 matrix $A = \begin{bmatrix} 1 & -1 \\ -1 & -1 \end{bmatrix}$ is $A^{-1} = \begin{bmatrix} \frac{1}{2} & -\frac{1}{2} \\ -\frac{1}{2} & -\frac{1}{2} \end{bmatrix}$,

   because $\begin{bmatrix} 1 & -1 \\ -1 & -1 \end{bmatrix} \begin{bmatrix} \frac{1}{2} & -\frac{1}{2} \\ -\frac{1}{2} & -\frac{1}{2} \end{bmatrix} = \begin{bmatrix} 1 & 0 \\ 0 & 1 \end{bmatrix}$ $\quad AA^{-1} = I$

   and $\begin{bmatrix} \frac{1}{2} & -\frac{1}{2} \\ -\frac{1}{2} & -\frac{1}{2} \end{bmatrix} \begin{bmatrix} 1 & -1 \\ -1 & -1 \end{bmatrix} = \begin{bmatrix} 1 & 0 \\ 0 & 1 \end{bmatrix}$. $\quad A^{-1}A = I$

Notes

1. It is possible to show that if A and B are square matrices with $AB = I$, then it must also be true that $BA = I$. In other words, once we have checked that $AB = I$, we know that B is the inverse of A. The second check, that $BA = I$, is unnecessary.
2. If B is the inverse of A, then we can also say that A is the inverse of B (why?). Thus, we sometimes refer to such a pair of matrices as an **inverse pair** of matrices. ■

EXAMPLE 1 Singular Matrix

Can $A = \begin{bmatrix} 1 & 1 \\ 0 & 0 \end{bmatrix}$ have an inverse?

Solution No. To see why not, notice that both entries in the second row of AB will be 0, no matter what B is. So AB cannot equal I, no matter what B is. Hence, A is singular.

➡ **Before we go on...** If you think about it, you can write down many similar examples of singular matrices. There is only one number with no multiplicative inverse (0), but there are many matrices having no inverses. ■

Finding the Inverse of a Square Matrix

Q: *In the box, it was stated that the inverse of* $\begin{bmatrix} 1 & -1 \\ -1 & -1 \end{bmatrix}$ *is* $\begin{bmatrix} \frac{1}{2} & -\frac{1}{2} \\ -\frac{1}{2} & -\frac{1}{2} \end{bmatrix}$. *How was that obtained?*

A: We can think of the problem of finding A^{-1} as a problem of finding four unknowns, the four unknown entries of A^{-1}:

$$A^{-1} = \begin{bmatrix} x & y \\ z & w \end{bmatrix}.$$

These unknowns must satisfy the equation $AA^{-1} = I$, or

$$\begin{bmatrix} 1 & -1 \\ -1 & -1 \end{bmatrix}\begin{bmatrix} x & y \\ z & w \end{bmatrix} = \begin{bmatrix} 1 & 0 \\ 0 & 1 \end{bmatrix}.$$

If we were to try to find the first column of A^{-1}, consisting of x and z, we would have to solve

$$\begin{bmatrix} 1 & -1 \\ -1 & -1 \end{bmatrix}\begin{bmatrix} x \\ z \end{bmatrix} = \begin{bmatrix} 1 \\ 0 \end{bmatrix}$$

or

$$\begin{aligned} x - z &= 1 \\ -x - z &= 0. \end{aligned}$$

To solve this system by Gauss-Jordan reduction, we would row-reduce the augmented matrix, which is A with the column $\begin{bmatrix} 1 \\ 0 \end{bmatrix}$ adjoined.

$$\left[\begin{array}{cc|c} 1 & -1 & 1 \\ -1 & -1 & 0 \end{array}\right] \to \left[\begin{array}{cc|c} 1 & 0 & x \\ 0 & 1 & z \end{array}\right]$$

To find the second column of A^{-1} we would similarly row-reduce the augmented matrix obtained by tacking on to A the second column of the identity matrix.

$$\left[\begin{array}{cc|c} 1 & -1 & 0 \\ -1 & -1 & 1 \end{array}\right] \to \left[\begin{array}{cc|c} 1 & 0 & y \\ 0 & 1 & w \end{array}\right]$$

The row operations used in doing these two reductions would be exactly the same. We could do both reductions simultaneously by "doubly augmenting" A, putting both columns of the identity matrix to the right of A.

$$\left[\begin{array}{cc|cc} 1 & -1 & 1 & 0 \\ -1 & -1 & 0 & 1 \end{array}\right] \to \left[\begin{array}{cc|cc} 1 & 0 & x & y \\ 0 & 1 & z & w \end{array}\right]$$

We carry out this reduction in the following example.

EXAMPLE 2 Computing Matrix Inverse

Find the inverse of each matrix.

a. $P = \begin{bmatrix} 1 & -1 \\ -1 & -1 \end{bmatrix}$ **b.** $Q = \begin{bmatrix} 1 & 0 & 1 \\ 2 & -2 & -1 \\ 3 & 0 & 0 \end{bmatrix}$

Solution

a. As described above, we put the matrix P on the left and the identity matrix I on the right to get a 2×4 matrix.

$$\underbrace{\left[\begin{array}{cc|cc} 1 & -1 & 1 & 0 \\ -1 & -1 & 0 & 1 \end{array}\right]}_{P \qquad I}$$

We now row-reduce the whole matrix:

$$\begin{bmatrix} 1 & -1 & 1 & 0 \\ -1 & -1 & 0 & 1 \end{bmatrix} \begin{matrix} \\ R_2 + R_1 \end{matrix} \rightarrow \begin{bmatrix} 1 & -1 & 1 & 0 \\ 0 & -2 & 1 & 1 \end{bmatrix} \begin{matrix} 2R_1 - R_2 \\ \\ \end{matrix} \rightarrow$$

$$\begin{bmatrix} 2 & 0 & 1 & -1 \\ 0 & -2 & 1 & 1 \end{bmatrix} \begin{matrix} \frac{1}{2}R_1 \\ -\frac{1}{2}R_2 \end{matrix} \rightarrow \left[\begin{array}{cc|cc} 1 & 0 & \frac{1}{2} & -\frac{1}{2} \\ 0 & 1 & -\frac{1}{2} & -\frac{1}{2} \end{array}\right].$$

I P^{-1}

We have now solved the systems of linear equations that define the entries of P^{-1}. Thus,

$$P^{-1} = \begin{bmatrix} \frac{1}{2} & -\frac{1}{2} \\ -\frac{1}{2} & -\frac{1}{2} \end{bmatrix}.$$

b. The procedure to find the inverse of a 3×3 matrix (or larger) is just the same as for a 2×2 matrix. We place Q on the left and the identity matrix (now 3×3) on the right, and reduce.

Q I

$$\left[\begin{array}{ccc|ccc} 1 & 0 & 1 & 1 & 0 & 0 \\ 2 & -2 & -1 & 0 & 1 & 0 \\ 3 & 0 & 0 & 0 & 0 & 1 \end{array}\right] \begin{matrix} \\ R_2 - 2R_1 \\ R_3 - 3R_1 \end{matrix} \rightarrow \begin{bmatrix} 1 & 0 & 1 & 1 & 0 & 0 \\ 0 & -2 & -3 & -2 & 1 & 0 \\ 0 & 0 & -3 & -3 & 0 & 1 \end{bmatrix} \begin{matrix} 3R_1 + R_3 \\ R_2 - R_3 \\ \\ \end{matrix} \rightarrow$$

$$\begin{bmatrix} 3 & 0 & 0 & 0 & 0 & 1 \\ 0 & -2 & 0 & 1 & 1 & -1 \\ 0 & 0 & -3 & -3 & 0 & 1 \end{bmatrix} \begin{matrix} \frac{1}{3}R_1 \\ -\frac{1}{2}R_2 \\ -\frac{1}{3}R_3 \end{matrix} \rightarrow \left[\begin{array}{ccc|ccc} 1 & 0 & 0 & 0 & 0 & \frac{1}{3} \\ 0 & 1 & 0 & -\frac{1}{2} & -\frac{1}{2} & \frac{1}{2} \\ 0 & 0 & 1 & 1 & 0 & -\frac{1}{3} \end{array}\right].$$

I Q^{-1}

Thus,

$$Q^{-1} = \begin{bmatrix} 0 & 0 & \frac{1}{3} \\ -\frac{1}{2} & -\frac{1}{2} & \frac{1}{2} \\ 1 & 0 & -\frac{1}{3} \end{bmatrix}.$$

We have already checked that P^{-1} is the inverse of P. You should also check that Q^{-1} is the inverse of Q.

using Technology

Here is an outline on the use of technology to invert the matrix in Example 2(b) (see the Technology Guides at the end of the chapter for additional details on using a TI-83/84 Plus or a spreadsheet):

TI-83/84 Plus

Enter the matrix [A] using MATRX; EDIT

Home Screen: [A] X^{-1} ENTER

[More details on page 242.]

Spreadsheet

Enter the matrix in a convenient block of cells (e.g., A1–C3).

Highlight a 3×3 block for the answer.

Type =MINVERSE(A1:C3)

Press Control+Shift+Enter.

[More details on page 245.]

Website

www.WanerMath.com

Student Website

→ On Line Utilities

→ Matrix Algebra Tool

Enter Q as shown:

```
Enter your matrices here.
Q = [1, 0, 1
2, -2,-1
3, 0, 0]
```

Type Q^-1 or Q^(-1) in the formula box and press "Compute".

The method we used in Example 2 can be summarized as follows:

Inverting an $n \times n$ Matrix

In order to determine whether an $n \times n$ matrix A is invertible or not, and to find A^{-1} if it does exist, follow this procedure:

1. Write down the $n \times 2n$ matrix $[A \mid I]$ (this is A with the $n \times n$ identity matrix set next to it).
2. Row-reduce $[A \mid I]$.
3. If the reduced form is $[I \mid B]$ (i.e., has the identity matrix in the left part), then A is invertible and $B = A^{-1}$. If you cannot obtain I in the left part, then A is singular. (See Example 3.)

Although there is a general formula for the inverse of a matrix, it is not a simple one. In fact, using the formula for anything larger than a 3×3 matrix is so inefficient that the row-reduction procedure is the method of choice even for computers. However, the general formula is very simple for the special case of 2×2 matrices:

Formula for the Inverse of a 2 × 2 Matrix

The inverse of a 2×2 matrix is

$$\begin{bmatrix} a & b \\ c & d \end{bmatrix}^{-1} = \frac{1}{ad - bc}\begin{bmatrix} d & -b \\ -c & a \end{bmatrix}, \quad \text{provided } ad - bc \neq 0.$$

If the quantity $ad - bc$ is zero, then the matrix is singular (noninvertible). The quantity $ad - bc$ is called the **determinant** of the matrix $\begin{bmatrix} a & b \\ c & d \end{bmatrix}$.

Quick Examples

1. $$\begin{bmatrix} 1 & 2 \\ 3 & 4 \end{bmatrix}^{-1} = \frac{1}{(1)(4) - (2)(3)}\begin{bmatrix} 4 & -2 \\ -3 & 1 \end{bmatrix} = -\frac{1}{2}\begin{bmatrix} 4 & -2 \\ -3 & 1 \end{bmatrix}$$
$$= \begin{bmatrix} -2 & 1 \\ \frac{3}{2} & -\frac{1}{2} \end{bmatrix}$$

2. $\begin{bmatrix} 1 & -1 \\ 2 & -2 \end{bmatrix}$ has determinant $ad - bc = (1)(-2) - (-1)(2) = 0$ and so is singular.

The formula for the inverse of a 2×2 matrix can be obtained using the technique of row reduction. (See the Communication and Reasoning Exercises associated with this section.)

As we have mentioned earlier, not every square matrix has an inverse, as we see in the next example.

EXAMPLE 3 Singular 3 × 3 Matrix

Find the inverse of the matrix $S = \begin{bmatrix} 1 & 1 & 2 \\ -2 & 0 & 4 \\ 3 & 1 & -2 \end{bmatrix}$, if it exists.

Solution We proceed as before.

$$\overset{S \qquad\qquad I}{\left[\begin{array}{rrr|rrr} 1 & 1 & 2 & 1 & 0 & 0 \\ -2 & 0 & 4 & 0 & 1 & 0 \\ 3 & 1 & -2 & 0 & 0 & 1 \end{array}\right]} \begin{array}{l} \\ R_2 + 2R_1 \\ R_3 - 3R_1 \end{array} \rightarrow \begin{bmatrix} 1 & 1 & 2 & 1 & 0 & 0 \\ 0 & 2 & 8 & 2 & 1 & 0 \\ 0 & -2 & -8 & -3 & 0 & 1 \end{bmatrix} \begin{array}{l} 2R_1 - R_2 \\ \\ R_3 + R_2 \end{array}$$

$$\rightarrow \begin{bmatrix} 2 & 0 & -4 & 0 & -1 & 0 \\ 0 & 2 & 8 & 2 & 1 & 0 \\ 0 & 0 & 0 & -1 & 1 & 1 \end{bmatrix}$$

We stopped here, even though the reduction is incomplete, because there is *no hope* of getting the identity on the left-hand side. Completing the row reduction will not change the three zeros in the bottom row. So what is wrong? Nothing. As in Example 1, we have here a singular matrix. Any square matrix that, after row reduction, winds up with a row of zeros is singular. (See Exercise 77.)

Before we go on... In practice, deciding whether a given matrix is invertible or singular is easy: Simply try to find its inverse. If the process works, then the matrix is invertible, and we get its inverse. If the process fails, then the matrix is singular. If you try to invert a singular matrix using a spreadsheet, calculator, or computer program, you should get an error. Sometimes, instead of an error, you will get a spurious answer due to round-off errors in the device. ■

Using the Inverse to Solve a System of *n* Linear Equations in *n* Unknowns

Having used systems of equations and row reduction to find matrix inverses, we will now use matrix inverses to solve systems of equations. Recall that, at the end of the previous section, we saw that a system of linear equations could be written in the form

$$AX = B,$$

where A is the coefficient matrix, X is the column matrix of unknowns, and B is the column matrix of right-hand sides. Now suppose that there are as many unknowns as equations, so that A is a square matrix, and suppose that A is invertible. The object is to solve for the matrix X of unknowns, so we multiply both sides of the equation by the inverse A^{-1} of A, getting

$$A^{-1}AX = A^{-1}B.$$

Notice that we put A^{-1} on the left on both sides of the equation. Order matters when multiplying matrices, so we have to be careful to do the same thing to both sides of the equation. But now $A^{-1}A = I$, so we can rewrite the last equation as

$$IX = A^{-1}B.$$

Also, $IX = X$ (I being the identity matrix), so we really have

$$X = A^{-1}B,$$

and we have solved for X!

Moreover, we have shown that, if A is invertible and $AX = B$, then the only *possible* solution is $X = A^{-1}B$. We should check that $A^{-1}B$ is actually a solution by substituting back into the original equation.

$$AX = A(A^{-1}B) = (AA^{-1})B = IB = B$$

Thus, $X = A^{-1}B$ is a solution and is the only solution. Therefore, if A is invertible, $AX = B$ has exactly one solution.

On the other hand, if $AX = B$ has no solutions or has infinitely many solutions, we can conclude that A is not invertible (why?). To summarize:

Solving the Matrix Equation *AX* = *B*

If A is an invertible matrix, then the matrix equation $AX = B$ has the unique solution

$$X = A^{-1}B.$$

Quick Example

The system of linear equations

$$\begin{aligned} 2x \quad\;\; + z &= 9 \\ 2x + y - z &= 6 \\ 3x + y - z &= 9 \end{aligned}$$

can be written as $AX = B$, where

$$A = \begin{bmatrix} 2 & 0 & 1 \\ 2 & 1 & -1 \\ 3 & 1 & -1 \end{bmatrix}, X = \begin{bmatrix} x \\ y \\ z \end{bmatrix} \quad \text{and} \quad B = \begin{bmatrix} 9 \\ 6 \\ 9 \end{bmatrix}.$$

The matrix A is invertible with inverse

$$A^{-1} = \begin{bmatrix} 0 & -1 & 1 \\ 1 & 5 & -4 \\ 1 & 2 & -2 \end{bmatrix}.$$

You should check this.

Thus,

$$X = A^{-1}B = \begin{bmatrix} 0 & -1 & 1 \\ 1 & 5 & -4 \\ 1 & 2 & -2 \end{bmatrix} \begin{bmatrix} 9 \\ 6 \\ 9 \end{bmatrix} = \begin{bmatrix} 3 \\ 3 \\ 3 \end{bmatrix}$$

so that $(x, y, z) = (3, 3, 3)$ is the (unique) solution to the system.

using Technology

Here is an outline on the use of technology to do the calculations in Example 4 (see the Technology Guides at the end of the chapter for additional details on using a TI-83/84 Plus or a spreadsheet):

TI-83/84 Plus

Enter the four matrices `[A]`, `[B]`, `[C]`, `[D]` using MATRX; EDIT

Home Screen: `[A]`$^{-1}$`*[B]`, `[A]`$^{-1}$`*[C]`, `[A]`$^{-1}$`*[D]`

[More details on page 242.]

Spreadsheet

Enter the matrices A and B in convenient blocks of cells (e.g., A1–C3, E1–E3).

Highlight a 3×1 block for the answer.

Type `=MMULT(MINVERSE(A1:C3),E1:E3)`

Press Control+Shift+Enter.

[More details on page 246.]

Website

www.WanerMath.com

Student Website

→ On Line Utilities

→ Matrix Algebra Tool

Enter `A^-1*B`, `A^-1*C`, and `A^-1*D` to compute these products after entering the matrices A, B, C, and D as shown earlier.

EXAMPLE 4 Solving Systems of Equations Using an Inverse

Solve the following three systems of equations.

a. $\begin{aligned} 2x \quad\;\; + z &= 1 \\ 2x + y - z &= 1 \\ 3x + y - z &= 1 \end{aligned}$ **b.** $\begin{aligned} 2x \quad\;\; + z &= 0 \\ 2x + y - z &= 1 \\ 3x + y - z &= 2 \end{aligned}$ **c.** $\begin{aligned} 2x \quad\;\; + z &= 0 \\ 2x + y - z &= 0 \\ 3x + y - z &= 0 \end{aligned}$

Solution We *could* go ahead and row-reduce all three augmented matrices as we did in Chapter 3, but this would require a lot of work. Notice that the coefficients are the same in all three systems. In other words, we can write the three systems in matrix form as

a. $AX = B$ **b.** $AX = C$ **c.** $AX = D$

where the matrix A is the same in all three cases:

$$A = \begin{bmatrix} 2 & 0 & 1 \\ 2 & 1 & -1 \\ 3 & 1 & -1 \end{bmatrix}.$$

Now the solutions to these systems are

a. $X = A^{-1}B$ **b.** $X = A^{-1}C$ **c.** $X = A^{-1}D$

so the main work is the calculation of the single matrix A^{-1}, which we have already noted (Quick Example on the previous page) is

$$A^{-1} = \begin{bmatrix} 0 & -1 & 1 \\ 1 & 5 & -4 \\ 1 & 2 & -2 \end{bmatrix}.$$

Thus, the three solutions are:

a. $X = A^{-1}B = \begin{bmatrix} 0 & -1 & 1 \\ 1 & 5 & -4 \\ 1 & 2 & -2 \end{bmatrix} \begin{bmatrix} 1 \\ 1 \\ 1 \end{bmatrix} = \begin{bmatrix} 0 \\ 2 \\ 1 \end{bmatrix}$

b. $X = A^{-1}C = \begin{bmatrix} 0 & -1 & 1 \\ 1 & 5 & -4 \\ 1 & 2 & -2 \end{bmatrix} \begin{bmatrix} 0 \\ 1 \\ 2 \end{bmatrix} = \begin{bmatrix} 1 \\ -3 \\ -2 \end{bmatrix}$

c. $X = A^{-1}D = \begin{bmatrix} 0 & -1 & 1 \\ 1 & 5 & -4 \\ 1 & 2 & -2 \end{bmatrix} \begin{bmatrix} 0 \\ 0 \\ 0 \end{bmatrix} = \begin{bmatrix} 0 \\ 0 \\ 0 \end{bmatrix}$

➡ **Before we go on...** We have been speaking of *the* inverse of a matrix A. Is there only one? It is not hard to prove that a matrix A cannot have more than one inverse: If B and C were both inverses of A, then

$B = BI$	Property of the identity
$= B(AC)$	Because C is an inverse of A
$= (BA)C$	Associative law
$= IC$	Because B is an inverse of A
$= C.$	Property of the identity

In other words, if B and C were both inverses of A, then B and C would have to be equal. ■

FAQs

Which Method to Use in Solving a System

Q: *Now we have two methods to solve a system of linear equations $AX = B$: (1) Compute $X = A^{-1}B$, or (2) row-reduce the augmented matrix. Which is the better method?*

A: Each method has its advantages and disadvantages. Method (1), as we have seen, is very efficient when you must solve several systems of equations with the same coefficients, but it works only when the coefficient matrix is *square* (meaning that you have the same number of equations as unknowns) *and invertible* (meaning that there is a unique solution). The row-reduction method will work for all systems. Moreover, for all but the smallest systems, the most efficient way to find A^{-1} is to use row reduction. Thus, in practice, the two methods are essentially the same when both apply.

4.3 EXERCISES

Access end-of-section exercises online at **www.webassign.net** ENHANCED WebAssign

4.4 Game Theory

It frequently happens that you are faced with making a decision or choosing a best strategy from several possible choices. For instance, you might need to decide whether to invest in stocks or bonds, whether to cut prices of the product you sell, or what offensive play to use in a football game. In these examples, the result depends on something you cannot control. In the first case, your success depends on the future behavior of the economy. In the second case, it depends in part on whether your competitors also cut prices, and in the third case, it depends on the defensive strategy chosen by the opposing team.

We can model situations like these using **game theory**. We represent the various options and payoffs in a matrix and can then calculate the best single strategy or combination of strategies using matrix algebra and other techniques.

Game theory is very new compared with most of the mathematics you learn. It was invented in the 1920s by the noted mathematicians Emile Borel (1871–1956) and John von Neumann (1903–1957). Game theory's connection with linear programming was discovered even more recently, in 1947, by von Neumann, and further advances were made by the mathematician John Nash* (1928–), for which he received the 1994 Nobel Prize for Economics.

* Nash's turbulent life is the subject of the biography *A Beautiful Mind* by Sylvia Nasar (Simon & Schuster, 1998). The 2001 Academy Award-winning movie of the same title is a somewhat fictionalized account.

The Payoff Matrix and Expected Payoff

We have probably all played the simple game "Rock, Paper, Scissors" at some time in our lives. It goes as follows: There are two players—let us call them A and B—and at each turn, both players produce, by a gesture of the hand, either paper, a pair of scissors, or a rock. Rock beats scissors (since a rock can crush scissors), but is beaten by paper (since a rock can be covered by paper), while scissors beat paper (since scissors can cut paper). The round is a draw if both A and B show the same item. We could turn this into a betting game if, at each turn, we require the loser to pay the winner 1¢. For instance, if A shows a rock and B shows paper, then A pays B 1¢.

Rock, Paper, Scissors is an example of a **two-person zero-sum game**. It is called a zero-sum game because each player's loss is equal to the other player's gain.† We can represent this game by a matrix, called the **payoff matrix**.

† An example of a *nonzero-sum game* would be one in which the government taxed the earnings of the winner. In that case the winner's gain would be less than the loser's loss.

$$\begin{array}{cc} & \mathbf{B} \\ & \begin{array}{ccc} r & p & s \end{array} \\ \mathbf{A} \begin{array}{c} r \\ p \\ s \end{array} & \begin{bmatrix} 0 & -1 & 1 \\ 1 & 0 & -1 \\ -1 & 1 & 0 \end{bmatrix} \end{array} \quad \text{or just} \quad P = \begin{bmatrix} 0 & -1 & 1 \\ 1 & 0 & -1 \\ -1 & 1 & 0 \end{bmatrix}$$

if we choose to omit the labels. In the payoff matrix, Player A's options, or **moves**, are listed on the left, while Player B's options are listed on top. We think of A as playing the rows and B as playing the columns. Positive numbers indicate a win for the row player, while negative numbers indicate a loss for the row player. Thus, for example, the p, s entry represents the outcome if A plays p (paper) and B plays s (scissors). In this event, B wins, and the -1 entry there indicates that A loses 1¢. (If that entry were -2 instead, it would have meant that A loses 2¢.)

Two-Person Zero-Sum Game

A **two-person zero-sum game** is one in which one player's loss equals the other's gain. We assume that the outcome is determined by each player's choice from among a fixed, finite set of moves. If Player A has m moves to choose from and Player B has n, we can represent the game using the **payoff matrix**, the $m \times n$ matrix showing the result of each possible pair of choices of moves.

In each round of the game, the way a player chooses a move is called a **strategy**. A player using a **pure strategy** makes the same move each round of the game. For example, if a player in the above game chooses to play scissors at each turn, then that player is using the pure strategy s. A player using a **mixed strategy** chooses each move a certain percentage of the time in a random fashion; for instance, Player A might choose to play p 50% of the time, and each of s and r 25% of the time.

Our ultimate goal is to be able to determine which strategy is best for each player to use. To do that, we need to know how to evaluate strategies. The fundamental calculation we need is that of the **expected payoff** resulting from a pair of strategies. Let's look at a simple example.

EXAMPLE 1 Expected Payoff

Consider the following game:

$$\begin{array}{cc} & \mathbf{B} \\ & \begin{array}{cc} a & b \end{array} \\ \mathbf{A} \begin{array}{c} p \\ q \end{array} & \begin{bmatrix} 3 & -1 \\ -2 & 3 \end{bmatrix}. \end{array}$$

Player A decides to pick moves at random, choosing to play p 75% of the time and q 25% of the time. Player B also picks moves at random, choosing a 20% of the time and b 80% of the time. On average, how much does A expect to win or lose?

Solution Suppose they play the game 100 times. Each time they play there are four possible outcomes:

Case 1: *A plays p, B plays a.*
Because A plays p only 75% of the time and B plays a only 20% of the time, we expect this case to occur $0.75 \times 0.20 = 0.15$, or 15% of the time, or 15 times out of 100. Each time this happens, A gains 3 points, so we get a contribution of $15 \times 3 = 45$ points to A's total winnings.

Case 2: *A plays p, B plays b.*
Because A plays p only 75% of the time and B plays b only 80% of the time, we expect this case to occur $0.75 \times 0.80 = 0.60$, or 60 times out of 100. Each time this happens, A loses 1 point, so we get a contribution of $60 \times -1 = -60$ to A's total winnings.

Case 3: *A plays q, B plays a.*
This case occurs $0.25 \times 0.20 = 0.05$, or 5 out of 100 times, with a loss of 2 points to A each time, giving a contribution of $5 \times -2 = -10$ to A's total winnings.

Case 4: *A plays q, B plays b.*
This case occurs $0.25 \times 0.80 = 0.20$, or 20 out of 100 times, with a gain of 3 points to A each time, giving a contribution of $20 \times 3 = 60$ to A's total winnings.

Summing to get A's total winnings and then dividing by the number of times the game is played gives the average value of

$$(45 - 60 - 10 + 60)/100 = 0.35$$

so that A can expect to win an average of 0.35 points per play of the game. We call 0.35 the **expected payoff** of the game resulting from these particular strategies for A and B.

This calculation was somewhat tedious and it would only get worse if A and B had many moves to choose from. There is a far more convenient way of doing exactly the same calculation, using matrix multiplication: We start by representing the player's strategies as matrices. For reasons to become clear in a moment, we record A's strategy as a row matrix:

$$R = [0.75 \quad 0.25].$$

We record B's strategy as a column matrix:

$$C = \begin{bmatrix} 0.20 \\ 0.80 \end{bmatrix}.$$

(We will sometimes write column vectors using transpose notation, writing, for example, $[0.20 \quad 0.80]^T$ for the column above, to save space.) Now: *The expected payoff is the matrix product RPC, where P is the payoff matrix!*

$$\text{Expected payoff} = RPC = [0.75 \quad 0.25]\begin{bmatrix} 3 & -1 \\ -2 & 3 \end{bmatrix}\begin{bmatrix} 0.20 \\ 0.80 \end{bmatrix}$$

$$= [1.75 \quad 0]\begin{bmatrix} 0.20 \\ 0.80 \end{bmatrix} = [0.35].$$

Why does this work? Write out the arithmetic involved in the matrix product RPC to see what we calculated:

$$[0.75 \times 3 + 0.25 \times (-2)] \times 0.20 + [0.75 \times (-1) + 0.25 \times 3] \times 0.80$$
$$= 0.75 \times 3 \times 0.20 + 0.25 \times (-2) \times 0.20 + 0.75 \times (-1) \times 0.80 + 0.25 \times 3 \times 0.80$$
$$= \text{Case 1} + \text{Case 3} + \text{Case 2} + \text{Case 4.}$$

So, the matrix product does all at once the various cases we considered above.

using Technology

The use of technology becomes indispensable when we need to do several calculations or when the matrices involved are big. See the technology note accompanying Example 3 in Section 4.2 on page 199 for instructions on multiplying matrices using a TI-83/84 Plus, a spreadsheet, and the Matrix Algebra Tool at the Website.

To summarize what we just saw:

The Expected Payoff resulting from Mixed Strategies *R* and *C*

The **expected payoff of a game resulting from given mixed strategies** is the average payoff that occurs if the game is played a large number of times with the row and column players using the given strategies.

To compute the expected payoff resulting from mixed strategies R and C:

1. Write the row player's mixed strategy as a row matrix R.
2. Write the column player's mixed strategy as a column matrix C.
3. Calculate the product RPC, where P is the payoff matrix. This product is a 1×1 matrix whose entry is the expected payoff e.

Quick Example

Consider "Rock, Paper, Scissors."

$$\begin{array}{cc} & \mathbf{B} \\ & \begin{array}{ccc} r & p & s \end{array} \\ \mathbf{A}\ \begin{array}{c} r \\ p \\ s \end{array} & \begin{bmatrix} 0 & -1 & 1 \\ 1 & 0 & -1 \\ -1 & 1 & 0 \end{bmatrix} \end{array}$$

Suppose that the row player plays *rock* half the time and each of the other two strategies a quarter of the time, and the column player always plays *paper*. We write

$$R = \begin{bmatrix} \frac{1}{2} & \frac{1}{4} & \frac{1}{4} \end{bmatrix}, \quad C = \begin{bmatrix} 0 \\ 1 \\ 0 \end{bmatrix}.$$

So,

$$e = RPC = \begin{bmatrix} \frac{1}{2} & \frac{1}{4} & \frac{1}{4} \end{bmatrix} \begin{bmatrix} 0 & -1 & 1 \\ 1 & 0 & -1 \\ -1 & 1 & 0 \end{bmatrix} \begin{bmatrix} 0 \\ 1 \\ 0 \end{bmatrix}$$

$$= \begin{bmatrix} \frac{1}{2} & \frac{1}{4} & \frac{1}{4} \end{bmatrix} \begin{bmatrix} -1 \\ 0 \\ 1 \end{bmatrix} = -\frac{1}{4}.$$

Thus, player A can expect to lose an average of once every four plays.

Solving a Game

Now that we know how to evaluate particular strategies, we want to find the *best* strategy. The next example takes us another step toward that goal.

EXAMPLE 2 Television Ratings Wars

Commercial TV station RTV and cultural station CTV are competing for viewers in the Tuesday prime-time 9–10 PM time slot. RTV is trying to decide whether to show a sitcom, a docudrama, a reality show, or a movie, while CTV is thinking about either a nature documentary, a symphony concert, a ballet, or an opera. A television rating company estimates the payoffs for the various alternatives as follows. (Each point indicates a shift of 1,000 viewers from one channel to the other; thus, for instance, –2 indicates a shift of 2,000 viewers from RTV to CTV.)

		CTV			
		Nature Doc.	**Symphony**	**Ballet**	**Opera**
RTV	**Sitcom**	2	1	–2	2
	Docudrama	–1	1	–1	2
	Reality Show	–2	0	0	1
	Movie	3	1	–1	1

a. If RTV notices that CTV is showing nature documentaries half the time and symphonies the other half, what would RTV's best strategy be, and how many viewers would it gain if it followed this strategy?

b. If, on the other hand, CTV notices that RTV is showing docudramas half the time and reality shows the other half, what would CTV's best strategy be, and how many viewers would it gain or lose if it followed this strategy?

Solution

a. We are given the matrix of the game, P, in the table above, and we are given CTV's strategy $C = [0.50 \;\; 0.50 \;\; 0 \;\; 0]^T$. We are not given RTV's strategy R. To say that RTV is looking for its best strategy is to say that it wants the resulting expected payoff $e = RPC$ to be as high as possible. So, we take $R = [x \;\; y \;\; z \;\; t]$ and look for values for x, y, z, and t that make RPC as high as possible. First, we calculate e in terms of these unknowns:

$$e = RPC = [x \;\; y \;\; z \;\; t]\begin{bmatrix} 2 & 1 & -2 & 2 \\ -1 & 1 & -1 & 2 \\ -2 & 0 & 0 & 1 \\ 3 & 1 & -1 & 1 \end{bmatrix}\begin{bmatrix} 0.50 \\ 0.50 \\ 0 \\ 0 \end{bmatrix}$$

$$= [x \;\; y \;\; z \;\; t]\begin{bmatrix} 1.5 \\ 0 \\ -1 \\ 2 \end{bmatrix} = 1.5x - z + 2t.$$

Now, the unknowns x, y, z, and t must be nonnegative and add up to 1 (why?). Because t has the largest coefficient, 2, we'll get the best result by making it as large as possible, namely, $t = 1$, leaving $x = y = z = 0$. Thus, RTV's best strategy is $R = [0 \;\; 0 \;\; 0 \;\; 1]$. In other words, RTV should use the pure strategy of showing a movie every Tuesday evening. If it does so, the expected payoff will be

$$e = 1.5(0) - 0 + 2(1) = 2$$

so RTV can expect to gain 2,000 viewers.

b. Here, we are given $R = [0 \;\; 0.50 \;\; 0.50 \;\; 0]$ and are not given CTV's strategy C, so this time we take $C = [x \;\; y \;\; z \;\; t]^T$ and calculate the resulting expected payoff e:

$$e = RPC = [0 \;\; 0.50 \;\; 0.50 \;\; 0]\begin{bmatrix} 2 & 1 & -2 & 2 \\ -1 & 1 & -1 & 2 \\ -2 & 0 & 0 & 1 \\ 3 & 1 & -1 & 1 \end{bmatrix}\begin{bmatrix} x \\ y \\ z \\ t \end{bmatrix}$$

$$= [-1.5 \;\; 0.5 \;\; -0.5 \;\; 1.5]\begin{bmatrix} x \\ y \\ z \\ t \end{bmatrix} = -1.5x + 0.5y - 0.5z + 1.5t.$$

Now, CTV wants e to be as *low* as possible (why?). Because x has the largest negative coefficient, CTV would like it to be as large as possible: $x = 1$, so the rest of the unknowns must be zero. Thus, CTV's best strategy is $C = [1 \;\; 0 \;\; 0 \;\; 0]^T$; that is, show a nature documentary every night. If it does so, the expected payoff will be

$$e = -1.5(1) + 0.5(0) - 0.5(0) + 1.5(0) = -1.5.$$

So CTV can expect to gain 1,500 viewers.

This example illustrates the fact that, no matter what mixed strategy one player selects, the other player can choose an appropriate *pure* counterstrategy in order to maximize its gain. How does this affect what decisions you should make as one of the players? If you were on the board of directors of RTV, you might reason as follows: Since for every mixed strategy you try, CTV can find a best counterstrategy (as in part (b)), it is in your company's best interest to select a mixed strategy that *minimizes* the effect of CTV's best counterstrategy. This is called the **minimax criterion**.

Minimax Criterion

A player using the **minimax criterion** chooses a strategy that, among all possible strategies, minimizes the effect of the other player's best counterstrategy. That is, an optimal (best) strategy according to the minimax criterion is one that minimizes the maximum damage the opponent can cause.

This criterion assumes that your opponent is determined to win. More precisely, it assumes the following.

Fundamental Principle of Game Theory

Each player tries to use its best possible strategy, and assumes that the other player is doing the same.

This principle is not always followed by every player. For example, one of the players may be nature and may choose its move at random, with no particular purpose in mind. In such a case, criteria other than the minimax criterion may be more appropriate. For example, there is the "maximax" criterion, which maximizes the maximum possible payoff (also known as the "reckless" strategy), or the criterion that seeks to minimize "regret" (the difference between the payoff you get and the payoff you *would have gotten* if you had known beforehand what was going to happen).*
But, we shall assume here the fundamental principle and try to find optimal strategies under the minimax criterion.

* See *Location in Space: Theoretical Perspectives in Economic Geography,* 3rd Edition, by Peter Dicken and Peter E. Lloyd, HarperCollins Publishers, 1990, p. 276.

Finding the optimal strategy is called **solving the game**. In general, solving a game can be done using linear programming, as we shall see in the next chapter. However, we can solve 2 × 2 games "by hand," as we shall see in the next example. First, we notice that some large games can be reduced to smaller games.

Consider the game in the preceding example, which had the following matrix:

$$P = \begin{bmatrix} 2 & 1 & -2 & 2 \\ -1 & 1 & -1 & 2 \\ -2 & 0 & 0 & 1 \\ 3 & 1 & -1 & 1 \end{bmatrix}.$$

Compare the second and third columns through the eyes of the column player, CTV. Every payoff in the third column is as good as or better, from CTV's point of view, than the corresponding entry in the second column. Thus, no matter what RTV does, CTV will do better showing a ballet (third column) than a symphony (second column). We say that the third column **dominates** the second column. As far as CTV is concerned, we might as well forget about symphonies entirely, so we remove the second column. Similarly, the third column dominates the fourth, so we can remove the fourth column, too. This gives us a smaller game to work with:

$$P = \begin{bmatrix} 2 & -2 \\ -1 & -1 \\ -2 & 0 \\ 3 & -1 \end{bmatrix}.$$

Now compare the first and last rows. Every payoff in the last row is larger than the corresponding payoff in the first row, so the last row is always better to RTV. Again, we say that the last row dominates the first row, and we can discard the first row. Similarly, the last row dominates the second row, so we discard the second row as well. This reduces us to the following game:

$$P = \begin{bmatrix} -2 & 0 \\ 3 & -1 \end{bmatrix}.$$

In this matrix, neither row dominates the other and neither column dominates the other. So, this is as far as we can go with this line of argument. We call this **reduction by dominance**.

Reduction by Dominance

One *row* **dominates** another if every entry in the former is greater than or equal to the corresponding entry in the latter. Put another way, one row dominates another if it is always at least as good for the row player.

One *column* dominates another if every entry in the former is less than or equal to the corresponding entry in the latter. Put another way, one column dominates another if it is always at least as good for the column player.

Procedure for Reducing by Dominance:

1. Check whether there is any row in the (remaining) matrix that is dominated by another row. Remove all dominated rows.
2. Check whether there is any column in the (remaining) matrix that is dominated by another column. Remove all dominated columns.
3. Repeat steps 1 and 2 until there are no dominated rows or columns.

Let us now go back to the "television ratings wars" example and see how we can solve a game using the minimax criterion once we are down to a 2×2 payoff matrix.

EXAMPLE 3 Solving a 2×2 Game

Continuing the preceding example:

a. Find the optimal strategy for RTV.

b. Find the optimal strategy for CTV.

c. Find the expected payoff of the game if RTV and CTV use their optimal strategies.

Solution As in the text, we begin by reducing the game by dominance, which brings us down to the following 2×2 game:

		CTV	
		Nature Doc.	**Ballet**
RTV	**Reality Show**	−2	0
	Movie	3	−1

a. Now let's find RTV's optimal strategy. Because we don't yet know what it is, we write down a general strategy:

$$R = [x \quad y].$$

Because $x + y = 1$, we can replace y by $1 - x$:

$$R = [x \quad 1 - x].$$

We know that CTV's best counterstrategy to R will be a pure strategy (see the discussion after Example 2), so let's compute the expected payoff that results from each of CTV's possible pure strategies:

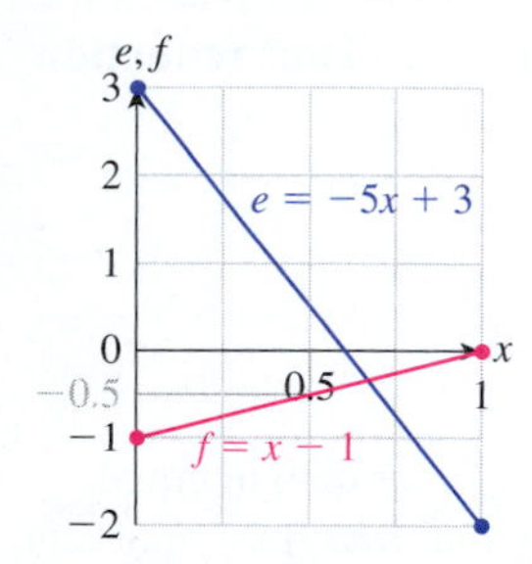

Figure 1

$$e = [x \quad 1 - x]\begin{bmatrix} -2 & 0 \\ 3 & -1 \end{bmatrix}\begin{bmatrix} 1 \\ 0 \end{bmatrix}$$
$$= (-2)x + 3(1 - x) = -5x + 3$$

$$f = [x \quad 1 - x]\begin{bmatrix} -2 & 0 \\ 3 & -1 \end{bmatrix}\begin{bmatrix} 0 \\ 1 \end{bmatrix}$$
$$= 0x - (1 - x) = x - 1.$$

Because both e and f depend on x, we can graph them as in Figure 1.

If, for instance, RTV happened to choose $x = 0.5$, then the expected payoffs resulting from CTV's two pure strategies are $e = -5(1/2) + 3 = 1/2$ and $f = 1/2 - 1 = -1/2$. The worst outcome for RTV is the lower of the two, f, and this will be true wherever the graph of f is below the graph of e. On the other hand, if RTV chose $x = 1$, the graph of e would be lower and the worst possible expected value would be $e = -5(1) + 3 = -2$. Since RTV can choose x to be any value between 0 and 1, the worst possible outcomes are those shown by the colored portion of the graph in Figure 2.

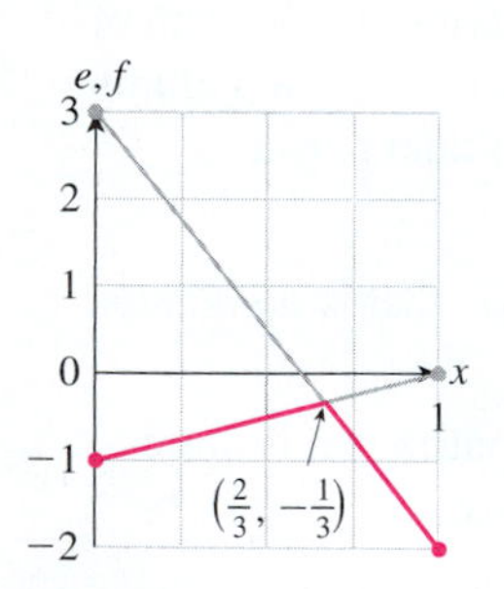

Figure 2

Because RTV is trying to make the worst possible outcome as large as possible (that is, to minimize damages), it is seeking the point on the colored portion of the graph that is highest. This is the intersection point of the two lines. To calculate its coordinates, it's easiest to equate the two functions of x:

$$-5x + 3 = x - 1,$$
$$-6x = -4,$$
or $$x = \frac{2}{3}.$$

The e (or f) coordinate is then obtained by substituting $x = 2/3$ into the expression for e (or f), giving:

$$e = -5\left(\frac{2}{3}\right) + 3$$
$$= -\frac{1}{3}.$$

We conclude that RTV's best strategy is to take $x = 2/3$, giving an expected value of $-1/3$. In other words, RTV's optimal mixed strategy is:

$$R = \begin{bmatrix} \frac{2}{3} & \frac{1}{3} \end{bmatrix}.$$

Going back to the original game, RTV should show reality shows 2/3 of the time and movies 1/3 of the time. It should not bother showing any sitcoms or docudramas. It expects to lose, on average, 333 viewers to CTV, but all of its other options are worse.

b. To find CTV's optimal strategy, we must reverse roles and start by writing its unknown strategy as follows:

$$C = \begin{bmatrix} x \\ 1 - x \end{bmatrix}.$$

We calculate the expected payoffs for the two pure row strategies:

$$e = [1 \quad 0]\begin{bmatrix} -2 & 0 \\ 3 & -1 \end{bmatrix}\begin{bmatrix} x \\ 1 - x \end{bmatrix}$$

$$= -2x$$

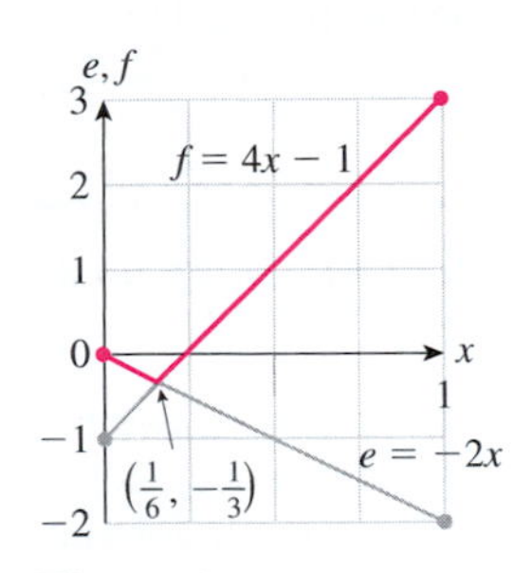

Figure 3

and

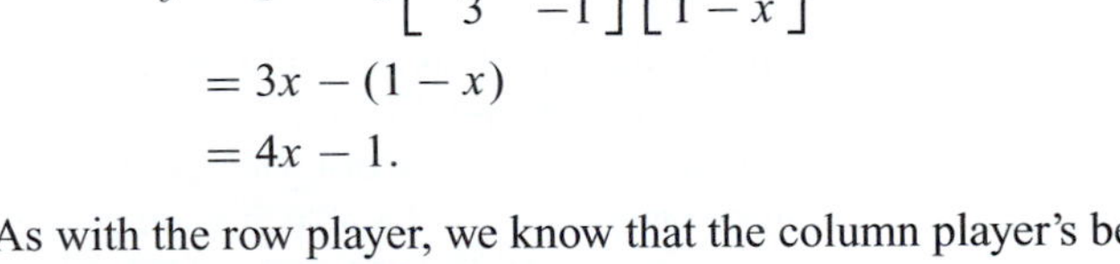

$$f = [0 \quad 1]\begin{bmatrix} -2 & 0 \\ 3 & -1 \end{bmatrix}\begin{bmatrix} x \\ 1 - x \end{bmatrix}$$

$$= 3x - (1 - x)$$

$$= 4x - 1.$$

As with the row player, we know that the column player's best strategy will correspond to the intersection of the graphs of e and f (Figure 3). (Why is the upper edge colored, rather than the lower edge?) The graphs intersect when

$$-2x = 4x - 1$$

or

$$x = \frac{1}{6}.$$

The corresponding value of e (or f) is

$$e = -2\left(\frac{1}{6}\right) = -\frac{1}{3}.$$

Thus, CTV's optimal mixed strategy is $\begin{bmatrix} \frac{1}{6} & \frac{5}{6} \end{bmatrix}^T$ and the expected payoff is $-1/3$. So, CTV should show nature documentaries 1/6 of the time and ballets 5/6 of the time. It should not bother to show symphonies or operas. It expects to gain, on average, 333 viewers from RTV.

c. We can now calculate the expected payoff as usual, using the optimal strategies we found in parts (a) and (b).

$$e = RPC$$

$$= \begin{bmatrix} \frac{2}{3} & \frac{1}{3} \end{bmatrix}\begin{bmatrix} -2 & 0 \\ 3 & -1 \end{bmatrix}\begin{bmatrix} \frac{1}{6} \\ \frac{5}{6} \end{bmatrix}$$

$$= -\frac{1}{3}$$

➡ **Before we go on...** In Example 3, it is no accident that the expected payoff resulting from the optimal strategies equals the expected payoff we found in (a) and (b). If we call the expected payoff resulting from the optimal strategies the **expected value of the game**, the row player's optimal strategy guarantees an expected payoff no smaller than the expected value, while the column player's optimal strategy guarantees an expected payoff no larger. Together they force the average payoff to be the expected value of the game. ■

Expected Value of a Game

The **expected value of a game** is its expected payoff that results when the row and column players use their optimal (minimax) strategies. By using its optimal strategy, the row player guarantees an expected payoff no lower than the expected value of the game, no matter what the column player does. Similarly, by using its optimal strategy, the column player guarantees an expected payoff no higher than the expected value of the game, no matter what the row player does.

Q: *What about games that don't reduce to 2×2 matrices? Are these solved in a similar way?*

A: The method illustrated in the previous example cannot easily be generalized to solve bigger games (i.e., games that cannot be reduced to 2×2 matrices); solving even a 2×3 game using this approach would require us to consider graphs in three dimensions. To be able to solve games of arbitrary size, we need to wait until the next chapter (Section 5.5), where we describe a method for solving a game, using the simplex method, that works for all payoff matrices.

Strictly Determined Games

Although we haven't yet discussed how to solve general $m \times n$ games, there are certain kinds of games that can be solved quite simply regardless of size, as illustrated by the following example.

EXAMPLE 4 Strictly Determined Game

Solve the following game:

$$\begin{array}{cc} & \mathbf{B} \\ & \begin{array}{ccc} p & q & r \end{array} \\ \mathbf{A} \begin{array}{c} s \\ t \\ u \end{array} & \begin{bmatrix} -4 & -3 & 3 \\ 2 & -1 & -2 \\ 1 & 0 & 2 \end{bmatrix} \end{array}.$$

Solution If we look carefully at this matrix, we see that no row dominates another and no column dominates another, so we can't reduce it. Nor do we know how to solve a 3×3 game, so it looks as if we're stuck. However, there is a way to understand this particular game. With the minimax criterion in mind, let's begin by considering the worst possible outcomes for the row player for each possible move. We do this by circling the smallest payoff in each row, the **row minima**:

$$\begin{array}{cc} & \mathbf{B} \\ & \begin{array}{ccc} p & q & r \end{array} \quad \text{Row minima} \\ \mathbf{A} \begin{array}{c} s \\ t \\ u \end{array} & \begin{bmatrix} \circled{-4} & -3 & 3 \\ 2 & -1 & \circled{-2} \\ 1 & \circled{0} & 2 \end{bmatrix} \quad \begin{array}{l} -4 \\ -2 \\ 0 \leftarrow \text{(largest)} \end{array} \end{array}$$

So, for example, if A plays move s, the worst possible outcome is to lose 4. Player A takes the least risk by using move u, which has the largest row minimum.

We do the same thing for the column player, remembering that smaller payoffs are better for B and larger payoffs worse. We draw a box around the largest payoff in each column, the **column maxima**:

$$\begin{array}{cc} & \mathbf{B} \\ & \begin{array}{ccc} p & q & r \end{array} \\ \mathbf{A} \begin{array}{c} s \\ t \\ u \end{array} & \begin{bmatrix} -4 & -3 & \boxed{3} \\ \boxed{2} & -1 & -2 \\ 1 & \boxed{0} & 2 \end{bmatrix}. \\ \text{Column maxima} & \begin{array}{ccc} 2 & 0 & 3 \\ & \uparrow & \\ & \text{(smallest)} & \end{array} \end{array}$$

Player B takes the least risk by using move q, which has the smallest column maximum.

Now put the circles and boxes together:

$$\begin{array}{cc} & \mathbf{B} \\ & \begin{array}{ccc} p & q & r \end{array} \\ \mathbf{A} \begin{array}{c} s \\ t \\ u \end{array} & \begin{bmatrix} \circled{-4} & -3 & \boxed{3} \\ \boxed{2} & -1 & \circled{-2} \\ 1 & \circled{\boxed{0}} & 2 \end{bmatrix}. \end{array}$$

Notice that the uq entry is both circled and boxed: It is both a row minimum and a column maximum. We call such an entry a **saddle point.**

Now we claim that the optimal strategy for A is to always play u while the optimal strategy for B is to always play q. By playing u, A guarantees that the payoff will be 0 or higher, no matter what B does, so the expected value of the game has to be *at least* 0. On the other hand, by playing q, B guarantees that the payoff will be 0 or less, so the expected value of the game has to be *no more than* 0. Combining these facts, we conclude that the expected value of the game must be exactly 0, and A and B have no strategies that could do any better for them than the pure strategies u and q.

➡ **Before we go on...** You should consider what happens in an example like the television rating wars game of Example 3. In that game, the largest row minimum is -1 while the smallest column maximum is 0; there is no saddle point. The row player can force a payoff of at least -1 by playing a pure strategy (always showing movies, for example), but can do better, forcing an expected payoff of $-1/3$, by playing a mixed strategy as we saw in Example 3. Similarly, the column player can force the payoff to be 0 or less with a pure strategy, but can do better, forcing an expected payoff of $-1/3$, with a mixed strategy. Only when there is a saddle point will pure strategies be optimal. ■

Strictly Determined Game

A **saddle point** is a payoff v that is simultaneously a row minimum and a column maximum (both boxed and circled in our approach). If a game has a saddle point, the corresponding row and column strategies are the optimal ones, the expected value of the game is the payoff v, and we say that the game is **strictly determined**.

using Technology

Website
www.WanerMath.com
To play a strictly determined game against your computer, and also to automatically reduce by dominance and solve arbitrary games up to 5 × 5, follow
Student Website
→ On Line Utilities
→ Game Theory Tool

FAQs

Solving a Game

Q: *We've seen several ways of trying to solve a game. What should I do and in what order?*

A: Here are the steps you should take when trying to solve a game:

1. Reduce by dominance. This should always be your first step.
2. If you were able to reduce to a 1 × 1 game, you're done. The optimal strategies are the corresponding pure strategies, as they dominate all the others.
3. Look for a saddle point in the reduced game. If it has one, the game is strictly determined, and the corresponding pure strategies are optimal.
4. If your reduced game is 2 × 2 and has no saddle point, use the method of Example 3 to find the optimal mixed strategies.
5. If your reduced game is larger than 2 × 2 and has no saddle point, you have to use linear programming to solve it, but that will have to wait until the following chapter.

4.4 EXERCISES

Access end-of-section exercises online at **www.webassign.net**

4.5 Input-Output Models

In this section we look at an application of matrix algebra developed by Wassily Leontief (1906–1999) in the middle of the twentieth century. In 1973, he won the Nobel Prize in Economics for this work. The application involves analyzing national and regional economies by looking at how various parts of the economy interrelate. We'll work out some of the details by looking at a simple scenario.

First, we can think of the economy of a country or a region as being composed of various **sectors**, or groups of one or more industries. Typical sectors are the manufacturing sector, the utilities sector, and the agricultural sector. To introduce the basic concepts, we shall consider two specific sectors: the coal-mining sector (Sector 1) and the electric utilities sector (Sector 2). Both produce a commodity: The coal-mining sector produces coal, and the electric utilities sector produces electricity. We measure these products by their dollar value. By **one unit** of a product, we mean $1 worth of that product.

Here is the scenario.

1. To produce one unit ($1 worth) of coal, assume that the coal-mining sector uses 50¢ worth of coal (to power mining machinery, say) and 10¢ worth of electricity.
2. To produce one unit ($1 worth) of electricity, assume that the electric utilities sector uses 25¢ worth of coal and 25¢ worth of electricity.

These are *internal* usage figures. In addition to this, assume that there is an *external* demand (from the rest of the economy) of 7,000 units ($7,000 worth) of coal and 14,000 units ($14,000 worth) of electricity over a specific time period (one year, say). Our basic question is: How much should each of the two sectors supply in order to meet both internal and external demand?

The key to answering this question is to set up equations of the form:

Total supply = Total demand.

The unknowns, the values we are seeking, are

x_1 = the total supply (in units) from Sector 1 (coal) and
x_2 = the total supply (in units) from Sector 2 (electricity).

Our equations then take the following form:

Total supply from Sector 1 = Total demand for Sector 1 products

$$x_1 = \underset{\substack{\uparrow \\ \text{Coal required by Sector 1}}}{0.50x_1} + \underset{\substack{\uparrow \\ \text{Coal required by Sector 2}}}{0.25x_2} + \underset{\substack{\uparrow \\ \text{External demand for coal}}}{7{,}000}$$

Total supply from Sector 2 = Total demand for Sector 2 products

$$x_2 = \underset{\substack{\uparrow \\ \text{Electricity required by Sector 1}}}{0.10x_1} + \underset{\substack{\uparrow \\ \text{Electricity required by Sector 2}}}{0.25x_2} + \underset{\substack{\uparrow \\ \text{External demand for electricity}}}{14{,}000.}$$

This is a system of two linear equations in two unknowns:

$$x_1 = 0.50x_1 + 0.25x_2 + 7{,}000$$
$$x_2 = 0.10x_1 + 0.25x_2 + 14{,}000.$$

We can rewrite this system of equations in matrix form as follows:

$$\underbrace{\begin{bmatrix} x_1 \\ x_2 \end{bmatrix}}_{\text{Production}} = \underbrace{\begin{bmatrix} 0.50 & 0.25 \\ 0.10 & 0.25 \end{bmatrix}\begin{bmatrix} x_1 \\ x_2 \end{bmatrix}}_{\text{Internal demand}} + \underbrace{\begin{bmatrix} 7{,}000 \\ 14{,}000 \end{bmatrix}}_{\text{External demand}}.$$

In symbols,

$$X = AX + D.$$

Here,

$$X = \begin{bmatrix} x_1 \\ x_2 \end{bmatrix}$$

is called the **production vector**. Its entries are the amounts produced by the two sectors. The matrix

$$D = \begin{bmatrix} 7{,}000 \\ 14{,}000 \end{bmatrix}$$

is called the **external demand** vector, and

$$A = \begin{bmatrix} 0.50 & 0.25 \\ 0.10 & 0.25 \end{bmatrix} \quad \text{Organization: } \begin{bmatrix} 1 \to 1 & 1 \to 2 \\ 2 \to 1 & 2 \to 2 \end{bmatrix}$$

is called the **technology matrix**. The entries of the technology matrix have the following meanings:

a_{11} = units of Sector 1 needed to produce one unit of Sector 1
a_{12} = units of Sector 1 needed to produce one unit of Sector 2
a_{21} = units of Sector 2 needed to produce one unit of Sector 1
a_{22} = units of Sector 2 needed to produce one unit of Sector 2.

Now that we have the matrix equation

$$X = AX + D$$

we can solve it as follows. First, subtract AX from both sides:

$$X - AX = D.$$

Because $X = IX$, where I is the 2×2 identity matrix, we can rewrite this as

$$IX - AX = D.$$

Now factor out X:

$$(I - A)X = D.$$

If we multiply both sides by the inverse of $(I - A)$, we get the solution

$$X = (I - A)^{-1}D.$$

Input-Output Model

In an input-output model, an economy (or part of one) is divided into n **sectors**. We then record the $n \times n$ **technology matrix** A, whose ijth entry is the number of units from Sector i used in producing one unit from Sector j (in symbols, "$i \rightarrow j$"). To meet an **external demand** of D, the economy must produce X, where X is the **production vector**. These are related by the equations

$$X = AX + D$$

or

$$X = (I - A)^{-1}D. \qquad \text{Provided } (I - A) \text{ is invertible}$$

Quick Example

In the previous scenario, $A = \begin{bmatrix} 0.50 & 0.25 \\ 0.10 & 0.25 \end{bmatrix}$, $X = \begin{bmatrix} x_1 \\ x_2 \end{bmatrix}$, and $D = \begin{bmatrix} 7{,}000 \\ 14{,}000 \end{bmatrix}$.

The solution is

$$X = (I - A)^{-1}D$$

$$\begin{bmatrix} x_1 \\ x_2 \end{bmatrix} = \left(\begin{bmatrix} 1 & 0 \\ 0 & 1 \end{bmatrix} - \begin{bmatrix} 0.50 & 0.25 \\ 0.10 & 0.25 \end{bmatrix} \right)^{-1} \begin{bmatrix} 7{,}000 \\ 14{,}000 \end{bmatrix} \qquad \text{Calculate } I - A.$$

$$= \begin{bmatrix} 0.50 & -0.25 \\ -0.10 & 0.75 \end{bmatrix}^{-1} \begin{bmatrix} 7{,}000 \\ 14{,}000 \end{bmatrix} \qquad \text{Calculate } (I - A)^{-1}.$$

$$= \begin{bmatrix} \frac{15}{7} & \frac{5}{7} \\ \frac{2}{7} & \frac{10}{7} \end{bmatrix} \begin{bmatrix} 7{,}000 \\ 14{,}000 \end{bmatrix}$$

$$= \begin{bmatrix} 25{,}000 \\ 22{,}000 \end{bmatrix}.$$

In other words, to meet the demand, the economy must produce \$25,000 worth of coal and \$22,000 worth of electricity.

The next example uses actual data from the U.S. economy (we have rounded the figures to make the computations less complicated). It is rare to find input-output data already packaged for you as a technology matrix. Instead, the data commonly found in statistical sources come in the form of "input-output tables," from which we will have to construct the technology matrix.

EXAMPLE 1 Petroleum and Natural Gas

Consider two sectors of the U.S. economy: crude petroleum and natural gas (*crude*) and petroleum refining and related industries (*refining*). According to government figures,[1] in 1998 the crude sector used $27,000 million worth of its own products and $750 million worth of the products of the refining sector to produce $87,000 million worth of goods (crude oil and natural gas). The refining sector in the same year used $59,000 million worth of the products of the crude sector and $15,000 million worth of its own products to produce $140,000 million worth of goods (refined oil and the like). What was the technology matrix for these two sectors? What was left over from each of these sectors for use by other parts of the economy or for export?

Solution First, for convenience, we record the given data in the form of a table, called the **input-output table**. (All figures are in millions of dollars.)

	To	**Crude**	**Refining**
From	**Crude**	27,000	59,000
	Refining	750	15,000
	Total Output	87,000	140,000

The entries in the top portion are arranged in the same way as those of the technology matrix: The *ij*th entry represents the number of units of Sector i that went to Sector j. Thus, for instance, the 59,000 million entry in the 1, 2 position represents the number of units of Sector 1, crude, that were used by Sector 2, refining. ("From the side, to the top.")

We now construct the technology matrix. The technology matrix has entries a_{ij} = units of Sector i used to produce *one* unit of Sector j. Thus,

a_{11} = units of crude to produce one unit of crude. We are told that 27,000 million units of crude were used to produce 87,000 million units of crude. Thus, to produce *one* unit of crude, $27{,}000/87{,}000 \approx 0.31$ units of crude were used, and so $a_{11} \approx 0.31$. (We have rounded this value to two significant digits; further digits are not reliable due to rounding of the original data.)

a_{12} = units of crude to produce one unit of refined:
$a_{12} = 59{,}000/140{,}000 \approx 0.42$

a_{21} = units of refined to produce one unit of crude:
$a_{21} = 750/87{,}000 \approx 0.0086$

a_{22} = units of refined to produce one unit of refined:
$a_{22} = 15{,}000/140{,}000 \approx 0.11$.

[1]The data have been rounded to two significant digits. Source: *Survey of Current Business*, December, 2001, U.S. Department of Commerce. The *Survey of Current Business* and the input-output tables themselves are available at the Web site of the Department of Commerce's Bureau of Economic Analysis (www.bea.gov).

This gives the technology matrix

$$A = \begin{bmatrix} 0.31 & 0.42 \\ 0.0086 & 0.11 \end{bmatrix}. \qquad \text{Technology matrix}$$

In short *we obtained the technology matrix from the input-output table by dividing the Sector 1 column by the Sector 1 total, and the Sector 2 column by the Sector 2 total.*

Now we also know the total output from each sector, so *we have already been given the production vector:*

$$X = \begin{bmatrix} 87{,}000 \\ 140{,}000 \end{bmatrix}. \qquad \text{Production vector}$$

What we are asked for is the external demand vector D, the amount available for the outside economy. To find D, we use the equation

$$X = AX + D, \qquad \text{Relationship of } X, A, \text{ and } D$$

where, this time, we are given A and X, and must solve for D. Solving for D gives

$$\begin{aligned} D &= X - AX \\ &= \begin{bmatrix} 87{,}000 \\ 140{,}000 \end{bmatrix} - \begin{bmatrix} 0.31 & 0.42 \\ 0.0086 & 0.11 \end{bmatrix} \begin{bmatrix} 87{,}000 \\ 140{,}000 \end{bmatrix} \\ &\approx \begin{bmatrix} 87{,}000 \\ 140{,}000 \end{bmatrix} - \begin{bmatrix} 86{,}000 \\ 16{,}000 \end{bmatrix} = \begin{bmatrix} 1{,}000 \\ 124{,}000 \end{bmatrix}. \qquad \text{We rounded to 2 digits.*} \end{aligned}$$

* Why?

The first number, $1,000 million, is the amount produced by the crude sector that is available to be used by other parts of the economy or to be exported. (In fact, because something has to happen to all that crude petroleum and natural gas, this is the amount actually used or exported, where use can include stockpiling.) The second number, $124,000 million, represents the amount produced by the refining sector that is available to be used by other parts of the economy or to be exported.

Note that we could have calculated D more simply from the input-output table. The internal use of units from the crude sector was the sum of the outputs from that sector:

$$27{,}000 + 59{,}000 = 86{,}000.$$

Because 87,000 units were actually produced by the sector, that left a surplus of $87{,}000 - 86{,}000 = 1{,}000$ units for export. We could compute the surplus from the refining sector similarly. (The two calculations actually come out slightly different, because we rounded the intermediate results.) The calculation in Example 2 on page 229 cannot be done as trivially, however.

using Technology

See the Technology Guides at the end of the chapter to see how to compute the technology matrix and the external demand vector in Example 1 using a TI-83/84 Plus or a spreadsheet.

Input-Output Table

National economic data are often given in the form of an **input-output table.** The ijth entry in the top portion of the table is the number of units that go from Sector i to Sector j. The "Total outputs" are the total numbers of units produced by each sector. We obtain the technology matrix from the input-output table by dividing the Sector 1 column by the Sector 1 total, the Sector 2 column by the Sector 2 total, and so on.

* The production of skateboards required skateboards due to the fact that skateboard workers tend to commute to work on (what else?) skateboards!

Quick Example

Input-Output Table:

	To	Skateboards	Wood
From	**Skateboards**	20,000*	0
	Wood	100,000	500,000
	Total Output	200,000	5,000,000

Technology Matrix:

$$A = \begin{bmatrix} \frac{20{,}000}{200{,}000} & \frac{0}{5{,}000{,}000} \\ \frac{100{,}000}{200{,}000} & \frac{500{,}000}{5{,}000{,}000} \end{bmatrix} = \begin{bmatrix} 0.1 & 0 \\ 0.5 & 0.1 \end{bmatrix}.$$

EXAMPLE 2 Rising Demand

Suppose that external demand for refined petroleum rises to \$200,000 million, but the demand for crude remains \$1,000 million (as in Example 1). How do the production levels of the two sectors considered in Example 1 have to change?

Solution We are being told that now

$$D = \begin{bmatrix} 1{,}000 \\ 200{,}000 \end{bmatrix}$$

and we are asked to find X. Remember that we can calculate X from the formula

$$X = (I - A)^{-1}D.$$

Now

$$I - A = \begin{bmatrix} 1 & 0 \\ 0 & 1 \end{bmatrix} - \begin{bmatrix} 0.31 & 0.42 \\ 0.0086 & 0.11 \end{bmatrix} = \begin{bmatrix} 0.69 & -0.42 \\ -0.0086 & 0.89 \end{bmatrix}.$$

We take the inverse using our favorite technique and find that, to four significant digits,†

† Because A is accurate to two digits, we should use more than two significant digits in intermediate calculations so as not to lose additional accuracy. We must, of course, round the final answer to two digits.

$$(I - A)^{-1} \approx \begin{bmatrix} 1.458 & 0.6880 \\ 0.01409 & 1.130 \end{bmatrix}.$$

Now we can compute X:

$$X = (I - A)^{-1}D = \begin{bmatrix} 1.458 & 0.6880 \\ 0.01409 & 1.130 \end{bmatrix} \begin{bmatrix} 1{,}000 \\ 200{,}000 \end{bmatrix} \approx \begin{bmatrix} 140{,}000 \\ 230{,}000 \end{bmatrix}.$$

(As in Example 1, we have rounded all the entries in the answer to two significant digits.) Comparing this vector to the production vector used in Example 1, we see that production in the crude sector has to increase from \$87,000 million to \$140,000 million, while production in the refining sector has to increase from \$140,000 million to \$230,000 million.

Note Using the matrix $(I - A)^{-1}$, we have a slightly different way of solving Example 2. We are asking for the effect on production of a *change* in the final demand of 0 for crude and \$200,000 – 124,000 = \$76,000 million for refined products. If we multiply $(I - A)^{-1}$ by the matrix representing this *change,* we obtain

$$\begin{bmatrix} 1.458 & 0.6880 \\ 0.01409 & 1.130 \end{bmatrix} \begin{bmatrix} 0 \\ 76,000 \end{bmatrix} \approx \begin{bmatrix} 53,000 \\ 90,000 \end{bmatrix}.$$

$(I - A)^{-1} \times$ Change in Demand = Change in Production

We see the changes required in production: an increase of \$53,000 million in the crude sector and an increase of \$90,000 million in the refining sector.

Notice that the increase in external demand for the products of the refining sector requires the crude sector to increase production as well, even though there is no increase in the *external* demand for its products. The reason is that, in order to increase production, the refining sector needs to use more crude oil, so that the *internal* demand for crude oil goes up. The inverse matrix $(I - A)^{-1}$ takes these **indirect effects** into account in a nice way.

By replacing the \$76,000 by \$1 in the computation we just did, we see that a \$1 increase in external demand for refined products will require an increase in production of \$0.6880 in the crude sector, as well as an increase in production of \$1.130 in the refining sector. This is how we interpret the entries in $(I - A)^{-1}$, and this is why it is useful to look at this matrix inverse rather than just solve $(I - A)X = D$ for X using, say, Gauss-Jordan reduction. Looking at $(I - A)^{-1}$, we can also find the effects of an increase of \$1 in external demand for crude: an increase in production of \$1.458 in the crude sector and an increase of \$0.01409 in the refining sector.

Here are some questions to think about: Why are the diagonal entries of $(I - A)^{-1}$ (slightly) larger than 1? Why is the entry in the lower left so small compared to the others? ■

Interpreting $(I - A)^{-1}$: Indirect Effects

If A is the technology matrix, then the ijth entry of $(I - A)^{-1}$ is the change in the number of units Sector i must produce in order to meet a one-unit increase in external demand for Sector j products. To meet a rising external demand, the necessary change in production for each sector is given by

$$\text{Change in production} = (I - A)^{-1}D^+,$$

where D^+ is the change in external demand.

Quick Example

Take Sector 1 to be skateboards, and Sector 2 to be wood, and assume that

$$(I - A)^{-1} = \begin{bmatrix} 1.1 & 0 \\ 0.6 & 1.1 \end{bmatrix}.$$

Then

$a_{11} = 1.1 =$ number of additional units of skateboards that must be produced to meet a one-unit increase in the demand for skateboards (Why is this number larger than 1?)

$a_{12} = 0 =$ number of additional units of skateboards that must be produced to meet a one-unit increase in the demand for wood (Why is this number 0?)

$a_{21} = 0.6 =$ number of additional units of wood that must be produced to meet a one-unit increase in the demand for skateboards

$a_{22} = 1.1 =$ number of additional units of wood that must be produced to meet a one-unit increase in the demand for wood.

To meet an increase in external demand of 100 skateboards and 400 units of wood, the necessary change in production is

$$(I - A)^{-1}D^{+} = \begin{bmatrix} 1.1 & 0 \\ 0.6 & 1.1 \end{bmatrix} \begin{bmatrix} 100 \\ 400 \end{bmatrix} = \begin{bmatrix} 110 \\ 500 \end{bmatrix}$$

so 110 additional skateboards and 500 additional units of wood will need to be produced.

In the preceding examples, we used only two sectors of the economy. The data used in Examples 1 and 2 were taken from an input-output table published by the U.S. Department of Commerce, in which the whole U.S. economy was broken down into 85 sectors. This in turn was a simplified version of a model in which the economy was broken into about 500 sectors. Obviously, computers are required to make a realistic input-output analysis possible. Many governments collect and publish input-output data as part of their national planning. The United Nations collects these data and publishes collections of national statistics. The United Nations also has a useful set of links to government statistics at the following URL:

www.un.org/Depts/unsd/sd_natstat.htm

using Technology

Technology can be used to do the computations in Example 3. Here is an outline for part (a) (see the Technology Guides at the end of the chapter for additional details on using a TI-83/84 Plus or a spreadsheet):

TI-83/84 Plus
Enter the matrices A and D using `MATRX`; EDIT.
(For A type the entries as quotients: Column entry/ Column total.) Home Screen: `(identity(4)-[A])^-1[D]`
[More details on page 243.]

Spreadsheet
Obtain the technology matrix as in Example 1.
Insert the 4×4 identity matrix and then use `MINVERSE` and `MMULT` to compute $(I - A)^{-1}D$.
[More details on page 247.]

Website
www.WanerMath.com
Student Website
→ On Line Utilities
→ Matrix Algebra Tool
Enter the matrices A and D. (For A type the entries as quotients: Column entry/Column total.)
Type `(I-A)^(-1)*D` in the formula box and press "Compute".

EXAMPLE 3 Kenya Economy

Consider four sectors of the economy of Kenya[2]: (1) the traditional economy, (2) agriculture, (3) manufacture of metal products and machinery, and (4) wholesale and retail trade. The input-output table for these four sectors for 1976 looks like this (all numbers are thousands of K£):

	To	1	2	3	4
From	1	8,600	0	0	0
	2	0	20,000	24	0
	3	1,500	530	15,000	660
	4	810	8,500	5,800	2,900
Total Output		87,000	530,000	110,000	180,000

Suppose that external demand for agriculture increased by K£50,000,000 and that external demand for metal products and machinery increased by K£10,000,000. How would production in these four sectors have to change to meet this rising demand?

[2]Figures are rounded. Source: *Input-Output Tables for Kenya 1976,* Central Bureau of Statistics of the Ministry of Economic Planning and Community Affairs, Kenya.

Solution To find the change in production necessary to meet the rising demand, we need to use the formula

$$\text{Change in production} = (I - A)^{-1}D^{+},$$

where A is the technology matrix and D^{+} is the change in demand:

$$D^{+} = \begin{bmatrix} 0 \\ 50{,}000 \\ 10{,}000 \\ 0 \end{bmatrix}.$$

With entries shown rounded to two significant digits, the matrix A is

$$A = \begin{bmatrix} 0.099 & 0 & 0 & 0 \\ 0 & 0.038 & 0.00022 & 0 \\ 0.017 & 0.001 & 0.14 & 0.0037 \\ 0.0093 & 0.016 & 0.053 & 0.016 \end{bmatrix}.$$

Entries shown are rounded to two significant digits.

The next calculation is best done using technology:

$$\text{Change in production} = (I - A)^{-1}D^{+} = \begin{bmatrix} 0 \\ 52{,}000 \\ 12{,}000 \\ 1{,}500 \end{bmatrix}.$$

Entries shown are rounded to two significant digits.

Looking at this result, we see that the changes in external demand will leave the traditional economy unaffected, production in agriculture will rise by K£52 million, production in the manufacture of metal products and machinery will rise by K£12 million, and activity in wholesale and retail trade will rise by K£1.5 million.

Before we go on... Can you see why the traditional economy was unaffected in Example 3? Although it takes inputs from other parts of the economy, it is not itself an input to any other part. In other words, there is no intermediate demand for the products of the traditional economy coming from any other part of the economy, and so an increase in production in any other sector of the economy will require no increase from the traditional economy. On the other hand, the wholesale and retail trade sector does provide input to the agriculture and manufacturing sectors, so increases in those sectors do require an increase in the trade sector.

One more point: If you calculate $(I - A)^{-1}$ you will notice how small the off-diagonal entries are. This says that increases in each sector have relatively small effects on the other sectors. We say that these sectors are **loosely coupled**. Regional economies, where many products are destined to be shipped out to the rest of the country, tend to show this phenomenon even more strongly. Notice in Example 2 that those two sectors are **strongly coupled**, because a rise in demand for refined products requires a comparable rise in the production of crude. ■

4.5 EXERCISES

Access end-of-section exercises online at **www.webassign.net** ENHANCED WebAssign

CHAPTER 4 REVIEW

KEY CONCEPTS

Website www.WanerMath.com
Go to the Website at www.WanerMath.com to find a comprehensive and interactive Web-based summary of Chapter 4.

4.1 Matrix Addition and Scalar Multiplication

$m \times n$ matrix, dimensions, entries *p. 188*
Referring to the entries of a matrix *p. 188*
Matrix equality *p. 189*
Row, column, and square matrices *p. 190*
Addition and subtraction of matrices *p. 190*
Scalar multiplication *p. 192*
Properties of matrix addition and scalar multiplication *p. 193*
The transpose of a matrix *p. 194*
Properties of transposition *p. 194*

4.2 Matrix Multiplication

Multiplying a row by a column *p. 195*
Linear equation as a matrix equation *p. 197*
The product of two matrices: general case *p. 198*
Identity matrix *p. 201*
Properties of matrix addition and multiplication *p. 203*
Properties of transposition and multiplication *p. 203*
A system of linear equations can be written as a single matrix equation *p. 204*

4.3 Matrix Inversion

The inverse of a matrix, singular matrix *p. 206*
Procedure for finding the inverse of a matrix *p. 206*
Formula for the inverse of a 2×2 matrix; determinant of a 2×2 matrix *p. 209*
Using an inverse matrix to solve a system of equations *p. 210*

4.4 Game Theory

Two-person zero-sum game, payoff matrix *p. 214*
A strategy specifies how a player chooses a move *p. 214*
The expected payoff of a game for given mixed strategies R and C *p. 215*
An optimal strategy, according to the minimax criterion, is one that minimizes the maximum damage your opponent can cause you. *p. 218*
The Fundamental Principle of Game Theory *p. 218*
Procedure for reducing by dominance *p. 219*
Procedure for solving a 2×2 game *p. 219*
The expected value of a game is its expected payoff when the players use their optimal strategies *p. 222*
A strictly determined game is one with a saddle point *p. 224*
Steps to follow in solving a game *p. 224*

4.5 Input-Output Models

An input-output model divides an economy into sectors. The technology matrix records the interactions of these sectors and allows us to relate external demand to the production vector. *p. 226*
Procedure for finding a technology matrix from an input-output table *p. 229*
The entries of $(I - A)^{-1}$ *p. 230*

REVIEW EXERCISES

For Exercises 1–10, let

$$A = \begin{bmatrix} 1 & 2 & 3 \\ 4 & 5 & 6 \end{bmatrix}, \quad B = \begin{bmatrix} 1 & -1 \\ 0 & 1 \end{bmatrix},$$

$$C = \begin{bmatrix} -1 & 0 \\ 1 & 1 \\ 0 & 1 \end{bmatrix}, \text{ and } D = \begin{bmatrix} -3 & -2 & -1 \\ 1 & 2 & 3 \end{bmatrix}.$$

Determine whether each expression is defined, and if it is, evaluate it.

1. $A + B$ **2.** $A - D$
3. $2A^T + C$ **4.** AB
5. $A^T B$ **6.** A^2
7. B^2 **8.** B^3
9. $AC + B$ **10.** $CD + B$

In Exercises 11–16, find the inverse of the given matrix or determine that the matrix is singular.

11. $\begin{bmatrix} 1 & -1 \\ 0 & 1 \end{bmatrix}$ **12.** $\begin{bmatrix} 1 & 2 \\ 0 & 0 \end{bmatrix}$

13. $\begin{bmatrix} 1 & 2 & 3 \\ 0 & 4 & 1 \\ 0 & 0 & 1 \end{bmatrix}$ **14.** $\begin{bmatrix} 1 & 2 & 3 & 4 \\ 1 & 3 & 4 & 2 \\ 0 & 1 & 2 & 3 \\ 0 & 0 & 1 & 2 \end{bmatrix}$

15. $\begin{bmatrix} 1 & 2 & 3 & 4 \\ 2 & 3 & 3 & 3 \\ 0 & 1 & 2 & 3 \\ 0 & 0 & 1 & 2 \end{bmatrix}$ **16.** $\begin{bmatrix} 0 & 1 & 0 & 0 \\ 1 & 0 & 0 & 0 \\ 0 & 0 & 0 & 1 \\ 0 & 0 & 1 & 0 \end{bmatrix}$

In Exercises 17–20, write the given system of linear equations as a matrix equation, and solve by inverting the coefficient matrix.

17. $\begin{aligned} x + 2y &= 0 \\ 3x + 4y &= 2 \end{aligned}$

18. $\begin{aligned} x + y + z &= 3 \\ y + 2z &= 4 \\ y - z &= 1 \end{aligned}$

19. $\begin{aligned} x + y + z &= 2 \\ x + 2y + z &= 3 \\ x + y + 2z &= 1 \end{aligned}$

20. $\begin{aligned} x + y \qquad &= 0 \\ y + z \qquad &= 1 \\ z + w &= 0 \\ x \qquad - w &= 3 \end{aligned}$

In each of Exercises 21–24, solve the game with the given payoff matrix and give the expected value of the game.

21. $P = \begin{bmatrix} 2 & 1 & 3 & 2 \\ -1 & 0 & -2 & 1 \\ 2 & 0 & 1 & 3 \end{bmatrix}$ **22.** $P = \begin{bmatrix} 3 & -3 & -2 \\ -1 & 3 & 0 \\ 2 & 2 & 1 \end{bmatrix}$

23. $P = \begin{bmatrix} -1 & -3 & -2 \\ -1 & 3 & 0 \\ 3 & 3 & -1 \end{bmatrix}$ **24.** $P = \begin{bmatrix} 1 & 4 & 3 & 3 \\ 0 & -1 & 2 & 3 \\ 2 & 0 & -1 & 2 \end{bmatrix}$

In each of Exercises 25–28, find the production vector X corresponding to the given technology matrix A and external demand vector D.

25. $A = \begin{bmatrix} 0.3 & 0.1 \\ 0 & 0.3 \end{bmatrix}, D = \begin{bmatrix} 700 \\ 490 \end{bmatrix}$

26. $A = \begin{bmatrix} 0.7 & 0.1 \\ 0.1 & 0.7 \end{bmatrix}, D = \begin{bmatrix} 1{,}000 \\ 2{,}000 \end{bmatrix}$

27. $A = \begin{bmatrix} 0.2 & 0.2 & 0.2 \\ 0 & 0.2 & 0.2 \\ 0 & 0 & 0.2 \end{bmatrix}, D = \begin{bmatrix} 32{,}000 \\ 16{,}000 \\ 8{,}000 \end{bmatrix}$

28. $A = \begin{bmatrix} 0.5 & 0.1 & 0 \\ 0.1 & 0.5 & 0.1 \\ 0 & 0.1 & 0.5 \end{bmatrix}, D = \begin{bmatrix} 23{,}000 \\ 46{,}000 \\ 23{,}000 \end{bmatrix}$

APPLICATIONS: OHaganBooks.com

It is now July 1 and online sales of romance, science fiction, and horror novels at OHaganBooks.com were disappointingly slow over the past month. Exercises 29–34 are based on the following tables:

Inventory of books in stock on June 1 at the OHaganBooks.com warehouses in Texas and Nevada:

Books in Stock (June 1)

	Romance	Sci Fi	Horror
Texas	2,500	4,000	3,000
Nevada	1,500	3,000	1,000

Online sales during June:

June Sales

	Romance	Sci Fi	Horror
Texas	300	500	100
Nevada	100	600	200

New books purchased each month:

Monthly Purchases

	Romance	Sci Fi	Horror
Texas	400	400	300
Nevada	200	400	300

July Sales (Projected)

	Romance	Sci Fi	Horror
Texas	280	550	100
Nevada	50	500	120

29. ***Inventory*** Use matrix algebra to compute the inventory at each warehouse at the end of June.

30. ***Inventory*** Use matrix algebra to compute the change in inventory at each warehouse during June.

31. ***Inventory*** Assuming that sales continue at the level projected for July for the next few months, write down a matrix equation showing the inventory N at each warehouse x months after July 1. How many months from now will OHaganBooks.com run out of Sci Fi novels at the Nevada warehouse?

32. ***Inventory*** Assuming that sales continue at the level projected for July for the next few months, write down a matrix equation showing the change in inventory N at each warehouse x months after July 1. How many months from now will OHaganBooks.com have 1,000 more horror novels in stock in Texas than currently?

33. ***Revenue*** It is now the end of July and OHaganBooks.com's e-commerce manager bursts into the CEO's office. "I thought you might want to know, John, that our sales figures are exactly what I projected a month ago. Is that good market analysis or what?" OHaganBooks.com has charged an average of \$5 for romance novels, \$6 for science fiction novels, and \$5.50 for horror novels. Use the projected July sales figures from above and matrix arithmetic to compute the total revenue OHaganBooks.com earned at each warehouse in July.

34. ***Cost*** OHaganBooks.com pays an average of \$2 for romance novels, \$3.50 for science fiction novels, and \$1.50 for horror novels. Use this information together with the monthly purchasing information to compute the monthly purchasing cost.

Acting on a "tip" from Marjory Duffin, John O'Hagan decided that his company should invest a significant sum in shares of Duffin House Publishers (DHP) and Duffin Subprime Ventures (DSV). Exercises 35–38 are based on the following table, which shows what information John was able to piece together later, after some of the records had been deleted by an angry student intern.

Date	Number of Shares: DHP	Price per Share: DHP	Number of Shares: DSV	Price per Share: DSV
July 1	?	\$20	?	\$10
August 1	?	\$10	?	\$20
September 1	?	\$5	?	\$40
Total	5,000		7,000	

35. ***Investments*** Over the three months shown, the company invested a total of $50,000 in DHP stock, and, on August 15, was paid dividends of 10¢ per share held on that date, for a total of $300. Use matrix inversion to determine how many shares of DHP OHaganBooks.com purchased on each of the three dates shown.

36. ***Investments*** Over the three months shown, the company invested a total of $150,000 in DSV stock, and, on July 15, was paid dividends of 20¢ per share held on that date, for a total of $600. Use matrix inversion to determine how many shares of DSV OHaganBooks.com purchased on each of the three dates shown.

37. ***Investments*** (Refer to Exercise 35.) On October 1, the shares of DHP purchased on July 1 were sold at $3 per share. The remaining shares were sold one month later at $1 per share. Use matrix algebra to determine the total loss (taking into account the dividends paid on August 15) incurred as a result of the Duffin stock debacle.

38. ***Investments*** (Refer to Exercise 36.) On September 15, DSV announced a two-for-one stock split, so that each share originally purchased was converted into two shares. On October 1, the company paid an additional special dividend of 10¢ per share. On October 15, OHaganBooks.com sold 3,000 shares at $20 per share. One week later the subprime market crashed, Duffin Subprime Ventures declared bankruptcy (after awarding its fund manager a $10 million bonus), and DSV stock became worthless. Use matrix algebra to determine the total loss (taking into account the dividends paid on July 15) incurred as a result of the Duffin stock debacle.

OHaganBooks.com has two main competitors: JungleBooks.com and FarmerBooks.com, and no other competitors of any significance on the horizon. Exercises 39–42 are based on the following table, which shows the movement of customers during July.[3] *(Thus, for instance, the first row tells us that 80% of OHaganBooks.com's customers remained loyal, 10% of them went to JungleBooks.com, and the remaining 10% went to FarmerBooks.com.)*

	To OHagan	To Jungle	To Farmer
From OHagan	0.8	0.1	0.1
From Jungle	0.4	0.6	0
From Farmer	0.2	0	0.8

At the beginning of July, OHaganBooks.com had an estimated 2,000 customers, while its two competitors had 4,000 each.

39. ***Competition*** Set up the July 1 customer numbers in a row matrix, and use matrix arithmetic to estimate the number of customers each company has at the end of July.

40. ***Competition*** Assuming the July trends continue in August, predict the number of customers each company will have at the end of August.

[3]By a "customer" of one of the three e-commerce sites, we mean someone who purchases more at that site than at either of the two competitors.

41. ***Competition*** Assuming the July trends continue in August, why is it not possible for a customer of FarmerBooks.com on July 1 to have ended up as a JungleBooks.com customer two months later without having ever been an OHaganBooks.com customer?

42. ***Competition*** Name one or more important factors that the model we have used does not take into account.

Publisher Marjory Duffin reveals that JungleBooks may be launching a promotional scheme in which it will offer either two books for the price of one, or three books for the price of two (Marjory can't quite seem to remember which, and is not certain whether they will go with the scheme at all). John O'Hagan's marketing advisers Flood and O'Lara seem to have different ideas as to how to respond. Flood suggests that the company counter by offering three books for the price of one, while O'Lara suggests that it offer instead a free copy of the Finite Mathematics Student Solutions Manual *with every purchase. After a careful analysis, O'Hagan comes up with the following payoff matrix, where the payoffs represent the number of customers, in thousands, he expects to gain from JungleBooks.*

		JungleBooks		
		No Promo	2 for Price of 1	3 for Price of 2
	No Promo	0	−60	−40
O'Hagan	**3 for Price of 1**	30	20	10
	Finite Math	20	0	15

Use the above information in Exercises 43–48.

43. ***Competition*** After a very expensive dinner at an exclusive restaurant, Marjory suddenly "remembers" that the JungleBooks CEO mentioned to her (at a less expensive restaurant) that there is only a 20% chance JungleBooks will launch a "2 for the price of 1" promotion, and a 40% chance that it will launch a "3 for the price of 2" promotion. What should OHaganBooks.com do in view of this information, and what will the expected effect be on its customer base?

44. ***Competition*** JungleBooks CEO François Dubois has been told by someone with personal ties to OHaganBooks.com staff that OHaganBooks is in fact 80% certain to opt for the "3 for the price of 1" option and will certainly go with one of the two possible promos. What should JungleBooks.com do in view of this information, and what will the expected effect be on its customer base?

45. ***Corporate Spies*** One of John O'Hagan's trusted marketing advisers has, without knowing it, accidentally sent him a copy of the following e-mail:

> To: René, JungleBooks Marketing Department
>
> From: O'Lara
>
> Hey René somehow the CEO here says he has learned that there is a 20% chance that you will opt for the "2 for the price of 1" promotion, and a 40% chance that you will opt for the "3 for the price of 2" promotion. Thought you might want to know. So when do I get my "commission"? —Jim

What will each company do in view of this new information, and what will the expected effect be on its customer base?

46. ***More Corporate Spies*** The next day, everything changes: OHaganBooks.com's mole at JungleBooks, Davíde DuPont, is found unconscious in the coffee room at JungleBooks.com headquarters, clutching in his hand the following correspondence he had apparently received moments earlier:

To: Davíde
From: John O'Hagan
Subject: Re: Urgent Information
Davíde: This information is much appreciated and definitely changes my plans—J
>
>To: John O
>From: Davíde
>Subject: Urgent Information
>Thought you might want to know that JungleBooks
>thinks you are 80% likely to opt for 3 for 1 and 20%
>likely to opt for Finite Math.
>—D

What will each company do in view of this information, and what will the expected effect be on its customer base?

47. ***Things Unravel*** John O'Hagan is about to go with the option chosen in Exercise 45 when he hears word about an exposé in the *Publisher Enquirer* on corporate spying in the two companies. Each company now knows that no information about the other's intentions can be trusted. Now what should OHaganBooks.com do, and how many customers should it expect to gain or lose?

48. ***Competition*** It is now apparent as a result of the *Publisher Enquirer* exposé that, not only can each company make no assumptions about the strategies the other might be using, but the payoff matrix they have been using is wrong: A crack investigative reporter at the *Enquirer* publishes the following revised matrix:

$$P = \begin{bmatrix} 0 & -60 & -40 \\ 30 & 20 & 10 \\ 20 & 0 & 20 \end{bmatrix}.$$

Now what should JungleBooks.com do, and how many customers should it expect to gain or lose?

Some of the books sold by OHaganBooks.com are printed at Bruno Mills, Inc., a combined paper mill and printing company. Exercises 49–52 are based on the following typical monthly input-output table for Bruno Mills' paper and book printing sectors.

	To	Paper	Books
From	**Paper**	\$20,000	\$50,000
	Books	2,000	5,000
	Total Output	200,000	100,000

49. ***Production*** Find the technology matrix for Bruno Mills' paper and book printing sectors.

50. ***Production*** Compute $(I - A)^{-1}$. What is the significance of the $(1, 2)$-entry?

51. ***Production*** Approximately \$1,700 worth of the books sold each month by OHaganBooks.com are printed at Bruno Mills, Inc., and OHaganBooks.com uses approximately \$170 worth of Bruno Mills' paper products each month. What is the total value of paper and books that must be produced by Bruno Mills, Inc. in order to meet demand from OHaganBooks.com?

52. ***Production*** Currently, Bruno Mills, Inc. has a monthly capacity of \$500,000 of paper products and \$200,000 of books. What level of external demand would cause Bruno to meet the capacity for both products?

Case Study: Projecting Market Share

GOH CHAI HIN/AFP/Newscom

You are the sales director at *Selular*, a cellphone provider, and things are not looking good for your company: The recently launched *iClone* competitor is beginning to chip away at Selular's market share. Particularly disturbing are rumors of fierce brand loyalty by iClone customers, with several bloggers suggesting that iClone retains close to 100% of their customers. Worse, you will shortly be presenting a sales report to the board of directors, and the CEO has "suggested" that your report include 2-, 5-, and 10-year projections of Selular's market share given the recent impact on the market by iClone, and also a "worst-case" scenario projecting what would happen if iClone customers are so loyal that none of them ever switch services.

The sales department has conducted two market surveys, taken one quarter apart, of the major cellphone providers, which are *iClone*, *Selular*, *AB&C*, and some smaller companies lumped together as "Other," and has given you the data shown in Figure 4, which shows the percentages of subscribers who switched from one service to another during the quarter.*

* If you go on to study probability theory in Chapter 7, you will see how this scenario can be interpreted as a Markov system, and you will revisit the analysis below in that context.

The percentages in the figure and market shares below reflect actual data for several cellphone services in the United States during a single quarter of 2003. (See the exercises for Section 7.7.)

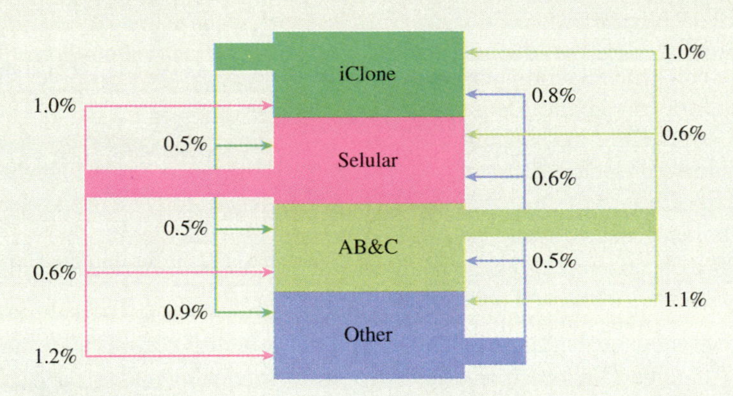

Figure 4

For example, by the end of the quarter 0.5% of iClone's customers had switched to Selular, 0.5% to AB&C, and 0.9% to Other. The current market shares are as follows: iClone: 29.7%, Selular: 19.3%, AB&C: 18.1%, Other: 32.9%.

"This is simple," you tell yourself. "Since I know the current market shares and percentage movements over one quarter, I can easily calculate the market shares next quarter (assuming the percentages that switch services remain the same), then repeat the calculation for the following quarter, and so on, until I get the long-term prediction I am seeking." So you begin your calculations by computing the market shares next quarter. First you note that, since a total of $0.5 + 0.5 + 0.9 = 1.9\%$ of iClone users switched to other brands, the rest, 98.1%, stayed with iClone. Similarly, 97.2% of Selular users, 97.3% of AB&C, and 98.1% of Other stayed with their respective brands. Then you compute:

iClone share after one quarter

$$= 98.1\% \text{ of iClone} + 1.0\% \text{ of Selular} + 1.0\% \text{ of AB\&C} + 0.8\% \text{ of Other}$$
$$= (0.981)(0.297) + (0.010)(0.193) + (0.010)(0.181) + (0.008)(0.329)$$
$$= 0.297729, \text{ or } 29.7729\%.$$

Similarly,

Selular share:

$$= (0.005)(0.297) + (0.972)(0.193) + (0.006)(0.181) + (0.006)(0.329)$$
$$= 0.192141$$

AB&C share:

$$= (0.005)(0.297) + (0.006)(0.193) + (0.973)(0.181) + (0.005)(0.329)$$
$$= 0.180401$$

Other share:

$$= (0.009)(0.297) + (0.012)(0.193) + (0.011)(0.181) + (0.981)(0.329)$$
$$= 0.329729.$$

So, you project the market shares after one quarter to be: iClone: 29.7729%, Selular: 19.2141%, AB&C: 18.0401%, Other: 32.9729%. You now begin to realize that repeating this kind of calculation for the large number of quarters required for long-term projections will be tedious. You call in your student intern (who happens to be a mathematics major) to see if she can help. After taking one look at the calculations, she makes the observation that all you have really done is compute the product of two matrices:

$$[0.297729 \quad 0.192141 \quad 0.180401 \quad 0.329729] = [0.297 \quad 0.193 \quad 0.181 \quad 0.329] \begin{bmatrix} .981 & .005 & .005 & .009 \\ .010 & .972 & .006 & .012 \\ .010 & .006 & .973 & .011 \\ .008 & .006 & .005 & .981 \end{bmatrix}$$

Market shares after one quarter = Market shares at start of quarter $\times A$

The 4×4 matrix A is organized as follows:

		To			
		iClone	**Selular**	**AB&C**	**Other**
From	**iClone**	.981	.005	.005	.009
	Selular	.010	.972	.006	.012
	AB&C	.010	.006	.973	.011
	Other	.008	.006	.005	.981

Since the market shares one quarter later can be obtained from the shares at the start of the quarter, you realize that you can now obtain the shares *two* quarters later by multiplying the result by A—and at this point you start using technology (such as the Matrix Algebra Tool at www.WanerMath.com) to continue the calculation:

Market shares after two quarters = Market shares after one quarter $\times A$

$$= [0.297729 \quad 0.192141 \quad 0.180401 \quad 0.329729] \begin{bmatrix} .981 & .005 & .005 & .009 \\ .010 & .972 & .006 & .012 \\ .010 & .006 & .973 & .011 \\ .008 & .006 & .005 & .981 \end{bmatrix}$$

$$\approx [0.298435 \quad 0.191310 \quad 0.179820 \quad 0.330434].$$

Although the use of matrices has simplified your work, continually multiplying the result by A over and over again to get the market shares for successive months is still tedious. Would it not be possible to get, say, the market share after 10 years (40 quarters) with a single calculation? To explore this, you decide to use symbols for the various market shares:

m_0 = Starting market shares = [0.297 0.193 0.181 0.329]

m_1 = Market shares after 1 quarter
= [0.297729 0.192141 0.180401 0.329729]

m_2 = Market shares after 2 quarters

...

m_n = Market share after n quarters.

You then rewrite the relationships above as

$m_1 = m_0 A$ Shares after one quarter = Shares at start of quarter $\times A$

$m_2 = m_1 A$ Shares after two quarters = Shares after one quarter $\times A$

On an impulse, you substitute the first equation in the second:

$$m_2 = m_1 A = (m_0 A)A = m_0 A^2.$$

Continuing,

$$m_3 = m_2 A = (m_0 A^2)A = m_0 A^3$$
$$\ldots$$
$$m_n = m_0 A^n,$$

which is exactly the formula you need! You can now obtain the 2-, 5-, and 10-year projections each in a single step (with the aid of technology):

2-year projection: $m_8 = m_0 A^8$

$$= [0.297 \quad 0.193 \quad 0.181 \quad 0.329] \begin{bmatrix} .981 & .005 & .005 & .009 \\ .010 & .972 & .006 & .012 \\ .010 & .006 & .973 & .011 \\ .008 & .006 & .005 & .981 \end{bmatrix}^8$$

$$\approx [0.302237 \quad 0.186875 \quad 0.176691 \quad 0.334198]$$

5-year projection: $m_{20} = m_0 A^{20} \approx [0.307992 \quad 0.180290 \quad 0.171938 \quad 0.33978]$

10-year projection: $m_{40} = m_0 A^{40} \approx [0.313857 \quad 0.173809 \quad 0.167074 \quad 0.34526]$.

In particular, Selular's market shares are projected to decline slightly: 2-year projection: 18.7%, 5-year projection: 18.0%, 10-year projection: 17.4%.

You now move on to the "worst-case" scenario, which you represent by assuming 100% loyalty by iClone users with the remaining percentages staying the same (Figure 5).

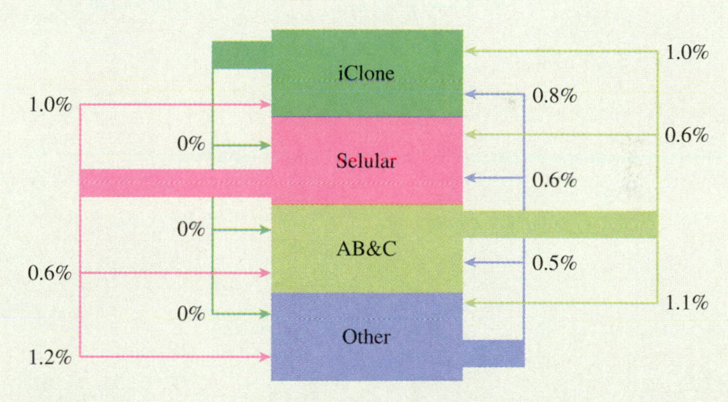

Figure 5

The matrix A corresponding to this diagram is $\begin{bmatrix} 1 & 0 & 0 & 0 \\ .010 & .972 & .006 & .012 \\ .010 & .006 & .973 & .011 \\ .008 & .006 & .005 & .981 \end{bmatrix}$, and you find:

2-year projection: $m_8 = m_0 A^8 \approx [0.346335 \quad 0.175352 \quad 0.165200 \quad 0.313113]$

5-year projection: $m_{20} = m_0 A^{20} \approx [0.413747 \quad 0.152940 \quad 0.144776 \quad 0.288536]$

10-year projection: $m_{40} = m_0 A^{40} \approx [0.510719 \quad 0.123553 \quad 0.117449 \quad 0.248279]$.

Thus, in the worst-case scenario, Selular's market shares are projected to decline more rapidly: 2-year projection: 17.5%, 5-year projection: 15.3%, 10-year projection: 12.4%.

EXERCISES

1. Project Selular's market share 20 and 30 years from now based on the original data shown in Figure 4. (Round all figures to the nearest 0.1%.)
2. Using the original data, what can you say about each company's share in 60 and 80 years (assuming current trends continue)? (Round all figures to the nearest 0.1%.)
3. Obtain a sequence of projections several hundreds of years into the future using the scenarios in both Figure 4 and Figure 5. What do you notice?
4. Compute a sequence of larger and larger powers of the matrix A for both scenarios with all figures rounded to four decimal places. What do you notice?
5. Suppose the trends noted in the market survey were in place for several quarters *before* the current quarter. Using the scenario in Figure 4, determine what the companies' market shares were one quarter before the present and one year before the present (to the nearest 0.1%). Do the same for the scenario in Figure 5.
6. If A is the matrix from the scenario in Figure 4, compute A^{-1}. In light of the preceding exercise, what do the entries in A^{-1} mean?

TI-83/84 Plus Technology Guide

Section 4.1

Example 2 (page 191) The A-Plus auto parts store chain has two outlets, one in Vancouver and one in Quebec. Among other things, it sells wiper blades, windshield cleaning fluid, and floor mats. The monthly sales of these items at the two stores for two months are given in the following tables:

January Sales

	Vancouver	Quebec
Wiper Blades	20	15
Cleaning Fluid (bottles)	10	12
Floor Mats	8	4

February Sales

	Vancouver	Quebec
Wiper Blades	23	12
Cleaning Fluid (bottles)	8	12
Floor Mats	4	5

Use matrix arithmetic to calculate the change in sales of each product in each store from January to February.

Solution with Technology

On the TI-83/84 Plus, matrices are referred to as [A], [B], and so on through [J]. To enter a matrix, press MATRX to bring up the matrix menu, select EDIT, select a matrix, and press ENTER. Then enter the dimensions of the matrix followed by its entries. When you want to use a matrix, press MATRX, select the matrix and press ENTER.

On the TI-83/84 Plus, adding matrices is similar to adding numbers. The sum of the matrices [A] and [B] is [A]+[B]; their difference, of course, is [A]-[B]. As in the text, for this example,

1. Create two matrices, [J] and [F].
2. Compute their difference, [F]-[J] using [F]-[J]→[D].

```
MATRIX[J] 3 ×2
[ 20   15 ]
[ 10   12 ]
[ 8    4  ]
3,2=4
```

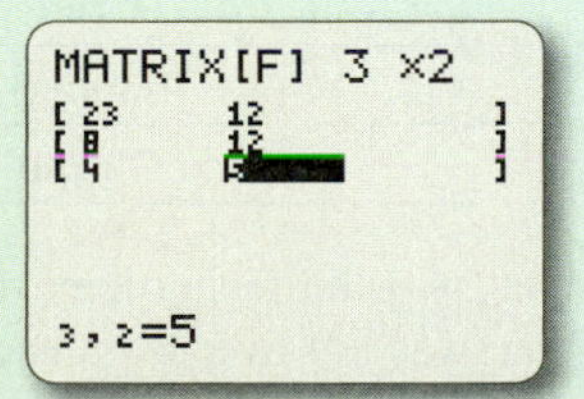

```
[F]-[J]→[D]
     [[3   -3]
      [-2  0 ]
      [-4  1 ]]
```

Note that we have stored the difference in the matrix [D] in case we need it for later use.

Section 4.2

Example 3(a) (page 199) Calculate

$$\begin{bmatrix} 2 & 0 & -1 & 3 \\ 1 & -1 & 2 & -2 \end{bmatrix} \begin{bmatrix} 1 & 1 & -8 \\ 1 & 0 & 0 \\ 0 & 5 & 2 \\ -2 & 8 & -1 \end{bmatrix}.$$

Solution with Technology

On the TI-83/84 Plus, the format for multiplying matrices is the same as for multiplying numbers: [A][B] or [A]*[B] will give the product. We enter the matrices and then multiply them. (Note that, while editing, you can see only three columns of [A] at a time.)

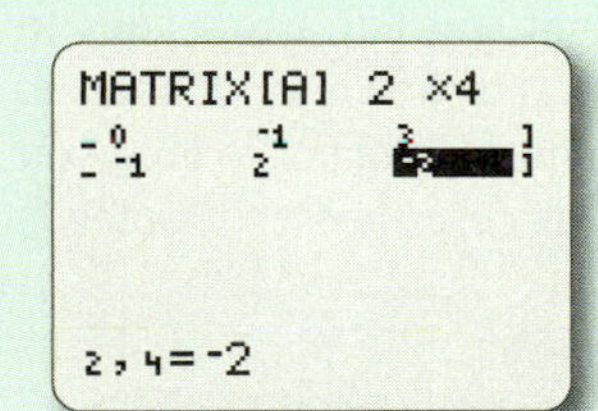

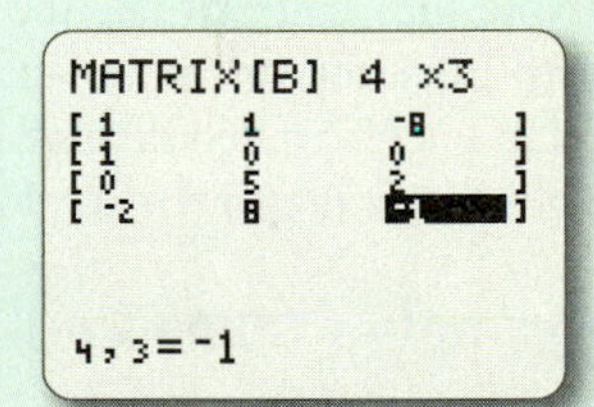

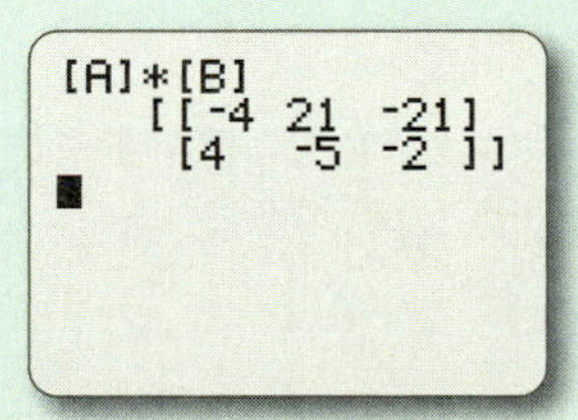

Note that if you try to multiply two matrices whose product is not defined, you will get the error "DIM MISMATCH" (dimension mismatch).

Example 6 (page 202)—Identity Matrix On the TI-83/84 Plus, the function `identity(n)` (in the MATRX MATH menu) returns the $n \times n$ identity matrix.

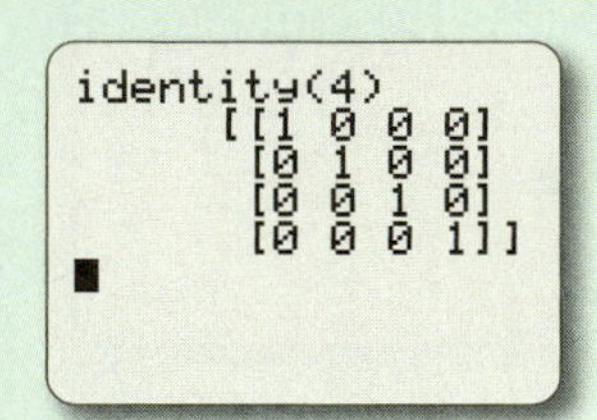

Section 4.3

Example 2(b) (page 207)

Find the inverse of

$$Q = \begin{bmatrix} 1 & 0 & 1 \\ 2 & -2 & -1 \\ 3 & 0 & 0 \end{bmatrix}.$$

Solution with Technology

On a TI-83/84 Plus, you can invert the square matrix [A] by entering [A] $\boxed{X^{-1}}$ ENTER.

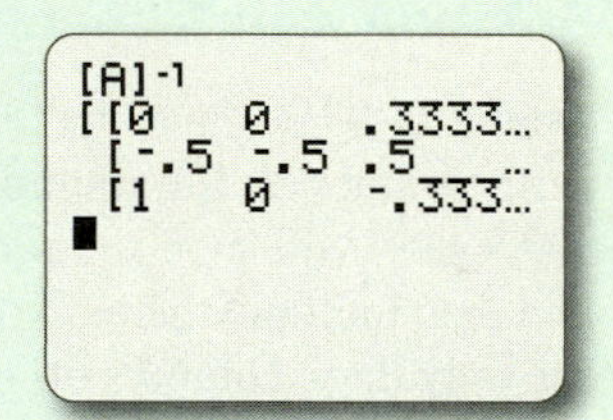

(Note that you can use the right and left arrow keys to scroll the answer, which cannot be shown on the screen all at once.) You could also use the calculator to help you go through the row reduction, as described in Chapter 2.

Example 4 (page 211)

Solve the following three systems of equations.

a. $2x \quad + z = 1$
$2x + y - z = 1$
$3x + y - z = 1$

b. $2x \quad + z = 0$
$2x + y - z = 1$
$3x + y - z = 2$

c. $2x \quad + z = 0$
$2x + y - z = 0$
$3x + y - z = 0$

Solution with Technology

1. Enter the four matrices A, B, C, and D:

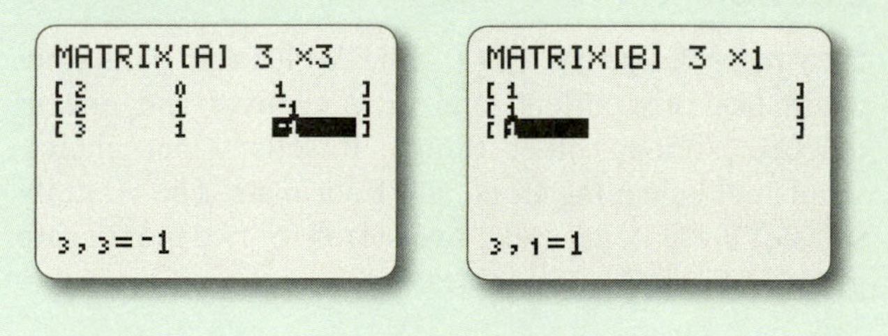

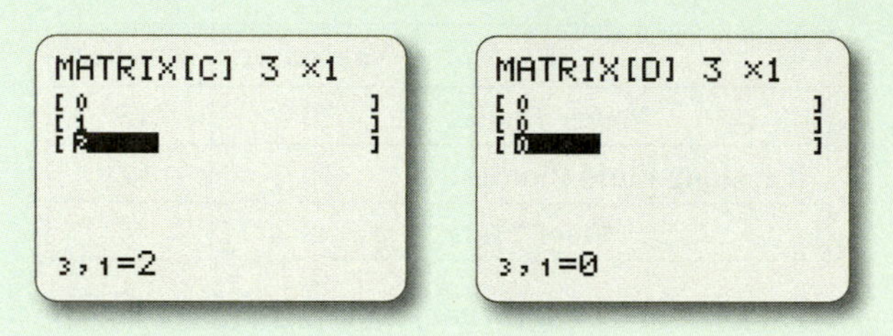

2. Compute the solutions $A^{-1}B$, $A^{-1}C$, and $A^{-1}D$:

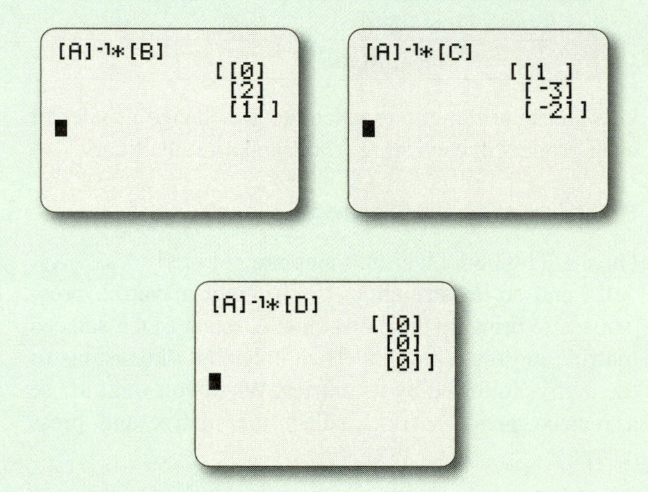

Section 4.5

Example 1 (page 227) Recall that the input-output table in Example 1 looks like this:

	To	**Crude**	**Refining**
From	**Crude**	27,000	59,000
	Refining	750	15,000
	Total Output	87,000	140,000

What was the technology matrix for these two sectors? What was left over from each of these sectors for use by other parts of the economy or for export?

Solution with Technology

There are several ways to use these data to create the technology matrix in your TI-83/84 Plus. For small matrices like this, the most straightforward is to use the matrix editor, where you can give each entry as the appropriate quotient:

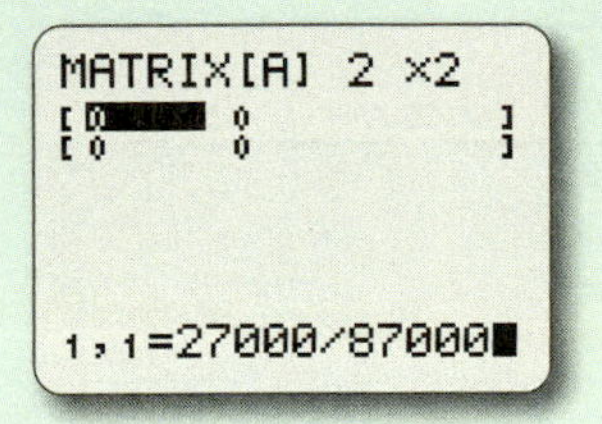

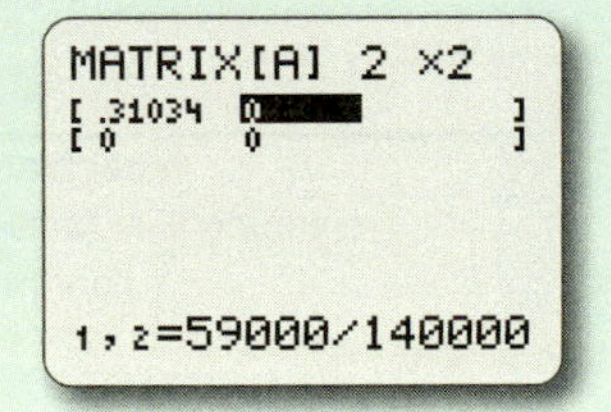

and so on. Once we have the technology matrix [A] and the production vector [B] (remember that we can't use [X] as a matrix name), we can calculate the external demand vector:

```
[B]-[A]*[B]
      [[1000   ]
       [124250]]
```

Of course, when interpreting these numbers, we must remember to round to two significant digits, because our original data were accurate to only that many digits.

Here is an alternative way to calculate the technology matrix that may be better for examples with more sectors.

1. Begin by entering the columns of the input-output table as lists in the list editor (STAT EDIT) (see below left).

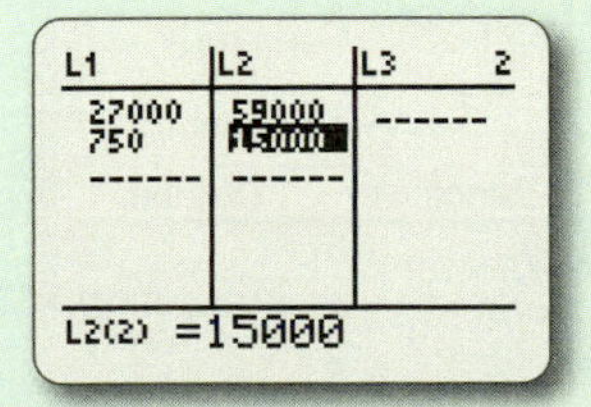

2. We now want to divide each column by the total output of its sector and assemble the results into a matrix. We can do this using the `List ▶matr` function (under the MATRX MATH menu) shown on the right above.

Example 3 (page 231) Consider four sectors of the economy of Kenya[4]: (1) the traditional economy, (2) agriculture, (3) manufacture of metal products and machinery, and (4) wholesale and retail trade. The input-output table for these four sectors for 1976 looks like this (all numbers are thousands of K£):

	To	1	2	3	4
From	1	8,600	0	0	0
	2	0	20,000	24	0
	3	1,500	530	15,000	660
	4	810	8,500	5,800	2,900
Total Output		87,000	530,000	110,000	180,000

Suppose that external demand for agriculture increased by K£50,000,000 and that external demand for metal products and machinery increased by K£10,000,000. How would production in these four sectors have to change to meet this rising demand?

Solution with Technology

1. Enter the technology matrices A as `[A]` and D^+ as `[D]` using one of the techniques above:

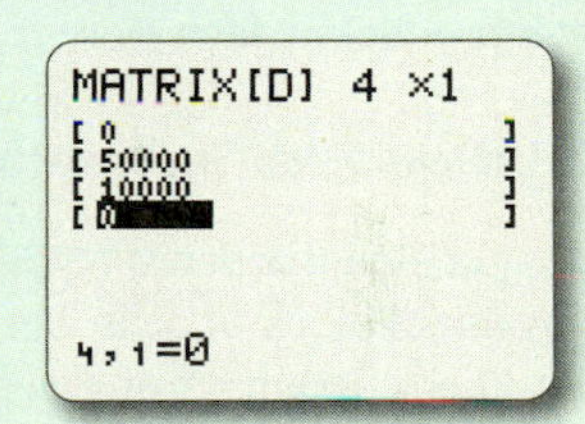

2. You can then compute the change in production with the formula: `(identity(4)-[A])`$^{-1}$ `[D]`.

```
(identity(4)-[A]
)-1[D]
 [[0           ]
  [51963.42476]
  [11645.36129]
  [1471.104956]]
```

[4]Figures are rounded. Source: *Input-Output Tables for Kenya 1976*, Central Bureau of Statistics of the Ministry of Economic Planning and Community Affairs, Kenya.

TECHNOLOGY GUIDE

TECHNOLOGY GUIDE

SPREADSHEET Technology Guide

Section 4.1

Example 2 (page 191) The A-Plus auto parts store chain has two outlets, one in Vancouver and one in Quebec. Among other things, it sells wiper blades, windshield cleaning fluid, and floor mats. The monthly sales of these items at the two stores for two months are given in the following tables:

January Sales

	Vancouver	Quebec
Wiper Blades	20	15
Cleaning Fluid (bottles)	10	12
Floor Mats	8	4

February Sales

	Vancouver	Quebec
Wiper Blades	23	12
Cleaning Fluid (bottles)	8	12
Floor Mats	4	5

Use matrix arithmetic to calculate the change in sales of each product in each store from January to February.

Solution with Technology

To enter a matrix in a spreadsheet, we put its entries in any convenient block of cells. For example, the matrix A in the first Quick Example of this section might look like this:

	A	B	C
1	2	0	1
2	33	-22	0

Spreadsheets refer to such blocks of data as **arrays**, which it can handle in much the same way as it handles single cells of data. For instance, when typing a formula, just as clicking on a cell creates a reference to that cell, selecting a whole array of cells will create a reference to that array. An array is referred to using an **array range** consisting of the top-left and bottom-right cell coordinates, separated by a colon. For example, the array range A1:C2 refers to the 2×3 matrix above, with top-left corner A1 and bottom-right corner C2.

1. To add or subtract two matrices in a spreadsheet, first input their entries in two separate arrays in the spreadsheet (we've also added labels as in the previous tables, which you might do if you wanted to save the spreadsheet for later use):

	A	B	C
1	January Sales		
2		Vancouver	Quebec
3	wiper blades	20	15
4	cleaning fluid (bottles)	10	12
5	floor mats	8	4
6			
7	February Sales		
8		Vancouver	Quebec
9	wiper blades	23	12
10	cleaning fluid (bottles)	8	12
11	floor mats	4	5

2. Select (highlight) a block of the same size (3×2 in this case) where you would like the answer, $F - J$, to appear, enter the formula `=B9:C11-B3:C5`, and then type Control+Shift+Enter. The easiest way to do this is as follows:

- Highlight cells B15:C17. *Where you want the answer to appear*
- Type "=".
- Highlight the matrix ***F***. *Cells B9 through C11*
- Type "-".
- Highlight the matrix ***J***. *Cells B3 through C5*
- Press Control+Shift+Enter. *Not just Enter*

	A	B	C
1	January Sales		
2		Vancouver	Quebec
3	wiper blades	20	15
4	cleaning fluid (bottles)	10	12
5	floor mats	8	4
6			
7	February Sales		
8		Vancouver	Quebec
9	wiper blades	23	12
10	cleaning fluid (bottles)	8	12
11	floor mats	4	5
12			
13	Change in Sales		
14		Vancouver	Quebec
15	wiper blades	=B9:C11-B3:C5	
16	cleaning fluid (bottles)		
17	floor mats		

Typing Control+Shift+Enter (instead of Enter) tells the spreadsheet that your formula is an *array formula,* one that returns a matrix rather than a single number.[5] Once entered, the formula bar will show the formula you entered enclosed in "curly braces," indicating that it is an array formula. Note that you must use Control+Shift+Enter to delete any array you create: Select the block you wish to delete and press Delete followed by Control+Shift+Enter.

Section 4.2

Example 3(a) (page 199) Calculate the product

$$\begin{bmatrix} 2 & 0 & -1 & 3 \\ 1 & -1 & 2 & -2 \end{bmatrix} \begin{bmatrix} 1 & 1 & -8 \\ 1 & 0 & 0 \\ 0 & 5 & 2 \\ -2 & 8 & -1 \end{bmatrix}.$$

Solution with Technology

In a spreadsheet, the function we use for matrix multiplication is `MMULT`. (Ordinary multiplication, *, will *not* work.)

1. Enter the two matrices as shown in the spreadsheet and highlight a block where you want the answer to appear. (Note that it should have the correct dimensions for the product: 2 × 3.)

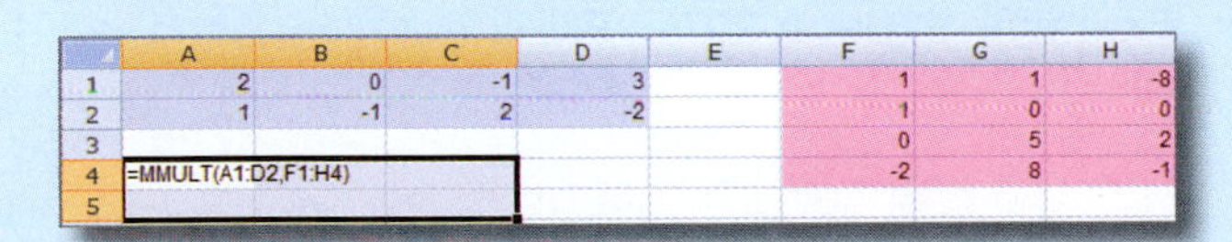

	A	B	C	D	E	F	G	H
1	2	0	-1	3		1	1	-8
2	1	-1	2	-2		1	0	0
3						0	5	2
4	=MMULT(A1:D2,F1:H4)					-2	8	-1
5								

2. Enter the formula `=MMULT(A1:D2,F1:H4)` (using the mouse to avoid typing the array ranges if you like) and press Control+Shift+Enter. The product will appear in the region you highlighted. If you try to multiply two matrices whose product is not defined, you will get the error "#VALUE!".

	A	B	C	D	E	F	G	H
1	2	0	-1	3		1	1	-8
2	1	-1	2	-2		1	0	0
3						0	5	2
4	-4	21	-21			-2	8	-1
5	4	-5	-2					

Example 6—Identity Matrix (page 202) There is no spreadsheet function that returns an identity matrix. If you need a small identity matrix, it's simplest to just enter the 1s and 0s by hand. If you need a large identity matrix, here is one way to get it quickly.

1. Say we want a 4 × 4 identity matrix in the cells B1:E4. Enter the following formula in cell B1:

```
=IF(ROW(B1)-ROW($B$1)
=COLUMN(B1)-COLUMN($B$1),1,0)
```

	A	B	C	D	E
1		=IF(ROW(B1)-ROW($B$1)=COLUMN(B1)-COLUMN($B$1),1,0)			
2					
3					
4					

2. Press Enter, then copy cell B1 to cells B1:E4. The formula will return 1s along the diagonal of the matrix and 0s elsewhere, giving you the identity matrix. Why does this formula work?

	A	B	C	D	E
1		1	0	0	0
2		0	1	0	0
3		0	0	1	0
4		0	0	0	1

Section 4.3

Example 2(b) (page 207) Find the inverse of

$$Q = \begin{bmatrix} 1 & 0 & 1 \\ 2 & -2 & -1 \\ 3 & 0 & 0 \end{bmatrix}.$$

Solution with Technology

In a spreadsheet, the function `MINVERSE` computes the inverse of a matrix.

1. Enter Q somewhere convenient, for example, in cells A1:C3.
2. Choose the block where you would like the inverse to appear, highlight the whole block.
3. Enter the formula `=MINVERSE(A1:C3)` and press Control+Shift+Enter.

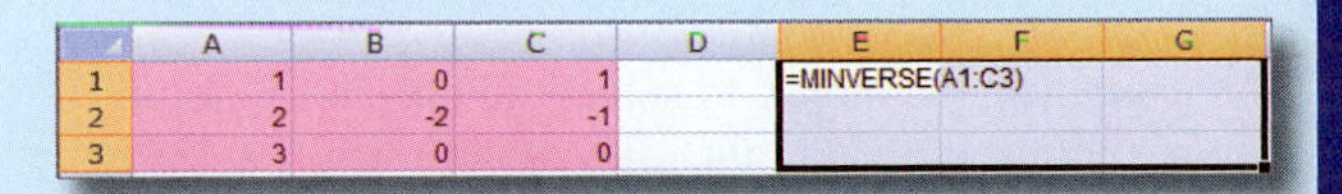

	A	B	C	D	E	F	G
1	1	0	1		=MINVERSE(A1:C3)		
2	2	-2	-1				
3	3	0	0				

[5]Note that on a Mac, Command-Enter has the same effect as Control+Shift+Enter.

The inverse will appear in the region you highlighted. (To convert the answer to fractions, format the cells as fractions.)

If a matrix is singular, a spreadsheet will register an error by showing `#NUM!` in each cell.

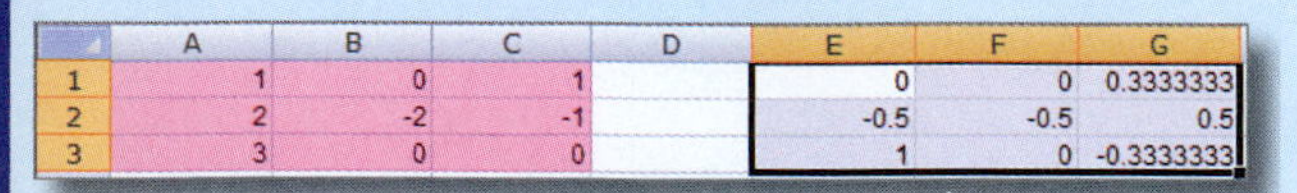

	A	B	C	D	E	F	G
1	1	0	1		0	0	0.3333333
2	2	-2	-1		-0.5	-0.5	0.5
3	3	0	0		1	0	-0.3333333

Although the spreadsheet appears to invert the matrix in one step, it is going through the procedure in the text or some variation of it to find the inverse. Of course, you could also use the spreadsheet to help you go through the row reduction, just as in Chapter 3.

Example 4 (page 211) Solve the following three systems of equations.

a. $2x \quad\quad + z = 1$
$2x + y - z = 1$
$3x + y - z = 1$

b. $2x \quad\quad + z = 0$
$2x + y - z = 1$
$3x + y - z = 2$

c. $2x \quad\quad + z = 0$
$2x + y - z = 0$
$3x + y - z = 0$

Solution with Technology

Spreadsheets instantly update calculated results every time the contents of a cell are changed. We can take advantage of this to solve the three systems of equations given above using the same worksheet as follows.

1. Enter the matrices A and B from the matrix equation $AX = B$.

2. Select a 3×1 block of cells for the matrix X.

3. The Excel formula we can use to calculate X is:

```
=MMULT(MINVERSE(A1:C3),E1:E3)
```

$A^{-1}B$

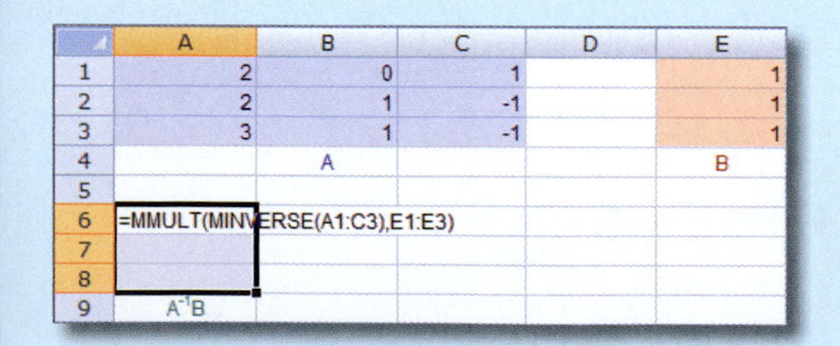

	A	B	C	D	E
1	2	0	1		1
2	2	1	-1		1
3	3	1	-1		1
4		A			B
5					
6	=MMULT(MINVERSE(A1:C3),E1:E3)				
7					
8					
9	A⁻¹B				

(As usual, use the mouse to select the ranges for A and B while typing the formula, and don't forget to press Control+Shift+Enter.) Having obtained the solution to part (a), you can now simply modify the entries in Column E to see the solutions for parts (b) and (c).

Note Your spreadsheet for part (a) may look like this:

	A	B	C	D	E
1	2	0	1		1
2	2	1	-1		1
3	3	1	-1		1
4		A			B
5					
6	1.11022E-16				
7	2				
8	1				
9	A⁻¹B				

What is that strange number doing in cell A6? "`E-16`" represents "$\times 10^{-16}$", so the entry is really

$$1.11022 \times 10^{-16} = 0.000\,000\,000\,000\,000\,111022 \approx 0.$$

Mathematically, it is supposed to be *exactly* zero (see the solution to part (a) in the text) but Excel made a small error in computing the inverse of A, resulting in this spurious value. Note, however, that it is accurate (agrees with zero) to 15 decimal places! In practice, when we see numbers arise in matrix calculations that are far smaller than all the other entries, we can usually assume they are supposed to be zero. ■

Section 4.5

Example 1 (page 227) Recall that the input-output table in Example 1 looks like this:

	To	**Crude**	**Refining**
From	**Crude**	27,000	59,000
	Refining	750	15,000
	Total Output	87,000	140,000

What was the technology matrix for these two sectors? What was left over from each of these sectors for use by other parts of the economy or for export?

Solution with Technology

1. Enter the input-output table in a spreadsheet:

	A	B	C	D
1			To	
2			Crude	Refining
3	From	Crude	27000	59000
4		Refining	750	15000
5		Total Output	87000	140000

2. To obtain the technology matrix, we divide each column by the total output of its sector:

	A	B	C	D
1			To	
2			Crude	Refining
3	From	Crude	27000	59000
4		Refining	750	15000
5		Total Output	87000	140000
6				
7			=C3/C$5	
8				

	A	B	C	D
1			To	
2			Crude	Refining
3	From	Crude	27000	59000
4		Refining	750	15000
5		Total Output	87000	140000
6				
7			0.3103448	0.4214286
8			0.0086207	0.1071429

The formula =C3/C$5 is copied into the shaded 2×2 block shown above. (The $ sign in front of the 5 forces the program to always divide by the total in Row 5 even when the formula is copied from Row 7 to Row 8.) The result is the technology matrix shown in the bottom screenshot above.

3. Using the techniques discussed in the second section, we can now compute $D = X - AX$ to find the demand vector.

Example 3 (page 231) Consider four sectors of the economy of Kenya[6]: (1) the traditional economy, (2) agriculture, (3) manufacture of metal products and machinery, and (4) wholesale and retail trade. The input-output table for these four sectors for 1976 looks like this (all numbers are thousands of K£):

To		1	2	3	4
From	1	8,600	0	0	0
	2	0	20,000	24	0
	3	1,500	530	15,000	660
	4	810	8,500	5,800	2,900
Total Output		87,000	530,000	110,000	180,000

Suppose that external demand for agriculture increased by K£50,000,000 and that external demand for metal products and machinery increased by K£10,000,000. How would production in these four sectors have to change to meet this rising demand?

Solution with Technology

1. Enter the input-output table in the spreadsheet.
2. Compute the technology matrix by dividing each column by the column total.
3. Insert the identity matrix I in preparation for the next step.

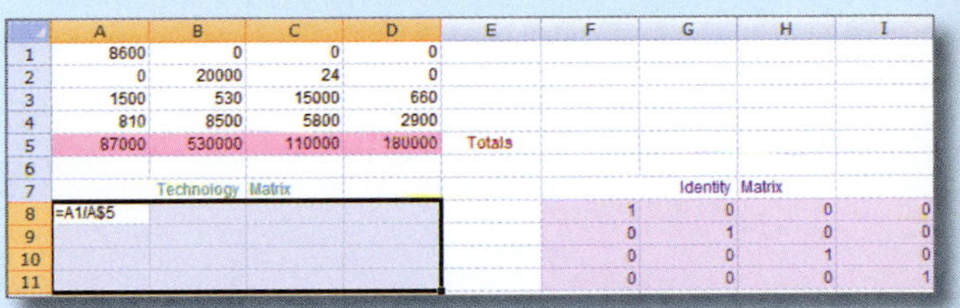

	A	B	C	D	E	F	G	H	I
1	8600	0	0	0					
2	0	20000	24	0					
3	1500	530	15000	660					
4	810	8500	5800	2900					
5	87000	530000	110000	180000	Totals				
6									
7		Technology	Matrix				Identity	Matrix	
8	=A1/A$5					1	0	0	0
9						0	1	0	0
10						0	0	1	0
11						0	0	0	1

4. To see how each sector reacts to rising external demand, you must calculate the inverse matrix $(I - A)^{-1}$, as shown below. (Remember to use Control+Shift+Enter each time.)

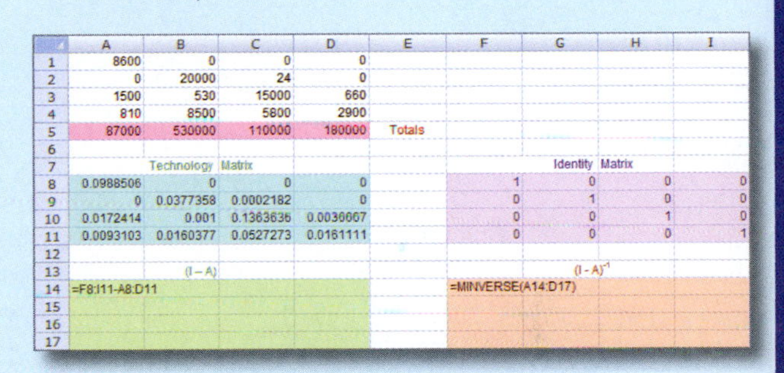

	A	B	C	D	E	F	G	H	I
1	8600	0	0	0					
2	0	20000	24	0					
3	1500	530	15000	660					
4	810	8500	5800	2900					
5	87000	530000	110000	180000	Totals				
6									
7		Technology	Matrix				Identity	Matrix	
8	0.0988506	0	0	0		1	0	0	0
9	0	0.0377358	0.0002182	0		0	1	0	0
10	0.0172414	0.001	0.1363636	0.0036667		0	0	1	0
11	0.0093103	0.0160377	0.0527273	0.0161111		0	0	0	1
12									
13		(I – A)					(I - A)^-1		
14	=F8:I11-A8:D11					=MINVERSE(A14:D17)			
15									
16									
17									

5. To compute $(I - A)^{-1}D^{+}$, enter D^{+} as a column and use the MMULT operation. (See Example 3 in Section 4.2.)

	A	B	C	D	E	F	G	H	I
13		(I – A)					(I - A)^-1		
14	0.9011494	0	0	0		1.1096939	0	0	0
15	0	0.9622642	-0.0002182	0		5.034E-06	1.039216	0.0002626	9.786E-07
16	-0.0172414	-0.001	0.8636364	-0.0036667		0.0222032	0.0012755	1.1581586	0.0043161
17	-0.0093103	-0.0160377	-0.0527273	0.9838889		0.0116908	0.0170079	0.0620708	1.0166062
18									
19	D		(I - A)^-1 D						
20	0		0						
21	50000		51963.425						
22	10000		11645.361						
23	0		1471.105						

Note Here is one of the beauties of spreadsheet programs: Once you are done with the calculation, you can use the spreadsheet as a template for any 4×4 input-output table by just changing the entries of the input-output matrix and/or D^{+}. The rest of the computation will then be done automatically as the spreadsheet is updated. In other words, you can use it to do your homework! ■

[6]Figures are rounded. Source: *Input-Output Tables for Kenya 1976*, Central Bureau of Statistics of the Ministry of Economic Planning and Community Affairs, Kenya.

5

Linear Programming

Website
www.WanerMath.com
At the Website you will find:

- Section-by-section tutorials, including game tutorials with randomized quizzes
- A detailed chapter summary
- A true/false quiz
- Additional review exercises
- A linear programming grapher
- A pivot and Gauss-Jordan tool
- A simplex method tool

Case Study The Diet Problem

The Galaxy Nutrition health-food mega-store chain provides free online nutritional advice and support to its customers. As Web site technical consultant, you are planning to construct an interactive Web page to assist customers prepare a diet tailored to their nutritional and budgetary requirements. Ideally, the customer would select foods to consider and specify nutritional and/or budgetary constraints, and the tool should return the optimal diet meeting those requirements. You would also like the Web page to allow the customer to decide whether, for instance, to find the cheapest possible diet meeting the requirements, the diet with the lowest number of calories, or the diet with the least total carbohydrates. **How do you go about constructing such a Web page?**

Image Studios/UpperCut Images/Getty Images

Introduction

In this chapter we begin to look at one of the most important types of problems for business and the sciences: finding the largest or smallest possible value of some quantity (such as profit or cost) under certain constraints (such as limited resources). We call such problems **optimization** problems because we are trying to find the best, or optimum, value. The optimization problems we look at in this chapter involve linear functions only and are known as **linear programming** (LP) problems. One of the main purposes of calculus, which you may study later, is to solve nonlinear optimization problems.

Linear programming problems involving only two unknowns can usually be solved by a graphical method that we discuss in Sections 5.1 and 5.2. When there are three or more unknowns, we must use an algebraic method, as we had to do for systems of linear equations. The method we use is called the **simplex method**. Invented in 1947 by George B. Dantzig* (1914–2005), the simplex method is still the most commonly used technique to solve LP problems in real applications, from finance to the computation of trajectories for guided missiles.

* Dantzig is the real-life source of the story of the student who, walking in late to a math class, copies down two problems on the board, thinking they're homework. After much hard work he hands in the solutions, only to discover that he's just solved two famous unsolved problems. This actually happened to Dantzig in graduate school in 1939.[1]

The simplex method can be used for hand calculations when the numbers are fairly small and the unknowns are few. Practical problems often involve large numbers and many unknowns, however. Problems such as routing telephone calls or airplane flights, or allocating resources in a manufacturing process can involve tens of thousands of unknowns. Solving such problems by hand is obviously impractical, and so computers are regularly used. Although computer programs most often use the simplex method, mathematicians are always seeking faster methods. The first radically different method of solving LP problems was the **ellipsoid algorithm** published in 1979 by the Soviet mathematician Leonid G. Khachiyan[2] (1952–2005). In 1984, Narendra Karmarkar (1957–), a researcher at Bell Labs, created a more efficient method now known as **Karmarkar's algorithm**. Although these methods (and others since developed) can be shown to be faster than the simplex method in the worst cases, it seems to be true that the simplex method is still the fastest in the applications that arise in practice.

Calculators and spreadsheets are very useful aids in the simplex method. In practice, software packages do most of the work, so you can think of what we teach you here as a peek inside a "black box." What the software cannot do for you is convert a real situation into a mathematical problem, so the most important lessons to get out of this chapter are (1) how to recognize and set up a linear programming problem, and (2) how to interpret the results.

5.1 Graphing Linear Inequalities

By the end of the next section, we will be solving linear programming (LP) problems with two unknowns. We use inequalities to describe the constraints in a problem, such as limitations on resources. Recall the basic notation for inequalities.

[1]Sources: D. J. Albers, and C. Reid, "An Interview of George B. Dantzig: The Father of Linear Programming," *College Math. Journal,* v. 17 (1986), pp. 293–314. Quoted and discussed in the context of the urban legends it inspired at www.snopes.com/college/homework/unsolvable.asp.

[2]Dantzig and Khachiyan died approximately two weeks apart in 2005. The *New York Times* ran their obituaries together on May 23, 2005.

Non-Strict Inequalities

$a \le b$ means that a **is less than or equal to** b.

$a \ge b$ means that a **is greater than or equal to** b.

Quick Examples

$3 \le 99$, $-2 \le -2$, $0 \le 3$

$3 \ge 3$, $1.78 \ge 1.76$, $\frac{1}{3} \ge \frac{1}{4}$

There are also the inequalities $<$ and $>$, called **strict** inequalities because they do not permit equality. We do not use them in this chapter.

Following are some of the basic rules for manipulating inequalities. Although we illustrate all of them with the inequality $\le$, they apply equally well to inequalities with $\ge$ and to the strict inequalities $<$ and $>$.

Rules for Manipulating Inequalities

Rule	Quick Examples
1. The same quantity can be added to or subtracted from both sides of an inequality: If $x \le y$, then $x + a \le y + a$ for any real number a.	$x \le y$ implies $x - 4 \le y - 4$
2. Both sides of an inequality can be multiplied or divided by a positive constant: If $x \le y$ and a is positive, then $ax \le ay$.	$x \le y$ implies $3x \le 3y$
3. Both sides of an inequality can be multiplied or divided by a negative constant if the inequality is *reversed*: If $x \le y$ and a is negative, then $ax \ge ay$.	$x \le y$ implies $-3x \ge -3y$
4. The left and right sides of an inequality can be switched if the inequality is *reversed*: If $x \le y$, then $y \ge x$; if $y \ge x$, then $x \le y$.	$3x \ge 5y$ implies $5y \le 3x$

Here are the particular kinds of inequalities in which we're interested:

Linear Inequalities and Solving Inequalities

An **inequality in the unknown** x is the statement that one expression involving x is less than or equal to (or greater than or equal to) another. Similarly, we can have an **inequality in** x **and** y, which involves expressions that contain x and y; an **inequality in** x**,** y**, and** z; and so on. A **linear inequality** in one or more unknowns is an inequality of the form

$ax \le b$ (or $ax \ge b$), *a and b real constants*

$ax + by \le c$ (or $ax + by \ge c$), *a, b, and c real constants*

$ax + by + cz \le d$, *a, b, c, and d real constants*

$ax + by + cz + dw \le e$, *a, b, c, d, and e real constants*

and so on.

Quick Examples

$2x + 8 \geq 89$	Linear inequality in x
$2x^3 \leq x^3 + y$	Nonlinear inequality in x and y
$3x - 2y \geq 8$	Linear inequality in x and y
$x^2 + y^2 \leq 19z$	Nonlinear inequality in x, y, and z
$3x - 2y + 4z \leq 0$	Linear inequality in x, y, and z

A **solution** of an inequality in the unknown x is a value for x that makes the inequality true. For example, $2x + 8 \geq 89$ has a solution $x = 50$ because $2(50) + 8 \geq 89$. Of course, it has many other solutions as well. Similarly, a solution of an inequality in x and y is a pair of values (x, y) making the inequality true. For example, $(5, 1)$ is a solution of $3x - 2y \geq 8$ because $3(5) - 2(1) \geq 8$. To **solve** an inequality is to find the set of *all* solutions.

Solving Linear Inequalities in Two Variables

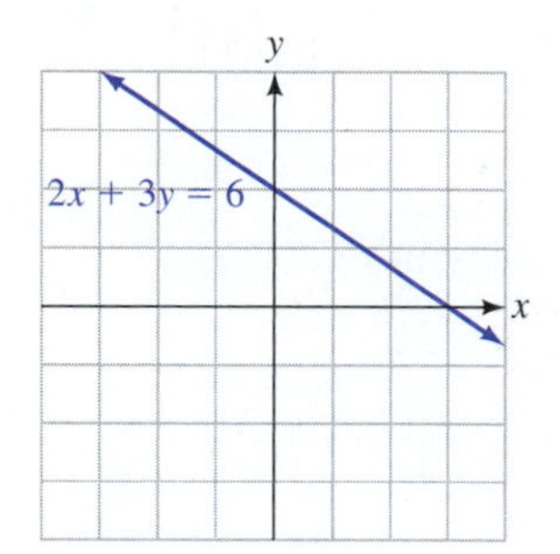

Figure 1

Our first goal is to solve linear inequalities in two variables—that is, inequalities of the form $ax + by \leq c$. As an example, let's solve

$$2x + 3y \leq 6.$$

We already know how to solve the *equation* $2x + 3y = 6$. As we saw in Chapter 1, the solution of this equation may be pictured as the set of all points (x, y) on the straight-line graph of the equation. This straight line has x-intercept 3 (obtained by putting $y = 0$ in the equation) and y-intercept 2 (obtained by putting $x = 0$ in the equation) and is shown in Figure 1.

Notice that, if (x, y) is any point on the line, then x and y not only satisfy the *equation* $2x + 3y = 6$, but they also satisfy the *inequality* $2x + 3y \leq 6$, because being equal to 6 qualifies as being less than or equal to 6.

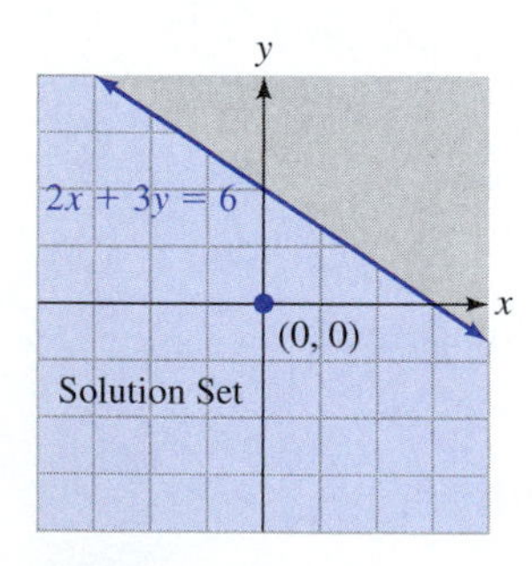

Figure 2

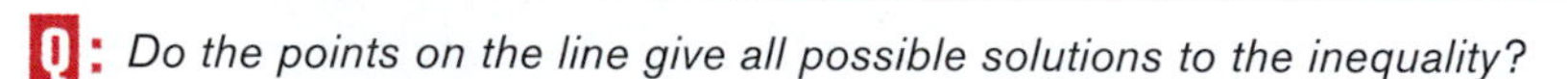

Q: *Do the points on the line give all possible solutions to the inequality?*

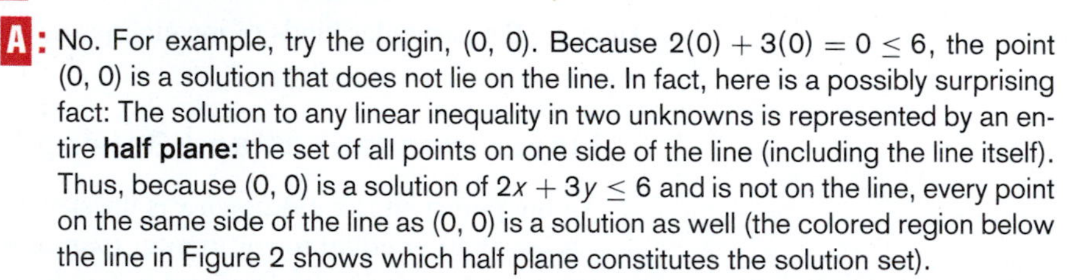

A: No. For example, try the origin, (0, 0). Because $2(0) + 3(0) = 0 \leq 6$, the point (0, 0) is a solution that does not lie on the line. In fact, here is a possibly surprising fact: The solution to any linear inequality in two unknowns is represented by an entire **half plane:** the set of all points on one side of the line (including the line itself). Thus, because (0, 0) is a solution of $2x + 3y \leq 6$ and is not on the line, every point on the same side of the line as (0, 0) is a solution as well (the colored region below the line in Figure 2 shows which half plane constitutes the solution set).

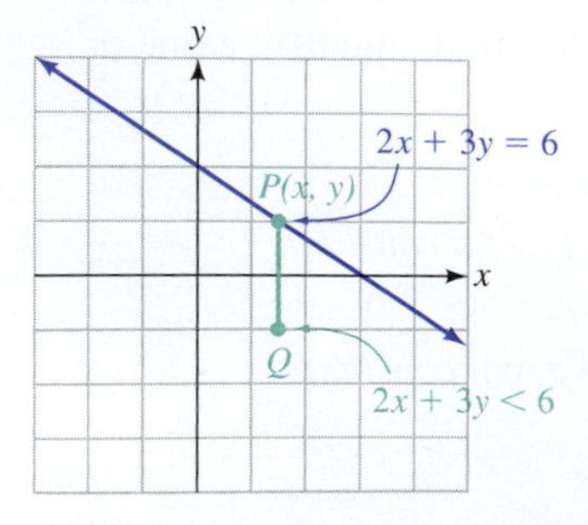

Figure 3

To see why the solution set of $2x + 3y \leq 6$ is the entire half plane shown in Figure 2, start with any point P on the line $2x + 3y = 6$. We already know that P is a solution of $2x + 3y \leq 6$. If we choose any point Q directly below P, the x-coordinate of Q will be the same as that of P, and the y-coordinate will be smaller. So the value of $2x + 3y$ at Q will be smaller than the value at P, which is 6. Thus, $2x + 3y < 6$ at Q, and so Q is another solution of the inequality. (See Figure 3.) In other words, *every point beneath the line is a solution of* $2x + 3y \leq 6$.

On the other hand, any point above the line is directly above a point on the line, and so $2x + 3y > 6$ for such a point. Thus, *no point above the line is a solution of* $2x + 3y \leq 6$.

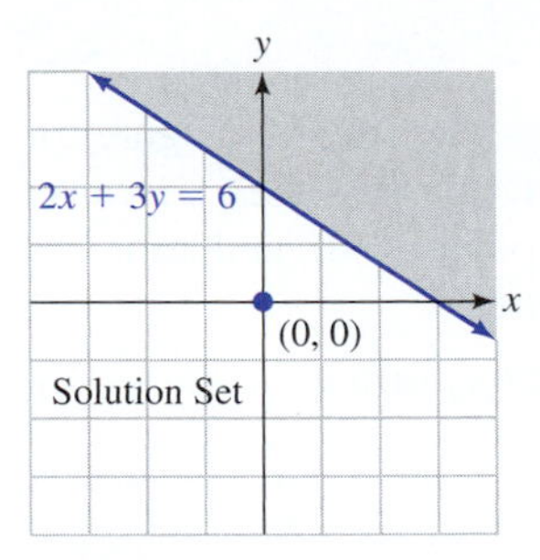

Figure 4

The same kind of argument can be used to show that the solution set of every inequality of the form $ax + by \le c$ or $ax + by \ge c$ consists of the half plane above or below the line $ax + by = c$. The "test-point" procedure we describe below gives us an easy method for deciding whether the solution set includes the region above or below the corresponding line.

Now we are going to do something that will appear backward at first (but makes it simpler to sketch sets of solutions of *systems* of linear inequalities). For our standard drawing of the region of solutions of $2x + 3y \le 6$, we are going to *shade only the part that we do not want and leave the solution region blank.* Think of covering over or "blocking out" the unwanted points, leaving those that we do want in full view (but remember that the points on the boundary line are also points that we want). The result is Figure 4. The reason we do this should become clear in Example 2.

Sketching the Region Represented by a Linear Inequality in Two Variables

1. Sketch the straight line obtained by replacing the given inequality with an equality.
2. Choose a test point not on the line; (0, 0) is a good choice if the line does not pass through the origin.
3. If the test point satisfies the inequality, then the set of solutions is the entire region on the same side of the line as the test point. Otherwise, it is the region on the other side of the line. In either case, shade (block out) the side that does *not* contain the solutions, leaving the solution set unshaded.

Quick Example

Here are the three steps used to graph the inequality $x + 2y \ge 5$:

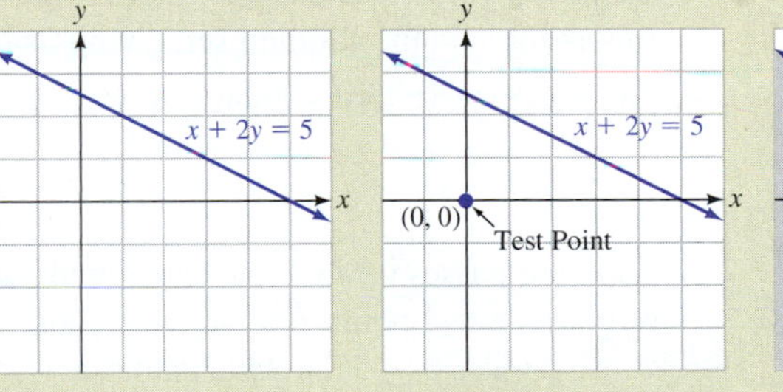

1. Sketch the line $x + 2y = 5$.
2. Test the point (0, 0) $0 + 2(0) \not\ge 5$. Inequality is not satisfied.
3. Because the inequality is not satisfied, shade the region containing the test point.

EXAMPLE 1 Graphing Single Inequalities

Sketch the regions determined by each of the following inequalities:

a. $3x - 2y \le 6$ **b.** $6x \le 12 + 4y$ **c.** $x \le -1$ **d.** $y \ge 0$ **e.** $x \ge 3y$

using Technology

Technology can be used to graph inequalities. Here is an outline (see the Technology Guides at the end of the chapter for additional details on using a TI-83/84 Plus or a spreadsheet):

TI-83/84 Plus

Solve the inequality for y and enter the resulting function of x; for example,

```
Y1=-(2/3)*X+2.
```

Position the cursor on the icon to the left of Y_1 and press ENTER until you see the kind of shading desired (above or below the line). [More details on page 308.]

Spreadsheet

Solve the inequality for y and create a scattergraph using two points on the line. Then use the drawing palette to create a polygon to provide the shading.
[More details on page 310.]

Website

www.WanerMath.com

→ On Line Utilities

→ Linear Programming Grapher

Type "graph" and enter one or more inequalities (each one on a new line) as shown:

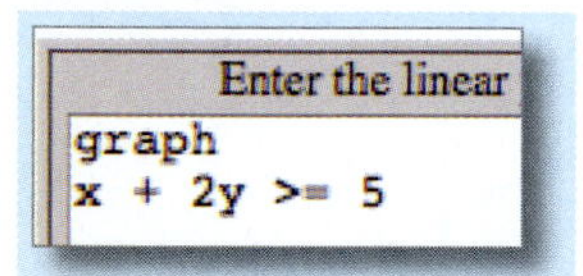

Adjust the graph window settings, and press "Graph".

Solution

a. The boundary line $3x - 2y = 6$ has x-intercept 2 and y-intercept -3 (Figure 5). We use (0, 0) as a test point (because it is not on the line). Because $3(0) - 2(0) \leq 6$, the inequality is satisfied by the test point (0, 0), and so it lies inside the solution set. The solution set is shown in Figure 5.

b. The given inequality, $6x \leq 12 + 4y$, can be rewritten in the form $ax + by \leq c$ by subtracting $4y$ from both sides:

$$6x - 4y \leq 12.$$

Dividing both sides by 2 gives the inequality $3x - 2y \leq 6$, which we considered in part (a). Now, *applying the rules for manipulating inequalities does not affect the set of solutions*. Thus, the inequality $6x \leq 12 + 4y$ has the same set of solutions as $3x - 2y \leq 6$. (See Figure 5.)

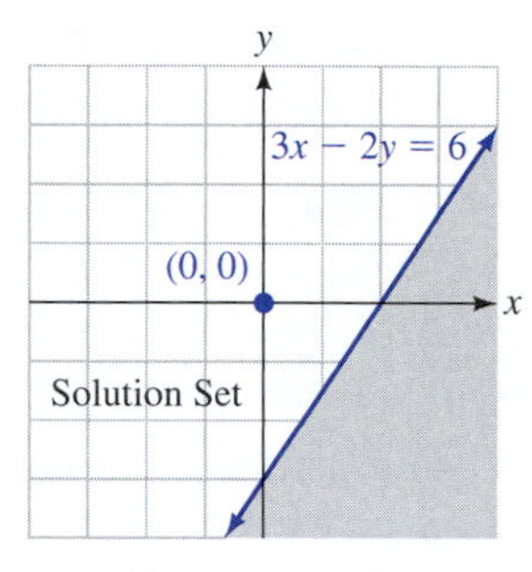

Figure 5

Figure 6

c. The region $x \leq -1$ has as boundary the vertical line $x = -1$. The test point (0, 0) is not in the solution set, as shown in Figure 6.

d. The region $y \geq 0$ has as boundary the horizontal line $y = 0$ (that is, the x-axis). We cannot use (0, 0) for the test point because it lies on the boundary line. Instead, we choose a convenient point not on the line $y = 0$—say, (0, 1). Because $1 \geq 0$, this point is in the solution set, giving us the region shown in Figure 7.

e. The line $x \geq 3y$ has as boundary the line $x = 3y$ or, solving for y,

$$y = \frac{1}{3}x.$$

This line passes through the origin with slope 1/3, so again we cannot choose the origin as a test point. Instead, we choose (0, 1). Substituting these coordinates in $x \geq 3y$ gives $0 \geq 3(1)$, which is false, so (0, 1) is not in the solution set, as shown in Figure 8.

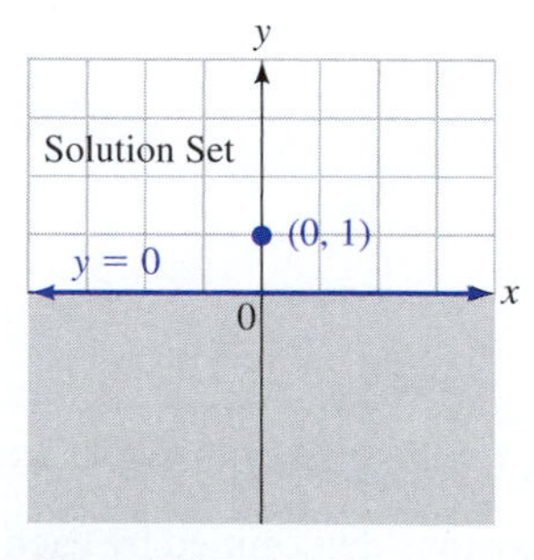

Figure 7

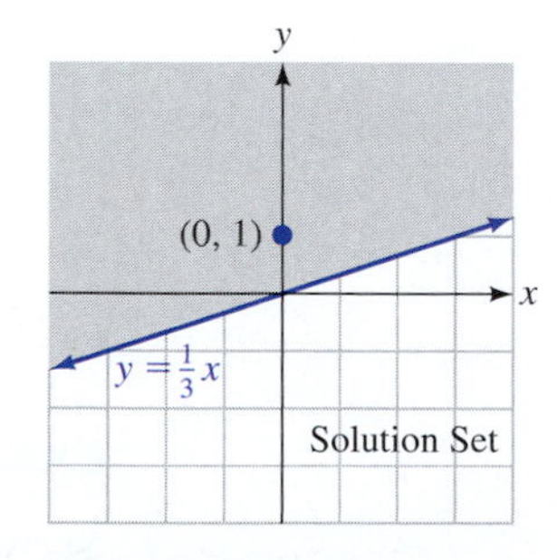

Figure 8

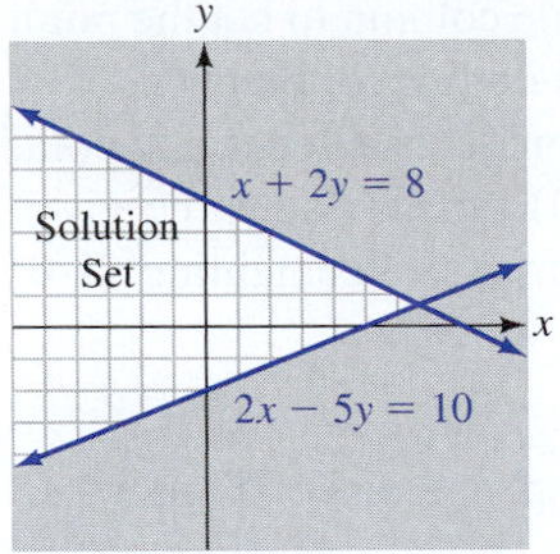

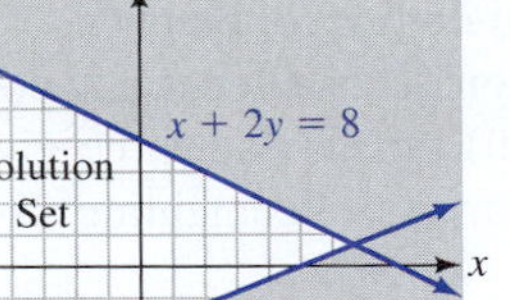

Figure 9

* Although these graphs are quite easy to do by hand, the more lines we have to graph the more difficult it becomes to get everything in the right place, and this is where graphing technology can become important. This is especially true when, for instance, three or more lines intersect in points that are very close together and hard to distinguish in hand-drawn graphs.

EXAMPLE 2 Graphing Simultaneous Inequalities

Sketch the region of points that satisfy both inequalities:

$$2x - 5y \leq 10$$
$$x + 2y \leq 8.$$

Solution Each inequality has a solution set that is a half plane. If a point is to satisfy *both* inequalities, it must lie in both sets of solutions. Put another way, if we cover the points that are not solutions to $2x - 5y \leq 10$ and then also cover the points that are not solutions to $x + 2y \leq 8$, the points that remain uncovered must be the points we want, those that are solutions to both inequalities. The result is shown in Figure 9, where the unshaded region is the set of solutions.*

As a check, we can look at points in various regions in Figure 9. For example, our graph shows that (0, 0) should satisfy both inequalities, and it does:

$$2(0) - 5(0) = 0 \leq 10 \quad ✔$$
$$0 + 2(0) = 0 \leq 8. \quad ✔$$

On the other hand, (0, 5) should fail to satisfy one of the inequalities:

$$2(0) - 5(5) = -25 \leq 10 \quad ✔$$
$$0 + 2(5) = 10 > 8 \quad ✗$$

One more: (5, −1) should fail one of the inequalities:

$$2(5) - 5(-1) = 15 > 10 \quad ✗$$
$$5 + 2(-1) = 3 \leq 8. \quad ✔$$

EXAMPLE 3 Corner Points

Sketch the region of solutions of the following system of inequalities and list the coordinates of all the corner points.

$$3x - 2y \leq 6$$
$$x + y \geq -5$$
$$y \leq 4$$

Figure 10

Solution Shading the regions that we do not want leaves us with the triangle shown in Figure 10. We label the corner points A, B, and C as shown.

Each of these corner points lies at the intersection of two of the bounding lines. So, to find the coordinates of each corner point, we need to solve the system of equations given by the two lines. To do this systematically, we make the following table:

Point	*Lines through Point*	*Coordinates*
A	$y = 4$ $x + y = -5$	$(-9, 4)$
B	$y = 4$ $3x - 2y = 6$	$\left(\frac{14}{3}, 4\right)$
C	$x + y = -5$ $3x - 2y = 6$	$\left(-\frac{4}{5}, -\frac{21}{5}\right)$

Technology Note Using the trace feature makes it easy to locate corner points graphically. Remember to zoom in for additional accuracy when appropriate. Of course, you can also use technology to help solve the systems of equations, as we discussed in Chapter 3.

Here, we have solved each system of equations in the middle column to get the point on the right, using the techniques of Chapter 2. You should do this for practice.*

As a partial check that we have drawn the correct region, let us choose any point in its interior—say, (0, 0). We can easily check that (0, 0) satisfies all three given inequalities. It follows that all of the points in the triangular region containing (0, 0) are also solutions.

Take another look at the regions of solutions in Examples 2 and 3 (Figure 11).

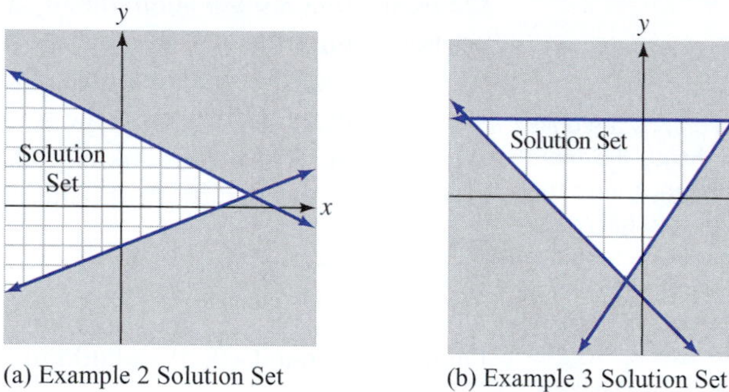

(a) Example 2 Solution Set

(b) Example 3 Solution Set

Figure 11

Notice that the solution set in Figure 11(a) extends infinitely far to the left, whereas the one in Figure 11(b) is completely enclosed by a boundary. Sets that are completely enclosed are called **bounded**, and sets that extend infinitely in one or more directions are **unbounded**. For example, all the solution sets in Example 1 are unbounded.

EXAMPLE 4 Resource Allocation

Socaccio Pistachio Inc. makes two types of pistachio nuts: Dazzling Red and Organic. Pistachio nuts require food color and salt, and the following table shows the amount of food color and salt required for a 1-kilogram batch of pistachios, as well as the total amount of these ingredients available each day.

	Dazzling Red	*Organic*	*Total Available*
Food Color (g)	2	1	20
Salt (g)	10	20	220

Use a graph to show the possible numbers of batches of each type of pistachio Socaccio can produce each day. This region (the solution set of a system of inequalities) is called the **feasible region**.

Solution As we did in Chapter 3, we start by identifying the unknowns: Let x be the number of batches of Dazzling Red manufactured per day and let y be the number of batches of Organic manufactured each day.

Now, because of our experience with systems of linear equations, we are tempted to say: For food color $2x + y = 20$ and for salt, $10x + 20y = 220$. However, no one is saying that Socaccio has to use all available ingredients; the company might choose to use fewer than the total available amounts if this proves more profitable. Thus, $2x + y$ can be anything *up to a total of* 20. In other words,

$$2x + y \le 20.$$

Similarly,

$$10x + 20y \le 220.$$

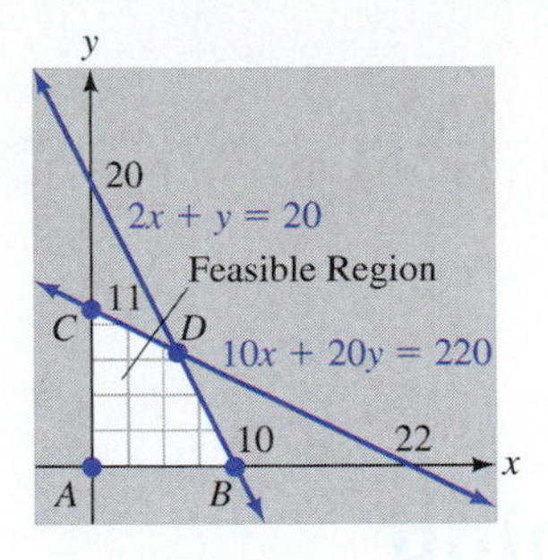

Figure 12

There are two more restrictions not explicitly mentioned: Neither x nor y can be negative. (The company cannot produce a negative number of batches of nuts.) Therefore, we have the additional restrictions

$$x \geq 0 \quad y \geq 0.$$

These two inequalities tell us that the feasible region (solution set) is restricted to the first quadrant, because in the other quadrants, either x or y or both x and y are negative. So instead of shading out all other quadrants, we can simply restrict our drawing to the first quadrant.

The (bounded) feasible region shown in Figure 12 is a graphical representation of the limitations the company faces.

Before we go on... Every point in the feasible region in Example 4 represents a value for x and a value for y that do not violate any of the company's restrictions. For example, the point (5, 6) lies well inside the region, so the company can produce five batches of Dazzling Red nuts and six batches of Organic without exceeding the limitations on ingredients [that is, $2(5) + 6 = 16 \leq 20$ and $10(5) + 20(6) = 170 \leq 220$]. The corner points A, B, C, and D are significant if the company wishes to realize the greatest profit, as we will see in Section 5.2. We can find the corners as in the following table:

Point	*Lines through Point*	*Coordinates*
A		(0, 0)
B		(10, 0)
C		(0, 11)
D	$2x + y = 20$ $10x + 20y = 220$	(6, 8)

(We have not listed the lines through the first three corners because their coordinates can be read easily from the graph.) Points on the line segment DB represent use of all the food color (because the segment lies on the line $2x + y = 20$), and points on the line segment CD represent use of all the salt (because the segment lies on the line $10x + 20y = 220$). Note that the point D is the only solution that uses all of both ingredients. ■

FAQs

Recognizing Whether to Use a Linear Inequality or a Linear Equation

Q: *How do I know whether to model a situation by a linear inequality like $3x + 2y \leq 10$ or by a linear equation like $3x + 2y = 10$?*

A: Here are some key phrases to look for: *at most, up to, no more than, at least, or more, exactly*. Suppose, for instance, that nuts cost 3¢, bolts cost 2¢, x is the number of nuts you can buy, and y is the number of bolts you can buy.

- If you have *up to* 10¢ to spend, then $3x + 2y \leq 10$.
- If you must spend *exactly* 10¢, then $3x + 2y = 10$.
- If you plan to spend *at least* 10¢, then $3x + 2y \geq 10$.

The use of inequalities to model a situation is often more realistic than the use of equations; for instance, one cannot always expect to exactly fill all orders, spend the exact amount of one's budget, or keep a plant operating at exactly 100% capacity.

5.1 EXERCISES

Access end-of-section exercises online at **www.webassign.net** 

5.2 Solving Linear Programming Problems Graphically

As we saw in Example 4 in Section 5.1, in some scenarios the possibilities are restricted by a system of linear inequalities. In that example, it would be natural to ask which of the various possibilities gives the company the largest profit. This is a kind of problem known as a *linear programming problem* (commonly referred to as an LP problem).

Linear Programming (LP) Problems

A **linear programming problem** in two unknowns x and y is one in which we are to find the maximum or minimum value of a linear expression

$$ax + by$$

called the **objective function**, subject to a number of linear **constraints** of the form

$$cx + dy \le e \quad \text{or} \quad cx + dy \ge e.$$

The largest or smallest value of the objective function is called the **optimal value**, and a pair of values of x and y that gives the optimal value constitutes an **optimal solution**.

Quick Example

$$\begin{aligned} \text{Maximize} \quad & p = x + y && \text{Objective function} \\ \text{subject to} \quad & \left.\begin{aligned} x + 2y &\le 12 \\ 2x + y &\le 12 \\ x \ge 0, y &\ge 0. \end{aligned}\right\} && \text{Constraints} \end{aligned}$$

See Example 1 for a method of solving this LP problem (that is, finding an optimal solution and value).

The set of points (x, y) satisfying all the constraints is the **feasible region** for the problem. Our methods of solving LP problems rely on the following facts:

Fundamental Theorem of Linear Programming

- If an LP problem has optimal solutions, then at least one of these solutions occurs at a corner point of the feasible region.
- Linear programming problems with bounded, nonempty feasible regions always have optimal solutions.

Let's see how we can use this to solve an LP problem, and then we'll discuss why it's true.

EXAMPLE 1 Solving an LP Problem

$$\begin{aligned} \text{Maximize} \quad & p = x + y \\ \text{subject to} \quad & x + 2y \leq 12 \\ & 2x + y \leq 12 \\ & x \geq 0, y \geq 0. \end{aligned}$$

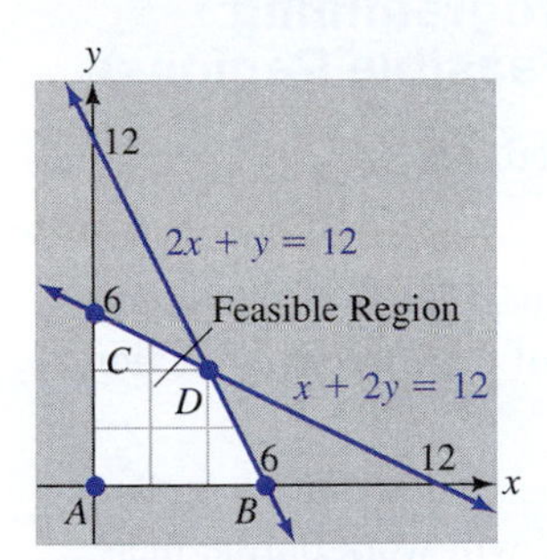

Figure 13

Solution We begin by drawing the feasible region for the problem. We do this using the techniques of Section 5.1, and we get Figure 13.

Each **feasible point** (point in the feasible region) gives an x and a y satisfying the constraints. The question now is, which of these points gives the largest value of the objective function $p = x + y$? The Fundamental Theorem of Linear Programming tells us that the largest value must occur at one (or more) of the corners of the feasible region. In the following table, we list the coordinates of each corner point and we compute the value of the objective function at each corner.

Corner Point	*Lines through Point*	*Coordinates*	$p = x + y$
A		(0, 0)	0
B		(6, 0)	6
C		(0, 6)	6
D	$x + 2y = 12$ $2x + y = 12$	(4, 4)	8

Now we simply pick the one that gives the largest value for p, which is D. Therefore, the optimal value of p is 8, and an optimal solution is (4, 4).

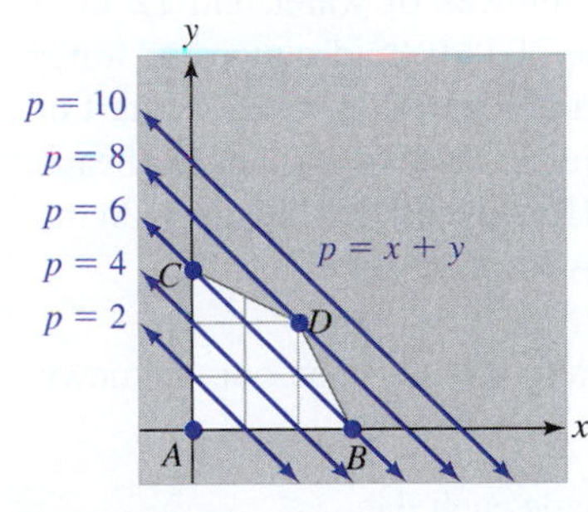

Figure 14

Now we owe you an explanation of why one of the corner points should be an optimal solution. The question is, which point in the feasible region gives the largest possible value of $p = x + y$?

Consider first an easier question: Which points result in a *particular value* of p? For example, which points result in $p = 2$? These would be the points on the line $x + y = 2$, which is the line labeled $p = 2$ in Figure 14.

Now suppose we want to know which points make $p = 4$: These would be the points on the line $x + y = 4$, which is the line labeled $p = 4$ in Figure 14. Notice that this line is parallel to but higher than the line $p = 2$. (If p represented profit in an application, we would call these **isoprofit lines**, or **constant-profit lines**.) Imagine moving this line up or down in the picture. As we move the line down, we see smaller values of p, and as we move it up, we see larger values. Several more of these lines are drawn in Figure 14. Look, in particular, at the line labeled $p = 10$. This line does not meet the feasible region, meaning that no feasible point makes p as large as 10. Starting with the line $p = 2$, as we move the line up, increasing p, there will be a last line that meets the feasible region. In the figure it is clear that this is the line $p = 8$, and this meets the feasible region in only one point, which is the corner point D. Therefore, D gives the greatest value of p of all feasible points.

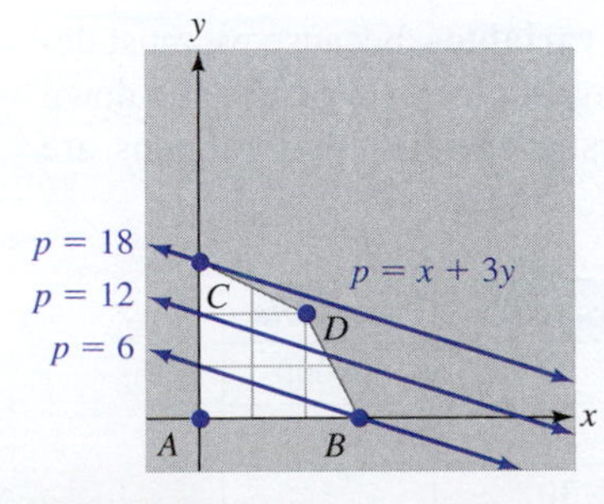

Figure 15

If we had been asked to maximize some other objective function, such as $p = x + 3y$, then the optimal solution might be different. Figure 15 shows some of the isoprofit lines for this objective function. This time, the last point that is hit as p

increases is C, not D. This tells us that the optimal solution is (0, 6), giving the optimal value $p = 18$.

This discussion should convince you that the optimal value in an LP problem will always occur at one of the corner points. By the way, it is possible for the optimal value to occur at *two* corner points and at all points along an edge connecting them. (Do you see why?) We will see this in Example 3(b).

Here is a summary of the method we have just been using.

Graphical Method for Solving Linear Programming Problems in Two Unknowns (Bounded Feasible Regions)

1. Graph the feasible region and check that it is bounded.
2. Compute the coordinates of the corner points.
3. Substitute the coordinates of the corner points into the objective function to see which gives the maximum (or minimum) value of the objective function.
4. Any such corner point is an optimal solution.

Note If the feasible region is unbounded, this method will work only if there are optimal solutions; otherwise, it will not work. We will show you a method for deciding this on page 264. ■

APPLICATIONS

EXAMPLE 2 Resource Allocation

Acme Baby Foods mixes two strengths of apple juice. One quart of Beginner's juice is made from 30 fluid ounces of water and 2 fluid ounces of apple juice concentrate. One quart of Advanced juice is made from 20 fluid ounces of water and 12 fluid ounces of concentrate. Every day Acme has available 30,000 fluid ounces of water and 3,600 fluid ounces of concentrate. Acme makes a profit of 20¢ on each quart of Beginner's juice and 30¢ on each quart of Advanced juice. How many quarts of each should Acme make each day to get the largest profit? How would this change if Acme made a profit of 40¢ on Beginner's juice and 20¢ on Advanced juice?

Solution Looking at the question that we are asked, we see that our unknown quantities are

$x =$ number of quarts of Beginner's juice made each day
$y =$ number of quarts of Advanced juice made each day.

(In this context, x and y are often called the **decision variables**, because we must decide what their values should be in order to get the largest profit.) We can write down the data given in the form of a table (the numbers in the first two columns are amounts per quart of juice):

	Beginner's, x	*Advanced, y*	*Available*
Water (ounces)	30	20	30,000
Concentrate (ounces)	2	12	3,600
Profit (¢)	20	30	

Because nothing in the problem says that Acme must use up all the water or concentrate, just that it can use no more than what is available, the first two rows of the table give us two inequalities:

$$30x + 20y \le 30{,}000$$
$$2x + 12y \le 3{,}600.$$

Dividing the first inequality by 10 and the second by 2 gives

$$3x + 2y \le 3{,}000$$
$$x + 6y \le 1{,}800.$$

We also have that $x \ge 0$ and $y \ge 0$ because Acme can't make a negative amount of juice. To finish setting up the problem, we are asked to maximize the profit, which is

$$p = 20x + 30y. \qquad \text{Expressed in ¢}$$

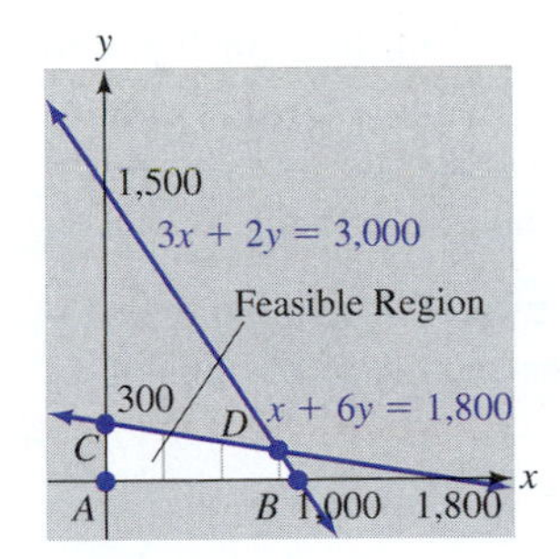

Figure 16

This gives us our LP problem:

$$\begin{aligned} \text{Maximize} \quad & p = 20x + 30y \\ \text{subject to} \quad & 3x + 2y \le 3{,}000 \\ & x + 6y \le 1{,}800 \\ & x \ge 0, y \ge 0. \end{aligned}$$

The (bounded) feasible region is shown in Figure 16.

The corners and the values of the objective function are listed in the following table:

Point	*Lines through Point*	*Coordinates*	$p = 20x + 30y$
A		(0, 0)	0
B		(1,000, 0)	20,000
C		(0, 300)	9,000
D	$3x + 2y = 3{,}000$ $x + 6y = 1{,}800$	(900, 150)	22,500

We are seeking to maximize the objective function p, so we look for corner points that give the maximum value for p. Because the maximum occurs at the point D, we conclude that the (only) optimal solution occurs at D. Thus, the company should make 900 quarts of Beginner's juice and 150 quarts of Advanced juice, for a largest possible profit of 22,500¢, or $225.

If, instead, the company made a profit of 40¢ on each quart of Beginner's juice and 20¢ on each quart of Advanced juice, then we would have $p = 40x + 20y$. This gives the following table:

Point	*Lines through Point*	*Coordinates*	$p = 40x + 20y$
A		(0, 0)	0
B		(1,000, 0)	40,000
C		(0, 300)	6,000
D	$3x + 2y = 3{,}000$ $x + 6y = 1{,}800$	(900, 150)	39,000

We can see that, in this case, Acme should make 1,000 quarts of Beginner's juice and no Advanced juice, for a largest possible profit of 40,000¢, or $400.

Before we go on... Notice that, in the first version of the problem in Example 2, the company used all the water and juice concentrate:

Water: $30(900) + 20(150) = 30{,}000$

Concentrate: $2(900) + 12(150) = 3{,}600.$

In the second version, it used all the water but not all the concentrate:

Water: $30(100) + 20(0) = 30{,}000$

Concentrate: $2(100) + 12(0) = 200 < 3{,}600.$ ■

EXAMPLE 3 Investments

The Solid Trust Savings & Loan Company has set aside \$25 million for loans to home buyers. Its policy is to allocate at least \$10 million annually for luxury condominiums. A government housing development grant it receives requires, however, that at least one third of its total loans be allocated to low-income housing.

a. Solid Trust's return on condominiums is 12% and its return on low-income housing is 10%. How much should the company allocate for each type of housing to maximize its total return?

b. Redo part (a), assuming that the return is 12% on both condominiums and low-income housing.

Solution

a. We first identify the unknowns: Let x be the annual amount (in millions of dollars) allocated to luxury condominiums and let y be the annual amount allocated to low-income housing.

We now look at the constraints. The first constraint is mentioned in the first sentence: The total the company can invest is \$25 million. Thus,

$$x + y \leq 25.$$

(The company is not required to invest all of the \$25 million; rather, it can invest *up to* \$25 million.) Next, the company has allocated at least \$10 million to condos. Rephrasing this in terms of the unknowns, we get

The amount allocated to condos is at least \$10 *million.*

The phrase "is at least" means $\geq$. Thus, we obtain a second constraint:

$$x \geq 10.$$

The third constraint is that at least one third of the total financing must be for low-income housing. Rephrasing this, we say:

The amount allocated to low-income housing is at least one third of the total.

Because the total investment will be $x + y$, we get

$$y \geq \frac{1}{3}(x + y).$$

We put this in the standard form of a linear inequality as follows:

$$3y \geq x + y \qquad \text{Multiply both sides by 3.}$$

$$-x + 2y \geq 0. \qquad \text{Subtract } x + y \text{ from both sides.}$$

There are no further constraints.

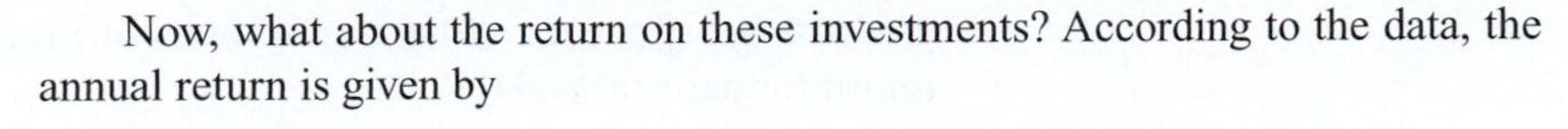

Now, what about the return on these investments? According to the data, the annual return is given by

$$p = 0.12x + 0.10y.$$

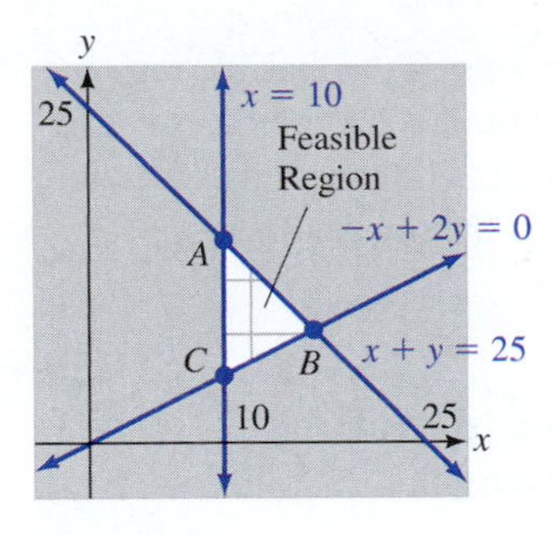

Figure 17

We want to make this quantity p as large as possible. In other words, we want to

$$\begin{aligned} \text{Maximize} \quad & p = 0.12x + 0.10y \\ \text{subject to} \quad & x + y \le 25 \\ & x \ge 10 \\ & -x + 2y \ge 0 \\ & x \ge 0, y \ge 0. \end{aligned}$$

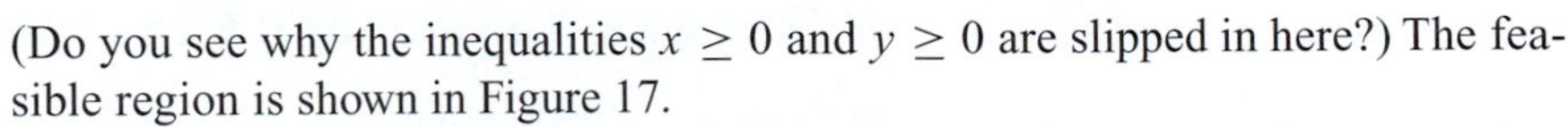

(Do you see why the inequalities $x \ge 0$ and $y \ge 0$ are slipped in here?) The feasible region is shown in Figure 17.

We now make a table that gives the return on investment at each corner point:

Point	*Lines through Point*	*Coordinates*	$p = 0.12x + 0.10y$
A	$x = 10$ $x + y = 25$	(10, 15)	2.7
B	$x + y = 25$ $-x + 2y = 0$	(50/3, 25/3)	2.833
C	$x = 10$ $-x + 2y = 0$	(10, 5)	1.7

From the table, we see that the values of x and y that maximize the return are $x = 50/3$ and $y = 25/3$, which give a total return of \$2.833 million. In other words, the most profitable course of action is to invest \$16.667 million in loans for condominiums and \$8.333 million in loans for low-income housing, giving a maximum annual return of \$2.833 million.

b. The LP problem is the same as for part (a) except for the objective function:

$$\begin{aligned} \text{Maximize} \quad & p = 0.12x + 0.12y \\ \text{subject to} \quad & x + y \le 25 \\ & x \ge 10 \\ & -x + 2y \ge 0 \\ & x \ge 0, y \ge 0. \end{aligned}$$

Here are the values of p at the three corners:

Point	*Coordinates*	$p = 0.12x + 0.12y$
A	(10, 15)	3
B	(50/3, 25/3)	3
C	(10, 5)	1.8

Looking at the table, we see that a curious thing has happened: We get the same maximum annual return at both A and B. Thus, we could choose either option to maximize the annual return. In fact, any point along the line segment AB will yield an annual return of \$3 million. For example, the point (12, 13) lies on the line segment AB and also yields an annual revenue of \$3 million. This happens because the "isoreturn" lines are parallel to that edge.

Before we go on... What breakdowns of investments would lead to the *lowest* return for parts (a) and (b)? ■

The preceding examples all had bounded feasible regions. If the feasible region is unbounded, then, *provided there are optimal solutions,* the fundamental theorem of linear programming guarantees that the above method will work. The following procedure determines whether or not optimal solutions exist and finds them when they do.

Solving Linear Programming Problems in Two Unknowns (Unbounded Feasible Regions)

If the feasible region of an LP problem is unbounded, proceed as follows:

1. Draw a rectangle large enough so that all the corner points are inside the rectangle (and not on its boundary):

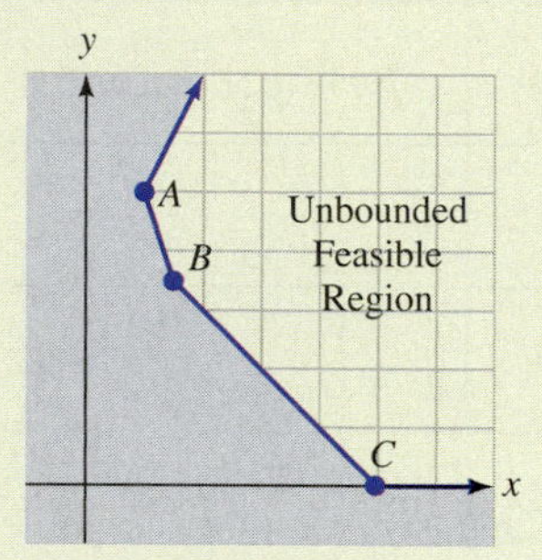

Corner points: A, B, C

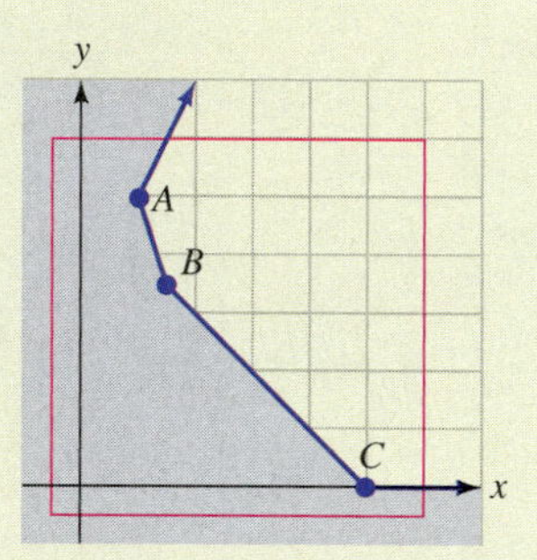

Corner points inside the rectangle

2. Shade the outside of the rectangle so as to define a new bounded feasible region, and locate the new corner points:

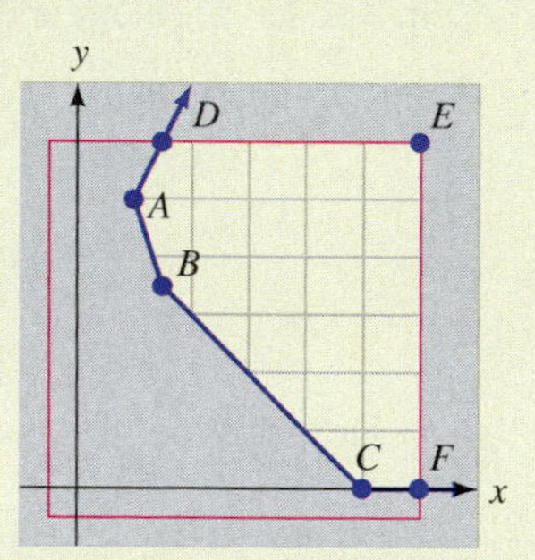

New corner points: D, E, and F

3. Obtain the optimal solutions using this bounded feasible region.
4. If any optimal solutions occur at one of the original corner points (A, B, and C in the figure), then the LP problem has that corner point as an optimal solution. Otherwise, the LP problem has no optimal solutions. When the latter occurs, we say that the **objective function is unbounded**, because it can assume arbitrarily large (positive or negative) values.

In the next two examples, we work with unbounded feasible regions.

EXAMPLE 4 Cost

You are the manager of a small store that specializes in hats, sunglasses, and other accessories. You are considering a sales promotion of a new line of hats and sunglasses. You will offer the sunglasses only to those who purchase two or more hats, so you will sell at least twice as many hats as pairs of sunglasses. Moreover, your supplier tells you that, due to seasonal demand, your order of sunglasses cannot exceed 100 pairs. To ensure that the sale items fill out the large display you have set aside, you estimate that you should order at least 210 items in all.

a. Assume that you will lose \$3 on every hat and \$2 on every pair of sunglasses sold. Given the constraints above, how many hats and pairs of sunglasses should you order to lose the least amount of money in the sales promotion?

b. Suppose instead that you lose \$1 on every hat sold but make a profit of \$5 on every pair of sunglasses sold. How many hats and pairs of sunglasses should you order to make the largest profit in the sales promotion?

c. Now suppose that you make a profit of \$1 on every hat sold but lose \$5 on every pair of sunglasses sold. How many hats and pairs of sunglasses should you order to make the largest profit in the sales promotion?

Solution

a. The unknowns are:

$$x = \text{number of hats you order}$$
$$y = \text{number of pairs of sunglasses you order.}$$

The objective is to minimize the total loss:

$$c = 3x + 2y.$$

Now for the constraints. The requirement that you will sell at least twice as many hats as sunglasses can be rephrased as:

The number of hats is at least twice the number of pairs of sunglasses,

or

$$x \geq 2y$$

which, in standard form, is

$$x - 2y \geq 0.$$

Next, your order of sunglasses cannot exceed 100 pairs, so

$$y \leq 100.$$

Finally, you would like to sell at least 210 items in all, giving

$$x + y \geq 210.$$

Thus, the LP problem is the following:

$$\begin{aligned} \text{Minimize} \quad & c = 3x + 2y \\ \text{subject to} \quad & x - 2y \geq 0 \\ & y \leq 100 \\ & x + y \geq 210 \\ & x \geq 0, y \geq 0. \end{aligned}$$

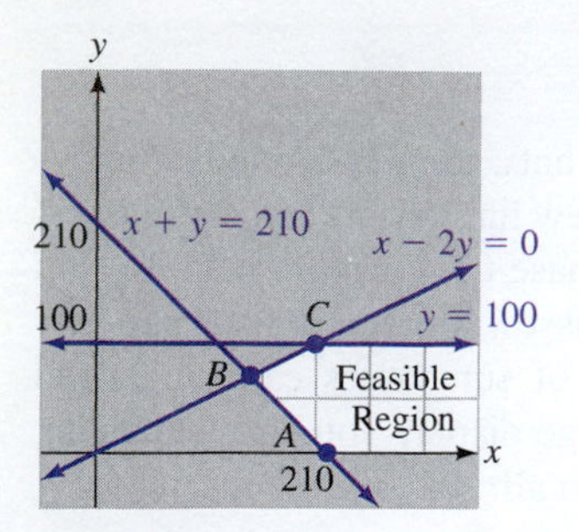

Figure 18

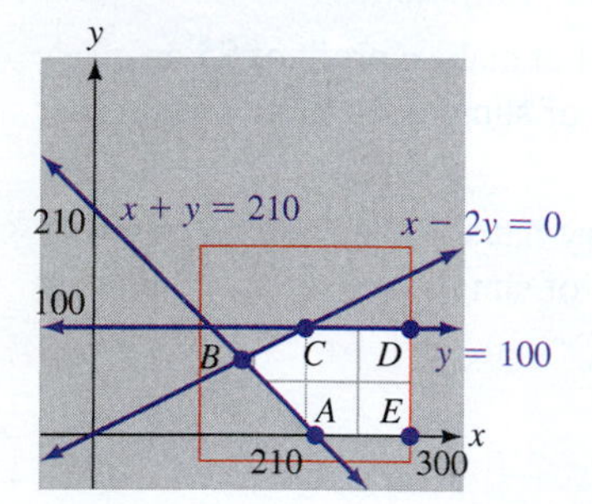

Figure 19

The feasible region is shown in Figure 18. This region is unbounded, so there is no guarantee that there are any optimal solutions. Following the procedure described above, we enclose the corner points in a rectangle as shown in Figure 19. (There are many infinitely many possible rectangles we could have used. We chose one that gives convenient coordinates for the new corners.)

We now list all the corners of this bounded region along with the corresponding values of the objective function c:

Point	*Lines through Point*	*Coordinates*	$c = 3x + 2y$ **($)**
A		(210, 0)	630
B	$x + y = 210$ $x - 2y = 0$	(140, 70)	560
C	$x - 2y = 0$ $y = 100$	(200, 100)	800
D		(300, 100)	1,100
E		(300, 0)	900

The corner point that gives the minimum value of the objective function c is B. Because B is one of the corner points of the original feasible region, we conclude that our linear programming problem has an optimal solution at B. Thus, the combination that gives the smallest loss is 140 hats and 70 pairs of sunglasses.

b. The LP problem is the following:

$$\begin{aligned} \text{Maximize} \quad & p = -x + 5y \\ \text{subject to} \quad & x - 2y \ge 0 \\ & y \le 100 \\ & x + y \ge 210 \\ & x \ge 0, y \ge 0. \end{aligned}$$

Because most of the work is already done for us in part (a), all we need to do is change the objective function in the table that lists the corner points:

Point	*Lines through Point*	*Coordinates*	$p = -x + 5y$ **($)**
A		(210, 0)	−210
B	$x + y = 210$ $x - 2y = 0$	(140, 70)	210
C	$x - 2y = 0$ $y = 100$	(200, 100)	300
D		(300, 100)	200
E		(300, 0)	−300

The corner point that gives the maximum value of the objective function p is C. Because C is one of the corner points of the original feasible region, we conclude that our LP problem has an optimal solution at C. Thus, the combination that gives the largest profit ($300) is 200 hats and 100 pairs of sunglasses.

c. The objective function is now $p = x - 5y$, which is the negative of the objective function used in part (b). Thus, the table of values of p is the same as in part (b), except that it has opposite signs in the p column. This time we find that the maximum value of p occurs at E. However, E is not a corner point of the original feasible region, so the LP problem has no optimal solution. Referring to Figure 18, we can make the objective p as large as we like by choosing a point far to the right in the unbounded feasible region. Thus, the objective function is unbounded; that is, it is possible to make an arbitrarily large profit.

EXAMPLE 5 Resource Allocation

You are composing a very avant-garde ballade for violins and bassoons. In your ballade, each violinist plays a total of two notes and each bassoonist only one note. To make your ballade long enough, you decide that it should contain at least 200 instrumental notes. Furthermore, after playing the requisite two notes, each violinist will sing one soprano note, while each bassoonist will sing three soprano notes.* To make the ballade sufficiently interesting, you have decided on a minimum of 300 soprano notes. To give your composition a sense of balance, you wish to have no more than three times as many bassoonists as violinists. Violinists charge $200 per performance and bassoonists $400 per performance. How many of each should your ballade call for in order to minimize personnel costs?

* Whether or not these musicians are capable of singing decent soprano notes will be left to chance. You reason that a few bad notes will add character to the ballade.

Solution First, the unknowns are x = number of violinists and y = number of bassoonists. The constraint on the number of instrumental notes implies that

$$2x + y \geq 200$$

because the total number is to be *at least* 200. Similarly, the constraint on the number of soprano notes is

$$x + 3y \geq 300.$$

The next one is a little tricky. As usual, we reword it in terms of the quantities x and y.

The number of bassoonists should be no more than three times the number of violinists.

Thus, $y \leq 3x$

or $3x - y \geq 0$.

Finally, the total cost per performance will be

$$c = 200x + 400y.$$

We wish to minimize total cost. So, our linear programming problem is as follows:

$$\begin{aligned} \text{Minimize} \quad & c = 200x + 400y \\ \text{subject to} \quad & 2x + y \geq 200 \\ & x + 3y \geq 300 \\ & 3x - y \geq 0 \\ & x \geq 0, y \geq 0. \end{aligned}$$

We get the feasible region shown in Figure 20.† The feasible region is unbounded, and so we add a convenient rectangle as before (Figure 21).

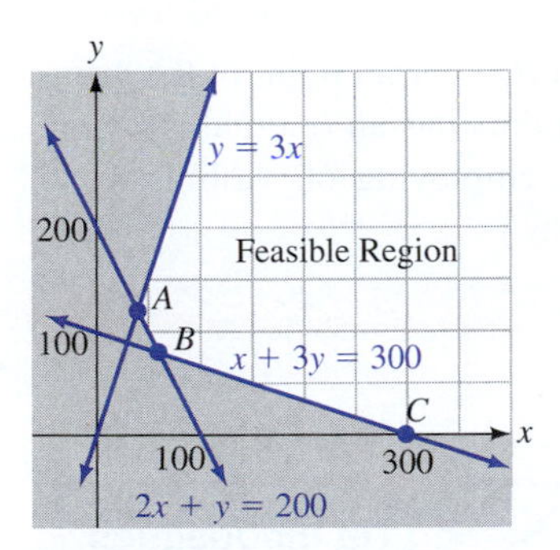

Figure 20

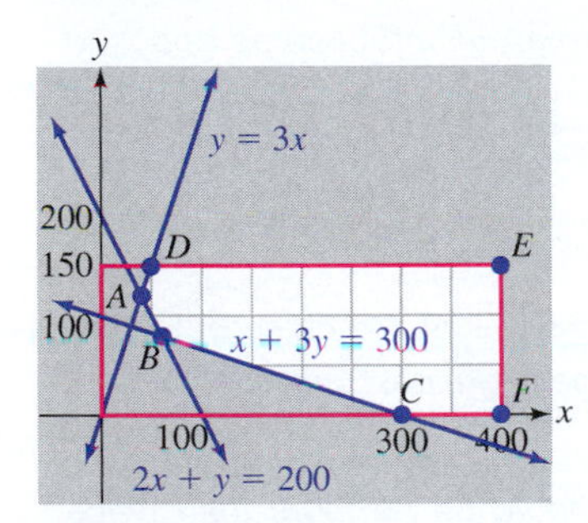

Figure 21

† In Figure 20 you can see how graphing technology would help in determining the corner points: Unless you are very confident in the accuracy of your sketch, how do you know that the line $y = 3x$ falls to the left of the point B? If it were to fall to the right, then B would not be a corner point and the solution would be different. You could (and should) check that B satisfies the inequality $3x - y \geq 0$, so that the line falls to the left of B as shown. However, if you use a graphing calculator or computer, you can be fairly confident of the picture produced without doing further calculations.

Point	*Lines through Point*	*Coordinates*	$c = 200x + 400y$
A	$2x + y = 200$ $3x - y = 0$	(40, 120)	56,000
B	$2x + y = 200$ $x + 3y = 300$	(60, 80)	44,000
C		(300, 0)	60,000
D	$3x - y = 0$ $y = 150$	(50, 150)	70,000
E		(400, 150)	140,000
F		(400, 0)	80,000

From the table we see that the minimum cost occurs at B, a corner point of the original feasible region. The linear programming problem thus has an optimal solution, and the minimum cost is \$44,000 per performance, employing 60 violinists and 80 bassoonists. (Quite a wasteful ballade, one might say.)

using Technology

Website
www.WanerMath.com
For an online utility that does everything (solves linear programming problems with two unknowns and even draws the feasible region), go to the Website and follow

→ On Line Utilities
→ Linear Programming Grapher.

FAQs

Recognizing a Linear Programming Problem, Setting Up Inequalities, and Dealing with Unbounded Regions

Q: *How do I recognize when an application leads to an LP problem as opposed to a system of linear equations?*

A: Here are some cues that suggest an LP problem:

- Key phrases suggesting inequalities rather than equalities, like *at most, up to, no more than, at least*, and *or more*.
- A quantity that is being maximized or minimized (this will be the objective). Key phrases are *maximum, minimum, most, least, largest, greatest, smallest, as large as possible*, and *as small as possible*.

Q: *How do I deal with tricky phrases like "there should be no more than twice as many nuts as bolts" or "at least 50% of the total should be bolts"?*

A: The easiest way to deal with phrases like this is to use the technique we discussed in Chapter 3: reword the phrases using "the number of . . . ", as in

The number of nuts (x) is no more than twice the number of bolts (y) $\quad x \le 2y$
The number of bolts is at least 50% of the total $\quad y \ge 0.50(x + y)$

Q: *Do I always have to add a rectangle to deal with unbounded regions?*

A: Under some circumstances, you can tell right away whether optimal solutions exist, even when the feasible region is unbounded.

Note that the following apply only when we have the constraints $x \ge 0$ and $y \ge 0$.

1. If you are minimizing $c = ax + by$ with a and b nonnegative, then optimal solutions always exist. (Examples 4(a) and 5 are of this type.)
2. If you are maximizing $p = ax + by$ with a and b nonnegative (and not both zero), then there is no optimal solution unless the feasible region is bounded.

Do you see why statements (1) and (2) are true?

5.2 EXERCISES

Access end-of-section exercises online at **www.webassign.net** 

5.3 The Simplex Method: Solving Standard Maximization Problems

The method discussed in Section 5.2 works quite well for LP problems in two unknowns, but what about three or more unknowns? Because we need an axis for each unknown, we would need to draw graphs in three dimensions (where we have x-, y-, and z-coordinates) to deal with problems in three unknowns, and we would have to draw in hyperspace to answer questions involving four or more unknowns. Given the state of technology as this book is being written, we can't easily do this. So we need another method for solving LP problems that will work for any number of unknowns. One such method, called the **simplex method**, has been the method of choice since it was invented by George Dantzig in 1947. (See the Introduction to this chapter for more about Dantzig.) To illustrate it best, we first use it to solve only so-called standard maximization problems.

General Linear Programming Problem

A **linear programming problem in n unknowns** $x_1, x_2, \ldots, x_n$ is one in which we are to find the maximum or minimum value of a linear **objective function**

$$a_1x_1 + a_2x_2 + \cdots + a_nx_n,$$

where $a_1, a_2, \ldots, a_n$ are numbers, subject to a number of linear **constraints** of the form

$$b_1x_1 + b_2x_2 + \cdots + b_nx_n \leq c \quad \text{or} \quad b_1x_1 + b_2x_2 + \cdots + b_nx_n \geq c,$$

where $b_1, b_2, \ldots, b_n, c$ are numbers.

Standard Maximization Problem

A **standard maximization problem** is an LP problem in which we are required to *maximize* (not minimize) an objective function of the form

$$p = a_1x_1 + a_2x_2 + \cdots + a_nx_n$$

subject to the constraints

$$x_1 \geq 0, x_2 \geq 0, \ldots, x_n \geq 0$$

and further constraints of the form

$$b_1x_1 + b_2x_2 + \cdots + b_nx_n \leq c$$

with c *nonnegative*. It is important that the inequality here be $\leq$, *not* $=$ or $\geq$.

Note As in the chapter on linear equations, we will almost always use $x, y, z, \ldots$ for the unknowns. Subscripted variables $x_1, x_2, \ldots$ are very useful names when you start running out of letters of the alphabet, but we should not find ourselves in that predicament. ■

Quick Examples

1. Maximize $p = 2x - 3y + 3z$
subject to
$$2x + z \le 7$$
$$-x + 3y - 6z \le 6$$
$$x \ge 0, y \ge 0, z \ge 0.$$
This is a standard maximization problem.

2. Maximize $p = 2x_1 + x_2 - x_3 + x_4$
subject to
$$x_1 - 2x_2 + x_4 \le 0$$
$$3x_1 \le 1$$
$$x_2 + x_3 \le 2$$
$$x_1 \ge 0, x_2 \ge 0, x_3 \ge 0, x_4 \ge 0.$$
This is a standard maximization problem.

3. Maximize $p = 2x - 3y + 3z$
subject to
$$2x + z \ge 7$$
$$-x + 3y - 6z \le 6$$
$$x \ge 0, y \ge 0, z \ge 0.$$
This is *not* a standard maximization problem.

The inequality $2x + z \ge 7$ cannot be written in the required form. If we reverse the inequality by multiplying both sides by -1, we get $-2x - z \le -7$, but a negative value on the right side is not allowed.

The idea behind the simplex method is this: In any linear programming problem, there is a feasible region. If there are only two unknowns, we can draw the region; if there are three unknowns, it is a solid region in space; and if there are four or more unknowns, it is an abstract higher-dimensional region. But it is a faceted region with corners (think of a diamond), and it is at one of these corners that we will find the optimal solution. Geometrically, what the simplex method does is to start at the corner where all the unknowns are 0 (possible because we are talking of standard maximization problems) and then walk around the region, from corner to adjacent corner, always increasing the value of the objective function, until the best corner is found. In practice, we will visit only a small number of the corners before finding the right one. Algebraically, as we are about to see, this walking around is accomplished by matrix manipulations of the same sort as those used in the chapter on systems of linear equations.

We describe the method while working through an example.

EXAMPLE 1 Meet the Simplex Method

Maximize $p = 3x + 2y + z$
subject to
$$2x + 2y + z \le 10$$
$$x + 2y + 3z \le 15$$
$$x \ge 0,\ y \ge 0,\ z \ge 0.$$

Solution

Step 1 ***Convert to a system of linear equations.*** The inequalities $2x + 2y + z \le 10$ and $x + 2y + 3z \le 15$ are less convenient than equations. Look at the first inequality. It says that the left-hand side, $2x + 2y + z$, must have some positive number (or zero) *added to it* if it is to equal 10. Because we don't yet know what x, y, and z are, we are not yet sure what number to add to the left-hand side. So we invent a new unknown, $s \geq 0$, called a **slack variable**, to "take up the slack," so that

$$2x + 2y + z + s = 10.$$

Turning to the next inequality, $x + 2y + 3z \leq 15$, we now add a slack variable to its left-hand side, to get it up to the value of the right-hand side. We might have to add a different number than we did the last time, so we use a new slack variable, $t \geq 0$, and obtain

$$x + 2y + 3z + t = 15.$$

Use a different slack variable for each constraint.

Now we write the system of equations we have (including the one that defines the objective function) in standard form.

$$\begin{aligned} 2x + 2y + z + s \phantom{{}+t+p} &= 10 \\ x + 2y + 3z \phantom{{}+s} + t \phantom{{}+p} &= 15 \\ -3x - 2y - z \phantom{{}+s+t} + p &= 0 \end{aligned}$$

Note three things: First, all the variables are neatly aligned in columns, as they were in Chapter 3. Second, in rewriting the objective function $p = 3x + 2y + z$, we have left the coefficient of p as $+1$ and brought the other variables over to the same side of the equation as p. This will be our standard procedure from now on. *Don't* write $3x + 2y + z - p = 0$ (even though it means the same thing) because the negative coefficients will be important in later steps. Third, the above system of equations has fewer equations than unknowns, and hence cannot have a unique solution.

Step 2 ***Set up the initial tableau.*** We represent our system of equations by the following table (which is simply the augmented matrix in disguise), called **the initial tableau**:

	x	y	z	s	t	p	
	2	2	1	1	0	0	10
	1	2	3	0	1	0	15
	−3	−2	−1	0	0	1	0

The labels along the top keep track of which columns belong to which variables.

Now notice a peculiar thing. If we rewrite the matrix using the variables s, t, and p first, we get the matrix

$$\begin{array}{c} \begin{matrix} s & t & p & x & y & z & \end{matrix} \\ \begin{bmatrix} 1 & 0 & 0 & 2 & 2 & 1 & 10 \\ 0 & 1 & 0 & 1 & 2 & 3 & 15 \\ 0 & 0 & 1 & -3 & -2 & -1 & 0 \end{bmatrix}, \end{array}$$

Matrix with s, t, and p columns first

which is already in reduced form. We can therefore read off the general solution (see Section 3.2) to our system of equations as

$$\begin{aligned} s &= 10 - 2x - 2y - z \\ t &= 15 - x - 2y - 3z \\ p &= 0 + 3x + 2y + z \end{aligned}$$

x, y, z arbitrary.

Thus, we get a whole family of solutions, one for each choice of x, y, and z. One possible choice is to set x, y, and z all equal to 0. This gives the particular solution

$$s = 10, t = 15, p = 0, x = 0, y = 0, z = 0.$$

Set $x = y = z = 0$ above.

This solution is called the **basic solution** associated with the tableau. The variables s and t are called the **active** variables, and x, y, and z are the **inactive** variables. (Other terms used are **basic** and **nonbasic** variables.)

We can obtain the basic solution directly from the tableau as follows.

- The active variables correspond to the cleared columns (columns with only one nonzero entry).
- The values of the active variables are calculated as shown below.
- All other variables are inactive, and set equal to zero.

Inactive $x = 0$	Inactive $y = 0$	Inactive $z = 0$	Active $s = \frac{10}{1}$	Active $t = \frac{15}{1}$	Active $p = \frac{0}{1}$	
x	y	z	s	t	p	
2	2	1	1	0	0	**10**
1	2	3	0	1	0	**15**
−3	−2	−1	0	0	1	**0**

As an additional aid to recognizing which variables are active and which are inactive, we label each row with the name of the corresponding active variable. Thus, the complete initial tableau looks like this.

	x	y	z	s	t	p	
s	2	2	1	1	0	0	10
t	1	2	3	0	1	0	15
p	−3	−2	−1	0	0	1	0

This basic solution represents our starting position $x = y = z = 0$ in the feasible region in xyz space.

We now need to move to another corner point. To do so, we choose a pivot* in one of the first three columns of the tableau and clear its column. Then we will get a different basic solution, which corresponds to another corner point. Thus, in order to move from corner point to corner point, all we have to do is choose suitable pivots and clear columns in the usual manner.

* Also see Section 3.2 for a discussion of pivots and pivoting.

The next two steps give the procedure for choosing the pivot.

Step 3 ***Select the pivot column*** (the column that contains the pivot we are seeking).

Selecting the Pivot Column

Choose the negative number with the largest magnitude on the left-hand side of the bottom row (that is, don't consider the last number in the bottom row). Its column is the pivot column. (If there are two or more candidates, choose any one.) If all the numbers on the left-hand side of the bottom row are zero or positive, then we are done, and the basic solution is the optimal solution.

Simple enough. The most negative number in the bottom row is −3, so we choose the x column as the pivot column:

	x	y	z	s	t	p	
s	**2**	2	1	1	0	0	10
t	**1**	2	3	0	1	0	15
p	**−3**	−2	−1	0	0	1	0
	↑ Pivot column						

Q: *Why choose the pivot column this way?*

A: The variable labeling the pivot column is going to be increased from 0 to something positive. In the equation $p = 3x + 2y + z$, the fastest way to increase p is to increase x because p would increase by 3 units for every 1-unit increase in x. (If we chose to increase y, then p would increase by only 2 units for every 1-unit increase in y, and if we increased z instead, p would grow even more slowly.) In short, choosing the pivot column this way makes it likely that we'll increase p as much as possible.

Step 4 ***Select the pivot in the pivot column.***

Selecting the Pivot

1. The pivot must always be a positive number. (This rules out zeros and negative numbers, such as the −3 in the bottom row.)
2. For each positive entry b in the pivot column, compute the ratio a/b, where a is the number in the rightmost column in that row. We call this a **test ratio**.
3. Of these ratios, choose the smallest one. (If there are two or more candidates, choose any one.) The corresponding number b is the pivot.

In our example, the test ratio in the first row is $10/2 = 5$, and the test ratio in the second row is $15/1 = 15$. Here, 5 is the smallest, so the 2 in the upper left is our pivot.

	x	y	z	s	t	p		Test Ratios
s	[2]	2	1	1	0	0	10	$10/2 = 5$
t	1	2	3	0	1	0	15	$15/1 = 15$
p	−3	−2	−1	0	0	1	0	

Q: *Why select the pivot this way?*

A: The rule given above guarantees that, after pivoting, all variables will be nonnegative in the basic solution. In other words, it guarantees that we will remain in the feasible region. We will explain further after finishing this example.

Step 5 ***Use the pivot to clear the column in the normal manner and then relabel the pivot row with the label from the pivot column.*** It is important to follow the exact prescription described in Section 3.2 for formulating the row operations:

$$aR_c \pm bR_p.$$

a and *b* both positive

↑ Row to change (R_c) ↑ Pivot row (R_p)

All entries in the last column should remain nonnegative after pivoting. Furthermore, because the x column (and no longer the s column) will be cleared, x will become an active variable. In other words, the s on the left of the pivot will be replaced by x. We call s the **departing**, or **exiting variable** and x the **entering variable** for this step.

Entering variable ↓

Departing variable →

	x	y	z	s	t	p		
s	[2]	2	1	1	0	0	10	
t	1	2	3	0	1	0	15	$2R_2 - R_1$
p	−3	−2	−1	0	0	1	0	$2R_3 + 3R_1$

This gives

	x	y	z	s	t	p	
x	2	2	1	1	0	0	10
t	0	2	5	−1	2	0	20
p	0	2	1	3	0	2	30

This is the second tableau.

Step 6 ***Go to Step 3.*** But wait! According to Step 3, we are finished because there are no negative numbers in the bottom row. Thus, we can read off the answer. Remember, though, that the solution for x, the first active variable, is not just $x = 10$, but is $x = 10/2 = 5$ because the pivot has not been reduced to a 1. Similarly, $t = 20/2 = 10$ and $p = 30/2 = 15$. All the other variables are zero because they are inactive. Thus, the solution is as follows: p has a maximum value of 15, and this occurs when $x = 5$, $y = 0$, and $z = 0$. (The slack variables then have the values $s = 0$ and $t = 10$.)

Q: *Why can we stop when there are no negative numbers in the bottom row? Why does this tableau give an optimal solution?*

A: The bottom row corresponds to the equation $2y + z + 3s + 2p = 30$, or

$$p = 15 - y - \frac{1}{2}z - \frac{3}{2}s.$$

Think of this as part of the general solution to our original system of equations, with y, z, and s as the parameters. Because these variables must be nonnegative, *the largest possible value of p in any feasible solution of the system comes when all three of the parameters are 0.* Thus, the current basic solution must be an optimal solution.*

* Calculators or spreadsheets could obviously be a big help in the calculations here, just as in Chapter 3. We'll say more about that after the next couple of examples.

We owe some further explanation for Step 4 of the simplex method. After Step 3, we knew that x would be the entering variable, and we needed to choose the departing variable. In the next basic solution, x was to have some positive value and we wanted this value to be as large as possible (to make p as large as possible) without making any other variables negative. Look again at the equations written in Step 2:

$$s = 10 - 2x - 2y - z$$
$$t = 15 - x - 2y - 3z.$$

We needed to make either s or t into an inactive variable and hence zero. Also, y and z were to remain inactive. If we had made s inactive, then we would have had $0 = 10 - 2x$, so $x = 10/2 = 5$. This would have made $t = 15 - 5 = 10$, which would be fine. On the other hand, if we had made t inactive, then we would have had $0 = 15 - x$, so $x = 15$, and this would have made $s = 10 - 2 \cdot 15 = -20$, which would *not* be fine, because slack variables must be nonnegative. In other words, we had a choice of making $x = 10/2 = 5$ or $x = 15/1 = 15$, but making x larger than 5 would have made another variable negative. We were thus compelled to choose the smaller ratio, 5, and make s the departing variable. Of course, we do not have to think it through this way every time. We just use the rule stated in Step 4. (For a graphical explanation, see Example 3.)

EXAMPLE 2 Simplex Method

Find the maximum value of $p = 12x + 15y + 5z$, subject to the constraints:

$$\begin{aligned} 2x + 2y + z &\le 8 \\ x + 4y - 3z &\le 12 \\ x \ge 0,\ y \ge 0,\ z &\ge 0. \end{aligned}$$

Solution Following Step 1, we introduce slack variables and rewrite the constraints and objective function in standard form:

$$\begin{aligned} 2x + 2y + z + s \qquad &= 8 \\ x + 4y - 3z \quad + t \quad &= 12 \\ -12x - 15y - 5z \qquad + p &= 0. \end{aligned}$$

We now follow with Step 2, setting up the initial tableau:

	x	y	z	s	t	p	
s	2	2	1	1	0	0	8
t	1	4	−3	0	1	0	12
p	−12	−15	−5	0	0	1	0

For Step 3, we select the column over the negative number with the largest magnitude in the bottom row, which is the y column. For Step 4, finding the pivot, we see that the test ratios are $8/2$ and $12/4$, the smallest being $12/4 = 3$. So we select the pivot in the t row and clear its column:

	x	y	z	s	t	p		
s	2	2	1	1	0	0	8	$2R_1 - R_2$
t	1	[4]	−3	0	1	0	12	
p	−12	−15	−5	0	0	1	0	$4R_3 + 15R_2$

The departing variable is t and the entering variable is y. This gives the second tableau.

	x	y	z	s	t	p	
s	3	0	5	2	−1	0	4
y	1	4	−3	0	1	0	12
p	−33	0	−65	0	15	4	180

We now go back to Step 3. Because we still have negative numbers in the bottom row, we choose the one with the largest magnitude (which is −65), and thus our pivot column is the z column. Because negative numbers can't be pivots, the only possible choice for the pivot is the 5. (We need not compute the test ratios because there would only be one from which to choose.) We now clear this column, remembering to take care of the departing and entering variables.

	x	y	z	s	t	p		
s	3	0	[5]	2	−1	0	4	
y	1	4	−3	0	1	0	12	$5R_2 + 3R_1$
p	−33	0	−65	0	15	4	180	$R_3 + 13R_1$

This gives

	x	y	z	s	t	p	
z	3	0	5	2	-1	0	4
y	14	20	0	6	2	0	72
p	6	0	0	26	2	4	232

Notice how the value of p keeps climbing: It started at 0 in the first tableau, went up to $180/4 = 45$ in the second, and is currently at $232/4 = 58$. Because there are no more negative numbers in the bottom row, we are done and can write down the solution: p has a maximum value of $232/4 = 58$, and this occurs when

$$x = 0$$

$$y = \frac{72}{20} = \frac{18}{5} \quad \text{and}$$

$$z = \frac{4}{5}.$$

The slack variables are both zero.

As a partial check on our answer, we can substitute these values into the objective function and the constraints:

$$58 = 12(0) + 15(18/5) + 5(4/5) \quad ✔$$

$$2(0) + 2(18/5) + (4/5) = 8 \leq 8 \quad ✔$$

$$0 + 4(18/5) - 3(4/5) = 12 \leq 12. \quad ✔$$

We say that this is only a partial check, because it shows only that our solution is feasible and that we have correctly calculated p. It does not show that we have the optimal solution. This check will *usually* catch any arithmetic mistakes we make, but it is not foolproof.

APPLICATIONS

In the next example (further exploits of Acme Baby Foods—compare Example 2 in Section 2) we show how the simplex method relates to the graphical method.

EXAMPLE 3 Resource Allocation

Acme Baby Foods makes two puddings, vanilla and chocolate. Each serving of vanilla pudding requires 2 teaspoons of sugar and 25 fluid ounces of water, and each serving of chocolate pudding requires 3 teaspoons of sugar and 15 fluid ounces of water. Acme has available each day 3,600 teaspoons of sugar and 22,500 fluid ounces of water. Acme makes no more than 600 servings of vanilla pudding because that is all that it can sell each day. If Acme makes a profit of 10¢ on each serving of vanilla pudding and 7¢ on each serving of chocolate, how many servings of each should it make to maximize its profit?

Solution We first identify the unknowns. Let

$x =$ the number of servings of vanilla pudding

$y =$ the number of servings of chocolate pudding.

The objective function is the profit $p = 10x + 7y$, which we need to maximize. For the constraints, we start with the fact that Acme will make no more than 600 servings of vanilla: $x \leq 600$. We can put the remaining data in a table as follows:

	Vanilla	*Chocolate*	*Total Available*
Sugar (teaspoons)	2	3	3,600
Water (ounces)	25	15	22,500

Because Acme can use no more sugar and water than is available, we get the two constraints:

$$2x + 3y \leq 3{,}600$$
$$25x + 15y \leq 22{,}500. \quad \text{Note that all the terms are divisible by 5.}$$

Thus our linear programming problem is this:

$$\begin{aligned} \text{Maximize} \quad & p = 10x + 7y \\ \text{subject to} \quad & x \leq 600 \\ & 2x + 3y \leq 3{,}600 \\ & 5x + 3y \leq 4{,}500 \quad \text{We divided } 25x + 15y \leq 22{,}500 \text{ by 5.} \\ & x \geq 0, \; y \geq 0. \end{aligned}$$

Next, we introduce the slack variables and set up the initial tableau.

$$\begin{aligned} x \quad + s \quad\quad\quad\quad & = 600 \\ 2x + 3y \quad + t \quad\quad & = 3{,}600 \\ 5x + 3y \quad\quad + u \quad & = 4{,}500 \\ -10x - 7y \quad\quad\quad + p & = 0 \end{aligned}$$

Note that we have had to introduce a third slack variable, u. There need to be as many slack variables as there are constraints (other than those of the $x \geq 0$ variety).

Q: *What do the slack variables say about Acme puddings?*

A: The first slack variable, *s*, represents the number you must add to the number of servings of vanilla pudding actually made to obtain the maximum of 600 servings. The second slack variable, *t*, represents the amount of sugar that is left over once the puddings are made, and *u* represents the amount of water left over.

We now use the simplex method to solve the problem:

	x	y	s	t	u	p		
s	[1]	0	1	0	0	0	600	
t	2	3	0	1	0	0	3,600	$R_2 - 2R_1$
u	5	3	0	0	1	0	4,500	$R_3 - 5R_1$
p	−10	−7	0	0	0	1	0	$R_4 + 10R_1$

	x	y	s	t	u	p		
x	1	0	1	0	0	0	600	
t	0	3	−2	1	0	0	2,400	$R_2 - R_3$
u	0	[3]	−5	0	1	0	1,500	
p	0	−7	10	0	0	1	6,000	$3R_4 + 7R_3$

using Technology

See the Technology Guides at the end of the chapter for a discussion of using a TI-83/84 Plus to help with the simplex method in Example 3. Or, go to the Website at www.WanerMath.com and follow the path

→ On Line Utilities

→ Pivot and Gauss-Jordan Tool

to find a utility that allows you to avoid doing the calculations in each pivot step: Just highlight the entry you wish to use as a pivot, and press "Pivot on Selection."

	x	y	s	t	u	p		
x	1	0	1	0	0	0	600	$3R_1 - R_2$
t	0	0	[3]	1	−1	0	900	
y	0	3	−5	0	1	0	1,500	$3R_3 + 5R_2$
p	0	0	−5	0	7	3	28,500	$3R_4 + 5R_2$

	x	y	s	t	u	p	
x	3	0	0	−1	1	0	900
s	0	0	3	1	−1	0	900
y	0	9	0	5	−2	0	9,000
p	0	0	0	5	16	9	90,000

Thus, the solution is as follows: The maximum value of p is $90{,}000/9 = 10{,}000$¢ $= \$100$, which occurs when $x = 900/3 = 300$, and $y = 9{,}000/9 = 1{,}000$. (The slack variables are $s = 900/3 = 300$ and $t = u = 0$.)

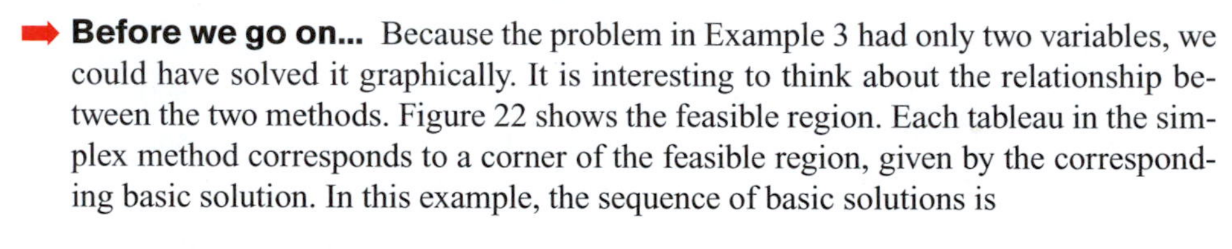

Before we go on... Because the problem in Example 3 had only two variables, we could have solved it graphically. It is interesting to think about the relationship between the two methods. Figure 22 shows the feasible region. Each tableau in the simplex method corresponds to a corner of the feasible region, given by the corresponding basic solution. In this example, the sequence of basic solutions is

$$(x, y) = (0, 0), (600, 0), (600, 500), (300, 1{,}000).$$

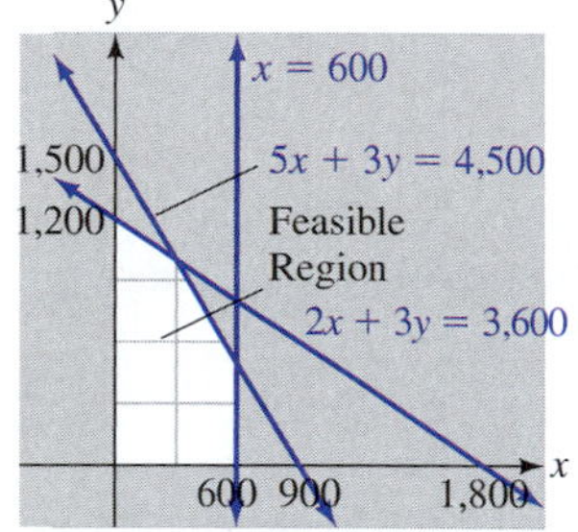

Figure 22

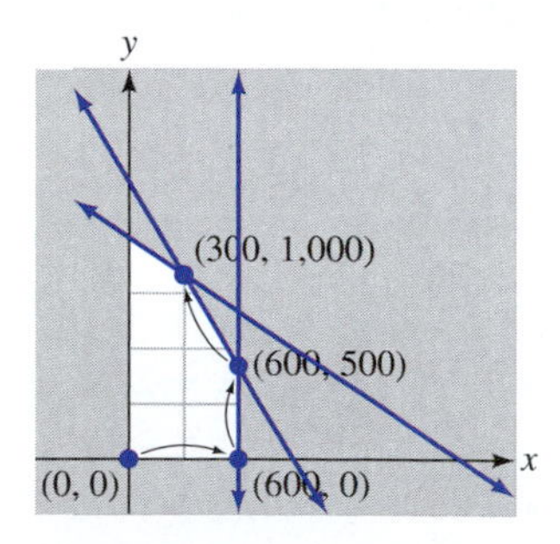

Figure 23

This is the sequence of corners shown in Figure 23. In general, we can think of the simplex method as walking from corner to corner of the feasible region, until it locates the optimal solution. In problems with many variables and many constraints, the simplex method usually visits only a small fraction of the total number of corners.

We can also explain again, in a different way, the reason we use the test ratios when choosing the pivot. For example, when choosing the first pivot we had to choose among the test ratios 600, 1,800, and 900 (look at the first tableau). In Figure 22, you can see that those are the three x-intercepts of the lines that bound the feasible region. If we had chosen 1,800 or 900, we would have jumped along the x-axis to a point outside of the feasible region, which we do not want to do. In general, the test ratios measure the distance from the current corner to the constraint lines, and we must choose the smallest such distance to avoid crossing any of them into the unfeasible region.

It is also interesting in an application like this to think about the values of the slack variables. We said above that s is the difference between the maximum 600 servings of vanilla that might be made and the number that is actually made. In the optimal solution, $s = 300$, which says that 300 fewer servings of vanilla were made than the maximum possible. Similarly, t was the amount of sugar left over. In the optimal solution, $t = 0$, which tells us that all of the available sugar is used. Finally, $u = 0$, so all of the available water is used as well. ■

 using Technology

 **Website**
www.WanerMath.com
On the
→ On Line Utilities
page you will find utilities that automate the simplex method to varying extents:

- Pivot and Gauss-Jordan Tool (Pivots and does row operations)
- Simplex Method Tool (Solves entire LP problems; shows all tableaux)

Summary: The Simplex Method for Standard Maximization Problems

To solve a standard maximization problem using the simplex method, we take the following steps:

1. Convert to a system of equations by introducing **slack variables** to turn the constraints into equations and by rewriting the objective function in standard form.
2. Write down the initial **tableau**.
3. Select the pivot column: Choose the negative number with the largest magnitude in the left-hand side of the bottom row. Its column is the pivot column. (If there are two or more candidates, choose any one.) If all the numbers in the left-hand side of the bottom row are zero or positive, then we are finished, and the basic solution maximizes the objective function. (See below for the basic solution.)
4. Select the pivot in the pivot column: The pivot must always be a positive number. For each positive entry b in the pivot column, compute the ratio a/b, where a is the number in the last column in that row. Of these **test ratios**, choose the smallest one. (If there are two or more candidates, choose any one.) The corresponding number b is the pivot.
5. Use the pivot to clear the column in the normal manner (taking care to follow the exact prescription for formulating the row operations described in Chapter 3) and then relabel the pivot row with the label from the pivot column. The variable originally labeling the pivot row is the **departing**, or **exiting**, **variable**, and the variable labeling the column is the **entering variable**.
6. Go to Step 3.

To get the **basic solution** corresponding to any tableau in the simplex method, set to zero all variables that do not appear as row labels. The value of a variable that does appear as a row label (an **active variable**) is the number in the rightmost column in that row divided by the number in that row in the column labeled by the same variable.

FAQs

Troubleshooting the Simplex Method

Q: *What if there is no candidate for the pivot in the pivot column? For example, what do we do with a tableau like the following?*

	x	y	z	s	t	p	
z	0	0	5	2	0	0	4
y	−8	20	0	6	5	0	72
p	−20	0	0	26	15	4	232

A: Here, the pivot column is the x column, but there is no suitable entry for a pivot (because zeros and negative numbers can't be pivots). This happens when the feasible region is unbounded and there is also no optimal solution. In other words, p can be made as large as we like without violating the constraints.

Q: *What should we do if there is a negative number in the rightmost column?*

A: A negative number will not appear above the bottom row in the rightmost column if we follow the procedure correctly. (The bottom right entry is allowed to be negative if the objective takes on negative values as in a negative profit, or loss.) Following are the most likely errors leading to this situation:

- The pivot was chosen incorrectly. (Don't forget to choose the *smallest* test ratio.) When this mistake is made, one or more of the variables will be negative in the corresponding basic solution.
- The row operation instruction was written backward or performed backward (for example, instead of $R_2 - R_1$, it was $R_1 - R_2$). This mistake can be corrected by multiplying the row by −1.
- An arithmetic error occurred. (We all make those annoying errors from time to time.)

Q: *What about zeros in the rightmost column?*

A: Zeros are permissible in the rightmost column. For example, the constraint $x - y \leq 0$ will lead to a zero in the rightmost column.*

* When there are zeros in the rightmost column there is a potential problem of *cycling*, where a sequence of pivots brings you back to a tableau you already considered, with no change in the objective function. You can usually break out of a cycle by choosing a different pivot. This problem should not arise in the exercises associated with this section.

Q: *What happens if we choose a pivot column other than the one with the most negative number in the bottom row?*

A: There is no harm in doing this as long as we choose the pivot in that column using the smallest test ratio. All it might do is slow the whole calculation by adding extra steps.

One last suggestion: If it is possible to do a simplification step (dividing a row by a positive number) *at any stage*, we should do so. As we saw in Chapter 3, this can help prevent the numbers from getting out of hand.

5.3 EXERCISES

Access end-of-section exercises online at **www.webassign.net** ENHANCED WebAssign

5.4 The Simplex Method: Solving General Linear Programming Problems

As we saw in Section 5.2, not all LP problems are standard maximization problems. We might have constraints like $2x + 3y \geq 4$ or perhaps $2x + 3y = 4$. Or, we might have to minimize, rather than maximize, the objective function. General problems like this are almost as easy to deal with as the standard kind: There is a modification of the simplex method that works very nicely. The best way to illustrate it is by means of examples. First, we discuss nonstandard maximization problems.

Nonstandard Maximization Problems

EXAMPLE 1 Maximizing with Mixed Constraints

$$\begin{aligned}\text{Maximize}\quad & p = 4x + 12y + 6z\\ \text{subject to}\quad & x + y + z \le 100\\ & 4x + 10y + 7z \le 480\\ & x + y + z \ge 60\\ & x \ge 0, y \ge 0, z \ge 0.\end{aligned}$$

Solution We begin by turning the first two inequalities into equations as usual because they have the standard form. We get

$$\begin{aligned} x + y + z + s \quad &= 100\\ 4x + 10y + 7z \quad + t &= 480.\end{aligned}$$

We are tempted to use a slack variable for the third inequality, $x + y + z \ge 60$, but *adding* something positive to the left-hand side will not make it equal to the right: It will get even bigger. To make it equal to 60, we must *subtract* some nonnegative number. We will call this number u (because we have already used s and t) and refer to u as a **surplus variable** rather than a slack variable. Thus, we write

$$x + y + z - u = 60.$$

Continuing with the setup, we have

$$\begin{aligned} x + y + z + s \qquad\qquad &= 100\\ 4x + 10y + 7z \quad + t \qquad &= 480\\ x + y + z \qquad\qquad - u \quad &= 60\\ -4x - 12y - 6z \qquad\qquad\quad + p &= 0.\end{aligned}$$

This leads to the initial tableau:

	x	y	z	s	t	u	p	
s	1	1	1	1	0	0	0	100
t	4	10	7	0	1	0	0	480
$*u$	1	1	1	0	0	−1	0	60
p	−4	−12	−6	0	0	0	1	0

We put a star next to the third row because the basic solution corresponding to this tableau is

$$x = y = z = 0, s = 100, t = 480, u = 60/(-1) = -60.$$

Several things are wrong here. First, the values $x = y = z = 0$ do not satisfy the third inequality $x + y + z \ge 60$. Thus, this basic solution is *not feasible*. Second—and this is really the same problem—the surplus variable u is negative, whereas we said that it should be nonnegative. The star next to the row labeled u alerts us to the fact that the present basic solution is not feasible and that the problem is located in the starred row, where the active variable u is negative.

Whenever an active variable is negative, we star the corresponding row.

In setting up the initial tableau, we star those rows coming from $\ge$ inequalities.

The simplex method as described in the preceding section assumed that we began in the feasible region, but now we do not. Our first task is to get ourselves into the feasible region. In practice, we can think of this as getting rid of the stars on the rows. Once we get into the feasible region, we go back to the method of the preceding section.

There are several ways to get into the feasible region. The method we have chosen is one of the simplest to state and carry out. (We will see why this method works at the end of the example.)

The Simplex Method for General Linear Programming Problems

Star all rows that give a negative value for the associated active variable (except for the objective variable, which is allowed to be negative). If there are starred rows, you will need to begin with Phase I.

Phase I: Getting into the Feasible Region (Getting Rid of the Stars)
In the first starred row, find the largest positive number. Use test ratios as in Section 5.3 to find the pivot in that column (exclude the bottom row), and then pivot on that entry. (If the lowest ratio occurs both in a starred row and an unstarred row, pivot in a starred row rather than the unstarred one.) Check to see which rows should now be starred. Repeat until no starred rows remain, and then go on to Phase II.

Phase II: Use the Simplex Method for Standard Maximization Problems
If there are any negative entries on the left side of the bottom row after Phase I, use the method described in the preceding section.

Because there is a starred row, we need to use Phase I. The largest positive number in the starred row is 1, which occurs three times. Arbitrarily select the first, which is in the first column. In that column, the smallest test ratio happens to be given by the 1 in the u row, so this is our first pivot.

Pivot column ↓

	x	y	z	s	t	u	p		
s	1	1	1	1	0	0	0	100	$R_1 - R_3$
t	4	10	7	0	1	0	0	480	$R_2 - 4R_3$
$*u$	[1]	1	1	0	0	−1	0	60	
p	−4	−12	−6	0	0	0	1	0	$R_4 + 4R_3$

This gives

	x	y	z	s	t	u	p	
s	0	0	0	1	0	1	0	40
t	0	6	3	0	1	4	0	240
x	1	1	1	0	0	−1	0	60
p	0	−8	−2	0	0	−4	1	240

Notice that we removed the star from row 3. To see why, look at the basic solution given by this tableau:

$$x = 60, y = 0, z = 0, s = 40, t = 240, u = 0.$$

None of the variables is negative anymore, so there are no rows to star. The basic solution is therefore feasible—it satisfies all the constraints.

Now that there are no more stars, we have completed Phase I, so we proceed to Phase II, which is just the method of the preceding section.

	x	y	z	s	t	u	p		
s	0	0	0	1	0	1	0	40	
t	0	6	3	0	1	4	0	240	
x	1	1	1	0	0	−1	0	60	$6R_3 - R_2$
p	0	−8	−2	0	0	−4	1	240	$3R_4 + 4R_2$

	x	y	z	s	t	u	p	
s	0	0	0	1	0	1	0	40
y	0	6	3	0	1	4	0	240
x	6	0	3	0	−1	−10	0	120
p	0	0	6	0	4	4	3	1,680

And we are finished. Thus the solution is

$$p = 1{,}680/3 = 560, x = 120/6 = 20, y = 240/6 = 40, z = 0.$$

The slack and surplus variables are

$$s = 40, t = 0, u = 0.$$

Before we go on... We owe you an explanation of why this method works. When we perform a pivot in Phase I, one of two things will happen. As in Example 1, we may pivot in a starred row. In that case, the negative active variable in that row will become inactive (hence zero) and some other variable will be made active with a positive value because we are pivoting on a positive entry. Thus, at least one star will be eliminated. (We will not introduce any new stars because pivoting on the entry with the smallest test ratio will keep all nonnegative variables nonnegative.)

The second possibility is that we may pivot on some row other than a starred row. Choosing the pivot via test ratios again guarantees that no new starred rows are created. A little bit of algebra shows that the value of the negative variable in the first starred row must increase toward zero. (Choosing the *largest* positive entry in the starred row will make it a little more likely that we will increase the value of that variable as much as possible; the rationale for choosing the largest entry is the same as that for choosing the most negative entry in the bottom row during Phase II.) Repeating this procedure as necessary, the value of the variable must eventually become zero or positive, assuming that there are feasible solutions to begin with.

So, one way or the other, we can eventually get rid of all of the stars. ■

Here is an example that begins with two starred rows.

EXAMPLE 2 More Mixed Constraints

$$\begin{aligned} \text{Maximize} \quad & p = 2x + y \\ \text{subject to} \quad & x + y \geq 35 \\ & x + 2y \leq 60 \\ & 2x + y \geq 60 \\ & x \leq 25 \\ & x \geq 0, y \geq 0. \end{aligned}$$

Solution We introduce slack and surplus variables, and write down the initial tableau:

$$
\begin{aligned}
x + y - s \quad &= 35 \\
x + 2y \quad + t \quad &= 60 \\
2x + y \quad - u \quad &= 60 \\
x \quad + v \quad &= 25 \\
-2x - y \quad + p &= 0.
\end{aligned}
$$

	x	y	s	t	u	v	p	
*s	1	1	−1	0	0	0	0	35
t	1	2	0	1	0	0	0	60
*u	2	1	0	0	−1	0	0	60
v	1	0	0	0	0	1	0	25
p	−2	−1	0	0	0	0	1	0

We locate the largest positive entry in the first starred row (row 1). There are two to choose from (both 1s); let's choose the one in the x column. The entry with the smallest test ratio in that column is the 1 in the v row, so that is the entry we use as the pivot:

Pivot column ↓ (x)

	x	y	s	t	u	v	p		
*s	1	1	−1	0	0	0	0	35	$R_1 - R_4$
t	1	2	0	1	0	0	0	60	$R_2 - R_4$
*u	2	1	0	0	−1	0	0	60	$R_3 - 2R_4$
v	[1]	0	0	0	0	1	0	25	
p	−2	−1	0	0	0	0	1	0	$R_5 + 2R_4$

	x	y	s	t	u	v	p	
*s	0	1	−1	0	0	−1	0	10
t	0	2	0	1	0	−1	0	35
*u	0	1	0	0	−1	−2	0	10
x	1	0	0	0	0	1	0	25
p	0	−1	0	0	0	2	1	50

Notice that both stars are still there because the basic solutions for s and u remain negative (but less so). The only positive entry in the first starred row is the 1 in the y column, and that entry also has the smallest test ratio in its column. (Actually, it is tied with the 1 in the u column, so we could choose either one.)

	x	y	s	t	u	v	p		
*s	0	[1]	−1	0	0	−1	0	10	
t	0	2	0	1	0	−1	0	35	$R_2 - 2R_1$
*u	0	1	0	0	−1	−2	0	10	$R_3 - R_1$
x	1	0	0	0	0	1	0	25	
p	0	−1	0	0	0	2	1	50	$R_5 + R_1$

	x	y	s	t	u	v	p	
y	0	1	−1	0	0	−1	0	10
t	0	0	2	1	0	1	0	15
u	0	0	1	0	−1	−1	0	0
x	1	0	0	0	0	1	0	25
p	0	0	−1	0	0	1	1	60

The basic solution is $x = 25$, $y = 10$, $s = 0$, $t = 15$, $u = 0/(-1) = 0$, and $v = 0$. Because there are no negative variables left (even u has become 0), we are in the feasible region, so we can go on to Phase II, shown next. (Filling in the instructions for the row operations is an exercise.)

	x	y	s	t	u	v	p	
y	0	1	−1	0	0	−1	0	10
t	0	0	2	1	0	1	0	15
u	0	0	[1]	0	−1	−1	0	0
x	1	0	0	0	0	1	0	25
p	0	0	−1	0	0	1	1	60

	x	y	s	t	u	v	p	
y	0	1	0	0	−1	−2	0	10
t	0	0	0	1	[2]	3	0	15
s	0	0	1	0	−1	−1	0	0
x	1	0	0	0	0	1	0	25
p	0	0	0	0	−1	0	1	60

	x	y	s	t	u	v	p	
y	0	2	0	1	0	−1	0	35
u	0	0	0	1	2	3	0	15
s	0	0	2	1	0	1	0	15
x	1	0	0	0	0	1	0	25
p	0	0	0	1	0	3	2	135

The optimal solution is

$$x = 25,\ y = 35/2 = 17.5,\ p = 135/2 = 67.5 \quad (s = 7.5,\ t = 0,\ u = 7.5).$$

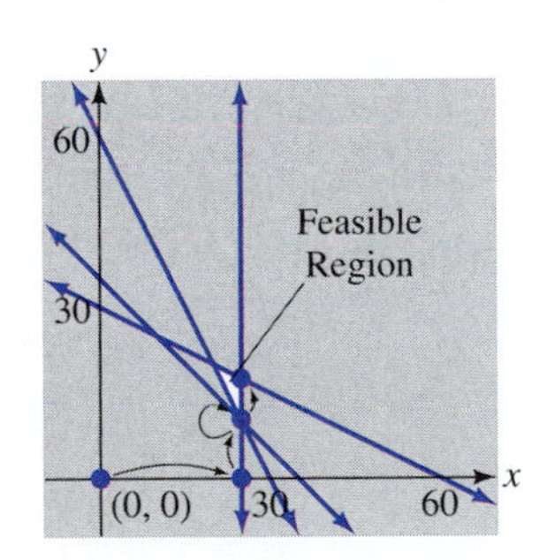

Figure 24

➡ **Before we go on...** Because Example 2 had only two unknowns, we can picture the sequence of basic solutions on the graph of the feasible region. This is shown in Figure 24.

You can see that there was no way to jump from (0, 0) in the initial tableau directly into the feasible region because the first jump must be along an axis. (Why?) Also notice that the third jump did not move at all. To which step of the simplex method does this correspond? ■

Minimization Problems

Now that we know how to deal with nonstandard constraints, we consider **minimization** problems, problems in which we have to minimize, rather than maximize, the objective function. The idea is to *convert a minimization problem into a maximization problem*, which we can then solve as usual.

Suppose, for instance, we want to minimize $c = 10x - 30y$ subject to some constraints. The technique is as follows: Define a new variable p by taking p to be the negative of c, so that $p = -c$. Then, the larger we make p, the smaller c becomes. For example, if we can make p increase from -10 to -5, then c will decrease from 10 to 5. So, if we are looking for the smallest value of c, we might as well look for the largest value of p instead. More concisely,

Minimizing c is the same as maximizing $p = -c$.

Now because $c = 10x - 30y$, we have $p = -10x + 30y$, and the requirement that we "minimize $c = 10x - 30y$" is now replaced by "maximize $p = -10x + 30y$."

Minimization Problems

We convert a minimization problem into a maximization problem by taking the negative of the objective function. All the constraints remain unchanged.

Quick Example

Minimization Problem	→	**Maximization Problem**
Minimize $c = 10x - 30y$		Maximize $p = -10x + 30y$
subject to $2x + y \le 160$		subject to $2x + y \le 160$
$x + 3y \ge 120$		$x + 3y \ge 120$
$x \ge 0, y \ge 0.$		$x \ge 0, y \ge 0.$

EXAMPLE 3 Purchasing

You are in charge of ordering furniture for your company's new headquarters. You need to buy at least 200 tables, 500 chairs, and 300 computer desks. Wall-to-Wall Furniture (WWF) is offering a package of 20 tables, 25 chairs, and 18 computer desks for \$2,000, whereas rival Acme Furniture (AF) is offering a package of 10 tables, 50 chairs, and 24 computer desks for \$3,000. How many packages should you order from each company to minimize your total cost?

Solution The unknowns here are

x = number of packages ordered from WWF
y = number of packages ordered from AF.

We can put the information about the various kinds of furniture in a table:

	WWF	*AF*	*Needed*
Tables	20	10	200
Chairs	25	50	500
Computer Desks	18	24	300
Cost (\$)	2,000	3,000	

From this table we get the following LP problem:

$$\begin{aligned} \text{Minimize} \quad & c = 2{,}000x + 3{,}000y \\ \text{subject to} \quad & 20x + 10y \ge 200 \\ & 25x + 50y \ge 500 \\ & 18x + 24y \ge 300 \\ & x \ge 0, y \ge 0. \end{aligned}$$

Before we start solving this problem, notice that all the inequalities may be simplified. The first is divisible by 10, the second by 25, and the third by 6. (However, this affects the meaning of the surplus variables; see Before we go on... on page 288.) Dividing gives the following simpler problem:

$$\begin{aligned} \text{Minimize} \quad & c = 2{,}000x + 3{,}000y \\ \text{subject to} \quad & 2x + y \ge 20 \\ & x + 2y \ge 20 \\ & 3x + 4y \ge 50 \\ & x \ge 0, y \ge 0. \end{aligned}$$

Following the discussion that preceded this example, we convert to a maximization problem:

$$\begin{aligned} \text{Maximize} \quad & p = -2{,}000x - 3{,}000y \\ \text{subject to} \quad & 2x + y \geq 20 \\ & x + 2y \geq 20 \\ & 3x + 4y \geq 50 \\ & x \geq 0, y \geq 0. \end{aligned}$$

We introduce surplus variables.

$$\begin{aligned} 2x + y - s &= 20 \\ x + 2y - t &= 20 \\ 3x + 4y - u &= 50 \\ 2{,}000x + 3{,}000y + p &= 0. \end{aligned}$$

The initial tableau is then

	x	y	s	t	u	p	
$*s$	2	1	−1	0	0	0	20
$*t$	1	2	0	−1	0	0	20
$*u$	3	4	0	0	−1	0	50
p	2,000	3,000	0	0	0	1	0

The largest entry in the first starred row is the 2 in the upper left, which happens to give the smallest test ratio in its column.

	x	y	s	t	u	p		
$*s$	[2]	1	−1	0	0	0	20	
$*t$	1	2	0	−1	0	0	20	$2R_2 - R_1$
$*u$	3	4	0	0	−1	0	50	$2R_3 - 3R_1$
p	2,000	3,000	0	0	0	1	0	$R_4 - 1{,}000R_1$

	x	y	s	t	u	p		
x	2	1	−1	0	0	0	20	$3R_1 - R_2$
$*t$	0	[3]	1	−2	0	0	20	
$*u$	0	5	3	0	−2	0	40	$3R_3 - 5R_2$
p	0	2,000	1,000	0	0	1	−20,000	$3R_4 - 2{,}000R_2$

	x	y	s	t	u	p		
x	6	0	−4	2	0	0	40	$5R_1 - R_3$
y	0	3	1	−2	0	0	20	$5R_2 + R_3$
$*u$	0	0	4	[10]	−6	0	20	
p	0	0	1,000	4,000	0	3	−100,000	$R_4 - 400R_3$

	x	y	s	t	u	p		
x	30	0	−24	0	6	0	180	$R_1/6$
y	0	15	9	0	−6	0	120	$R_2/3$
t	0	0	4	10	−6	0	20	$R_3/2$
p	0	0	−600	0	2,400	3	−108,000	$R_4/3$

This completes Phase I. We are not yet at the optimal solution, so after performing the simplifications indicated we proceed with Phase II.

	x	y	s	t	u	p		
x	5	0	−4	0	1	0	30	$R_1 + 2R_3$
y	0	5	3	0	−2	0	40	$2R_2 - 3R_3$
t	0	0	2	5	−3	0	10	
p	0	0	−200	0	800	1	−36,000	$R_4 + 100R_3$

	x	y	s	t	u	p	
x	5	0	0	10	−5	0	50
y	0	10	0	−15	5	0	50
s	0	0	2	5	−3	0	10
p	0	0	0	500	500	1	−35,000

The optimal solution is

$x = 50/5 = 10$, $y = 50/10 = 5$, $p = -35{,}000$, so $c = 35{,}000$
$(s = 5, t = 0, u = 0)$.

You should buy 10 packages from Wall-to-Wall Furniture and 5 from Acme Furniture, for a minimum cost of $35,000.

Before we go on... The surplus variables in the preceding example represent pieces of furniture over and above the minimum requirements. The order you place will give you 50 extra tables ($s = 5$, but s was introduced after we divided the first inequality by 10, so the actual surplus is $10 \times 5 = 50$), the correct number of chairs ($t = 0$), and the correct number of computer desks ($u = 0$). ■

The preceding LP problem is an example of a **standard minimization problem**—in a sense the opposite of a standard maximization problem: We are *minimizing* an objective function, where all the constraints have the form $Ax + By + Cz + \cdots \geq N$. We will discuss standard minimization problems more fully in Section 5.5, as well as another method of solving them.

FAQs

When to Switch to Phase II, Equality Constraints, and Troubleshooting

Q: *How do I know when to switch to Phase II?*

A: After each step, check the basic solution for starred rows. You are not ready to proceed with Phase II until all the stars are gone.

Q: *How do I deal with an equality constraint, such as* $2x + 7y - z = 90$?

A: Although we haven't given examples of equality constraints, they can be treated by the following trick: Replace an equality by two inequalities. For example, replace the equality $2x + 7y - z = 90$ by the two inequalities $2x + 7y - z \leq 90$ and $2x + 7y - z \geq 90$. A little thought will convince you that these two inequalities amount to the same thing as the original equality!

Q: *What happens if it is impossible to choose a pivot using the instructions in Phase I?*

A: In that case, the LP problem has no solution. In fact, the feasible region is empty. If it is impossible to choose a pivot in Phase II, then the feasible region is unbounded and there is no optimal solution.

5.4 EXERCISES

Access end-of-section exercises online at **www.webassign.net**

5.5 The Simplex Method and Duality

We mentioned **standard minimization problems** in the last section. These problems have the following form.

Standard Minimization Problem

A **standard minimization problem** is an LP problem in which we are required to *minimize* (not maximize) a linear objective function

$$c = as + bt + cu + \cdots$$

of the variables $s, t, u, \ldots$ (in this section, we will always use the letters $s, t, u, \ldots$ for the unknowns in a standard minimization problem) subject to the constraints

$$s \geq 0, t \geq 0, u \geq 0, \ldots$$

and further constraints of the form

$$As + Bt + Cu + \cdots \geq N$$

where $A, B, C, \ldots$ and N are numbers with N nonnegative.

A **standard linear programming problem** is an LP problem that is either a standard maximization problem or a standard minimization problem. An LP problem satisfies the **nonnegative objective condition** if all the coefficients in the objective function are nonnegative.

Standard Minimization and Maximization Problems

1. Minimize $c = 2s + 3t + 3u$

subject to
$$2s + u \geq 10$$
$$s + 3t - 6u \geq 5$$
$$s \geq 0, t \geq 0, u \geq 0.$$

This is a standard minimization problem satisfying the nonnegative objective condition.

2. Maximize $p = 2x + 3y + 3z$
subject to
$$2x + z \le 7$$
$$x + 3y - 6z \le 6$$
$$x \ge 0, y \ge 0, z \ge 0.$$

This is a standard maximization problem satisfying the nonnegative objective condition.

3. Minimize $c = 2s - 3t + 3u$
subject to
$$2s + u \ge 10$$
$$s + 3t - 6u \ge 5$$
$$s \ge 0, t \ge 0, u \ge 0.$$

This is a standard minimization problem that does *not* satisfy the nonnegative objective condition.

We saw a way of solving minimization problems in Section 5.4, but a mathematically elegant relationship between maximization and minimization problems gives us another way of solving minimization problems that satisfy the nonnegative objective condition. This relationship is called **duality**.

To describe duality, we must first represent an LP problem by a matrix. This matrix is *not* the first tableau but something simpler: Pretend you forgot all about slack variables and also forgot to change the signs of the objective function.* As an example, consider the following two standard† problems:

* Forgetting these things is exactly what happens to many students under test conditions!

† Although duality does not require the problems to be standard, it does require them to be written in so-called *standard form*: In the case of a maximization problem all constraints need to be (re)written using $\le$, while for a minimization problem all constraints need to be (re)written using $\ge$. It is least confusing to stick with standard problems, which is what we will do in this section.

Problem 1

Maximize $p = 20x + 20y + 50z$
subject to
$$2x + y + 3z \le 2{,}000$$
$$x + 2y + 4z \le 3{,}000$$
$$x \ge 0, y \ge 0, z \ge 0.$$

We represent this problem by the matrix

$$\left[\begin{array}{ccc|c} 2 & 1 & 3 & 2{,}000 \\ 1 & 2 & 4 & 3{,}000 \\ \hline 20 & 20 & 50 & 0 \end{array}\right]. \quad \begin{array}{l}\text{Constraint 1}\\ \text{Constraint 2}\\ \text{Objective}\end{array}$$

Notice that the coefficients of the objective function go in the bottom row, and we place a zero in the bottom right corner.

Problem 2 (from Example 3 in Section 5.4)

Minimize $c = 2{,}000s + 3{,}000t$
subject to
$$2s + t \ge 20$$
$$s + 2t \ge 20$$
$$3s + 4t \ge 50$$
$$s \ge 0, t \ge 0.$$

Problem 2 is represented by

$$\left[\begin{array}{cc|c} 2 & 1 & 20 \\ 1 & 2 & 20 \\ 3 & 4 & 50 \\ \hline 2{,}000 & 3{,}000 & 0 \end{array}\right]. \quad \begin{array}{l}\text{Constraint 1}\\ \text{Constraint 2}\\ \text{Constraint 3}\\ \text{Objective}\end{array}$$

These two problems are related: The matrix for Problem 1 is the transpose of the matrix for Problem 2. (Recall that the transpose of a matrix is obtained by writing its rows as columns; see Section 4.1.) When we have a pair of LP problems related in this way, we say that the two are *dual* LP problems.

Dual Linear Programming Problems

Two LP problems, one a maximization and one a minimization problem, are **dual** if the matrix that represents one is the transpose of the matrix that represents the other.

Finding the Dual of a Given Problem

Given an LP problem, we find its dual as follows:

1. Represent the problem as a matrix (see previous page).
2. Take the transpose of the matrix.
3. Write down the dual, which is the LP problem corresponding to the new matrix. If the original problem was a maximization problem, its dual will be a minimization problem, and vice versa.

The original problem is called the **primal problem**, and its dual is referred to as the **dual problem**.

Quick Example

Primal problem

Minimize $c = s + 2t$
subject to $5s + 2t \geq 60$
$3s + 4t \geq 80$
$s + t \geq 20$
$s \geq 0, t \geq 0.$

$\xrightarrow{1}$ $\begin{bmatrix} 5 & 2 & 60 \\ 3 & 4 & 80 \\ 1 & 1 & 20 \\ 1 & 2 & 0 \end{bmatrix}$

$\xrightarrow{2}$ $\begin{bmatrix} 5 & 3 & 1 & 1 \\ 2 & 4 & 1 & 2 \\ 60 & 80 & 20 & 0 \end{bmatrix}$

$\xrightarrow{3}$ Dual problem

Maximize $p = 60x + 80y + 20z$
subject to $5x + 3y + z \leq 1$
$2x + 4y + z \leq 2$
$x \geq 0, y \geq 0, z \geq 0.$

The following theorem justifies what we have been doing, and says that solving the dual problem of an LP problem is equivalent to solving the original problem.

Fundamental Theorem of Duality

a. If an LP problem has an optimal solution, then so does its dual. Moreover, the primal problem and the dual problem have the same optimal value for their objective functions.

b. Contained in the final tableau of the simplex method applied to an LP problem is the solution to its dual problem: It is given by the bottom entries in the columns associated with the slack variables, divided by the entry under the objective variable.

The theorem* gives us an alternative way of solving minimization problems that satisfy the nonnegative objective condition. Let's illustrate by solving Problem 2 above.

* The proof of the theorem is beyond the scope of this book but can be found in a textbook devoted to linear programming, like *Linear Programming* by Vašek Chvátal (San Francisco: W. H. Freeman and Co., 1983), which has a particularly well-motivated discussion.

EXAMPLE 1 Solving by Duality

Minimize $c = 2{,}000s + 3{,}000t$
subject to $2s + t \geq 20$
$s + 2t \geq 20$
$3s + 4t \geq 50$
$s \geq 0, t \geq 0.$

Solution

Step 1 ***Find the dual problem.*** Write the primal problem in matrix form and take the transpose:

$$\begin{bmatrix} 2 & 1 & 20 \\ 1 & 2 & 20 \\ 3 & 4 & 50 \\ 2{,}000 & 3{,}000 & 0 \end{bmatrix} \to \begin{bmatrix} 2 & 1 & 3 & 2{,}000 \\ 1 & 2 & 4 & 3{,}000 \\ 20 & 20 & 50 & 0 \end{bmatrix}.$$

The dual problem is:

$$\begin{aligned} \text{Maximize} \quad & p = 20x + 20y + 50z \\ \text{subject to} \quad & 2x + y + 3z \le 2{,}000 \\ & x + 2y + 4z \le 3{,}000 \\ & x \ge 0, y \ge 0, z \ge 0. \end{aligned}$$

Step 2 ***Use the simplex method to solve the dual problem.*** Because we have a standard maximization problem, we do not have to worry about Phase I but go straight to Phase II.

	x	y	z	s	t	p	
s	2	1	[3]	1	0	0	2,000
t	1	2	4	0	1	0	3,000
p	−20	−20	−50	0	0	1	0

	x	y	z	s	t	p	
z	2	1	3	1	0	0	2,000
t	−5	[2]	0	−4	3	0	1,000
p	40	−10	0	50	0	3	100,000

	x	y	z	s	t	p	
z	9	0	6	6	−3	0	3,000
y	−5	2	0	−4	3	0	1,000
p	15	0	0	30	15	3	105,000

Note that the maximum value of the objective function is $p = 105{,}000/3 = 35{,}000$. By the theorem, this is also the optimal value of c in the primal problem!

Step 3 ***Read off the solution to the primal problem by dividing the bottom entries in the columns associated with the slack variables by the entry in the p column.*** Here is the final tableau again with the entries in question highlighted.

	x	y	z	s	t	p	
z	9	0	6	6	−3	0	3,000
y	−5	2	0	−4	3	0	1,000
p	15	0	0	**30**	**15**	**3**	**105,000**

The solution to the primal problem is

$$s = 30/3 = 10, t = 15/3 = 5, c = 105{,}000/3 = 35{,}000.$$

(Compare this with the method we used to solve Example 3 in the preceding section. Which method seems more efficient?)

➡ **Before we go on...** Can you now see the reason for using the variable names $s, t, u, \dots$ in standard minimization problems? ■

Q: *Is the theorem also useful for solving problems that do not satisfy the nonnegative objective condition?*

A: Consider a standard minimization problem that does not satisfy the nonnegative objective condition, such as

$$\begin{aligned} \text{Minimize} \quad & c = 2s - t \\ \text{subject to} \quad & 2s + 3t \geq 2 \\ & s + 2t \geq 2 \\ & s \geq 0, t \geq 0. \end{aligned}$$

Its dual would be

$$\begin{aligned} \text{Maximize} \quad & p = 2x + 2y \\ \text{subject to} \quad & 2x + y \leq 2 \\ & 3x + 2y \leq -1 \\ & x \geq 0, y \geq 0. \end{aligned}$$

This is not a standard maximization problem because the right-hand side of the second constraint is negative. In general, if a problem does not satisfy the nonnegative objective condition, its dual is not standard. Therefore, to solve the dual by the simplex method will require using Phase I as well as Phase II, and we may as well just solve the primal problem that way to begin with. Thus, duality helps us solve problems only when the primal problem satisfies the nonnegative objective condition.

In many economic applications, the solution to the dual problem also gives us useful information about the primal problem, as we will see in the following example.

EXAMPLE 2 Shadow Costs

You are trying to decide how many vitamin pills to take. SuperV brand vitamin pills each contain 2 milligrams of vitamin X, 1 milligram of vitamin Y, and 1 milligram of vitamin Z. Topper brand vitamin pills each contain 1 milligram of vitamin X, 1 milligram of vitamin Y, and 2 milligrams of vitamin Z. You want to take enough pills daily to get at least 12 milligrams of vitamin X, 10 milligrams of vitamin Y, and 12 milligrams of vitamin Z. However, SuperV pills cost 4¢ each and Toppers cost 3¢ each, and you would like to minimize the total cost of your daily dosage. How many of each brand of pill should you take? How would changing your daily vitamin requirements affect your minimum cost?

Solution This is a straightforward minimization problem. The unknowns are

s = number of SuperV brand pills
t = number of Topper brand pills.

The linear programming problem is

$$\begin{aligned} \text{Minimize} \quad & c = 4s + 3t \\ \text{subject to} \quad & 2s + t \geq 12 \\ & s + t \geq 10 \\ & s + 2t \geq 12 \\ & s \geq 0, t \geq 0. \end{aligned}$$

We solve this problem by using the simplex method on its dual, which is

$$\begin{aligned} \text{Maximize} \quad & p = 12x + 10y + 12z \\ \text{subject to} \quad & 2x + y + z \le 4 \\ & x + y + 2z \le 3 \\ & x \ge 0, y \ge 0, z \ge 0. \end{aligned}$$

After pivoting three times, we arrive at the final tableau:

	x	y	z	s	t	p	
x	6	0	−6	6	−6	0	6
y	0	1	3	−1	2	0	2
p	0	0	6	2	8	1	32

Therefore, the answer to the original problem is that you should take two SuperV vitamin pills and eight Toppers at a cost of 32¢ per day.

Now, the key to answering the last question, which asks you to determine how changing your daily vitamin requirements would affect your minimum cost, is to look at the solution to the dual problem. From the tableau we see that $x = 1$, $y = 2$, and $z = 0$. To see what x, y, and z might tell us about the original problem, let's look at their units. In the inequality $2x + y + z \le 4$, the coefficient 2 of x has units "mg of vitamin X/SuperV pill," and the 4 on the right-hand side has units "¢/SuperV pill." For $2x$ to have the same units as the 4 on the right-hand side, x must have units "¢/mg of vitamin X." Similarly, y must have units "¢/mg of vitamin Y" and z must have units "¢/mg of vitamin Z." One can show (although we will not do it here) that x gives the amount that would be added to the minimum cost for each increase* of 1 milligram of vitamin X in our daily requirement. For example, if we were to increase our requirement from 12 milligrams to 14 milligrams, an increase of 2 milligrams, the minimum cost would change by $2x = 2$¢, from 32¢ to 34¢. (Try it; you'll end up taking four SuperV pills and six Toppers.) Similarly, each increase of 1 milligram of vitamin Y in the requirements would increase the cost by $y = 2$¢. These costs are called the **marginal costs** or the **shadow costs** of the vitamins.

* To be scrupulously correct, this works only for relatively small changes in the requirements, not necessarily for very large ones.

What about $z = 0$? The shadow cost of vitamin Z is 0¢/mg, meaning that you can increase your requirement of vitamin Z without changing your cost. In fact, the solution $s = 2$ and $t = 8$ provides you with 18 milligrams of vitamin Z, so you can increase the required amount of vitamin Z up to 18 milligrams without changing the solution at all.

We can also interpret the shadow costs as the effective cost to you of each milligram of each vitamin in the optimal solution. You are paying 1¢/milligram of vitamin X, 2¢/milligram of vitamin Y, and getting the vitamin Z for free. This gives a total cost of $1 \times 12 + 2 \times 10 + 0 \times 12 = 32$¢, as we know. Again, if you change your requirements slightly, these are the amounts you will pay per milligram of each vitamin.

Game Theory

We return to a topic we discussed in Section 4.4: solving two-person zero-sum games. In that section, we described how to solve games that could be reduced to 2×2 games or smaller. It turns out that we can solve larger games using linear programming and duality. We summarize the procedure, work through an example, and then discuss why it works.

Solving a Matrix Game

Step 1 Reduce the payoff matrix by dominance.

Step 2 Add a fixed number k to each of the entries so that they all become nonnegative and no column is all zero.

Step 3 Write 1s to the right of and below the matrix, and then write down the associated standard maximization problem. Solve this primal problem using the simplex method.

Step 4 Find the optimal strategies and the expected value as follows:

Column Strategy

1. Express the solution to the primal problem as a column vector.
2. Normalize by dividing each entry of the solution vector by p (which is also the sum of all the entries).
3. Insert zeros in positions corresponding to the columns deleted during reduction.

Row Strategy

1. Express the solution to the dual problem as a row vector.
2. Normalize by dividing each entry by p, which will once again be the sum of all the entries.
3. Insert zeros in positions corresponding to the rows deleted during reduction.

Value of the Game

$$e = \frac{1}{p} - k$$

EXAMPLE 3 Restaurant Inspector

You manage two restaurants, Tender Steaks Inn (TSI) and Break for a Steak (BFS). Even though you run the establishments impeccably, the Department of Health has been sending inspectors to your restaurants on a daily basis and fining you for minor infractions. You've found that you can head off a fine if you're present, but you can cover only one restaurant at a time. The Department of Health, on the other hand, has two inspectors, who sometimes visit the same restaurant and sometimes split up, one to each restaurant. The average fines you have been getting are shown in the following matrix.

Health Inspectors

		Both at BFS	*Both at TSI*	*One at Each*
You go to	**TSI**	$8,000	0	$2,000
	BFS	0	$10,000	$4,000

How should you choose which restaurant to visit to minimize your expected fine?

Solution This matrix is not quite the payoff matrix because fines, being penalties, should be negative payoffs. Thus, the payoff matrix is the following:

$$P = \begin{bmatrix} -8{,}000 & 0 & -2{,}000 \\ 0 & -10{,}000 & -4{,}000 \end{bmatrix}.$$

We follow the steps above to solve the game using the simplex method.

Step 1 There are no dominated rows or columns, so this game does not reduce.

Step 2 We add $k = 10{,}000$ to each entry so that none are negative, getting the following new matrix (with no zero column):

$$\begin{bmatrix} 2{,}000 & 10{,}000 & 8{,}000 \\ 10{,}000 & 0 & 6{,}000 \end{bmatrix}.$$

Step 3 We write 1s to the right and below this matrix:

$$\begin{bmatrix} 2{,}000 & 10{,}000 & 8{,}000 & 1 \\ 10{,}000 & 0 & 6{,}000 & 1 \\ 1 & 1 & 1 & 0 \end{bmatrix}.$$

The corresponding standard maximization problem is the following:

$$\begin{aligned} \text{Maximize} \quad & p = x + y + z \\ \text{subject to} \quad & 2{,}000x + 10{,}000y + 8{,}000z \le 1 \\ & 10{,}000x \qquad\qquad\quad + 6{,}000z \le 1 \\ & x \ge 0, y \ge 0, z \ge 0. \end{aligned}$$

Step 4 We use the simplex method to solve this problem. After pivoting twice, we arrive at the final tableau:

	x	y	z	s	t	p	
y	0	50,000	34,000	5	-1	0	4
x	10,000	0	6,000	0	1	0	1
p	0	0	14,000	5	4	50,000	9

Column Strategy The solution to the primal problem is

$$\begin{bmatrix} x \\ y \\ z \end{bmatrix} = \begin{bmatrix} \frac{1}{10{,}000} \\ \frac{4}{50{,}000} \\ 0 \end{bmatrix}.$$

We divide each entry by $p = 9/50{,}000$, which is also the sum of the entries. This gives the optimal column strategy:

$$C = \begin{bmatrix} \frac{5}{9} \\ \frac{4}{9} \\ 0 \end{bmatrix}.$$

Thus, the inspectors' optimal strategy is to stick together, visiting BFS with probability $5/9$ and TSI with probability $4/9$.

Row Strategy The solution to the dual problem is

$$[s \quad t] = \begin{bmatrix} \frac{5}{50{,}000} & \frac{4}{50{,}000} \end{bmatrix}.$$

Once again, we divide by $p = 9/50{,}000$ to find the optimal row strategy:

$$R = \begin{bmatrix} \frac{5}{9} & \frac{4}{9} \end{bmatrix}.$$

Thus, you should visit TSI with probability 5/9 and BFS with probability 4/9.

Value of the Game Your expected average fine is

$$e = \frac{1}{p} - k = \frac{50{,}000}{9} - 10{,}000 = -\frac{40{,}000}{9} \approx -\$4{,}444.$$

Before we go on... We owe you an explanation of why the procedure we used in Example 3 works. The main point is to understand how we turn a game into a linear programming problem. It's not hard to see that adding a fixed number k to all the payoffs will change only the payoff, increasing it by k, and not change the optimal strategies. So let's pick up Example 3 from the point where we were considering the following game:

$$P = \begin{bmatrix} 2{,}000 & 10{,}000 & 8{,}000 \\ 10{,}000 & 0 & 6{,}000 \end{bmatrix}.$$

We are looking for the optimal strategies R and C for the row and column players, respectively; if e is the value of the game, we will have $e = RPC$. Let's concentrate first on the column player's strategy $C = [u \;\; v \;\; w]^T$, where u, v, and w are the unknowns we want to find. Because e is the value of the game, if the column player uses the optimal strategy C and the row player uses any old strategy S, the expected value with these strategies has to be e or better for the column player, so $SPC \le e$. Let's write that out for two particular choices of S. First, consider $S = [1 \quad 0]$:

$$[1 \quad 0]\begin{bmatrix} 2{,}000 & 10{,}000 & 8{,}000 \\ 10{,}000 & 0 & 6{,}000 \end{bmatrix}\begin{bmatrix} u \\ v \\ w \end{bmatrix} \le e.$$

Multiplied out, this gives

$$2{,}000u + 10{,}000v + 8{,}000w \le e.$$

Next, do the same thing for $S = [0 \quad 1]$:

$$[0 \quad 1]\begin{bmatrix} 2{,}000 & 10{,}000 & 8{,}000 \\ 10{,}000 & 0 & 6{,}000 \end{bmatrix}\begin{bmatrix} u \\ v \\ w \end{bmatrix} \le e$$

$$10{,}000u + 6{,}000w \le e.$$

It turns out that if these two inequalities are true, then $SPC \le e$ for any S at all, which is what the column player wants. These are starting to look like constraints in a linear programming problem, but the e appearing on the right is in the way. We get around this by dividing by e, which we know to be positive because all of the payoffs are nonnegative and no column is all zero (so the column player can't force the value of the game to be 0; here is where we need these assumptions). We get the following inequalities:

$$2{,}000\left(\frac{u}{e}\right) + 10{,}000\left(\frac{v}{e}\right) + 8{,}000\left(\frac{w}{e}\right) \le 1$$

$$10{,}000\left(\frac{u}{e}\right) \qquad\qquad + 6{,}000\left(\frac{w}{e}\right) \le 1.$$

Now we're getting somewhere. To make these look even more like linear constraints, we replace our unknowns u, v, and w with new unknowns, $x = u/e$, $y = v/e$, and $z = w/e$. Our inequalities then become.

$$\begin{aligned} 2{,}000x + 10{,}000y + 8{,}000z &\le 1 \\ 10{,}000x \qquad\qquad + 6{,}000z &\le 1. \end{aligned}$$

What about an objective function? From the point of view of the column player, the objective is to find a strategy that will minimize the expected value e. In order to write e in terms of our new variables x, y, and z, we use the fact that our original variables, being the entries in the column strategy, have to add up to 1: $u + v + w = 1$. Dividing by e gives

$$\frac{u}{e} + \frac{v}{e} + \frac{w}{e} = \frac{1}{e}$$

or

$$x + y + z = \frac{1}{e}.$$

Now we notice that, if we *maximize* $p = x + y + z = 1/e$, it will have the effect of minimizing e, which is what we want. So, we get the following linear programming problem:

$$\begin{aligned} \text{Maximize} \quad & p = x + y + z \\ \text{subject to} \quad & 2{,}000x + 10{,}000y + 8{,}000z \le 1 \\ & 10{,}000x \qquad\qquad + 6{,}000z \le 1 \\ & x \ge 0, y \ge 0, z \ge 0. \end{aligned}$$

Why can we say that x, y, and z should all be nonnegative? Because the unknowns u, v, w, and e must all be nonnegative.

So now, if we solve this linear programming problem to find x, y, z, and p, we can find the column player's optimal strategy by computing $u = xe = x/p$, $v = y/p$, and $w = z/p$. Moreover, the value of the game is $e = 1/p$. (If we added k to all the payoffs, we should now adjust by subtracting k again to find the correct value of the game.)

Turning now to the row player's strategy, if we repeat the above type of argument from the row player's viewpoint, we'll end up with the following linear programming problem to solve:

$$\begin{aligned} \text{Minimize} \quad & c = s + t \\ \text{subject to} \quad & 2{,}000s + 10{,}000t \ge 1 \\ & 10{,}000s \qquad\qquad \ge 1 \\ & 8{,}000s + 6{,}000t \ge 1 \\ & s \ge 0, t \ge 0. \end{aligned}$$

This is, of course, the dual to the problem we solved to find the column player's strategy, so we know that we can read its solution off of the same final tableau. The optimal value of c will be the same as the value of p, so $c = 1/e$ also. The entries in the optimal row strategy will be s/c and t/c. ■

FAQs

When to Use Duality

Q: *Given a minimization problem, when should I use duality, and when should I use the two-phase method in Section 5.4?*

A: If the original problem satisfies the nonnegative objective condition (none of the coefficients in the objective function are negative), then you can use duality to convert the problem to a standard maximization one, which can be solved with the one-phase method. If the original problem does not satisfy the nonnegative objective condition, then dualizing results in a nonstandard LP problem, so dualizing may not be worthwhile.

Q: *When is it absolutely necessary to use duality?*

A: Never. Duality gives us an efficient but not necessary alternative for solving standard minimization problems.

5.5 EXERCISES

Access end-of-section exercises online at **www.webassign.net**

CHAPTER 5 REVIEW

KEY CONCEPTS

 Website www.WanerMath.com

Go to the Website at www.WanerMath.com to find a comprehensive and interactive Web-based summary of Chapter 5.

5.1 Graphing Linear Inequalities

5.2 Solving Linear Programming Problems Graphically

5.3 The Simplex Method: Solving Standard Maximization Problems

5.4 The Simplex Method: Solving General Linear Programming Problems

5.5 The Simplex Method and Duality

REVIEW EXERCISES

In each of Exercises 1–4, sketch the region corresponding to the given inequalities, say whether it is bounded, and give the coordinates of all corner points.

1. $2x - 3y \le 12$

2. $x \le 2y$

3. $x + 2y \le 20$
$3x + 2y \le 30$
$x \ge 0, y \ge 0$

4. $3x + 2y \ge 6$
$2x - 3y \le 6$
$3x - 2y \ge 0$
$x \ge 0, y \ge 0$

In each of Exercises 5–8, solve the given linear programming problem graphically.

5. Maximize $p = 2x + y$
subject to $3x + y \le 30$
$x + y \le 12$
$x + 3y \le 30$
$x \ge 0, y \ge 0.$

6. Maximize $p = 2x + 3y$
subject to $x + y \ge 10$
$2x + y \ge 12$
$x + y \le 20$
$x \ge 0, y \ge 0.$

7. Minimize $c = 2x + y$
subject to $3x + y \ge 30$
$x + 2y \ge 20$
$2x - y \ge 0$
$x \ge 0, y \ge 0.$

8. Minimize $c = 3x + y$
subject to $3x + 2y \ge 6$
$2x - 3y \le 0$
$3x - 2y \ge 0$
$x \ge 0, y \ge 0.$

In each of Exercises 9–18, solve the given linear programming problem using the simplex method. If no optimal solution exists, indicate whether the feasible region is empty or the objective function is unbounded.

9. Maximize $p = x + y + 2z$
subject to $x + 2y + 2z \le 60$
$2x + y + 3z \le 60$
$x \ge 0, y \ge 0, z \ge 0.$

10. Maximize $p = x + y + 2z$
subject to $x + 2y + 2z \le 60$
$2x + y + 3z \le 60$
$x + 3y + 6z \le 60$
$x \ge 0, y \ge 0, z \ge 0.$

11. Maximize $p = x + y + 3z$
subject to $x + y + z \ge 100$
$y + z \le 80$
$x + z \le 80$
$x \ge 0, y \ge 0, z \ge 0.$

12. Maximize $p = 2x + y$
subject to $x + 2y \ge 12$
$2x + y \le 12$
$x + y \le 5$
$x \ge 0, y \ge 0.$

13. Minimize $c = x + 2y + 3z$
subject to
$3x + 2y + z \geq 60$
$2x + y + 3z \geq 60$
$x \geq 0, y \geq 0, z \geq 0.$

14. Minimize $c = 5x + 4y + 3z$
subject to
$x + y + 4z \geq 30$
$2x + y + 3z \geq 60$
$x \geq 0, y \geq 0, z \geq 0.$

15. T Minimize $c = x - 2y + 4z$
subject to
$3x + 2y - z \geq 10$
$2x + y + 3z \geq 20$
$x + 3y - 2z \geq 30$
$x \geq 0, y \geq 0, z \geq 0.$

16. T Minimize $c = x + y - z$
subject to
$3x + 2y + z \geq 60$
$2x + y + 3z \geq 60$
$x + 3y + 2z \geq 60$
$x \geq 0, y \geq 0, z \geq 0.$

17. Minimize $c = x + y + z + w$
subject to
$x + y \geq 30$
$x + z \geq 20$
$x + y - w \leq 10$
$y + z - w \leq 10$
$x \geq 0, y \geq 0, z \geq 0, w \geq 0.$

18. Minimize $c = 4x + y + z + w$
subject to
$x + y \geq 30$
$y - z \leq 20$
$z - w \leq 10$
$x \geq 0, y \geq 0, z \geq 0, w \geq 0.$

In each of Exercises 19–22, solve the given linear programming problem using duality.

19. Minimize $c = 2x + y$
subject to
$3x + 2y \geq 60$
$2x + y \geq 60$
$x + 3y \geq 60$
$x \geq 0, y \geq 0.$

20. Minimize $c = 2x + y + 2z$
subject to
$3x + 2y + z \geq 100$
$2x + y + 3z \geq 200$
$x \geq 0, y \geq 0, z \geq 0.$

21. Minimize $c = 2x + y$
subject to
$3x + 2y \geq 10$
$2x - y \leq 30$
$x + 3y \geq 60$
$x \geq 0, y \geq 0.$

22. Minimize $c = 2x + y + 2z$
subject to
$3x - 2y + z \geq 100$
$2x + y - 3z \leq 200$
$x \geq 0, y \geq 0, z \geq 0.$

In each of Exercises 23–26, solve the game with the given payoff matrix.

23. $P = \begin{bmatrix} -1 & 2 & -1 \\ 1 & -2 & 1 \\ 3 & -1 & 0 \end{bmatrix}$ **24.** $P = \begin{bmatrix} -3 & 0 & 1 \\ -4 & 0 & 0 \\ 0 & -1 & -2 \end{bmatrix}$

25. $P = \begin{bmatrix} -3 & -2 & 3 \\ 1 & 0 & 0 \\ -2 & 2 & 1 \end{bmatrix}$ **26.** $P = \begin{bmatrix} -4 & -2 & -3 \\ 1 & -3 & -2 \\ -3 & 1 & -4 \end{bmatrix}$

Exercises 27–30 are adapted from the Actuarial Exam on Operations Research.

27. You are given the following linear programming problem:

Minimize $c = x + 2y$
subject to
$-2x + y \geq 1$
$x - 2y \geq 1$
$x \geq 0, y \geq 0.$

Which of the following is true?

(A) The problem has no feasible solutions.
(B) The objective function is unbounded.
(C) The problem has optimal solutions.

28. Repeat the preceding exercise with the following linear programming problem:

Maximize $p = x + y$
subject to
$-2x + y \leq 1$
$x - 2y \leq 2$
$x \geq 0, y \geq 0.$

29. Determine the optimal value of the objective function. You are given the following linear programming problem.

Maximize $Z = x_1 + 4x_2 + 2x_3 - 10$
subject to
$4x_1 + x_2 + x_3 \leq 45$
$-x_1 + x_2 + 2x_3 \leq 0$
$x_1, x_2, x_3 \geq 0.$

30. Determine the optimal value of the objective function. You are given the following linear programming problem.

Minimize $Z = x_1 + 4x_2 + 2x_3 + x_4 + 40$
subject to
$4x_1 + x_2 + x_3 \leq 45$
$-x_1 + 2x_2 + x_4 \geq 40$
$x_1, x_2, x_3 \geq 0.$

APPLICATIONS: OHaganBooks.com

In Exercises 31–34, you are the buyer for OHaganBooks.com and are considering increasing stocks of romance and horror novels at the new OHaganBooks.com warehouse in Texas. You have offers from several publishers: Duffin House, Higgins Press, McPhearson Imprints, and O'Conell Books. Duffin offers a package of 5 horror novels and 5 romance novels for \$50, Higgins offers a package of 5 horror and 10 romance novels for \$80, McPhearson offers a package of 10 horror novels and 5 romance novels for \$80, and

O'Conell offers a package of 10 horror novels and 10 romance novels for $90.

31. How many packages should you purchase from Duffin House and Higgins Press to obtain at least 4,000 horror novels and 6,000 romance novels at minimum cost? What is the minimum cost?

32. How many packages should you purchase from McPhearson Imprints and O'Conell Books to obtain at least 5,000 horror novels and 4,000 romance novels at minimum cost? What is the minimum cost?

33. Refer to the scenario in Exercise 31. As it turns out, John O'Hagan promised Marjory Duffin that OHaganBooks.com would buy at least 20 percent more packages from Duffin as from Higgins, but you still want to obtain at least 4,000 horror novels and 6,000 romance novels at minimum cost.

a. *Without solving the problem,* say which of the following statements are possible:

(A) The cost will stay the same.
(B) The cost will increase.
(C) The cost will decrease.
(D) It will be impossible to meet all the conditions.
(E) The cost will become unbounded.

b. If you wish to meet all the requirements at minimum cost, how many packages should you purchase from each publisher? What is the minimum cost?

34. Refer to Exercise 32. You are about to place the order meeting the requirements of Exercise 32 when you are told that you can order no more than a total of 500 packages, and that at least half of the packages should be from McPhearson. Explain why this is impossible by referring to the feasible region for Exercise 32.

35. ***Investments*** Marjory Duffin's portfolio manager has suggested two high-yielding stocks: European Emerald Emporium (EEE) and Royal Ruby Retailers (RRR).[3] EEE shares cost $50, yield 4.5% in dividends, and have a risk index of 2.0 per share. RRR shares cost $55, yield 5% in dividends, and have a risk index of 3.0 per share. Marjory has up to $12,100 to invest and would like to earn at least $550 in dividends. How many shares of each stock should she purchase to meet her requirements and minimize the total risk index for her portfolio? What is the minimum total risk index?

36. ***Investments*** Marjory Duffin's other portfolio manager has suggested another two high-yielding stocks: Countrynarrow Mortgages (CNM) and Scotland Subprime (SS).[4] CNM shares cost $40, yield 5.5% in dividends, and have a risk index of 1.0 per share. SS shares cost $25, yield 7.5% in dividends, and have a risk index of 1.5 per share. Marjory can invest up to $30,000 in these stocks and would like to earn at least $1,650 in dividends. How many shares of each stock should she purchase in order to meet her requirements and minimize the total risk index for her portfolio?

[3]RRR and EEE happen to be, respectively, the ticker symbols of RSC Holdings (an equipment rental provider) and Evergreen Energy Inc. (an environmentally friendly energy technology company) and thus have nothing to do with rubies and emeralds.

[4]CNM is actually the ticker symbol of Carnegie Wave, whereas SS is not the ticker symbol of any U.S.-based company we are aware of.

37. ***Resource Allocation*** Billy-Sean O'Hagan has joined the Physics Society at Suburban State University, and the group is planning to raise money to support the dying space program by making and selling umbrellas. The society intends to make three models: the Sprinkle, the Storm, and the Hurricane. The amounts of cloth, metal, and wood used in making each model are given in this table:

	Sprinkle	*Storm*	*Hurricane*	*Total Available*
Cloth (sq. yd)	1	2	2	600
Metal (lbs)	2	1	3	600
Wood (lbs)	1	3	6	600
Profit ($)	1	1	2	

The table also shows the amounts of each material available in a given day and the profits to be made from each model. How many of each model should the society make in order to maximize its profit?

38. ***Profit*** Duffin House, which is now the largest publisher of books sold at the OHaganBooks.com site, prints three kinds of books: paperback, quality paperback, and hardcover. The amounts of paper, ink, and time on the presses required for each kind of book are given in this table:

	Paperback	*Quality Paperback*	*Hardcover*	*Total Available*
Paper (pounds)	3	2	1	6,000
Ink (gallons)	2	1	3	6,000
Time (minutes)	10	10	10	22,000
Profit ($)	1	2	3	

The table also lists the total amounts of paper, ink, and time available in a given day and the profits made on each kind of book. How many of each kind of book should Duffin print to maximize profit?

39. ***Purchases*** You are just about to place book orders from Duffin and Higgins (see Exercise 31) when everything changes: Duffin House informs you that, due to a global romance crisis, its packages now each will contain 5 horror novels but only 2 romance novels and still cost $50 per package. Packages from Higgins will now contain 10 of each type of novel, but now cost $150 per package. Ewing Books enters the fray and offers its own package of 5 horror and 5 romance novels for $100. The sales manager now tells you that at least 50% of the packages must come from Higgins Press and, as before, you want to obtain at least 4,000 horror novels and 6,000 romance novels at minimum cost. Taking all of this into account, how many packages should you purchase from each publisher? What is the minimum cost?

40. ***Purchases*** You are about to place book orders from McPhearson and O'Conell (see Exercise 32) when you get an e-mail from McPhearson Imprints saying that, sorry, but they have stopped publishing romance novels due to the global romance crisis and can now offer only packages of 10 horror novels for \$50. O'Conell is still offering packages of 10 horror novels and 10 romance novels for \$90, and now the United States Treasury, in an attempt to bolster the floundering romance industry, is offering its own package of 20 romance novels for \$120. Furthermore, Congress, in approving this measure, has passed legislation dictating that at least two thirds of the packages in every order must come from the U.S. Treasury. As before, you wish to obtain at least 5,000 horror novels and 4,000 romance novels at minimum cost. Taking all of this into account, how many packages should you purchase from each supplier? What is the minimum cost?

41. ***Degree Requirements*** During his lunch break, John O'Hagan decides to devote some time to assisting his son Billy-Sean, who continues to have a terrible time planning his college course schedule. The latest Bulletin of Suburban State University claims to have added new flexibility to its course requirements, but it remains as complicated as ever. It reads as follows:

> *All candidates for the degree of Bachelor of Arts at SSU must take at least 120 credits from the Sciences, Fine Arts, Liberal Arts, and Mathematics combined, including at least as many Science credits as Fine Arts credits, and at most twice as many Mathematics credits as Science credits, but with Liberal Arts credits exceeding Mathematics credits by no more than one third of the number of Fine Arts credits.*

Science and fine arts credits cost \$300 each, and liberal arts and mathematics credits cost \$200 each. John would like to have Billy-Sean meet all the requirements at a minimum total cost.

a. Set up (without solving) the associated linear programming problem.

b. Use technology to determine how many of each type of credit Billy-Sean should take. What will the total cost be?

42. ***Degree Requirements*** No sooner had the "new and flexible" course requirement been released than the English Department again pressured the University Senate to include their vaunted "Verbal Expression" component in place of the fine arts requirement in all programs (including the sciences):

> *All candidates for the degree of Bachelor of Science at SSU must take at least 120 credits from the Liberal Arts, Sciences, Verbal Expression, and Mathematics, including at most as many Science credits as Liberal Arts credits, and at least twice as many Verbal Expression credits as Science credits and Liberal Arts credits combined, with Liberal Arts credits exceeding Mathematics credits by at least a quarter of the number of Verbal Expression credits.*

Science credits cost \$300 each, while each credit in the remaining subjects now costs \$400. John would like to have Billy-Sean meet all the requirements at a minimum total cost.

a. Set up (without solving) the associated linear programming problem.

b. Use technology to determine how many of each type of credit Billy-Sean should take. What will the total cost be?

43. ***Shipping*** On the same day that the sales department at Duffin House received an order for 600 packages from the OHaganBooks.com Texas headquarters, it received an additional order for 200 packages from FantasyBooks.com, based in California. Duffin House has warehouses in New York and Illinois. The New York warehouse has 600 packages in stock, but the Illinois warehouse is closing down and has only 300 packages in stock. Shipping costs per package of books are as follows: New York to Texas: \$20; New York to California: \$50; Illinois to Texas: \$30; Illinois to California: \$40. What is the lowest total shipping cost for which Duffin House can fill the orders? How many packages should be sent from each warehouse to each online bookstore at a minimum shipping cost?

44. ***Transportation Scheduling*** Duffin House is about to start a promotional blitz for its new book, *Advanced String Theory for the Liberal Arts*. The company has 25 salespeople stationed in Austin and 10 in San Diego, and would like to fly at least 15 to sales fairs in each of Houston and Cleveland. A round-trip plane flight from Austin to Houston costs \$200; from Austin to Cleveland costs \$150; from San Diego to Houston costs \$400; and from San Diego to Cleveland costs \$200. How many salespeople should the company fly from each of Austin and San Diego to each of Houston and Cleveland for the lowest total cost in air fare?

45. ***Marketing*** Marjory Duffin, head of Duffin House, reveals to John O'Hagan that FantasyBooks.com is considering several promotional schemes: It may offer two books for the price of one, three books for the price of two, or possibly a free copy of *Brain Surgery for Klutzes* with each order. OHaganBooks.com's marketing advisers Floody and O'Lara seem to have different ideas as to how to respond. Floody suggests offering *three* books for the price of one, while O'Lara suggests instead offering a free copy of the *Finite Mathematics Student Solutions Manual* with every purchase. After a careful analysis, O'Hagan comes up with the following payoff matrix, where the payoffs represent the number of customers, in thousands, O'Hagan expects to gain from FantasyBooks.com.

		FantasyBooks.com			
		No Promo	2 for Price of 1	3 for Price of 2	Brain Surgery
OHaganBooks.com	No Promo	0	−60	−40	10
	3 for Price of 1	30	20	10	15
	Finite Math	20	0	15	10

Find the optimal strategies for both companies and the expected shift in customers.

46. ***Study Techniques*** Billy-Sean's friend Pat from college has been spending all of his time in fraternity activities, and thus knows absolutely nothing about any of the three topics on tomorrow's math test. He has turned to Billy-Sean for advice as to how to spend his "all-nighter." The table on the right shows the scores Pat could expect to earn if the entire test were to be in a specific subject. (Because he knows no linear programming or matrix algebra, the table shows, for instance, that studying game theory all night will not be much use in preparing him for this topic.)

	Test		
Pat's Strategies ↓	**Game Theory**	**Linear Programming**	**Matrix Algebra**
Study Game Theory	30	0	20
Study Linear Programming	0	70	0
Study Matrix Algebra	0	0	70

What percentage of the night should Pat spend on each topic, assuming the principles of game theory, and what score can he expect to get?

Case Study: The Diet Problem

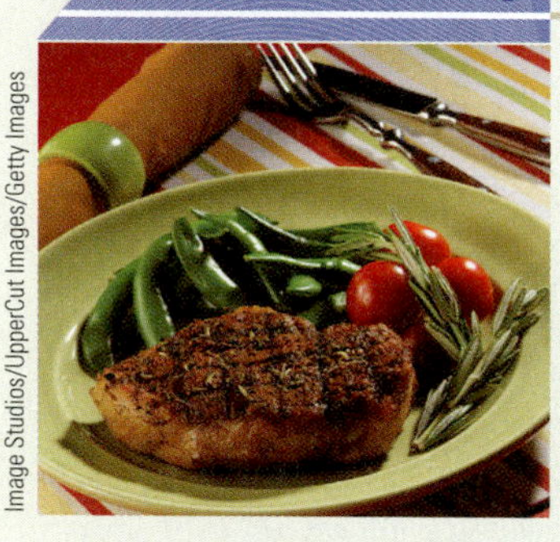

Image Studios/UpperCut Images/Getty Images

The Galaxy Nutrition health-food mega-store chain provides free online nutritional advice and support to its customers. As Web site technical consultant, you are planning to construct an interactive Web page to assist customers in preparing a diet tailored to their nutritional and budgetary requirements. Ideally, the customer would select foods to consider and specify nutritional and/or budgetary constraints, and the tool should return the optimal diet meeting those requirements. You would also like the Web page to allow the customer to decide whether, for instance, to find the cheapest possible diet meeting the requirements, the diet with the lowest number of calories, or the diet with the least total carbohydrates.

After doing a little research, you notice that the kind of problem you are trying is solve is quite well known and referred to as the *diet problem*, and that solving the diet problem is a famous example of linear programming. Indeed, there are already some online pages that solve versions of the problem that minimize total cost, so you have adequate information to assist you as you plan the page.*

* See, for instance, the Diet Problem Demo at the NEOS Wiki: www.neos-guide.org/NEOS/index.php/Diet_Problem_Demo

You decide to start on a relatively small scale, starting with a program that uses a list of 10 foods, and minimizes either total caloric intake or total cost and satisfies a small list of requirements. Following is a small part of a table of nutritional information from the demo at the NEOS Wiki (all the values shown are for a single serving) as well as approximate minimum daily requirements:

	Price per Serving	*Calories*	*Total Fat g*	*Carbs g*	*Dietary Fiber g*	*Protein g*	*Vit C IU*
Tofu	\$0.31	88.2	5.5	2.2	1.4	9.4	0.1
Roast Chicken	\$0.84	277.4	10.8	0	0	42.2	0
Spaghetti w/Sauce	\$0.78	358.2	12.3	58.3	11.6	8.2	27.9
Tomato	\$0.27	25.8	0.4	5.7	1.4	1.0	23.5
Oranges	\$0.15	61.6	0.2	15.4	3.1	1.2	69.7
Wheat Bread	\$0.05	65.0	1.0	12.4	1.3	2.2	0
Cheddar Cheese	\$0.25	112.7	9.3	0.4	0	7.0	0
Oatmeal	\$0.82	145.1	2.3	25.3	4.0	6.1	0
Peanut Butter	\$0.07	188.5	16.0	6.9	2.1	7.7	0
White Tuna in Water	\$0.69	115.6	2.1	0	0	22.7	0
Minimum Requirements		2,200	20	80	25	60	90

Source: www.neos-guide.org/NEOS/index.php/Diet_Problem_Demo

Now you get to work. As always, you start by identifying the unknowns. Since the output of the Web page will consist of a recommended diet, the unknowns should logically be the number of servings of each item of food selected by the user. In your first trial run, you decide to include all the 10 food items listed, so you take

$$x_1 = \text{Number of servings of tofu}$$
$$x_2 = \text{Number of servings of roast chicken}$$
$$\vdots$$
$$x_{10} = \text{Number of servings of white tuna in water.}$$

You now set up a linear programming problem for two sample scenarios:

Scenario 1 (Minimum Cost): Satisfy all minimum nutritional requirements at a minimum cost. Here the linear programming problem is:

Minimize

$$c = 0.31x_1 + 0.84x_2 + 0.78x_3 + 0.27x_4 + 0.15x_5 + 0.05x_6 + 0.25x_7 + 0.82x_8 + 0.07x_9 + 0.69x_{10}$$

subject to

$$88.2x_1 + 277.4x_2 + 358.2x_3 + 25.8x_4 + 61.6x_5 + 65x_6 + 112.7x_7 + 145.1x_8 + 188.5x_9 + 115.6x_{10} \ge 2{,}200$$
$$5.5x_1 + 10.8x_2 + 12.3x_3 + 0.4x_4 + 0.2x_5 + 1x_6 + 9.3x_7 + 2.3x_8 + 16x_9 + 2.1x_{10} \ge 20$$
$$2.2x_1 + 58.3x_3 + 5.7x_4 + 15.4x_5 + 12.4x_6 + 0.4x_7 + 25.3x_8 + 6.9x_9 \ge 80$$
$$1.4x_1 + 11.6x_3 + 1.4x_4 + 3.1x_5 + 1.3x_6 + 4x_8 + 2.1x_9 \ge 25$$
$$9.4x_1 + 42.2x_2 + 8.2x_3 + 1x_4 + 1.2x_5 + 2.2x_6 + 7x_7 + 6.1x_8 + 7.7x_9 + 22.7x_{10} \ge 60$$
$$0.1x_1 + 27.9x_3 + 23.5x_4 + 69.7x_5 \ge 90.$$

This is clearly the kind of linear programming problem no one in their right mind would like to do by hand (solving it requires 16 tableaus!) so you decide to use the online simplex method tool at the Website (www.WanerMath.com → Student Web Site → On Line Utilities → Simplex Method Tool).

Here is a picture of the input, entered almost exactly as written above (you need to enter each constraint on a new line, and `"Minimize c = 0.31x1 + ··· Subject to"` must be typed on a single line):

Type your linear programming problem below. (Press "Example" to see how to set it up.)

```
Minimize  c =
0.31x1+0.84x2+0.78x3+0.27x4+0.15x5+0.05x6+0.25x7+0.82x8+0.07x9+0.69x10 Subject to
88.2x1+277.4x2+358.2x3+25.8x4+61.6x5+65x6+112.7x7+145.1x8+188.5x9+115.6x10 >= 2200
5.5x1+10.8x2+12.3x3+0.4x4+0.2x5+1x6+9.3x7+2.3x8+16x9+2.1x10 >= 20
2.2x1+58.3x3+5.7x4+15.4x5+12.4x6+0.4x7+25.3x8+6.9x9 >= 80
1.4x1+11.6x3+1.4x4+3.1x5+1.3x6+4x8+2.1x9 >= 25
9.4x1+42.2x2+8.2x3+1x4+1.2x5+2.2x6+7x7+6.1x8+7.7x9+22.7x10 >= 60
0.1x1+0x2+27.9x3+23.5x4+69.7x5 >= 90
```

Clicking "Solve" results in the following solution:

$$c = 0.981126; x_1 = 0, x_2 = 0, x_3 = 0, x_4 = 0, x_5 = 1.29125, x_6 = 0, x_7 = 0, x_8 = 0, x_9 = 11.2491, x_{10} = 0.$$

This means that you can satisfy all the daily requirements for less than $1 on a diet of 1.3 servings of orange juice and 11.2 servings of peanut butter! Although you enjoy peanut butter, 11.2 servings seems a little over the top, so you modify the LP problem by adding a new constraint (which also suggests to you that some kind of flexibility needs to be built into the site to allow users to set limits on the number of servings of any one item):

$$x_9 \le 3.$$

This new constraint results in the following solution:

$$c = 1.59981; x_1 = 0, x_2 = 0, x_3 = 0, x_4 = 0, x_5 = 1.29125, x_6 = 23.9224, x_7 = 0, x_8 = 0, x_9 = 3, x_{10} = 0.$$

Because wheat bread is cheap and, in large enough quantities, supplies ample protein, the program has now substituted 23.9 servings of wheat bread for the missing peanut butter for a total cost of $1.60.

Unfettered, you now add

$$x_6 \le 4$$

and obtain the following spaghetti, bread, and peanut butter diet for $3.40 per day:

$$c = 3.40305; x_1 = 0, x_2 = 0, x_3 = 3.83724, x_4 = 0, x_5 = 0, x_6 = 4, x_7 = 0, x_8 = 0, x_9 = 3, x_{10} = 0.$$

Scenario 2 (Minimum Calories): Minimize total calories and satisfy all minimum nutritional requirements (except for caloric intake).

Here, the linear programming problem is

Minimize

$$c = 88.2x_1 + 277.4x_2 + 358.2x_3 + 25.8x_4 + 61.6x_5 + 65x_6 + 112.7x_7 + 145.1x_8 + 188.5x_9 + 115.6x_{10}$$

subject to

$$5.5x_1 + 10.8x_2 + 12.3x_3 + 0.4x_4 + 0.2x_5 + 1x_6 + 9.3x_7 + 2.3x_8 + 16x_9 + 2.1x_{10} \ge 20$$

$$2.2x_1 + 58.3x_3 + 5.7x_4 + 15.4x_5 + 12.4x_6 + 0.4x_7 + 25.3x_8 + 6.9x_9 \ge 80$$

$$1.4x_1 + 11.6x_3 + 1.4x_4 + 3.1x_5 + 1.3x_6 + 4x_8 + 2.1x_9 \ge 25$$

$$9.4x_1 + 42.2x_2 + 8.2x_3 + 1x_4 + 1.2x_5 + 2.2x_6 + 7x_7 + 6.1x_8 + 7.7x_9 + 22.7x_{10} \ge 60$$

$$0.1x_1 + 27.9x_3 + 23.5x_4 + 69.7x_5 \ge 90.$$

You obtain the following 716-calorie tofu, tomato, and tuna diet:

$$x_1 = 2.07232, x_2 = 0, x_3 = 0, x_4 = 15.7848, x_5 = 0, x_6 = 0, x_7 = 0, x_8 = 0, x_9 = 0, x_{10} = 1.08966.$$

As 16 servings of tomatoes seems a little over the top, you add the new constraint $x_4 \le 3$ and obtain a 783-calorie tofu, tomato, orange, and tuna diet:

$$x_1 = 2.81682, x_2 = 0, x_3 = 0, x_4 = 3, x_5 = 5.43756, x_6 = 0, x_7 = 0, x_8 = 0, x_9 = 0, x_{10} = 1.05713.$$

What the trial runs have shown you is that your Web site will need to allow the user to set reasonable upper bounds for the number of servings of each kind of food considered. You now get to work writing the algorithm, which appears here:

www.WanerMath.com → Student Web Site → On Line Utilities → Diet Problem Solver

EXERCISES

1. Briefly explain why roast chicken, which supplies protein more cheaply than either tofu or tuna, does not appear in the optimal solution in either scenario.
2. Consider the optimal solution obtained in Scenario 1 when peanut butter and bread were restricted. Experiment on the Simplex Method Tool by increasing the protein requirement 10 grams at a time until chicken appears in the optimal diet. At what level of protein does the addition of chicken first become necessary?
3. What constraints would you add for a person who wants to eat at most two servings of chicken a day and is allergic to tomatoes and peanut butter? What is the resulting diet for Scenario 2?
4. What is the linear programming problem for someone who wants as much protein as possible at a cost of no more than $6 per day with no more than 50 g of carbohydrates per day assuming they want to satisfy the minimum requirements for all the remaining nutrients? What is the resulting diet?
5. Is it possible to obtain a diet with no bread or peanut butter in Scenario 1 costing less than $4 per day?

TI-83/84 Plus Technology Guide

Section 5.1

Some calculators, including the TI-83/84 Plus, will shade one side of a graph, but you need to tell the calculator which side to shade. For instance, to obtain the solution set of $2x + 3y \leq 6$ shown in Figure 4:

1. Solve the corresponding equation $2x + 3y = 6$ for y and use the input shown below:

```
Plot1 Plot2 Plot3
◥Y1■-(2/3)*X+2
\Y2=
\Y3=
\Y4=
\Y5=
\Y6=
\Y7=
```

2. The icon to the left of "Y_1" tells the calculator to shade above the line. You can cycle through the various shading options by positioning the cursor to the left of Y_1 and pressing ENTER until you see the one you want. Here's what the graph will look like:

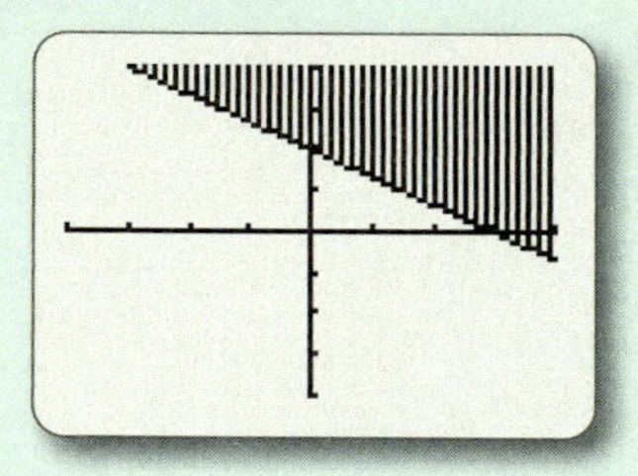

Section 5.3

Example 3 (page 276) The Acme Baby Foods example in the text leads to the following linear programming problem:

$$\begin{aligned} \text{Maximize} \quad & p = 10x + 7y \\ \text{subject to} \quad & x \leq 600 \\ & 2x + 3y \leq 3{,}600 \\ & 5x + 3y \leq 4{,}500 \\ & x \geq 0,\ y \geq 0. \end{aligned}$$

Solve it using technology.

Solution with Technology

When we introduce slack variables, we get the following system of equations:

$$\begin{aligned} x \quad\quad + s \quad\quad\quad\quad &= 600 \\ 2x + 3y \quad + t \quad\quad\quad &= 3{,}600 \\ 5x + 3y \quad\quad + u \quad &= 4{,}500 \\ -10x - 7y \quad\quad\quad + p &= 0. \end{aligned}$$

We use the PIVOT program for the TI-83/84 Plus to help with the simplex method. This program is available at the Website by following

Everything for Finite Math → Math Tools for Chapter 5.

Because the calculator handles decimals as easily as integers, there is no need to avoid them, except perhaps to save limited screen space. If we don't need to avoid decimals, we can use the traditional Gauss-Jordan method (see the discussion at the end of Section 3.2): After selecting your pivot, and prior to clearing the pivot column, *divide the pivot row by the value of the pivot, thereby turning the pivot into a 1.*

The main drawback to using the TI-83/84 Plus is that we can't label the rows and columns. We can mentally label them as we go, but we can do without labels entirely if we wish. We begin by entering the initial tableau as the matrix [A]. (Another drawback to using the TI-83/84 Plus is that it can't show the whole tableau at once. Here and below we show tableaux across several screens. Use the TI-83/84 Plus's arrow keys to scroll a matrix left and right so you can see all of it.)

```
[A]
[[1   0  1 0 0 ...
 [2   3  0 1 0 ...
 [5   3  0 0 1 ...
 [-10 -7 0 0 0 ...
```

```
[A]
... 1 0 0 0 600 ]
... 0 1 0 0 3600]
... 0 0 1 0 4500]
... 0 0 0 1 0   ]]
```

The following is the sequence of tableaux we get while using the simplex method with the help of the PIVOT program.

```
?1
ROW 1
COLUMN 1
[[1 0   1  0 0 0...
 [0 3  -2  1 0 0...
 [0 3  -5  0 1 0...
 [0 -7 10  0 0 1...
```

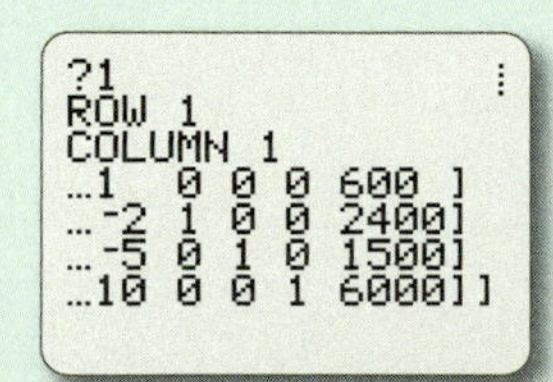

```
?1
ROW 1
COLUMN 1
...1  0 0 0 600 ]
...-2 1 0 0 2400]
...-5 0 1 0 1500]
...10 0 0 1 6000]]
```

After determining that the next pivot is in the third row and second column, we divide the third row by the pivot, 3, and then pivot:

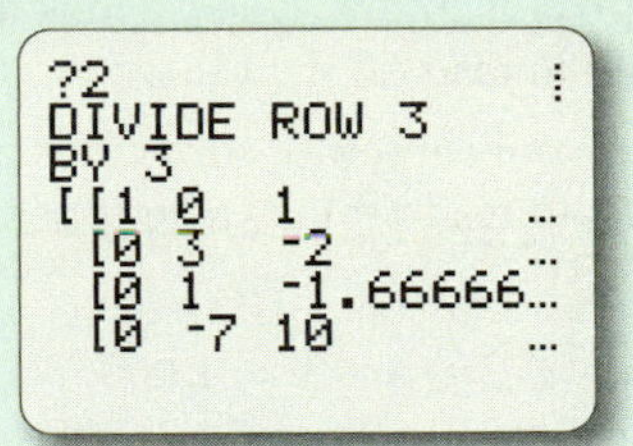


```
?2
DIVIDE ROW 3
BY 3
...     0 0   ...
...     1 0   ...
...667 0 .3333333...
...     0 0   ...
```

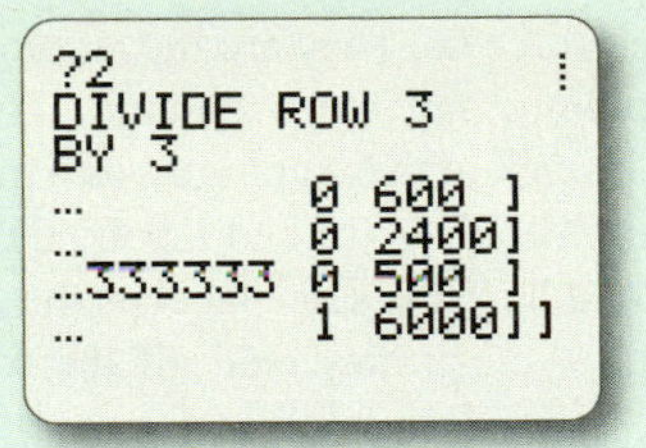


```
?1
ROW 3
COLUMN 2
[[1 0 1 ...
 [0 0 3 ...
 [0 1 -1.666666...
 [0 0 -1.666666...
```

```
?1
ROW 3
COLUMN 2
...     0 0   ...
...     1 -1  ...
...667 0 .3333333...
...667 0 2.333333...
```

```
?1
ROW 3
COLUMN 2
...        0 600 ]
...        0 900 ]
...333333 0 500 ]
...333333 1 9500]]
```

The next pivot is the 3 in the second row, third column. We divide its row by 3 and pivot:

```
?2
DIVIDE ROW 2
BY 3
[[1 0 1 ...
 [0 0 1 ...
 [0 1 -1.666666...
 [0 0 -1.666666...
```

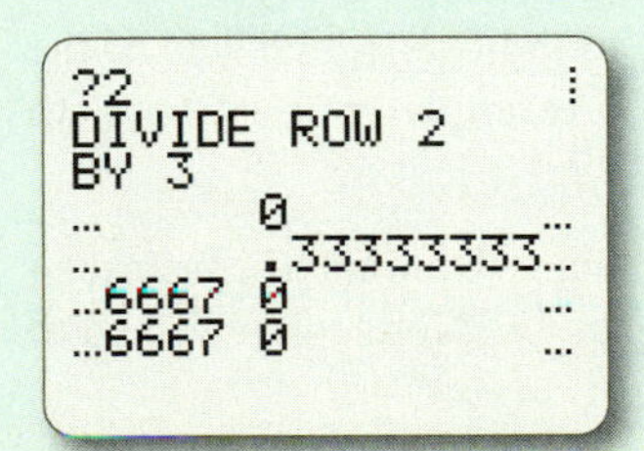


```
?2
DIVIDE ROW 2
BY 3
...     0       ...
...33 -.333333333...
...    .333333333...
...   2.333333333...
```

```
?2
DIVIDE ROW 2
BY 3
...        0 600 ]
...333333 0 300 ]
...33333  0 500 ]
...33333  1 9500]]
```

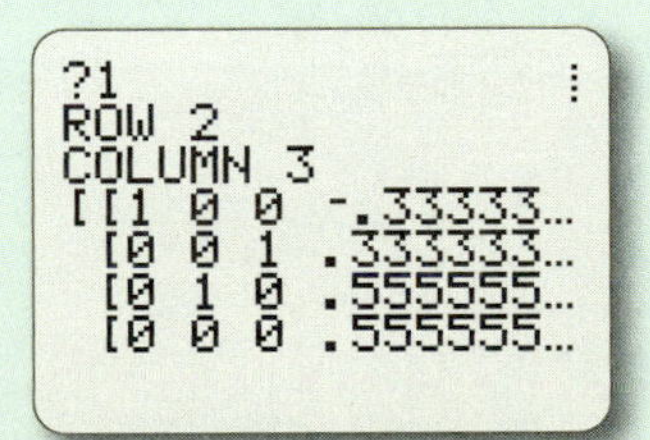

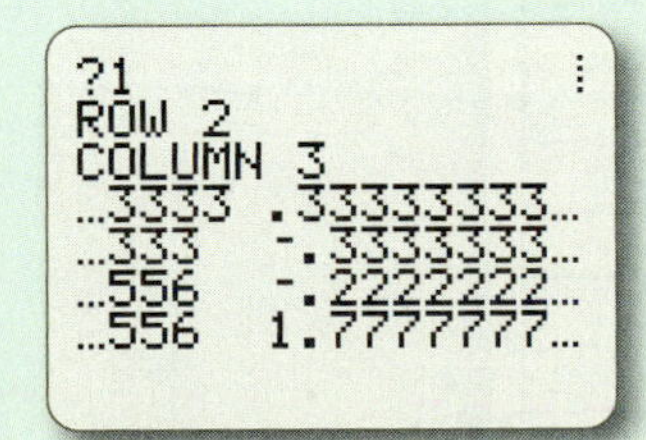

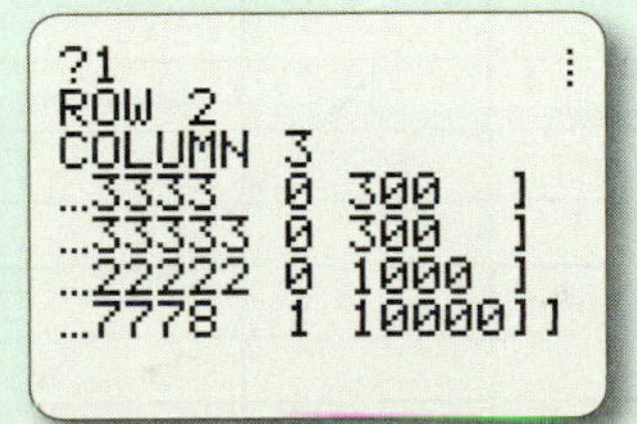

There are no negative numbers in the bottom row, so we're finished. How do we read off the optimal solutions if we don't have labels, though? Look at the columns containing one 1 and three 0s. They are the x column, the y column, the s column, and the p column. Think of the 1 that appears in each of these columns as a pivot whose column has been cleared. If we had labels, the row containing a pivot would have the same label as the column containing that pivot. We can now read off the solution as follows:

x column: The pivot is in the first row, so row 1 would have been labeled with x. We look at the rightmost column to read off the value $x = 300$.

y column: The pivot is in row 3, so we look at the rightmost column to read off the value $y = 1,000$.

s column: The pivot is in row 2, so we look at the rightmost column to read off the value $s = 300$.

p column: The pivot is in row 3, so we look at the rightmost column to read off the value $p = 10,000$.

Thus, the maximum value of p is $10,000¢ = \$100$, which occurs when $x = 300$ and $y = 1,000$. The values of the slack variables are $s = 300$ and $t = u = 0$. (Look at the t and u columns to see that they must be inactive.)

SPREADSHEET Technology Guide

Section 5.1

Excel is not a particularly good tool for graphing linear inequalities because it cannot easily shade one side of a line. One solution available in Excel, for instance, is to use the "error bar" feature to indicate which side of the line *should* be shaded. For example, here is how we might graph the inequality $2x + 3y \le 6$.

1. Create a scatter graph using two points to construct a line segment (as in Chapter 1). (Notice that we had to solve the equation $2x + 3y = 6$ for y.)

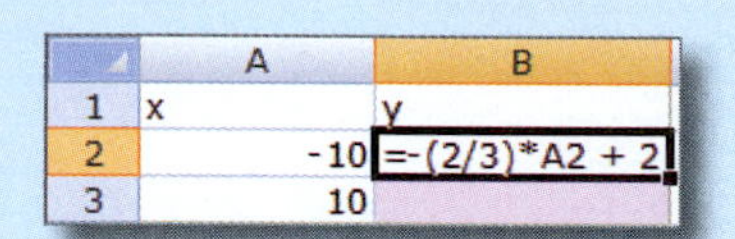

	A	B
1	x	y
2	-10	=-(2/3)*A2 + 2
3	10	

2. Double-click on the line segment, and use the "X-Error Bars" feature to obtain a diagram similar to the one on the left below, where the error bars indicate the direction of shading.

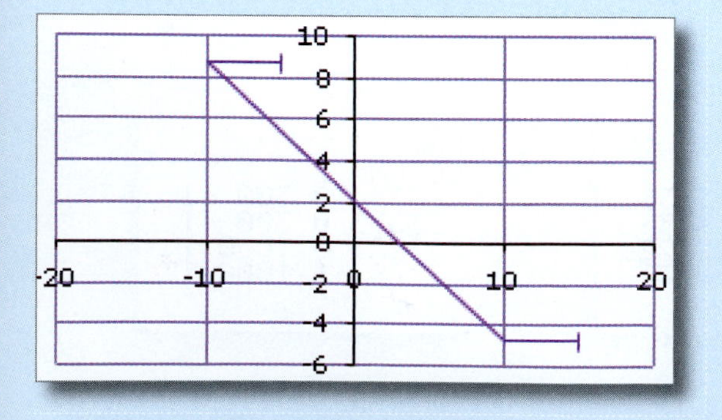

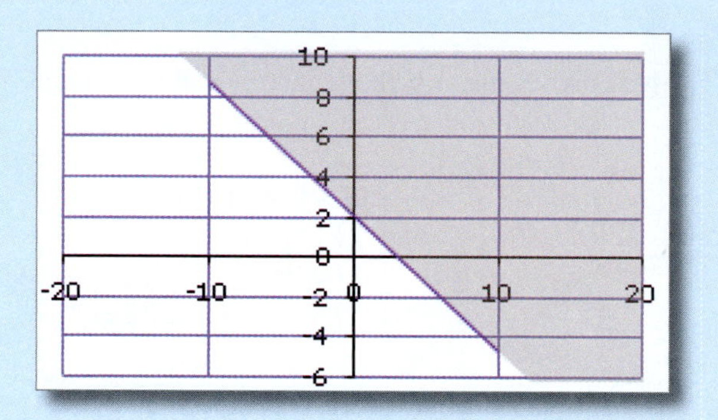

Alternatively, you can use the Drawing Palette to create a polygon with a semi-transparent fill, as shown above.

6

Sets and Counting

Website

www.WanerMath.com

At the Website you will find:

- Section-by-section tutorials, including game tutorials with randomized quizzes
- A detailed chapter summary
- A true/false quiz
- A utility to compute factorials, permutations, and combinations
- Additional review exercises

Case Study Designing a Puzzle

As Product Design Manager for Cerebral Toys, Inc., you are constantly on the lookout for ideas for intellectually stimulating yet inexpensive toys. Your design team recently came up with an idea for a puzzle consisting of a number of plastic cubes. Each cube will have two faces colored red, two white, and two blue, and there will be exactly two cubes with each possible configuration of colors. The goal of the puzzle is to seek out the matching pairs, thereby enhancing a child's geometric intuition and three-dimensional manipulation skills. **If the kit is to include every possible configuration of colors, how many cubes will the kit contain?**

Image Source/Getty Images

Introduction

The theory of sets is the foundation for most of mathematics. It also has direct applications—for example, in searching computer databases. We will use set theory extensively in the chapter on probability, and thus much of this chapter revolves around the idea of a **set of outcomes** of a procedure such as rolling a pair of dice or choosing names from a database. Also important in probability is the theory of **counting** the number of elements in a set, which is called **combinatorics**.

Counting elements is not a trivial proposition; for example, the betting game Lotto (used in many state lotteries) has you pick six numbers from some range—say, 1–55. If your six numbers match the six numbers chosen in the "official drawing," you win the top prize. How many Lotto tickets would you need to buy to guarantee that you will win? That is, how many Lotto tickets are possible? By the end of this chapter, we will be able to answer these questions.

6.1 Sets and Set Operations

In this section we introduce some of the basic ideas of set theory. Some of the examples and applications we see here are derived from the theory of probability and will recur throughout the rest of this chapter and the next.

Sets

Sets and Elements

A **set** is a collection of items, referred to as the **elements** of the set.

Visualizing a Set

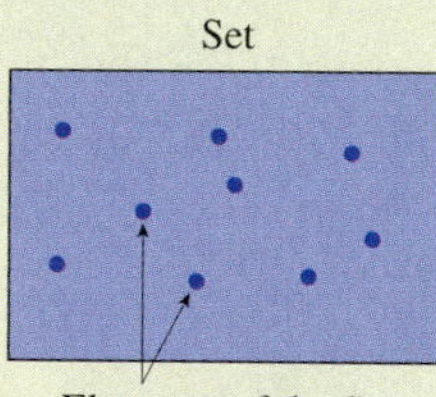

We usually use a capital letter to name a set and braces to enclose the elements of a set.

$x \in A$ means that x **is an element of** the set A. If x is not an element of A, we write $x \notin A$.

$B = A$ means that A and B have the same elements. The order in which the elements are listed does not matter.

$B \subseteq A$ means that B is a **subset** of A; every element of B is also an element of A.

$B \subset A$ means that B is a **proper subset** of A: $B \subseteq A$, but $B \neq A$.

Quick Examples

$W = \{\text{Amazon, eBay, Apple}\}$
$N = \{1, 2, 3, \ldots\}$

Amazon $\in W$ (W as above)
Microsoft $\notin W$ $\quad 2 \in N$

$\{5, -9, 1, 3\} = \{-9, 1, 3, 5\}$
$\{1, 2, 3, 4\} \neq \{1, 2, 3, 6\}$

$\{\text{eBay, Apple}\} \subseteq W$
$\{1, 2, 3, 4\} \subseteq \{1, 2, 3, 4\}$

$\{\text{eBay, Apple}\} \subset W$
$\{1, 2, 3\} \subset \{1, 2, 3, 4\}$
$\{1, 2, 3\} \subset N$ (N as above)

$\emptyset$ is the **empty set**, the set containing no elements. It is a subset of every set.	$\emptyset \subseteq W$ $\emptyset \subset W$
A **finite** set has finitely many elements. An **infinite** set does not have finitely many elements.	$W = \{\text{Amazon, eBay, Apple}\}$ is a finite set. $N = \{1, 2, 3, \ldots\}$ is an infinite set.

One type of set we'll use often is the **set of outcomes** of some activity or experiment. For example, if we toss a coin and observe which side faces up, there are two possible outcomes, heads (H) and tails (T). The set of outcomes of tossing a coin once can be written

$$S = \{\text{H, T}\}.$$

As another example, suppose we roll a die that has faces numbered 1 through 6, as usual, and observe which number faces up. The set of outcomes *could* be represented as

S = {⚀, ⚁, ⚂, ⚃, ⚄, ⚅}.

However, we can much more easily write

$$S = \{1, 2, 3, 4, 5, 6\}.$$

EXAMPLE 1 Two Dice: Distinguishable vs. Indistinguishable

a. Suppose we have two dice that we can distinguish in some way—say, one is green and one is red. If we roll both dice, what is the set of outcomes?

b. Describe the set of outcomes if the dice are indistinguishable.

Solution

a. A systematic way of laying out the set of outcomes for a distinguishable pair of dice is shown in Figure 1.

Figure 1

In the first row all the green dice show a 1, in the second row a 2, in the third row a 3, and so on. Similarly, in the first column all the red dice show a 1, in the second column a 2, and so on. The diagonal pairs (top left to bottom right) show all the "doubles." Using the picture as a guide, we can write the set of 36 outcomes as follows.

$$S = \left\{ \begin{array}{llllll} (1,1), & (1,2), & (1,3), & (1,4), & (1,5), & (1,6), \\ (2,1), & (2,2), & (2,3), & (2,4), & (2,5), & (2,6), \\ (3,1), & (3,2), & (3,3), & (3,4), & (3,5), & (3,6), \\ (4,1), & (4,2), & (4,3), & (4,4), & (4,5), & (4,6), \\ (5,1), & (5,2), & (5,3), & (5,4), & (5,5), & (5,6), \\ (6,1), & (6,2), & (6,3), & (6,4), & (6,5), & (6,6) \end{array} \right\}$$

Distinguishable dice

Notice that S is also the set of outcomes if we roll a single die twice, if we take the first number in each pair to be the outcome of the first roll and the second number the outcome of the second roll.

b. If the dice are truly indistinguishable, we will have no way of knowing which die is which once they are rolled. Think of placing two identical dice in a closed box and then shaking the box. When we look inside afterward, there is no way to tell which die is which. (If we make a small marking on one of the dice or somehow keep track of it as it bounces around, we are *distinguishing* the dice.) We regard two dice as **indistinguishable** if we make no attempt to distinguish them. Thus, for example, the two different outcomes (1, 3) and (3, 1) from part (a) would represent the same outcome in part (b) (one die shows a 3 and the other a 1). Because the set of outcomes should contain each outcome only once, we can remove (3, 1). Following this approach gives the following smaller set of outcomes:

$$S = \left\{ \begin{array}{llllll} (1,1), & (1,2), & (1,3), & (1,4), & (1,5), & (1,6), \\ & (2,2), & (2,3), & (2,4), & (2,5), & (2,6), \\ & & (3,3), & (3,4), & (3,5), & (3,6), \\ & & & (4,4), & (4,5), & (4,6), \\ & & & & (5,5), & (5,6), \\ & & & & & (6,6) \end{array} \right\}.$$

Indistinguishable dice

EXAMPLE 2 Set-Builder Notation

Let $B = \{0, 2, 4, 6, 8\}$. B is the set of all nonnegative* even integers less than 10. If we don't want to list the individual elements of B, we can instead use "set-builder notation," and write

$$B = \{n \mid n \text{ is a nonnegative even integer less than } 10\}.$$

This is read "*B is the set of all n such that n is a nonnegative even integer less than 10.*" Here is the correspondence between the words and the symbols:

B | is | the set of | all n | such that | n is a nonnegative even integer less than 10

$$B = \{n \mid n \text{ is a nonnegative even integer less than } 10\}.$$

* Note that the **nonnegative** integers *include* 0, whereas the **positive** integers *exclude* 0.

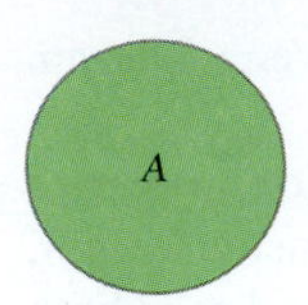

Figure 2

Venn Diagrams

We can visualize sets and relations between sets using **Venn diagrams**. In a Venn diagram, we represent a set as a region, often a disk (Figure 2).

The elements of A are the points inside the region. The following Venn diagrams illustrate the relations we've discussed so far.

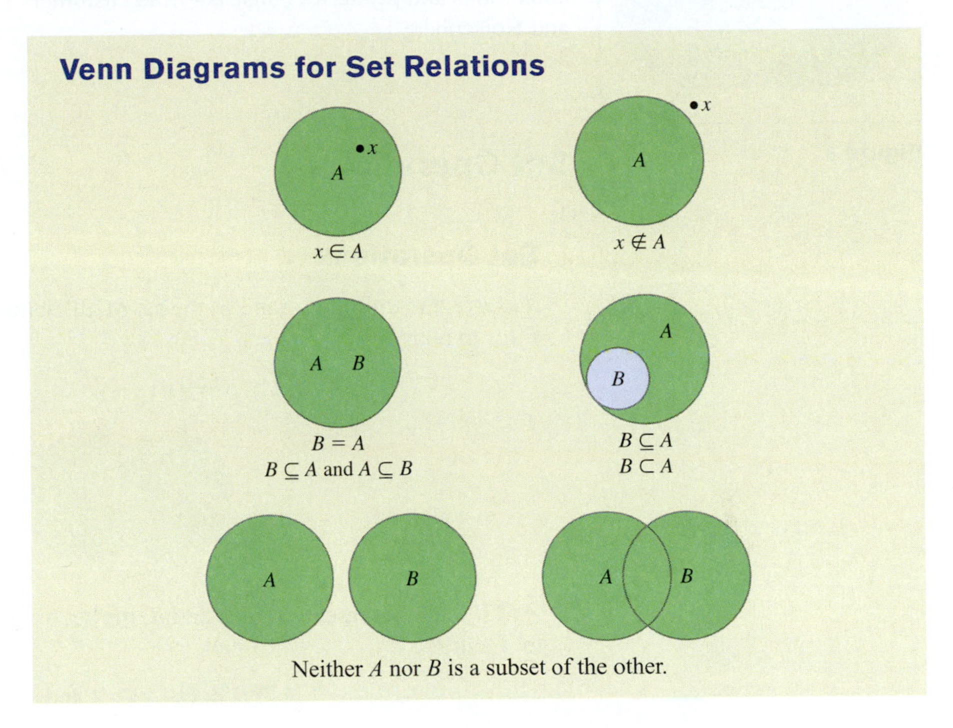

Note Although the diagram for $B \subseteq A$ suggests a proper subset, it is customary to use the same diagram for both subsets and proper subsets. ■

EXAMPLE 3 Customer Interests

NobelBooks.com (a fierce competitor of OHaganBooks.com) maintains a database of customers and the types of books they have purchased. In the company's database is the set of customers

$$S = \{\text{Einstein, Bohr, Millikan, Heisenberg, Schrödinger, Dirac}\}.$$

A search of the database for customers who have purchased cookbooks yields the subset

$$A = \{\text{Einstein, Bohr, Heisenberg, Dirac}\}.$$

Another search, this time for customers who have purchased mysteries, yields the subset

$$B = \{\text{Bohr, Heisenberg, Schrödinger}\}.$$

NobelBooks.com wants to promote a new combination mystery/cookbook, and wants to target two subsets of customers: those who have purchased either cookbooks or mysteries (or both) and, for additional promotions, those who have purchased both cookbooks and mysteries. Name the customers in each of these subsets.

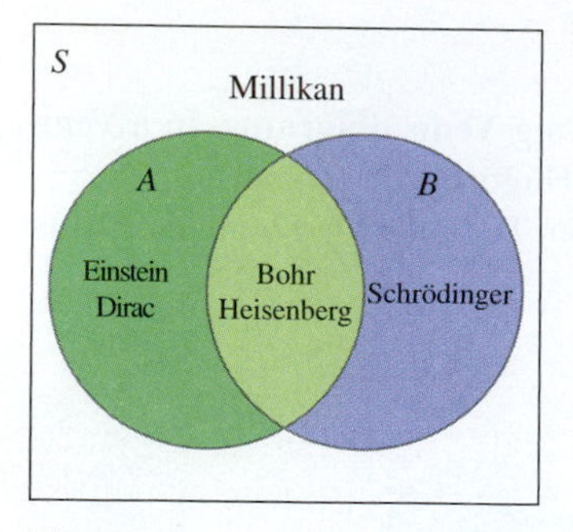

Figure 3

Solution We can picture the database and the two subsets using the Venn diagram in Figure 3.

The set of customers who have purchased either cookbooks or mysteries (or both) consists of the customers who are in A or B or both: Einstein, Bohr, Heisenberg, Schrödinger, and Dirac. The set of customers who have purchased both cookbooks and mysteries consists of the customers in the overlap of A and B, Bohr and Heisenberg.

Set Operations

Set Operations

$A \cup B$ is the **union** of A and B, the set of all elements that are either in A or in B (or in both).

$$A \cup B = \{x \mid x \in A \text{ or } x \in B\}$$

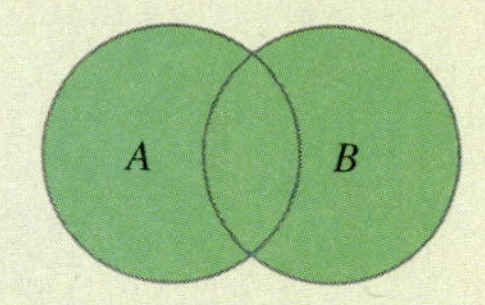

$A \cap B$ is the **intersection** of A and B, the set of all elements that are common to A and B.

$$A \cap B = \{x \mid x \in A \text{ and } x \in B\}$$

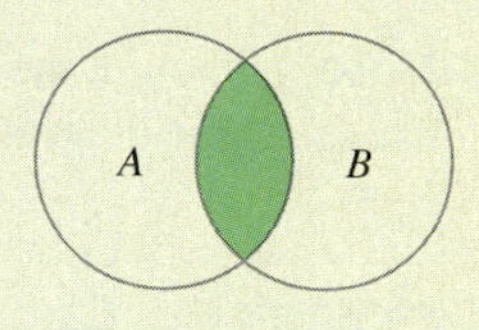

Logical Equivalents

Union: For an element to be in $A \cup B$, it must be in A **or** in B.
Intersection: For an element to be in $A \cap B$, it must be in A **and** in B.

Quick Examples

If $A = \{\text{a, b, c, d}\}$ and $B = \{\text{c, d, e, f}\}$, then

$A \cup B = \{\text{a, b, c, d, e, f}\}$

$A \cap B = \{\text{c, d}\}$.

Note Mathematicians always use "or" in its *inclusive* sense: one thing or another *or both.* ■

There is one other operation we use, called the **complement** of a set A, which, roughly speaking, is the set of things *not* in A.

Q: *Why only "roughly"? Why not just form the set of things not in A?*

A: This would amount to assuming that there is a set of *all things*. (It would be the complement of the empty set.) Although tempting, talking about entities such as the "set of all things" leads to paradoxes.* Instead, we first need to fix a set *S of all objects under consideration,* or the *universe of discourse*, which we generally call the **universal set** for the discussion. For example, when we search the Web, we take S to be the set of all Web pages. When talking about integers, we take S to be the set of all integers. In other words, our choice of universal set depends on the context. The complement of a set $A \subseteq S$ is then the set of *things in S* that are not in A.

* The most famous such paradox is called "Russell's Paradox," after the mathematical logician (and philosopher and pacifist) Bertrand Russell. It goes like this: If there were a set of all things, then there would also be a (smaller) set of all sets. Call it S. Now, because S is the *set* of *all* sets, it must contain itself as a member. In other words, $S \in S$. Let P be the subset of S consisting of all sets that are *not* members of themselves. Now we pose the following question: Is P a member of itself? If it is, then, because it is the set of all sets that are *not* members of themselves, it is not a member of itself. On the other hand, if it is *not* a member of itself, then it qualifies as an element of P. In other words, it *is* a member of itself! Because neither can be true, something is wrong. What is wrong is the assumption that there is such a thing as the set of all sets or the set of all things.

Complement

If S is the universal set and $A \subseteq S$, then A' is the **complement** of A (in S), the set of all elements of S not in A.

$$A' = \{x \in S \mid x \notin A\}$$ = Green Region Below

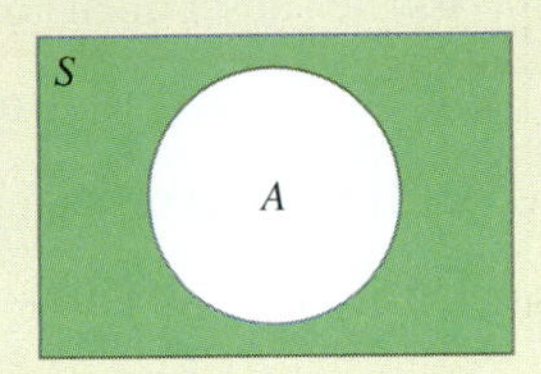

Logical Equivalent

For an element to be in A', it must be in S but **not** in A.

Quick Example

If $S = \{a, b, c, d, e, f, g\}$ and $A = \{a, b, c, d\}$, then

$A' = \{e, f, g\}$.

In the following example we use set operations to describe the sets we found in Example 3, as well as some others.

Joe Hawkins Photography/Alamy

EXAMPLE 4 Customer Interests

NobelBooks.com maintains a database of customers and the types of books they have purchased. In the company's database is the set of customers

$S = \{$Einstein, Bohr, Millikan, Heisenberg, Schrödinger, Dirac$\}$.

A search of the database for customers who have purchased cookbooks yields the subset

$A = \{$Einstein, Bohr, Heisenberg, Dirac$\}$.

Another search, this time for customers who have purchased mysteries, yields the subset

$B = \{$Bohr, Heisenberg, Schrödinger$\}$.

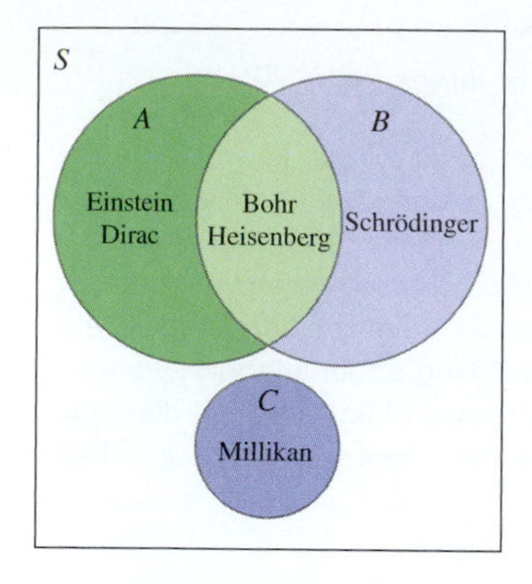

A third search, for customers who had registered with the site but not used their first-time customer discount, yields the subset

$$C = \{\text{Millikan}\}.$$

Use set operations to describe the following subsets:

a. The subset of customers who have purchased either cookbooks or mysteries

b. The subset of customers who have purchased both cookbooks and mysteries

c. The subset of customers who have not purchased cookbooks

d. The subset of customers who have purchased cookbooks but have not used their first-time customer discount

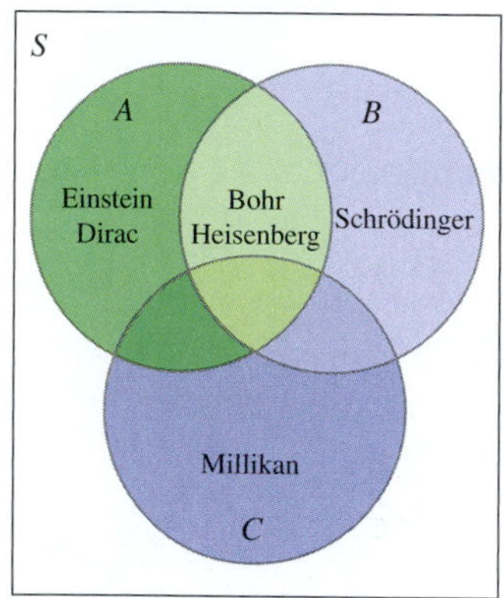

Figure 4

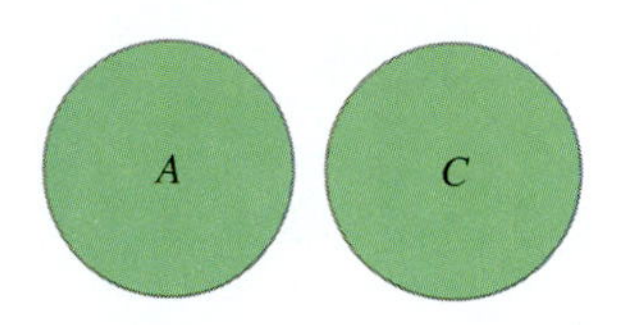

Figure 5

Solution Figure 4 shows two alternative Venn diagram representations of the database. Although the second version shows C overlapping A and B, the placement of the names inside shows that there are no customers in those overlaps.

a. The subset of customers who have bought either cookbooks *or* mysteries is

$$A \cup B = \{\text{Einstein, Bohr, Heisenberg, Schrödinger, Dirac}\}.$$

b. The subset of customers who have bought both cookbooks *and* mysteries is

$$A \cap \text{B} = \{\text{Bohr, Heisenberg}\}.$$

c. The subset of customers who have *not* bought cookbooks is

$$A' = \{\text{Millikan, Schrödinger}\}.$$

Note that, for the universal set, we are using the set S of all customers in the database.

d. The subset of customers who have bought cookbooks but have not used their first-time purchase discount is the empty set

$$A \cap C = \emptyset.$$

When the intersection of two sets is empty, we say that the two sets are **disjoint**. In a Venn diagram, disjoint sets are drawn as regions that don't overlap, as in Figure 5.*

* People new to set theory sometimes find it strange to consider the empty set a valid set. Here is one of the times where it is very useful to do so. If we did not, we would have to say that $A \cap C$ was defined only when A and C had something in common. Having to deal with the fact that this set operation was not always defined would quickly get tiresome.

Before we go on... Computer databases and the Web can be searched using so-called "Boolean searches." These are search requests using "and," "or," and "not." Using "and" gives the intersection of separate searches, using "or" gives the union, and using "not" gives the complement. In the next section we'll see how Web search engines allow such searches. ■

Cartesian Product

There is one more set operation we need to discuss.

Cartesian Product

The **Cartesian product** of two sets, A and B, is the set of all ordered pairs (a, b) with $a \in A$ and $b \in B$.

$$A \times B = \{(a, b) \mid a \in A \text{ and } b \in B\}$$

In words, $A \times B$ is the set of all ordered pairs whose first component is in A and whose second component is in B.

Quick Examples

1. If $A = \{a, b\}$ and $B = \{1, 2, 3\}$, then

$$A \times B = \{(a, 1), (a, 2), (a, 3), (b, 1), (b, 2), (b, 3)\}.$$

Visualizing A × B

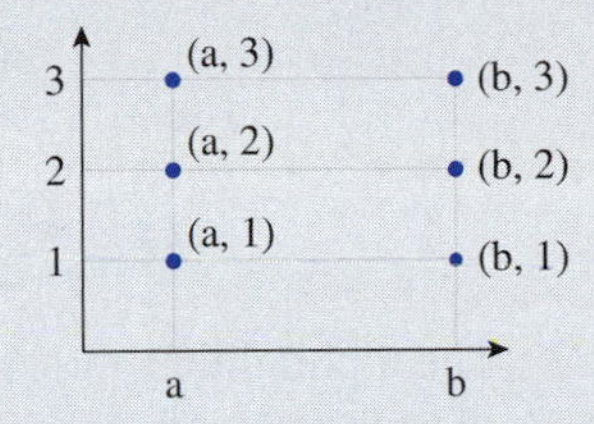

2. If $S = \{H, T\}$, then

$$S \times S = \{(H, H), (H, T), (T, H), (T, T)\}.$$

In other words, if S is the set of outcomes of tossing a coin once, then $S \times S$ is the set of outcomes of tossing a coin twice.

3. If $S = \{1, 2, 3, 4, 5, 6\}$, then

$$S \times S = \left\{ \begin{matrix} (1,1), & (1,2), & (1,3), & (1,4), & (1,5), & (1,6), \\ (2,1), & (2,2), & (2,3), & (2,4), & (2,5), & (2,6), \\ (3,1), & (3,2), & (3,3), & (3,4), & (3,5), & (3,6), \\ (4,1), & (4,2), & (4,3), & (4,4), & (4,5), & (4,6), \\ (5,1), & (5,2), & (5,3), & (5,4), & (5,5), & (5,6), \\ (6,1), & (6,2), & (6,3), & (6,4), & (6,5), & (6,6) \end{matrix} \right\}.$$

In other words, if S is the set of outcomes of rolling a die once, then $S \times S$ is the set of outcomes of rolling a die twice (or rolling two distinguishable dice).

4. If $A = \{\text{red, yellow}\}$ and $B = \{\text{Mustang, Firebird}\}$, then

$A \times B = \{$(red, Mustang), (red, Firebird), (yellow, Mustang), (yellow, Firebird)$\}$ which we might also write as

$A \times B = \{$red Mustang, red Firebird, yellow Mustang, yellow Firebird$\}$.

EXAMPLE 5 Representing Cartesian Products

The manager of an automobile dealership has collected data on the number of pre-owned Acura, Infiniti, Lexus, and Mercedes cars the dealership has from the 2009, 2010, and 2011 model years. In entering this information on a spreadsheet, the manager would like to have each spreadsheet cell represent a particular year and make. Describe this set of cells.

Solution Because each cell represents a year and a make, we can think of the cell as a pair (year, make), as in (2009, Acura). Thus, the set of cells can be thought of as a Cartesian product:

$$Y = \{2009, 2010, 2011\} \qquad \text{Year of car}$$

$$M = \{\text{Acura, Infiniti, Lexus, Mercedes}\} \qquad \text{Make of car}$$

$$Y \times M = \left\{ \begin{matrix} (2009, \text{Acura}) & (2009, \text{Infiniti}) & (2009, \text{Lexus}) & (2009, \text{Mercedes}) \\ (2010, \text{Acura}) & (2010, \text{Infiniti}) & (2010, \text{Lexus}) & (2010, \text{Mercedes}) \\ (2011, \text{Acura}) & (2011, \text{Infiniti}) & (2011, \text{Lexus}) & (2011, \text{Mercedes}) \end{matrix} \right\}. \qquad \text{Cells}$$

Thus, the manager might arrange the spreadsheet as follows:

	A	B	C	D	E
1		**Acura**	**Infiniti**	**Lexus**	**Mercedes**
2	**2009**	(2009 Acura)	(2009 Infiniti)	(2009 Lexus)	(2009 Mercedes)
3	**2010**	(2010 Acura)	(2010 Infiniti)	(2010 Lexus)	(2010 Mercedes)
4	**2011**	(2011 Acura)	(2011 Infiniti)	(2011 Lexus)	(2011 Mercedes)

The highlighting shows the 12 cells to be filled in, representing the numbers of cars of each year and make. For example, in cell B2 should go the number of 2009 Acuras the dealership has.

Before we go on... The arrangement in the spreadsheet in Example 5 is consistent with the matrix notation in Chapter 2. We could also have used the elements of Y as column labels along the top and the elements of M as row labels down the side. Along those lines, we can also visualize the Cartesian product $Y \times M$ as a set of points in the xy-plane ("Cartesian plane") as shown in Figure 6.

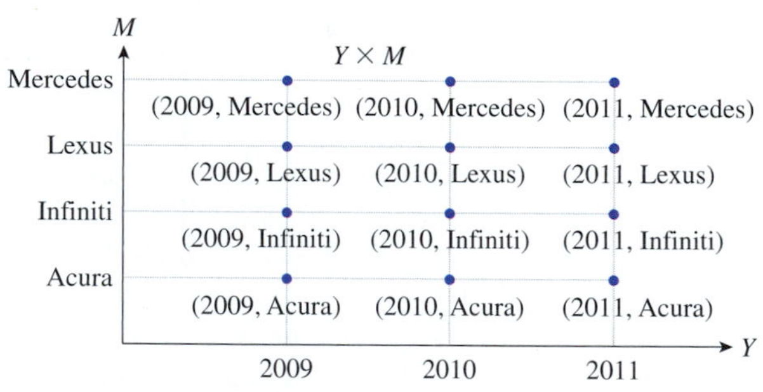

Figure 6

FAQs

The Many Meanings of "And"

Q: *Suppose A is the set of actors and B is the set of all baseball players. Then the set of all actors* and *baseball players is $A \cap B$, right?*

A: Wrong. The fact that the word "and" appears in the description of a set does not always mean that the set is an intersection; the word "and" can mean different things in different contexts. $A \cap B$ refers to the set of elements that are in both A and B, hence to actors who are also baseball players. On the other hand, the set of all actors and baseball players is the set of people who are either actors or baseball players (or both), which is $A \cup B$. We can use the word "and" to describe both sets:

$A \cap B = \{$people who are both actors *and* baseball players$\}$

$A \cup B = \{$people who are actors *or* baseball players$\}$

$= \{$all actors *and* baseball players$\}$.

6.1 EXERCISES

Access end-of-section exercises online at **www.webassign.net** 

6.2 Cardinality

In this section, we begin to look at a deceptively simple idea: the size of a set, which we call its **cardinality**.

Cardinality

If A is a finite set, then its **cardinality** is

$$n(A) = \text{number of elements in } A.$$

Visualizing Cardinality

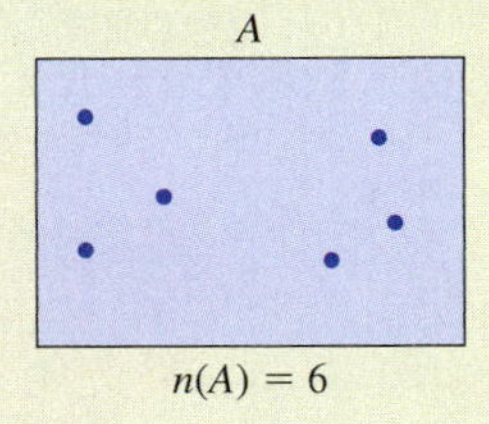

Quick Examples

1. Let $S = \{\text{a, b, c}\}$. Then $n(S) = 3$.
2. Let S be the set of outcomes when two distinguishable dice are rolled. Then $n(S) = 36$ (see Example 1 in Section 6.1).
3. $n(\emptyset) = 0$ because the empty set has no elements.

Counting the elements in a small, simple set is straightforward. To count the elements in a large, complicated set, we try to describe the set as built of simpler sets using the set operations. We then need to know how to calculate the number of elements in, for example, a union, based on the number of elements in the simpler sets whose union we are taking.

The Cardinality of a Union

How can we calculate $n(A \cup B)$ if we know $n(A)$ and $n(B)$? Our first guess might be that $n(A \cup B)$ is $n(A) + n(B)$. But consider a simple example. Let

$$A = \{\text{a, b, c}\}$$

and

$$B = \{\text{b, c, d}\}.$$

Then $A \cup B = \{a, b, c, d\}$ so $n(A \cup B) = 4$, but $n(A) + n(B) = 3 + 3 = 6$. The calculation $n(A) + n(B)$ gives the wrong answer because the elements b and c are counted twice, once for being in A and again for being in B. To correct for this overcounting, we need to subtract the number of elements that get counted twice, which is the number of elements that A and B have in common, or $n(A \cap B) = 2$ in this case. So, we get the right number for $n(A \cup B)$ from the following calculation:

$$n(A) + n(B) - n(A \cap B) = 3 + 3 - 2 = 4.$$

This argument leads to the following general formula.

Cardinality of a Union

If A and B are finite sets, then

$$n(A \cup B) = n(A) + n(B) - n(A \cap B).$$

In particular, if A and B are disjoint (meaning that $A \cap B = \emptyset$), then

$$n(A \cup B) = n(A) + n(B).$$

(When A and B are disjoint we say that $A \cup B$ is a **disjoint union**.)

Visualizing Cardinality of a Union

Disjoint Sets

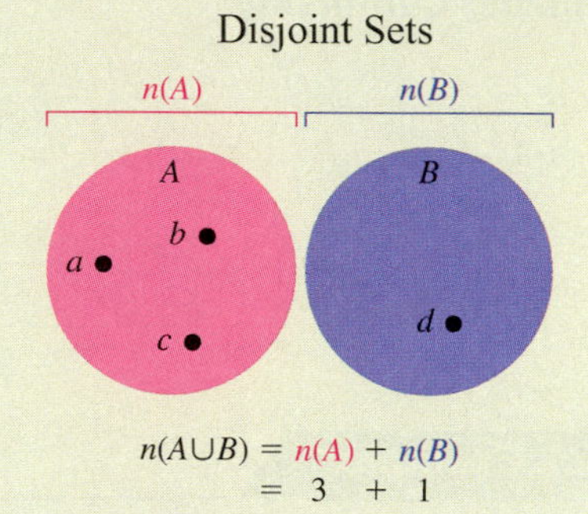

$n(A \cup B) = n(A) + n(B)$
$= 3 + 1$

Not Disjoint

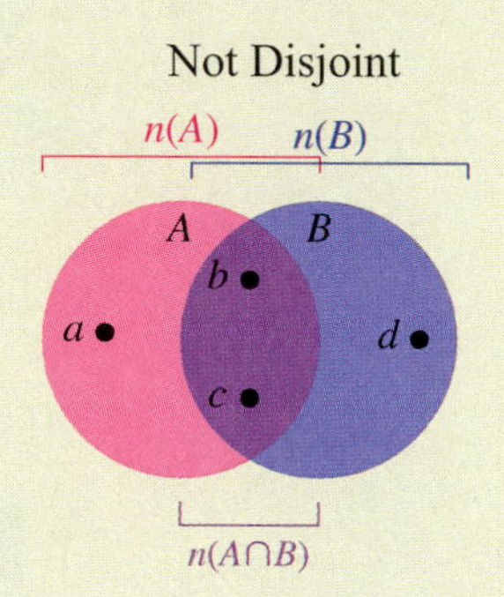

$n(A \cup B) = n(A) + n(B) - n(A \cap B)$
$= 3 + 3 - 2$

Quick Examples

1. If $A = \{a, b, c, d\}$ and $B = \{b, c, d, e, f\}$, then

$$n(A \cup B) = n(A) + n(B) - n(A \cap B) = 4 + 5 - 3 = 6.$$

In fact, $A \cup B = \{a, b, c, d, e, f\}$.

2. If $A = \{a, b, c\}$ and $B = \{d, e, f\}$, then $A \cap B = \emptyset$, so

$$n(A \cup B) = n(A) + n(B) = 3 + 3 = 6.$$

digitallife/Alamy

EXAMPLE 1 Web Searches

In October 2011, a search on Bing™ for "Lt. William Burrows" yielded 29 Web sites containing that phrase, and a search for "Romulan probe" yielded 17 sites. A search for sites containing both phrases yielded 4 Web sites. How many Web sites contained either "Lt. William Burrows," "Romulan probe," or both?

Solution Let A be the set of sites containing "Lt. William Burrows" and let B be the set of sites containing "Romulan probe." We are told that

$$n(A) = 29$$
$$n(B) = 17$$
$$n(A \cap B) = 4. \quad \text{"Lt. William Burrows" AND "Romulan probe"}$$

The formula for the cardinality of the union tells us that

$$n(A \cup B) = n(A) + n(B) - n(A \cap B) = 29 + 17 - 4 = 42.$$

So, 42 sites in the Bing database contained one or both of the phrases "Lt. William Burrows" or "Romulan probe."

Before we go on... Each search engine has a different way of specifying a search for a union or an intersection. At Bing or Google™, you can use "OR" for the union and "AND" for the intersection.

Although the formula $n(A \cup B) = n(A) + n(B) - n(A \cap B)$ always holds mathematically, you may sometimes find that, in an actual search, the numbers don't add up. Google, for example, appears not to adhere strictly to the search rule you enter, but instead presents results that it thinks you want.* ■

* For example, in October 2011, a search on Google gave the following results:
"Romulan probe": 471 results
"Abraham Lincoln is born on Earth": 8 results
"Romulan probe" OR "Abraham Lincoln is born on Earth": 1,600 results!

Q: *Is there a similar formula for $n(A \cap B)$?*

A: The formula for the cardinality of a union can also be thought of as a formula for the cardinality of an intersection. We can solve for $n(A \cap B)$ to get

$$n(A \cap B) = n(A) + n(B) - n(A \cup B).$$

In fact, we can think of this formula as an equation relating four quantities. If we know any three of them, we can use the equation to find the fourth. (See Example 2 on next page).

Q: *Is there a similar formula for $n(A')$?*

A: We can get a formula for the cardinality of a complement as follows: If S is our universal set and $A \subseteq S$, then S is the disjoint union of A and its complement. That is,

$$S = A \cup A' \quad \text{and} \quad A \cap A' = \emptyset.$$

Applying the cardinality formula for a disjoint union, we get

$$n(S) = n(A) + n(A').$$

We can then solve for $n(A')$ or for $n(A)$ to get the formulas shown below:

Cardinality of a Complement

If S is a finite universal set and A is a subset of S, then

$$n(A') = n(S) - n(A)$$

and

$$n(A) = n(S) - n(A').$$

Visualizing Cardinality of a Complement

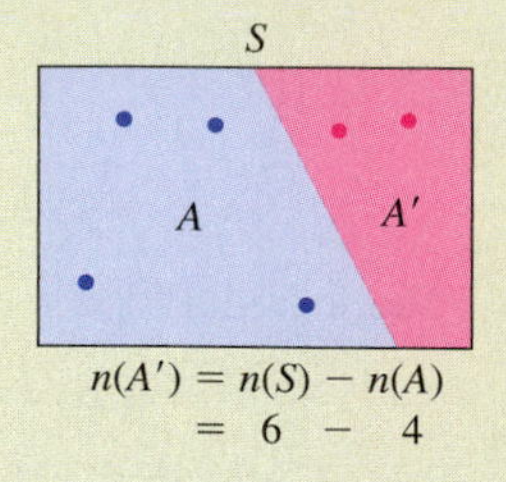

Quick Example

If $S = \{a, b, c, d, e, f\}$ and $A = \{a, b, c, d\}$, then

$$n(A') = n(S) - n(A) = 6 - 4 = 2.$$

In fact, $A' = \{e, f\}$.

EXAMPLE 2 Cookbooks

In November 2011, a search at Amazon.com found 132,000 books on cooking.[1] Of these, 20,000 were on regional cooking, 5,000 were on vegetarian cooking, and 24,000 were on either regional or vegetarian cooking (or both). How many of these books were not on both regional and vegetarian cooking?

Solution Let S be the set of all 132,000 books on cooking, let A be the set of books on regional cooking, and let B be the set of books on vegetarian cooking. We wish to find the size of the complement of the set of books on both regional and vegetarian cooking—that is, $n((A \cap B)')$. Using the formula for the cardinality of a complement, we have

$$n((A \cap B)') = n(S) - n(A \cap B) = 132{,}000 - n(A \cap B).$$

To find $n(A \cap B)$, we use the formula for the cardinality of a union:

$$n(A \cup B) = n(A) + n(B) - n(A \cap B).$$

Substituting the values we were given, we find

$$24{,}000 = 20{,}000 + 5{,}000 - n(A \cap B),$$

which we can solve to get

$$n(A \cap B) = 1{,}000.$$

Therefore,

$$n((A \cap B)') = 132{,}000 - n(A \cap B) = 132{,}000 - 1{,}000 = 131{,}000.$$

So, 131,000 of the books on cooking were not on both regional and vegetarian cooking.

---

[1] Precisely, it found that many books under the subject "Cooking, Food & Wine." Regional cooking falls under "Regional & International." Figures are rounded to the nearest 1,000.

EXAMPLE 3 iPods, iPhones, and iPads

The following table shows sales, in millions of units, of iPods®, iPhones®, and iPads® in the last three quarters of 2011.[2]

	iPods (A)	iPhones (B)	iPads (C)	Total
2011 Q2 (U)	9.0	18.7	4.7	32.4
2011 Q3 (V)	7.5	20.3	9.3	37.1
2011 Q4 (W)	6.6	17.1	11.1	34.8
Total	23.1	56.1	25.1	104.3

Let S be the set of all these iPods, iPhones, and iPads, and label the sets representing the sales in each row and column as shown (so that, for example, A is the set of all iPods sold during the three quarters). Describe the following sets and compute their cardinality:

a. U' **b.** $A \cap U'$ **c.** $(A \cap U)'$ **d.** $C \cup U$

Solution Because all the figures are stated in millions of units, we'll give our calculations and results in millions of units as well.

a. U' is the set of all items not sold in the second quarter of 2011. To compute its cardinality, we could add the totals for all the other quarters listed in the rightmost column:

$$n(U') = n(V) + n(W) = 37.1 + 34.8 = 71.9 \text{ million items.}$$

Alternatively, we can use the formula for the cardinality of a complement (referring again to the totals in the table):

$$n(U') = n(S) - n(U) = 104.3 - 32.4 = 71.9 \text{ million items.}$$

b. $A \cap U'$ is the intersection of the set of all iPods and the set of all items not sold in 2011 Q2. In other words, it is the set of all iPods not sold in the second quarter of 2011. Here is the table with the corresponding sets A and U' shaded ($A \cap U'$ is the overlap):

	iPods (A)	iPhones (B)	iPads (C)	Total
2011 Q2 (U)	9.0	18.7	4.7	32.4
2011 Q3 (V)	7.5	20.3	9.3	37.1
2011 Q4 (W)	6.6	17.1	11.1	34.8
Total	23.1	56.1	25.1	104.3

From the table:

$$n(A \cap U') = 7.5 + 6.6 = 14.1 \text{ million items.}$$

[2]Figures are rounded to one decimal place. Source: Apple quarterly press releases (www.investor.apple.com).

c. $A \cap U$ is the set of all iPods sold in 2011 Q2, and so $(A \cap U)'$ is the set of all items remaining if we exclude iPods sold in 2011 Q2:

	iPods (A)	iPhones (B)	iPads (C)	Total
2011 Q2 (U)	9.0	18.7	4.7	32.4
2011 Q3 (V)	7.5	20.3	9.3	37.1
2011 Q4 (W)	6.6	17.1	11.1	34.8
Total	23.1	56.1	25.1	104.3

From the formula for the cardinality of a complement:

$$n((A \cap U)') = n(S) - n(A \cap U)$$
$$= 104.3 - 9.0 = 95.3 \text{ million items.}$$

d. $C \cup U$ is the set of items that were either iPads or sold in 2011 Q2:

	iPods (A)	iPhones (B)	iPads (C)	Total
2011 Q2 (U)	9.0	18.7	4.7	32.4
2011 Q3 (V)	7.5	20.3	9.3	37.1
2011 Q4 (W)	6.6	17.1	11.1	34.8
Total	23.1	56.1	25.1	104.3

To compute it, we can use the formula for the cardinality of a union:

$$n(C \cup U) = n(C) + n(U) - n(C \cap U)$$
$$= 25.1 + 32.4 - 4.7 = 52.8 \text{ million items.}$$

To determine the cardinality of a union of three or more sets, like $n(A \cup B \cup C)$, we can think of $A \cup B \cup C$ as a union of two sets, $(A \cup B)$ and C, and then analyze each piece using the techniques we already have. Alternatively, there are formulas for the cardinalities of unions of any number of sets, but these formulas get more and more complicated as the number of sets grows. In many applications, like the following example, we can use Venn diagrams instead.

EXAMPLE 4 Reading Lists

A survey of 300 college students found that 100 had read *War and Peace,* 120 had read *Crime and Punishment,* and 100 had read *The Brothers Karamazov*. It also found that 40 had read only *War and Peace,* 70 had read *War and Peace* but not *The Brothers Karamazov,* and 80 had read *The Brothers Karamazov* but not *Crime and Punishment*. Only 10 had read all three novels. How many had read none of these three novels?

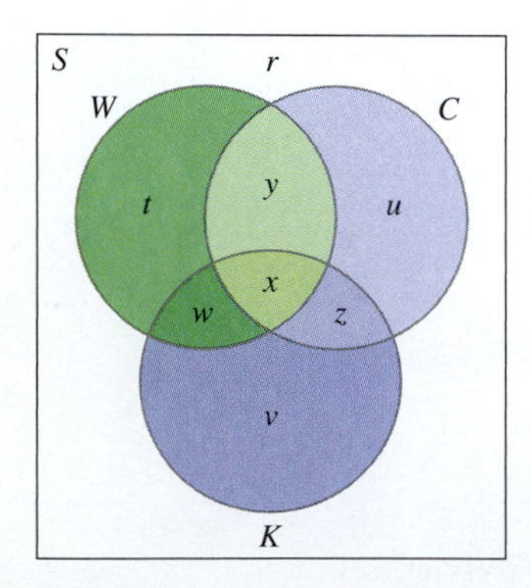

Figure 7

Solution There are four sets mentioned in the problem: the universe S consisting of the 300 students surveyed, the set W of students who had read *War and Peace,* the set C of students who had read *Crime and Punishment,* and the set K of students who had read *The Brothers Karamazov*. Figure 7 shows a Venn diagram representing these sets.

We have put labels in the various regions of the diagram to represent the number of students in each region. For example, x represents the number of students in

$W \cap C \cap K$, which is the number of students who have read all three novels. We are told that this number is 10, so

$$x = 10.$$

(You should draw the diagram for yourself and fill in the numbers as we go along.) We are also told that 40 students had read only *War and Peace,* so

$$t = 40.$$

We are given none of the remaining regions directly. However, because 70 had read *War and Peace* but not *The Brothers Karamazov,* we see that t and y must add up to 70. Because we already know that $t = 40$, it follows that $y = 30$. Further, because a total of 100 students had read *War and Peace,* we have

$$x + y + t + w = 100.$$

Substituting the known values of x, y, and t gives

$$10 + 30 + 40 + w = 100$$

so $w = 20$. Because 80 students had read *The Brothers Karamazov* but not *Crime and Punishment,* we see that $v + w = 80$, so $v = 60$ (because we know $w = 20$). We can now calculate z using the fact that a total of 100 students had read *The Brothers Karamazov:*

$$60 + 20 + 10 + z = 100$$

giving $z = 10$. Similarly, we can now get u using the fact that 120 students had read *Crime and Punishment:*

$$10 + 30 + 10 + u = 120,$$

giving $u = 70$. Of the 300 students surveyed, we've now found $x + y + z + w + t + u + v = 240$. This leaves

$$r = 60$$

who had read none of the three novels.

The Cardinality of a Cartesian Product

We've covered all the operations except Cartesian product. To find a formula for $n(A \times B)$, consider the following simple example:

$$A = \{\text{H, T}\}$$
$$B = \{1, 2, 3, 4, 5, 6\}$$

so that

$$A \times B = \{\text{H1, H2, H3, H4, H5, H6, T1, T2, T3, T4, T5, T6}\}.$$

As we saw in Example 5 in Section 6.1, the elements of $A \times B$ can be arranged in a table or spreadsheet with $n(A) = 2$ rows and $n(B) = 6$ elements in each row.

	A	B	C	D	E	F	G
1		**1**	**2**	**3**	**4**	**5**	**6**
2	**H**	H1	H2	H3	H4	H5	H6
3	**T**	T1	T2	T3	T4	T5	T6

In a region with 2 rows and 6 columns, there are $2 \times 6 = 12$ cells. So,

$$n(A \times B) = n(A)n(B)$$

in this case. There is nothing particularly special about this example, however, and that formula holds true in general.

Cardinality of a Cartesian Product

If A and B are finite sets, then

$$n(A \times B) = n(A)n(B).$$

Quick Example

If $A = \{a, b, c\}$ and $B = \{x, y, z, w\}$, then

$$n(A \times B) = n(A)n(B) = 3 \times 4 = 12.$$

EXAMPLE 5 Coin Tosses

a. If we toss a coin twice and observe the sequence of heads and tails, how many possible outcomes are there?

b. If we toss a coin three times, how many possible outcomes are there?

c. If we toss a coin ten times, how many possible outcomes are there?

Solution

a. Let $A = \{H, T\}$ be the set of possible outcomes when a coin is tossed once. The set of outcomes when a coin is tossed twice is $A \times A$, which has

$$n(A \times A) = n(A)n(A) = 2 \times 2 = 4$$

possible outcomes.

b. When a coin is tossed three times, we can think of the set of outcomes as the product of the set of outcomes for the first two tosses, which is $A \times A$, and the set of outcomes for the third toss, which is just A. The set of outcomes for the three tosses is then $(A \times A) \times A$, which we usually write as $A \times A \times A$ or A^3. The number of outcomes is

$$n((A \times A) \times A) = n(A \times A)n(A) = (2 \times 2) \times 2 = 8.$$

c. Considering the result of part (b), we can easily see that the set of outcomes here is $A^{10} = A \times A \times \cdots \times A$ (10 copies of A), or the set of ordered sequences of ten Hs and Ts. It's also easy to see that

$$n(A^{10}) = [n(A)]^{10} = 2^{10} = 1{,}024.$$

➡ **Before we go on...** We can start to see the power of these formulas for cardinality. In Example 5, we were able to calculate that there are 1,024 possible outcomes when we toss a coin 10 times without writing out all 1,024 possibilities and counting them. ■

6.2 EXERCISES

Access end-of-section exercises online at **www.webassign.net**

6.3 Decision Algorithms: The Addition and Multiplication Principles

Let's start with a really simple example. You walk into an ice cream parlor and find that you can choose between ice cream, of which there are 15 flavors, and frozen yogurt, of which there are 5 flavors. How many different selections can you make? Clearly, you have $15 + 5 = 20$ different desserts from which to choose. Mathematically, this is an example of the formula for the cardinality of a disjoint union: If we let A be the set of ice creams you can choose from, and B the set of frozen yogurts, then $A \cap B = \emptyset$ and we want $n(A \cup B)$. But the formula for the cardinality of a disjoint union is $n(A \cup B) = n(A) + n(B)$, which gives $15 + 5 = 20$ in this case.

This example illustrates a very useful general principle.

Addition Principle

When choosing among r disjoint alternatives, suppose that

alternative 1 has n_1 possible outcomes,

alternative 2 has n_2 possible outcomes,

...

alternative r has n_r possible outcomes,

with no two of these outcomes the same. Then there are a total of $n_1 + n_2 + \cdots + n_r$ possible outcomes.

Quick Example

At a restaurant you can choose among 8 chicken dishes, 10 beef dishes, 4 seafood dishes, and 12 vegetarian dishes. This gives a total of $8 + 10 + 4 + 12 = 34$ different dishes to choose from.

Here is another simple example. In that ice cream parlor, not only can you choose from 15 flavors of ice cream, but you can also choose from 3 different sizes of cone. How many different ice cream cones can you select from? This time, we want to choose both a flavor and a size, or, in other words, a pair (flavor, size). Therefore, if we let A again be the set of ice cream flavors and now let C be the set of cone sizes, the pair we want to choose is an element of $A \times C$, the Cartesian product. To find the number of choices we have, we use the formula for the cardinality of a Cartesian product: $n(A \times C) = n(A)n(C)$. In this case, we get $15 \times 3 = 45$ different ice cream cones we can select.

This example illustrates another general principle.

Multiplication Principle

When making a sequence of choices with r steps, suppose that

step 1 has n_1 possible outcomes

step 2 has n_2 possible outcomes

...

step r has n_r possible outcomes

* See Example 3 for a case in which different sequences of choices can lead to the same outcome, with the result that the multiplication principle does not apply.

and that each sequence of choices results in a distinct outcome.* Then there are a total of $n_1 \times n_2 \times \cdots \times n_r$ possible outcomes.

Quick Example

At a restaurant you can choose among 5 appetizers, 34 main dishes, and 10 desserts. This gives a total of $5 \times 34 \times 10 = 1{,}700$ different meals (each including one appetizer, one main dish, and one dessert) from which you can choose.

Things get more interesting when we have to use the addition and multiplication principles in tandem.

EXAMPLE 1 Desserts

You walk into an ice cream parlor and find that you can choose between ice cream, of which there are 15 flavors, and frozen yogurt, of which there are 5 flavors. In addition, you can choose among 3 different sizes of cones for your ice cream or 2 different sizes of cups for your yogurt. How many different desserts can you choose from?

Solution It helps to think about a definite procedure for deciding which dessert you will choose. Here is one we can use:

Alternative 1: An ice cream cone
- **Step 1** Choose a flavor.
- **Step 2** Choose a size.

Alternative 2: A cup of frozen yogurt
- **Step 1** Choose a flavor.
- **Step 2** Choose a size.

That is, we can choose between alternative 1 and alternative 2. If we choose alternative 1, we have a sequence of two choices to make: flavor and size. The same is true of alternative 2. We shall call a procedure in which we make a sequence of decisions a **decision algorithm.**† Once we have a decision algorithm, we can use the addition and multiplication principles to count the number of possible outcomes.

† An algorithm is a procedure with definite rules for what to do at every step.

Alternative 1: An ice cream cone
- **Step 1** Choose a flavor: 15 choices.
- **Step 2** Choose a size: 3 choices.

There are $15 \times 3 = 45$ possible choices in alternative 1. Multiplication Principle

Alternative 2: A cup of frozen yogurt
- **Step 1** Choose a flavor: 5 choices.
- **Step 2** Choose a size: 2 choices.

There are $5 \times 2 = 10$ possible choices in alternative 2. Multiplication Principle

So, there are $45 + 10 = 55$ possible choices of desserts. Addition Principle

➡ **Before we go on...** Decision algorithms can be illustrated by **decision trees**. To simplify the picture, suppose we had fewer choices in Example 1—say, only two choices of ice cream flavor: vanilla and chocolate, and two choices of yogurt flavor: banana and raspberry. This gives us a total of $2 \times 3 + 2 \times 2 = 10$ possible desserts.

We can illustrate the decisions we need to make when choosing what to buy in the diagram in Figure 8, called a decision tree.

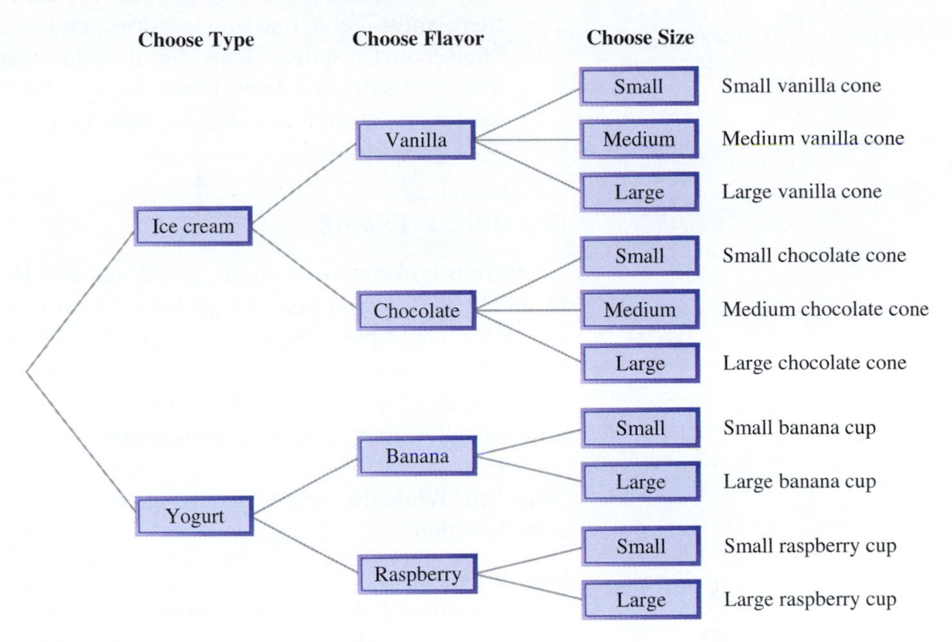

Figure 8

We do not use decision trees much in this chapter because, while they provide a good way of thinking about decision algorithms, they're not really practical for counting large sets. Similar diagrams will be very useful, however, in the chapter on probability. ■

Decision Algorithm

A **decision algorithm** is a procedure in which we make a sequence of decisions. We can use decision algorithms to determine the number of possible items by pretending we are *designing* such an item (for example, an ice-cream cone) and listing the decisions or choices we should make at each stage of the process.

Quick Example

An iPod is available in two sizes. The larger size comes in two colors and the smaller size (the Mini) comes in four colors. A decision algorithm for "designing" an iPod is:

Alternative 1: Select Large:

Step 1 Choose a color: Two choices.
(So, there are two choices for alternative 1.)

Alternative 2: Select a Mini:

Step 1 Choose a color: Four choices.
(So, there are four choices for alternative 2.)

Thus, there are $2 + 4 = 6$ possible choices of iPods.

Caution

For a decision algorithm to give the correct number of possible items, there must be a one-to-one correspondence between sequences of choices and resulting items. So, it is necessary that each sequence of choices results in a distinct item. In other words, *changing one or more choices must result in a different item.* (See Example 3.)

EXAMPLE 2 Exams

An exam is broken into two parts, Part A and Part B, both of which you are required to do. In Part A you can choose between answering 10 true-false questions or answering 4 multiple-choice questions, each of which has 5 answers to choose from. In Part B you can choose between answering 8 true-false questions and answering 5 multiple-choice questions, each of which has 4 answers to choose from. How many different collections of answers are possible?

Solution While deciding what answers to write down, we use the following decision algorithm:

Step 1 Do Part A.

Alternative 1: Answer the 10 true-false questions.

Steps 1–10 Choose true or false for each question: 2 choices each.
There are $2 \times 2 \times \cdots \times 2 = 2^{10} = 1{,}024$ choices in alternative 1.

Alternative 2: Answer the 4 multiple-choice questions.

Steps 1–4 Choose one answer for each question: 5 choices each.
There are $5 \times 5 \times 5 \times 5 = 5^4 = 625$ choices in alternative 2.

$1{,}024 + 625 = 1{,}649$ choices in step 1

Step 2 Do Part B.

Alternative 1: Answer the 8 true-false questions: 2 choices each.
$2^8 = 256$ choices in alternative 1

Alternative 2: Answer the 5 multiple-choice questions, 4 choices each.
$4^5 = 1{,}024$ choices in alternative 2

$256 + 1{,}024 = 1{,}280$ choices in step 2

There are $1{,}649 \times 1{,}280 = 2{,}110{,}720$ different collections of answers possible.

The next example illustrates the need to select your decision algorithm with care.

EXAMPLE 3 Scrabble®

You are playing Scrabble and have the following letters to work with: k, e, r, e. Because you are losing the game, you would like to use all your letters to make a single word, but you can't think of any four-letter words using all these letters. In desperation, you decide to list *all* the four-letter sequences possible to see if there are any valid words among them. How large is your list?

Solution It may first occur to you to try the following decision algorithm.

Step 1 Select the first letter: 4 choices.
Step 2 Select the second letter: 3 choices.

Step 3 Select the third letter: 2 choices.
Step 4 Select the last letter: 1 choice.

This gives $4 \times 3 \times 2 \times 1 = 24$ choices. However, something is wrong with the algorithm.

Q: *What is wrong with this decision algorithm?*

A: We didn't take into account the fact that there are two "e"s;* different decisions in Steps 1–4 can produce the same sequence. Suppose, for example, that we selected the first "e" in Step 1, the second "e" in Step 2, and then the "k" and the "r." This would produce the sequence "eekr." If we selected the *second* "e" in Step 1, the *first* "e" in Step 2, and then the "k" and "r," we would obtain the *same* sequence: "eekr." In other words, the decision algorithm produces two copies of the sequence "eekr" in the associated decision tree. (In fact, it produces two copies of each possible sequence of the letters.) In short, different sequences of choices produce the same result, violating the requirement in the note of caution that precedes Example 2:

For a decision algorithm to be valid, each sequence of choices must produce a different result.

* Consider the following extreme case: If all four letters were "e", then there would be only a single sequence: "eeee" and not the 24 predicted by the decision algorithm.

Because our original algorithm is not valid, we need a new one. Here is a strategy that works nicely for this example. Imagine, as before, that we are going to construct a sequence of four letters. This time we are going to imagine that we have a sequence of four empty slots: □□□□, and instead of selecting letters to fill the slots from left to right, we are going to select *slots* in which to place each of the letters. Remember that we have to use these letters: k, e, r, e. We proceed as follows, leaving the "e"s until last.

Step 1 Select an empty slot for the k: 4 choices. (e.g., □□[*k*]□)
Step 2 Select an empty slot for the r: 3 choices. (e.g., [*r*]□[*k*]□)
Step 3 Place the "e"s in the remaining two slots: 1 choice!

Thus the multiplication principle yields $4 \times 3 \times 1 = 12$ choices.

➡ **Before we go on...** You should try constructing a decision tree for Example 3, and you will see that each sequence of four letters is produced exactly once when we use the correct (second) decision algorithm. ■

FAQs

Creating and Testing a Decision Algorithm

Q: *How do I set up a decision algorithm to count how many items there are in a given scenario?*

A: Pretend that you are *designing* such an item (for example, pretend that you are designing an ice-cream cone) and come up with a step-by-step procedure for doing so, listing the decisions you should make at each stage.

Q: *Once I have my decision algorithm, how do I check if it is valid?*

A: Ask yourself the following question: "Is it possible to get the same item (the exact same ice-cream cone, say) by making different decisions when applying the algorithm?" If the answer is "yes," then your decision algorithm is invalid. Otherwise, it is valid.

6.3 EXERCISES

Access end-of-section exercises online at **www.webassign.net** 

6.4 Permutations and Combinations

Certain classes of counting problems come up frequently, and it is useful to develop formulas to deal with them.

EXAMPLE 1 Casting

Ms. Birkitt, the English teacher at Brakpan Girls High School, wanted to stage a production of R. B. Sheridan's play, *The School for Scandal.* The casting was going well until she was left with five unfilled characters and five seniors who were yet to be assigned roles. The characters were Lady Sneerwell, Lady Teazle, Mrs. Candour, Maria, and Snake; while the unassigned seniors were April, May, June, Julia, and Augusta. How many possible assignments are there?

Solution To decide on a specific assignment, we use the following algorithm:

Step 1 Choose a senior to play Lady Sneerwell: 5 choices.

Step 2 Choose one of the remaining seniors to play Lady Teazle: 4 choices.

Step 3 Choose one of the now remaining seniors to play Mrs. Candour: 3 choices.

Step 4 Choose one of the now remaining seniors to play Maria: 2 choices.

Step 5 Choose the remaining senior to play Snake: 1 choice.

Thus, there are $5 \times 4 \times 3 \times 2 \times 1 = 120$ possible assignments of seniors to roles.

What the situation in Example 1 has in common with many others is that we start with a set—here the set of seniors—and we want to know how many ways we can put the elements of that set in order in a list. In this example, an ordered list of the five seniors, for instance

1. May

2. Augusta

3. June

4. Julia

5. April

corresponds to a particular casting:

Cast	
Lady Sneerwell	May
Lady Teazle	Augusta
Mrs. Candour	June
Maria	Julia
Snake	April

We call an ordered list of items a **permutation** of those items.

If we have n items, how many permutations of those items are possible? We can use a decision algorithm similar to the one we used in the preceding example to select a permutation.

Step 1 Select the first item: n choices.
Step 2 Select the second item: $n - 1$ choices.
Step 3 Select the third item: $n - 2$ choices.
. . .
Step $n - 1$ Select the next-to-last item: 2 choices.
Step n Select the last item: 1 choice.
Thus, there are $n \times (n - 1) \times (n - 2) \times \cdots \times 2 \times 1$ possible permutations. We call this number ***n* factorial**, which we write as $n!$.

using Technology

Technology can be used to compute factorials. For instance, to compute 5!:

TI-83/84 Plus
Home screen: `5!`
(To obtain the `!` symbol, press MATH, choose PRB, and select `4:!`.)

Spreadsheet
Use the formula
`=FACT(5)`

Website
www.WanerMath.com
On the Main Page enter `5!` and press "Calculate".

Permutations

A **permutation of *n* items** is an ordered list of those items. The number of possible permutations of n items is given by ***n* factorial**, which is

$$n! = n \times (n - 1) \times (n - 2) \times \cdots \times 2 \times 1$$

for n a positive integer, and

$$0! = 1.$$

Visualizing Permutations

Permutations of three colors in a flag:

$3! = 3 \times 2 \times 1 = 6$ Possible Flags

Quick Examples

1. The number of permutations of five items is $5! = 5 \times 4 \times 3 \times 2 \times 1 = 120$.
2. The number of ways four CDs can be played in sequence is $4! = 4 \times 3 \times 2 \times 1 = 24$.
3. The number of ways three cars can be matched with three drivers is $3! = 6$.

Sometimes, instead of constructing an ordered list of *all* the items of a set, we might want to construct a list of only *some* of the items, as in the next example.

EXAMPLE 2 Corporations[3]

At the end of 2008, the 10 largest companies (by market capitalization) listed on the New York Stock Exchange were, in alphabetical order, **AT&T Inc.**, **Berkshire Hathaway Inc.**, **Chevron Corporation**, **China Mobile Ltd.** (ADR), **Exxon Mobil Corporation**, **Johnson & Johnson**, **PetroChina Company Limited** (ADR), **Royal Dutch Shell plc** (ADR), **The Procter & Gamble Company**, and **Wal-Mart Stores, Inc.** You would like to apply to six of these

[3]Source: New York Stock Exchange Web site (www.nyse.com).

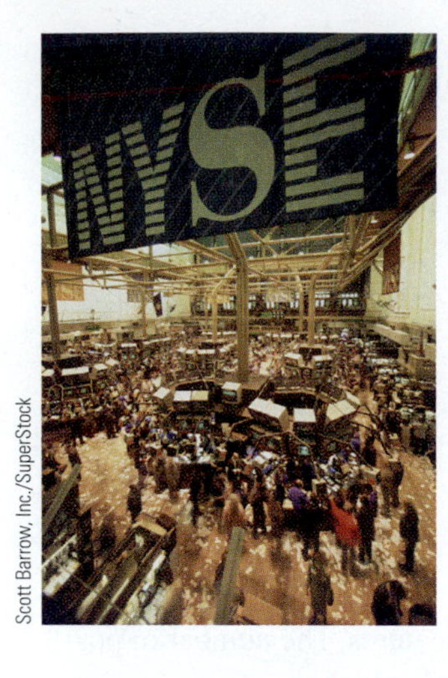

Scott Barrow, Inc./SuperStock

companies for a job and you would like to list them in order of job preference. How many such ordered lists are possible?

Solution We want to count ordered lists, but we can't use the permutation formula because we don't want all 10 companies in the list, just 6 of them. So we fall back to a decision algorithm.

Step 1 Choose the first company: 10 choices.

Step 2 Choose the second company: 9 choices.

Step 3 Choose the third one: 8 choices;

Step 4 Choose the fourth one: 7 choices.

Step 5 Choose the fifth one: 6 choices.

Step 6 Choose the sixth one: 5 choices.

Thus, there are $10 \times 9 \times 8 \times 7 \times 6 \times 5 = 151{,}200$ possible lists of 6. We call this number the **number of permutations of 6 items chosen from 10**, or the **number of permutations of 10 items taken 6 at a time**.

Before we go on... We wrote the answer as the product $10 \times 9 \times 8 \times 7 \times 6 \times 5$. But it is useful to notice that we can write this number in a more compact way:

$$10 \times 9 \times 8 \times 7 \times 6 \times 5 = \frac{10 \times 9 \times 8 \times 7 \times 6 \times 5 \times 4 \times 3 \times 2 \times 1}{4 \times 3 \times 2 \times 1} = \frac{10!}{4!} = \frac{10!}{(10-6)!}$$

■

So, we can generalize our definition of permutation to allow for the case in which we use only some of the items, not all. Check that, if $r = n$ below, this is the same definition we gave above.

using Technology

Technology can be used to compute permutations. For instance, to compute $P(6, 2)$:

TI-83/84 Plus

Home screen: `6 nPr 2`

(To obtain the `nPr` symbol, press `MATH`, choose `PRB`, and select `2:nPr`.)

Spreadsheet

Use the formula

`=PERMUT(6,2)`

Website

www.WanerMath.com

On the Main Page enter

`perm(6,2)`

and press "Calculate".

Permutations of *n* items taken *r* at a time

A **permutation of *n* items taken *r* at a time** is an ordered list of r items chosen from a set of n items. The number of permutations of n items taken r at a time is given by

$$P(n, r) = n \times (n-1) \times (n-2) \times \cdots \times (n-r+1)$$

(there are r terms multiplied together). We can also write

$$P(n, r) = \frac{n!}{(n-r)!}.$$

Quick Example

The number of permutations of six items taken two at a time is

$$P(6, 2) = 6 \times 5 = 30$$

which we could also calculate as

$$P(6, 2) = \frac{6!}{(6-2)!} = \frac{6!}{4!} = \frac{720}{24} = 30.$$

What if we don't care about the order of the items we're choosing? Consider the following example:

EXAMPLE 3 Corporations

Suppose we simply wanted to pick two of the 10 companies listed in Example 2 to apply to, without regard to order. How many possible choices do we have? What if we wanted to choose six to apply to, without regard to order?

Solution Our first guess might be $P(10, 2) = 10 \times 9 = 90$. However, that is the number of *ordered lists* of two companies. We said that we don't care which is first and which second. For example, we consider the list

1. AT&T **2.** Johnson & Johnson

to be the same as

1. Johnson & Johnson **2.** AT&T.

Because every set of two companies occurs twice in the 90 lists, once in one order and again in the reverse order, we would count every set of two twice. Thus, there are $90/2 = 45$ possible choices of two companies.

Now, if we wish to pick six companies, again we might start with $P(10, 6) = 151{,}200$. But now, every set of six companies appears as many times as there are different orders in which they could be listed. Six things can be listed in $6! = 720$ different orders, so the number of ways of choosing six companies is $151{,}200/720 = 210$.

In Example 3 we were concerned with counting not the number of ordered lists, but the number of *unordered sets* of companies. For ordered lists we used the word *permutation*; for unordered sets we use the word **combination**.

Permutations and Combinations

A **permutation** of n items taken r at a time is an *ordered list* of r items chosen from n. A **combination** of n items taken r at a time is an *unordered set* of r items chosen from n.

Visualizing

1.
2.
3.

Permutation Combination

Note Because lists are usually understood to be ordered, when we refer to a list of items, we will always mean an *ordered* list. Similarly, because sets are understood to be unordered, when we refer to a set of items we will always mean an *unordered* set. In short:

Lists are ordered. Sets are unordered. ■

Quick Example

There are six permutations of the three letters a, b, c taken two at a time:

1. a, b; **2.** b, a; **3.** a, c; **4.** c, a; **5.** b, c; **6.** c, b.

There are six lists containing two of the letters a, b, c.

There are three combinations of the three letters a, b, c taken two at a time:

1. {a, b}; **2.** {a, c}; **3.** {b, c}.

There are three sets containing two of the letters a, b, c.

How do we count the number of possible combinations of n items taken r at a time? We generalize the calculation done in Example 3. The number of permutations is $P(n, r)$, but each set of r items occurs $r!$ times because this is the number of ways in which those r items can be ordered. So, the number of combinations is $P(n, r)/r!$.

Combinations of *n* items taken *r* at a time

The number of **combinations of *n* items taken *r* at a time** is given by

$$C(n, r) = \frac{P(n, r)}{r!} = \frac{n \times (n-1) \times (n-2) \times \cdots \times (n-r+1)}{r!}.$$

We can also write

$$C(n, r) = \frac{n!}{r!(n-r)!}.$$

Quick Examples

1. The number of combinations of six items taken two at a time is

$$C(6, 2) = \frac{6 \times 5}{2 \times 1} = 15,$$

which we can also calculate as

$$C(6, 2) = \frac{6!}{2!(6-2)!} = \frac{6!}{2!4!} = \frac{720}{2 \times 24} = 15.$$

2. The number of sets of four marbles chosen from six is

$$C(6, 4) = \frac{6 \times 5 \times 4 \times 3}{4 \times 3 \times 2 \times 1} = 15.$$

Note There are other common notations for $C(n, r)$. Calculators often have ${}_nC_r$. In mathematics we often write $\binom{n}{r}$ which is also known as a **binomial coefficient**. Because $C(n, r)$ is the number of ways of choosing a set of r items from n, it is often read "n choose r." ■

EXAMPLE 4 Calculating Combinations

Calculate **a.** $C(11, 3)$ **b.** $C(11, 8)$

Solution The easiest way to calculate $C(n, r)$ by hand is to use the first formula above:

$$C(n, r) = \frac{P(n, r)}{r!}$$
$$= \frac{n \times (n-1) \times (n-2) \times \cdots \times (n-r+1)}{r \times (r-1) \times (r-2) \times \cdots \times 1}.$$

using Technology

Technology can be used to compute combinations. For instance, to compute C(11, 8):

TI-83/84 Plus

Home screen: `11 nCr 8`
(To obtain the `nCr` symbol, press `MATH`, choose `PRB`, and select `3:nCr`.)

Spreadsheet

Use the formula
`=COMBIN(11,8)`

Website

www.WanerMath.com
On the Main Page enter
`comb(11,8)`
and press "Calculate".

Both the numerator and the denominator have r factors, so we can begin with n/r and then continue multiplying by decreasing numbers on the top and the bottom until we hit 1 in the denominator. When calculating, it helps to cancel common factors from the numerator and denominator before doing the multiplication in either one.

a. $C(11, 3) = \dfrac{11 \times 10 \times 9}{3 \times 2 \times 1} = \dfrac{11 \times \overset{5}{\cancel{10}} \times \overset{3}{\cancel{9}}}{\cancel{3} \times \cancel{2} \times 1} = 165$

b. $C(11, 8) = \dfrac{11 \times 10 \times 9 \times \cancel{8} \times \cancel{7} \times \cancel{6} \times \cancel{5} \times \cancel{4}}{\cancel{8} \times \cancel{7} \times \cancel{6} \times \cancel{5} \times \cancel{4} \times 3 \times 2 \times 1}$

$= \dfrac{11 \times \overset{5}{\cancel{10}} \times \overset{3}{\cancel{9}}}{\cancel{3} \times \cancel{2} \times 1} = 165$

Before we go on... It is no coincidence that the answers for parts (a) and (b) of Example 4 are the same. Consider what each represents. $C(11, 3)$ is the number of ways of choosing 3 items from 11—for example, electing 3 trustees from a slate of 11. Electing those 3 is the same as choosing the 8 who *do not* get elected. Thus, there are exactly as many ways to choose 3 items from 11 as there are ways to choose 8 items from 11. So, $C(11, 3) = C(11, 8)$. In general,

$$C(n, r) = C(n, n - r).$$

We can also see this equality by using the formula

$$C(n, r) = \frac{n!}{r!(n - r)!}.$$

If we substitute $n - r$ for r, we get exactly the same formula.

Use the equality $C(n, r) = C(n, n - r)$ to make your calculations easier. Choose the one with the smaller denominator to begin with. ■

EXAMPLE 5 Calculating Combinations

Calculate **a.** $C(11, 11)$ **b.** $C(11, 0)$.

Solution

a. $C(11, 11) = \dfrac{11 \times 10 \times 9 \times 8 \times 7 \times 6 \times 5 \times 4 \times 3 \times 2 \times 1}{11 \times 10 \times 9 \times 8 \times 7 \times 6 \times 5 \times 4 \times 3 \times 2 \times 1} = 1$

b. What do we do with that 0? What does it mean to multiply 0 numbers together? We know from above that $C(11, 0) = C(11, 11)$, so we must have $C(11, 0) = 1$. How does this fit with the formulas? Go back to the calculation of $C(11, 11)$:

$$1 = C(11, 11) = \frac{11!}{11!(11 - 11)!} = \frac{11!}{11!0!}.$$

This equality is true only if we agree that $0! = 1$, which we do. Then

$$C(11, 0) = \frac{11!}{0!11!} = 1.$$

Before we go on... There is nothing special about 11 in the calculation in Example 5. In general,

$$C(n, n) = C(n, 0) = 1.$$

After all, there is only one way to choose n items out of n: Choose them all. Similarly, there is only one way to choose 0 items out of n: Choose none of them. ■

Now for a few more complicated examples that illustrate the applications of the counting techniques we've discussed.

EXAMPLE 6 Lotto

In the betting game Lotto, used in many state lotteries, you choose six different numbers in the range 1–55 (the upper number varies). The order in which you choose them is irrelevant. If your six numbers match the six numbers chosen in the "official drawing," you win the top prize. If Lotto tickets cost \$1 for two sets of numbers and you decide to buy tickets that cover every possible combination, thereby guaranteeing that you will win the top prize, how much money will you have to spend?

Solution We first need to know how many sets of numbers are possible. Because order does not matter, we are asking for the number of combinations of 55 numbers taken 6 at a time. This is

$$C(55, 6) = \frac{55 \times 54 \times 53 \times 52 \times 51 \times 50}{6 \times 5 \times 4 \times 3 \times 2 \times 1} = 28{,}989{,}675.$$

Because \$1 buys you two of these, you need to spend $28{,}989{,}675/2 = \$14{,}494{,}838$ (rounding up to the nearest dollar) to be assured of a win!

Before we go on... The calculation in Example 6 shows that you should not bother buying all these tickets if the winning prize is less than about \$14.5 million. Even if the prize is higher, you need to account for the fact that many people will play and the prize may end up split among several winners, not to mention the impracticality of filling out millions of betting slips. ■

EXAMPLE 7 Marbles

A bag contains three red marbles, three blue ones, three green ones, and two yellow ones (all distinguishable from one another).

a. How many sets of four marbles are possible?

b. How many sets of four are there such that each one is a different color?

c. How many sets of four are there in which at least two are red?

d. How many sets of four are there in which none are red, but at least one is green?

Solution

a. We simply need to find the number of ways of choosing 4 marbles out of 11, which is

$C(11, 4) = 330$ possible sets of 4 marbles

b. We use a decision algorithm for choosing such a set of marbles.

Step 1 Choose one red one from the three red ones: $C(3, 1) = 3$ choices.

Step 2 Choose one blue one from the three blue ones: $C(3, 1) = 3$ choices.

Step 3 Choose one green one from the three green ones: $C(3, 1) = 3$ choices.
Step 4 Choose one yellow one from the two yellow ones: $C(2, 1) = 2$ choices.

This gives a total of $3 \times 3 \times 3 \times 2 = 54$ possible sets.

c. We need another decision algorithm. To say that at least two must be red means that either two are red or three are red (with a total of three red ones). In other words, we have two *alternatives*.

Alternative 1: Exactly two red marbles
Step 1 Choose two red ones: $C(3, 2) = 3$ choices.
Step 2 Choose two nonred ones. There are eight of these, so we get $C(8, 2) = 28$ possible choices.
Thus, the total number of choices for this alternative is $3 \times 28 = 84$.

Alternative 2: Exactly three red marbles.
Step 1 Choose the three red ones: $C(3, 3) = 1$ choice.
Step 2 Choose one nonred one: $C(8, 1) = 8$ choices.
Thus, the total number of choices for this alternative is $1 \times 8 = 8$.

By the addition principle, we get a total of $84 + 8 = 92$ sets.

d. The phrase "at least one green" tells us that we again have some alternatives.

Alternative 1: One green marble
Step 1 Choose one green marble from the three: $C(3, 1) = 3$ choices.
Step 2 Choose three nongreen, nonred marbles: $C(5, 3) = 10$ choices.
Thus, the total number of choices for alternative 1 is $3 \times 10 = 30$.

Alternative 2: Two green marbles
Step 1 Choose two green marbles from the three: $C(3, 2) = 3$ choices.
Step 2 Choose two nongreen, nonred marbles: $C(5, 2) = 10$ choices.
Thus, the total number of choices for alternative 2 is $3 \times 10 = 30$.

Alternative 3: Three green marbles
Step 1 Choose three green marbles from the three: $C(3, 3) = 1$ choice.
Step 2 Choose one nongreen, nonred marble: $C(5, 1) = 5$ choices.
Thus, the total number of choices for alternative 3 is $1 \times 5 = 5$.

The addition principle now tells us that the number of sets of four marbles with none red, but at least one green, is $30 + 30 + 5 = 65$.

➡ **Before we go on...** Here is an easier way to answer part (d) of Example 7. First, the total number of sets having *no* red marbles is $C(8, 4) = 70$. Next, of those, the number containing no green marbles is $C(5, 4) = 5$. This leaves $70 - 5 = 65$ sets that contain no red marbles but have at least one green marble. (We have really used here the formula for the cardinality of the complement of a set.) ■

The last example concerns poker hands. For those unfamiliar with playing cards, here is a short description. A standard deck consists of 52 playing cards. Each card is in one of 13 denominations: ace, 2, 3, 4, 5, 6, 7, 8, 9, 10, jack (J), queen (Q), and king (K), and in one of four suits: hearts (♥), diamonds (♦), clubs (♣), and spades (♠). Thus, for instance, the jack of spades, J♠, refers to the denomination of jack in the suit of spades. The entire deck of cards is thus

A♥	2♥	3♥	4♥	5♥	6♥	7♥	8♥	9♥	10♥	J♥	Q♥	K♥
A♦	2♦	3♦	4♦	5♦	6♦	7♦	8♦	9♦	10♦	J♦	Q♦	K♦
A♣	2♣	3♣	4♣	5♣	6♣	7♣	8♣	9♣	10♣	J♣	Q♣	K♣
A♠	2♠	3♠	4♠	5♠	6♠	7♠	8♠	9♠	10♠	J♠	Q♠	K♠

EXAMPLE 8 Poker Hands

In the card game poker, a hand consists of a set of 5 cards from a standard deck of 52. A **full house** is a hand consisting of three cards of one denomination ("three of a kind"—e.g., three 10s) and two of another ("two of a kind"—e.g., two queens). Here is an example of a full house: 10♣, 10♦, 10♠, Q♥, Q♣.

a. How many different poker hands are there?

b. How many different full houses are there that contain three 10s and two queens?

c. How many different full houses are there altogether?

Solution

a. Because the order of the cards doesn't matter, we simply need to know the number of ways of choosing a set of 5 cards out of 52, which is

$C(52, 5) = 2{,}598{,}960$ hands.

b. Here is a decision algorithm for choosing a full house with three 10s and two queens.

Step 1 Choose three 10s. Because there are four 10s to choose from, we have $C(4, 3) = 4$ choices.

Step 2 Choose two queens: $C(4, 2) = 6$ choices.

Thus, there are $4 \times 6 = 24$ possible full houses with three 10s and two queens.

c. Here is a decision algorithm for choosing a full house.

Step 1 Choose a denomination for the three of a kind; 13 choices.

Step 2 Choose three cards of that denomination. Because there are four cards of each denomination (one for each suit), we get $C(4, 3) = 4$ choices.

Step 3 Choose a different denomination for the two of a kind. There are only 12 denominations left, so we have 12 choices.

Step 4 Choose two of that denomination: $C(4, 2) = 6$ choices.

Thus, by the multiplication principle, there are a total of $13 \times 4 \times 12 \times 6 = 3{,}744$ possible full houses.

FAQs

Recognizing When to Use Permutations or Combinations

Q: *How can I tell whether a given application calls for permutations or combinations?*

A: Decide whether the application calls for ordered lists (as in situations where order is implied) or for unordered sets (as in situations where order is not relevant). Ordered lists are permutations, whereas unordered sets are combinations.

6.4 EXERCISES

Access end-of-section exercises online at **www.webassign.net**

CHAPTER 6 REVIEW

KEY CONCEPTS

Website www.WanerMath.com

Go to the Website at www.WanerMath.com to find a comprehensive and interactive Web-based summary of Chapter 6, along with section-by-section tutorials.

6.1 Sets and Set Operations

Sets, elements, subsets, proper subsets, empty set, finite and infinite sets *pp. 312 & 313*

Visualizing sets and relations between sets using Venn diagrams *p. 315*

Union: $A \cup B = \{x \mid x \in A \text{ or } x \in B\}$ *p. 316*

Intersection: $A \cap B = \{x \mid x \in A \text{ and } x \in B\}$ *p. 316*

Universal sets, complements *p. 317*

Disjoint sets: $A \cap B = \emptyset$ *p. 318*

Cartesian product: $A \times B = \{(a, b) \mid a \in A \text{ and } b \in B\}$ *p. 318*

6.2 Cardinality

Cardinality: $n(A) =$ number of elements in A. *p. 321*

If A and B are finite sets, then $n(A \cup B) = n(A) + n(B) - n(A \cap B)$. *p. 322*

If A and B are disjoint finite sets, then $n(A \cup B) = n(A) + n(B)$. In this case, we say that $A \cup B$ is a **disjoint union**. *p. 322*

If S is a finite universal set and A is a subset of S, then $n(A') = n(S) - n(A)$ and $n(A) = n(S) - n(A')$. *p. 323*

If A and B are finite sets, then $n(A \times B) = n(A)n(B)$. *p. 328*

6.3 Decision Algorithms: The Addition and Multiplication Principles

Addition principle *p. 329*

Multiplication principle *p. 329*

Decision algorithm: a procedure for making a sequence of decisions to choose an element of a set *p. 331*

6.4 Permutations and Combinations

n factorial: $n! = n \times (n-1) \times (n-2) \times \cdots \times 2 \times 1$ *p. 335*

Permutation of n items taken r at a time:

$P(n, r) = \dfrac{n!}{(n-r)!}$ *p. 336*

Combination of n items taken r at a time:

$C(n, r) = P(n, r)/r! = \dfrac{n!}{r!(n-r)!}$ *p. 338*

Using the equality $C(n, r) = C(n, n-r)$ to simplify calculations of combinations *p. 339*

REVIEW EXERCISES

In each of Exercises 1–5, list the elements of the given set.

1. The set N of all negative integers greater than or equal to -3

2. The set of all outcomes of tossing a coin five times

3. The set of all outcomes of tossing two distinguishable dice such that the numbers are different

4. The sets $(A \cap B) \cup C$ and $A \cap (B \cup C)$, where $A = \{1,2,3,4,5\}$, $B = \{3, 4, 5\}$, and $C = \{1, 2, 5, 6, 7\}$

5. The sets $A \cup B'$ and $A \times B'$, where $A = \{\text{a, b}\}$, $B = \{\text{b, c}\}$, and $S = \{\text{a, b, c, d}\}$

In each of Exercises 6–10, write the indicated set in terms of the given sets.

6. S: the set of all customers; A: the set of all customers who owe money; B: the set of all customers who owe at least \$1,000. The set of all customers who owe money but owe less than \$1,000

7. A: the set of outcomes when a day in August is selected; B: the set of outcomes when a time of day is selected. The set of outcomes when a day in August and a time of that day are selected

8. S: the set of outcomes when two dice are rolled; E: those outcomes in which at most one die shows an even number, F: those outcomes in which the sum of the numbers is 7. The set of outcomes in which both dice show an even number or sum to seven

9. S: the set of all integers; P: the set of all positive integers; E: the set of all even integers; Q: the set of all integers that are perfect squares ($Q = \{0, 1, 4, 9, 16, 25, \ldots\}$). The set of all integers that are not positive odd perfect squares

10. S: the set of all integers; N: the set of all negative integers; E: the set of all even integers; T: the set of all integers that are multiples of 3 ($T = \{0, 3, -3, 6, -6, 9, -9, \ldots\}$). The set of all even integers that are neither negative nor multiples of three

In each of Exercises 11–14, give a formula for the cardinality rule or rules needed to answer the question, and then give the solution.

11. You have read 150 of the 400 novels in your home, but your sister Roslyn has read 200, of which only 50 are novels you have read as well. How many have neither of you read?

12. There are 32 students in categories A and B combined; 24 are in A and 24 are in B. How many are in both A and B?

13. You roll two dice, one red and one green. Losing combinations are doubles (both dice show the same number) and outcomes in which the green die shows an odd number and the red die shows an even number. The other combinations are winning ones. How many winning combinations are there?

14. The *Apple iMac* used to come in three models, each with five colors to choose from. How many combinations were possible?

Recall that a poker hand consists of 5 cards from a standard deck of 52. In each of Exercises 15–18, find the number of different poker hands of the specified type. Leave your answer in terms of combinations.

15. Two of a kind with no aces

16. A full house with either two kings and three queens or two queens and three kings

17. Straight Flush (five cards of the same suit with consecutive denominations: A, 2, 3, 4, 5 up through 10, J, Q, K, A)

18. Three of a kind with no aces

In each of Exercises 19–24, consider a bag containing four red marbles, two green ones, one transparent one, three yellow ones, and two orange ones.

19. How many possible sets of five marbles are there in which all of them are red or green?

20. How many possible sets of five marbles are there in which none of them are red or green?

21. How many sets of five marbles include all the red ones?

22. How many sets of five marbles do not include all the red ones?

23. How many sets of five marbles include at least two yellow ones?

24. How many sets of five marbles include at most one of the red ones but no yellow ones?

APPLICATIONS: OHaganBooks.com

Inventories *OHaganBooks.com currently operates three warehouses: one in Washington, one in California, and the new one in Texas. Exercises 25–30 are based on the following table, which shows the book inventories at each warehouse:*

	Sci Fi	Horror	Romance	Other	Total
Washington	10,000	12,000	12,000	30,000	64,000
California	8,000	12,000	6,000	16,000	42,000
Texas	15,000	15,000	20,000	44,000	94,000
Total	33,000	39,000	38,000	90,000	200,000

Take the first letter of each category to represent the corresponding set of books; for instance, S is the set of sci fi books in stock, W is the set of books in the Washington warehouse, and so on. In each exercise describe the given set in words and compute its cardinality.

25. $S \cup T$

26. $H \cap C$

27. $C \cup S'$

28. $(R \cap T) \cup H$

29. $R \cap (T \cup H)$

30. $(S \cap W) \cup (H \cap C')$

Customers *OHaganBooks.com has two main competitors: JungleBooks.com and FarmerBooks.com. At the beginning of August, OHaganBooks.com had 3,500 customers. Of these, a total of 2,000 customers were shared with JungleBooks.com and 1,500 with FarmerBooks.com. Furthermore, 1,000 customers were shared with both. JungleBooks.com has a total of 3,600 customers, FarmerBooks.com has 3,400, and they share 1,100 customers between them. Use these data for Exercises 31–36.*

31. How many of all these customers are exclusive OHaganBooks.com customers?

32. How many customers of the other two companies are not customers of OHaganBooks.com?

33. Which of the three companies has the largest number of exclusive customers?

34. Which of the three companies has the smallest number of exclusive customers?

35. OHaganBooks.com is interested in merging with one of its two competitors. Which merger would give it the largest combined customer base, and how large would that be?

36. Referring to the preceding exercise, which merger would give OHaganBooks.com the largest *exclusive* customer base, and how large would that be?

Online IDs *As the customer base at OHaganBooks.com grows, software manager Ruth Nabarro is thinking of introducing identity codes for all the online customers. Help her with the questions in Exercises 37–40.*

37. If she uses 3-letter codes, how many different customers can be identified?

38. If she uses codes with three different letters, how many different customers can be identified?

39. It appears that Nabarro has finally settled on codes consisting of two letters followed by two digits. For technical reasons, the letters must be different and the first digit cannot be a zero. How many different customers can be identified?

40. O'Hagan sends Nabarro the following memo:

```
To: Ruth Nabarro, Software Manager
From: John O'Hagan, CEO
Subject: Customer Identity Codes

I have read your proposal for the customer ID codes. However, due to our ambitious expansion plans, I would like our system software to allow for at least 500,000 customers. Please adjust your proposal accordingly.
```

Nabarro is determined to have a sequence of letters followed by some digits, and, for reasons too complicated to explain, there cannot be more than two letters, the letters must be different, the digits must all be different, and the first digit cannot be a zero. What is the form of the shortest code she can use to satisfy the CEO, and how many different customers can be identified?

Degree Requirements *After an exhausting day at the office, John O'Hagan returns home and finds himself having to assist his son Billy-Sean, who continues to have a terrible time planning his first-year college course schedule. The latest Bulletin of Suburban State University reads as follows:*

> *All candidates for the degree of Bachelor of Arts at SSU must take, in their first year, at least 10 courses in the Sciences, Fine Arts, Liberal Arts, and Mathematics combined, of which at least 2 must be in each of the Sciences and Fine Arts, and exactly 3 must be in each of the Liberal Arts and Mathematics.*

Help him with the answers to Exercises 41–43.

41. If the Bulletin lists exactly five first-year-level science courses and six first-year-level courses in each of the other categories, how many course combinations are possible that meet the minimum requirements?

42. Reading through the course descriptions in the bulletin a second time, John O'Hagan notices that Calculus I (listed as one of the mathematics courses) is a required course for many of the other courses, and so decides that it would be best if Billy-Sean included Calculus I. Further, two of the Fine Arts courses cannot both be taken in the first year. How many course combinations are possible that meet the minimum requirements and include Calculus I?

43. To complicate things further, in addition to the requirement in Exercise 42, Physics II has Physics I as a prerequisite (both are listed as first-year science courses, but it is not necessary to take both). How many course combinations are possible that include Calculus I and meet the minimum requirements?

Case Study: Designing a Puzzle

Image Source/Getty Images

As Product Design Manager for Cerebral Toys Inc., you are constantly on the lookout for ideas for intellectually stimulating yet inexpensive toys. You recently received the following memo from Felix Frost, the developmental psychologist on your design team.

> To: Felicia
> From: Felix
> Subject: Crazy Cubes
>
> We've hit on an excellent idea for a new educational puzzle (which we are calling "Crazy Cubes" until Marketing comes up with a better name). Basically, Crazy Cubes will consist of a set of plastic cubes. Two faces of each cube will be colored red, two will be colored blue, and two white, and there will be exactly two cubes with each possible configuration of colors. The goal of the puzzle is to seek out the matching pairs, thereby enhancing a child's geometric intuition and three-dimensional manipulation skills. The kit will include every possible configuration of colors. We are, however, a little stumped on the following question: How many cubes will the kit contain? In other words, how many possible ways can one color the faces of a cube so that two faces are red, two are blue, and two are white?

Looking at the problem, you reason that the following three-step decision algorithm ought to suffice:

Step 1 Choose a pair of faces to color red; $C(6, 2) = 15$ choices.

Step 2 Choose a pair of faces to color blue; $C(4, 2) = 6$ choices.

Step 3 Choose a pair of faces to color white; $C(2, 2) = 1$ choice.

This algorithm appears to give a total of $15 \times 6 \times 1 = 90$ possible cubes. However, before sending your reply to Felix, you realize that something is wrong, because there are different choices that result in the same cube. To describe some of these choices, imagine a cube oriented so that four of its faces are facing the four compass directions (Figure 9). Consider choice 1, with the top and bottom faces blue, north and south faces white, and east and west faces red; and choice 2, with the top and bottom faces blue,

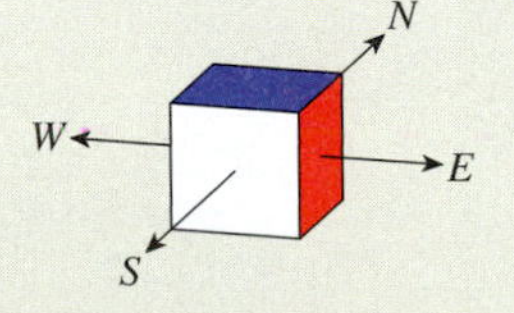

Figure 9

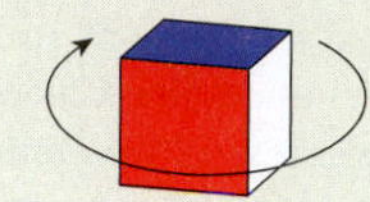

Figure 10

north and south faces red, and east and west faces white. These cubes are actually the same, as you see by rotating the second cube 90 degrees (Figure 10).

You therefore decide that you need a more sophisticated decision algorithm. Here is one that works.

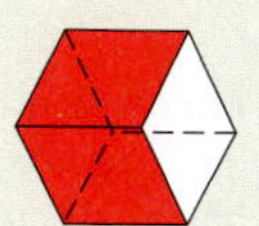

Figure 11

Alternative 1: Faces with the same color opposite each other. Place one of the blue faces down. Then the top face is also blue. The cube must look like the one drawn in Figure 10. Thus there is only one choice here.

Alternative 2: Red faces opposite each other and the other colors on adjacent pairs of faces. Again there is only one choice, as you can see by putting the red faces on the top and bottom and then rotating.

Alternative 3: White faces opposite each other and the other colors on adjacent pairs of faces; one possibility.

Alternative 4: Blue faces opposite each other and the other colors on adjacent pairs of faces; one possibility.

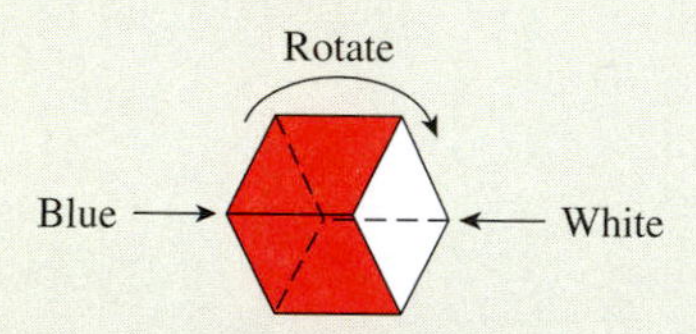

Figure 12

Alternative 5: Faces with the same color adjacent to each other. Look at the cube so that the edge common to the two red faces is facing you and horizontal (Figure 11). Then the faces on the left and right must be of different colors because they are opposite each other. Assume that the face on the right is white. (If it's blue, then rotate the die with the red edge still facing you to move it there, as in Figure 12.) This leaves two choices for the other white face, on the upper or the lower of the two back faces. This alternative gives two choices.

It follows that there are $1 + 1 + 1 + 1 + 2 = 6$ choices. Because the Crazy Cubes kit will feature two of each cube, the kit will require 12 different cubes.*

* There is a beautiful way of calculating this and similar numbers, called Pólya enumeration, but it requires a discussion of topics well outside the scope of this book. Take this as a hint that counting techniques can use some of the most sophisticated mathematics.

EXERCISES

In all of the following exercises, there are three colors to choose from: red, white, and blue.

1. In order to enlarge the kit, Felix suggests including two each of two-colored cubes (using two of the colors red, white, and blue) with three faces one color and three another. How many additional cubes will be required?

2. If Felix now suggests adding two each of cubes with two faces one color, one face another color, and three faces the third color, how many additional cubes will be required?

3. Felix changes his mind and suggests the kit use tetrahedral blocks with two colors instead (see the figure). How many of these would be required?

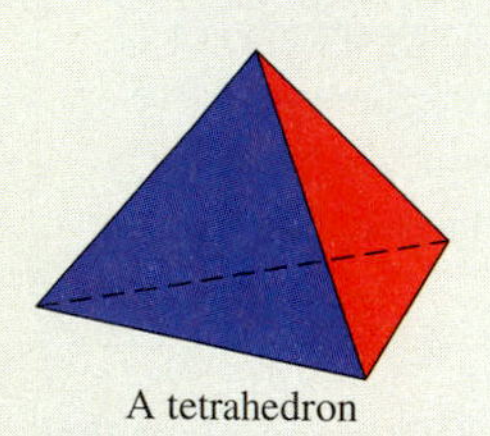

A tetrahedron

4. Once Felix finds the answer to the preceding exercise, he decides to go back to the cube idea, but this time insists that all possible combinations of up to three colors should be included. (For instance, some cubes will be all one color, others will be two colors.) How many cubes should the kit contain?

7

Probability

Website

www.WanerMath.com

At the Website you will find:

- Section-by-section tutorials, including game tutorials with randomized quizzes
- A detailed chapter summary
- A true/false quiz
- A Markov system simulation and matrix algebra tool
- Additional review exercises

Case Study The Monty Hall Problem

On the game show *Let's Make a Deal*, you are shown three doors, A, B, and C, and behind one of them is the Big Prize. After you select one of them—say, door A—to make things more interesting the host (Monty Hall) opens one of the other doors—say, door B—revealing that the Big Prize is not there. He then offers you the opportunity to change your selection to the remaining door, door C. Should you switch or stick with your original guess? **Does it make any difference?**

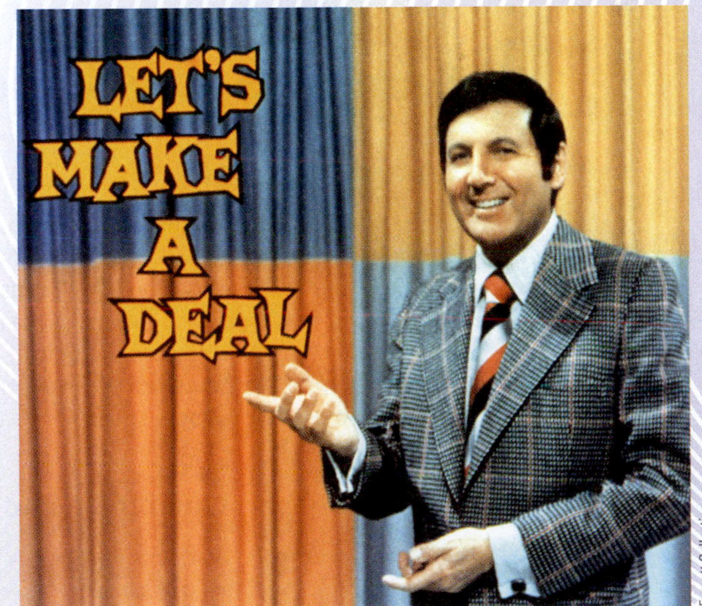

Everett Collection

Introduction

What is the probability of winning the lottery twice? What are the chances that a college athlete whose drug test is positive for steroid use is actually using steroids? You are playing poker and have been dealt two iacks. What is the likelihood that one of the next three cards you are dealt will also be a jack? These are all questions about probability.

Understanding probability is important in many fields, ranging from risk management in business through hypothesis testing in psychology to quantum mechanics in physics. Historically, the theory of probability arose in the sixteenth and seventeenth centuries from attempts by mathematicians such as Gerolamo Cardano, Pierre de Fermat, Blaise Pascal, and Christiaan Huygens to understand games of chance. Andrey Nikolaevich Kolmogorov set forth the foundations of modern probability theory in his 1933 book *Foundations of the Theory of Probability.*

The goal of this chapter is to familiarize you with the basic concepts of modern probability theory and to give you a working knowledge that you can apply in a variety of situations. In the first two sections, the emphasis is on translating real-life situations into the language of sample spaces, events, and probability. Once we have mastered the language of probability, we spend the rest of the chapter studying some of its theory and applications. The last section gives an interesting application of both probability and matrix arithmetic.

7.1 Sample Spaces and Events

Sample Spaces

At the beginning of a football game, to ensure fairness, the referee tosses a coin to decide who will get the ball first. When the ref tosses the coin and observes which side faces up, there are two possible results: heads (H) and tails (T). These are the *only* possible results, ignoring the (remote) possibility that the coin lands on its edge. The act of tossing the coin is an example of an **experiment**. The two possible results, H and T, are possible **outcomes** of the experiment, and the set $S = \{H, T\}$ of all possible outcomes is the **sample space** for the experiment.

Experiments, Outcomes, and Sample Spaces

An **experiment** is an occurrence with a result, or **outcome**, that is uncertain before the experiment takes place. The set of all possible outcomes is called the **sample space** for the experiment.

Quick Examples

1. *Experiment:* Flip a coin and observe the side facing up.
   Outcomes: H, T
   Sample Space: $S = \{H, T\}$
2. *Experiment:* Select a student in your class.
   Outcomes: The students in your class
   Sample Space: The set of students in your class

3. *Experiment:* Select a student in your class and observe the color of his or her hair.
   Outcomes: red, black, brown, blond, green, . . .
   Sample Space: {red, black, brown, blond, green, . . .}
4. *Experiment:* Cast a die and observe the number facing up.
   Outcomes: 1, 2, 3, 4, 5, 6
   Sample Space: $S = \{1, 2, 3, 4, 5, 6\}$
5. *Experiment:* Cast two distinguishable dice (see Example 1(a) of Section 6.1) and observe the numbers facing up.
   Outcomes: (1, 1), (1, 2), . . . , (6, 6) (36 outcomes)

$$\text{Sample Space: } S = \begin{Bmatrix} (1,1), & (1,2), & (1,3), & (1,4), & (1,5), & (1,6), \\ (2,1), & (2,2), & (2,3), & (2,4), & (2,5), & (2,6), \\ (3,1), & (3,2), & (3,3), & (3,4), & (3,5), & (3,6), \\ (4,1), & (4,2), & (4,3), & (4,4), & (4,5), & (4,6), \\ (5,1), & (5,2), & (5,3), & (5,4), & (5,5), & (5,6), \\ (6,1), & (6,2), & (6,3), & (6,4), & (6,5), & (6,6) \end{Bmatrix}$$

$$n(S) = \text{the number of outcomes in } S = 36$$

6. *Experiment:* Cast two indistinguishable dice (see Example 1(b) of Section 6.1) and observe the numbers facing up.
   Outcomes: (1, 1), (1, 2), . . . , (6, 6) (21 outcomes)

$$\text{Sample Space: } S = \begin{Bmatrix} (1,1), & (1,2), & (1,3), & (1,4), & (1,5), & (1,6), \\ & (2,2), & (2,3), & (2,4), & (2,5), & (2,6), \\ & & (3,3), & (3,4), & (3,5), & (3,6), \\ & & & (4,4), & (4,5), & (4,6), \\ & & & & (5,5), & (5,6), \\ & & & & & (6,6) \end{Bmatrix}$$

$$n(S) = 21$$

7. *Experiment:* Cast two dice and observe the *sum* of the numbers facing up.
   Outcomes: 2, 3, 4, 5, 6, 7, 8, 9, 10, 11, 12
   Sample Space: $S = \{2, 3, 4, 5, 6, 7, 8, 9, 10, 11, 12\}$
8. *Experiment:* Choose 2 cars (without regard to order) at random from a fleet of 10.
   Outcomes: Collections of 2 cars chosen from 10
   Sample Space: The set of all collections of 2 cars chosen from 10

$$n(S) = C(10, 2) = 45$$

The following example introduces a sample space that we'll use in several other examples.

EXAMPLE 1 School and Work

In a survey conducted by the Bureau of Labor Statistics,[1] the high school graduating class of 2010 was divided into those who went on to college and those who did not. Those who went on to college were further divided into those who went to 2-year colleges and those who went to 4-year colleges. All graduates were also asked whether they were working or not. Find the sample space for the experiment "Select a member of the high school graduating class of 2010 and classify his or her subsequent school and work activity."

Solution The tree in Figure 1 shows the various possibilities.

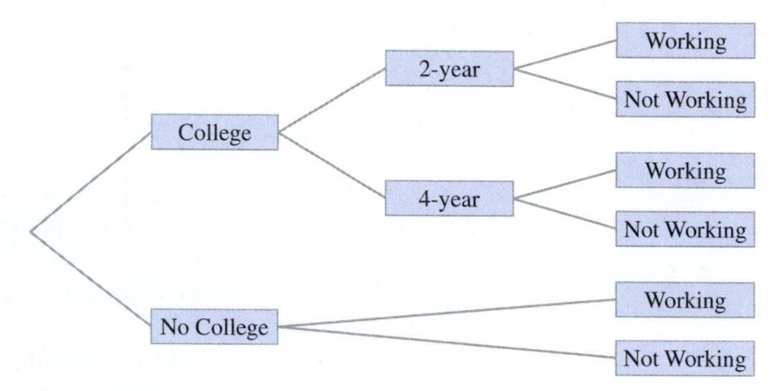

Figure 1

The sample space is

$$S = \{\text{2-year college \& working, 2-year college \& not working,}\\ \text{4-year college \& working, 4-year college \& not working,}\\ \text{no college \& working, no college \& not working}\}.$$

Events

In Example 1, suppose we are interested in the event that a 2010 high school graduate was working. In mathematical language, we are interested in the *subset* of the sample space consisting of all outcomes in which the graduate was working.

Events

Given a sample space S, an **event** E is a subset of S. The outcomes in E are called the **favorable** outcomes. We say that E **occurs** in a particular experiment if the outcome of that experiment is one of the elements of E—that is, if the outcome of the experiment is favorable.

[1] "College Enrollment and Work Activity of High School Graduates," U.S. Bureau of Labor Statistics, www.bls.gov/news.release/hsgec.htm.

Visualizing an Event

In the following figure, the favorable outcomes (events in E) are shown in green.

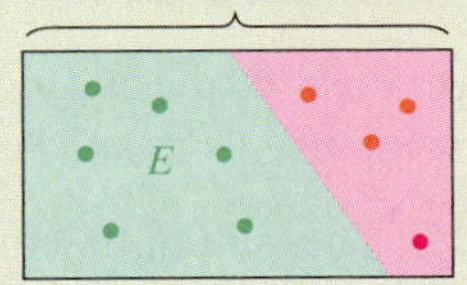

Quick Examples

1. *Experiment:* Roll a die and observe the number facing up.

   $S = \{1, 2, 3, 4, 5, 6\}$

   Event: E: The number observed is odd.

   $E = \{1, 3, 5\}$

2. *Experiment:* Roll two distinguishable dice and observe the numbers facing up.

   $S = \{(1, 1), (1, 2), \dots, (6, 6)\}$

   Event: F: The dice show the same number.

   $F = \{(1, 1), (2, 2), (3, 3), (4, 4), (5, 5), (6, 6)\}$

3. *Experiment:* Roll two distinguishable dice and observe the numbers facing up.

   $S = \{(1, 1), (1, 2), \dots, (6, 6)\}$

   Event: G: The sum of the numbers is 1.

   $G = \emptyset$ There are no favorable outcomes.

4. *Experiment:* Select a city beginning with "J."
   Event: E: The city is Johannesburg.

   $E = \{\text{Johannesburg}\}$ An event can consist of a single outcome.

5. *Experiment:* Roll a die and observe the number facing up.
   Event: E: The number observed is either even or odd.

   $E = S = \{1, 2, 3, 4, 5, 6\}$ An event can consist of all possible outcomes.

6. *Experiment:* Select a student in your class.
   Event: E: The student has red hair.

   $E = \{\text{red-haired students in your class}\}$

7. *Experiment:* Draw a hand of 2 cards from a deck of 52.
   Event: H: Both cards are diamonds.

   H is the set of all hands of 2 cards chosen from 52 such that both cards are diamonds.

Here are some more examples of events.

EXAMPLE 2 Dice

We roll a red die and a green die and observe the numbers facing up. Describe the following events as subsets of the sample space.

a. E: The sum of the numbers showing is 6.

b. F: The sum of the numbers showing is 2.

Solution Here (again) is the sample space for the experiment of throwing two dice.

$$S = \left\{ \begin{array}{llllll} (1,1), & (1,2), & (1,3), & (1,4), & (1,5), & (1,6), \\ (2,1), & (2,2), & (2,3), & (2,4), & (2,5), & (2,6), \\ (3,1), & (3,2), & (3,3), & (3,4), & (3,5), & (3,6), \\ (4,1), & (4,2), & (4,3), & (4,4), & (4,5), & (4,6), \\ (5,1), & (5,2), & (5,3), & (5,4), & (5,5), & (5,6), \\ (6,1), & (6,2), & (6,3), & (6,4), & (6,5), & (6,6) \end{array} \right\}$$

a. In mathematical language, E is the subset of S that consists of all those outcomes in which the sum of the numbers showing is 6. Here is the sample space once again, with the outcomes in question shown in color:

$$S = \left\{ \begin{array}{llllll} (1,1), & (1,2), & (1,3), & (1,4), & (1,5), & (1,6), \\ (2,1), & (2,2), & (2,3), & (2,4), & (2,5), & (2,6), \\ (3,1), & (3,2), & (3,3), & (3,4), & (3,5), & (3,6), \\ (4,1), & (4,2), & (4,3), & (4,4), & (4,5), & (4,6), \\ (5,1), & (5,2), & (5,3), & (5,4), & (5,5), & (5,6), \\ (6,1), & (6,2), & (6,3), & (6,4), & (6,5), & (6,6) \end{array} \right\}$$

Thus, $E = \{(1, 5), (2, 4), (3, 3), (4, 2), (5, 1)\}$.

b. The only outcome in which the numbers showing add to 2 is (1, 1). Thus,

$$F = \{(1, 1)\}.$$

EXAMPLE 3 School and Work

Let S be the sample space of Example 1. List the elements in the following events:

a. The event E that a 2010 high school graduate was working.

b. The event F that a 2010 high school graduate was not going to a 2-year college.

Solution

a. We had this sample space:

$S = \{$2-year college & working, two-year college & not working,
4-year college & working, four-year college & not working,
no college & working, no college & not working$\}$.

We are asked for the event that a graduate was working. Whenever we encounter a phrase involving "the event that . . . ," we mentally translate this into mathematical language by changing the wording.

Replace the phrase "the event that . . ." by the phrase "the subset of the sample space consisting of all outcomes in which . . . ."

Thus we are interested in the subset of the sample space consisting of all outcomes in which the graduate was working. This gives

$E = \{$2-year college & working, 4-year college & working, no college & working$\}$.

The outcomes in E are illustrated by the shaded cells in Figure 2.

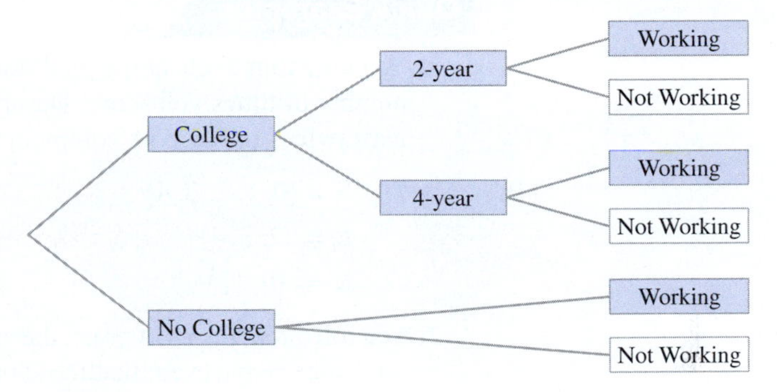

Figure 2

b. We are looking for the event that a graduate was not going to a 2-year college; that is, the subset of the sample space consisting of all outcomes in which the graduate was not going to a 2-year college. Thus,

$F = \{$4-year college & working, 4-year college & not working, no college & working, no college & not working$\}$.

The outcomes in F are illustrated by the shaded cells in Figure 3.

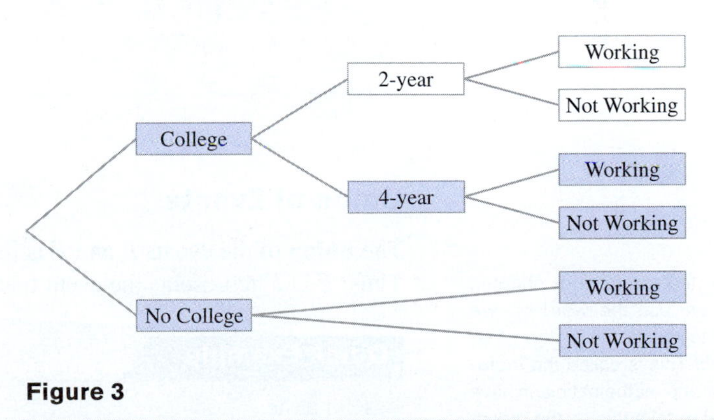

Figure 3

Complement, Union, and Intersection of Events

Events may often be described in terms of other events, using set operations such as complement, union, and intersection.

Complement of an Event

The **complement** of an event E is the set of outcomes not in E. Thus, the complement of E represents the event that E *does not occur*.

Visualizing the Complement

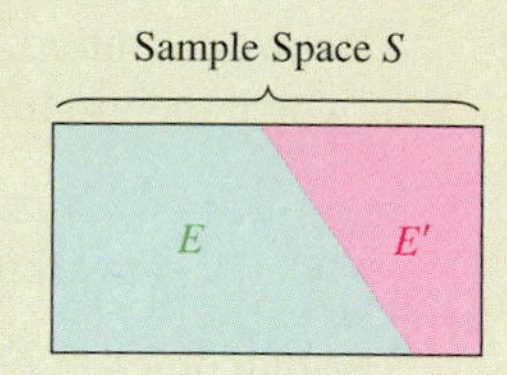

Quick Examples

1. You take four shots at the goal during a soccer game and record the number of times you score. Describe the event that you score at least twice, and also its complement.

$S = \{0, 1, 2, 3, 4\}$ Set of outcomes

$E = \{2, 3, 4\}$ Event that you score at least twice

$E' = \{0, 1\}$ Event that you do not score at least twice

2. You roll a red die and a green die and observe the two numbers facing up. Describe the event that the sum of the numbers is not 6.

$S = \{(1, 1), (1, 2), \dots, (6, 6)\}$

$F = \{(1, 5), (2, 4), (3, 3), (4, 2), (5, 1)\}$ Sum of numbers is 6.

$$F' = \left\{\begin{array}{cccccc} (1,1), & (1,2), & (1,3), & (1,4), & & (1,6), \\ (2,1), & (2,2), & (2,3), & & (2,5), & (2,6), \\ (3,1), & (3,2), & & (3,4), & (3,5), & (3,6), \\ (4,1), & & (4,3), & (4,4), & (4,5), & (4,6), \\ & (5,2), & (5,3), & (5,4), & (5,5), & (5,6), \\ (6,1), & (6,2), & (6,3), & (6,4), & (6,5), & (6,6) \end{array}\right\}$$

Sum of numbers is not 6.

Union of Events

The **union** of the events E and F is the set of all outcomes in E or F (or both). Thus, $E \cup F$ represents the event that E occurs *or* F occurs (or both).*

* As in the preceding chapter, when we use the word *or*, we agree to mean one or the other *or both*. This is called the **inclusive or** and mathematicians have agreed to take this as the meaning of *or* to avoid confusion.

Quick Example

Roll a die.

E: The outcome is a 5; $E = \{5\}$.

F: The outcome is an even number; $F = \{2, 4, 6\}$.

$E \cup F$: The outcome is either a 5 *or* an even number;
$E \cup F = \{2, 4, 5, 6\}$.

Intersection of Events

The **intersection** of the events E and F is the set of all outcomes common to E and F. Thus, $E \cap F$ represents the event that both E *and* F occur.

Quick Example

Roll two dice; one red and one green.

E: The red die is 2.

F: The green die is odd.

$E \cap F$: The red die is 2 and the green die is odd;

$$E \cap F = \{(2, 1), (2, 3), (2, 5)\}.$$

EXAMPLE 4 Weather

Let R be the event that it will rain tomorrow, let P be the event that it will be pleasant, let C be the event that it will be cold, and let H be the event that it will be hot.

a. Express in words: $R \cap P'$, $R \cup (P \cap C)$.

b. Express in symbols: Tomorrow will be either a pleasant day or a cold and rainy day; it will not, however, be hot.

Solution The key here is to remember that intersection corresponds to *and* and union to *or*.

a. $R \cap P'$ is the event that it will rain *and* it will not be pleasant.
$R \cup (P \cap C)$ is the event that either it will rain, or it will be pleasant and cold.

b. If we rephrase the given statement using *and* and *or* we get "Tomorrow will be either a pleasant day or a cold and rainy day, and it will not be hot."

$$[P \cup (C \cap R)] \cap H' \qquad \text{Pleasant, or cold and rainy, and not hot.}$$

The nuances of the English language play an important role in this formulation. For instance, the effect of the pause (comma) after "rainy day" suggests placing the preceding clause $P \cup (C \cap R)$ in parentheses. In addition, the phrase "cold and rainy" suggests that C and R should be grouped together in their own parentheses.

EXAMPLE 5 iPods, iPhones, and iPads

(Compare Example 3 in Section 6.2.) The following table shows sales, in millions of units, of iPods®, iPhones®, and iPads® in the last three quarters of 2011.[2]

	iPods (A)	iPhones (B)	iPads (C)	Total
2011 Q2 (U)	9.0	18.7	4.7	32.4
2011 Q3 (V)	7.5	20.3	9.3	37.1
2011 Q4 (W)	6.6	17.1	11.1	34.8
Total	23.1	56.1	25.1	104.3

[2]Figures are rounded to one decimal place. Source: Apple quarterly press releases (www.investor.apple.com).

Consider the experiment in which a device is selected at random from those represented in the table. Let A be the event that it was an iPod, let C be the event that it was an iPad, and let U be the event that it was sold in the second quarter of 2011. Describe the following events and compute their cardinality:

a. U' **b.** $A \cap U'$ **c.** $C \cup U$

Solution Before we answer the questions, note that S is the set of all items represented in the table, so S has a total of 104.3 million outcomes.

a. U' is the event that the item was not sold in the second quarter of 2011. Its cardinality is

$$n(U') = n(S) - n(U) = 104.3 - 32.4 = 71.9 \text{ million items.}$$

b. $A \cap U'$ is the event that it was an iPod not sold in the second quarter of 2011.

	iPods (A)	iPhones (B)	iPads (C)	Total
2011 Q2 (U)	9.0	18.7	4.7	32.4
2011 Q3 (V)	7.5	20.3	9.3	37.1
2011 Q4 (W)	6.6	17.1	11.1	34.8
Total	23.1	56.1	25.1	104.3

Referring to the table, we find

$$n(A \cap U') = 7.5 + 6.6 = 14.1 \text{ million items.}$$

c. $C \cup U$ is the event that either it was an iPad or it was sold in the second quarter of 2011:

	iPods (A)	iPhones (B)	iPads (C)	Total
2011 Q2 (U)	9.0	18.7	4.7	32.4
2011 Q3 (V)	7.5	20.3	9.3	37.1
2011 Q4 (W)	6.6	17.1	11.1	34.8
Total	23.1	56.1	25.1	104.3

To compute its cardinality, we can use the formula for the cardinality of a union:

$$n(C \cup U) = n(C) + n(U) - n(C \cap U)$$
$$= 25.1 + 32.4 - 4.7 = 52.8 \text{ million items.}$$

The case where $E \cap F$ is empty is interesting, and we give it a name.

Mutually Exclusive Events

If E and F are events, then E and F are said to be **disjoint** or **mutually exclusive** if $E \cap F$ is empty. (Hence, they have no outcomes in common.)

Visualizing Mutually Exclusive Events

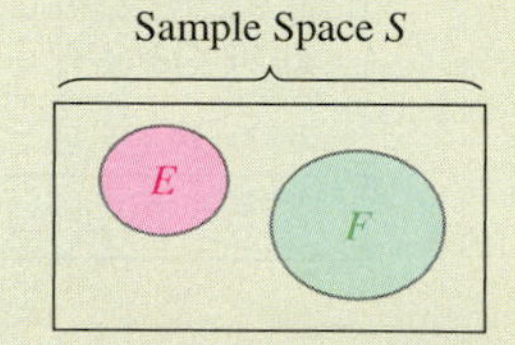

Interpretation
It is impossible for mutually exclusive events to occur simultaneously.

Quick Examples

In each of the following examples, E and F are mutually exclusive events.

1. Roll a die and observe the number facing up. E: The outcome is even; F: The outcome is odd.

$$E = \{2, 4, 6\}, F = \{1, 3, 5\}$$

2. Toss a coin three times and record the sequence of heads and tails. E: All three tosses land the same way up, F: One toss shows heads and the other two show tails.

$$E = \{\text{HHH, TTT}\}, F = \{\text{HTT, THT, TTH}\}$$

3. Observe tomorrow's weather. E: It is raining; F: There is not a cloud in the sky.

FAQs

Specifying the Sample Space

Q: *How do I determine the sample space in a given application?*

A: Strictly speaking, an experiment should include a description of what kinds of objects are in the sample space, as in:

Cast a die and observe the number facing up.
Sample space: the possible numbers facing up, {1, 2, 3, 4, 5, 6}.

Choose a person at random and record her Social Security number and whether she is blonde.
Sample space: pairs (9-digit number, Y/N).

However, in many of the scenarios discussed in this chapter and the next, an experiment is specified more vaguely, as in "Select a student in your class." In cases like this, the nature of the sample space should be determined from the context. For example, if the discussion is about grade-point averages and gender, the sample space can be taken to consist of pairs (grade-point average, M/F).

7.1 EXERCISES

Access end-of-section exercises online at **www.webassign.net**

7.2 Relative Frequency

Suppose you have a coin that you think is not fair and you would like to determine the likelihood that heads will come up when it is tossed. You could estimate this likelihood by tossing the coin a large number of times and counting the number of times heads comes up. Suppose, for instance, that in 100 tosses of the coin, heads comes up 58 times. The fraction of times that heads comes up, $58/100 = .58$, is the **relative frequency**, or **estimated probability** of heads coming up when the coin is tossed. In other words, saying that the relative frequency of heads coming up is .58 is the same as saying that heads came up 58% of the time in your series of experiments.

Now let's think about this example in terms of sample spaces and events. First of all, there is an experiment that has been repeated $N = 100$ times: Toss the coin and observe the side facing up. The sample space for this experiment is $S = \{\text{H, T}\}$. Also, there is an event E in which we are interested: the event that heads comes up, which is $E = \{\text{H}\}$. The number of times E has occurred, or the **frequency** of E, is $fr(E) = 58$. The relative frequency of the event E is then

$$P(E) = \frac{fr(E)}{N} \qquad \frac{\text{Frequency of event } E}{\text{Number of repetitions } N}$$

$$= \frac{58}{100} = .58.$$

Notes

1. The relative frequency gives us an *estimate* of the likelihood that heads will come up when that particular coin is tossed. This is why statisticians often use the alternative term *estimated probability* to describe it.
2. The larger the number of times the experiment is performed, the more accurate an estimate we expect this estimated probability to be. ■

Relative Frequency

When an experiment is performed a number of times, the **relative frequency** or **estimated probability** of an event E is the fraction of times that the event E occurs. If the experiment is performed N times and the event E occurs $fr(E)$ times, then the relative frequency is given by

$$P(E) = \frac{fr(E)}{N}. \qquad \text{Fraction of times } E \text{ occurs}$$

The number $fr(E)$ is called the **frequency** of E. N, the number of times that the experiment is performed, is called the number of **trials** or the **sample size**. If E consists of a single outcome s, then we refer to $P(E)$ as the relative frequency or estimated probability of the outcome s, and we write $P(s)$.

Visualizing Relative Frequency

● ● ● ● ○
● ○ ○ ● ○

$$P(E) = \frac{fr(E)}{N} = \frac{4}{10} = .4$$

The collection of the estimated probabilities of *all* the outcomes is the **relative frequency distribution** or **estimated probability distribution**.

Quick Examples

1. **Experiment:** Roll a pair of dice and add the numbers that face up.

   Event: E: The sum is 5.

   If the experiment is repeated 100 times and E occurs on 10 of the rolls, then the relative frequency of E is

   $$P(E) = \frac{fr(E)}{N} = \frac{10}{100} = .10.$$

2. If 10 rolls of a single die resulted in the outcomes 2, 1, 4, 4, 5, 6, 1, 2, 2, 1, then the associated relative frequency distribution is shown in the following table:

Outcome	1	2	3	4	5	6
Rel. Frequency	.3	.3	0	.2	.1	.1

3. **Experiment:** Note the cloud conditions on a particular day in April.

   If the experiment is repeated a number of times, and it is clear 20% of those times, partly cloudy 30% of those times, and overcast the rest of those times, then the relative frequency distribution is:

Outcome	Clear	Partly Cloudy	Overcast
Rel. Frequency	.20	.30	.50

EXAMPLE 1 Sales of Hybrid Vehicles

In a survey of 250 hybrid vehicles sold in the United States, 125 were Toyota Prii,* 30 were Honda Civics, 20 were Toyota Camrys, 15 were Ford Escapes, and the rest were other makes.[3] What is the relative frequency that a hybrid vehicle sold in the United States is not a Toyota Camry?

* The official plural form of Prius according to Toyota.

Solution The experiment consists of choosing a hybrid vehicle sold in the United States and determining its make. The sample space suggested by the information given is

$$S = \{\text{Toyota Prius, Honda Civic, Toyota Camry, Ford Escape, Other}\}$$

and we are interested in the event

$$E = \{\text{Toyota Prius, Honda Civic, Ford Escape, Other}\}.$$

The sample size is $N = 250$, of which 20 were Toyota Camrys. Thus, the frequency of E is $fr(E) = 250 - 20 = 230$ and the relative frequency of E is

$$P(E) = \frac{fr(E)}{N} = \frac{230}{250} = .92.$$

[3] The proportions are based on approximate actual cumulative sales through October 2011 (www.wikipedia.com).

➡ **Before we go on...** In Example 1, you might ask how accurate the estimate of .92 is or how well it reflects *all* of the hybrid vehicles sold in the United States absent any information about national sales figures. The field of statistics provides the tools needed to say to what extent this estimated probability can be trusted. ■

EXAMPLE 2 Auctions on eBay

The following chart shows the results of a survey of the bid prices for 50 paintings on eBay with the highest number of bids.[4]

Bid Price	\$0–\$9.99	\$10–\$49.99	\$50–\$99.99	$\geq$ \$100
Frequency	6	23	15	6

Consider the experiment in which a painting is chosen and the bid price is observed.

a. Find the relative frequency distribution.

b. Find the relative frequency that a painting in the survey had a bid price of less than \$50.

Solution

a. The following table shows the relative frequency of each outcome, which we find by dividing each frequency by the sum $N = 50$:

Bid Price	\$0–\$9.99	\$10–\$49.99	\$50–\$99.99	$\geq$ \$100
Rel. Frequency	$\frac{6}{50} = .12$	$\frac{23}{50} = .46$	$\frac{15}{50} = .30$	$\frac{6}{50} = .12$

b. Method 1: Computing Directly

$E = \{\$0\text{–}\$9.99, \$10\text{–}\$49.99\}$

Thus,

$$P(E) = \frac{fr(E)}{N} = \frac{6+23}{50} = \frac{29}{50} = .58.$$

Method 2: Using the Relative Frequency Distribution

Notice that we can obtain the same answer from the distribution in part (a) by simply adding the relative frequencies of the outcomes in E:

$$P(E) = .12 + .46 = .58.$$

[4]In the category "Art—Direct from Artist" on November 14, 2011 (www.eBay.com).

Q: *Why did we get the same result in part (b) of Example 2 by simply adding the relative frequencies of the outcomes in E?*

A: The reason can be seen by doing the calculation in the first method a slightly different way:

$$P(E) = \frac{fr(E)}{N} = \frac{6+23}{50}$$

$$= \frac{6}{50} + \frac{23}{50}. \quad \text{Sum of rel. frequencies of the individual outcomes}$$

This property of relative frequency distributions is discussed below.

Following are some important properties of estimated probability that we can observe in Example 2.

Some Properties of Relative Frequency Distributions

Let $S = \{s_1, s_2, \ldots, s_n\}$ be a sample space and let $P(s_i)$ be the relative frequency of the event $\{s_i\}$. Then

1. $0 \leq P(s_i) \leq 1$

2. $P(s_1) + P(s_2) + \cdots + P(s_n) = 1$

3. If $E = \{e_1, e_2, \ldots, e_r\}$, then $P(E) = P(e_1) + P(e_2) + \cdots + P(e_r)$.

In words:

1. The relative frequency of each outcome is a number between 0 and 1 (inclusive).

2. The relative frequencies of all the outcomes add up to 1.

3. The relative frequency of an event E is the sum of the relative frequencies of the individual outcomes in E.

Relative Frequency and Increasing Sample Size

using Technology

See the Technology Guides at the end of the chapter to see how to use a TI-83/84 Plus or a spreadsheet to simulate experiments.

A "fair" coin is one that is as likely to come up heads as it is to come up tails. In other words, we expect heads to come up 50% of the time if we toss such a coin many times. Put more precisely, we expect the relative frequency to approach .5 as the number of trials gets larger. Figure 4 shows how the relative frequency behaved for one sequence of coin tosses. For each N we have plotted what fraction of times the coin came up heads in the first N tosses.

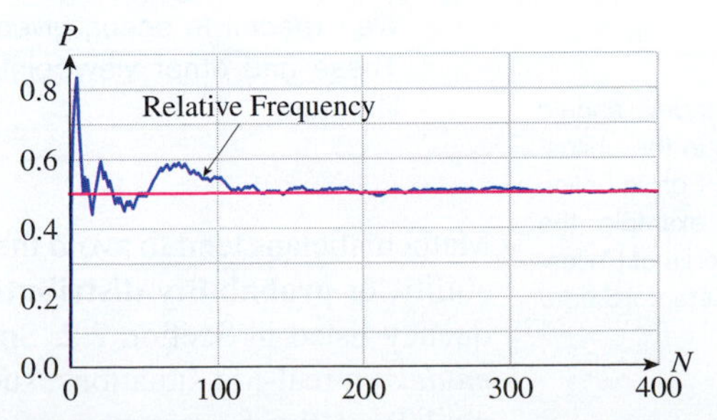

Figure 4

Notice that the relative frequency graph meanders as N increases, sometimes getting closer to .5, and sometimes drifting away again. However, the graph tends to meander within smaller and smaller distances of .5 as N increases.*

* This can be made more precise by the concept of "limit" used in calculus.

In general, this is how relative frequency seems to behave; as N gets large, the relative frequency appears to approach some fixed value. Some refer to this value as the "actual" probability, whereas others point out that there are difficulties with this this notion. For instance, how can we actually determine this limit to any accuracy by experiment? How exactly is the experiment conducted? Technical and philosophical issues aside, the relative frequencies do approach a fixed value and, in the next section, we will talk about how we use probability models to predict this limiting value.

7.2 EXERCISES

Access end-of-section exercises online at **www.webassign.net**

7.3 Probability and Probability Models

It is understandable if you are a little uncomfortable with using relative frequency as the estimated probability because it does not always agree with what you intuitively feel to be true. For instance, if you toss a fair coin (one as likely to come up heads as tails) 100 times and heads happen to come up 62 times, the experiment seems to suggest that the probability of heads is .62, even though you *know* that the "actual" probability is .50 (because the coin is fair).

Q: *So what do we mean by "actual" probability?*

A: There are various philosophical views as to exactly what we should mean by "actual" probability. For example, (finite) *frequentists* say that there is no such thing as "actual probability"—all we should really talk about is what we can actually measure, the relative frequency. *Propensitists* say that the actual probability p of an event is a (often physical) property of the event that makes its relative frequency tend to p in the long run; that is, p will be the limiting value of the relative frequency as the number of trials in a repeated experiment gets larger and larger. (See Figure 4 in the preceding section.) *Bayesians*, on the other hand, argue that the actual probability of an event is the degree to which we *expect* it to occur, given our knowledge about the nature of the experiment. These and other viewpoints have been debated in considerable depth in the literature.*

* The interested reader should consult references in the philosophy of probability. For an online summary, see, for example, the Stanford Encyclopedia of Philosophy (http://plato.stanford.edu/contents.html).

Mathematicians tend to avoid the whole debate, and talk instead about *abstract* probability, or **probability distributions**, based purely on the properties of relative frequency listed in Section 7.2. Specific probability distributions can then be used as *models* in real-life situations, such as flipping a coin or tossing a die, to predict (or model) relative frequency.

Probability Distribution; Probability

(Compare with the properties of relative frequency distributions in Section 7.2.) A (finite) **probability distribution** is an assignment of a number $P(s_i)$, the **probability of s_i**, to each outcome of a finite sample space $S = \{s_1, s_2, \dots, s_n\}$. The probabilities must satisfy

1. $0 \leq P(s_i) \leq 1$

and

2. $P(s_1) + P(s_2) + \cdots + P(s_n) = 1$.

We find the **probability of an event E**, written $P(E)$, by adding up the probabilities of the outcomes in E.

If $P(E) = 0$, we call E an **impossible event**. The empty event $\emptyset$ is always impossible, since *something* must happen.

Quick Examples

1. All the examples of estimated probability distributions in Section 7.2 are examples of probability distributions. (See page 361.)
2. Let us take $S = \{\text{H, T}\}$ and make the assignments $P(\text{H}) = .5$, $P(\text{T}) = .5$. Because these numbers are between 0 and 1 and add to 1, they specify a probability distribution.
3. In Quick Example 2, we can instead make the assignments $P(\text{H}) = .2$, $P(\text{T}) = .8$. Because these numbers are between 0 and 1 and add to 1, they, too, specify a probability distribution.
4. With $S = \{\text{H, T}\}$ again, we could also take $P(\text{H}) = 1$, $P(\text{T}) = 0$, so that $\{\text{T}\}$ is an impossible event.
5. The following table gives a probability distribution for the sample space $S = \{1, 2, 3, 4, 5, 6\}$.

Outcome	1	2	3	4	5	6
Probability	.3	.3	0	.1	.2	.1

It follows that

$$P(\{1, 6\}) = .3 + .1 = .4$$
$$P(\{2, 3\}) = .3 + 0 = .3$$
$$P(3) = 0. \qquad \{3\} \text{ is an impossible event.}$$

The above Quick Examples included models for the experiments of flipping fair and unfair coins. In general:

Probability Models

A **probability model** for a particular experiment is a probability distribution that predicts the relative frequency of each outcome if the experiment is performed a large number of times (see Figure 4 at the end of the preceding section).* Just as we think of relative frequency as *estimated probability*, we can think of modeled probability as *theoretical probability*.

* Just how large is a "large number of times"? That depends on the nature of the experiment. For example, if you toss a fair coin 100 times, then the relative frequency of heads will be between .45 and .55 about 73% of the time. If an outcome is extremely unlikely (such as winning the lotto), you might need to repeat the experiment billions or trillions of times before the relative frequency approaches any specific number.

Quick Examples

1. **Fair Coin Model:** (See Quick Example 2 on the previous page.) Flip a fair coin and observe the side that faces up. Because we expect that heads is as likely to come up as tails, we model this experiment with the probability distribution specified by $S = \{H, T\}$, $P(H) = .5$, $P(T) = .5$. Figure 4 on p. 361 suggests that the relative frequency of heads approaches .5 as the number of coin tosses gets large, so the fair coin model predicts the relative frequency for a large number of coin tosses quite well.
2. **Unfair Coin Model:** (See Quick Example 3 on the previous page.) Take $S = \{H, T\}$ and $P(H) = .2$, $P(T) = .8$. We can think of this distribution as a model for the experiment of flipping an unfair coin that is four times as likely to land with tails uppermost than heads.
3. **Fair Die Model:** Roll a fair die and observe the uppermost number. Because we expect to roll each specific number one sixth of the time, we model the experiment with the probability distribution specified by $S = \{1, 2, 3, 4, 5, 6\}$, $P(1) = 1/6$, $P(2) = 1/6, \ldots, P(6) = 1/6$. This model predicts, for example, that the relative frequency of throwing a 5 approaches $1/6$ as the number of times you roll the die gets large.
4. Roll a pair of fair dice (recall that there are a total of 36 outcomes if the dice are distinguishable). Then an appropriate model of the experiment has

$$S = \left\{ \begin{array}{cccccc} (1,1), & (1,2), & (1,3), & (1,4), & (1,5), & (1,6), \\ (2,1), & (2,2), & (2,3), & (2,4), & (2,5), & (2,6), \\ (3,1), & (3,2), & (3,3), & (3,4), & (3,5), & (3,6), \\ (4,1), & (4,2), & (4,3), & (4,4), & (4,5), & (4,6), \\ (5,1), & (5,2), & (5,3), & (5,4), & (5,5), & (5,6), \\ (6,1), & (6,2), & (6,3), & (6,4), & (6,5), & (6,6) \end{array} \right\}$$

   with each outcome being assigned a probability of $1/36$.
5. In the experiment in Quick Example 4, take E to be the event that the sum of the numbers that face up is 5, so

$$E = \{(1,4), (2,3), (3,2), (4,1)\}.$$

   By the properties of probability distributions,

$$P(E) = \frac{1}{36} + \frac{1}{36} + \frac{1}{36} + \frac{1}{36} = \frac{4}{36} = \frac{1}{9}.$$

Notice that, in all of the Quick Examples above except for the unfair coin, all the outcomes are equally likely, and each outcome s has a probability of

$$P(s) = \frac{1}{\text{Total number of outcomes}} = \frac{1}{n(S)}.$$

More generally, in the last Quick Example we saw that adding the probabilities of the individual outcomes in an event E amounted to computing the ratio (Number of favorable outcomes)/(Total number of outcomes):

$$P(E) = \frac{\text{Number of favorable outcomes}}{\text{Total number of outcomes}} = \frac{n(E)}{n(S)}.$$

Probability Model for Equally Likely Outcomes

In an experiment in which all outcomes are equally likely, we model the experiment by taking the probability of an event E to be

$$P(E) = \frac{\text{Number of favorable outcomes}}{\text{Total number of outcomes}} = \frac{n(E)}{n(S)}.$$

Visualizing Probability for Equally Likely Outcomes

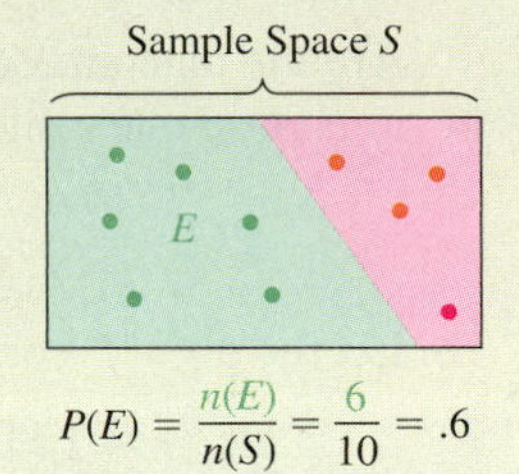

$$P(E) = \frac{n(E)}{n(S)} = \frac{6}{10} = .6$$

Note

Remember that this formula will work *only* when the outcomes are equally likely. If, for example, a die is *weighted*, then the outcomes may not be equally likely, and the formula above will not give an appropriate probability model. ■

Quick Examples

1. Toss a fair coin three times, so $S = \{\text{HHH, HHT, HTH, HTT, THH, THT, TTH, TTT}\}$. The probability that we throw exactly two heads is

$$P(E) = \frac{n(E)}{n(S)} = \frac{3}{8}.$$

There are eight equally likely outcomes and $E = \{\text{HHT, HTH, THH}\}$.

2. Roll a pair of fair dice. The probability that we roll a double (both dice show the same number) is

$$P(E) = \frac{n(E)}{n(S)} = \frac{6}{36} = \frac{1}{6}.$$

$E = \{(1, 1), (2, 2), (3, 3), (4, 4), (5, 5), (6, 6)\}$

3. Randomly choose a person from a class of 40, in which 6 have red hair. If E is the event that a randomly selected person in the class has red hair, then

$$P(E) = \frac{n(E)}{n(S)} = \frac{6}{40} = .15.$$

EXAMPLE 1 Sales of Hybrid Vehicles

(Compare Example 1 in Section 7.2.) A total of 1.9 million hybrid vehicles had been sold in the United States through October of 2011. Of these, 955,000 were Toyota Prii, 205,000 were Honda Civics, 170,000 were Toyota Camrys, 105,000 were Ford Escapes, and the rest were other makes.[5]

a. What is the probability that a randomly selected hybrid vehicle sold in the United States was either a Toyota Prius or a Honda Civic?

b. What is the probability that a randomly selected hybrid vehicle sold in the United States was not a Toyota Camry?

[5]Source for sales data: www.wikipedia.com.

Solution

a. The experiment suggested by the question consists of randomly choosing a hybrid vehicle sold in the United States and determining its make. We are interested in the event E that the hybrid vehicle was either a Toyota Prius or a Honda Civic. So,

$$S = \text{the set of hybrid vehicles sold; } n(S) = 1{,}900{,}000$$

$$E = \text{the set of Toyota Prii and Honda Civics sold;}$$
$$n(E) = 955{,}000 + 205{,}000 = 1{,}160{,}000.$$

Are the outcomes equally likely in this experiment? Yes, because we are as likely to choose one vehicle as another. Thus,

$$P(E) = \frac{n(E)}{n(S)} = \frac{1{,}160{,}000}{1{,}900{,}000} \approx .61.$$

b. Let the event F consist of those hybrid vehicles sold that were not Toyota Camrys.

$$n(F) = 1{,}900{,}000 - 170{,}000 = 1{,}730{,}000$$

Hence,

$$P(F) = \frac{n(F)}{n(S)} = \frac{1{,}730{,}000}{1{,}900{,}000} \approx .91.$$

Q: *In Example 1 of Section 7.2 we had a similar example about hybrid vehicles, but we called the probabilities calculated there relative frequencies. Here they are probabilities. What is the difference?*

A: In Example 1 of Section 7.2, the data were based on the results of a survey, or sample, of only 250 hybrid vehicles (out of a total of about 1.9 million sold in the United States), and were therefore incomplete. (A statistician would say that we were given *sample data*.) It follows that any inference we draw from the 250 surveyed, such as the probability that a hybrid vehicle sold in the United States is not a Toyota Camry, is uncertain, and this is the cue that tells us that we are working with relative frequency, or estimated probability. Think of the survey as an experiment (choosing a hybrid vehicle) repeated 250 times—exactly the setting for estimated probability.

In Example 1 above, on the other hand, the data do not describe how *some* hybrid vehicle sales are broken down into the categories described, but they describe how *all 1.9 million* hybrid vehicle sales in the United States are broken down. (The statistician would say that we were given *population data* in this case, because the data describe the entire "population" of hybrid vehicles sold in the United States.)

EXAMPLE 2 Indistinguishable Dice

We recall from Section 7.1 that the sample space when we roll a pair of indistinguishable dice is

$$S = \left\{ \begin{array}{rrrrrr} (1,1), & (1,2), & (1,3), & (1,4), & (1,5), & (1,6), \\ & (2,2), & (2,3), & (2,4), & (2,5), & (2,6), \\ & & (3,3), & (3,4), & (3,5), & (3,6), \\ & & & (4,4), & (4,5), & (4,6), \\ & & & & (5,5), & (5,6), \\ & & & & & (6,6) \end{array} \right\}.$$

Construct a probability model for this experiment.

Solution Because there are 21 outcomes, it is tempting to say that the probability of each outcome should be taken to be 1/21. However, the outcomes are not all equally likely. For instance, the outcome (2, 3) is twice as likely as (2, 2), because (2, 3) can occur in two ways (it corresponds to the event {(2, 3), (3, 2)} for distinguishable dice). For purposes of calculating probability, it is easiest to use calculations for distinguishable dice.* Here are some examples.

* Note that any pair of real dice can be distinguished in principle because they possess slight differences, although we may regard them as indistinguishable by not attempting to distinguish them. Thus, the probabilities of events must be the same as for the corresponding events for distinguishable dice.

Outcome (indistinguishable dice)	(1, 1)	(1, 2)	(2, 2)	(1, 3)	(2, 3)	(3, 3)
Corresponding Event (distinguishable dice)	{(1, 1)}	{(1, 2), (2, 1)}	{(2, 2)}	{(1, 3), (3, 1)}	{(2, 3), (3, 2)}	{(3, 3)}
Probability	$\frac{1}{36}$	$\frac{2}{36} = \frac{1}{18}$	$\frac{1}{36}$	$\frac{2}{36} = \frac{1}{18}$	$\frac{2}{36} = \frac{1}{18}$	$\frac{1}{36}$

If we continue this process for all 21 outcomes, we will find that they add to 1. Figure 5 illustrates the complete probability distribution:

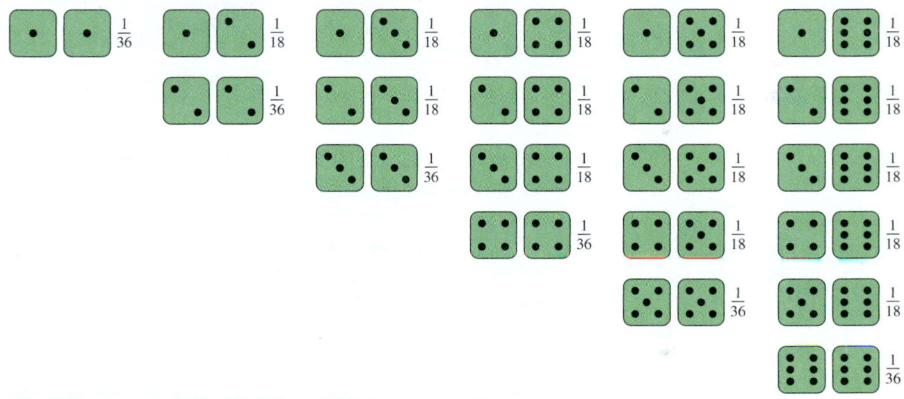

Figure 5

EXAMPLE 3 Weighted Dice

In order to impress your friends with your die-rolling skills, you have surreptitiously weighted your die in such a way that 6 is three times as likely to come up as any one of the other numbers. (All the other outcomes are equally likely.) Obtain a probability distribution for a roll of the die and use it to calculate the probability of an even number coming up.

Solution Let us label our unknowns (there appear to be two of them):

x = probability of rolling a 6

y = probability of rolling any one of the other numbers

We are first told that "6 is three times as likely to come up as any one of the other numbers." If we rephrase this in terms of our unknown probabilities we get, "the probability of rolling a 6 is three times the probability of rolling any one of the other numbers." In symbols,

$$x = 3y.$$

We must also use a piece of information not given to us, but one we know must be true: The sum of the probabilities of all the outcomes is 1:

$$x + y + y + y + y + y = 1$$

or

$$x + 5y = 1.$$

We now have two linear equations in two unknowns, and we solve for x and y. Substituting the first equation ($x = 3y$) in the second ($x + 5y = 1$) gives

$$8y = 1$$

or

$$y = \frac{1}{8}.$$

To get x, we substitute the value of y back into either equation and find

$$x = \frac{3}{8}.$$

Thus, the probability model we seek is the one shown in the following table.

Outcome	1	2	3	4	5	6
Probability	$\frac{1}{8}$	$\frac{1}{8}$	$\frac{1}{8}$	$\frac{1}{8}$	$\frac{1}{8}$	$\frac{3}{8}$

We can use the distribution to calculate the probability of an even number coming up by adding the probabilities of the favorable outcomes.

$$P(\{2, 4, 6\}) = \frac{1}{8} + \frac{1}{8} + \frac{3}{8} = \frac{5}{8}.$$

Thus there is a $5/8 = .625$ chance that an even number will come up.

Before we go on... We should check that the probability distribution in Example 3 satisfies the requirements: 6 is indeed three times as likely to come up as any other number. Also, the probabilities we calculated do add up to 1:

$$\frac{1}{8} + \frac{1}{8} + \frac{1}{8} + \frac{1}{8} + \frac{1}{8} + \frac{3}{8} = 1.$$ ■

Probability of Unions, Intersections, and Complements

So far, all we know about computing the probability of an event E is that $P(E)$ is the sum of the probabilities of the individual outcomes in E. Suppose, though, that we do not know the probabilities of the individual outcomes in E but we do know that $E = A \cup B$, where we happen to know $P(A)$ and $P(B)$. How do we compute the probability of $A \cup B$? We might be tempted to say that $P(A \cup B)$ is $P(A) + P(B)$, but let us look at an example using the probability distribution in Quick Example 5 at the beginning of this section:

Outcome	1	2	3	4	5	6
Probability	.3	.3	0	.1	.2	.1

For A let us take the event $\{2, 4, 5\}$, and for B let us take $\{2, 4, 6\}$. $A \cup B$ is then the event $\{2, 4, 5, 6\}$. We know that we can find the probabilities $P(A)$, $P(B)$, and $P(A \cup B)$ by adding the probabilities of all the outcomes in these events, so

$$P(A) = P(\{2, 4, 5\}) = .3 + .1 + .2 = .6$$
$$P(B) = P(\{2, 4, 6\}) = .3 + .1 + .1 = .5, \text{ and}$$
$$P(A \cup B) = P(\{2, 4, 5, 6\}) = .3 + .1 + .2 + .1 = .7.$$

Our first guess was wrong: $P(A \cup B) \neq P(A) + P(B)$. Notice, however, that the outcomes in $A \cap B$ are counted twice in computing $P(A) + P(B)$, but only once in computing $P(A \cup B)$:

$$P(A) + P(B) = P(\{2, 4, 5\}) + P(\{2, 4, 6\}) \qquad A \cap B = \{2, 4\}$$
$$= (.3 + .1 + .2) + (.3 + .1 + .1) \qquad P(A \cap B) \text{ counted twice}$$
$$= 1.1$$

whereas

$$P(A \cup B) = P(\{2, 4, 5, 6\}) = .3 + .1 + .2 + .1 \qquad P(A \cap B) \text{ counted once}$$
$$= .7.$$

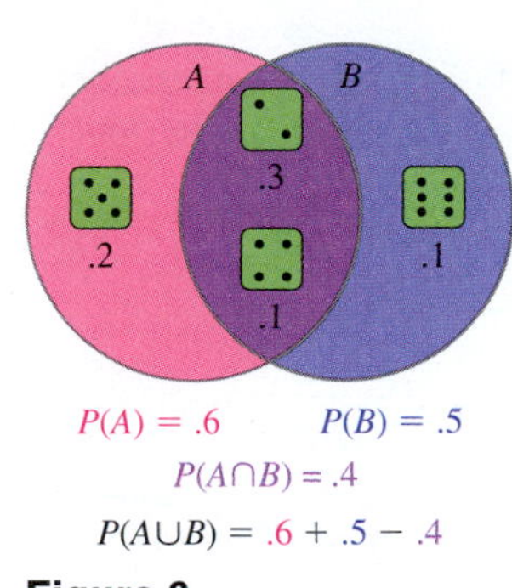

Figure 6

Thus, if we take $P(A) + P(B)$ and then subtract the surplus $P(A \cap B)$, we get $P(A \cup B)$. In symbols,

$$P(A \cup B) = P(A) + P(B) - P(A \cap B)$$
$$.7 = .6 + .5 - .4$$

(see Figure 6). We call this formula the **addition principle**. One more thing: Notice that our original guess $P(A \cup B) = P(A) + P(B)$ would have worked if we had chosen A and B with no outcomes in common; that is, if $A \cap B = \emptyset$. When $A \cap B = \emptyset$, recall that we say that A and B are mutually exclusive.

Addition Principle

If A and B are any two events, then

$$P(A \cup B) = P(A) + P(B) - P(A \cap B).$$

Visualizing the Addition Principle

In the figure, the area of the union is obtained by adding the areas of A and B and then subtracting the overlap (because it is counted twice when we add the areas).

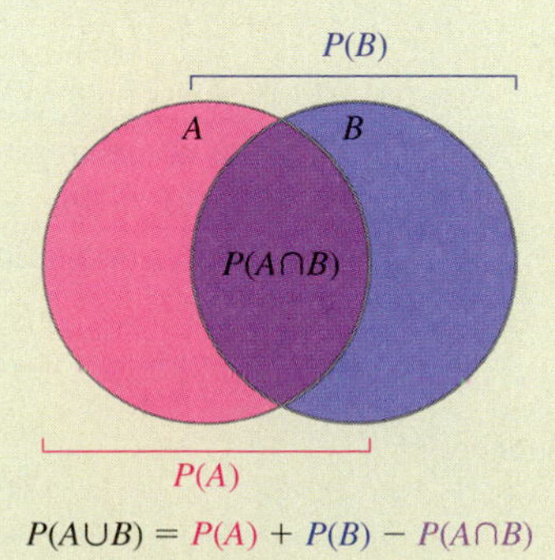

Addition Principle for Mutually Exclusive Events

If $A \cap B = \emptyset$, we say that A and B are **mutually exclusive**, and we have

$$P(A \cup B) = P(A) + P(B).$$ Because $P(A \cap B) = 0$

Visualizing the Addition Principle for Mutually Exclusive Events

If A and B do not overlap, then the area of the union is obtained by adding the areas of A and B.

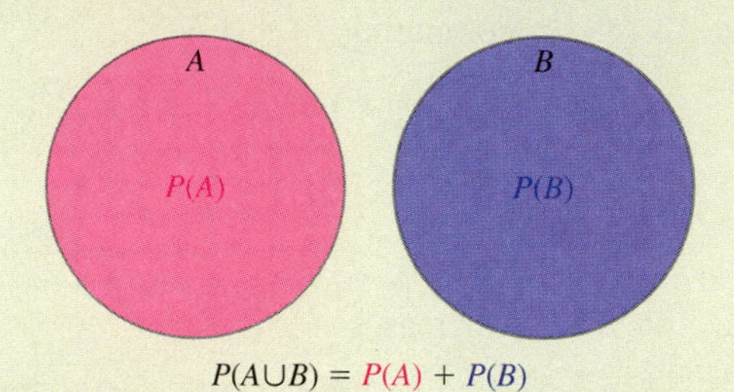

$P(A \cup B) = P(A) + P(B)$

This holds true also for more than two events: If $A_1, A_2, \ldots, A_n$ are mutually exclusive events (that is, the intersection of every pair of them is empty), then

$$P(A_1 \cup A_2 \cup \cdots \cup A_n) = P(A_1) + P(A_2) + \cdots + P(A_n).$$ Addition principle for many mutually exclusive events

Quick Examples

1. There is a 10% chance of rain (R) tomorrow, a 20% chance of high winds (W), and a 5% chance of both. The probability of either rain or high winds (or both) is

$$P(R \cup W) = P(R) + P(W) - P(R \cap W)$$
$$= .10 + .20 - .05 = .25.$$

2. The probability that you will be in Cairo at 6:00 am tomorrow (C) is .3, while the probability that you will be in Alexandria at 6:00 am tomorrow (A) is .2. Thus, the probability that you will be either in Cairo or Alexandria at 6:00 am tomorrow is

$$P(C \cup A) = P(C) + P(A)$$ A and C are mutually exclusive.
$$= .3 + .2 = .5.$$

3. When a pair of fair dice is rolled, the probability of the numbers that face up adding to 7 is 6/36, the probability of their adding to 8 is 5/36, and the probability of their adding to 9 is 4/36. Thus, the probability of the numbers adding to 7, 8, or 9 is

$$P(\{7\} \cup \{8\} \cup \{9\}) = P(7) + P(8) + P(9)$$
$$= \frac{6}{36} + \frac{5}{36} + \frac{4}{36} = \frac{15}{36} = \frac{5}{12}.$$ The events are mutually exclusive.*

* The sum of the numbers that face up cannot equal two different numbers at the same time.

EXAMPLE 4 School and Work

A survey[6] conducted by the Bureau of Labor Statistics found that 68% of the high school graduating class of 2010 went on to college the following year, while 42% of the class was working. Furthermore, 92% were either in college or working, or both.

a. What percentage went on to college and work at the same time?

b. What percentage went on to college but not work?

Solution We can think of the experiment of choosing a member of the high school graduating class of 2010 at random. The sample space is the set of all these graduates.

a. We are given information about two events:

A: A graduate went on to college; $P(A) = .68$.

B: A graduate went on to work; $P(B) = .42$.

We are also told that $P(A \cup B) = .92$. We are asked for the probability that a graduate went on to both college and work, $P(A \cap B)$. To find $P(A \cap B)$, we take advantage of the fact that the formula

$$P(A \cup B) = P(A) + P(B) - P(A \cap B)$$

can be used to calculate any one of the four quantities that appear in it as long as we know the other three. Substituting the quantities we know, we get

$$.92 = .68 + .42 - P(A \cap B)$$

so

$$P(A \cap B) = .68 + .42 - .92 = .18.$$

Thus, 18% of the graduates went on to college and work at the same time.

b. We are asked for the probability of a new event:

C: A graduate went on to college but not work.

C is the part of A outside of $A \cap B$, so $C \cup (A \cap B) = A$, and C and $A \cap B$ are mutually exclusive. (See Figure 7.)

Thus, applying the addition principle, we have

$$P(C) + P(A \cap B) = P(A).$$

From part (a), we know that $P(A \cap B) = .18$, so

$$P(C) + .18 = .68$$

giving

$$P(C) = .50.$$

In other words, 50% of the graduates went on to college but not work.

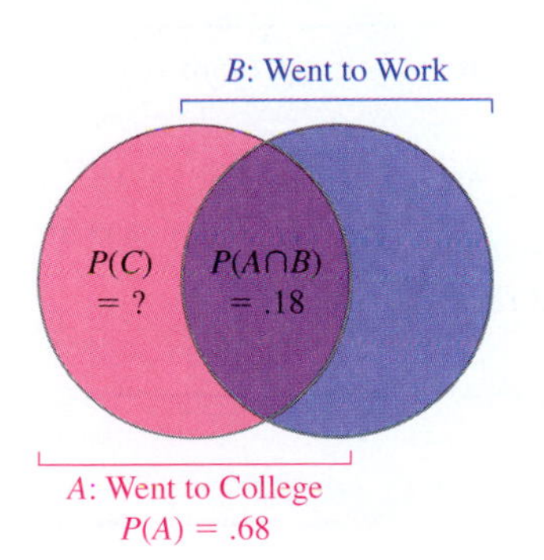

Figure 7

We can use the addition principle to deduce other useful properties of probability distributions.

[6]Source: "College Enrollment and Work Activity of High School Graduates," U.S. Bureau of Labor Statistics (www.bls.gov/news.release/hsgec.htm).

More Principles of Probability Distributions

The following rules hold for any sample space S and any event A:

$P(S) = 1$ — The probability of *something* happening is 1.

$P(\emptyset) = 0$ — The probability of *nothing* happening is 0.

$P(A') = 1 - P(A).$ — The probability of A *not* happening is 1 minus the probability of A.

Note

We can also write the third equation as

$$P(A) = 1 - P(A')$$

or

$$P(A) + P(A') = 1.$$

■

Visualizing the Rule for Complements

Think of A' as the portion of S outside of A. Adding the two areas gives the area of all of S, equal to 1.

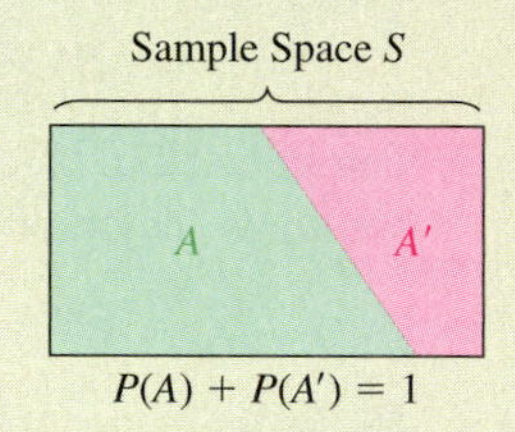

Quick Examples

1. There is a 10% chance of rain (R) tomorrow. Therefore, the probability that it will *not* rain is

$$P(R') = 1 - P(R) = 1 - .10 = .90.$$

2. The probability that Eric Ewing will score at least two goals is .6. Therefore, the probability that he will score at most one goal is $1 - .6 = .4$.

Q: *Can you persuade me that all of these principles are true?*

A: Let us take them one at a time.

We know that $S = \{s_1, s_2, \ldots, s_n\}$ is the set of all outcomes, and so

$$\begin{aligned} P(S) &= P(\{s_1, s_2, \ldots, s_n\}) \\ &= P(s_1) + P(s_2) + \cdots + P(s_n) \\ &= 1. \end{aligned}$$

We add the probabilities of the outcomes to obtain the probability of an event.

By the definition of a probability distribution

Now, note that $S \cap \emptyset = \emptyset$, so that S and $\emptyset$ are mutually exclusive. Applying the addition principle gives

$$P(S) = P(S \cup \emptyset) = P(S) + P(\emptyset).$$

Subtracting $P(S)$ from both sides gives $0 = P(\emptyset)$.

If A is any event in S, then we can write

$$S = A \cup A'$$

where A and A' are mutually exclusive. (Why?) Thus, by the addition principle,

$$P(S) = P(A) + P(A').$$

Because $P(S) = 1$, we get

$$1 = P(A) + P(A')$$

or $$P(A') = 1 - P(A).$$

EXAMPLE 5 Subprime Mortgages during the Housing Bubble

A home loan is either current, 30–59 days past due, 60–89 days past due, 90 or more days past due, in foreclosure, or repossessed by the lender. In November 2008, the probability that a randomly selected subprime home mortgage in California was not current was .51. The probability that a mortgage was not current, but neither in foreclosure nor repossessed, was .28.[7] Calculate the probabilities of the following events.

a. A California home mortgage was current.

b. A California home mortgage was in foreclosure or repossessed.

Solution

a. Let us write C for the event that a randomly selected subprime home mortgage in California was current. The event that the home mortgage was *not* current is its complement C', and we are given that $P(C') = .51$. We have

$$P(C) + P(C') = 1$$
$$P(C) + .51 = 1,$$

so $$P(C) = 1 - .51 = .49.$$

b. Take

F: A mortgage was in foreclosure or repossessed.

N: A mortgage was neither current, in foreclosure, nor repossessed.

We are given $P(N) = .28$. Further, the events F and N are mutually exclusive with union C', the set of all non-current mortgages. Hence,

$$P(C') = P(F) + P(N)$$
$$.51 = P(F) + .28$$

giving

$$P(F) = .51 - .28 = .23.$$

Thus, there was a 23% chance that a subprime home mortgage was either in foreclosure or repossessed.

---

[7]Source: Federal Reserve Bank of New York (www.newyorkfed.org/regional/subprime.html).

EXAMPLE 6 iPods, iPhones, and iPads

The following table shows sales, in millions of units, of iPods®, iPhones®, and iPads® in the last three quarters of 2011.[8]

	iPods (A)	iPhones (B)	iPads (C)	Total
2011 Q2 (U)	9.0	18.7	4.7	32.4
2011 Q3 (V)	7.5	20.3	9.3	37.1
2011 Q4 (W)	6.6	17.1	11.1	34.8
Total	23.1	56.1	25.1	104.3

If one of the items sold is selected at random, find the probabilities of the following events:

a. It is an iPod.

b. It was sold in the third quarter of 2011.

c. It is an iPod sold in the third quarter of 2011.

d. It either is an iPod or was sold in the third quarter of 2011.

e. It is not an iPod.

Solution Before we answer the questions, note that S is the set of all items represented in the table, so S has a total of 104.3 million outcomes.

a. Because an item is being selected at random, all the outcomes are equally likely. If A is the event that the selected item is an iPod, then

$$P(A) = \frac{n(A)}{n(S)} = \frac{23.1}{104.3} \approx .221.$$

The event A is represented by the pink shaded region in the table:

	iPods (A)	iPhones (B)	iPads (C)	Total
2011 Q2 (U)	9.0	18.7	4.7	32.4
2011 Q3 (V)	7.5	20.3	9.3	37.1
2011 Q4 (W)	6.6	17.1	11.1	34.8
Total	23.1	56.1	25.1	104.3

b. If V is the event that the selected item was sold in the third quarter of 2011, then

$$P(V) = \frac{n(V)}{n(S)} = \frac{37.1}{104.3} \approx .356.$$

[8]Figures are rounded to one decimal place. Source: Apple quarterly press releases (www.investor.apple.com).

In the table, V is represented as shown:

	iPods (A)	iPhones (B)	iPads (C)	Total
2011 Q2 (U)	9.0	18.7	4.7	32.4
2011 Q3 (V)	7.5	20.3	9.3	37.1
2011 Q4 (W)	6.6	17.1	11.1	34.8
Total	23.1	56.1	25.1	104.3

c. The event that the selected item is an iPod sold in the third quarter of 2011 is the event $A \cap V$.

$$P(A \cap V) = \frac{n(A \cap V)}{n(S)} = \frac{7.5}{104.3} \approx .072$$

In the table, $A \cap V$ is represented by the overlap of the regions representing A and V:

	iPods (A)	iPhones (B)	iPads (C)	Total
2011 Q2 (U)	9.0	18.7	4.7	32.4
2011 Q3 (V)	7.5	20.3	9.3	37.1
2011 Q4 (W)	6.6	17.1	11.1	34.8
Total	23.1	56.1	25.1	104.3

d. The event that the selected item either is an iPod or was sold in the third quarter of 2011 is the event $A \cup V$, and is represented by the pink shaded area in the table:

	iPods (A)	iPhones (B)	iPads (C)	Total
2011 Q2 (U)	9.0	18.7	4.7	32.4
2011 Q3 (V)	7.5	20.3	9.3	37.1
2011 Q4 (W)	6.6	17.1	11.1	34.8
Total	23.1	56.1	25.1	104.3

We can compute its probability in two ways:

1. Directly from the table:

$$P(A \cup V) = \frac{n(A \cup V)}{n(S)} = \frac{23.1 + 37.1 - 7.5}{104.3} \approx .505$$

2. Using the addition principle:

$$\begin{aligned} P(A \cup V) &= P(A) + P(V) - P(A \cap V) \\ &\approx .221 + .356 - .072 = .505 \end{aligned}$$

e. The event that the selected item is not an iPod is the event A'. Its probability may be computed using the formula for the probability of the complement:

$$P(A') = 1 - P(A) \approx 1 - .221 = .779.$$

FAQs

Distinguishing Probability from Relative Frequency

Q: *Relative frequency and modeled probability using equally likely outcomes have essentially the same formula: (Number of favorable outcomes)/(Total number of outcomes). How do I know whether a given probability is one or the other?*

A: Ask yourself this: Has the probability been arrived at experimentally, by performing a number of trials and counting the number of times the event occurred? If so, the probability is estimated; that is, relative frequency. If, on the other hand, the probability was computed by analyzing the experiment under consideration rather than by performing actual trials of the experiment, it is a probability model.

Q: *Out of every 100 homes, 22 have broadband Internet service. Thus, the probability that a house has broadband service is .22. Is this probability estimated (relative frequency) or theoretical (a probability model)?*

A: That depends on how the ratio 22 out of 100 was arrived at. If it is based on a poll of *all* homes, then the probability is theoretical. If it is based on a survey of only a *sample* of homes, it is estimated (see the Q/A following Example 1).

7.3 EXERCISES

Access end-of-section exercises online at **www.webassign.net**

7.4 Probability and Counting Techniques

We saw in the preceding section that, when all outcomes in a sample space are equally likely, we can use the following formula to model the probability of each event:

Modeling Probability: Equally Likely Outcomes

In an experiment in which all outcomes are equally likely, the probability of an event E is given by

$$P(E) = \frac{\text{Number of favorable outcomes}}{\text{Total number of outcomes}} = \frac{n(E)}{n(S)}.$$

This formula is simple, but calculating $n(E)$ and $n(S)$ may not be. In this section, we look at some examples in which we need to use the counting techniques discussed in Chapter 6.

EXAMPLE 1 Marbles

A bag contains four red marbles and two green ones. Upon seeing the bag, Suzan (who has compulsive marble-grabbing tendencies) sticks her hand in and grabs three at random. Find the probability that she will get both green marbles.

S is the set of *all* outcomes that can occur, and has nothing to do with having green marbles.

Solution According to the formula, we need to know these numbers:

- The number of elements in the sample space S.
- The number of elements in the event E.

First of all, what is the sample space? The sample space is the set of all possible outcomes, and each outcome consists of a set of three marbles (in Suzan's hand). So, the set of outcomes is the set of all sets of three marbles chosen from a total of six marbles (four red and two green). Thus,

$$n(S) = C(6, 3) = 20.$$

Now what about E? This is the event that Suzan gets both green marbles. We must *rephrase this as a subset of S* in order to deal with it: "E is the collection of sets of three marbles such that one is red and two are green." Thus, $n(E)$ is the *number* of such sets, which we determine using a decision algorithm.

Step 1 Choose a red marble: $C(4, 1) = 4$ possible outcomes.

Step 2 Choose the two green marbles: $C(2, 2) = 1$ possible outcome.

We get $n(E) = 4 \times 1 = 4$. Now,

$$P(E) = \frac{n(E)}{n(S)} = \frac{4}{20} = \frac{1}{5}.$$

Thus, there is a one in five chance of Suzan's getting both the green marbles.

EXAMPLE 2 Investment Lottery

After a down day on the stock market, you decide to ignore your broker's cautious advice and purchase three stocks at random from the six most active stocks listed on the New York Stock Exchange at the end of the day's trading.[9]

Company	*Symbol*	*Price*	*Change*
Bank of America	BAC	$5.80	−$1.00
PowerShares QQQ	QQQ	$55.83	−$1.34
Och-Ziff Capital Mgmt	OZM	$7.98	−$0.43
UBS AG Common Stock	UBS	$11.21	−$0.30
iShares Silver Trust	SLV	$30.64	−$2.18
Direxion Small Cap Bull	TNA	$42.80	−$1.92

Find the probabilities of the following events:

a. You purchase BAC and QQQ.

b. At most two of the stocks you purchase declined in value by more than $1.

[9]Most active stocks on November 17, 2011. Source: Yahoo! Finance (http://finance.yahoo.com).

Solution First, the sample space is the set of all collections of 3 stocks chosen from the 6. Thus,

$$n(S) = C(6, 3) = 20.$$

a. The event E of interest is the event that you purchase BAC and QQQ. Thus, E is the set of all groups of 3 stocks that include BAC and QQQ. Because there is only one more stock left to choose,

$$n(E) = C(4, 1) = 4.$$

We now have

$$P(E) = \frac{n(E)}{n(S)} = \frac{4}{20} = \frac{1}{5} = .2.$$

b. Let F be the event that at most two of the stocks you purchase declined in value by more than \$1. Thus, F is the set of all groups of three stocks of which at most two declined in value by more than \$1. To calculate $n(F)$, we use the following decision algorithm.

Alternative 1: None of the stocks declined in value by more than \$1.

Step 1 Choose three stocks that did not decline in value by more than \$1: $C(3, 3) = 1$ possibility.

Alternative 2: One of the stocks declined in value by more than \$1.

Step 1 Choose one stock that declined in value by more than \$1: $C(3, 1) = 3$ possibilities.

Step 2 Choose two stocks that did not decline in value by more than \$1: $C(3, 2) = 3$ possibilities.

This gives $3 \times 3 = 9$ possibilities for this alternative.

Alternative 3: Two of the stocks declined in value by more than \$1.

Step 1 Choose two stocks that declined in value by more than \$1: $C(3, 2) = 3$ possibilities.

Step 2 Choose one stock that did not decline in value by more than \$1: $C(3, 1) = 3$ possibilities.

This gives $3 \times 3 = 9$ possibilities for this alternative.

So, we have a total of $1 + 9 + 9 = 19$ possible outcomes. Thus,

$$n(F) = 19$$

and

$$P(F) = \frac{n(F)}{n(S)} = \frac{19}{20} = .95.$$

Before we go on... When counting the number of outcomes in an event, the calculation is sometimes easier if we look at the *complement* of that event. In the case of part (b) of Example 2, the complement of the event F is

> F': At least three of the stocks you purchase declined in value by more than \$1.

At most two of the stocks declined in value by more than \$1.

↕ *Complementary Events*

At least three of the stocks declined in value by more than \$1.

Because there are only three stocks in your portfolio, this is the same as the event that all three you purchase declined in value by more than \$1. The decision algorithm for $n(F')$ is far simpler:

Step 1 Choose three stocks that declined in value by more than \$1: $C(3, 3) = 1$ possibility.

So, $n(F') = 1$, giving

$$n(F) = n(S) - n(F') = 20 - 1 = 19$$

as we calculated above. ■

EXAMPLE 3 Poker Hands

You are dealt 5 cards from a well-shuffled standard deck of 52. Find the probability that you have a full house. (Recall that a full house consists of 3 cards of one denomination and 2 of another.)

Solution The sample space S is the set of all possible 5-card hands dealt from a deck of 52. Thus,

$$n(S) = C(52, 5) = 2{,}598{,}960.$$

If the deck is thoroughly shuffled, then each of these 5-card hands is equally likely. Now consider the event E, the set of all possible 5-card hands that constitute a full house. To calculate $n(E)$, we use a decision algorithm, which we show in the following compact form.

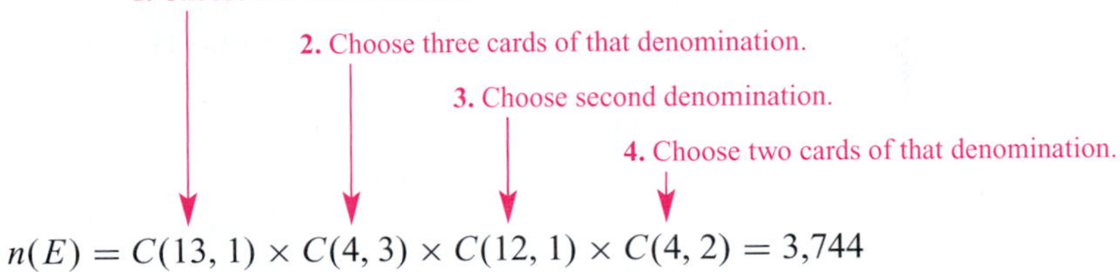

$$n(E) = C(13, 1) \times C(4, 3) \times C(12, 1) \times C(4, 2) = 3{,}744$$

Thus,

$$P(E) = \frac{n(E)}{n(S)} = \frac{3{,}744}{2{,}598{,}960} \approx .00144.$$

In other words, there is an approximately 0.144% chance that you will be dealt a full house.

EXAMPLE 4 More Poker Hands

You are playing poker, and you have been dealt the following hand:

J♠, J♦, J♥, 2♣, 10♠.

You decide to exchange the last two cards. The exchange works as follows: The two cards are discarded (not replaced in the deck), and you are dealt two new cards.

a. Find the probability that you end up with a full house.

b. Find the probability that you end up with four jacks.

c. What is the probability that you end up with either a full house or four jacks?

Solution

a. In order to get a full house, you must be dealt two of a kind. The sample space S is the set of all pairs of cards selected from what remains of the original deck of 52. You were dealt 5 cards originally, so there are $52 - 5 = 47$ cards left in the deck. Thus, $n(S) = C(47, 2) = 1{,}081$. The event E is the set of all pairs of cards that constitute two of a kind. Note that you cannot get two jacks because only one is left in the deck. Also, only three 2s and three 10s are left in the deck. We have

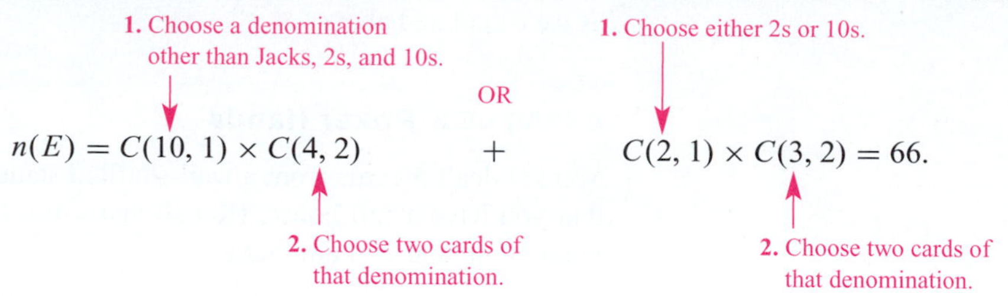

$$n(E) = C(10, 1) \times C(4, 2) + C(2, 1) \times C(3, 2) = 66.$$

Thus,

$$P(E) = \frac{n(E)}{n(S)} = \frac{66}{1{,}081} \approx .0611.$$

b. We have the same sample space as in part (a). Let F be the set of all pairs of cards that include the missing jack of clubs. So,

1. Choose the jack of clubs.
2. Choose 1 card from the remaining 46.

$$n(F) = C(1, 1) \times C(46, 1) = 46.$$

Thus,

$$P(F) = \frac{n(F)}{n(S)} = \frac{46}{1{,}081} \approx .0426.$$

c. We are asked to calculate the probability of the event $E \cup F$. From the addition principle, we have

$$P(E \cup F) = P(E) + P(F) - P(E \cap F).$$

Because $E \cap F$ means "E and F," $E \cap F$ is the event that the pair of cards you are dealt are two of a kind and include the jack of clubs. But this is impossible because only one jack is left. Thus $E \cap F = \emptyset$, and so $P(E \cap F) = 0$. This gives us

$$P(E \cup F) = P(E) + P(F) \approx .0611 + .0426 = .1037.$$

In other words, there is slightly better than a one in ten chance that you will wind up with either a full house or four of a kind, given the original hand.

Before we go on... A more accurate answer to part (c) of Example 4 is $(66 + 46)/1{,}081 \approx .1036$; we lost some accuracy in rounding the answers to parts (a) and (b). ■

EXAMPLE 5 Committees

The University Senate bylaws at Hofstra University state the following:[10]

> The Student Affairs Committee shall consist of one elected faculty senator, one faculty senator-at-large, one elected student senator, five student senators-at-large (including one from the graduate school), two delegates from the Student Government Association, the President of the Student Government Association or his/her designate, and the President of the Graduate Student Organization. It shall be chaired by the elected student senator on the Committee and it shall be advised by the Dean of Students or his/her designate.

[10] As of 2011. Source: Hofstra University Senate Bylaws.

You are an undergraduate student and, even though you are not an elected student senator, you would very much like to serve on the Student Affairs Committee. The senators-at-large as well as the Student Government delegates are chosen by means of a random drawing from a list of candidates. There are already 13 undergraduate candidates for the position of senator-at-large, and 6 candidates for Student Government delegates, and you have been offered a position on the Student Government Association by the President (who happens to be a good friend of yours), should you wish to join it. (This would make you ineligible for a senator-at-large position.) What should you do?

Solution You have two options. Option 1 is to include your name on the list of candidates for the senator-at-large position. Option 2 is to join the Student Government Association (SGA) and add your name to its list of candidates. Let us look at the two options separately.

Option 1: Add your name to the senator-at-large list.
This will result in a list of 14 undergraduates for 4 undergraduate positions. The sample space is the set of all possible outcomes of the random drawing. Each outcome consists of a set of 4 lucky students chosen from 14. Thus,

$$n(S) = C(14, 4) = 1{,}001.$$

We are interested in the probability that you are among the chosen four. Thus, E is the set of sets of four that include you.

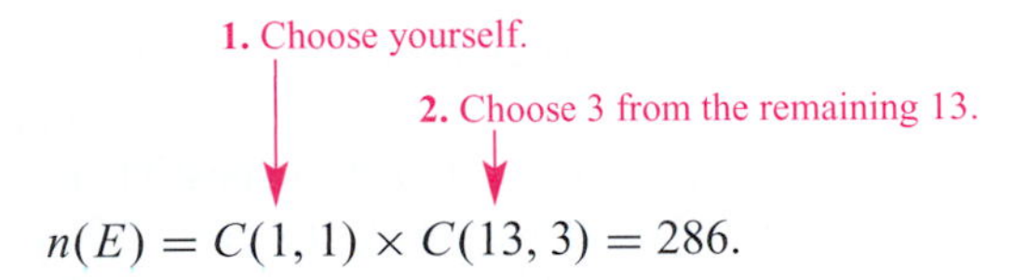

$$n(E) = C(1, 1) \times C(13, 3) = 286.$$

So,

$$P(E) = \frac{n(E)}{n(S)} = \frac{286}{1{,}001} = \frac{2}{7} \approx .2857.$$

Option 2: Join the SGA and add your name to its list.
This results in a list of seven candidates from which two are selected. For this case, the sample space consists of all sets of two chosen from seven, so

$$n(S) = C(7, 2) = 21$$

and

1. Choose yourself.
2. Choose one from the remaining six.

$$n(E) = C(1, 1) \times C(6, 1) = 6.$$

Thus,

$$P(E) = \frac{n(E)}{n(S)} = \frac{6}{21} = \frac{2}{7} \approx .2857.$$

In other words, the probability of being selected is exactly the same for Option 1 as it is for Option 2! Thus, you can choose either option, and you will have slightly less than a 29% chance of being selected.

7.4 EXERCISES

Access end-of-section exercises online at **www.webassign.net** 

7.5 Conditional Probability and Independence

Cyber Video Games, Inc., ran a television ad in advance of the release of its latest game, "Ultimate Hockey." As Cyber Video's director of marketing, you would like to assess the ad's effectiveness, so you ask your market research team to survey video game players. The results of its survey of 2,000 video game players are summarized in the following table:

	Saw Ad	Did Not See Ad	Total
Purchased Game	100	200	300
Did Not Purchase Game	200	1,500	1,700
Total	300	1,700	2,000

The market research team concludes in its report that the ad is highly persuasive, and recommends using the company that produced the ad for future projects.

But wait, how could the ad possibly have been persuasive? Only 100 people who saw the ad purchased the game, while 200 people purchased the game without seeing the ad at all! At first glance, it looks as though potential customers are being *put off* by the ad. But let us analyze the figures a little more carefully.

First, let us restrict attention to those players who saw the ad (first column of data: "Saw Ad") and compute the estimated probability that a player *who saw the ad* purchased Ultimate Hockey.

	Saw Ad
Purchased Game	100
Did Not Purchase Game	200
Total	300

To compute this probability, we calculate

Probability that someone who saw the ad purchased the game

$$= \frac{\text{Number of people who saw the ad and bought the game}}{\text{Total number of people who saw the ad}} = \frac{100}{300} \approx .33.$$

In other words, 33% of game players who saw the ad went ahead and purchased the game. Let us compare this with the corresponding probability for those players who did *not* see the ad (second column of data "Did Not See Ad"):

	Did Not See Ad
Purchased Game	200
Did Not Purchase Game	1,500
Total	1,700

Probability that someone who did not see the ad purchased the game

$$= \frac{\text{Number of people who did not see the ad and bought the game}}{\text{Total number of people who did not see the ad}} = \frac{200}{1{,}700} \approx .12.$$

In other words, only 12% of game players who did not see the ad purchased the game, whereas 33% of those who *did* see the ad purchased the game. Thus, it appears that the ad *was* highly persuasive.

Here's some terminology. In this example there were two related events of importance:

A: A video game player purchased Ultimate Hockey.

B: A video game player saw the ad.

The first probability we computed was the estimated probability that a video game player purchased Ultimate Hockey *given that* he or she saw the ad. We call the latter probability the (estimated) **probability of A, given B**, and we write it as $P(A \mid B)$. We call $P(A \mid B)$ a **conditional probability**—it is the probability of A under the condition that B occurred. Put another way, it is the probability of A occurring if the sample space is reduced to just those outcomes in B.

$$P(\text{Purchased game } \textit{given that} \text{ saw the ad}) = P(A \mid B) \approx .33$$

The second probability we computed was the estimated probability that a video game player purchased Ultimate Hockey *given that* he or she did not see the ad, or the **probability of A, given B'**.

$$P(\text{Purchased game } \textit{given that} \text{ did not see the ad}) = P(A \mid B') \approx .12$$

Calculating Conditional Probabilities

How do we calculate conditional probabilities? In the example above we used the ratio

$$P(A \mid B) = \frac{\text{Number of people who saw the ad and bought the game}}{\text{Total number of people who saw the ad}}.$$

The numerator is the frequency of $A \cap B$, and the denominator is the frequency of B:

$$P(A \mid B) = \frac{fr(A \cap B)}{fr(B)}.$$

Now, we can write this formula in another way:

$$P(A \mid B) = \frac{fr(A \cap B)}{fr(B)} = \frac{fr(A \cap B)/N}{fr(B)/N} = \frac{P(A \cap B)}{P(B)}.$$

We therefore have the following definition, which applies to general probability distributions.

Conditional Probability

If A and B are events with $P(B) \neq 0$, then the probability of A given B is

$$P(A \mid B) = \frac{P(A \cap B)}{P(B)}.$$

Visualizing Conditional Probability

In the figure, $P(A \mid B)$ is represented by the fraction of B that is covered by A.

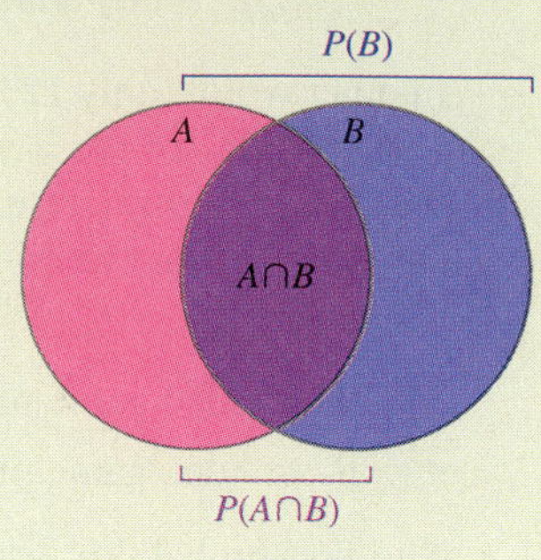

$$P(A|B) = \frac{P(A \cap B)}{P(B)}$$

Quick Examples

1. If there is a 50% chance of rain (R) and a 10% chance of both rain and lightning (L), then the probability of lightning, given that it rains, is

$$P(L \mid R) = \frac{P(L \cap R)}{P(R)} = \frac{.10}{.50} = .20.$$

Here are two more ways to express the result:

- If it rains, the probability of lightning is .20.
- Assuming that it rains, there is a 20% chance of lightning.

2. Referring to the Cyber Video data at the beginning of this section, the probability that a video game player did not purchase the game (A'), given that she did not see the ad (B'), is

$$P(A' \mid B') = \frac{P(A' \cap B')}{P(B')} = \frac{1{,}500/2{,}000}{1{,}700/2{,}000} = \frac{15}{17} \approx .88.$$

Q: *Returning to the video game sales survey, how do we compute the* ordinary probability *of A, not "given" anything?*

A: We look at the event A that a randomly chosen game player purchased Ultimate Hockey *regardless of whether or not he or she saw the ad.* In the "Purchased Game" row we see that a total of 300 people purchased the game out of a total of 2,000 surveyed. Thus, the (estimated) probability of A is

$$P(A) = \frac{fr(A)}{N} = \frac{300}{2{,}000} = .15.$$

We sometimes refer to $P(A)$ as the **unconditional** probability of A to distinguish it from conditional probabilities like $P(A|B)$ and $P(A|B')$.

Now, let's see some more examples involving conditional probabilities.

EXAMPLE 1 Dice

If you roll a fair die twice and observe the numbers that face up, find the probability that the sum of the numbers is 8, given that the first number is 3.

Solution We begin by recalling that the sample space when we roll a fair die twice is the set $S = \{(1, 1), (1, 2), \ldots, (6, 6)\}$ containing the 36 different equally likely outcomes.

The two events under consideration are

A: The sum of the numbers is 8.

B: The first number is 3.

We also need

$A \cap B$: The sum of the numbers is 8 and the first number is 3.

But this can only happen in one way: $A \cap B = \{(3, 5)\}$. From the formula, then,

$$P(A \mid B) = \frac{P(A \cap B)}{P(B)} = \frac{1/36}{6/36} = \frac{1}{6}.$$

➡ **Before we go on...** There is another way to think about Example 1. When we say that the first number is 3, we are restricting the sample space to the six outcomes (3, 1), (3, 2), . . . , (3, 6), all still equally likely. Of these six, only one has a sum of 8, so the probability of the sum being 8, given that the first number is 3, is 1/6. ■

Notes

1. Remember that, in the expression $P(A \mid B)$, A is the event whose probability you want, given that you know the event B has occurred.
2. From the formula, notice that $P(A \mid B)$ is not defined if $P(B) = 0$. Could $P(A \mid B)$ make any sense if the event B were impossible? ■

EXAMPLE 2 School and Work

A survey[11] of the high school graduating class of 2010, conducted by the Bureau of Labor Statistics, found that, if a graduate went on to college, there was a 40% chance that he or she would work at the same time. On the other hand, there was a 68% chance that a randomly selected graduate would go on to college. What is the probability that a graduate went to college and work at the same time?

Solution To understand what the question asks and what information is given, it is helpful to rephrase everything using the standard wording "*the probability that* ___ " and "*the probability that* ___ *given that* ___." Now we have, "The probability that a

[11]Source: "College Enrollment and Work Activity of High School Graduates," U.S. Bureau of Labor Statistics (www.bls.gov/news.release/hsgec.htm).

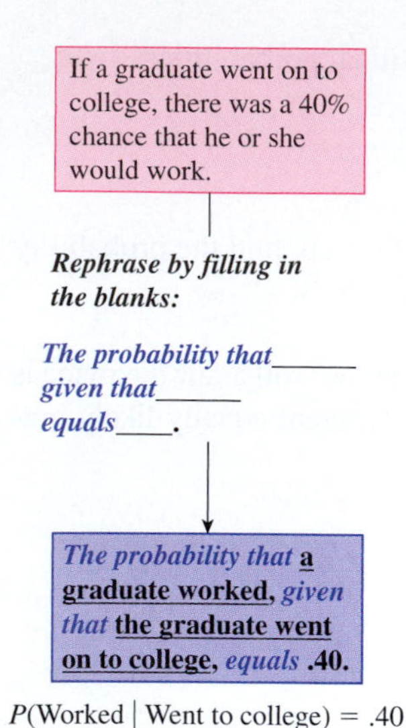

P(Worked | Went to college) = .40

Figure 8

graduate worked, given that the graduate went on to college, equals .40. (See Figure 8.) The probability that a graduate went on to college is .68." The events in question are as follows:

W: A high school graduate went on to work.

C: A high school graduate went on to college.

From our rephrasing of the question we can write:

$$P(W \mid C) = .40. \qquad P(C) = .68. \qquad \text{Find } P(W \cap C).$$

The definition

$$P(W \mid C) = \frac{P(W \cap C)}{P(C)}$$

can be used to find $P(W \cap C)$:

$$\begin{aligned} P(W \cap C) &= P(W \mid C)P(C) \\ &= (.40)(.68) \approx .27. \end{aligned}$$

Thus there is a 27% chance that a member of the high school graduating class of 2010 went on to college and work at the same time.

The Multiplication Principle and Trees

In Example 2, we saw that the formula

$$P(A \mid B) = \frac{P(A \cap B)}{P(B)}$$

can be used to calculate $P(A \cap B)$ if we rewrite the formula in the following form, known as the **multiplication principle for conditional probability**:

Multiplication Principle for Conditional Probability

If A and B are events, then

$$P(A \cap B) = P(A \mid B)P(B).$$

Quick Example

If there is a 50% chance of rain (R) and a 20% chance of a lightning (L) if it rains, then the probability of both rain and lightning is

$$P(R \cap L) = P(L \mid R)P(R) = (.20)(.50) = .10.$$

The multiplication principle is often used in conjunction with **tree diagrams**. Let's return to Cyber Video Games, Inc., and its television ad campaign. Its marketing survey was concerned with the following events:

A: A video game player purchased Ultimate Hockey.

B: A video game player saw the ad.

We can illustrate the various possibilities by means of the two-stage "tree" shown in Figure 9.

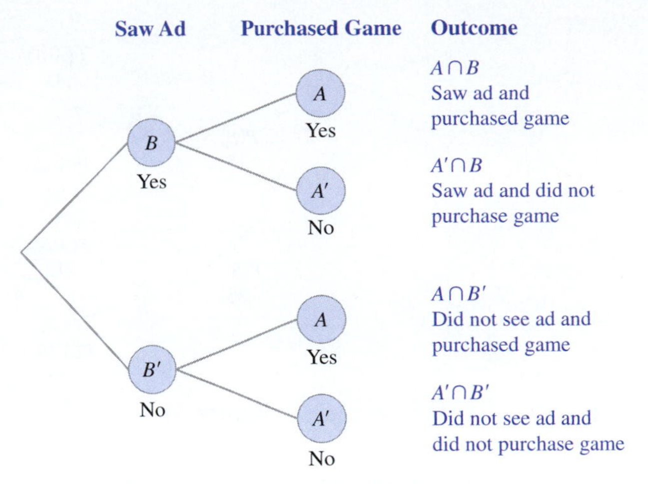

Figure 9

Consider the outcome $A \cap B$. To get there from the starting position on the left, we must first travel up to the B node. (In other words, B must occur.) Then we must travel up the branch from the B node to the A node. We are now going to associate a probability with each branch of the tree: the probability of traveling along that branch *given that we have gotten to its beginning node.* For instance, the probability of traveling up the branch from the starting position to the B node is $P(B) = 300/2{,}000 = .15$ (see the data in the survey). The probability of going up the branch from the B node to the A node is the probability that A occurs, given that B has occurred. In other words, it is the *conditional* probability $P(A \mid B) \approx .33$. (We calculated this probability at the beginning of the section.) The probability of the outcome $A \cap B$ can then be computed using the multiplication principle:

$$P(A \cap B) = P(B)P(A \mid B) \approx (.15)(.33) \approx .05.$$

In other words, *to obtain the probability of the outcome $A \cap B$, we multiply the probabilities on the branches leading to that outcome* (Figure 10).

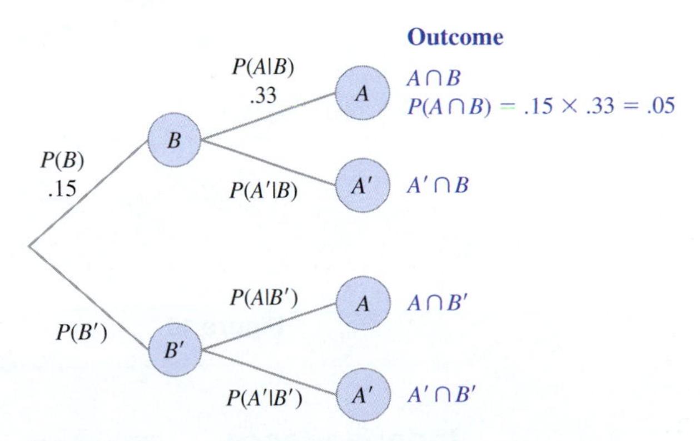

Figure 10

The same argument holds for the remaining three outcomes, and we can use the table given at the beginning of this section to calculate all the conditional probabilities shown in Figure 11.

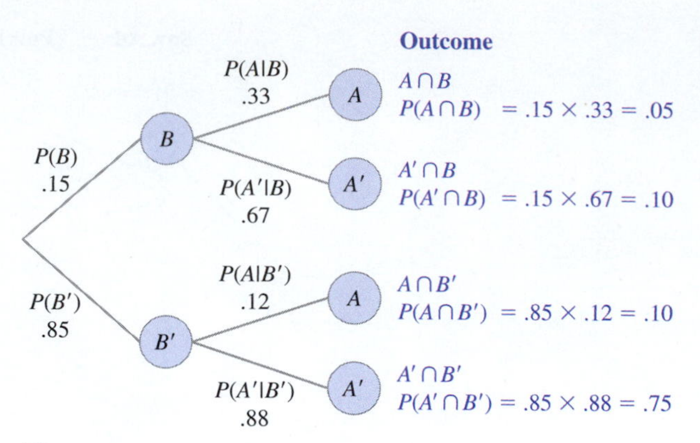

Figure 11

Note The sum of the probabilities on the branches leaving any node is always 1 (why?). This observation often speeds things up because after we have labeled one branch (or, all but one, if a node has more than two branches leaving it), we can easily label the remaining one. ■

EXAMPLE 3 Unfair Coins

An experiment consists of tossing two coins. The first coin is fair, while the second coin is twice as likely to land with heads facing up as it is with tails facing up. Draw a tree diagram to illustrate all the possible outcomes, and use the multiplication principle to compute the probabilities of all the outcomes.

Solution A quick calculation shows that the probability distribution for the second coin is $P(\text{H}) = 2/3$ and $P(\text{T}) = 1/3$. (How did we get that?) Figure 12 shows the tree diagram and the calculations of the probabilities of the outcomes.

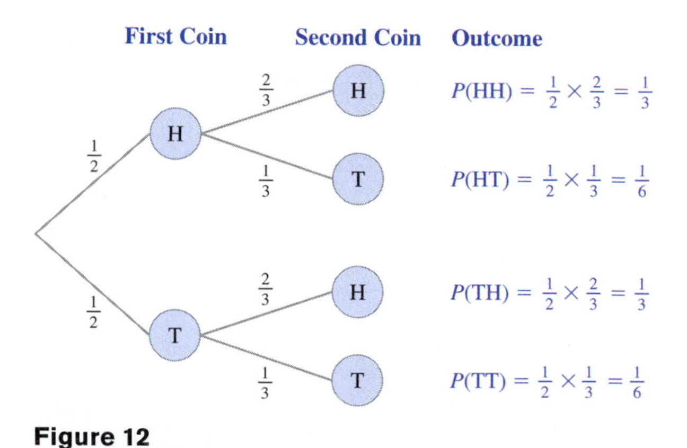

Figure 12

Independence

Let us go back once again to Cyber Video Games, Inc., and its ad campaign. How did we assess the ad's effectiveness? We considered the following events.

A: A video game player purchased Ultimate Hockey.

B: A video game player saw the ad.

We used the survey data to calculate $P(A)$, the probability that a video game player purchased Ultimate Hockey, and $P(A \mid B)$, the probability that a video game player *who saw the ad* purchased Ultimate Hockey. When these probabilities are compared, one of three things can happen.

Case 1 $P(A \mid B) > P(A)$
This is what the survey data actually showed: A video game player was more likely to purchase Ultimate Hockey if he or she saw the ad. This indicates that the ad is effective; seeing the ad had a positive effect on a player's decision to purchase the game.

Case 2 $P(A \mid B) < P(A)$
If this had happened, then a video game player would have been *less* likely to purchase Ultimate Hockey if he or she saw the ad. This would have indicated that the ad had "backfired"; it had, for some reason, put potential customers off. In this case, just as in the first case, the event B would have had an effect—a negative one—on the event A.

Case 3 $P(A \mid B) = P(A)$
In this case seeing the ad would have had absolutely no effect on a potential customer's buying Ultimate Hockey. Put another way, the probability of A occurring *does not depend* on whether B occurred or not. We say in a case like this that the events A and B are **independent**.

In general, we say that two events A and B are independent if $P(A \mid B) = P(A)$. When this happens, we have

$$P(A) = P(A \mid B) = \frac{P(A \cap B)}{P(B)}$$

so

$$P(A \cap B) = P(A)P(B).$$

* We shall only discuss the independence of two events in cases where their probabilities are both nonzero.

Conversely, if $P(A \cap B) = P(A)P(B)$, then, assuming $P(B) \neq 0$,* $P(A) = P(A \cap B)/P(B) = P(A \mid B)$. Thus, saying that $P(A) = P(A \mid B)$ is the same as saying that $P(A \cap B) = P(A)P(B)$. Also, we can switch A and B in this last formula and conclude that saying that $P(A \cap B) = P(A)P(B)$ is the same as saying that $P(B \mid A) = P(B)$.

Independent Events

The events A and B are **independent** if

$$P(A \cap B) = P(A)P(B).$$

Equivalent formulas (assuming neither A nor B is impossible) are

$$P(A \mid B) = P(A)$$

and $$P(B \mid A) = P(B).$$

If two events A and B are not independent, then they are **dependent**.

The property $P(A \cap B) = P(A)P(B)$ can be extended to three or more independent events. If, for example, A, B, and C are three mutually independent events (that is, each one of them is independent of each of the other two and of their intersection), then, among other things,

$$P(A \cap B \cap C) = P(A)P(B)P(C).$$

Quick Examples

1. If A and B are independent, and if A has a probability of .2 and B has a probability of .3, then $A \cap B$ has a probability of $(.2)(.3) = .06$.
2. Let us assume that the phase of the moon has no effect on whether or not my newspaper is delivered. The probability of a full moon (M) on a randomly selected day is about .034, and the probability that my newspaper will be delivered (D) on the random day is .20. Therefore, the probability that it is a full moon and my paper is delivered is

$$P(M \cap D) = P(M)P(D) = (.034)(.20) = .0068.$$

Testing for Independence

To check whether two events A and B are independent, we compute $P(A)$, $P(B)$, and $P(A \cap B)$. If $P(A \cap B) = P(A)P(B)$, the events are independent; otherwise, they are dependent. Sometimes it is obvious that two events, by their nature, are independent, so a test is not necessary. For example, the event that a die you roll comes up 1 is clearly independent of whether or not a coin you toss comes up heads.

Quick Examples

1. Roll two distinguishable dice (one red, one green) and observe the numbers that face up.

A: The red die is even; $P(A) = \frac{18}{36} = \frac{1}{2}$.

B: The dice have the same parity*; $P(B) = \frac{18}{36} = \frac{1}{2}$.

$A \cap B$: Both dice are even; $P(A \cap B) = \frac{9}{36} = \frac{1}{4}$.

$P(A \cap B) = P(A)P(B)$, and so A and B are independent.

2. Roll two distinguishable dice and observe the numbers that face up.

A: The sum of the numbers is 6; $P(A) = \frac{5}{36}$.

B: Both numbers are odd; $P(B) = \frac{9}{36} = \frac{1}{4}$.

$A \cap B$: The sum is 6, and both are odd; $P(A \cap B) = \frac{3}{36} = \frac{1}{12}$.

$P(A \cap B) \neq P(A)P(B)$, and so A and B are dependent.

* Two numbers have the **same parity** if both are even or both are odd. Otherwise, they have **opposite parity**.

EXAMPLE 4 Weather Prediction

According to the weather service, there is a 50% chance of rain in New York and a 30% chance of rain in Honolulu. Assuming that New York's weather is independent of Honolulu's, find the probability that it will rain in at least one of these cities.

Solution We take A to be the event that it will rain in New York and B to be the event that it will rain in Honolulu. We are asked to find the probability of $A \cup B$, the event that it will rain in at least one of the two cities. We use the addition principle:

$$P(A \cup B) = P(A) + P(B) - P(A \cap B).$$

We know that $P(A) = .5$ and $P(B) = .3$. But what about $P(A \cap B)$? Because the events A and B are independent, we can compute

$$\begin{aligned} P(A \cap B) &= P(A)P(B) \\ &= (.5)(.3) = .15. \end{aligned}$$

Thus,

$$\begin{aligned} P(A \cup B) &= P(A) + P(B) - P(A \cap B) \\ &= .5 + .3 - .15 \\ &= .65. \end{aligned}$$

So, there is a 65% chance that it will rain either in New York or in Honolulu (or in both).

EXAMPLE 5 Roulette

You are playing roulette and have decided to leave all 10 of your $1 chips on black for five consecutive rounds, hoping for a sequence of five blacks which, according to the rules, will leave you with $320. There is a 50% chance of black coming up on each spin, ignoring the complicating factor of zero or double zero. What is the probability that you will be successful?

Solution Because the roulette wheel has no memory, each spin is independent of the others. Thus, if A_1 is the event that black comes up the first time, A_2 the event that it comes up the second time, and so on, then

$$P(A_1 \cap A_2 \cap A_3 \cap A_4 \cap A_5) = P(A_1)P(A_2)P(A_3)P(A_4)P(A_5) = \left(\frac{1}{2}\right)^5 = \frac{1}{32}.$$

The next example is a version of a well known "brain teaser" that forces one to think carefully about conditional probability.

EXAMPLE 6 Legal Argument

A man was arrested for attempting to smuggle a bomb on board an airplane. During the subsequent trial, his lawyer claimed that, by means of a simple argument, she would prove beyond a shadow of a doubt that her client was not only innocent of any crime, but was in fact contributing to the safety of the other passengers on the flight. This was her eloquent argument: "Your Honor, first of all, my client had absolutely no intention of setting off the bomb. As the record clearly shows, the detonator was unarmed when he was apprehended. In addition—and your Honor is certainly aware of this—there is a small but definite possibility that there will be a bomb on any given flight. On the other hand, the chances of there being *two* bombs on a flight are so remote as to be negligible. There is in fact no record of this having *ever* occurred. Thus, because my client had already brought one bomb on board (with no intention of setting it off) and because we have seen that the chances of there being a second bomb on board were vanishingly remote, it follows that the flight was far safer as a result of his action! I rest my case." This argument was so elegant in its simplicity that the judge acquitted the defendant. Where is the flaw in the argument? (Think about this for a while before reading the solution.)

Solution The lawyer has cleverly confused the phrases "two bombs on board" and "a second bomb on board." To pinpoint the flaw, let us take B to be the event that there is one bomb on board a given flight, and let A be the event that there are two independent bombs on board. Let us assume for argument's sake that $P(B) = 1/1{,}000{,}000 = .000\,001$. Then the probability of the event A is

$$(.000\,001)(.000\,001) = .000\,000\,000\,001.$$

This *is* vanishingly small, as the lawyer contended. It was at this point that the lawyer used a clever maneuver: She assumed in concluding her argument that the probability of having two bombs on board was the same as the probability of having a *second* bomb on board. But to say that there is a *second* bomb on board is to imply that there already is one bomb on board. This is therefore a *conditional* event: the event that there are two bombs on board, *given that there is already one bomb on board.* Thus, the probability that there is a second bomb on board is the probability that there are two bombs on board, given that there is already one bomb on board, which is

$$P(A \mid B) = \frac{P(A \cap B)}{P(B)} = \frac{.000\,000\,000\,001}{.000\,001} = .000\,001.$$

In other words, it is the same as the probability of there being a single bomb on board to begin with! Thus the man's carrying the bomb onto the plane did not improve the flight's safety at all.*

* If we want to be picky, there was a *slight* decrease in the probability of a second bomb because there was one less seat for a potential second bomb bearer to occupy. In terms of our analysis, this is saying that the event of one passenger with a bomb and the event of a second passenger with a bomb are not completely independent.

FAQ

Probability of what given what?

Q: *How do I tell if a statement in an application is talking about conditional probability or unconditional probability? And if it is talking about conditional probability, how do I determine what to use as A and B in P(A|B)?*

A: Look carefully at the wording of the statement. If there is some kind of qualification or restriction to a smaller set than the entire sample space, then it is probably talking about conditional probability, as in the following examples:

60% of veterans vote Republican while 40% of the entire voting population vote Republican.
Here the sample space can be taken to be the entire voting population.
Reworded (see Example 2): *The probability of voting Republican (R) is 60% given that the person is a veteran (V); P(R|V) = .60, whereas the probability of voting Republican is .40: P(R) = .40.*

The likelihood of being injured if in an accident is 80% for a driver not wearing a seatbelt but it is 50% for all drivers.
Here, the sample space can be taken to be the set of drivers involved in an accident—these are the only drivers discussed.
Reworded: *The probability of a driver being injured (I) is .80 given that the driver is not wearing a seatbelt (B); P(I|B) = .80 whereas, for all drivers, the probability of being injured is .50: P(I) = .50.*

7.5 EXERCISES

Access end-of-section exercises online at **www.webassign.net**

7.6 Bayes' Theorem and Applications

Should schools test their athletes for drug use? A problem with drug testing is that there are always false positive results, so one can never be certain that an athlete who tests positive is in fact using drugs. Here is a typical scenario.

EXAMPLE 1 Steroids Testing

Gamma Chemicals advertises its anabolic steroid detection test as being 95% effective at detecting steroid use, meaning that it will show a positive result on 95% of all anabolic steroid users. It also states that its test has a false positive rate of 6%. This means that the probability of a nonuser testing positive is .06. Estimating that about 10% of its athletes are using anabolic steroids, Enormous State University (ESU) begins testing its football players. The quarterback, Hugo V. Huge, tests positive and is promptly dropped from the team. Hugo claims that he is not using anabolic steroids. How confident can we be that he is not telling the truth?

Solution There are two events of interest here: the event T that a person tests positive, and the event A that the person tested uses anabolic steroids. Here are the probabilities we are given:

$$P(T \mid A) = .95$$
$$P(T \mid A') = .06$$
$$P(A) = .10$$

We are asked to find $P(A \mid T)$, the probability that someone who tests positive is using anabolic steroids. We can use a tree diagram to calculate $P(A \mid T)$. The trick to setting up the tree diagram is to use as the first branching the events with *unconditional* probabilities we know. Because the only unconditional probability we are given is $P(A)$, we use A and A' as our first branching (Figure 13).

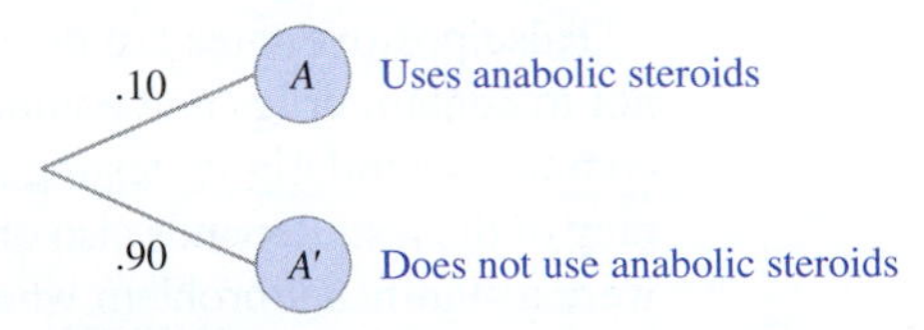

Figure 13

For the second branching, we use the outcomes of the drug test: positive (T) or negative (T'). The probabilities on these branches are conditional probabilities because they depend on whether or not an athlete uses steroids. (See Figure 14.) (We fill in the probabilities that are not supplied by remembering that the sum of the probabilities on the branches leaving any node must be 1.)

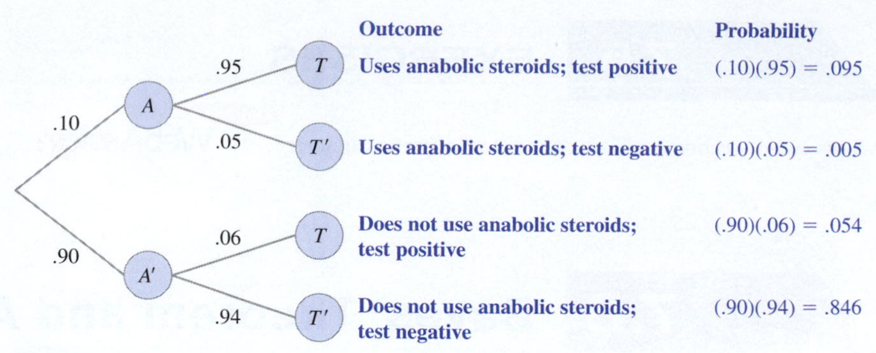

Figure 14

We can now calculate the probability we are asked to find:

$$P(A \mid T) = \frac{P(A \cap T)}{P(T)} = \frac{P(\text{Uses anabolic steroids and tests positive})}{P(\text{Tests positive})}$$

$$= \frac{P(\text{Using } A \text{ and } T \text{ branches})}{\text{Sum of } P(\text{Using branches ending in } T)}.$$

From the tree diagram, we see that $P(A \cap T) = .095$. To calculate $P(T)$, the probability of testing positive, notice that there are two outcomes on the tree diagram that reflect a positive test result. The probabilities of these events are .095 and .054. Because these two events are mutually exclusive (an athlete either uses steroids or does not, but not both), the probability of a test being positive (ignoring whether or not steroids are used) is the sum of these probabilities, .149. Thus,

$$P(A \mid T) = \frac{.095}{.095 + .054} = \frac{.095}{.149} \approx .64.$$

Thus there is a 64% chance that a randomly selected athlete who tests positive, like Hugo, is using steroids. In other words, we can be 64% confident that Hugo is lying.

Before we go on... Note that the correct answer in Example 1 is 64%, *not* the 94% we might suspect from the test's false positive rating. In fact, we can't answer the question asked without knowing the percentage of athletes who actually use steroids. For instance, if *no* athletes at all use steroids, then Hugo must be telling the truth, and so the test result has no significance whatsoever. On the other hand, if *all* athletes use steroids, then Hugo is definitely lying, regardless of the outcome of the test.

False positive rates are determined by testing a large number of samples known not to contain drugs and computing estimated probabilities. False negative rates are computed similarly by testing samples known to contain drugs. However, the accuracy of the tests depends also on the skill of those administering them. False positives were a significant problem when drug testing started to become common, with estimates of false positive rates for common immunoassay tests ranging from 10% to 30% on the high end,* but the accuracy has improved since then. Because of the possibility of false positive results, positive immunoassay tests need to be confirmed by the more expensive and much more reliable gas chromatograph/mass spectrometry (GC/MS) test. See also the National Collegiate Athletic Association's (NCAA) Drug-Testing Program Handbook, available at www.ncaa.org. (The section on Institutional Drug Testing addresses the problem of false positives.) ■

* *Drug Testing in the Workplace,* ACLU Briefing Paper, 1996.

Bayes' Theorem

The calculation we used to answer the question in Example 1 can be recast as a formula known as **Bayes' theorem**. Figure 15 shows a general form of the tree we used in Example 1.

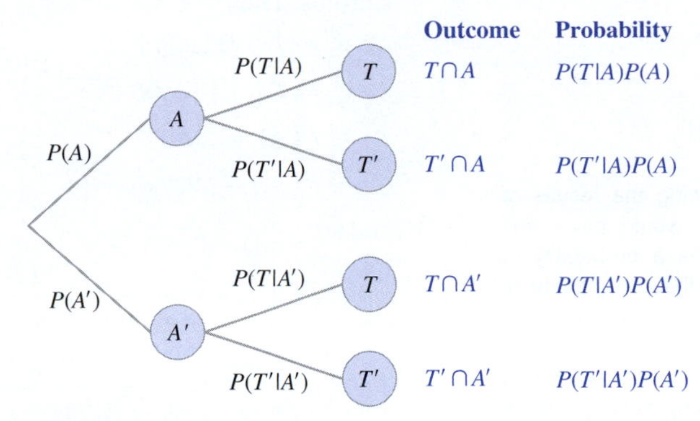

Figure 15

We first calculated

$$P(A \mid T) = \frac{P(A \cap T)}{P(T)}$$

as follows. We first calculated the numerator $P(A \cap T)$ using the multiplication principle:

$$P(A \cap T) = P(T \mid A)P(A).$$

We then calculated the denominator $P(T)$ by using the addition principle for mutually exclusive events together with the multiplication principle:

$$\begin{aligned} P(T) &= P(A \cap T) + P(A' \cap T) \\ &= P(T \mid A)P(A) + P(T \mid A')P(A'). \end{aligned}$$

Substituting gives

$$P(A \mid T) = \frac{P(T \mid A)P(A)}{P(T \mid A)P(A) + P(T \mid A')P(A')}.$$

This is the short form of Bayes' theorem.

Bayes' Theorem (Short Form)

If A and T are events, then

Bayes' Formula

$$P(A \mid T) = \frac{P(T \mid A)P(A)}{P(T \mid A)P(A) + P(T \mid A')P(A')}.$$

Using a Tree

$$P(A \mid T) = \frac{P(\text{Using } A \text{ and } T \text{ branches})}{\text{Sum of } P(\text{Using branches ending in } T)}$$

Quick Example

Let us calculate the probability that an athlete from Example 1 who tests positive is actually using steroids if only 5% of ESU athletes are using steroids. Thus,

$$P(T \mid A) = .95$$
$$P(T \mid A') = .06$$
$$P(A) = .05$$
$$P(A') = .95$$

and so

$$P(A \mid T) = \frac{P(T \mid A)P(A)}{P(T \mid A)P(A) + P(T \mid A')P(A')}$$
$$= \frac{(.95)(.05)}{(.95)(.05) + (.06)(.95)} \approx .45.$$

In other words, it is actually more likely that such an athlete does *not* use steroids than he does.*

* Without knowing the results of the test, we would have said that there was a probability of $P(A) = 0.05$ that the athlete is using steroids. The positive test result raises the probability to $P(A \mid T) = 0.45$, but the test gives too many false positives for us to be any more than 45% certain that the athlete is actually using steroids.

Remembering the Formula

Although the formula looks complicated at first sight, it is not hard to remember if you notice the pattern. Or, you could re-derive it yourself by thinking of the tree diagram.

The next example illustrates that we can use either a tree diagram or the Bayes' theorem formula.

EXAMPLE 2 Lie Detectors

The Sherlock Lie Detector Company manufactures the latest in lie detectors, and the Count-Your-Pennies (CYP) store chain is eager to use them to screen its employees for theft. Sherlock's advertising claims that the test misses a lie only once in every 100 instances. On the other hand, an analysis by a consumer group reveals 20% of people who are telling the truth fail the test anyway.† The local police department estimates that 1 out of every 200 employees has engaged in theft. When the CYP store first screened its employees, the test indicated Mrs. Prudence V. Good was lying when she claimed that she had never stolen from CYP. What is the probability that she was lying and had in fact stolen from the store?

† The reason for this is that many people show physical signs of distress when asked accusatory questions. Many people are nervous around police officers even if they have done nothing wrong.

Solution We are asked for the probability that Mrs. Good was lying, and in the preceding sentence we are told that the lie detector test showed her to be lying. So, we are looking for a conditional probability: the probability that she is lying, given that the lie detector test is positive. Now we can start to give names to the events:

L: A subject is lying.

T: The test is positive (indicated that the subject was lying).

We are looking for $P(L \mid T)$. We know that 1 out of every 200 employees engages in theft; let us assume that no employee admits to theft while taking a lie detector test, so the probability $P(L)$ that a test subject is lying is $1/200$. We also know the false negative and false positive rates $P(T' \mid L)$ and $P(T \mid L')$.

Using a Tree Diagram
Figure 16 shows the tree diagram.

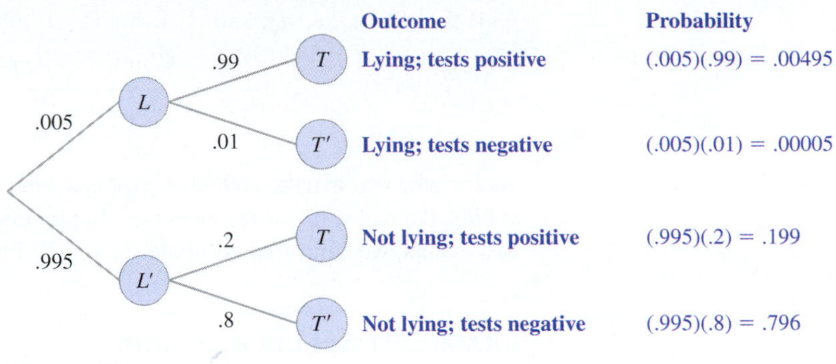

Figure 16

We see that

$$P(L \mid T) = \frac{P(\text{Using } L \text{ and } T \text{ branches})}{\text{Sum of } P(\text{Using branches ending in } T)}$$

$$= \frac{.00495}{.00495 + .199} \approx .024.$$

This means that there was only a 2.4% chance that poor Mrs. Good was lying and had stolen from the store!

Using Bayes' Theorem
We have

$$P(L) = .005$$
$$P(T' \mid L) = .01, \text{ from which we obtain}$$
$$P(T \mid L) = .99$$
$$P(T \mid L') = .2$$

and so

$$P(L \mid T) = \frac{P(T \mid L)P(L)}{P(T \mid L)P(L) + P(T \mid L')P(L')} = \frac{(.99)(.005)}{(.99)(.005) + (.2)(.995)} \approx .024.$$

Expanded Form of Bayes' Theorem

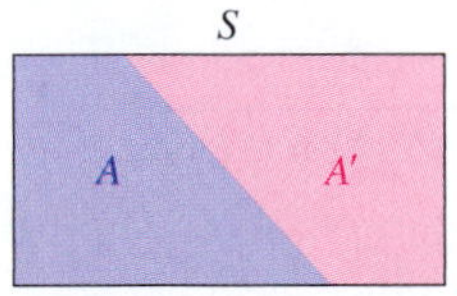

A and A' form a partition of S.

Figure 17

We have seen the "short form" of Bayes' theorem. What is the "long form?" To motivate an expanded form of Bayes' theorem, look again at the formula we've been using:

$$P(A \mid T) = \frac{P(T \mid A)P(A)}{P(T \mid A)P(A) + P(T \mid A')P(A')}.$$

The events A and A' form a **partition** of the sample space S; that is, their union is the whole of S and their intersection is empty (Figure 17).

The expanded form of Bayes' theorem applies to a partition of S into three or more events, as shown in Figure 18.

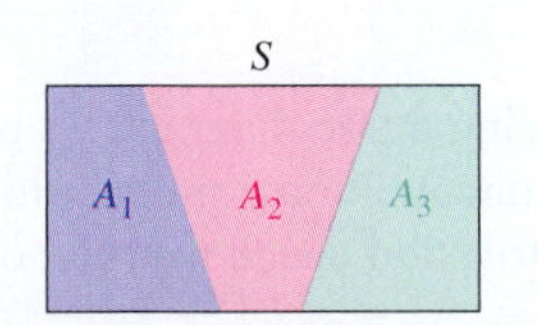

A_1, A_2, and A_3 form a partition of S.

Figure 18

By saying that the events A_1, A_2, and A_3 form a partition of S, we mean that their union is the whole of S and the intersection of any two of them is empty, as in the figure. When we have a partition into three events as shown, the formula gives us $P(A_1 \mid T)$ in terms of $P(T \mid A_1)$, $P(T \mid A_2)$, $P(T \mid A_3)$, $P(A_1)$, $P(A_2)$, and $P(A_3)$.

Bayes' Theorem (Expanded Form)

If the events A_1, A_2, and A_3 form a partition of the sample space S, then

$$P(A_1 \mid T) = \frac{P(T \mid A_1)P(A_1)}{P(T \mid A_1)P(A_1) + P(T \mid A_2)P(A_2) + P(T \mid A_3)P(A_3)}$$

As for why this is true, and what happens when we have a partition into *four or more* events, we will wait for the exercises. In practice, as was the case with a partition into two events, we can often compute $P(A_1 \mid T)$ by constructing a tree diagram.

EXAMPLE 3 School and Work

A survey[12] conducted by the Bureau of Labor Statistics found that approximately 27% of the high school graduating class of 2010 went on to a 2-year college, 41% went on to a 4-year college, and the remaining 32% did not go on to college. Of those who went on to a 2-year college, 52% worked at the same time, 32% of those going on to a 4-year college worked, and 78% of those who did not go on to college worked. What percentage of those working had not gone on to college?

Solution We can interpret these percentages as probabilities if we consider the experiment of choosing a member of the high school graduating class of 2010 at random. The events we are interested in are these:

R_1: A graduate went on to a 2-year college

R_2: A graduate went on to a 4-year college

R_3: A graduate did not go to college

A: A graduate went on to work.

The three events R_1, R_2, and R_3 partition the sample space of all graduates into three events. We are given the following probabilities:

$$P(R_1) = .27 \qquad P(R_2) = .41 \qquad P(R_3) = .32$$
$$P(A \mid R_1) = .52 \quad P(A \mid R_2) = .32 \quad P(A \mid R_3) = .78.$$

We are asked to find the probability that a graduate who went on to work did not go to college, so we are looking for $P(R_3 \mid A)$. Bayes' formula for these events is

$$P(R_3 \mid A) = \frac{P(A \mid R_3)P(R_3)}{P(A \mid R_1)P(R_1) + P(A \mid R_2)P(R_2) + P(A \mid R_3)P(R_3)}$$

$$= \frac{(.78)(.32)}{(.52)(.27) + (.32)(.41) + (.78)(.32)} \approx .48.$$

Thus we conclude that 48% of all those working had not gone on to college.

➡ **Before we go on...** We could also solve Example 3 using a tree diagram. As before, the first branching corresponds to the events with unconditional probabilities that we know: R_1, R_2, and R_3. You should complete the tree and check that you obtain the same result as above. ■

[12]Source: "College Enrollment and Work Activity of High School Graduates," U.S. Bureau of Labor Statistics (www.bls.gov/news.release/hsgec.htm).

7.6 EXERCISES

Access end-of-section exercises online at **www.webassign.net**

7.7 Markov Systems

Many real-life situations can be modeled by processes that pass from state to state with given probabilities. A simple example of such a **Markov system** is the fluctuation of a gambler's fortune as he or she continues to bet. Other examples come from the study of trends in the commercial world and the study of neural networks and artificial intelligence. The mathematics of Markov systems is an interesting combination of probability and matrix arithmetic.

Here is a basic example we shall use many times: A market analyst for Gamble Detergents is interested in whether consumers prefer powdered laundry detergents or liquid detergents. Two market surveys taken one year apart revealed that 20% of powdered detergent users had switched to liquid one year later, while the rest were still using powder. Only 10% of liquid detergent users had switched to powder one year later, with the rest still using liquid.

We analyze this example as follows: Every year a consumer may be in one of two possible **states**: He may be a powdered detergent user or a liquid detergent user. Let us number these states: A consumer is in state 1 if he uses powdered detergent and in state 2 if he uses liquid. There is a basic **time step** of one year. If a consumer happens to be in state 1 during a given year, then there is a probability of $20\% = .2$ (the chance that a randomly chosen powder user will switch to liquid) that he will be in state 2 the next year. We write

$$p_{12} = .2$$

to indicate that the probability of going *from* state 1 *to* state 2 in one time step is .2. The other 80% of the powder users are using powder the next year. We write

$$p_{11} = .8$$

to indicate that the probability of *staying* in state 1 from one year to the next is .8.* What if a consumer is in state 2? Then the probability of going to state 1 is given as $10\% = .1$, so the probability of remaining in state 2 is .9. Thus,

* Notice that these are actually *conditional* probabilities. For instance, p_{12} is the probability that the system (the consumer in this case) will go into state 2, *given that the system* (the consumer) *is in state 1*.

$$p_{21} = .1$$

and

$$p_{22} = .9.$$

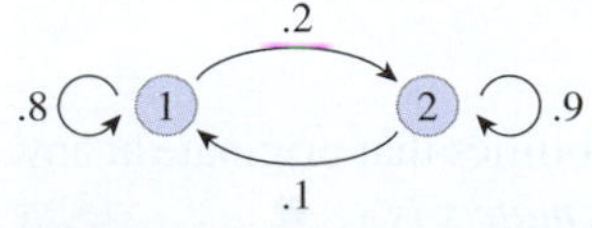

Figure 19

We can picture this system as in Figure 19, which shows the **state transition diagram** for this example. The numbers p_{ij}, which appear as labels on the arrows, are the **transition probabilities**.

Markov System, States, and Transition Probabilities

A **Markov system*** (or **Markov process** or **Markov chain**) is a system that can be in one of several specified **states**. There is specified a certain **time step**, and at each step the system will randomly change states or remain where it is. The probability of going from state i to state j is a fixed number p_{ij}, called the **transition probability**.

* Named after the Russian mathematician A.A. Markov (1856–1922), who first studied these "nondeterministic" processes.

Quick Example

The Markov system depicted in Figure 19 has two states: state 1 and state 2. The transition probabilities are as follows:

$$p_{11} = \text{Probability of going from state 1 to state 1} = .8$$
$$p_{12} = \text{Probability of going from state 1 to state 2} = .2$$
$$p_{21} = \text{Probability of going from state 2 to state 1} = .1$$
$$p_{22} = \text{Probability of going from state 2 to state 2} = .9.$$

Notice that, because the system must go somewhere at each time step, the transition probabilities originating at a particular state always add up to 1. For example, in the transition diagram above, when we add the probabilities originating at state 1, we get $.8 + .2 = 1$.

The transition probabilities may be conveniently arranged in a matrix.

Transition Matrix

The **transition matrix** associated with a given Markov system is the matrix P whose ijth entry is the transition probability p_{ij}, the transition probability of going *from* state i *to* state j. In other words, the entry in position ij is the *label on the arrow going from state i to state j* in a state transition diagram.

Thus, the transition matrix for a system with two states would be set up as follows:

$$\begin{array}{cc} & \textbf{To:} \\ & \begin{array}{cc} \mathbf{1} & \mathbf{2} \end{array} \\ \textbf{From:} \begin{array}{c} \mathbf{1} \\ \mathbf{2} \end{array} & \begin{bmatrix} p_{11} & p_{12} \\ p_{21} & p_{22} \end{bmatrix} . \end{array} \quad \begin{array}{l} \text{Arrows Originating in State 1} \\ \text{Arrows Originating in State 2} \end{array}$$

Quick Example

In the system pictured in Figure 19, the transition matrix is

$$P = \begin{bmatrix} .8 & .2 \\ .1 & .9 \end{bmatrix}.$$

Note Notice that because the sum of the transition probabilities that originate at any state is 1, *the sum of the entries in any row of a transition matrix is 1.* ■

Now, let's start doing some calculations.

EXAMPLE 1 Laundry Detergent Switching

Consider the Markov system found by Gamble Detergents at the beginning of this section. Suppose that 70% of consumers are now using powdered detergent, while the other 30% are using liquid.

a. What will be the distribution one year from now? (That is, what percentage will be using powdered and what percentage liquid detergent?)

b. Assuming that the probabilities remain the same, what will be the distribution two years from now? Three years from now?

Solution

a. First, let us think of the statement that 70% of consumers are using powdered detergent as telling us a probability: The probability that a randomly chosen consumer uses powdered detergent is .7. Similarly, the probability that a randomly chosen consumer uses liquid detergent is .3. We want to find the corresponding probabilities one year from now. To do this, consider the tree diagram in Figure 20.

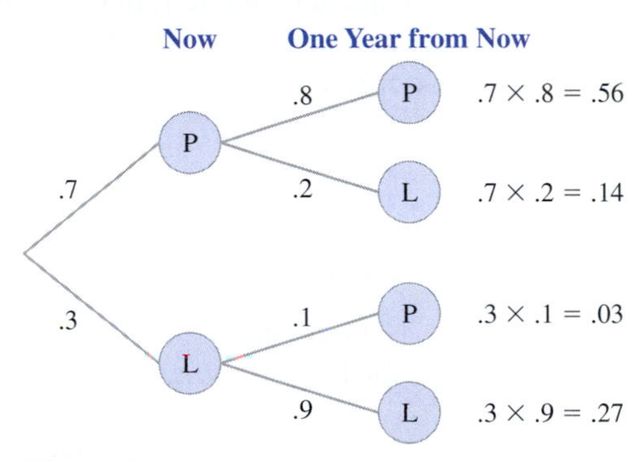

Figure 20

Technology can be used to compute the distribution vectors in Example 1:

TI-83/84 Plus
Define `[A]` as the initial distribution and `[B]` as the transition matrix. Then compute `[A]*[B]` for the distribution after 1 step. `Ans*[B]` gives the distribution after successive steps. [More details on page 415.]

Spreadsheet
Use the `MMULT` command to multiply the associated matrices. [More details on page 417.]

Website
www.WanerMath.com
Student Home
→ On Line Utilities
→ Matrix Algebra Tool
You can enter the transition matrix P and the initial distribution vector v in the input area, and compute the various distribution vectors using the following formulas:

`v*P`	1 step
`v*P^2`	2 steps
`v*P^3`	3 steps.

The first branching shows the probabilities now, while the second branching shows the (conditional) transition probabilities. So, if we want to know the probability that a consumer is using powdered detergent one year from now, it will be

$$\text{Probability of using powder after one year} = .7 \times .8 + .3 \times .1 = .59.$$

On the other hand, we have:

$$\text{Probability of using liquid after one year} = .7 \times .2 + .3 \times .9 = .41.$$

Now, here's the crucial point: *These are exactly the same calculations as in the matrix product*

$$\underset{\uparrow\ \text{Initial distribution}}{[.7 \quad .3]} \underset{\uparrow\ \text{Transition matrix}}{\begin{bmatrix} .8 & .2 \\ .1 & .9 \end{bmatrix}} = \underset{\uparrow\ \text{Distribution after 1 step}}{[.59 \quad .41]}.$$

Thus, to get the distribution of detergent users after one year, all we have to do is multiply the **initial distribution vector** [.7 .3] by the transition matrix P. The result is [.59 .41], the **distribution vector after one step**.

b. Now what about the distribution after *two* years? If we assume that the same fraction of consumers switch or stay put in the second year as in the first, we can simply repeat the calculation we did above, using the new distribution vector:

$$[.59 \quad .41]\begin{bmatrix} .8 & .2 \\ .1 & .9 \end{bmatrix} = [.513 \quad .487].$$

↑ Distribution after 1 step ↑ Transition matrix ↑ Distribution after 2 steps

Thus, after two years we can expect 51.3% of consumers to be using powdered detergent and 48.7% to be using liquid detergent. Similarly, after three years we have

$$[.513 \quad .487]\begin{bmatrix} .8 & .2 \\ .1 & .9 \end{bmatrix} = [.4591 \quad .5409].$$

So, after three years, 45.91% of consumers will be using powdered detergent and 54.09% will be using liquid. Slowly but surely, liquid detergent seems to be winning.

➡ **Before we go on...** Note that the sum of the entries is 1 in each of the distribution vectors in Example 1. In fact, these vectors are giving the probability distributions for each year of finding a randomly chosen consumer using either powdered or liquid detergent. A vector having nonnegative entries adding up to 1 is called a **probability vector**. ■

Distribution Vector after *m* Steps

A **distribution vector** is a probability vector giving the probability distribution for finding a Markov system in its various possible states. If v is a distribution vector, then the distribution vector one step later will be vP. The distribution m steps later will be

$$\text{Distribution after } m \text{ steps} = v \cdot P \cdot P \cdot \ldots \cdot P \ (m \text{ times}) = vP^m.$$

Quick Example

If $P = \begin{bmatrix} 0 & 1 \\ .5 & .5 \end{bmatrix}$ and $v = [.2 \quad .8]$, then we can calculate the following distribution vectors:

$$vP = [.2 \quad .8]\begin{bmatrix} 0 & 1 \\ .5 & .5 \end{bmatrix} = [.4 \quad .6]$$ Distribution after one step

$$vP^2 = (vP)P = [.4 \quad .6]\begin{bmatrix} 0 & 1 \\ .5 & .5 \end{bmatrix} = [.3 \quad .7]$$ Distribution after two steps

$$vP^3 = (vP^2)P = [.3 \quad .7]\begin{bmatrix} 0 & 1 \\ .5 & .5 \end{bmatrix} = [.35 \quad .65].$$ Distribution after three steps

What about the matrix P^m that appears above? Multiplying a distribution vector v times P^m gives us the distribution m steps later, so we can think of P^m as the m-step transition matrix. More explicitly, consider the following example.

EXAMPLE 2 Powers of the Transition Matrix

Continuing the example of detergent switching, suppose that a consumer is now using powdered detergent. What are the probabilities that the consumer will be using powdered or liquid detergent two years from now? What if the consumer is now using liquid detergent?

Solution To record the fact that we know that the consumer is using powdered detergent, we can take as our initial distribution vector $v = [1 \quad 0]$. To find the distribution two years from now, we compute vP^2. To make a point, we do the calculation slightly differently:

$$\begin{aligned} vP^2 &= [1 \quad 0]\begin{bmatrix} .8 & .2 \\ .1 & .9 \end{bmatrix}\begin{bmatrix} .8 & .2 \\ .1 & .9 \end{bmatrix} \\ &= [1 \quad 0]\begin{bmatrix} .66 & .34 \\ .17 & .83 \end{bmatrix} \\ &= [.66 \quad .34]. \end{aligned}$$

So, the probability that our consumer is using powdered detergent two years from now is .66, while the probability of using liquid detergent is .34. The point to notice is that these are the entries in the first row of P^2. Similarly, if we consider a consumer now using liquid detergent, we should take the initial distribution vector to be $v = [0 \quad 1]$ and compute

$$vP^2 = [0 \quad 1]\begin{bmatrix} .66 & .34 \\ .17 & .83 \end{bmatrix} = [.17 \quad .83].$$

Thus, the bottom row gives the probabilities that a consumer, now using liquid detergent, will be using either powdered or liquid detergent two years from now.

In other words, the ijth entry of P^2 gives the probability that a consumer, starting in state i, will be in state j after two time steps.

What is true in Example 2 for two time steps is true for any number of time steps:

Powers of the Transition Matrix

P^m $(m = 1, 2, 3, \ldots)$ is the ***m*-step transition matrix**. The ijth entry in P^m is the probability of a transition from state i to state j in m steps.

Quick Example

If $P = \begin{bmatrix} 0 & 1 \\ .5 & .5 \end{bmatrix}$ then

$$P^2 = P \cdot P = \begin{bmatrix} .5 & .5 \\ .25 & .75 \end{bmatrix} \qquad \text{2-step transition matrix}$$

$$P^3 = P \cdot P^2 = \begin{bmatrix} .25 & .75 \\ .375 & .625 \end{bmatrix}. \qquad \text{3-step transition matrix}$$

The probability of going from state 1 to state 2 in two steps = (1, 2)-entry of $P^2 = .5$.

The probability of going from state 1 to state 2 in three steps = (1, 2)-entry of $P^3 = .75$.

What happens if we follow our laundry detergent–using consumers for many years?

EXAMPLE 3 Long-Term Behavior

Suppose that 70% of consumers are now using powdered detergent while the other 30% are using liquid. Assuming that the transition matrix remains valid the whole time, what will be the distribution 1, 2, 3, . . . , and 50 years later?

Solution Of course, to do this many matrix multiplications, we're best off using technology. We already did the first three calculations in an earlier example.

Distribution after 1 year: $[.7 \quad .3]\begin{bmatrix} .8 & .2 \\ .1 & .9 \end{bmatrix} = [.59 \quad .41].$

Distribution after 2 years: $[.59 \quad .41]\begin{bmatrix} .8 & .2 \\ .1 & .9 \end{bmatrix} = [.513 \quad .487].$

Distribution after 3 years: $[.513 \quad .487]\begin{bmatrix} .8 & .2 \\ .1 & .9 \end{bmatrix} = [.4591 \quad .5409].$

. . .

Distribution after 48 years: $[.33333335 \quad .66666665].$

Distribution after 49 years: $[.33333334 \quad .66666666].$

Distribution after 50 years: $[.33333334 \quad .66666666].$

Thus, the distribution after 50 years is approximately $[.33333334 \quad .66666666]$.

Something interesting seems to be happening in Example 3. The distribution seems to be getting closer and closer to

$$[.333333\ldots \quad .666666\ldots] = \begin{bmatrix} \frac{1}{3} & \frac{2}{3} \end{bmatrix}.$$

Let's call this distribution vector v_∞. Notice two things about v_∞:

- v_∞ is a probability vector.
- If we calculate $v_\infty P$, we find

$$v_\infty P = \begin{bmatrix} \frac{1}{3} & \frac{2}{3} \end{bmatrix}\begin{bmatrix} .8 & .2 \\ .1 & .9 \end{bmatrix} = \begin{bmatrix} \frac{1}{3} & \frac{2}{3} \end{bmatrix} = v_\infty.$$

In other words,

$$v_\infty P = v_\infty.$$

We call a probability vector v with the property that $vP = v$ a **steady-state (probability) vector**.

Q: *Where does the name steady-state vector come from?*

A: If $vP = v$, then v is a distribution that will not change from time step to time step. In the example above, because $[1/3 \quad 2/3]$ is a steady-state vector, if 1/3 of consumers use powdered detergent and 2/3 use liquid detergent one year, then the proportions will be the same the next year. Individual consumers may still switch from year to year, but as many will switch from powder to liquid as switch from liquid to powder, so the number using each will remain constant.

But how do we find a steady-state vector?

EXAMPLE 4 Calculating the Steady-State Vector

Calculate the steady-state probability vector for the transition matrix in the preceding examples:

$$P = \begin{bmatrix} .8 & .2 \\ .1 & .9 \end{bmatrix}.$$

Solution We are asked to find

$$v_\infty = [x \quad y].$$

This vector must satisfy the equation

$$v_\infty P = v_\infty$$

or

$$[x \quad y]\begin{bmatrix} .8 & .2 \\ .1 & .9 \end{bmatrix} = [x \quad y].$$

Doing the matrix multiplication gives

$$[.8x + .1y \quad .2x + .9y] = [x \quad y].$$

Equating corresponding entries gives

$$\begin{aligned} .8x + .1y &= x \\ .2x + .9y &= y \end{aligned}$$

or

$$\begin{aligned} -.2x + .1y &= 0 \\ .2x - .1y &= 0. \end{aligned}$$

Now these equations are really the same equation. (Do you see that?) There is one more thing we know, though: Because $[x \quad y]$ is a probability vector, its entries must add up to 1. This gives one more equation:

$$x + y = 1.$$

Taking this equation together with one of the two equations above gives us the following system:

$$\begin{aligned} x + \quad y &= 1 \\ -.2x + .1y &= 0. \end{aligned}$$

We now solve this system using any of the techniques we learned for solving systems of linear equations. We find that the solution is $x = 1/3$, and $y = 2/3$, so the steady-state vector is

$$v_\infty = [x \quad y] = \begin{bmatrix} \frac{1}{3} & \frac{2}{3} \end{bmatrix}$$

as suggested in Example 3.

using Technology

Technology can be used to compute the steady-state vector in Example 4.

TI-83/84 Plus
Define `[A]` as the coefficient matrix of the system of equations being solved, and `[B]` as the column matrix of the right-hand sides.
Then compute `[A]`$^{-1}$`[B]`
[More details on page 416.]

Spreadsheet
Use the `MMULT` and `MINVERSE` commands to solve the necessary system of equations.
[More details on page 418.]

Website
www.WanerMath.com
Student Home
→ On Line Utilities
→ Pivot and Gauss-Jordan Tool
You can use the Pivot and Gauss-Jordan Tool to solve the system of equations that gives you the steady-state vector.

The method we just used works for any size transition matrix and can be summarized as follows.

Calculating the Steady-State Distribution Vector

To calculate the steady-state probability vector for a Markov system with transition matrix P, we solve the system of equations given by

$$\begin{aligned} &x + y + z + \cdots = 1 \\ &[x \quad y \quad z \ \ldots]P = [x \quad y \quad z \ \ldots], \end{aligned}$$

where we use as many unknowns as there are states in the Markov system. The steady-state probability vector is then

$$v_\infty = [x \quad y \quad z \ \ldots].$$

Q: *Is there always a steady-state distribution vector?*

A: Yes, although the explanation why is more involved than we can give here.

Q: *In Example 3, we started with a distribution vector v and found that vP^m got closer and closer to v_∞ as m got larger. Does that always happen?*

A: It does if the Markov system is **regular**, as we define below, but may not for other kinds of systems. Again, we shall not prove this fact here.

Regular Markov Systems

A **regular** Markov system is one for which some power of its transition matrix P has no zero entries. If a Markov system is regular, then

1. It has a unique steady-state probability vector v_∞, and

2. If v is any probability vector whatsoever, then vP^m approaches v_∞ as m gets large. We say that the **long-term behavior** of the system is to have distribution (close to) v_∞.

Interpreting the Steady-State Vector

In a regular Markov system, the entries in the steady-state probability vector give the long-term probabilities that the system will be in the corresponding states, or the fractions of time one can expect to find the Markov system in the corresponding states.

Quick Examples

1. The system with transition matrix $P = \begin{bmatrix} .8 & .2 \\ .1 & .9 \end{bmatrix}$ is regular because $P(= P^1)$ has no zero entries.

2. The system with transition matrix $P = \begin{bmatrix} 0 & 1 \\ .5 & .5 \end{bmatrix}$ is regular because $P^2 = \begin{bmatrix} .5 & .5 \\ .25 & .75 \end{bmatrix}$ has no zero entries.

3. The system with transition matrix $P = \begin{bmatrix} 0 & 1 \\ 1 & 0 \end{bmatrix}$ is *not* regular: $P^2 = \begin{bmatrix} 1 & 0 \\ 0 & 1 \end{bmatrix}$ and $P^3 = P$ again, so the powers of P alternate between these two matrices. Thus, every power of P has zero entries. Although this system has a steady-state vector, namely $[.5 \quad .5]$, if we take $v = [1 \quad 0]$, then $vP = [0 \quad 1]$ and $vP^2 = v$, so the distribution vectors vP^m just alternate between these two vectors, not approaching v_∞.

We finish with one more example.

EXAMPLE 5 Gambler's Ruin

A timid gambler, armed with her annual bonus of \$20, decides to play roulette using the following scheme. At each spin of the wheel, she places \$10 on red. If red comes up, she wins an additional \$10; if black comes up, she loses her \$10. For the sake of simplicity, assume that she has a probability of $1/2$ of winning. (In the real game, the probability is slightly lower—a fact that many gamblers forget.) She keeps playing until she has either gotten up to \$30 or lost it all. In either case, she then packs up and leaves. Model this situation as a Markov system and find the associated transition matrix. What can we say about the long-term behavior of this system?

Solution We must first decide on the states of the system. A good choice is the gambler's financial state, the amount of money she has at any stage of the game. According to her rules, she can be broke, have \$10, \$20, or \$30. Thus, there are four states: $1 = \$0$, $2 = \$10$, $3 = \$20$, and $4 = \$30$. Because she bets \$10 each time, she moves down \$10 if she loses (with probability $1/2$) and up \$10 if she wins (with probability also $1/2$), until she reaches one of the extremes. The transition diagram is shown in Figure 21.

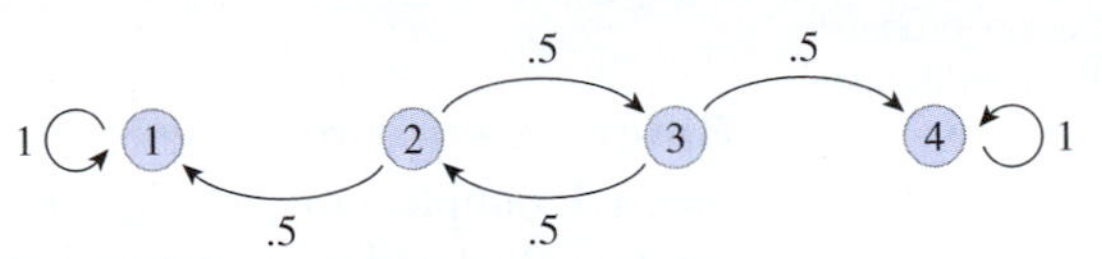

Figure 21

Note that once the system enters state 1 or state 4, it does not leave; with probability 1, it stays in the same state.* We call such states **absorbing states**. We can now write down the transition matrix:

* These states are like "Roach Motels" ("Roaches check in, but they don't check out")!

$$P = \begin{bmatrix} 1 & 0 & 0 & 0 \\ .5 & 0 & .5 & 0 \\ 0 & .5 & 0 & .5 \\ 0 & 0 & 0 & 1 \end{bmatrix}.$$

(Notice all the 0 entries, corresponding to possible transitions that we did not draw in the transition diagram because they have 0 probability of occurring. We usually leave out such arrows.) Is this system regular? Take a look at P^2:

$$P^2 = \begin{bmatrix} 1 & 0 & 0 & 0 \\ .5 & .25 & 0 & .25 \\ .25 & 0 & .25 & .5 \\ 0 & 0 & 0 & 1 \end{bmatrix}.$$

Notice that the first and last rows haven't changed. After two steps, there is still no chance of leaving states 1 or 4. In fact, no matter how many powers we take, no matter how many steps we look at, there will still be no way to leave either of those states, and the first and last rows will still have plenty of zeros. This system is not regular.

Nonetheless, we can try to find a steady-state probability vector. If we do this (and you should set up the system of linear equations and solve it), we find that there are infinitely many steady-state probability vectors, namely all vectors of the form $[x \;\; 0 \;\; 0 \;\; 1-x]$ for $0 \le x \le 1$. (You can check directly that these are all steady-state vectors.) As with a regular system, if we start with any distribution, the system will tend toward one of these steady-state vectors. In other words, eventually the gambler will either lose all her money or leave the table with \$30.

But which outcome is more likely, and with what probability? One way to approach this question is to try computing the distribution after many steps. The distribution that represents the gambler starting with \$20 is $v = [0 \;\; 0 \;\; 1 \;\; 0]$. Using technology, it's easy to compute vP^n for some large values of n:

$$vP^{10} \approx [.333008 \;\; 0 \;\; .000976 \;\; .666016]$$

$$vP^{50} \approx [.333333 \;\; 0 \;\; 0 \;\; .666667].$$

Website
www.WanerMath.com
Follow the path
Chapter 7
→ On Line Utilities
→ Markov Simulation
where you can enter the transition matrix and watch a visual simulation showing the system hopping from state to state. Run the simulation a number of times to get an experimental estimate of the time to absorption. (Take the average time to absorption over many runs.) Also, try varying the transition probabilities to study the effect on the time to absorption.

So, it looks like the probability that she will leave the table with \$30 is approximately 2/3, while the probability that she loses it all is 1/3.

What if she started with only \$10? Then our initial distribution would be $v = [0 \ 1 \ 0 \ 0]$ and

$$vP^{10} \approx [.666016 \quad .000976 \quad 0 \quad .333008]$$
$$vP^{50} \approx [.666667 \quad 0 \quad 0 \quad .333333].$$

So, this time, the probability of her losing everything is about 2/3 while the probability of her leaving with \$30 is 1/3. There is a way of calculating these probabilities exactly using matrix arithmetic; however, it would take us too far afield to describe it here.

➡ **Before we go on...** Another interesting question is, how long will it take the gambler in Example 5 to get to \$30 or lose it all? This is called the *time to absorption* and can be calculated using matrix arithmetic. ■

7.7 EXERCISES

Access end-of-section exercises online at **www.webassign.net** ENHANCED WebAssign

CHAPTER 7 REVIEW

KEY CONCEPTS

 Website www.WanerMath.com

Go to the Website at www.WanerMath.com to find a comprehensive and interactive Web-based summary of Chapter 7.

7.1 Sample Spaces and Events

Experiment, outcome, sample space *p. 348*
Event *p. 350*
The complement of an event *p. 354*
Unions of events *p. 354*
Intersections of events *p. 355*
Mutually exclusive events *p. 356*

7.2 Relative Frequency

Relative frequency or estimated probability *p. 358*
Relative frequency distribution *p. 358*
Properties of relative frequency distribution *p. 361*

7.3 Probability and Probability Models

Probability distribution, probability: $0 \leq P(s_i) \leq 1$ and $P(s_1) + \cdots + P(s_n) = 1$ *p. 363*
If $P(E) = 0$, we call E an impossible event *p. 363*
Probability models *p. 363*
Probability models for equally likely outcomes: $P(E) = n(E)/n(S)$ *p. 365*
Addition principle: $P(A \cup B) = P(A) + P(B) - P(A \cap B)$ *p. 369*
If A and B are mutually exclusive, then $P(A \cup B) = P(A) + P(B)$ *p. 370*
If S is the sample space, then $P(S) = 1$, $P(\emptyset) = 0$, and $P(A') = 1 - P(A)$ *p. 372*

7.4 Probability and Counting Techniques

Use counting techniques from Chapter 6 to calculate probability *p. 376*

7.5 Conditional Probability and Independence

Conditional probability: $P(A \mid B) = P(A \cap B)/P(B)$ *p. 383*
Multiplication principle for conditional probability: $P(A \cap B) = P(A \mid B)P(B)$ *p. 386*
Independent events: $P(A \cap B) = P(A)P(B)$ *p. 389*

7.6 Bayes' Theorem and Applications

Bayes' theorem (short form):

$$P(A \mid T) = \frac{P(T \mid A)P(A)}{P(T \mid A)P(A) + P(T \mid A')P(A')} \quad p.\ 395$$

Bayes' theorem (partition of sample space into three events):

$$P(A_1 \mid T) = \frac{P(T \mid A_1)P(A_1)}{P(T \mid A_1)P(A_1) + P(T \mid A_2)P(A_2) + P(T \mid A_3)P(A_3)}$$

p. 398

7.7 Markov Systems

Markov system, Markov process, states, transition probabilities *p. 400*
Transition matrix associated with a given Markov system *p. 400*
A vector having non-negative entries adding up to 1 is called a probability vector *p. 402*
A distribution vector is a probability vector giving the probability distribution for finding a Markov system in its various possible states *p. 402*
If v is a distribution vector, then the distribution vector one step later will be vP. The distribution after m steps will be vP^m *p. 402*
P^m is the m-step transition matrix. The ijth entry in P^m is the probability of a transition from state i to state j in m steps *p. 403*
A steady-state (probability) vector is a probability vector v such that $vP = v$ *p. 404*
Calculation of the steady-state distribution vector *p. 405*
A regular Markov system, long-term behavior of a regular Markov system *p. 406*
An absorbing state is one for which the probability of staying in the state is 1 (and the probability of leaving it for any other state is 0) *p. 407*

REVIEW EXERCISES

In each of Exercises 1–6, say how many elements are in the sample space S, list the elements of the given event E, and compute the probability of E.

1. Three coins are tossed; the result is one or more tails.

2. Four coins are tossed; the result is fewer heads then tails.

3. Two distinguishable dice are rolled; the numbers facing up add to 7.

4. Three distinguishable dice are rolled; the number facing up add to 5.

5. A die is weighted so that each of 2, 3, 4, and 5 is half as likely to come up as either 1 or 6; however, 2 comes up.

6. Two indistinguishable dice are rolled; the numbers facing up add to 7.

In each of Exercises 7–10, calculate the relative frequency $P(E)$.

7. Two coins are tossed 50 times, and two heads come up 12 times. E is the event that at least one tail comes up.

8. Ten stocks are selected at random from a portfolio. Seven of them have increased in value since their purchase, and the rest have decreased. Eight of them are Internet stocks and two of those have decreased in value. E is the event that a stock has either increased in value or is an Internet stock.

9. You have read 150 of the 400 novels in your home, but your sister Roslyn has read 200, of which only 50 are novels you have read as well. E is the event that a novel has been read by neither you nor your sister.

10. You roll two dice 10 times. Both dice show the same number 3 times, and on 2 rolls, exactly one number is odd. E is the event that the sum of the numbers is even.

In each of Exercises 11–14, calculate the probability $P(E)$.

11. There are 32 students in categories A and B combined. Some are in both, 24 are in A, and 24 are in B. E is the event that a randomly selected student (among the 32) is in both categories.

12. You roll two dice, one red and one green. Losing combinations are doubles (both dice showing the same number) and outcomes in which the green die shows an odd number and the red die shows an even number. The other combinations are winning ones. E is the event that you roll a winning combination.

13. The *jPlay* portable music/photo/video player and bottle opener comes in three models: A, B, and C, each with five colors to choose from, and there are equal numbers of each combination. E is the event that a randomly selected *jPlay* is either orange (one of the available colors), a Model A, or both.

14. The Heavy Weather Service predicts that for tomorrow there is a 50% chance of tornadoes, a 20% chance of a monsoon, and a 10% chance of both. What is the probability that we will be lucky tomorrow and encounter neither tornadoes nor a monsoon?

A bag contains four red marbles, two green ones, one transparent one, three yellow ones, and two orange ones. You select five at random. In each of Exercises 15–20, compute the probability of the given event.

15. You have selected all the red ones.

16. You have selected all the green ones.

17. All are different colors.

18. At least one is not red.

19. At least two are yellow.

20. None are yellow and at most one is red.

In each of Exercises 21–26, find the probability of being dealt the given type of 5-card hand from a standard deck of 52 cards. (None of these is a recognized poker hand.) Express your answer in terms of combinations.

21. **Kings and Queens:** Each of the five cards is either a king or a queen.

22. **Five Pictures:** Each card is a picture card (J, Q, K).

23. **Fives and Queens:** Three fives, the queen of spades, and one other queen.

24. **Prime Full House:** A full house (three cards of one denomination, two of another) with the face value of each card a prime number (Ace = 1, J = 11, Q = 12, K = 13).

25. **Full House of Commons:** A full house (three cards of one denomination, two of another) with no royal cards (that is, no J, Q, K, or Ace).

26. **Black Two Pair:** Five black cards (spades or clubs), two with one denomination, two with another, and one with a third.

Two dice, one green and one yellow, are rolled. In each of Exercises 27–32, find the conditional probability, and also say whether the indicated pair of events is independent.

27. The sum is 5, given that the green one is not 1 and the yellow one is 1.

28. The sum is 6, given that the green one is either 1 or 3 and the yellow one is 1.

29. The yellow one is 4, given that the green one is 4.

30. The yellow one is 5, given that the sum is 6.

31. The dice have the same parity, given that both of them are odd.

32. The sum is 7, given that the dice do not have the same parity.

A poll shows that half the consumers who use Brand A switched to Brand B the following year, while the other half stayed with Brand A. Three quarters of the Brand B users stayed with Brand B the following year, while the rest switched to Brand A. Use this information to answer Exercises 33–36.

33. Give the associated Markov transition matrix, with state 1 representing using Brand A and state 2 representing using Brand B.

34. Compute the associated two- and three-step transition matrices. What is the probability that a Brand A user will be using Brand B three years later?

35. If two thirds of consumers are presently using Brand A and one third are using Brand B, how are these consumers distributed in three years' time?

36. In the long term, what fraction of the time will a user spend using each of the two brands?

APPLICATIONS: OHaganBooks.com

OHaganBooks.com currently operates three warehouses: one in Washington, one in California, and the new one in Texas. Book inventories are shown in the following table.

	Sci Fi	Horror	Romance	Other	Total
Washington	10,000	12,000	12,000	30,000	64,000
California	8,000	12,000	6,000	16,000	42,000
Texas	15,000	15,000	20,000	44,000	94,000
Total	33,000	39,000	38,000	90,000	200,000

A book is selected at random. In each of Exercises 37–42, compute the probability of the given event.

37. That it is either a sci-fi book or stored in Texas (or both)

38. That it is a sci-fi book stored in Texas

39. That it is a sci-fi book, given that it is stored in Texas

40. That it was stored in Texas, given that it was a sci-fi book

41. That it was stored in Texas, given that it was not a sci-fi book

42. That it was not stored in Texas, given that it was a sci-fi book

In order to gauge the effectiveness of the OHaganBooks.com site, you recently commissioned a survey of online shoppers. According to the results, 2% of online shoppers visited OHaganBooks.com during a one-week period, while 5% of them visited at least one of OHaganBooks.com's two main competitors: JungleBooks.com and FarmerBooks.com. Use this information to answer Exercises 43–50.

43. What percentage of online shoppers never visited OHaganBooks.com?

44. What percentage of shoppers never visited either of OHaganBooks.com's main competitors?

45. Assuming that visiting OHaganBooks.com was independent of visiting a competitor, what percentage of online shoppers visited either OHaganBooks.com or a competitor?

46. Assuming that visiting OHaganBooks.com was independent of visiting a competitor, what percentage of online shoppers visited OHaganBooks.com but not a competitor?

47. Under the assumption of Exercise 45, what is the probability that an online shopper will visit none of the three sites during a week?

48. If no one who visited OHaganBooks.com ever visited any of the competitors, what is the probability that an online shopper will visit none of the three sites during a week?

49. Actually, the assumption in Exercise 45 is not what was found by the survey, because an online shopper visiting a competitor was in fact more likely to visit OHaganBooks.com than a randomly selected online shopper. Let H be the event that an online shopper visits OHaganBooks.com, and let C be the event that he visits a competitor. Which is greater: $P(H \cap C)$ or $P(H)P(C)$? Why?

50. What the survey found is that 25% of online shoppers who visited a competitor also visited OHaganBooks.com. Given this information, what percentage of online shoppers visited OHaganBooks.com and neither of its competitors?

51. ***Sales*** According to statistics gathered by OHaganBooks.com, 2% of online shoppers visited the OHaganBooks.com Web site during the course of a week, and 8% of those purchased books from the company. Further, 0.5% of online shoppers who did not visit the OHaganBooks.com Web site during the course of a week nonetheless purchased books from the company (through mail-order catalogs and other sites like LemmaZorn.com). What is the probability that an online shopper who purchased books from OHaganBooks.com during a given week visited the site?

52. ***Sales*** Repeat Exercise 51 in the event that 1% of online shoppers visited OHaganBooks.com during the week.

53. ***University Admissions*** In the year Billy-Sean O'Hagan applied to Suburban State U, 56% of in-state applicants were admitted, while only 15% of out-of-state applicants were admitted. Further, 72% of all applicants were in-state. What percentage of admitted applicants were in-state? (Round your answer to the nearest percentage point.)

54. ***University Admissions*** Billy-Sean O'Hagan had also applied to Gigantic State U, at which time 75% of all applicants were from the United States and 22% of those applicants were admitted. Also, 14% of the applicants from foreign countries were admitted. What percentage of admitted applicants were from the United States? (Round your answer to the nearest percentage point.)

As mentioned earlier, OHaganBooks.com has two main competitors, JungleBooks.com and FarmerBooks.com, and no other competitors of any significance. The following table shows the movement of customers during July.[13] *(Thus, for instance, the first row tells us that 80% of OHaganBooks.com's customers remained loyal, 10% of them went to JungleBooks.com and the remaining 10% to FarmerBooks.com.)*

		To		
		OHaganBooks	**JungleBooks**	**FarmerBooks**
From	**OHaganBooks**	80%	10%	10%
	JungleBooks	40%	60%	0%
	FarmerBooks	20%	0%	80%

At the beginning of July, OHaganBooks.com had an estimated market share of one fifth of all customers, while its two competitors had two fifths each. Use this information to answer Exercises 55–58.

55. Estimate the market shares each company had at the end of July.

56. Assuming the July trends continue in August, predict the market shares of each company at the end of August.

57. Name one or more important factors that the Markov model does not take into account.

58. Assuming the July trend were to continue indefinitely, predict the market share enjoyed by each of the three e-commerce sites.

[13]By a "customer" of one of the three e-commerce sites, we mean someone who purchases more at that site than at any of the two competitors' sites.

Case Study: The Monty Hall Problem

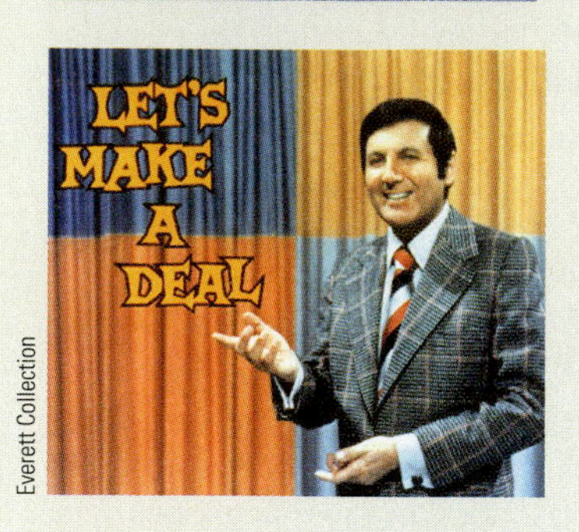

Everett Collection

* This problem caused quite a stir in late 1991 when this problem was discussed in Marilyn vos Savant's column in *Parade* magazine. Vos Savant gave the answer that you should switch. She received about 10,000 letters in response, most of them disagreeing with her, including several from mathematicians.

Here is a famous "paradox" that even mathematicians find counterintuitive. On the game show *Let's Make a Deal*, you are shown three doors, A, B, and C, and behind one of them is the Big Prize. After you select one of them–say, door A–to make things more interesting the host (Monty Hall), who knows what is behind each door, opens one of the other doors–say, door B–to reveal that the Big Prize is not there. He then offers you the opportunity to change your selection to the remaining door, door C. Should you switch or stick with your original guess? Does it make any difference?

Most people would say that the Big Prize is equally likely to be behind door A or door C, so there is no reason to switch.* In fact, this is wrong: The prize is more likely to be behind door C! There are several ways of seeing why this is so. Here is how you might work it out using Bayes' theorem.

Let A be the event that the Big Prize is behind door A, B the event that it is behind door B, and C the event that it is behind door C. Let F be the event that Monty has opened door B and revealed that the prize is not there. You wish to find $P(C \mid F)$ using Bayes' theorem. To use that formula you need to find $P(F \mid A)$ and $P(A)$ and similarly for B and C. Now, $P(A) = P(B) = P(C) = 1/3$ because at the outset, the prize is equally likely to be behind any of the doors. $P(F \mid A)$ is the probability that Monty will open door B if the prize is actually behind door A, and this is $1/2$ because we assume that he will choose either B or C randomly in this case. On the other hand, $P(F \mid B) = 0$, because he will never open the door that hides the prize. Also, $P(F \mid C) = 1$ because if the prize is behind door C, he must open door B to keep from revealing that the prize is behind door C. Therefore,

$$P(C \mid F) = \frac{P(F \mid C)P(C)}{P(F \mid A)P(A) + P(F \mid B)P(B) + P(F \mid C)P(C)}$$

$$= \frac{1 \cdot \frac{1}{3}}{\frac{1}{2} \cdot \frac{1}{3} + 0 \cdot \frac{1}{3} + 1 \cdot \frac{1}{3}} = \frac{2}{3}.$$

You conclude from this that you *should* switch to door C because it is more likely than door A to be hiding the Prize.

Here is a more elementary way you might work it out. Consider the tree diagram of possibilities shown in Figure 22. The top two branches of the tree give the cases in which the prize is behind door A, and there is a total probability of $1/3$ for that case. The remaining two branches with nonzero probabilities give the cases in which the prize is behind the door that you did not choose, and there is a total probability of $2/3$ for that case. Again, you conclude that you should switch your choice of doors because the one you did not choose is twice as likely as door A to be hiding the Big Prize.

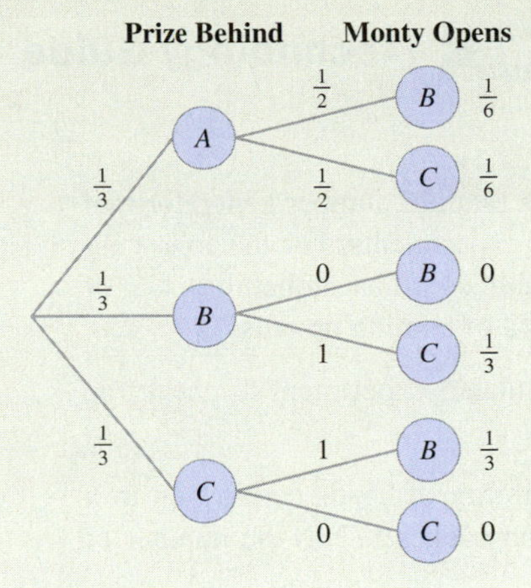

Figure 22

EXERCISES

1. The answer you came up with, to switch to the other door, depends on the strategy Monty Hall uses in picking the door to open. Suppose that he actually picks one of doors B and C at random, so that there is a chance that he will reveal the Big Prize. If he opens door B and it happens that the Prize is not there, should you switch or not?
2. What if you know that Monty's strategy is always to open door B if possible (i.e., it does not hide the Big Prize) after you choose A?
   a. If he opens door B, should you switch?
   b. If he opens door C, should you switch?
3. Repeat the analysis of the original game, but suppose that the game uses four doors instead of three (and still only one prize).
4. Repeat the analysis of the original game, but suppose that the game uses 1,000 doors instead of 3 (and still only one prize).

TI-83/84 Plus Technology Guide

Section 7.2

The TI-83/84 Plus has a random number generator that we can use to simulate experiments. For the following example, recall that a fair coin has probability 1/2 of coming up heads and 1/2 of coming up tails.

Example Use a simulated experiment to check the following.

a. The estimated probability of heads coming up in a toss of a fair coin approaches 1/2 as the number of trials gets large.

b. The estimated probability of heads coming up in two consecutive tosses of a fair coin approaches 1/4 as the number of trials gets large.[14]

Solution with Technology

a. Let us use 1 to represent heads and 0 to represent tails. We need to generate a list of **random binary digits** (0 or 1). One way to do this—and a method that works for most forms of technology—is to generate a random number between 0 and 1 and then round it to the nearest whole number, which will be either 0 or 1.

We generate random numbers on the TI-83/84 Plus using the "`rand`" function. To round the number X to the nearest whole number on the TI-83/84 Plus, follow [MATH] →NUM, select "`round`," and enter `round(X,0)`. This instruction rounds X to zero decimal places—that is, to the nearest whole number. Since we wish to round a random number we need to enter

`round(rand,0)` To obtain `rand`, follow [MATH] → `PRB`.

The result will be either 0 or 1. Each time you press [ENTER] you will now get another 0 or 1. The TI-83/84 Plus can also generate a random integer directly (without the need for rounding) through the instruction

`randInt(0,1)` To obtain `randInt`, follow [MATH] → `PRB`.

In general, the command `randInt(`m`,`n`)` generates a random integer in the range $[m, n]$. The following sequence of 100 random binary digits was produced using technology.[15]

0	1	0	0	1	1	0	1	0	0
0	1	0	0	0	0	0	0	1	0
1	1	0	0	0	1	0	0	1	1
1	1	1	0	1	0	0	0	1	0
1	1	1	1	1	1	1	0	0	1
1	0	1	1	1	0	0	1	1	0
0	1	0	1	1	1	0	1	1	1
1	0	0	0	0	0	0	1	1	1
1	1	1	1	0	0	1	1	1	0
1	1	1	0	1	1	0	1	0	0

If we use only the first row of data (corresponding to the first ten tosses), we find

$$P(\text{H}) = \frac{fr(1)}{N} = \frac{4}{10} = .4.$$

Using the first two rows ($N = 20$) gives

$$P(\text{H}) = \frac{fr(1)}{N} = \frac{6}{20} = .3.$$

Using all ten rows ($N = 100$) gives

$$P(\text{H}) = \frac{fr(1)}{N} = \frac{54}{100} = .54.$$

This is somewhat closer to the theoretical probability of 1/2 and supports our intuitive notion that the larger the number of trials, the more closely the estimated probability should approximate the theoretical value.[16]

b. We need to generate pairs of random binary digits and then check whether they are both 1s. Although the TI-83/84 Plus will generate a pair of random digits if you enter `round(rand(2),0)`, it would be a lot more convenient if the calculator could tell you right away whether both digits are 1s (corresponding to two consecutive heads in a coin toss). Here is a simple way of accomplishing this. Notice that if we *add* the two random binary digits, we obtain either 0, 1, or 2, telling us the number of heads that result from the two consecutive throws. Therefore, all we need to do is add the pairs of random digits and then count the number of times 2 comes up. A formula we can use is

```
randInt(0,1)+randInt(0,1)
```

[14]Since the set of outcomes of a pair of coin tosses is {HH, HT, TH, TT}, we expect HH to come up once in every four trials, on average.

[15]The instruction `randInt(0,1,100)→L1` will generate a list of 100 random 0s and 1s and store it in L_1, where it can be summed with Sum(L_1) (under [2ND] [LIST] →MATH).

[16]Do not expect this to happen every time. Compare, for example, $P(\text{H})$ for the first five rows and for all ten rows.

What would be even *more* convenient is if the result of the calculation would be either 0 or 1, with 1 signifying success (two consecutive heads) and 0 signifying failure. Then, we could simply add up all the results to obtain the number of times two heads occurred. To do this, we first divide the result of the previous calculation above by 2 (obtaining 0, .5, or 1, where now 1 signifies success) and then round *down* to an integer using a function called "int":

```
int(0.5*(randInt(0,1)+randInt(0,1)))
```

Following is the result of 100 such pairs of coin tosses, with 1 signifying success (two heads) and 0 signifying failure (all other outcomes). The last column records the number of successes in each row and the total number at the end.

1	1	0	0	0	0	0	0	0	0	2
0	1	0	0	0	0	0	1	0	1	3
0	1	0	0	1	1	0	0	0	1	4
0	0	0	0	0	0	0	0	1	0	1
0	1	0	0	1	0	0	1	0	0	3
1	0	1	0	0	0	0	0	0	0	2
0	0	0	0	0	0	0	0	0	1	1
0	1	1	1	1	0	0	0	0	1	5
1	1	0	1	0	0	1	1	0	0	5
0	0	0	0	0	0	0	0	1	0	1
										27

Now, as in part (a), we can compute estimated probabilities, with D standing for the outcome "two heads":

First 10 trials: $$P(D) = \frac{fr(1)}{N} = \frac{2}{10} = .2$$

First 20 trials: $$P(D) = \frac{fr(1)}{N} = \frac{5}{20} = .25$$

First 50 trials: $$P(D) = \frac{fr(1)}{N} = \frac{13}{50} = .26$$

100 trials: $$P(D) = \frac{fr(1)}{N} = \frac{27}{100} = .27.$$

Q: *What is happening with the data? The probabilities seem to be getting* less *accurate as N increases!*

A: Quite by chance, exactly 5 of the first 20 trials resulted in success, which matches the theoretical probability. Figure 23 shows an Excel plot of estimated probability versus N (for N a multiple of 10). Notice that, as N increases, the graph seems to meander within smaller distances of .25.

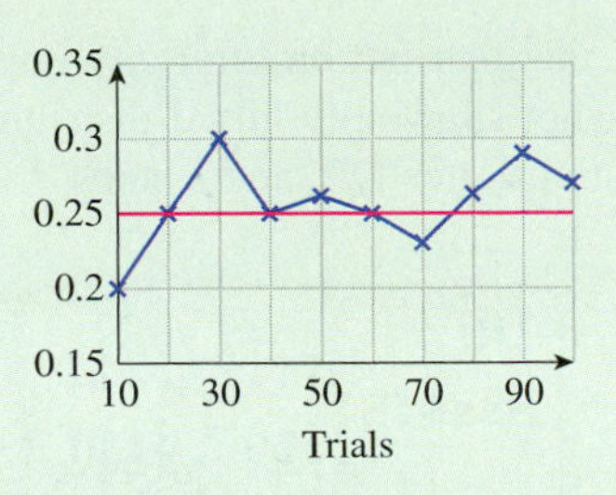

Figure 23

Q: *The previous techniques work fine for simulating coin tosses. What about rolls of a fair die, where we want outcomes between 1 and 6?*

A: We can simulate a roll of a die by generating a random integer in the range 1 through 6. The following formula accomplishes this:

```
1 + int(5.99999*rand).
```

(We used 5.99999 instead of 6 to avoid the outcome 7.)

Section 7.7

Example 1 (page 401) Consider the Markov system found by Gamble Detergents at the beginning of this section. Suppose that 70% of consumers are now using powdered detergent while the other 30% are using liquid. What will be the distribution one year from now? Two years from now? Three years from now?

Solution with Technology

In Chapter 4 we saw how to set up and multiply matrices. For this example, we can use the matrix editor to define [A] as the initial distribution and [B] as the transition matrix (remember that the only names we can use are [A] through [J]).

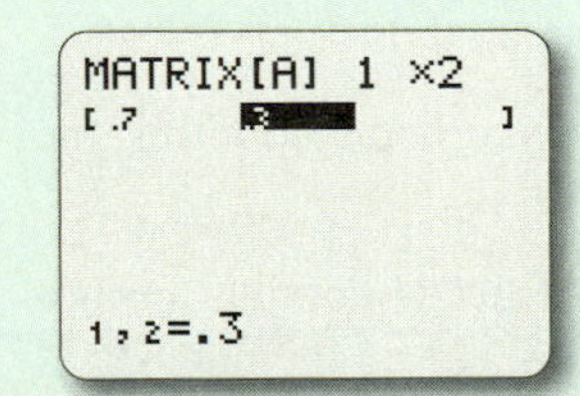

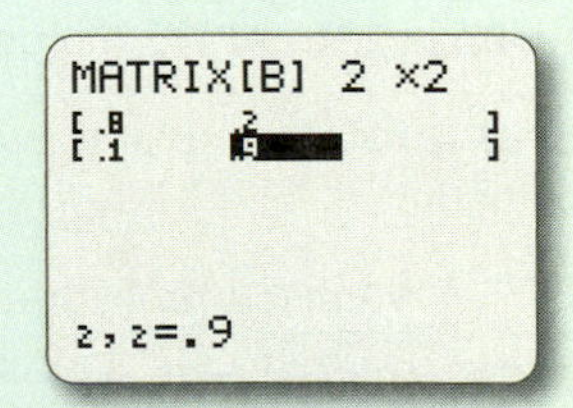

Entering [A] (obtained by pressing MATRX 1 ENTER) will show you the initial distribution. To obtain the distribution after 1 step, press X MATRX 2 ENTER, which has the effect of multiplying the previous answer by the transition matrix [B]. Now, just press ENTER repeatedly to continue multiplying by the transition matrix

TECHNOLOGY GUIDE

and obtain the distribution after any number of steps. The screenshot shows the initial distribution [A] and the distributions after 1, 2, and 3 steps.

```
[A]
        [[.7 .3]]
Ans*[B]
       [[.59 .41]]
     [[.513 .487]]
   [[.4591 .5409]]
```

Example 4 (page 405) Calculate the steady-state probability vector for the transition matrix in the preceding examples.

Solution with Technology

Finding the steady-state probability vector comes down to solving a system of equations. As discussed in Chapters 3 and 4, there are several ways to use a calculator to help. The most straightforward is to use matrix inversion to solve the matrix form of the system. In this case, as in the text, the system of equations we need to solve is

$$\begin{aligned} x + \quad y &= 1 \\ -.2x + .1y &= 0. \end{aligned}$$

We write this as the matrix equation $AX = B$ with

$$A = \begin{bmatrix} 1 & 1 \\ -.2 & .1 \end{bmatrix} \qquad B = \begin{bmatrix} 1 \\ 0 \end{bmatrix}.$$

To find $X = A^{-1}B$ using the TI-83/84 Plus, we first use the matrix editor to enter these matrices as [A] and [B], then compute [A]$^{-1}$[B] on the home screen.

```
[A]-1[B]
  [[.3333333333]
   [.6666666667]]
```

To convert the entries to fractions, we can follow this by the command.

➤ Frac MATH ENTER ENTER

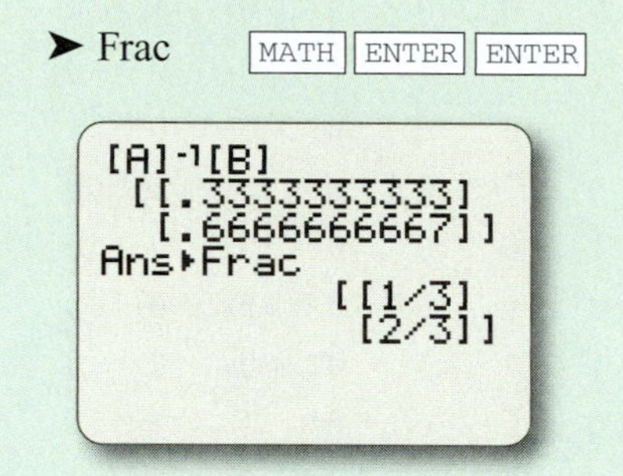

SPREADSHEET Technology Guide

Section 7.2

Spreadsheets have random number generators that we can use to simulate experiments. For the following example, recall that a fair coin has probability 1/2 of coming up heads and 1/2 of coming up tails.

Example Use a simulated experiment to check the following.

a. The estimated probability of heads coming up in a toss of a fair coin approaches 1/2 as the number of trials gets large.

b. The estimated probability of heads coming up in two consecutive tosses of a fair coin approaches 1/4 as the number of trials gets large.[17]

Solution with Technology

a. Let us use 1 to represent heads and 0 to represent tails. We need to generate a list of **random binary digits** (0 or 1). One way to do this—and a method that works for most forms of technology—is to generate a random number between 0 and 1 and then round it to the nearest whole number, which will be either 0 or 1.

In a spreadsheet, the formula RAND() gives a random number between 0 and 1.[18] The function ROUND(X,0) rounds X to zero decimal places—that is, to the nearest integer. Therefore, to obtain a random binary digit in any cell, just enter the following formula:

```
=ROUND(RAND(),0)
```

[17] Because the set of outcomes of a pair of coin tosses is {HH, HT, TH, TT}, we expect HH to come up once in every four trials, on average.

[18] The parentheses after RAND are necessary even though the function takes no arguments.

Spreadsheets can also generate a random integer directly (without the need for rounding) through the formula

```
=RANDBETWEEN(0,1)
```

To obtain a whole array of random numbers, just drag this formula into the cells you wish to use.

b. We need to generate pairs of random binary digits and then check whether they are both 1s. It would be convenient if the spreadsheet could tell you right away whether both digits are 1s (corresponding to two consecutive heads in a coin toss). Here is a simple way of accomplishing this. Notice that if we *add* two random binary digits, we obtain either 0, 1, or 2, telling us the number of heads that result from the two consecutive throws. Therefore, all we need to do is add pairs of random digits and then count the number of times 2 comes up. Formulas we can use are

```
=RANDBETWEEN(0,1)+RANDBETWEEN(0,1)
```

What would be even *more* convenient is if the result of the calculation would be either 0 or 1, with 1 signifying success (two consecutive heads) and 0 signifying failure. Then, we could simply add up all the results to obtain the number of times two heads occurred. To do this, we first divide the result of the calculation above by 2 (obtaining 0, .5, or 1, where now 1 signifies success) and then round *down* to an integer using a function called "int":

```
=INT(0.5*(RANDBETWEEN(0,1)+
 RANDBETWEEN(0,1)))
```

Following is the result of 100 such pairs of coin tosses, with 1 signifying success (two heads) and 0 signifying failure (all other outcomes). The last column records the number of successes in each row and the total number at the end.

1	1	0	0	0	0	0	0	0	0	2
0	1	0	0	0	0	0	1	0	1	3
0	1	0	0	1	1	0	0	0	1	4
0	0	0	0	0	0	0	0	1	0	1
0	1	0	0	1	0	0	1	0	0	3
1	0	1	0	0	0	0	0	0	0	2
0	0	0	0	0	0	0	0	0	1	1
0	1	1	1	1	0	0	0	0	1	5
1	1	0	1	0	0	1	1	0	0	5
0	0	0	0	0	0	0	0	1	0	1
										27

Now, as in part (a), we can compute estimated probabilities, with D standing for the outcome "two heads":

First 10 trials: $P(D) = \dfrac{fr(1)}{N} = \dfrac{2}{10} = .2$

First 20 trials: $P(D) = \dfrac{fr(1)}{N} = \dfrac{5}{20} = .25$

First 50 trials: $P(D) = \dfrac{fr(1)}{N} = \dfrac{13}{50} = .26$

100 trials: $P(D) = \dfrac{fr(1)}{N} = \dfrac{27}{100} = .27.$

Q: *What is happening with the data? The probabilities seem to be getting* less accurate *as N increases!*

A: Quite by chance, exactly 5 of the first 20 trials resulted in success, which matches the theoretical probability. Figure 24 shows an Excel plot of estimated probability versus N (for N a multiple of 10). Notice that, as N increases, the graph seems to meander within smaller distances of .25.

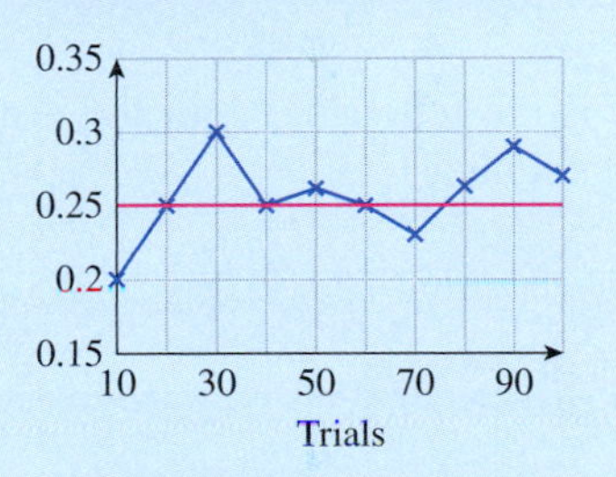

Figure 24

Q: *The previous techniques work fine for simulating coin tosses. What about rolls of a fair die, where we want outcomes between 1 and 6?*

A: We can simulate a roll of a die by generating a random integer in the range 1 through 6. The following formula accomplishes this:

```
=1 + INT(5.99999*RAND())
```

(We used 5.99999 instead of 6 to avoid the outcome 7.)

Section 7.7

Example 1 (page 401) Consider the Markov system found by Gamble Detergents at the beginning of this section. Suppose that 70% of consumers are now using powdered detergent while the other 30% are using liquid. What will be the distribution one year from now? Two years from now? Three years from now?

Solution with Technology

In your spreadsheet, enter the initial distribution vector in cells A1 and B1 and the transition matrix to the right of that, as shown.

	A	B	C	D	E
1	0.7	0.3		0.8	0.2
2				0.1	0.9

To calculate the distribution after one step, use the array formula

```
=MMULT(A1:B1,$D$1:$E$2)
```

The absolute cell references (dollar signs) ensure that the formula always refers to the same transition matrix, even if we copy it into other cells. To use the array formula, select cells A2 and B2, where the distribution vector will go, enter this formula, and then press Control+Shift+Enter.[19]

	A	B	C	D	E
1	0.7	0.3		0.8	0.2
2	=MMULT(A1:B1,$D$1:$E$2)			0.1	0.9

The result is the following, with the distribution after one step highlighted.

	A	B	C	D	E
1	0.7	0.3		0.8	0.2
2	0.59	0.41		0.1	0.9

To calculate the distribution after two steps, select cells A2 and B2 and drag the fill handle down to copy the formula to cells A3 and B3. Note that the formula now takes the vector in A2:B2 and multiplies it by the transition matrix to get the vector in A3:B3. To calculate several more steps, drag down as far as desired.

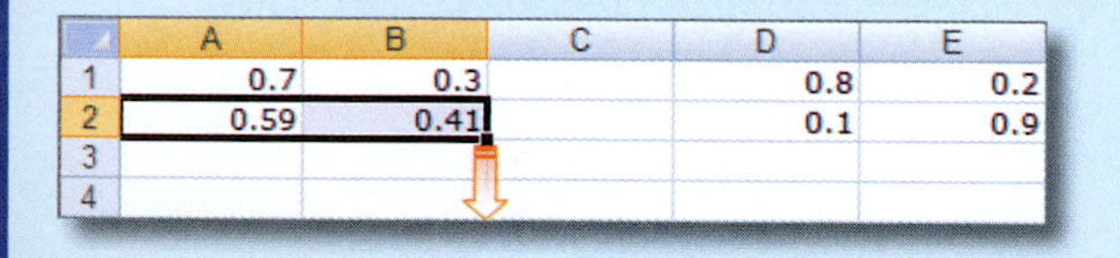

	A	B	C	D	E
1	0.7	0.3		0.8	0.2
2	0.59	0.41		0.1	0.9
3	0.513	0.487			
4	0.4591	0.5409			

Example 4 (page 405) Calculate the steady-state probability vector for the transition matrix in the preceding examples.

Solution with Technology

Finding the steady-state probability vector comes down to solving a system of equations. As discussed in Chapters 3 and 4, there are several ways to use Excel to help. The most straightforward is to use matrix inversion to solve the matrix form of the system. In this case, as in the text, the system of equations we need to solve is

$$\begin{aligned} x + \quad y &= 1 \\ -.2x + .1y &= 0. \end{aligned}$$

We write this as the matrix equation $AX = B$ with

$$A = \begin{bmatrix} 1 & 1 \\ -.2 & .1 \end{bmatrix} \qquad B = \begin{bmatrix} 1 \\ 0 \end{bmatrix}.$$

We enter A in cells A1:B2, B in cells D1:D2, and the formula for $X = A^{-1}B$ in a convenient location, say B4:B5.

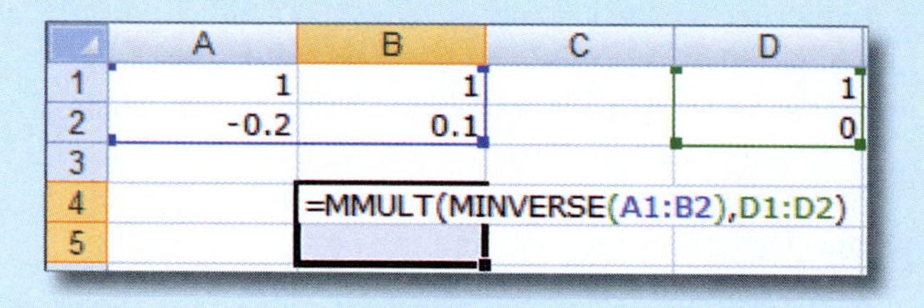

When we press Control+Shift+Enter we see the result:

	A	B	C	D
1	1	1		1
2	-0.2	0.1		0
3				
4		0.3333333		
5		0.6666667		

If we want to see the answer in fraction rather than decimal form, we format the cells as fractions.

	A	B	C	D
1	1	1		1
2	-0.2	0.1		0
3				
4		1/3		
5		2/3		

[19]On a Macintosh, you can also use Command+Enter.

8

Random Variables and Statistics

 **Website**

www.WanerMath.com

At the Website you will find:

- Section-by-section tutorials, including game tutorials with randomized quizzes
- A detailed chapter summary
- A true/false quiz
- Additional review exercises
- Histogram, Bernoulli trials, and normal distribution utilities
- The following optional extra sections:

  Sampling Distributions and the Central Limit Theorem

  Confidence Intervals

  Calculus and Statistics

Case Study Spotting Tax Fraud with Benford's Law

You are a tax fraud specialist working for the Internal Revenue Service (IRS), and you have just been handed a portion of the tax return from Colossal Conglomerate. The IRS suspects that the portion you were handed may be fraudulent and would like your opinion. **Is there any mathematical test, you wonder, that can point to a suspicious tax return based on nothing more than the numbers entered?**

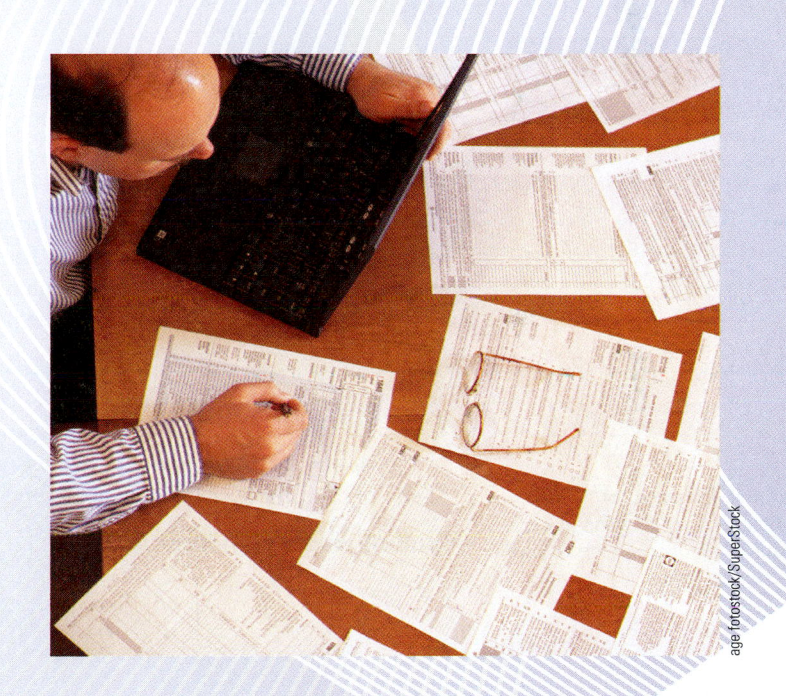

age fotostock/SuperStock

Introduction

Statistics is the branch of mathematics concerned with organizing, analyzing, and interpreting numerical data. For example, given the current annual incomes of 1,000 lawyers selected at random, you might wish to answer some questions: If I become a lawyer, what income am I likely to earn? Do lawyers' salaries vary widely? How widely?

To answer questions like these, it helps to begin by organizing the data in the form of tables or graphs. This is the topic of the first section of the chapter. The second section describes an important class of examples that are applicable to a wide range of situations, from tossing a coin to product testing.

Once the data are organized, the next step is to apply mathematical tools for analyzing the data and answering questions like those posed above. Numbers such as the **mean** and the **standard deviation** can be computed to reveal interesting facts about the data. These numbers can then be used to make predictions about future events.

The chapter ends with a section on one of the most important distributions in statistics, the **normal distribution**. This distribution describes many sets of data and also plays an important role in the underlying mathematical theory.

8.1 Random Variables and Distributions

Random Variables

In many experiments we can assign numerical values to the outcomes. For instance, if we roll a die, each outcome has a value from 1 through 6. If you select a lawyer and ascertain his or her annual income, the outcome is again a number. We call a rule that assigns a number to each outcome of an experiment a **random variable**.

Random Variable

A **random variable** X is a rule that assigns a number, or **value**, to each outcome in the sample space of an experiment.*

* In the language of functions (Chapter 1), a random variable is a *real-valued function* whose domain is the sample space.

Visualizing a Random Variable

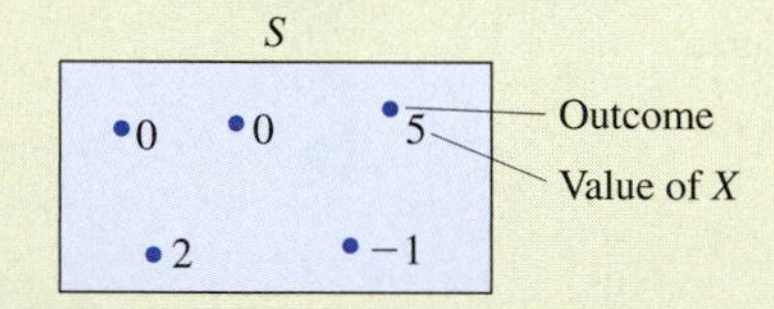

Quick Examples

1. Roll a die; X = the number facing up.
2. Select a mutual fund; X = the number of companies in the fund portfolio.
3. Select a computer; X = the number of gigabytes of memory it has.
4. Survey a group of 20 college students; X = the mean SAT.

Discrete and Continuous Random Variables

A **discrete** random variable can take on only specific, isolated numerical values, like the outcome of a roll of a die, or the number of dollars in a randomly chosen bank account. A **continuous** random variable, on the other hand, can take on any values within a continuum or an interval, like the temperature in Central Park, or the height of an athlete in centimeters. Discrete random variables that can take on only finitely many values (like the outcome of a roll of a die) are called **finite** random variables.

Quick Examples

Random Variable	Values	Type
1. Select a mutual fund; $X =$ the number of companies in the fund portfolio.	$\{1, 2, 3, \dots\}$	Discrete Infinite
2. Take five shots at the goal during a soccer match; $X =$ the number of times you score.	$\{0, 1, 2, 3, 4, 5\}$	Finite
3. Measure the length of an object; $X =$ its length in centimeters.	Any positive real number	Continuous
4. Roll a die until you get a 6; $X =$ the number of times you roll the die.	$\{1, 2, \dots\}$	Discrete Infinite
5. Bet a whole number of dollars in a race where the betting limit is \$100; $X =$ the amount you bet.	$\{0, 1, \dots, 100\}$	Finite
6. Bet a whole number of dollars in a race where there is no betting limit; $X =$ the amount you bet.	$\{0, 1, \dots, 100, 101, \dots\}$	Discrete Infinite

Notes

1. In Chapter 7, the only sample spaces we considered in detail were finite sample spaces. However, in general, sample spaces can be infinite, as in many of the experiments mentioned above.
2. There are some "borderline" situations. For instance, if X is the salary of a factory worker, then X is, strictly speaking, discrete. However, the values of X are so numerous and close together that in some applications it makes sense to model X as a continuous random variable. ■

For the moment, we shall consider only finite random variables.

EXAMPLE 1 Finite Random Variable

Let X be the number of heads that come up when a coin is tossed three times. List the value of X for each possible outcome. What are the possible values of X?

2 Heads ($X = 2$)

Solution First, we describe X as a random variable.

X is the rule that assigns to each outcome the number of heads that come up.

We take as the outcomes of this experiment all possible sequences of three heads and tails. Then, for instance, if the outcome is HTH, the value of X is 2. An easy way to list the values of X for all the outcomes is by means of a table.

Outcome	HHH	HHT	HTH	HTT	THH	THT	TTH	TTT
Value of X	3	2	2	1	2	1	1	0

Website
www.WanerMath.com
Go to the Chapter 8 Topic Summary to find an interactive simulation based on Example 1.

From the table, we also see that the possible values of X are 0, 1, 2, and 3.

Before we go on... Remember that X is just a rule we decide on. In Example 1, we could have taken X to be a different rule, such as the number of tails or perhaps the number of heads minus the number of tails. These different rules are examples of different random variables associated with the same experiment. ■

EXAMPLE 2 Stock Prices

You have purchased \$10,000 worth of stock in a biotech company whose newest arthritis drug is awaiting approval by the Food and Drug Administration (FDA). If the drug is approved this month, the value of the stock will double by the end of the month. If the drug is rejected this month, the stock's value will decline by 80%. If no decision is reached this month, its value will decline by 10%. Let X be the value of your investment at the end of this month. List the value of X for each possible outcome.

Solution There are three possible outcomes: the drug is approved this month, it is rejected this month, and no decision is reached. Once again, we express the random variable as a rule.

The random variable X is the rule that assigns to each outcome the value of your investment at the end of this month.

We can now tabulate the values of X as follows:

Outcome	Approved This Month	Rejected This Month	No Decision
Value of X	\$20,000	\$2,000	\$9,000

Probability Distribution of a Finite Random Variable

Given a random variable X, it is natural to look at certain *events*—for instance, the event that $X = 2$. By this, we mean the event consisting of all outcomes that have an assigned X-value of 2. Looking once again at the chart in Example 1, with X being the number of heads that face up when a coin is tossed three times, we find the following events:

The event that $X = 0$ is {TTT}.
The event that $X = 1$ is {HTT, THT, TTH}.
The event that $X = 2$ is {HHT, HTH, THH}.
The event that $X = 3$ is {HHH}.
The event that $X = 4$ is $\emptyset$. There are no outcomes with four heads.

Each of these events has a certain probability. For instance, the probability of the event that $X = 2$ is $3/8$ because the event in question consists of three of the eight possible (equally likely) outcomes. We shall abbreviate this by writing

$$P(X = 2) = \frac{3}{8}.$$ The probability that $X = 2$ is $3/8$.

Similarly,

$$P(X = 4) = 0.$$ The probability that $X = 4$ is 0.

When X is a finite random variable, the collection of the probabilities of X equaling each of its possible values is called the **probability distribution** of X. Because the probabilities in a probability distribution can be estimated or theoretical, we shall discuss both *estimated probability distributions* (or *relative frequency distributions*) and *theoretical (modeled) probability distributions* of random variables. (See the next two examples.)

Probability Distribution of a Finite Random Variable

If X is a finite random variable, with values $n_1, n_2, \ldots$ then its **probability distribution** lists the probabilities that $X = n_1$, $X = n_2$, ... The sum of these probabilities is always 1.

Visualizing the Probability Distribution of a Random Variable

If each outcome in S is equally likely, we get the probability distribution shown for the random variable X.

S

•0 •0 •5
•2 •−1

Probability Distribution of X

x	-1	0	2	5
$P(X = x)$	$\frac{1}{5} = .2$	$\frac{2}{5} = .4$	$\frac{1}{5} = .2$	$\frac{1}{5} = .2$

Here $P(X = x)$ means "the probability that the random variable X has the specific value x."

Quick Example

Roll a fair die; $X =$ the number facing up. Then, the probability that any specific value of X occurs is $\frac{1}{6}$. So, the probability distribution of X is the following (notice that the probabilities add up to 1):

x	1	2	3	4	5	6
$P(X = x)$	$\frac{1}{6}$	$\frac{1}{6}$	$\frac{1}{6}$	$\frac{1}{6}$	$\frac{1}{6}$	$\frac{1}{6}$

Using this probability distribution, we can calculate the probabilities of certain events; for instance:

$$P(X < 3) = \frac{1}{3}$$ The event that $X < 3$ is the event $\{1, 2\}$.

$$P(1 < X < 5) = \frac{1}{2}$$ The event that $1 < X < 5$ is the event $\{2, 3, 4\}$.

Note The distinction between X (uppercase) and x (lowercase) in the tables above is important; X stands for the random variable in question, whereas x stands for a specific *value* of X (so that x is always a number). Thus, if, say $x = 2$, then $P(X = x)$ means $P(X = 2)$, the probability that X is 2. Similarly, if Y is a random variable, then $P(Y = y)$ is the probability that Y has the specific value y. ■

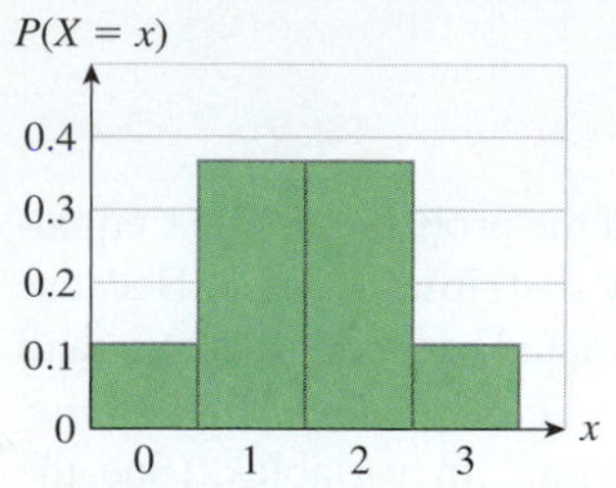

Figure 1

EXAMPLE 3 Probability Distribution

Let X be the number of heads that face up in three tosses of a coin. Give the probability distribution of X. What is the probability of throwing at least two heads?

Solution X is the random variable of Example 1, so its values are 0, 1, 2, and 3. The probability distribution of X is given in the following table:

x	0	1	2	3
$P(X = x)$	$\frac{1}{8}$	$\frac{3}{8}$	$\frac{3}{8}$	$\frac{1}{8}$

Notice that the probabilities add to 1, as we might expect. From the distribution, the probability of throwing at least two heads is

$$P(X \geq 2) = P(\{2, 3\}) = \frac{3}{8} + \frac{1}{8} = \frac{1}{2}.$$

We can use a bar graph to visualize a probability distribution. Figure 1 shows the bar graph for the probability distribution we obtained. Such a graph is sometimes called a **histogram**.

using Technology

Technology can be used to replicate the histogram in Example 3.

TI-83/84 Plus

STAT EDIT values of x in L_1 and probabilities in L_2

WINDOW $0 \leq X \leq 4$, $0 \leq Y \leq 0.5$, Xscl = 1

2ND Y= STAT PLOT on, Histogram icon, Xlist = L_1 Freq = L_2, then GRAPH

[More details on page 465.]

Spreadsheet

x-values and probabilities in Columns A and B.

Highlight probabilities (column B only) and insert a column chart. Right click on resulting graph and use Select Data to set the category labels to be the values of x.

[More details on page 467.]

Website

www.WanerMath.com

Student Home

→ On Line Utilities

→ Histogram Utility

Enter the x-values and probabilities as shown:

```
0, 1/8
1, 3/8
2, 3/8
3, 1/8
```

Make sure "Show histogram" is checked, and press "Results".

The online simulation for Example 1 in the Chapter 8 Summary also gives relative frequencies for the random variable of Example 3.

Before we go on... The probabilities in the table in Example 3 are *modeled* probabilities. To obtain a similar table of relative frequencies, we would have to repeatedly toss a coin three times and calculate the fraction of times we got 0, 1, 2, and 3 heads. ■

Note The table of probabilities in Example 3 looks like the probability distribution associated with an experiment, as we studied in Section 7.3. In fact, the probability distribution of a random variable is not really new. Consider the following experiment: toss three coins and count the number of heads. The associated probability distribution (as per Section 7.3) would be this:

Outcome	0	1	2	3
Probability	$\frac{1}{8}$	$\frac{3}{8}$	$\frac{3}{8}$	$\frac{1}{8}$

The difference is that in this chapter we are thinking of 0, 1, 2, and 3 not as the outcomes of the experiment, but as values of the random variable X. ■

EXAMPLE 4 Relative Frequency Distribution

The following table shows the (fictitious) income brackets of a sample of 1,000 lawyers in their first year out of law school.

Income Bracket	\$20,000–\$29,999	\$30,000–\$39,999	\$40,000–\$49,999	\$50,000–\$59,999	\$60,000–\$69,999	\$70,000–\$79,999	\$80,000–\$89,999
Number	20	80	230	400	170	70	30

Think of the experiment of choosing a first-year lawyer at random (all being equally likely) and assign to each lawyer the number X that is the midpoint of his or her income bracket. Find the relative frequency distribution of X.

Solution Statisticians refer to the income brackets as **measurement classes**. Because the first measurement class contains incomes that are at least \$20,000, but less than \$30,000, its midpoint is \$25,000.* Similarly the second measurement class has midpoint \$35,000, and so on. We can rewrite the table with the midpoints, as follows:

x	25,000	35,000	45,000	55,000	65,000	75,000	85,000
Frequency	20	80	230	400	170	70	30

We have used the term *frequency* rather than *number*, although it means the same thing. This table is called a **frequency table**. It is *almost* the relative frequency distribution for X, except that we must replace frequencies by relative frequencies. (We did this in calculating relative frequencies in the preceding chapter.) We start with the lowest measurement class. Because 20 of the 1,000 lawyers fall in this group, we have

$$P(X = 25{,}000) = \frac{20}{1{,}000} = .02.$$

We can calculate the remaining relative frequencies similarly to obtain the following distribution:

x	25,000	35,000	45,000	55,000	65,000	75,000	85,000
$P(X = x)$	.02	.08	.23	.40	.17	.07	.03

Note again the distinction between X and x: X stands for the random variable in question, whereas x stands for a specific value (25,000, 35,000, . . . , or 85,000) of X.

* One might argue that the midpoint should be $(20{,}000 + 29{,}999)/2 = 24{,}999.50$, but we round this to 25,000. So, technically we are using "rounded" midpoints of the measurement classes.

using Technology

Technology can be used to automate the calculations in Example 4. Here is an outline.

TI-83/84 Plus

STAT EDIT values of x in L_1 and frequencies in L_2.
Home screen: `L2/sum(L2)→L3`
[More details on page 465.]

Spreadsheet

Headings x, Fr, and $P(X = x)$ in A1–C1
x-values and frequencies in Columns A2–B8
`=B2/SUM(B:B)` in C2
Copy down column C.
[More details on page 468.]

Website

www.WanerMath.com
Student Home
→ On Line Utilities
→ Histogram Utility

Enter the x-values and frequencies as shown:

```
25000, 20
35000, 80
45000, 230
55000, 400
65000, 170
75000, 70
85000, 30
```

Make sure "Show probability distribution" is checked, and press "Results". The relative frequency distribution will appear at the bottom of the page.

EXAMPLE 5 Probability Distribution: Greenhouse Gases

The following table shows per capita emissions of greenhouse gases for the 30 countries with the highest per capita carbon dioxide emissions. (Emissions are rounded to the nearest 5 metric tons.)[1]

Country	*Per Capita Emissions (metric tons)*	*Country*	*Per Capita Emissions (metric tons)*
Qatar	70	Estonia	15
Kuwait	40	Faroe Islands	15
United Arab Emirates	40	Saudi Arabia	15
Luxembourg	25	Kazakhstan	15
Trinidad and Tobago	25	Gibraltar	15
Brunei	25	Finland	15
Bahrain	25	Oman	15
Netherlands Antilles	20	Singapore	10
Aruba	20	Palau	10
United States	20	Montserrat	10
Canada	20	Czech Republic	10
Norway	20	Equatorial Guinea	10
Australia	15	New Caledonia	10
Falkland Islands	15	Israel	10
Nauru	15	Russia	10

[1] Figures are based on 2004 data. Source: United Nations Millennium Development Goals Indicators (http://mdgs.un.org/unsd/mdg/Data.aspx).

Consider the experiment in which a country is selected at random from this list, and let X be the per capita carbon dioxide emissions for that country. Find the probability distribution of X and graph it with a histogram. Use the probability distribution to compute $P(X \geq 20)$ (the probability that X is 20 or more) and interpret the result.

Solution The values of X are the possible emissions figures, which we can take to be 0, 5, 10, 15, . . . , 70. In the table below, we first compute the frequency of each value of X by counting the number of countries that produce that per capita level of greenhouse gases. For instance, there are seven countries that have $X = 15$. Then, we divide each frequency by the sample size $N = 30$ to obtain the probabilities.*

* Even though we are using the term "frequency," we are really calculating *modeled* probability based on the assumption of equally likely outcomes. In this context, the frequencies are the number of favorable outcomes for each value of X. (See the Q&A discussion at the end of Section 7.3.)

x	0	5	10	15	20	25	30	35	40	45	50	55	60	65	70
Frequency	0	0	8	10	5	4	0	0	2	0	0	0	0	0	1
$P(X = x)$	0	0	$\frac{8}{30}$	$\frac{10}{30}$	$\frac{5}{30}$	$\frac{4}{30}$	0	0	$\frac{2}{30}$	0	0	0	0	0	$\frac{1}{30}$

Figure 2 shows the resulting histogram.

Finally, we compute $P(X \geq 20)$, the probability of the event that X has a value of 20 or more, which is the sum of the probabilities $P(X = 20)$, $P(X = 25)$, and so on. From the table, we obtain

$$P(X \geq 20) = \frac{5}{30} + \frac{4}{30} + \frac{2}{30} + \frac{1}{30} = \frac{12}{30} = .4.$$

Thus, there is a 40% chance that a country randomly selected from the given list produces 20 or more metric tons per capita of carbon dioxide.

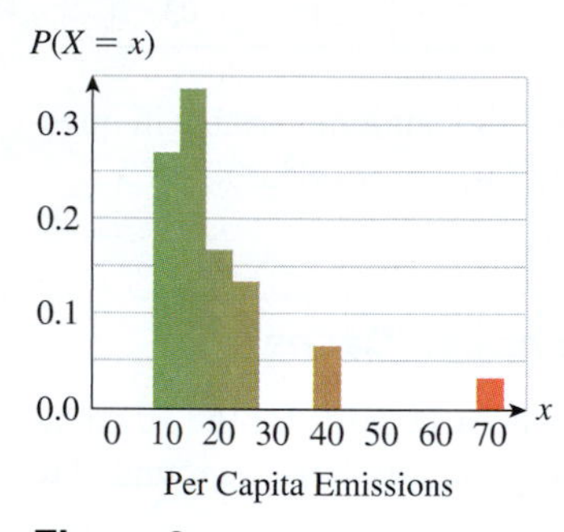

Figure 2

FAQs

Recognizing What to Use as a Random Variable and Deciding on Its Values

Q: *In an application, how, exactly, do I decide what to use as a random variable X?*

A: Be as systematic as possible: First, decide what the experiment is and what its sample space is. Then, based on what is asked for in the application, complete the following sentence: "X assigns ___ to each outcome." For instance, "X assigns the number of flavors to each packet of gummy bears selected," or "X assigns the average faculty salary to each college selected."

Q: *Once I have decided what X should be, how do I decide what values to assign it?*

A: Ask yourself: What are the conceivable values I could get for X? Then choose a collection of values that includes all of these. For instance, if X is the number of heads obtained when a coin is tossed five times, then the possible values of X are 0, 1, 2, 3, 4, and 5. If X is the average faculty salary in dollars, rounded to the nearest \$5,000, then possible values of X could be 20,000, 25,000, 30,000, and so on, up to the highest salary in your data.

8.1 EXERCISES

Access end-of-section exercises online at **www.webassign.net** ENHANCED WebAssign

8.2 Bernoulli Trials and Binomial Random Variables

Your electronic production plant produces video game joysticks. Unfortunately, quality control at the plant leaves much to be desired, and 10% of the joysticks the plant produces are defective. A large corporation has expressed interest in adopting your product for its new game console, and today an inspection team will be visiting to test video game joysticks as they come off the assembly line. If the team tests five joysticks, what is the probability that none will be defective? What is the probability that more than one will be defective?

In this scenario we are interested in the following, which is an example of a particular type of finite random variable called a **binomial random variable**: Think of the experiment as a sequence of five "trials" (in each trial the inspection team chooses one joystick at random and tests it) each with two possible outcomes: "success" (a defective joystick) and "failure" (a non-defective one).* If we now take X to be the number of successes (defective joysticks) the inspection team finds, we can recast the questions above as follows: Find $P(X = 0)$ and $P(X > 1)$.

* These are customary names for the two possible outcomes, and often do not indicate actual success or failure at anything. "Success" is the label we give the outcome of interest—in this case, finding a defective joystick.

† Jakob Bernoulli (1654–1705) was one of the pioneers of probability theory.

Bernoulli Trial

A **Bernoulli† trial** is an experiment that has two possible outcomes, called **success** and **failure**. If the probability of success is p, then the probability of failure is $q = 1 - p$.

Visualizing a Bernoulli Trial

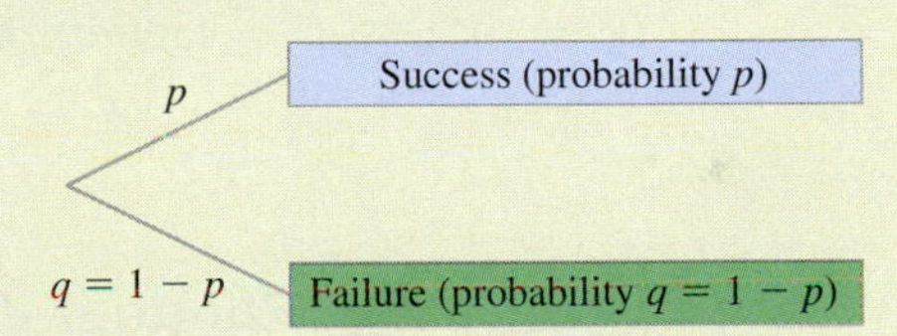

Tossing a coin three times is an example of a **sequence of independent Bernoulli trials**: a sequence of Bernoulli trials in which the outcomes in any one trial are independent (in the sense of the preceding chapter) of those in any other trial, and in which the probability of success is the same for all the trials.

Quick Examples

1. Roll a die, and take success to be the event that you roll a 6. Then $p = 1/6$ and $q = 5/6$. Rolling the die 10 times is then an example of a sequence of 10 independent Bernoulli trials.
2. Provide a property with flood insurance for 20 years, and take success to be the event that the property is flooded during a particular year. Observing whether or not the property is flooded each year for 20 years is then an example of 20 independent Bernoulli trials (assuming that the occurrence of flooding one year is independent of whether there was flooding in earlier years).

3. You know that 60% of all bond funds will depreciate in value next year. Take success to be the event that a randomly chosen fund depreciates next year. Then $p = .6$ and $q = .4$. Choosing five funds at random for your portfolio from a very large number of possible funds is, approximately,* an example of five independent Bernolli trials.

* Choosing a "loser" (a fund that will depreciate next year) slightly depletes the pool of "losers" and hence slightly decreases the probability of choosing another one. However, the fact that the pool of funds is very large means that this decrease is extremely small. Hence, p is very nearly constant.

4. Suppose that E is an event in an experiment with sample space S. Then we can think of the experiment as a Bernoulli trial with two outcomes; success if E occurs, and failure if E' occurs. The probability of success is then

$$p = P(E) \qquad \text{Success is the occurrence of } E.$$

and the probability of failure is

$$q = P(E') = 1 - P(E) = 1 - p. \qquad \text{Failure is the occurrence of } E'.$$

Repeating the experiment 30 times, say, is then an example of 30 independent Bernoulli trials.

Note Quick Example 4 above tells us that Bernoulli trials are not very special kinds of experiments; in fact, we are performing a Bernoulli trial every time we repeat *any* experiment and observe whether a specific event E occurs. Thinking of an experiment in this way amounts, mathematically, to thinking of $\{E, E'\}$ as our sample space (E = success, E' = failure). ■

Binomial Random Variable

A **binomial random variable** is one that counts the number of successes in a sequence of independent Bernoulli trials, where the number of trials is fixed.

Visualizing a Binomial Random Variable

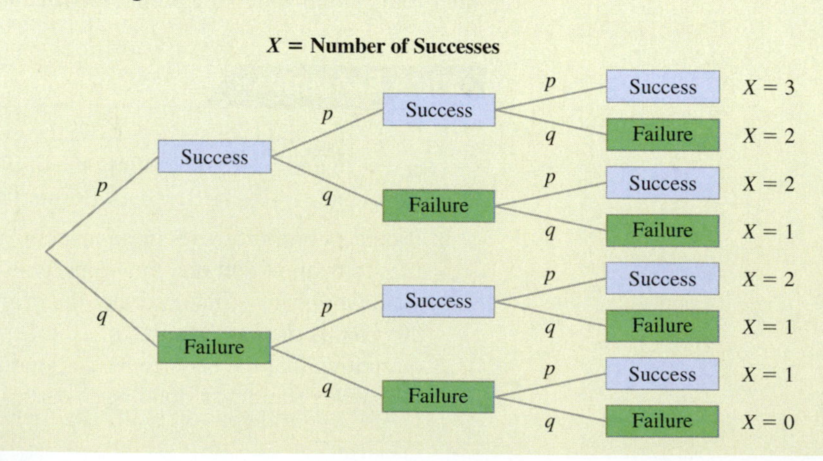

Quick Examples

1. Roll a die 10 times; X is the number of times you roll a 6.
2. Provide a property with flood insurance for 20 years; X is the number of years, during the 20-year period, during which the property is flooded (assuming that the occurrence of flooding one year is independent of whether there was flooding in earlier years).
3. You know that 60% of all bond funds will depreciate in value next year, and you randomly select four from a very large number of possible choices; X is the number of bond funds you hold that will depreciate next year. (X is approximately binomial; see the margin note on p. 428.)

EXAMPLE 1 Probability Distribution of a Binomial Random Variable

Suppose that we have a possibly unfair coin with the probability of heads p and the probability of tails $q = 1 - p$.

a. Let X be the number of heads you get in a sequence of five tosses. Find $P(X = 2)$.

b. Let X be the number of heads you get in a sequence of n tosses. Find $P(X = x)$.

Solution

a. We are looking for the probability of getting exactly two heads in a sequence of five tosses. Let's start with a simpler question: What is the probability that we will get the sequence HHTTT?

The probability that the first toss will come up heads is p.

The probability that the second toss will come up heads is also p.

The probability that the third toss will come up tails is q.

The probability that the fourth toss will come up tails is q.

The probability that the fifth toss will come up tails is q.

The probability that the first toss will be heads *and* the second will be heads *and* the third will be tails *and* the fourth will be tails *and* the fifth will be tails equals the probability of the *intersection* of these five events. Because these are independent events, the probability of the intersection is the product of the probabilities, which is

$$p \times p \times q \times q \times q = p^2q^3.$$

Now HHTTT is only one of several outcomes with two heads and three tails. Two others are HTHTT and TTTHH. How many such outcomes are there altogether? This is the number of "words" with two H's and three T's, and we know from Chapter 6 that the answer is $C(5, 2) = 10$.

Each of the 10 outcomes with two H's and three T's has the same probability: p^2q^3. (Why?) Thus, the probability of getting one of these 10 outcomes is the probability of the union of all these (mutually exclusive) events, and we saw in the preceding chapter that this is just the sum of the probabilities. In other words, the probability we are after is

$$\begin{aligned} P(X = 2) &= p^2q^3 + p^2q^3 + \cdots + p^2q^3 \qquad C(5, 2) \text{ times} \\ &= C(5, 2)p^2q^3. \end{aligned}$$

The structure of this formula is as follows:

$$P(X = 2) = C(5, 2)p^2q^3$$

Number of heads (exponent of p) Number of tails (exponent of q)

Number of tosses | Number of heads | Probability of heads | Probability of tails

b. What we did using the numbers 5 and 2 in part (a) works as well in general. For the general case, with n tosses and x heads, replace 5 with n and replace 2 with x to get:

$$P(X = x) = C(n, x)p^x q^{n-x}.$$

(Note that the coefficient of q is the number of tails, which is $n - x$.)

The calculation in Example 1 applies to any binomial random variable, so we can say the following:

Probability Distribution of Binomial Random Variables

If X is the number of successes in a sequence of n independent Bernoulli trials, then

$$P(X = x) = C(n, x)p^x q^{n-x},$$

where

n = number of trials
p = probability of success
q = probability of failure = $1 - p$.

Quick Example

If you roll a fair die five times, the probability of throwing exactly two 6s is

$$P(X = 2) = C(5, 2)\left(\frac{1}{6}\right)^2\left(\frac{5}{6}\right)^3 = 10 \times \frac{1}{36} \times \frac{125}{216} \approx .1608.$$

Here, we used $n = 5$ and $p = 1/6$, the probability of rolling a 6 on one roll of the die.

EXAMPLE 2 Aging

By 2030, the probability that a randomly chosen resident in the United States will be 65 years old or older is projected to be .2.[2]

a. What is the probability that, in a randomly selected sample of six U.S. residents, exactly four of them will be 65 or older?

b. If X is the number of people aged 65 or older in a sample of six, construct the probability distribution of X and plot its histogram.

[2] Source: U.S. Census Bureau, Decennial Census, Population Estimates and Projections (www.agingstats.gov/agingstatsdotnet/Main_Site/Data/2008_Documents/Population.aspx).

c. Compute $P(X \le 2)$.

d. Compute $P(X \ge 2)$.

Solution

a. The experiment is a sequence of Bernoulli trials; in each trial we select a person and ascertain his or her age. If we take "success" to mean selection of a person aged 65 or older, then the probability distribution is

$$P(X = x) = C(n, x)p^x q^{n-x}$$

where n = number of trials = 6

p = probability of success = .2

q = probability of failure = .8

So,

$$P(X = 4) = C(6, 4)(.2)^4(.8)^2$$
$$= 15 \times .0016 \times .64 = .01536$$

b. We have already computed $P(X = 4)$. Here are all the calculations:

$$P(X = 0) = C(6, 0)(.2)^0(.8)^6$$
$$= 1 \times 1 \times .262144 = .262144$$

$$P(X = 1) = C(6, 1)(.2)^1(.8)^5$$
$$= 6 \times .2 \times .32768 = .393216$$

$$P(X = 2) = C(6, 2)(.2)^2(.8)^4$$
$$= 15 \times .04 \times .4096 = .24576$$

$$P(X = 3) = C(6, 3)(.2)^3(.8)^3$$
$$= 20 \times .008 \times .512 = .08192$$

$$P(X = 4) = C(6, 4)(.2)^4(.8)^2$$
$$= 15 \times .0016 \times .64 = .01536$$

$$P(X = 5) = C(6, 5)(.2)^5(.8)^1$$
$$= 6 \times .00032 \times .8 = .001536$$

$$P(X = 6) = C(6, 6)(.2)^6(.8)^0$$
$$= 1 \times .000064 \times 1 = .000064.$$

$P(X = x)$

0.4

0.3

0.2

0.1

0

0 1 2 3 4 5 6 x

Figure 3

using Technology

Technology can be used to replicate the histogram in Example 2.

TI-83/84 Plus

Y= screen: `Y1=6 nCr X*0.2^X*0.8^(6-X)`

2ND TBLSET Set Indpt to Ask.

2ND TABLE Enter x-values 0, 1, . . . , 6.

[More details on page 465.]

The probability distribution is therefore as follows:

x	0	1	2	3	4	5	6
$P(X = x)$	.262144	.393216	.24576	.08192	.01536	.001536	.000064

Figure 3 shows its histogram.

c. $P(X \le 2)$—the probability that the number of people selected who are at least 65 years old is either 0, 1, or 2—is the probability of the union of these events and is thus the sum of the three probabilities:

$$P(X \le 2) = P(X = 0) + P(X = 1) + P(X = 2)$$
$$= .262144 + .393216 + .24576$$
$$= .90112.$$

Spreadsheet
Headings x and $P(X = x)$ in A1, B1
x-values 0, 1, . . . , 6 in A2–A8
`=BINOMDIST(A2,6,0,.2,0)` in B2
Copy down to B6.
[More details on page 468.]

 **Website**
www.WanerMath.com
Student Home
→ On Line Utilities
→ Binomial Distribution Utility
Enter $n = 6$ and $p = .2$ and press "Generate Distribution."

d. To compute $P(X \geq 2)$, we *could* compute the sum

$$P(X \geq 2) = P(X = 2) + P(X = 3) + P(X = 4) + P(X = 5) + P(X = 6)$$

but it is far easier to compute the probability of the complement of the event:

$$P(X < 2) = P(X = 0) + P(X = 1)$$
$$= .262144 + .393216 = .65536$$

and then subtract the answer from 1:

$$P(X \geq 2) = 1 - P(X < 2)$$
$$= 1 - .65536 = .34464.$$

FAQs

Terminology and Recognizing when to Use the Binomial Distribution

Q: *What is the difference between Bernoulli trials and a binomial random variable?*

A: A Bernoulli trial is a type of experiment, whereas a binomial random variable is the resulting kind of random variable. More precisely, if your experiment consists of performing a sequence of n Bernoulli trials (think of throwing a dart n times at random points on a dartboard hoping to hit the bull's eye), then the random variable X that counts the number of successes (the number of times you actually hit the bull's eye) is a binomial random variable.

Q: *How do I recognize when a situation gives a binomial random variable?*

A: Make sure that the experiment consists of a sequence of independent Bernoulli trials; that is, a sequence of a fixed number of trials of an experiment that has two outcomes, where the outcome of each trial does not depend on the outcomes in previous trials, and where the probability of success is the same for all the trials. For instance, repeatedly throwing a dart at a dartboard hoping to hit the bull's eye does not constitute a sequence of Bernoulli trials if you adjust your aim each time depending on the outcome of your previous attempt. This dart-throwing experiment can be modeled by a sequence of Bernoulli trials if you make no adjustments after each attempt and your aim does not improve (or deteriorate) with time.

8.2 EXERCISES

Access end-of-section exercises online at **www.webassign.net**

8.3 Measures of Central Tendency

Mean, Median, and Mode of a Set of Data

One day you decide to measure the popularity rating of your statistics instructor, Mr. Pelogrande. Ideally, you should poll all of Mr. Pelogrande's students, which is what statisticians would refer to as the **population**. However, it would be difficult to poll all the members of the population in question (Mr. Pelogrande teaches more than 400 students). Instead, you decide to survey 10 of his students, chosen at random, and ask them to rate Mr. Pelogrande on a scale of 0–100. The survey results in the following set of data:

60, 50, 55, 0, 100, 90, 40, 20, 40, 70.

Such a collection of data is called a **sample**, because the 10 people polled represent only a (small) sample of Mr. Pelogrande's students. We should think of the individual scores 60, 50, 55, . . . as values of a random variable: Choose one of Mr. Pelogrande's students at random and let X be the rating the student gives to Mr. Pelogrande.

How do we distill a single measurement, or **statistic**, from this sample that would describe Mr. Pelogrande's popularity? Perhaps the most commonly used statistic is the **average**, or **mean**, which is computed by adding the scores and dividing the sum by the number of scores in the sample:

$$\text{Sample Mean} = \frac{60 + 50 + 55 + 0 + 100 + 90 + 40 + 20 + 40 + 70}{10}$$

$$= \frac{525}{10} = 52.5.$$

We might then conclude, based on the sample, that Mr. Pelogrande's average popularity rating is about 52.5. The usual notation for the sample mean is $\bar{x}$, and the formula we use to compute it is

$$\bar{x} = \frac{x_1 + x_2 + \cdots + x_n}{n},$$

where $x_1, x_2, \ldots, x_n$ are the values of X in the sample.

A convenient way of writing the sum that appears in the numerator is to use **summation** or **sigma notation**. We write the sum $x_1 + x_2 + \cdots + x_n$ as

$$\sum_{i=1}^{n} x_i.$$

$\sum_{i=1}^{n}$ by itself stands for "the sum, from $i = 1$ to n."

$\sum_{i=1}^{n} x_i$ stands for "the sum of the x_i, from $i = 1$ to n."

We think of i as taking on the values 1, 2, . . . , n in turn, making x_i equal $x_1, x_2, \ldots, x_n$ in turn, and we then add up these values.

Sample and Mean

A **sample** is a sequence of values (or scores) of a random variable X. (The process of collecting such a sequence is sometimes called **sampling** X.) The **sample mean** is the average of the values, or **scores**, in the sample. To compute the sample mean, we use the following formula:

$$\bar{x} = \frac{x_1 + x_2 + \cdots + x_n}{n} = \frac{\sum_{i=1}^{n} x_i}{n}$$

or simply

$$\bar{x} = \frac{\sum_i x_i}{n}.$$

$\sum_i$ stands for "sum over all i"*

Here, n is the **sample size** (number of scores), and $x_1, x_2, \ldots, x_n$ are the individual values.

* In Section 1.4 we simply wrote $\sum x$ for the sum of all the x_i, but here we will use the subscripts to make it easier to interpret formulas in this and the next section.

If the sample $x_1, x_2, \ldots, x_n$ consists of all the values of X from the entire population* (for instance, the ratings given Mr. Pelogrande by *all* of his students), we refer to the mean as the **population mean**, and write it as μ (Greek "mu") instead of $\bar{x}$.

* When we talk about *populations*, the understanding is that the underlying experiment consists of selecting a member of a given population and ascertaining the value of X.

Visualizing the Mean

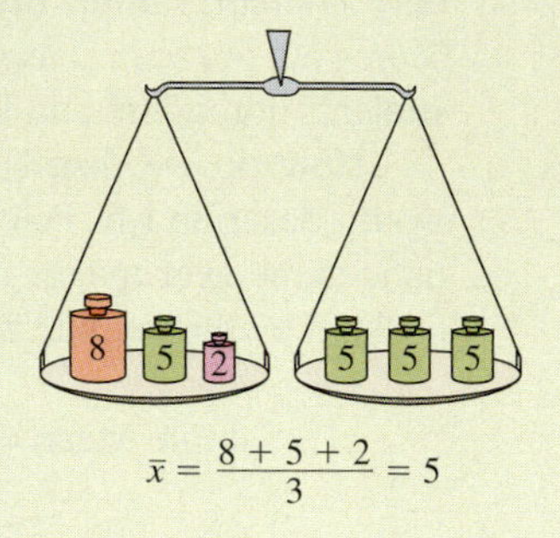

$$\bar{x} = \frac{8+5+2}{3} = 5$$

Quick Examples

1. The mean of the sample 1, 2, 3, 4, 5 is $\bar{x} = 3$.
2. The mean of the sample −1, 0, 2 is $\bar{x} = \dfrac{-1+0+2}{3} = \dfrac{1}{3}$.
3. The mean of the population −3, −3, 0, 0, 1 is

$$\mu = \frac{-3-3+0+0+1}{5} = -1.$$

Note: Sample Mean versus Population Mean

Determining a population mean can be difficult or even impossible. For instance, computing the mean household income for the United States would entail recording the income of every single household in the United States. Instead of attempting to do this, we usually use sample means instead. The larger the sample used, the more accurately we expect the sample mean to approximate the population mean. Estimating how accurately a sample mean based on a given sample size approximates the population mean is possible, but we will not go into that in this book. ■

The mean $\bar{x}$ is an attempt to describe where the "center" of the sample is. It is therefore called a **measure of central tendency**. There are two other common measures of central tendency: the "middle score," or **median**, and the "most frequent score," or **mode**. These are defined as follows.

Median and Mode

The **sample median** m is the middle score (in the case of an odd-size sample), or average of the two middle scores (in the case of an even-size sample) when the scores in a sample are arranged in ascending order.

A **sample mode** is a score that appears most often in the collection. (There may be more than one mode in a sample.)

Visualizing the Median and Mode

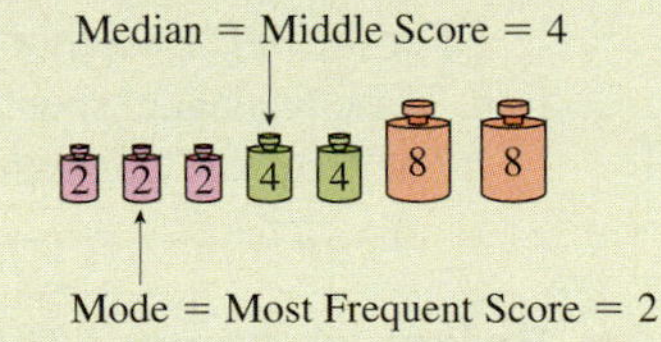

As before, we refer to the **population median** and **population mode** if the sample consists of the data from the entire population.

Quick Examples

1. The sample median of 2, –3, –1, 4, 2 is found by first arranging the scores in ascending order: –3, –1, 2, 2, 4 and then selecting the middle (third) score: $m = 2$. The sample mode is also 2 because this is the score that appears most often.
2. The sample 2, 5, 6, –1, 0, 6 has median $m = (2 + 5)/2 = 3.5$ and mode 6.

The mean tends to give more weight to scores that are further away from the center than does the median. For example, if you take the largest score in a collection of more than two numbers and make it larger, the mean will increase but the median will remain the same. For this reason the median is often preferred for collections that contain a wide range of scores. The mode can sometimes lie far from the center and is thus used less often as an indication of where the "center" of a sample lies.

EXAMPLE 1 Teenage Spending in the 1990s

A 10-year survey of spending patterns of U.S. teenagers in the 1990s yielded the following figures (in billions of dollars spent in a year):[3] 90, 90, 85, 80, 80, 80, 80, 85, 90, 100. Compute and interpret the mean, median, and mode, and illustrate the data on a graph.

Solution The *mean* is given by

$$\bar{x} = \frac{\sum_i x_i}{n}$$

$$= \frac{90 + 90 + 85 + 80 + 80 + 80 + 80 + 85 + 90 + 100}{10} = \frac{860}{10} = 86.$$

Thus, spending by teenagers averaged \$86 billion per year.

For the *median*, we arrange the sample data in ascending order:

80, 80, 80, 80, 85, 85, 90, 90, 90, 100.

We then take the average of the two middle scores:

$$m = \frac{85 + 85}{2} = 85.$$

[3]Spending figures are rounded, and cover the years 1988 through 1997. Source: Rand Youth Poll/Teen-Age Research Unlimited/*New York Times*, March 14, 1998, p. D1.

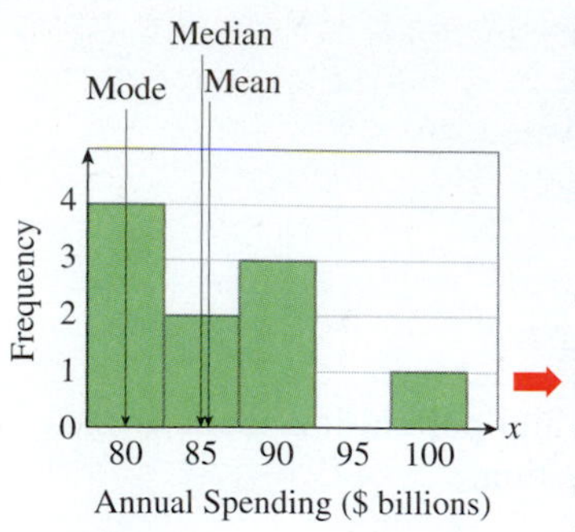

Figure 4

This means that in half the years in question, teenagers spent \$85 billion or less, and in half they spent \$85 billion or more.

For the *mode* we choose the score (or scores) that occurs most frequently: \$80 billion. Thus, teenagers spent \$80 billion per year more often than any other amount.

The frequency histogram in Figure 4 illustrates these three measures.

Before we go on... There is a nice geometric interpretation of the difference between the median and mode: The median line shown in Figure 4 divides the total area of the histogram into two equal pieces, whereas the mean line passes through its "center of gravity"; if you placed the histogram on a knife-edge along the mean line, it would balance. ■

Expected Value of a Finite Random Variable

Now, instead of looking at a sample of values of a given random variable, let us look at the probability distribution of the random variable itself and see if we can predict the sample mean without actually taking a sample. This prediction is what we call the *expected value* of the random variable.

EXAMPLE 2 Expected Value of a Random Variable

Suppose you roll a fair die a large number of times. What do you expect to be the average of the numbers that face up?

Solution Suppose we take a sample of n rolls of the die (where n is large). Because the probability of rolling a 1 is $1/6$, we would expect that we would roll a 1 one sixth of the time, or $n/6$ times. Similarly, each other number should also appear $n/6$ times. The frequency table should then look like this:

x	1	2	3	4	5	6
Number of Times x Is Rolled (frequency)	$\frac{n}{6}$	$\frac{n}{6}$	$\frac{n}{6}$	$\frac{n}{6}$	$\frac{n}{6}$	$\frac{n}{6}$

Note that we would not really expect the scores to be evenly distributed in practice, although for very large values of n we would expect the frequencies to vary only by a small percentage. To calculate the sample mean, we would add up all the scores and divide by the sample size. Now, the table tells us that there are $n/6$ ones, $n/6$ twos, $n/6$ threes, and so on, up to $n/6$ sixes. Adding these all up gives

$$\sum_i x_i = \frac{n}{6}\cdot 1 + \frac{n}{6}\cdot 2 + \frac{n}{6}\cdot 3 + \frac{n}{6}\cdot 4 + \frac{n}{6}\cdot 5 + \frac{n}{6}\cdot 6.$$

(Notice that we can obtain this number by multiplying the frequencies by the values of X and then adding.) Thus, the mean is

$$\begin{aligned}\bar{x} &= \frac{\sum_i x_i}{n} \\ &= \frac{\frac{n}{6}\cdot 1 + \frac{n}{6}\cdot 2 + \frac{n}{6}\cdot 3 + \frac{n}{6}\cdot 4 + \frac{n}{6}\cdot 5 + \frac{n}{6}\cdot 6}{n} \\ &= \frac{1}{6}\cdot 1 + \frac{1}{6}\cdot 2 + \frac{1}{6}\cdot 3 + \frac{1}{6}\cdot 4 + \frac{1}{6}\cdot 5 + \frac{1}{6}\cdot 6 \qquad \text{Divide top and bottom by } n. \\ &= 3.5.\end{aligned}$$

This is the average value we expect to get after a large number of rolls or, in short, the **expected value** of a roll of the die. More precisely, we say that this is the expected value of the random variable X whose value is the number we get by rolling a die. Notice that n, the number of rolls, does not appear in the expected value. In fact, we could redo the calculation more simply by dividing the frequencies in the table by n *before* adding. Doing this replaces the frequencies with the *probabilities*, $1/6$. That is, it *replaces the frequency distribution with the probability distribution.*

x	1	2	3	4	5	6
$P(X = x)$	$\frac{1}{6}$	$\frac{1}{6}$	$\frac{1}{6}$	$\frac{1}{6}$	$\frac{1}{6}$	$\frac{1}{6}$

The expected value of X is then the sum of the products $x \cdot P(X = x)$. This is how we shall compute it from now on.

Expected Value of a Finite Random Variable

If X is a finite random variable that takes on the values $x_1, x_2, \ldots, x_n$, then the **expected value** of X, written $E(X)$ or μ, is

$$\mu = E(X) = x_1 \cdot P(X = x_1) + x_2 \cdot P(X = x_2) + \cdots + x_n \cdot P(X = x_n)$$
$$= \sum_i x_i \cdot P(X = x_i).$$

In Words

To compute the expected value from the probability distribution of X, we multiply the values of X by their probabilities and add up the results.

Interpretation

We interpret the expected value of X as a *prediction* of the mean of a large random sample of measurements of X; in other words, it is what we "expect" the mean of a large number of scores to be. (The larger the sample, the more accurate this prediction will tend to be.)

Quick Example

If X has the distribution shown,

x	-1	0	4	5
$P(X = x)$	.3	.5	.1	.1

then $\mu = E(X) = -1(.3) + 0(.5) + 4(.1) + 5(.1) = -.3 + 0 + .4 + .5 = .6$.

EXAMPLE 3 Sports Injuries

According to historical data, the number of injuries that a member of the Enormous State University women's soccer team will sustain during a typical season is given by the following probability distribution table:

Injuries	0	1	2	3	4	5	6
Probability	.20	.20	.22	.20	.15	.01	.02

using Technology

Technology can be used to compute the expected values in Example 3.

TI-83/84 Plus

STAT EDIT values of X in L_1 and probabilities in L_2.
Home screen: `sum(L1*L2)`
[More details on page 466.]

Spreadsheet

Headings x and $P(X = x)$ in A1, B1.
x-values and probabilities in A2–B8.
`=A2*B2` in C2, copied down to B6.
`=SUM(C2:C8)` in C9.
[More details on page 468.]

Website

www.WanerMath.com
Student Home
→ On Line Utilities
→ Histogram Utility

Enter the x-values and frequencies as shown:

```
0, .2
1, .2
2, .22
3, .2
4, .15
5, .01
6, .02
```

Make sure "Show expected value and standard deviation" is checked, and press "Results." The results will appear at the bottom of the page.

If X denotes the number of injuries sustained by a player during one season, compute $E(X)$ and interpret the result.

Solution We can compute the expected value using the following tabular approach: Take the probability distribution table, add another row in which we compute the product $x P(X = x)$, and then add these products together.

x	0	1	2	3	4	5	6	
$P(X = x)$	.20	.20	.22	.20	.15	.01	.02	**Total:**
$xP(X = x)$	0	.20	.44	.60	.60	.05	.12	2.01

The total of the entries in the bottom row is the expected value. Thus,

$$E(X) = 2.01.$$

We interpret the result as follows: If many soccer players are observed for a season, we predict that the average number of injuries each will sustain is about two.

EXAMPLE 4 Roulette

A roulette wheel (of the kind used in the United States) has the numbers 1 through 36, 0 and 00. A bet on a single number pays 35 to 1. This means that if you place a $1 bet on a single number and win (your number comes up), you get your $1 back plus $35 (that is, you gain $35). If your number does not come up, you lose the $1 you bet. What is the expected gain from a $1 bet on a single number?

Solution The probability of winning is $1/38$, so the probability of losing is $37/38$. Let X be the gain from a $1 bet. X has two possible values: $X = -1$ if you lose and $X = 35$ if you win. $P(X = -1) = 37/38$ and $P(X = 35) = 1/38$. This probability distribution and the calculation of the expected value are given in the following table:

x	-1	35	
$P(X = x)$	$\frac{37}{38}$	$\frac{1}{38}$	**Total:**
$xP(X = x)$	$-\frac{37}{38}$	$\frac{35}{38}$	$-\frac{2}{38}$

So, we expect to average a small loss of $2/38 \approx \$0.0526$ on each spin of the wheel.

➡ **Before we go on...** Of course, you cannot actually lose the expected $0.0526 on one spin of the roulette wheel in Example 4. However, if you play many times, this is what you expect your *average* loss per bet to be. For example, if you played 100 times, you could expect to lose about $100 \times 0.0526 = \$5.26$. ■

A betting game in which the expected value is zero is called a **fair game**. For example, if you and I flip a coin, and I give you $1 each time it comes up heads but you give me $1 each time it comes up tails, then the game is fair. Over the long run, we expect to come out even. On the other hand, a game like roulette, in which the expected value is not zero, is **biased**. Most casino games are slightly biased in favor of

* Only rarely are games not biased in favor of the house. However, blackjack played without continuous shuffle machines can be beaten by card counting.

the house.* Thus, most gamblers will lose only a small amount and many gamblers will actually win something (and return to play some more). However, when the earnings are averaged over the huge numbers of people playing, the house is guaranteed to come out ahead. This is how casinos make (lots of) money.

Expected Value of a Binomial Random Variable

Suppose you guess all the answers to the questions on a multiple-choice test. What score can you expect to get? This scenario is an example of a sequence of Bernoulli trials (see the preceding section), and the number of correct guesses is therefore a binomial random variable whose expected value we wish to know. There is a simple formula for the expected value of a binomial random variable.

Expected Value of Binomial Random Variable

If X is the binomial random variable associated with n independent Bernoulli trials, each with probability p of success, then the expected value of X is

$$\mu = E(X) = np.$$

Quick Examples

1. If X is the number of successes in 20 Bernoulli trials with $p = .7$, then the expected number of successes is $\mu = E(X) = (20)(.7) = 14$.
2. If an event F in some experiment has $P(F) = .25$, the experiment is repeated 100 times, and X is the number of times F occurs, then $E(X) = (100)(.25) = 25$ is the number of times we expect F to occur.

Where does this formula come from? We *could* use the formula for expected value and compute the sum

$$E(X) = 0C(n, 0)p^0q^n + 1C(n, 1)p^1q^{n-1} + 2C(n, 2)p^2q^{n-2} + \cdots + nC(n, n)p^nq^0$$

directly (using the binomial theorem), but this is one of the many places in mathematics where a less direct approach is much easier. X is the number of successes in a sequence of n Bernoulli trials, each with probability p of success. Thus, p is the fraction of time we expect a success, so out of n trials we expect np successes. Because X counts successes, we expect the value of X to be np. (With a little more effort, this can be made into a formal proof that the sum above equals np.)

EXAMPLE 5 Guessing on an Exam

An exam has 50 multiple-choice questions, each having four choices. If a student randomly guesses on each question, how many correct answers can he or she expect to get?

Solution Each guess is a Bernoulli trial with probability of success 1 in 4, so $p = .25$. Thus, for a sequence of $n = 50$ trials,

$$\mu = E(X) = np = (50)(.25) = 12.5.$$

Thus, the student can expect to get about 12.5 correct answers.

Q: *Wait a minute. How can a student get a fraction of a correct answer?*

A: Remember that the expected value is the average number of correct answers a student will get if he or she guesses on a large number of such tests. Or, we can say that if many students use this strategy of guessing, they will average about 12.5 correct answers each.

Estimating the Expected Value from a Sample

It is not always possible to know the probability distribution of a random variable. For instance, if we take X to be the income of a randomly selected lawyer, we could not be expected to know the probability distribution of X. However, we can still obtain a good *estimate* of the expected value of X (the average income of all lawyers) by using the relative frequency distribution based on a large random sample.

EXAMPLE 6 Estimating an Expected Value

The following table shows the (fictitious) incomes of a random sample of 1,000 lawyers in the United States in their first year out of law school.

Income Bracket	\$20,000–\$29,999	\$30,000–\$39,999	\$40,000–\$49,999	\$50,000–\$59,999	\$60,000–\$69,999	\$70,000–\$79,999	\$80,000–\$89,999
Number	20	80	230	400	170	70	30

Estimate the average of the incomes of all lawyers in their first year out of law school.

Solution We first interpret the question in terms of a random variable. Let X be the income of a lawyer selected at random from among all currently practicing first-year lawyers in the United States. We are given a sample of 1,000 values of X, and we are asked to find the expected value of X. First, we use the midpoints of the income brackets to set up a relative frequency distribution for X:

x	25,000	35,000	45,000	55,000	65,000	75,000	85,000
$P(X = x)$	.02	.08	.23	.40	.17	.07	.03

Our estimate for $E(X)$ is then

$$E(X) = \sum_i x_i \cdot P(X = x_i)$$

$$= (25{,}000)(.02) + (35{,}000)(.08) + (45{,}000)(.23) + (55{,}000)(.40) + (65{,}000)(.17) + (75{,}000)(.07) + (85{,}000)(.03) = \$54{,}500.$$

Thus, $E(X)$ is approximately \$54,500. That is, the average income of all currently practicing first-year lawyers in the United States is approximately \$54,500.

FAQ

Recognizing when to Compute the Mean and when to Compute the Expected Value

Q: *When am I supposed to compute the mean (add the values of X and divide by n) and when am I supposed to use the expected value formula?*

A: The formula for the mean (adding and dividing by the number of observations) is used to compute the mean of a sequence of random scores, or sampled values of *X*. If, on the other hand, you are given the probability distribution for *X* (even if it is only an estimated probability distribution), then you need to use the expected value formula.

8.3 EXERCISES

Access end-of-section exercises online at **www.webassign.net**

8.4 Measures of Dispersion

Variance and Standard Deviation of a Set of Scores

Your grade on a recent midterm was 68%; the class average was 72%. How do you stand in comparison with the rest of the class? If the grades were widely scattered, then your grade may be close to the mean and a fair number of people may have done a lot worse than you (Figure 5a). If, on the other hand, almost all the grades were within a few points of the average, then your grade may not be much higher than the lowest grade in the class (Figure 5b).

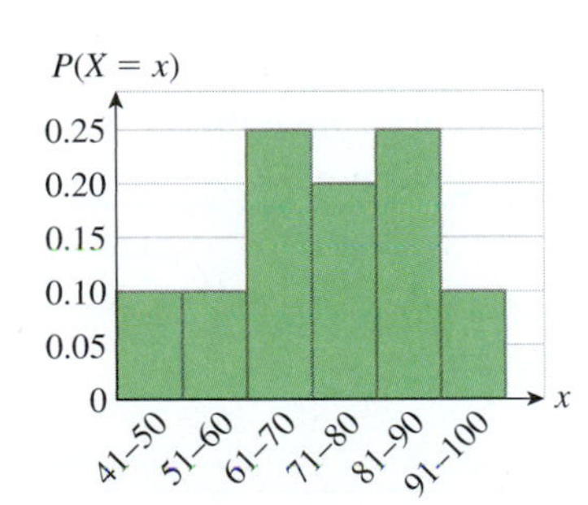

Figure 5(a)

Figure 5(b)

This scenario suggests that it would be useful to have a way of measuring not only the central tendency of a set of scores (mean, median, or mode) but also the amount of "scatter" or "dispersion" of the data.

If the scores in our set are $x_1, x_2, \dots, x_n$ and their mean is $\bar{x}$ (or μ in the case of a population mean), we are really interested in the distribution of the differences $x_i - \bar{x}$. We could compute the *average* of these differences, but this average will always be 0. (Why?) It is really the *sizes* of these differences that interest us, so we might try computing the average of the absolute values of the differences. This idea is reasonable, but it leads to technical difficulties that are avoided by a slightly different approach: The statistic we use is based on the average of the *squares* of the differences, as explained in the following definitions.

Population Variance and Standard Deviation

If the values $x_1, x_2, \dots, x_n$ are all the measurements of X in the entire population, then the **population variance** is given by

$$\sigma^2 = \frac{(x_1 - \mu)^2 + (x_2 - \mu)^2 + \cdots + (x_n - \mu)^2}{n} = \frac{\sum_{i=1}^{n} (x_i - \mu)^2}{n}.$$

(Remember that μ is the symbol we use for the *population* mean.) The **population standard deviation** is the square root of the population variance:

$$\sigma = \sqrt{\sigma^2}.$$

Sample Variance and Standard Deviation

The **sample variance** of a sample $x_1, x_2, \ldots, x_n$ of n values of X is given by

$$s^2 = \frac{(x_1 - \bar{x})^2 + (x_2 - \bar{x})^2 + \cdots + (x_n - \bar{x})^2}{n-1} = \frac{\sum_{i=1}^{n}(x_i - \bar{x})^2}{n-1}.$$

The **sample standard deviation** is the square root of the sample variance:

$$s = \sqrt{s^2}.$$

Visualizing Small and Large Variance

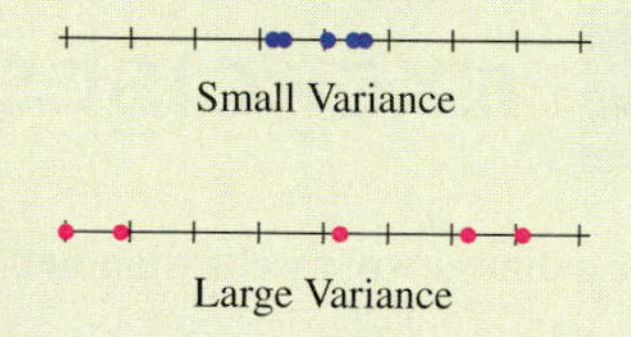

Quick Examples

1. The sample variance of the scores 1, 2, 3, 4, 5 is the sum of the squares of the differences between the scores and the mean $\bar{x} = 3$, divided by $n - 1 = 4$:

$$s^2 = \frac{(1-3)^2 + (2-3)^2 + (3-3)^2 + (4-3)^2 + (5-3)^2}{4}$$
$$= \frac{10}{4} = 2.5$$

so

$$s = \sqrt{2.5} \approx 1.58.$$

2. The population variance of the scores 1, 2, 3, 4, 5 is the sum of the squares of the differences between the scores and the mean $\mu = 3$, divided by $n = 5$:

$$\sigma^2 = \frac{(1-3)^2 + (2-3)^2 + (3-3)^2 + (4-3)^2 + (5-3)^2}{5} = \frac{10}{5} = 2$$

so

$$\sigma = \sqrt{2} \approx 1.41.$$

using Technology

Technology can be used to compute means and standard deviations as follows:

TI-83/84 Plus

STAT EDIT Enter the values $x_1, x_2, \ldots, x_n$ in L_1

STAT CALC `1-Var Stats` ENTER

s will appear as "Sx" and σ as "σx".

Spreadsheet

Enter the values $x_1, x_2, \ldots, x_n$ in column A

`=AVERAGE(A:A)` computes the mean.

`=STDEV(A:A)` computes s.

`=STDEVP(A:A)` computes σ.

Q: *The population variance is the average of the squares of the differences between the values and the mean. But why do we divide by n − 1 instead of n when calculating the sample variance?*

A: In real-life applications, we would like the variance we calculate from a sample to approximate the variance of the whole population. In statistical terms, we would like the expected value of the sample variance s^2 to be the same as the population variance σ^2. The sample variance s^2 as we have defined it is the "unbiased estimator" of the population variance σ^2 that accomplishes this task; if, instead, we divided by n in the formula for s^2, we would, on average, tend to underestimate the population variance. (See the online text on Sampling Distributions at the Website for further discussion of unbiased estimators.) Note that as the sample size gets larger and larger, the discrepancy between the formulas for s^2 and σ^2 becomes negligible; dividing by n gives almost the same answer as dividing by $n-1$. It is traditional, nonetheless, to use the sample variance in preference to the population variance when working with samples, and we do that here. In practice we should not try to draw conclusions about the entire population from samples so small that the difference between the two formulas matters. As one book puts it, "If the difference between n and $n-1$ ever matters to you, then you are probably up to no good anyway—e.g., trying to substantiate a questionable hypothesis with marginal data."*

* W. H. Press, S. A. Teukolsky, E. T. Vetterling, and B. P. Flannery, *Numerical Recipes: The Art of Scientific Computing*, Cambridge University Press, 2007.

Here's a simple example of calculating standard deviation.

EXAMPLE 1 Income

Following is a sample of the incomes (in thousands of dollars) of eight U.S. residents selected at random:[4]

$$50, 40, 60, 20, 90, 10, 30, 20.$$

Compute the sample mean and standard deviation, rounded to one decimal place. What percentage of the scores fall within one standard deviation of the mean? What percentage fall within two standard deviations of the mean?

Solution The sample mean is

$$\bar{x} = \frac{\sum_i x_i}{n} = \frac{50+40+60+20+90+10+30+20}{8} = \frac{320}{8} = 40.$$

The sample variance is

$$\begin{aligned} s^2 &= \frac{\sum_i (x_i - \bar{x})^2}{n-1} \\ &= \frac{1}{7}[(50-40)^2 + (40-40)^2 + (60-40)^2 + (20-40)^2 \\ &\qquad + (90-40)^2 + (10-40)^2 + (30-40)^2 + (20-40)^2] \\ &= \frac{1}{7}(100 + 0 + 400 + 400 + 2{,}500 + 900 + 100 + 400) \\ &= \frac{4{,}800}{7}. \end{aligned}$$

Thus, the sample standard deviation is

$$s = \sqrt{\frac{4{,}800}{7}} \approx 26.2. \qquad \text{Rounded to one decimal place}$$

To ask which scores fall "within one standard deviation of the mean" is to ask which scores fall in the interval $[\bar{x} - s, \bar{x} + s]$, or about $[40 - 26.2, 40 + 26.2] = [13.8, 66.2]$. Six out of the eight scores fall in this interval, so the percentage of scores that fall within one standard deviation of the mean is $6/8 = .75$, or 75%.

For two standard deviations, the interval in question is $[\bar{x} - 2s, \bar{x} + 2s] \approx [40 - 52.4, 40 + 52.4] = [-12.4, 92.4]$, which includes all of the scores. In other words, 100% of the scores fall within two standard deviations of the mean.

[4] The sample is roughly statistically representative of the actual income distribution in the U.S. in 2010 for incomes up to \$100,000. (See Exercise 35.) Source for income distribution: U.S. Census Bureau (www.census.gov).

Q: *In Example 1, 75% of the scores fell within one standard deviation of the mean and all of them fell within two standard deviations of the mean. Is this typical?*

A: Actually, the percentage of scores within a number of standard deviations of the mean depends a great deal on the way the scores are distributed. There are two useful methods for *estimating* the percentage of scores that fall within any number of standard deviations of the mean. The first method applies to any set of data and is due to P.L. Chebyshev (1821–1894), while the second applies to "nice" sets of data and is based on the "normal distribution," which we shall discuss in Section 5.

Chebyshev's Rule

For any set of data, the following statements are true:

At least $3/4$ of the scores fall within two standard deviations of the mean (within the interval $[\bar{x} - 2s, \bar{x} + 2s]$ for samples or $[\mu - 2\sigma, \mu + 2\sigma]$ for populations).

At least $8/9$ of the scores fall within three standard deviations of the mean (within the interval $[\bar{x} - 3s, \bar{x} + 3s]$ for samples or $[\mu - 3\sigma, \mu + 3\sigma]$ for populations).

At least $15/16$ of the scores fall within four standard deviations of the mean (within the interval $[\bar{x} - 4s, \bar{x} + 4s]$ for samples or $[\mu - 4\sigma, \mu + 4\sigma]$ for populations).

. . .

At least $1 - 1/k^2$ of the scores fall within k standard deviations of the mean (within the interval $[\bar{x} - ks, \bar{x} + ks]$ for samples or $[\mu - k\sigma, \mu + k\sigma]$ for populations).

Visualizing Chebyshev's Rule

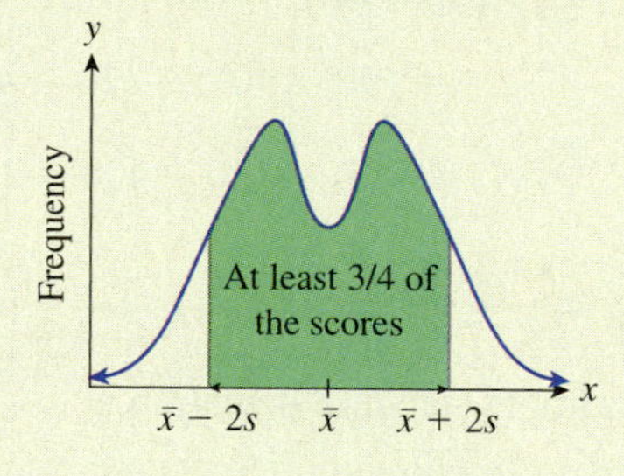

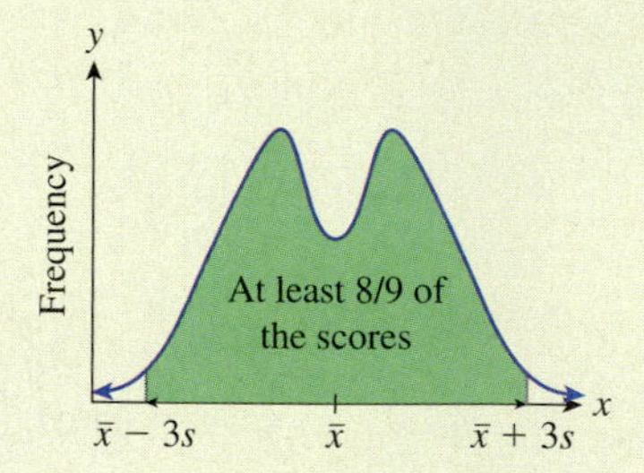

Empirical Rule*

For a set of data whose frequency distribution is bell shaped and symmetric (see Figure 6), the following is true:

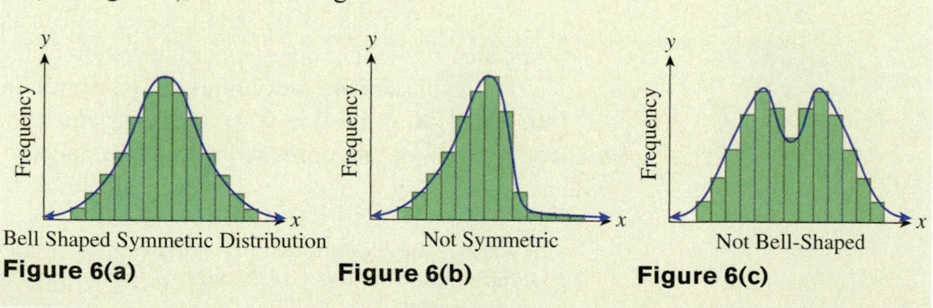

Figure 6(a) **Figure 6(b)** **Figure 6(c)**

* Unlike Chebyshev's rule, which is a precise theorem, the empirical rule is a "rule of thumb" that is intentionally vague about what exactly is meant by a "bell shaped distribution" and "approximately such-and-such %." (As a result, the rule is often stated differently in different textbooks.) We will see in Section 8.5 that if the distribution is a *normal* one, the empirical rule translates to a precise statement.

Approximately 68% of the scores fall within one standard deviation of the mean (within the interval $[\bar{x} - s, \bar{x} + s]$ for samples or $[\mu - \sigma, \mu + \sigma]$ for populations).

Approximately 95% of the scores fall within two standard deviations of the mean (within the interval $[\bar{x} - 2s, \bar{x} + 2s]$ for samples or $[\mu - 2\sigma, \mu + 2\sigma]$ for populations).

Approximately 99.7% of the scores fall within three standard deviations of the mean (within the interval $[\bar{x} - 3s, \bar{x} + 3s]$ for samples or $[\mu - 3\sigma, \mu + 3\sigma]$ for populations).

Visualizing the Empirical Rule

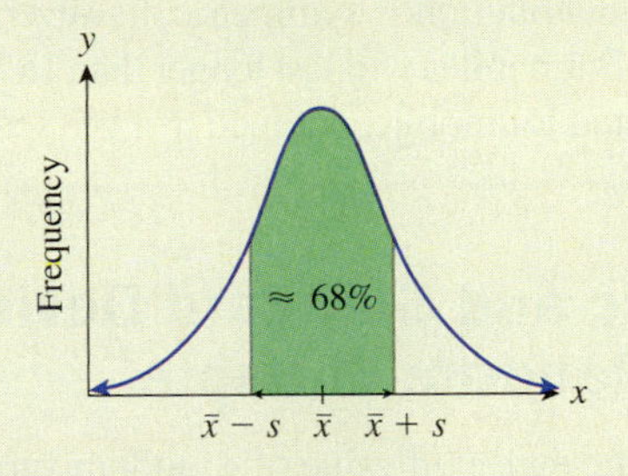

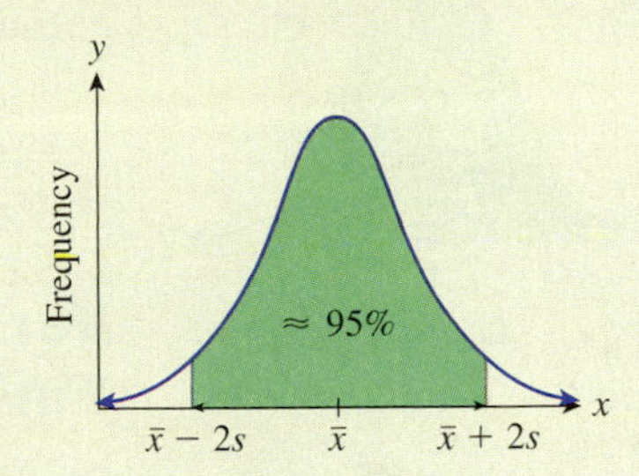

Quick Examples

1. If the mean of a sample is 20 with standard deviation $s = 2$, then at least 15/16, or 93.75% of the scores lie within four standard deviations of the mean—that is, in the interval [12, 28].
2. If the mean of a sample with a bell shaped symmetric distribution is 20 with standard deviation $s = 2$, then approximately 95% of the scores lie in the interval [16, 24].

The empirical rule could not be applied in Example 1. The distribution there is not symmetric (sketch it to see for yourself) and the fact that there were only 8 scores limits the accuracy further. The empirical rule is, however, accurate in distributions that are bell shaped and symmetric, even if not perfectly so. Chebyshev's rule, on the other hand, is always valid (and applies in Example 1 in particular) but tends to be "overcautious" and in practice underestimates how much of a distribution lies in a given interval.

EXAMPLE 2 Automobile Life

The average life span of a Batmobile is 9 years, with a standard deviation of 2 years. My own Batmobile lasted less than 3 years before being condemned to the bat-junkyard.

a. Without any further knowledge about the distribution of Batmobile life spans, what can one say about the percentage of Batmobiles that last less than 3 years?

b. Refine the answer in part (a), assuming that the distribution of Batmobile life spans is bell shaped and symmetric.

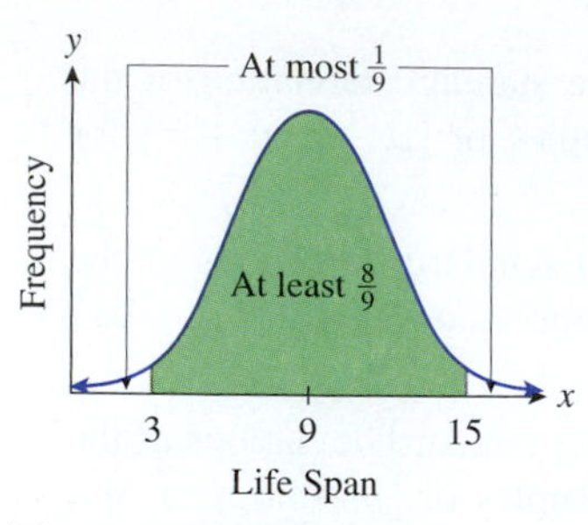

Figure 7

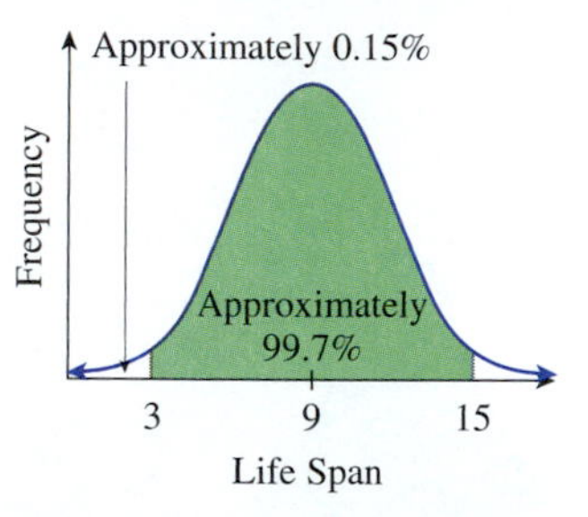

Figure 8

Solution

a. If we are given no further information about the distribution of Batmobile life spans, we need to use Chebyshev's rule. Because the life span of my Batmobile was more than 6 years (or three standard deviations) shorter than the mean, it lies outside the range $[\mu - 3\sigma, \mu + 3\sigma] = [3, 15]$. Because *at least* 8/9 of the life spans of all Batmobiles lie in this range, *at most* 1/9, or 11%, of the life spans lie outside this range (see Figure 7). Some of these, like the life span of my own Batmobile, are less than 3 years, while the rest are more than $\mu + 3\sigma = 15$ years.

b. Because we know more about the distribution now than we did in part (a), we can use the empirical rule and obtain sharper results. The empirical rule predicts that approximately 99.7% of the life spans of Batmobiles lie in the range $[\mu - 3\sigma, \mu + 3\sigma] = [3, 15]$. Thus, approximately $1 - 99.7\% = 0.3\%$ of them lie outside that range. Because the distribution is symmetric, however, more can be said: half of that 0.3%, or 0.15% of Batmobiles will last longer than 15 years, while the other 0.15% are, like my own ill-fated Batmobile, doomed to a life span of less than 3 years (see Figure 8).

Variance and Standard Deviation of a Finite Random Variable

Recall that the expected value of a random variable X is a prediction of the average of a large sample of values of X. Can we similarly predict the variance of a large sample? Suppose we have a sample $x_1, x_2, \ldots, x_n$. If n is large, the sample and population variances are essentially the same, so we concentrate on the population variance, which is the average of the numbers $(x_i - \bar{x})^2$. This average can be predicted using $E([X - \mu]^2)$ the expected value of $(X - \mu)^2$. In general, we make the following definition.

Variance and Standard Deviation of a Finite Random Variable

If X is a finite random variable taking on values $x_1, x_2, \ldots, x_n$, then the **variance** of X is

$$\begin{aligned}\sigma^2 &= E([X - \mu]^2)\\ &= (x_1 - \mu)^2 P(X = x_1) + (x_2 - \mu)^2 P(X = x_2) + \cdots + (x_n - \mu)^2 P(X = x_n)\\ &= \sum_i (x_i - \mu)^2 P(X = x_i).\end{aligned}$$

The **standard deviation** of X is then the square root of the variance:

$$\sigma = \sqrt{\sigma^2}.$$

To compute the variance from the probability distribution of X, first compute the expected value μ and then compute the expected value of $(X - \mu)^2$.

Quick Example

The following distribution has expected value $\mu = E(X) = 2$:

x	−1	2	3	10
$P(X = x)$	.3	.5	.1	.1

The variance of X is

$$\sigma^2 = (x_1 - \mu)^2 P(X = x_1) + (x_2 - \mu)^2 P(X = x_2) + \cdots + (x_n - \mu)^2 P(X = x_n)$$
$$= (-1 - 2)^2(.3) + (2 - 2)^2(.5) + (3 - 2)^2(.1) + (10 - 2)^2(.1) = 9.2.$$

The standard deviation of X is

$$\sigma = \sqrt{9.2} \approx 3.03.$$

Note We can interpret the variance of X as the number we expect to get for the variance of a large sample of values of X, and similarly for the standard deviation. ■

We can calculate the variance and standard deviation of a random variable using a tabular approach just as when we calculated the expected value in Example 3 in the preceding section.

using Technology

Technology can be used to automate the calculations of Example 3.

TI-83/84 Plus

STAT EDIT values of x in L_1, probabilities in L_2.
Home screen: `sum(L1*L2) → M`
Then `sum((L1-M)^2*L2)`
[More details on page 466.]

Spreadsheet

x-values in A2–A7, probabilities in B2–B7
`=A2*B2` in C2; copy down to C7.
`=SUM(C2:C7)` in C8
`=(A2-$C$8)^2*B2` in D2; copy down to D7.
`=SUM(D2:D7)` in D8
[More details on page 469.]

Website

www.WanerMath.com
Student Home
→ On Line Utilities
→ Histogram Utility
Enter the x-values and probabilities as shown:

```
10, .2
20, .2
30, .3
40, .1
50, .1
60, .1
```

Make sure "Show expected value and standard deviation" is checked, and press "Results".

EXAMPLE 3 Variance of a Random Variable

Compute the variance and standard deviation for the following probability distribution.

x	10	20	30	40	50	60
$P(X = x)$	.2	.2	.3	.1	.1	.1

Solution We first compute the expected value, μ, in the usual way:

x	10	20	30	40	50	60	
$P(X = x)$	.2	.2	.3	.1	.1	.1	
$xP(X = x)$	2	4	9	4	5	6	$\mu = 30$

Next, we add an extra three rows:

- a row for the differences $(x - \mu)$, which we get by subtracting μ from the values of X
- a row for the squares $(x - \mu)^2$, which we obtain by squaring the values immediately above
- a row for the products $(x - \mu)^2 P(X = x)$, which we obtain by multiplying the values in the second and the fifth rows.

x	10	20	30	40	50	60	
$P(X = x)$	.2	.2	.3	.1	.1	.1	
$xP(X = x)$	2	4	9	4	5	6	$\mu = 30$
$x - \mu$	−20	−10	0	10	20	30	
$(x - \mu)^2$	400	100	0	100	400	900	
$(x - \mu)^2P(X = x)$	80	20	0	10	40	90	$\sigma^2 = 240$

The sum of the values in the last row is the variance. The standard deviation is then the square root of the variance:

$$\sigma = \sqrt{240} \approx 15.49.$$

Note Chebyshev's rule and the empirical rule apply to random variables just as they apply to samples and populations, as we illustrate in the following example. ■

EXAMPLE 4 Internet Commerce

Your newly launched company, CyberPromo, Inc., sells computer games on the Internet.

a. Statistical research indicates that the lifespan of an Internet marketing company such as yours is symmetrically distributed with an expected value of 30 months and standard deviation of 4 months. Complete the following sentence:

There is (at least/at most/approximately)______ a _____ percent chance that CyberPromo will still be around for more than 3 years.

b. How would the answer to part (a) be affected if the distribution of lifespans was not known to be symmetric?

Solution

a. Do we use Chebyshev's rule or the empirical rule? Because the empirical rule requires that the distribution be both symmetric and bell shaped—not just symmetric—we cannot conclude that it applies here, so we are forced to use Chebyshev's rule instead.

Let X be the lifespan of an Internet commerce site. The expected value of X is 30 months, and the hoped-for lifespan of CyberPromo, Inc., is 36 months, which is 6 months, or $6/4 = 1.5$ standard deviations, above the mean. Chebyshev's rule tells us that X is within $k = 1.5$ standard deviations of the mean at least $1 - 1/k^2$ of the time; that is,

$$P(24 \leq X \leq 36) \geq 1 - \frac{1}{k^2} = 1 - \frac{1}{1.5^2} \approx .56.$$

In other words, at least 56% of all Internet marketing companies have life spans in the range of 24 to 36 months. Thus, *at most* 44 % have life spans outside this range. Because the distribution is symmetric, at most 22% have life spans longer than 36 months. Thus we can complete the sentence as follows:

There is at most a 22 percent chance that CyberPromo will still be around for more than 3 years.

b. If the given distribution was not known to be symmetric, how would this affect the answer? We saw above that regardless of whether the distribution is symmetric or not, at most 44% have lifespans outside the range 24 to 36 months. Because the distribution is not symmetric, we cannot conclude that at most half of the 44% have lifespans longer than 36 months, and all we can say is that *no more than 44% can possibly have life spans longer than 36 years.* In other words:

There is at most a 44 percent chance that CyberPromo will still be around for more than 3 years.

Variance and Standard Deviation of a Binomial Random Variable

We saw that there is an easy formula for the expected value of a binomial random variable: $\mu = np$, where n is the number of trials and p is the probability of success. Similarly, there is a simple formula for the variance and standard deviation.

Variance and Standard Deviation of a Binomial Random Variable

If X is a binomial random variable associated with n independent Bernoulli trials, each with probability p of success, then the variance and standard deviation of X are given by

$$\sigma^2 = npq \quad \text{and} \quad \sigma = \sqrt{npq}$$

where $q = 1 - p$ is the probability of failure.

Quick Example

If X is the number of successes in 20 Bernoulli trials with $p = .7$, then the standard deviation is $\sigma = \sqrt{npq} = \sqrt{(20)(.7)(.3)} \approx 2.05$.

For values of p near $1/2$ and large values of n, a binomial distribution is bell shaped and (nearly) symmetric, hence the empirical rule applies. One rule of thumb is that we can use the empirical rule when both $np \geq 10$ and $nq \geq 10$.*

* Remember that the empirical rule only gives an *estimate* of probabilities. In Section 8.5 we give a more accurate approximation that takes into account the fact that the binomial distribution is not continuous.

EXAMPLE 5 Internet Commerce

You have calculated that there is a 40% chance that a hit on your Web page results in a fee paid to your company CyberPromo, Inc. Your Web page receives 25 hits per day. Let X be the number of hits that result in payment of the fee ("successful hits").

a. What are the expected value and standard deviation of X?

b. Complete the following: On approximately 95 out of 100 days, I will get between ___ and ___ successful hits.

Solution

a. The random variable X is binomial with $n = 25$ and $p = .4$. To compute μ and σ, we use the formulas

$$\mu = np = (25)(.4) = 10 \text{ successful hits}$$

$$\sigma = \sqrt{npq} = \sqrt{(25)(.4)(.6)} \approx 2.45 \text{ hits.}$$

b. Because $np = 10 \geq 10$ and $nq = (25)(.6) = 15 \geq 10$, we can use the empirical rule, which tells us that there is an approximately 95% probability that the number of successful hits is within two standard deviations of the mean—that is, in the interval

$$[\mu - 2\sigma, \mu + 2\sigma] = [10 - 2(2.45), 10 + 2(2.45)] = [5.1, 14.9].$$

Thus, on approximately 95 out of 100 days, I will get between <u>5.1</u> and <u>14.9</u> successful hits.

FAQ

Recognizing when to Use the Empirical Rule or Chebyshev's Rule

Q: *How do I decide whether to use Chebyshev's rule or the empirical rule?*

A: Check to see whether the probability distribution you are considering is both symmetric and bell shaped. If so, you can use the empirical rule. If not, then you must use Chebyshev's rule. Thus, for instance, if the distribution is symmetric but not known to be bell shaped, you must use Chebyshev's rule.

8.4 EXERCISES

Access end-of-section exercises online at **www.webassign.net**

8.5 Normal Distributions

Continuous Random Variables

Figure 9 shows the probability distributions for the number of successes in sequences of 10 and 15 independent Bernoulli trials, each with probability of success $p = .5$.

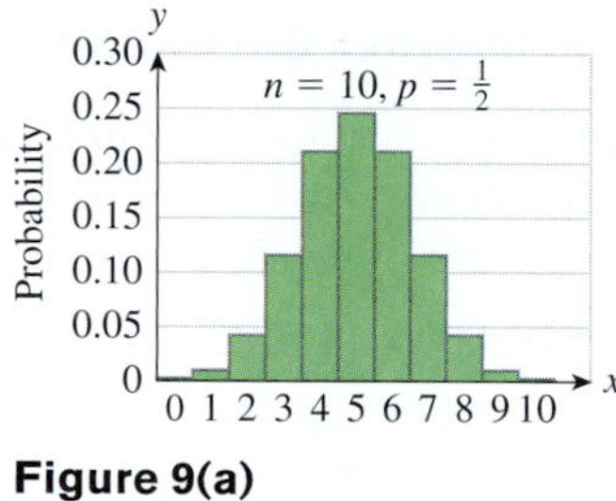

Figure 9(a)

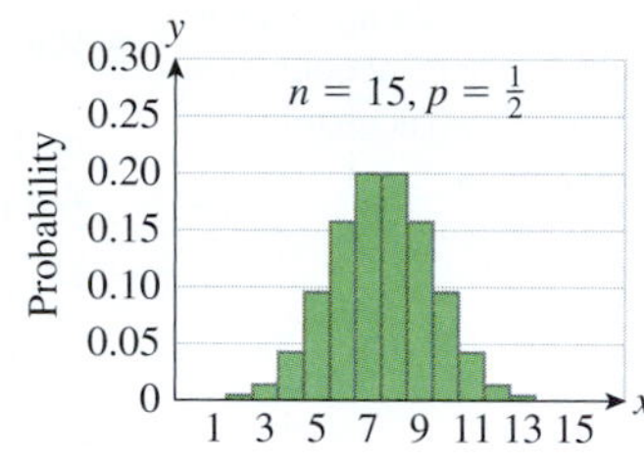

Figure 9(b)

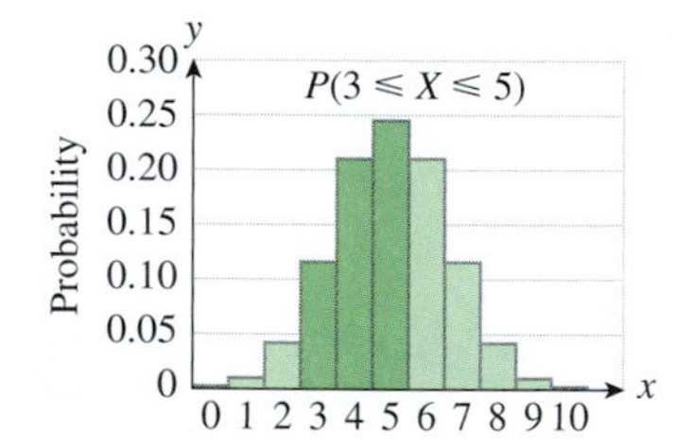

Figure 10

Because each column is 1 unit wide, its area is numerically equal to its height. Thus, the area of each rectangle can be interpreted as a probability. For example, in Figure 9(a) the area of the rectangle over $X = 3$ represents $P(X = 3)$. If we want to find $P(3 \le X \le 5)$, we can add up the areas of the three rectangles over 3, 4, and 5, shown shaded in Figure 10. Notice that if we add up the areas of *all* the rectangles in Figure 9(a), the total is 1 because $P(0 \le X \le 10) = 1$. We can summarize these observations.

Properties of the Probability Distribution Histogram

In a probability distribution histogram where each column is 1 unit wide:

- The total area enclosed by the histogram is 1 square unit.
- $P(a \le X \le b)$ is the area enclosed by the rectangles lying between and including $X = a$ and $X = b$.

This discussion is motivation for considering another kind of random variable, one whose probability distribution is specified not by a bar graph, as above, but by the graph of a function.

Continuous Random Variable; Probability Density Function

A **continuous random variable** X may take on any real value whatsoever. The probabilities $P(a \le X \le b)$ are defined by means of a **probability density function**, a function whose graph lies above the x-axis with the total area between the graph and the x-axis being 1. The probability $P(a \le X \le b)$ is defined to be the area enclosed by the curve, the x-axis, and the lines $x = a$ and $x = b$ (see Figure 11).

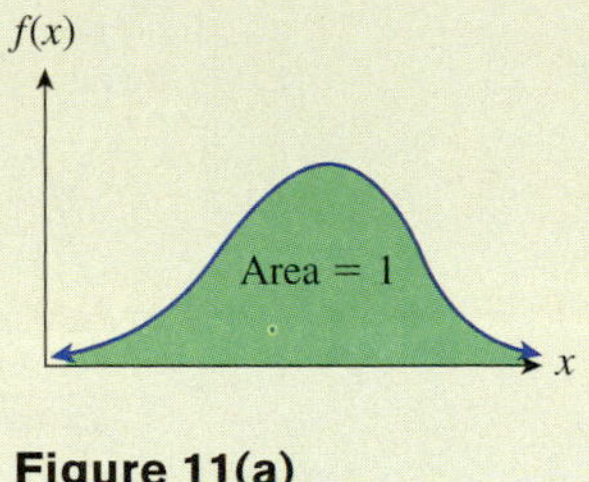

Figure 11(a)

Figure 11(b)

Notes

1. In Chapter 7, we defined probability distributions only for *finite* sample spaces. Because continuous random variables have infinite sample spaces, we need the definition above to give meaning to $P(a \le X \le b)$ if X is a continuous random variable.

2. If $a = b$, then $P(X = a) = P(a \le X \le a)$ is the area under the curve between the lines $x = a$ and $x = a$—no area at all! Thus, when X is a continuous random variable, $P(X = a) = 0$ for every value of a.

3. Whether we take the region in Figure 11(b) to include the boundary or not does not affect the area. The probability $P(a < X < b)$ is defined as the area strictly between the vertical lines $x = a$ and $x = b$, but is, of course, the same as $P(a \le X \le b)$, because the boundary contributes nothing to the area. When we are calculating probabilities associated with a continuous random variable,

$$P(a \le X \le b) = P(a < X \le b) = P(a \le X < b) = P(a < X < b). \quad \blacksquare$$

Normal Density Functions

Among all the possible probability density functions, there is an important class of functions called **normal density functions**, or **normal distributions**. The graph of a normal density function is bell shaped and symmetric, as the following figure shows. The formula for a normal density function is rather complicated looking:

$$f(x) = \frac{1}{\sigma\sqrt{2\pi}} e^{-\frac{(x-\mu)^2}{2\sigma^2}}.$$

The quantity μ is called the **mean** and can be any real number. The quantity σ is called the **standard deviation** and can be any positive real number. The number $e = 2.7182\ldots$ is a useful constant that shows up many places in mathematics, much as the constant π does. Finally, the constant $1/(\sigma\sqrt{2\pi})$ that appears in front is there to make the total area come out to be 1. We rarely use the actual formula in computations; instead, we use tables or technology.

Normal Density Function; Normal Distribution

A **normal density function**, or **normal distribution**, is a function of the form

$$f(x) = \frac{1}{\sigma\sqrt{2\pi}}e^{-\frac{(x-\mu)^2}{2\sigma^2}},$$

where μ is the mean and σ is the standard deviation. Its graph is bell shaped and symmetric, and has the following form:

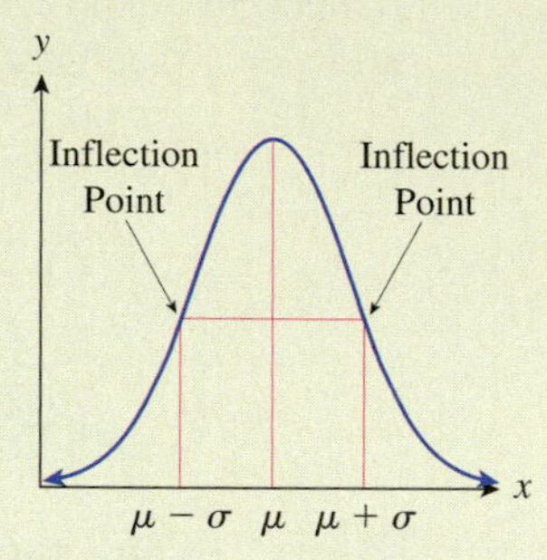

The "inflection points" are the points where the curve changes from bending in one direction to bending in another.*

* Pretend you were driving along the curve in a car. Then the points of inflection are the points where you would change the direction in which you are steering (from left to right or right to left).

Figure 12 shows the graphs of several normal density functions. The third of these has mean 0 and standard deviation 1, and is called the **standard normal distribution**. We use Z rather than X to refer to the standard normal variable.

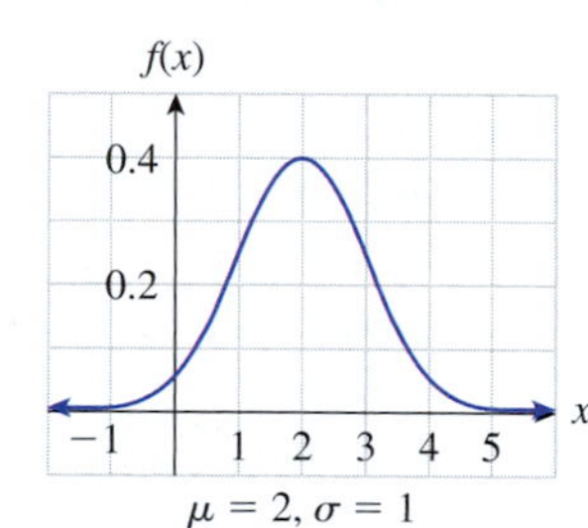

$\mu = 2, \sigma = 1$

Figure 12(a)

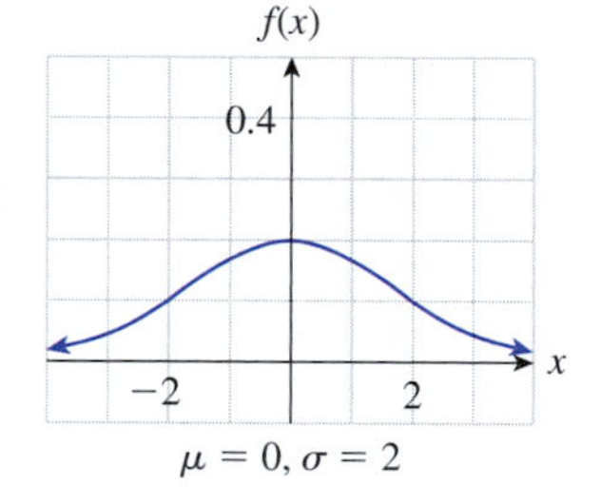

$\mu = 0, \sigma = 2$

Figure 12(b)

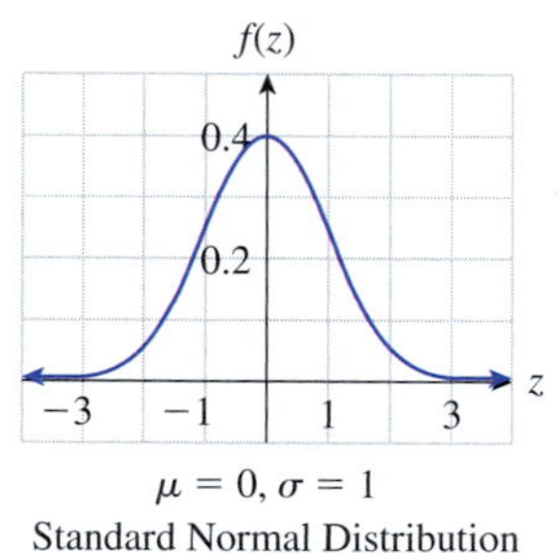

$\mu = 0, \sigma = 1$

Standard Normal Distribution

Figure 12(c)

using Technology

The graphs in Figure 12 can be drawn on a TI-83/84 Plus or the Website grapher.

TI-83/84 Plus

Figure 12(a):
```
Y1=normalpdf(x,2,1)
```
Figure 12(b):
```
Y1=normalpdf(x,0,2)
```
Figure 12(c):
```
Y1=normalpdf(x)
```

Website

www.WanerMath.com

Student Home

→ On Line Utilities

→ Function Evaluator and Grapher

Figure 12(a):
```
normalpdf(x,2,1)
```
Figure 12(b):
```
normalpdf(x,0,2)
```
Figure 12(c):
```
normalpdf(x)
```

Calculating Probabilities for the Standard Normal Distribution

The standard normal distribution has $\mu = 0$ and $\sigma = 1$. The corresponding variable is called the **standard normal variable**, which we always denote by Z. Recall that to calculate the probability $P(a \le Z \le b)$, we need to find the area under the distribution curve between the vertical lines $z = a$ and $z = b$. We can use the table in the Appendix to look up these areas, or we can use technology. Here is an example.

EXAMPLE 1 Standard Normal Distribution

Let Z be the standard normal variable. Calculate the following probabilities:

a. $P(0 \le Z \le 2.4)$ **b.** $P(0 \le Z \le 2.43)$

c. $P(-1.37 \le Z \le 2.43)$ **d.** $P(1.37 \le Z \le 2.43)$

Solution

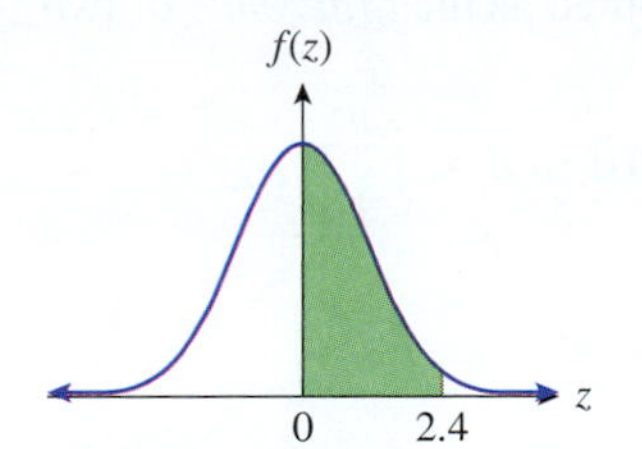

Figure 13

a. We are asking for the shaded area under the standard normal curve shown in Figure 13. We can find this area, correct to four decimal places, by looking at the table in the Appendix, which lists the area under the standard normal curve from $Z = 0$ to $Z = b$ for any value of b between 0 and 3.09. To use the table, write 2.4 as 2.40, and read the entry in the row labeled 2.4 and the column labeled 0.00 $(2.4 + 0.00 = 2.40)$. Here is the relevant portion of the table:

Z	**0.00**	**0.01**	**0.02**	**0.03**
2.3	.4893	.4896	.4898	.4901
→ **2.4**	.4918	.4920	.4922	.4925
2.5	.4938	.4940	.4941	.4943

Thus, $P(0 \leq Z \leq 2.40) = .4918$.

b. The area we require can be read from the same portion of the table shown above. Write 2.43 as $2.4 + 0.03$, and read the entry in the row labeled 2.4 and the column labeled 0.03:

Z	**0.00**	**0.01**	**0.02**	**0.03**
2.3	.4893	.4896	.4898	.4901
→ **2.4**	.4918	.4920	.4922	.4925
2.5	.4938	.4940	.4941	.4943

Thus, $P(0 \leq Z \leq 2.43) = .4925$.

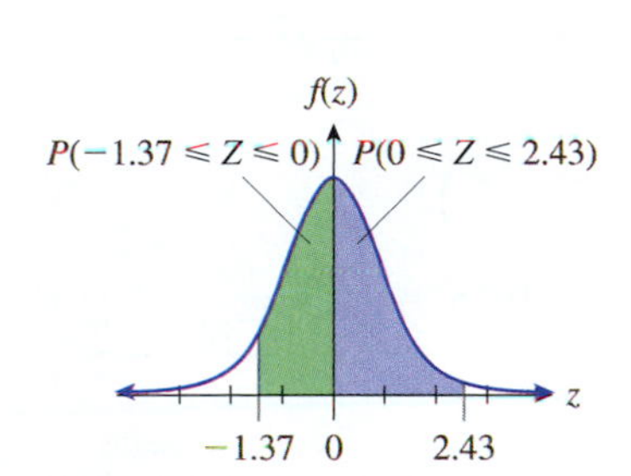

Figure 14

c. Here we cannot use the table directly because the range $-1.37 \leq Z \leq 2.43$ does not start at 0. But we can break the area up into two smaller areas that start or end at 0:

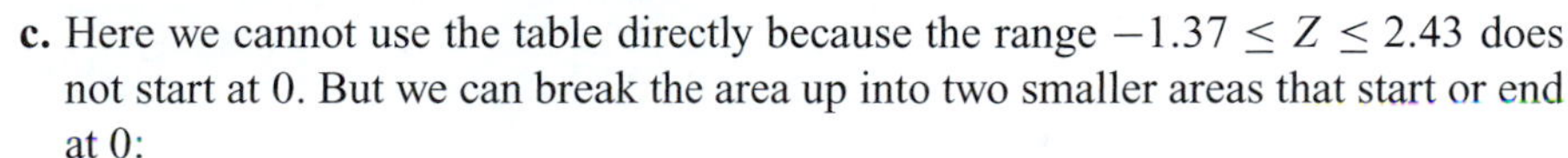

$$P(-1.37 \leq Z \leq 2.43) = P(-1.37 \leq Z \leq 0) + P(0 \leq Z \leq 2.43).$$

In terms of the graph, we are splitting the desired area into two smaller areas (Figure 14).

We already calculated the area of the right-hand piece in part (b):

$$P(0 \leq Z \leq 2.43) = .4925.$$

For the left-hand piece, the symmetry of the normal curve tells us that

$$P(-1.37 \leq Z \leq 0) = P(0 \leq Z \leq 1.37).$$

This we can find on the table. Look at the row labeled 1.3 and the column labeled .07, and read

$$P(-1.37 \leq Z \leq 0) = P(0 \leq Z \leq 1.37) = .4147.$$

Thus,

$$\begin{aligned} P(-1.37 \leq Z \leq 2.43) &= P(-1.37 \leq Z \leq 0) + P(0 \leq Z \leq 2.43) \\ &= .4147 + .4925 \\ &= .9072. \end{aligned}$$

using Technology

Technology can be used to calculate the probabilities in Example 1. For instance, the calculation for part (c) is as follows:

TI-83/84 Plus
Home Screen:
`normalcdf(−1.37, 2.43)`
(`normalcdf` is in [2ND] [VARS].)
[More details on page 466.]

Spreadsheet
```
=NORMSDIST(2.43)
-NORMSDIST(-1.37)
```
[More details on page 469.]

Website
www.WanerMath.com
Student Home
→ On Line Utilities
→ Normal Distribution Utility

Set up as shown and press "Calculate Probability".

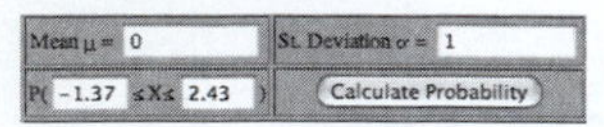

d. The range $1.37 \leq Z \leq 2.43$ does not contain 0, so we cannot use the technique of part (c). Instead, the corresponding area can be computed as the *difference* of two areas:

$$
\begin{aligned}
P(1.37 \leq Z \leq 2.43) &= P(0 \leq Z \leq 2.43) - P(0 \leq Z \leq 1.37) \\
&= .4925 - .4147 \\
&= .0778.
\end{aligned}
$$

Calculating Probabilities for Any Normal Distribution

Although we have tables to compute the area under the *standard* normal curve, there are no readily available tables for nonstandard distributions. For example, if $\mu = 2$ and $\sigma = 3$, then how would we calculate $P(0.5 \leq X \leq 3.2)$? The following conversion formula provides a method for doing so:

Standardizing a Normal Distribution

If X has a normal distribution with mean μ and standard deviation σ, and if Z is the standard normal variable, then

$$P(a \leq X \leq b) = P\left(\frac{a-\mu}{\sigma} \leq Z \leq \frac{b-\mu}{\sigma}\right).$$

Quick Example

If $\mu = 2$ and $\sigma = 3$, then

$$
\begin{aligned}
P(0.5 \leq X \leq 3.2) &= P\left(\frac{0.5-2}{3} \leq Z \leq \frac{3.2-2}{3}\right) \\
&= P(-0.5 \leq Z \leq 0.4) = .1915 + .1554 = .3469.
\end{aligned}
$$

To completely justify the above formula requires more mathematics than we shall discuss here. However, here is the main idea: If X is normal with mean μ and standard deviation σ, then $X - \mu$ is normal with mean 0 and standard deviation still σ, while $(X - \mu)/\sigma$ is normal with mean 0 and standard deviation 1. In other words, $(X - \mu)/\sigma = Z$. Therefore,

$$P(a \leq X \leq b) = P\left(\frac{a-\mu}{\sigma} \leq \frac{X-\mu}{\sigma} \leq \frac{b-\mu}{\sigma}\right) = P\left(\frac{a-\mu}{\sigma} \leq Z \leq \frac{b-\mu}{\sigma}\right).$$

EXAMPLE 2 Quality Control

Pressure gauges manufactured by Precision Corp. must be checked for accuracy before being placed on the market. To test a pressure gauge, a worker uses it to measure the pressure of a sample of compressed air known to be at a pressure of exactly 50 pounds per square inch. If the gauge reading is off by more than 1% (0.5 pounds), it is rejected. Assuming that the reading of a pressure gauge under these circumstances is a normal random variable with mean 50 and standard deviation 0.4, find the percentage of gauges rejected.

using Technology

Technology can be used to calculate the probability $P(49.5 \le X \le 50.5)$ in Example 2:

TI-83/84 Plus

Home Screen:

`normalcdf(49.5,50.5,50,0.4)`

(`normalcdf` is in [2ND] [VARS].)

[More details on page 467.]

Spreadsheet

```
=NORMDIST(50.5,50,0.4,1)
-NORMDIST(49.5,50,0.4,1)
```

[More details on page 470.]

Website

www.WanerMath.com

Student Home

→ On Line Utilities

→ Normal Distribution Utility

Set up as shown and press "Calculate Probability".

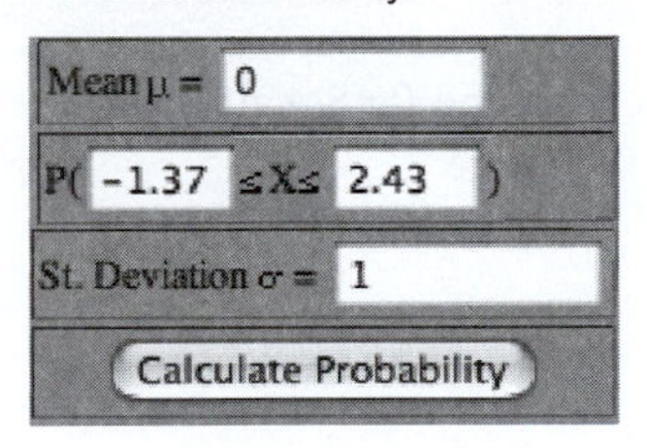

Solution If X is the reading of the gauge, then X has a normal distribution with $\mu = 50$ and $\sigma = 0.4$. We are asking for $P(X < 49.5 \text{ or } X > 50.5) = 1 - P(49.5 \le X \le 50.5)$. We calculate

$$\begin{aligned} P(49.5 \le X \le 50.5) &= P\left(\frac{49.5 - 50}{0.4} \le Z \le \frac{50.5 - 50}{0.4}\right) \quad \text{Standardize} \\ &= P(-1.25 \le Z \le 1.25) \\ &= 2 \cdot P(0 \le Z \le 1.25) \\ &= 2(.3944) = .7888. \end{aligned}$$

So, $$\begin{aligned} P(X < 49.5 \text{ or } X > 50.5) &= 1 - P(49.5 \le X \le 50.5) \\ &= 1 - .7888 = .2112. \end{aligned}$$

In other words, about 21% of the gauges will be rejected.

In many applications, we need to know the probability that a value of a normal random variable will lie within one standard deviation of the mean, or within two standard deviations, or within some number of standard deviations. To compute these probabilities, we first notice that, if X has a normal distribution with mean μ and standard deviation σ, then

$$P(\mu - k\sigma \le X \le \mu + k\sigma) = P(-k \le Z \le k)$$

by the standardizing formula. We can compute these probabilities for various values of k using the table in the Appendix, and we obtain the following results.

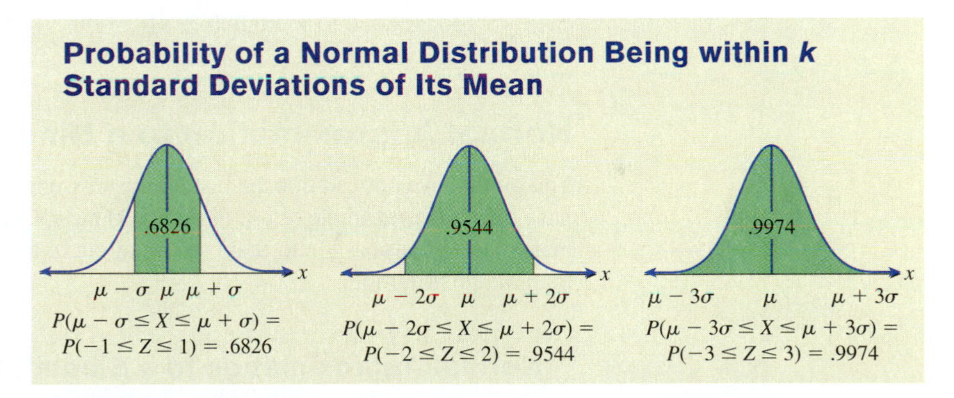

Now you can see where the empirical rule in Section 8.4 comes from! Notice also that the probabilities above are a good deal larger than the lower bounds given by Chebyshev's rule. Chebyshev's rule must work for distributions that are skew or any shape whatsoever.

EXAMPLE 3 Loans

The values of mortgage loans made by a certain bank one year were normally distributed with a mean of \$120,000 and a standard deviation of \$40,000.

a. What is the probability that a randomly selected mortgage loan was in the range of \$40,000–\$200,000?

b. You would like to state in your annual report that 50% of all mortgage loans were in a certain range with the mean in the center. What is that range?

Solution

a. We are asking for the probability that a loan was within two standard deviations (\$80,000) of the mean. By the calculation done previously, this probability is .9544.

b. We look for the k such that

$$P(120{,}000 - k \cdot 40{,}000 \leq X \leq 120{,}000 + k \cdot 40{,}000) = .5.$$

Because

$$P(120{,}000 - k \cdot 40{,}000 \leq X \leq 120{,}000 + k \cdot 40{,}000) = P(-k \leq Z \leq k)$$

we look in the Appendix to see for which k we have

$$P(0 \leq Z \leq k) = .25$$

so that $P(-k \leq Z \leq k) = .5$. That is, we look *inside* the table to see where 0.25 is, and find the corresponding k. We find

$$P(0 \leq Z \leq 0.67) = .2486$$

and $P(0 \leq Z \leq 0.68) = .2517$.

Therefore, the k we want is about halfway between 0.67 and 0.68, call it 0.675. This tells us that 50% of all mortgage loans were in the range

$$120{,}000 - 0.675 \cdot 40{,}000 = \$93{,}000$$

to $120{,}000 + 0.675 \cdot 40{,}000 = \$147{,}000$.

Normal Approximation to a Binomial Distribution

You might have noticed that the histograms of some of the binomial distributions we have drawn (for example, those in Figure 9) have a very rough bell shape. In fact, in many cases it is possible to draw a normal curve that closely approximates a given binomial distribution.

Normal Approximation to a Binomial Distribution

If X is the number of successes in a sequence of n independent Bernoulli trials, with probability p of success in each trial, and if the range of values of X within three standard deviations of the mean lies entirely within the range 0 to n (the possible values of X), then

$$P(a \leq X \leq b) \approx P(a - 0.5 \leq Y \leq b + 0.5)$$

where Y has a normal distribution with the same mean and standard deviation as X; that is, $\mu = np$ and $\sigma = \sqrt{npq}$, where $q = 1 - p$.

Notes

1. The condition that $0 \leq \mu - 3\sigma < \mu + 3\sigma \leq n$ is satisfied if n is sufficiently large and p is not too close to 0 or 1; it ensures that most of the normal curve lies in the range 0 to n.

2. In the formula $P(a \le X \le b) \approx P(a - 0.5 \le Y \le b + 0.5)$, we assume that a and b are integers. The use of $a - 0.5$ and $b + 0.5$ is called the **continuity correction**. To see that it is necessary, think about what would happen if you wanted to approximate, say, $P(X = 2) = P(2 \le X \le 2)$. Should the answer be 0? ■

Figures 15 and 16 show two binomial distributions with their normal approximations superimposed, and illustrate how closely the normal approximation fits the binomial distribution.

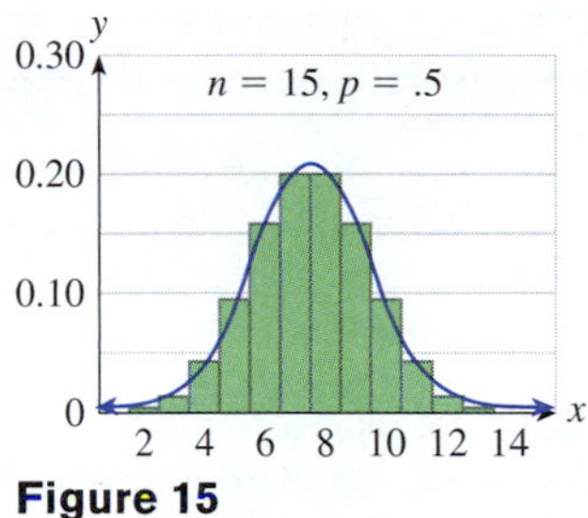

Figure 15

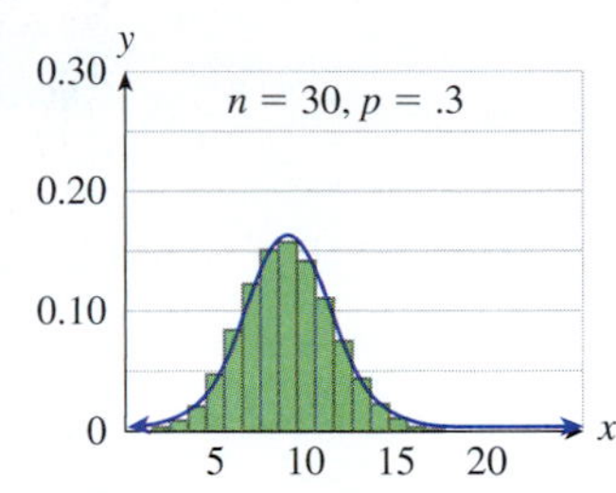

Figure 16

EXAMPLE 4 Coin Tosses

a. If you flip a fair coin 100 times, what is the probability of getting more than 55 heads or fewer than 45 heads?

b. What number of heads (out of 100) would make you suspect that the coin is not fair?

Solution

a. We are asking for

$$P(X < 45 \text{ or } X > 55) = 1 - P(45 \le X \le 55).$$

We *could* compute this by calculating

$$1 - [C(100, 45)(.5)^{45}(.5)^{55} + C(100, 46)(.5)^{46}(.5)^{54} + \cdots + C(100, 55)(.5)^{55}(.5)^{45}]$$

but we can much more easily *approximate* it by looking at a normal distribution with mean $\mu = 50$ and standard deviation $\sigma = \sqrt{(100)(.5)(.5)} = 5$. (Notice that three standard deviations above and below the mean is the range 35 to 65, which is well within the range of possible values for X, which is 0 to 100, so the approximation should be a good one.) Let Y have this normal distribution. Then

$$\begin{aligned} P(45 \le X \le 55) &\approx P(44.5 \le Y \le 55.5) \\ &= P(-1.1 \le Z \le 1.1) \\ &= .7286. \end{aligned}$$

Therefore,

$$P(X < 45 \text{ or } X > 55) \approx 1 - .7286 = .2714.$$

b. This is a deep question that touches on the concept of **statistical significance**: What evidence is strong enough to overturn a reasonable assumption (the assumption that the coin is fair)? Statisticians have developed sophisticated ways of answering this question, but we can look at one simple test now. Suppose we tossed a coin 100 times and got 66 heads. If the coin were fair, then $P(X > 65) \approx P(Y > 65.5) = P(Z > 3.1) \approx .001$. This is small enough to raise a reasonable doubt that the coin is fair. However, we should not be too surprised if we threw 56 heads because we can calculate $P(X > 55) \approx .1357$, which is not such a small probability. As we said, the actual tests of statistical significance are more sophisticated than this, but we shall not go into them.

FAQ

When to Subtract from .5 and when Not to

Q: *When computing probabilities like, say $P(Z \le 1.2)$, $P(Z \ge 1.2)$, or $P(1.2 \le Z \le 2.1)$ using a table, just looking up the given values (1.2, 2.1, or whatever) is not enough. Sometimes you have to subtract from .5, sometimes not. Is there a simple rule telling me what to do when?*

A: The simplest—and also most instructive—way of knowing what to do is to draw a picture of the standard normal curve, and shade in the area you are looking for. Drawing pictures also helps you come up with the following mechanical rules:

1. To compute $P(a \le Z \le b)$, look up the areas corresponding to $|a|$ and $|b|$ in the table. If a and b have opposite signs, add these areas. Otherwise, subtract the smaller area from the larger.
2. To compute $P(Z \le a)$, look up the area corresponding to $|a|$. If a is positive, add .5. Otherwise, subtract from .5.
3. To compute $P(Z \ge a)$, look up the area corresponding to $|a|$. If a is positive, subtract from .5. Otherwise, add .5.

8.5 EXERCISES

Access end-of-section exercises online at **www.webassign.net** ENHANCED WebAssign

CHAPTER 8 REVIEW

KEY CONCEPTS

Website www.WanerMath.com

Go to the Website at www.WanerMath.com to find a comprehensive and interactive Web-based summary of Chapter 8.

8.1 Random Variables and Distributions

Random variable; discrete vs continuous random variable *pp. 420–421*

Probability distribution of a finite random variable *p. 423*

Using measurement classes *p. 424*

8.2 Bernoulli Trials and Binomial Random Variables

Bernoulli trial; binomial random variable *pp. 427, 428*

Probability distribution of binomial random variable:
$P(X = x) = C(n, x)p^x q^{n-x}$ *p. 430*

8.3 Measures of Central Tendency

Sample, sample mean; population, population mean *pp. 433–434*

Sample median, sample mode *p. 434*

Expected value of a random variable:
$\mu = E(X) = \sum_i x_i \cdot P(X = x_i)$ *p. 437*

Expected value of a binomial random variable: $\mu = E(X) = np$ *p. 439*

8.4 Measures of Dispersion

Population variance:
$\sigma^2 = \frac{\sum_{i=1}^{n}(x_i - \mu)^2}{n}$ *p. 441*

Population standard deviation:
$\sigma = \sqrt{\sigma^2}$ *p. 442*

Sample variance:
$s^2 = \frac{\sum_{i=1}^{n}(x_i - \bar{x})^2}{n-1}$ *p. 442*

Sample standard deviation:
$s = \sqrt{s^2}$ *p. 442*

Chebyshev's rule *p. 444*

Empirical rule *p. 444*

Variance of a random variable:
$\sigma^2 = \sum_i (x_i - \mu)^2 P(X = x_i)$ *p. 446*

Standard deviation of X: $\sigma = \sqrt{\sigma^2}$ *p. 446*

Variance and standard deviation of a binomial random variable:
$\sigma^2 = npq, \sigma = \sqrt{npq}$ *p. 449*

8.5 Normal Distributions

Probability density function *p. 451*

Normal density function; normal distribution; standard normal distribution *p. 452*

Calculating probabilities based on the standard normal distribution *p. 452*

Standardizing a normal distribution *p. 454*

Calculating probabilities based on non-standard normal distributions *p. 454*

Normal approximation to a binomial distribution *p. 456*

REVIEW EXERCISES

In Exercises 1–6, find the probability distribution for the given random variable and draw a histogram.

1. A couple has two children; $X =$ the number of boys. (Assume an equal likelihood of a child being a boy or a girl.)

2. A couple has three children; $X =$ the number of girls. (Assume an equal likelihood of a child being a boy or a girl.)

3. A four-sided die (with sides numbered 1 through 4) is rolled twice in succession; $X =$ the sum of the two numbers.

4. 48.2% of *Xbox* players are in their teens, 38.6% are in their twenties, 11.6% are in their thirties, and the rest are in their forties; $X =$ age of an *Xbox* player. (Use the midpoints of the measurement classes.)

5. From a bin that contains 20 defective joysticks and 30 good ones, 3 are chosen at random; $X =$ the number of defective joysticks chosen. (Round all probabilities to four decimal places.)

6. Two dice are weighted so that each number 2, 3, 4, and 5 is half as likely to face up as each 1 and 6; $X =$ the number of 1s that face up when both are thrown.

7. Use any method to calculate the sample mean, median, and standard deviation of the following sample of scores: −1, 2, 0, 3, 6.

8. Use any method to calculate the sample mean, median, and standard deviation of the following sample of scores: 4, 4, 5, 6, 6.

9. Give an example of a sample of four scores with mean 1 and median 0. (Arrange them in ascending order.)

10. Give an example of a sample of six scores with sample standard deviation 0 and mean 2.

11. Give an example of a population of six scores with mean 0 and population standard deviation 1.

12. Give an example of a sample of five scores with mean 0 and sample standard deviation 1.

A die is constructed in such a way that rolling a 6 is twice as likely as rolling each other number. That die is rolled four times. Let X be the number of times a 6 is rolled. Evaluate the probabilities in Exercises 13–20.

13. $P(X = 1)$

14. $P(X = 3)$

15. The probability that 6 comes up at most twice

16. The probability that 6 comes up at most once

17. The probability that X is more than 3

18. The probability that X is at least 2

19. $P(1 \le X \le 3)$

20. $P(X \le 3)$

21. A couple has three children; $X =$ the number of girls. (Assume an equal likelihood of a child being a boy or a girl.) Find the expected value and standard deviation of X, and complete the following sentence with the smallest possible whole number: All values of X lie within ___ standard deviations of the expected value.

22. A couple has four children; $X =$ the number of boys. (Assume only a 25% chance of a child being a boy.) Find the expected value and standard deviation of X, and complete the following sentence with the smallest possible whole number: All values of X lie within ___ standard deviations of the expected value.

23. A random variable X has the following frequency distribution.

x	−3	−2	−1	0	1	2	3
$fr(X = x)$	1	2	3	4	3	2	1

Find the probability distribution, expected value, and standard deviation of X, and complete the following sentence: 87.5% (or 14/16) of the time, X is within ___ (round to one decimal place) standard deviations of the expected value.

24. A random variable X has the following frequency distribution.

x	−4	−2	0	2	4	6
$fr(X = x)$	3	3	4	5	3	2

Find the probability distribution, expected value, and standard deviation of X, and complete the following sentence: ___ percent of the values of X lie within one standard deviation of the expected value.

25. A random variable X has expected value $\mu = 100$ and standard deviation $\sigma = 16$. Use Chebyshev's rule to find an interval in which X is guaranteed to lie with a probability of at least 90%.

26. A random variable X has a symmetric distribution and an expected value $\mu = 200$ and standard deviation $\sigma = 5$. Use Chebyshev's rule to find a value that X is guaranteed to exceed with a probability of at most 10%.

27. A random variable X has a bell shaped, symmetric distribution, with expected value $\mu = 200$ and standard deviation $\sigma = 20$. The empirical rule tells us that X has a value greater than ___ approximately 0.15% of the time.

28. A random variable X has a bell shaped, symmetric distribution, with expected value $\mu = 100$ and standard deviation $\sigma = 30$. Use the empirical rule to give an interval in which X lies approximately 95% of the time.

In Exercises 29–34 the mean and standard deviation of a normal variable X are given. Find the indicated probability.

29. X is the standard normal variable Z; $P(0 \le X \le 1.5)$.

30. X is the standard normal variable Z; $P(X \le -1.5)$.

31. X is the standard normal variable Z; $P(|X| \ge 2.1)$.

32. $\mu = 100, \sigma = 16$; $P(80 \le X \le 120)$

33. $\mu = 0, \sigma = 2$; $P(X \le -1)$

34. $\mu = -1, \sigma = 0.5$; $P(X \ge 1)$

APPLICATIONS: OHaganBooks.com

Marketing *As a promotional gimmick, OHaganBooks.com has been selling copies of the Encyclopædia Galactica at an extremely low price that is changed each week at random in a nationally televised drawing. Exercises 35–40 are based on the following table, which summarizes the anticipated sales.*

Price	$5.50	$10	$12	$15
Frequency (weeks)	1	2	3	4
Weekly Sales	6,200	3,500	3,000	1,000

35. What is the expected value of the price of *Encyclopædia Galactica*?

36. What are the expected weekly sales of *Encyclopædia Galactica*?

37. What is the expected weekly revenue from sales of *Encyclopædia Galactica*? (Revenue = Price per copy sold × Number of copies sold.)

38. OHaganBooks.com originally paid Duffin House $20 per copy for the *Encyclopædia Galactica*. What is the expected weekly loss from sales of the encyclopædia? (Loss = Loss per copy sold × Number of copies sold.)

39. True or false? If X and Y are two random variables, then $E(XY) = E(X)E(Y)$ (the expected value of the product of two random variables is the product of the expected values). Support your claim by referring to the answers of Exercises 35, 36, and 37.

40. True or false? If X and Y are two random variables, then $E(X/Y) = E(X)/E(Y)$ (the expected value of the ratio of two random variables is the ratio of the expected values). Support your claim by referring to the answers of Exercises 36 and 38.

41. ***Online Sales*** The following table shows the number of online orders at OHaganBooks.com per million residents in 100 U.S. cities during one month:

Orders (per million residents)	1–2.9	3–4.9	5–6.9	7–8.9	9–10.9
Number of Cities	25	35	15	15	10

a. Let X be the number of orders per million residents in a randomly chosen U.S. city (use rounded midpoints of the given measurement classes). Construct the probability distribution for X and hence compute the expected value μ of X and standard deviation σ. (Round answers to four decimal places.)

b. What range of orders per million residents does the empirical rule predict from approximately 68% of all cities? Would you judge that the empirical rule applies? Why?

c. The actual percentage of cities from which you obtain between 3 and 8 orders per million residents is (choose the correct answer that gives the most specific information):
(A) Between 50% and 65% **(B)** At least 65%
(C) At least 50% **(D)** 57.5%

42. ***Pollen*** Marjory Duffin is planning a joint sales meeting with OHaganBooks.com in Atlanta at the end of March, but is extremely allergic to pollen, so she went online to find pollen counts for the period. The following table shows the results of her search:

Pollen Count	0–1.9	2–3.9	4–5.9	6–7.9	8–9.9	10–11.9
Number of Days	3	5	7	2	1	2

a. Let X be the pollen count on a given day (use rounded midpoints of the given measurement classes). Construct the probability distribution for X and hence compute the expected value μ of X and standard deviation σ. (Round answers to four decimal places.)

b. What range of pollen counts does the empirical rule predict on approximately 95% of the days? Would you judge that the empirical rule applies? Why?

c. The actual percentage of days on which the pollen count is between 2 and 7 is (choose the correct answer that gives the most specific information):

(A) Between 50% and 60% **(B)** At least 60%
(C) At most 70% **(D)** Between 60% and 70%

Mac vs. Windows *On average, 5% of all hits by Mac OS users and 10% of all hits by Windows users result in orders for books at OHaganBooks.com. Due to online promotional efforts, the site traffic is approximately 10 hits per hour by Mac OS users, and 20 hits per hour by Windows users. Compute the probabilities in Exercises 43–48. (Round all answers to three decimal places.)*

43. What is the probability that exactly three Windows users will order books in the next hour?

44. What is the probability that at most three Windows users will order books in the next hour?

45. What is the probability that exactly one Mac OS user and three Windows users will order books in the next hour?

46. What assumption must you make to justify your calculation in Exercise 45?

47. How many orders for books can OHaganBooks.com expect in the next hour from Mac OS users?

48. How many orders for books can OHaganBooks.com expect in the next hour from Windows users?

Online Cosmetics *OHaganBooks.com has launched a subsidiary, GnuYou.com, which sells beauty products online. Most products sold by GnuYou.com are skin creams and hair products. Exercises 49–52 are based on the following table, which shows monthly revenues earned through sales of these products. (Assume a normal distribution. Round all answers to three decimal places.)*

Product	*Skin Creams*	*Hair Products*
Mean Monthly Revenue	$38,000	$34,000
Standard Deviation	$21,000	$14,000

49. What is the probability that GnuYou.com will sell *at least* $50,000 worth of skin cream next month?

50. What is the probability that GnuYou.com will sell *at most* $50,000 worth of hair products next month?

51. What is the probability that GnuYou.com will sell less than $12,000 of skin creams next month?

52. What is the probability that GnuYou.com will sell less than $12,000 of hair products next month?

53. ***Intelligence*** Billy-Sean O'Hagan, now a senior at Suburban State University, has done exceptionally well and has just joined Mensa, a club for people with high IQs. Within Mensa is a group called the Three Sigma Club because their IQ scores are at least 3 standard deviations higher than the U.S. mean. Assuming a U.S. population of 313,000,000, how many people in the United States are qualified for the Three Sigma Club? (Round your answer to the nearest 1,000 people.)

54. ***Intelligence*** To join Mensa (not necessarily the Three Sigma Club), one needs an IQ of at least 132, corresponding to the top 2% of the population. Assuming that scores on this test are normally distributed with a mean of 100, what is the standard deviation? (Round your answer to the nearest whole number.)

55. ***Intelligence*** Based on the information given in Exercises 53 and 54, what score must Billy-Sean have to get into the Three Sigma Club? (Assume that IQ scores are normally distributed with a mean of 100, and use the rounded standard deviation.)

56. ***Intelligence*** Mensa allows the results of various standardized tests to be used to gain membership. Suppose that there was such a test with a mean of 500 on which one needed to score at least 600 to be in the top 2% of the population, hence eligible to join Mensa. What would Billy-Sean need to score on this test to get into the Three Sigma Club?

Case Study

Spotting Tax Fraud with Benford's Law[5]

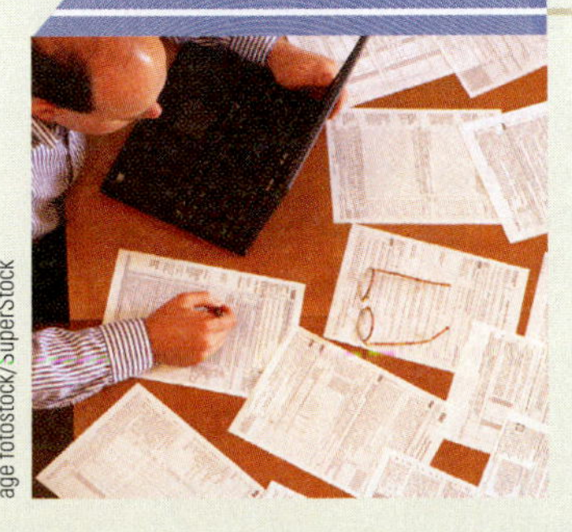
age fotostock/SuperStock

You are a tax fraud specialist working for the Internal Revenue Service (IRS), and you have just been handed a portion of the tax return from Colossal Conglomerate. The IRS suspects that the portion you were handed may be fraudulent, and would like your opinion. Is there any mathematical test, you wonder, that can point to a suspicious tax return based on nothing more than the numbers entered?

[5] The discussion is based on the article "Following Benford's Law, or Looking Out for No. 1" by Malcolm W. Browne, *New York Times*, August 4, 1998, p. F4. The use of Benford's Law in detecting tax evasion is discussed in a Ph.D. dissertation by Dr. Mark J. Nigrini (Southern Methodist University, Dallas).

You decide, on an impulse, to make a list of the first digits of all the numbers entered in the portion of the Colossal Conglomerate tax return (there are 625 of them). You reason that, if the tax return is an honest one, the first digits of the numbers should be uniformly distributed. More precisely, if the experiment consists of selecting a number at random from the tax return, and the random variable X is defined to be the first digit of the selected number, then X should have the following probability distribution:

x	1	2	3	4	5	6	7	8	9
$P(X = x)$	$\frac{1}{9}$	$\frac{1}{9}$	$\frac{1}{9}$	$\frac{1}{9}$	$\frac{1}{9}$	$\frac{1}{9}$	$\frac{1}{9}$	$\frac{1}{9}$	$\frac{1}{9}$

You then do a quick calculation based on this probability distribution and find an expected value of $E(X) = 5$. Next, you turn to the Colossal Conglomerate tax return data and calculate the relative frequency (estimated probability) of the actual numbers in the tax return. You find the following results.

Colossal Conglomerate Return

y	1	2	3	4	5	6	7	8	9
$P(Y = y)$	.29	.1	.04	.15	.31	.08	.01	.01	.01

It certainly does look suspicious! For one thing, the digits 1 and 5 seem to occur a lot more often than any of the other digits, and roughly three times what you predicted. Moreover, when you compute the expected value, you obtain $E(Y) = 3.48$, considerably lower than the value of 5 you predicted. Gotcha! you exclaim.

You are about to file a report recommending a detailed audit of Colossal Conglomerate when you recall an article you once read about first digits in lists of numbers. The article dealt with a remarkable discovery in 1938 by Dr. Frank Benford, a physicist at General Electric. What Dr. Benford noticed was that the pages of logarithm tables that listed numbers starting with the digits 1 and 2 tended to be more soiled and dog-eared than the pages that listed numbers starting with higher digits—say, 8. For some reason, numbers that start with low digits seemed more prevalent than numbers that start with high digits. He subsequently analyzed more than 20,000 sets of numbers, such as tables of baseball statistics, listings of widths of rivers, half-lives of radioactive elements, street addresses, and numbers in magazine articles. The result was always the same: Inexplicably, numbers that start with low digits tended to appear more frequently than those that start with high ones, with numbers beginning with the digit 1 most prevalent of all.[6] Moreover, the expected value of the first digit was not the expected 5, but 3.44.

Because the first digits in Colossal Conglomerate's return have an expected value of 3.48, very close to Benford's value, it might appear that your suspicion was groundless after all. (Back to the drawing board . . .)

Out of curiosity, you decide to investigate Benford's discovery more carefully. What you find is that Benford did more than simply observe a strange phenomenon in lists of numbers. He went further and derived the following formula for the probability distribution of first digits in lists of numbers:

$$P(X = x) = \log(1 + 1/x) \qquad (x = 1, 2, \ldots, 9).$$

[6] This does not apply to all lists of numbers. For instance, a list of randomly chosen numbers between 100 and 999 will have first digits uniformly distributed between 1 and 9.

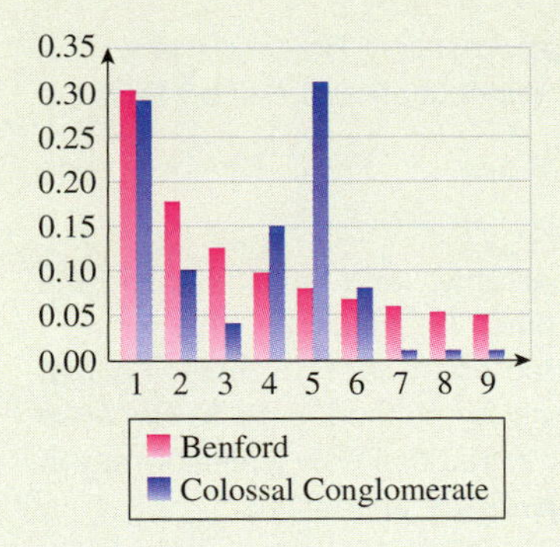

Figure 17

You compute these probabilities, and find the following distribution (the probabilities are all rounded, and thus do not add to exactly 1).

x	1	2	3	4	5	6	7	8	9
$P(X = x)$	.30	.18	.12	.10	.08	.07	.06	.05	.05

You then enter these data along with the Colossal Conglomerate tax return data in your spreadsheet program and obtain the graph shown in Figure 17.

The graph shows something awfully suspicious happening with the digit 5. The percentage of numbers in the Colossal Conglomerate return that begin with 5 far exceeds Benford's prediction that approximately 8% of all numbers should begin with 5.

Now it seems fairly clear that you are justified in recommending Colossal Conglomerate for an audit, after all.

Q: *Because no given set of data can reasonably be expected to satisfy Benford's law exactly, how can I be certain that the Colossal Conglomerate data is not simply due to chance?*

A: You can never be 100% certain. It is certainly conceivable that the tax figures just happen to result in the "abnormal" distribution in the Colossal Conglomerate tax return. However—and this is the subject of "inferential statistics"—there is a method for deciding whether you can be, say "95% certain" that the anomaly reflected in the data is not due to chance. To check, you must first compute a statistic that determines how far a given set of data deviates from satisfying a theoretical prediction (Benford's law, in this case). This statistic is called a **sum-of-squares error** and is given by the following formula (reminiscent of the variance):

$$\text{SSE} = n\left[\frac{[P(y_1) - P(x_1)]^2}{P(x_1)} + \frac{[P(y_2) - P(x_2)]^2}{P(x_2)} + \cdots + \frac{[P(y_9) - P(x_9)]^2}{P(x_9)}\right].$$

Here, n is the sample size: 625 in the case of Colossal Conglomerate. The quantities $P(x_i)$ are the theoretically predicted probabilities according to Benford's Law, and the $P(y_i)$ are the probabilities in the Colossal Conglomerate return. Notice that if the Colossal Conglomerate return probabilities had exactly matched the theoretically predicted probabilities, then SSE would have been zero. Notice also the effect of multiplying by the sample size n: The larger the sample, the more likely that the discrepancy between the $P(x_i)$ and the $P(y_i)$ is not due to chance. Substituting the numbers gives

$$\text{SSE} \approx 625\left[\frac{[.29 - .30]^2}{.30} + \frac{[.1 - .18]^2}{.18} + \cdots + \frac{[.01 - .05]^2}{.05}\right]$$
$$\approx 552.^7$$

Q: *The value of SSE does seem quite large. But how can I use this figure in my report? I would like to say something impressive, such as "Based on the portion of the Colossal Conglomerate tax return analyzed, one can be 95% certain that the figures are anomalous."*

A: The error SSE is used by statisticians to answer exactly such a question. What they would do is compare this figure to the largest SSE we would have expected to get

[7] If you use more accurate values for the probabilities in Benford's distribution, the value is approximately 560.

by chance in 95 out of 100 selections of data that *do* satisfy Benford's law. This "biggest error" is computed using a "chi-squared" distribution and can be found in Excel by entering

```
=CHIINV(0.05,8)
```

Here, the 0.05 is $1 - 0.95$, encoding the "95% certainty," and the 8 is called the "number of degrees of freedom" = number of outcomes (9) minus 1.

You now find, using Excel, that the chi-squared figure is 15.5, meaning that the largest SSE that you could have expected purely by chance is 15.5. Because Colossal Conglomerate's error is much larger at 552, you can now justifiably say in your report that there is a 95% certainty that the figures are anomalous.[8]

EXERCISES

Which of the following lists of data would you expect to follow Benford's law? If the answer is "no," give a reason.

1. Distances between cities in France, measured in kilometers

2. Distances between cities in France, measured in miles

3. The grades (0–100) in your math instructor's grade book

4. The Dow Jones averages for the past 100 years

5. Verbal SAT scores of college-bound high school seniors

6. Life spans of companies

T *Use technology to determine whether the given distribution of first digits fails, with 95% certainty, to follow Benford's law.*

7. Good Neighbor Inc.'s tax return ($n = 1{,}000$)

y	1	2	3	4	5	6	7	8	9
$P(Y = y)$	.31	.16	.13	.11	.07	.07	.05	.06	.04

8. Honest Growth Funds Stockholder Report ($n = 400$)

y	1	2	3	4	5	6	7	8	9
$P(Y = y)$	.28	.16	.1	.11	.07	.09	.05	.07	.07

[8] What this actually means is that, if you were to do a similar analysis on a large number of tax returns, and you designated as "not conforming to Benford's law" all of those whose value of SSE was larger than 15.5, you would be justified in 95% of the cases.

TI-83/84 Plus Technology Guide

Section 8.1

Example 3 (page 424) Let X be the number of heads that face up in three tosses of a coin. We obtained the following probability distribution of X in the text:

x	0	1	2	3
$P(X = x)$	$\frac{1}{8}$	$\frac{3}{8}$	$\frac{3}{8}$	$\frac{1}{8}$

Use technology to obtain the corresponding histogram.

Solution with Technology

1. In the TI-83/84 Plus, you can enter a list of probabilities as follows: press STAT, choose EDIT, and then press ENTER. Clear columns L_1 and L_2 if they are not already cleared. (Select the heading of a column and press CLEAR ENTER to clear it.) Enter the values of X in the column under L_1 (pressing ENTER after each entry) and enter the frequencies in the column under L_2.

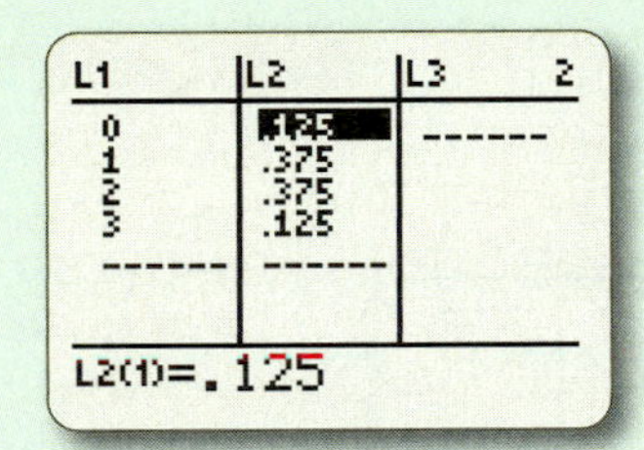

2. To graph the data as in Figure 1, first set the WINDOW to $0 \le X \le 4$, $0 \le Y \le 0.5$, and Xscl = 1 (the width of the bars). Then turn STAT PLOT on ([2nd] Y=), and configure it by selecting the histogram icon, setting Xlist = L_1 and Freq = L_2. Then hit GRAPH.

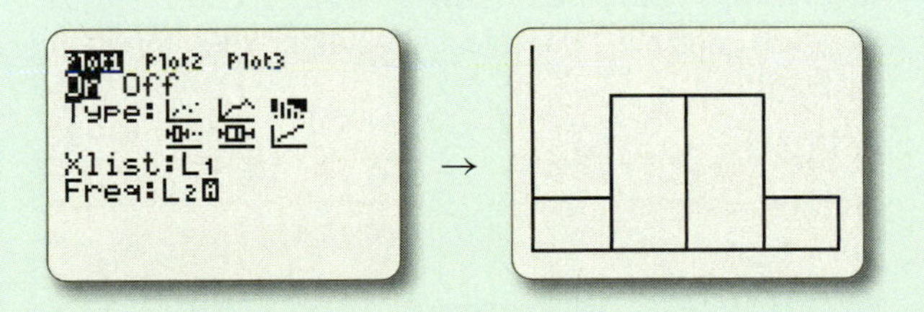

Example 4 (page 424) We obtained the following frequency table in the text:

x	25,000	35,000	45,000	55,000	65,000	75,000	85,000
Frequency	20	80	230	400	170	70	30

Find the probability distribution of X.

Solution with Technology

We need to divide each frequency by the sum. Although the computations in this example (dividing the seven frequencies by 1,000) are simple to do by hand, they could become tedious in general, so technology is helpful.

1. On the TI-83/84 Plus, press STAT, select EDIT, enter the values of X in the L_1 list, and enter the frequencies in the L_2 list as in Example 3 (below left).

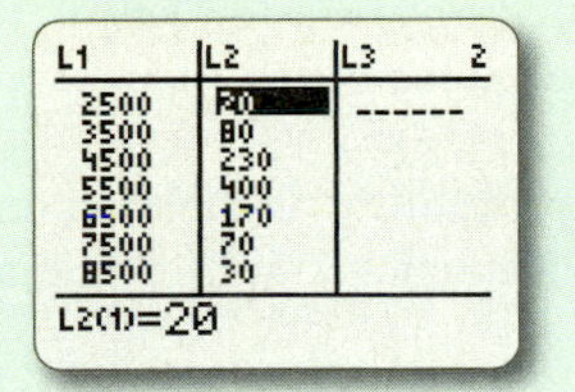

2. Then, on the home screen, enter

   $L_2/1,000 \rightarrow L_3$ L_2 is 2nd 2, L_3 is 2nd 3.

   or, better yet,

   $L_2/\text{sum}(L_2) \rightarrow L_3$ Sum is found in 2nd STAT, under MATH.

3. After pressing ENTER you can now go back to the STAT EDIT screen, and you will find the probabilities displayed in L_3 as shown above on the right.

Section 8.2

Example 2(b) (page 430) By 2030, the probability that a randomly chosen resident in the United States will be 65 years old or older is projected to be .2. If X is the number of people aged 65 or older in a sample of 6, construct the probability distribution of X.

Solution with Technology

In the "Y=" screen, you can enter the binomial distribution formula

```
Y1 = 6 nCr X*0.2^X*0.8^(6-X)
```

directly (to get nCr, press MATH and select PRB), and hit TABLE. You can then replicate the table in the text by choosing $X = 0, 1, \ldots, 6$ (use the TBLSET screen to set "Indpnt" to "Ask" if you have not already done so).

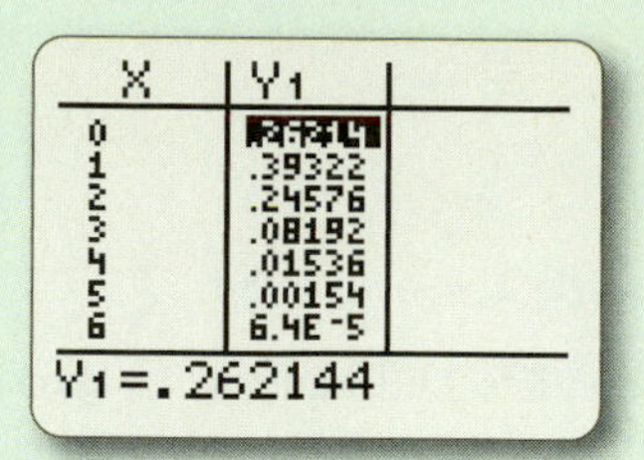

X	Y1
0	.26214
1	.39322
2	.24576
3	.08192
4	.01536
5	.00154
6	6.4E-5

Y1=.262144

The TI-83/84 Plus also has a built-in binomial distribution function that you can use in place of the explicit formula:

`Y1 = binompdf(6, 0.2, X)` Press [2nd] [VARS] [0].

The TI-83/84 Plus function `binomcdf` (directly following `binompdf`) gives the value of the *cumulative* distribution function, $P(0 \le X \le x)$.

To graph the resulting probability distribution on your calculator, follow the instructions for graphing a histogram in Section 8.1.

Section 8.3

Example 3 (page 437) According to historical data, the number of injuries that a member of the Enormous State University women's soccer team will sustain during a typical season is given by the following probability distribution table:

Injuries	0	1	2	3	4	5	6
Probability	.20	.20	.22	.20	.15	.01	.02

If X denotes the number of injuries sustained by a player during one season, compute $E(X)$.

Solution with Technology

To obtain the expected value of a probability distribution on the TI-83/84 Plus, press [STAT], select `EDIT`, and then press [ENTER], and enter the values of X in the L_1 list and the probabilities in the column in the L_2 list. Then, on the home screen, you can obtain the expected value as

sum (L_1 *L_2) — L_1 is [2nd] [1] L_2 is [2nd] [2]. Sum is found in [2nd] [STAT], under `MATH`

L1: 0 1 2 3 4 5 6 | L2: .2 .2 .22 .2 .15 .01 .02 | L3
L2(8) =

sum(L1*L2)
2.01

Section 8.4

Example 3 (page 447) Compute the variance and standard deviation for the following probability distribution.

x	10	20	30	40	50	60
$P(X = x)$	.2	.2	.3	.1	.1	.1

Solution with Technology

1. As in Example 3 in the preceding section, begin by entering the probability distribution of X into columns L_1 and L_2 in the `LIST` screen (press [STAT] and select `EDIT`). (See below left.)
2. Then, on the home screen, enter

   `sum(L1*L2)→M` — Stores the value of μ as M. Sum is found in [2nd] [STAT], under `MATH`.

3. To obtain the variance, enter the following.

   `sum((L1-M)^2*L2)` — Computation of $\sum(x-\mu)^2 P(X = x)$

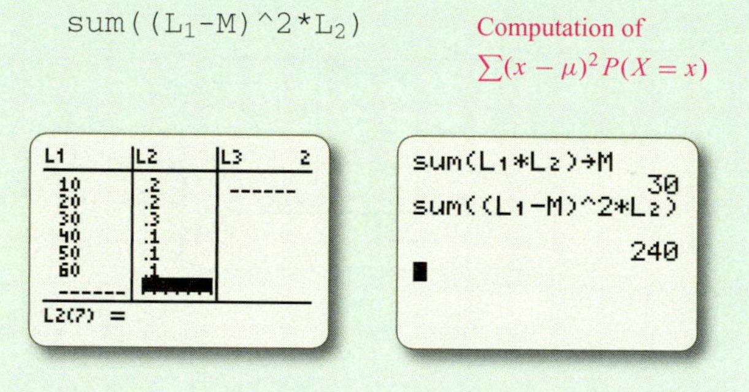

Section 8.5

Example 1(b), (c) (page 452) Let Z be the standard normal variable. Calculate the following probabilities.

b. $P(0 \le Z \le 2.43)$

c. $P(-1.37 \le Z \le 2.43)$

Solution with Technology

On the TI-83/84 Plus, press [2nd] [VARS] to obtain the selection of distribution functions. The first function, `normalpdf`, gives the values of the normal density function (whose graph is the normal curve). The second, `normalcdf`, gives $P(a \le Z \le b)$. For example, to compute $P(0 \le Z \le 2.43)$, enter

`normalcdf(0, 2.43)`

To compute $P(-1.37 \le Z \le 2.43)$, enter

```
normalcdf(-1.37, 2.43)
```

```
normalcdf(0,2.43
)
          .4924505885
normalcdf(-1.37,
2.43)
          .9071070809
```

Example 2 (page 454) Pressure gauges manufactured by Precision Corp. must be checked for accuracy before being placed on the market. To test a pressure gauge, a worker uses it to measure the pressure of a sample of compressed air known to be at a pressure of exactly 50 pounds per square inch. If the gauge reading is off by more than 1% (0.5 pounds), it is rejected. Assuming that the reading of a pressure gauge under these circumstances is a normal random variable with mean 50 and standard deviation 0.4, find the percentage of gauges rejected.

Solution with Technology

As seen in the text, we need to compute $1 - P(49.5 \le X \le 50.5)$ with $\mu = 50$ and $\sigma = 0.4$. On the TI-83/84 Plus, the built-in `normalcdf` function permits us to compute $P(a \le X \le b)$ for nonstandard normal distributions as well. The format is

normalcdf(a, b, μ, σ) $P(a \le X \le b)$

For example, we can compute $P(49.5 \le X \le 50.5)$ by entering

```
normalcdf(49.5, 50.5, 50, 0.4)
```

Then subtract it from 1 to obtain the answer:

```
normalcdf(49.5,5
0.5,50,.4)
          .7887003221
1-Ans
          .2112996779
```

SPREADSHEET Technology Guide

Section 8.1

Example 3 (page 424) Let X be the number of heads that face up in three tosses of a coin. We obtained the following probability distribution of X in the text:

x	0	1	2	3
$P(X = x)$	$\frac{1}{8}$	$\frac{3}{8}$	$\frac{3}{8}$	$\frac{1}{8}$

Use technology to obtain the corresponding histogram.

Solution with Technology

1. In your spreadsheet, enter the values of X in one column and the probabilities in another.

	A	B
1	x	P(X=x)
2	0	0.125
3	1	0.375
4	2	0.375
5	3	0.125

2. Next, select *only* the column of probabilities (B2–B5) and then choose Insert → Column Chart. The procedure for doing this depends heavily on the spreadsheet and platform you are using.

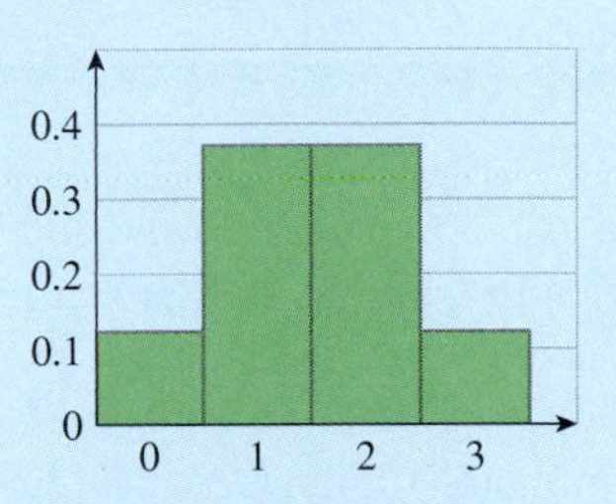

TECHNOLOGY GUIDE

Example 4 (page 424) We obtained the following frequency table in the text:

x	25,000	35,000	45,000	55,000	65,000	75,000	85,000
Frequency	20	80	230	400	170	70	30

Find the probability distribution of X.

Solution with Technology

We need to divide each frequency by the sum. Although the computations in this example (dividing the seven frequencies by 1,000) are simple to do by hand, they could become tedious in general, so technology is helpful. Spreadsheets manipulate lists with ease. Set up your spreadsheet as shown.

	A	B	C
1	x	Fr	P(X=x)
2	25000	20	=B2/SUM(B:B)
3	35000	80	
4	45000	230	
5	55000	400	
6	65000	170	
7	75000	70	
8	85000	30	

↓

	A	B	C
1	x	Fr	P(X=x)
2	25000	20	0.02
3	35000	80	0.08
4	45000	230	0.23
5	55000	400	0.4
6	65000	170	0.17
7	75000	70	0.07
8	85000	30	0.03

The formula `SUM(B:B)` gives the sum of all the numerical entries in Column B. You can now change the frequencies to see the effect on the probabilities. You can also add new values and frequencies to the list if you copy the formula in column C further down the column.

Section 8.2

Example 2(b) (page 430) By 2030, the probability that a randomly chosen resident in the United States will be 65 years old or older is projected to be .2. If X is the number of people aged 65 or older in a sample of 6, construct the probability distribution of X.

Solution with Technology

You can generate the binomial distribution as follows in your spreadsheet:

	A	B
1	x	P(X=x)
2	0	=BINOMDIST(A2,6,0.2,0)
3	1	
4	2	
5	3	
6	4	
7	5	
8	6	

↓

	A	B
1	x	P(X=x)
2	0	0.262144
3	1	0.393216
4	2	0.24576
5	3	0.08192
6	4	0.01536
7	5	0.001536
8	6	6.4E-05

The values of X are shown in column A, and the probabilities are computed in column B. The arguments of the `BINOMDIST` function are as follows:

`BINOMDIST`(x, n, p, Cumulative ($0 =$ no, $1 =$ yes)).

Setting the last argument to 0 (as shown) gives $P(X = x)$. Setting it to 1 gives $P(X \leq x)$.

To graph the resulting probability distribution using your spreadsheet, insert a bar chart as in Section 8.1.

Section 8.3

Example 3 (page 437) According to historical data, the number of injuries that a member of the Enormous State University women's soccer team will sustain during a typical season is given by the following probability distribution table:

Injuries	0	1	2	3	4	5	6
Probability	.20	.20	.22	.20	.15	.01	.02

If X denotes the number of injuries sustained by a player during one season, compute $E(X)$.

Solution with Technology

As the method we used suggests, the calculation of the expected value from the probability distribution is particularly easy to do using a spreadsheet program such as Excel.

The following worksheet shows one way to do it. (The first two columns contain the probability distribution of X; the quantities $xP(X=x)$ are summed in cell C9.)

	A	B	C
1	x	P(X=x)	x*P(X=x)
2	0	0.2	=A2*B2
3	1	0.2	
4	2	0.22	
5	3	0.2	
6	4	0.15	
7	5	0.01	
8	6	0.02	
9			=SUM(C2:C8)

↓

	A	B	C
1	x	P(X=x)	x*P(X=x)
2	0	0.2	0
3	1	0.2	0.2
4	2	0.22	0.44
5	3	0.2	0.6
6	4	0.15	0.6
7	5	0.01	0.05
8	6	0.02	0.12
9			2.01

An alternative is to use the SUMPRODUCT function: Once we enter the first two columns above, the formula

```
=SUMPRODUCT(A2:A8,B2:B8)
```

computes the sum of the products of corresponding entries in the columns, giving us the expected value.

Section 8.4

Example 3 (page 447) Compute the variance and standard deviation for the following probability distribution.

x	10	20	30	40	50	60
$P(X=x)$	.2	.2	.3	.1	.1	.1

Solution with Technology

As in Example 3 in the preceding section, begin by entering the probability distribution into columns A and B, and then proceed as shown:

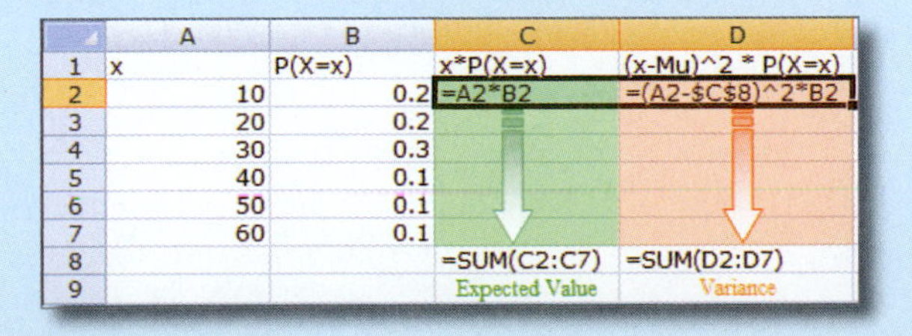

	A	B	C	D
1	x	P(X=x)	x*P(X=x)	(x-Mu)^2 * P(X=x)
2	10	0.2	=A2*B2	=(A2-$C$8)^2*B2
3	20	0.2		
4	30	0.3		
5	40	0.1		
6	50	0.1		
7	60	0.1		
8			=SUM(C2:C7)	=SUM(D2:D7)
9			Expected Value	Variance

The variance then appears in cell D8:

	A	B	C	D
1	x	P(X=x)	x*P(X=x)	(x-Mu)^2 * P(X=x)
2	10	0.2	2	80
3	20	0.2	4	20
4	30	0.3	9	0
5	40	0.1	4	10
6	50	0.1	5	40
7	60	0.1	6	90
8			30	240
9			Expected Value	Variance

Section 8.5

Example 1(b), (c) (page 452) Let Z be the standard normal variable. Calculate the following probabilities.

b. $P(0 \leq Z \leq 2.43)$

c. $P(-1.37 \leq Z \leq 2.43)$

Solution with Technology

In spreadsheets, the function `NORMSDIST` (Normal Standard Distribution) gives the area shown on the left in Figure 18. (Tables such as the one in the Appendix give the area shown on the right.)

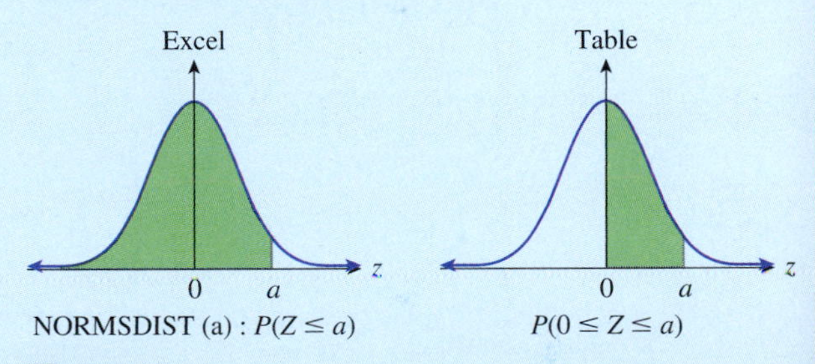

Figure 18

To compute a general area, $P(a \leq Z \leq b)$ in your spreadsheet, subtract the cumulative area to a from that to b:

`=NORMSDIST(b)-NORMSDIST(a)` $P(a \leq Z \leq b)$

In particular, to compute $P(0 \leq Z \leq 2.43)$, use

```
=NORMSDIST(2.43)-NORMSDIST(0)
```

and to compute $P(-1.37 \leq Z \leq 2.43)$, use

```
=NORMSDIST(2.43)-NORMSDIST(-1.37)
```

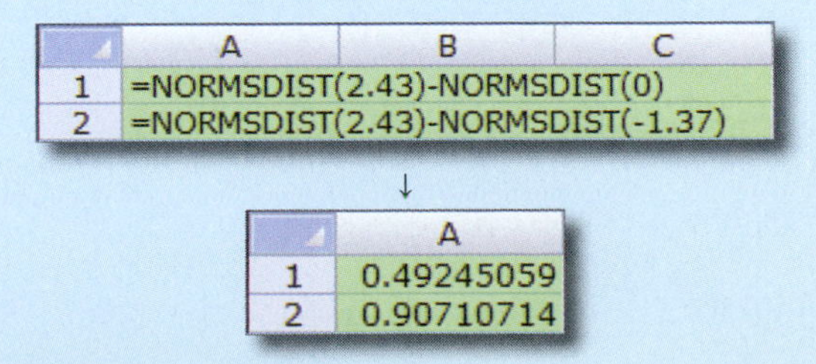

	A	B	C
1	=NORMSDIST(2.43)-NORMSDIST(0)		
2	=NORMSDIST(2.43)-NORMSDIST(-1.37)		

↓

	A
1	0.49245059
2	0.90710714

Example 2 (page 454) Pressure gauges manufactured by Precision Corp. must be checked for accuracy before being placed on the market. To test a pressure gauge, a worker uses it to measure the pressure of a sample of compressed air known to be at a pressure of exactly 50 pounds per square inch. If the gauge reading is off by more than 1% (0.5 pounds), it is rejected. Assuming that the reading of a pressure gauge under these circumstances is a normal random variable with mean 50 and standard deviation 0.4, find the percentage of gauges rejected.

Solution with Technology

In spreadsheets, we use the function NORMDIST instead of NORMSDIST. Its format is similar to NORMSDIST, but includes extra arguments as shown.

```
=NORMDIST(a, μ, σ, 1)
```
$P(X \le a)$

(The last argument, set to 1, tells the spreadsheet that we want the cumulative distribution.) To compute $P(a \le X \le b)$ we enter the following in any vacant cell:

```
=NORMDIST(b, μ, σ, 1)
  -NORMDIST(a, μ, σ, 1)
```
$P(a \le X \le b)$

For example, we can compute $P(49.5 \le X \le 50.5)$ by entering

```
=NORMDIST(50.5,50,0.4,1)
  -NORMDIST(49.5,50,0.4,1)
```

We then subtract it from 1 to obtain the answer:

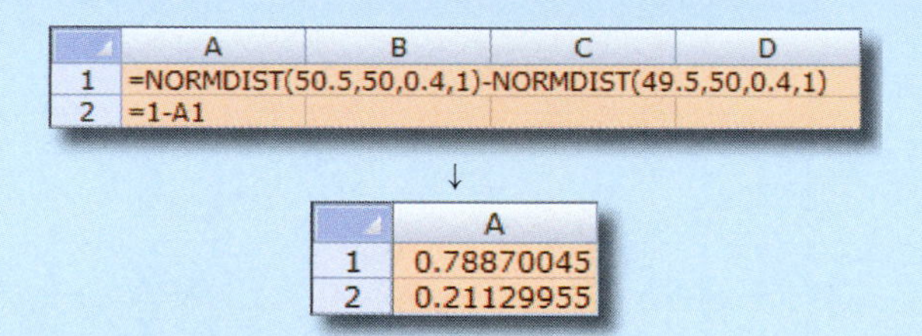

	A	B	C	D
1	=NORMDIST(50.5,50,0.4,1)-NORMDIST(49.5,50,0.4,1)			
2	=1-A1			

↓

	A
1	0.78870045
2	0.21129955

Appendix A Logic

Introduction

Logic is the underpinning of all reasoned argument. The ancient Greeks recognized its role in mathematics and philosophy, and studied it extensively. Aristotle, in his *Organon*, wrote the first systematic treatise on logic. His work had a heavy influence on philosophy, science and religion through the Middle Ages.

But Aristotle's logic was expressed in ordinary language, so was subject to the ambiguities of ordinary language. Philosophers came to want to express logic more formally and symbolically, more like the way that mathematics is written (Leibniz, in the 17th century, was probably the first to envision and call for such a formalism). It was with the publication in 1847 of G. Boole's *The Mathematical Analysis of Logic* and A. DeMorgan's *Formal Logic* that **symbolic logic** came into being, and logic became recognized as part of mathematics. Since Boole and DeMorgan, logic and mathematics have been inextricably intertwined. Logic is part of mathematics, but at the same time it is the language of mathematics.

The study of symbolic logic is usually broken into several parts. The first and most fundamental is the **propositional logic**. Built on top of this is the **predicate logic**, which is the language of mathematics. In this appendix we give an introduction to propositional logic.

A.1 Statements and Logical Operators

Propositional logic is the study of *propositions*. A **statement**, or **proposition**, is any declarative sentence which is either true (T) or false (F). We refer to T or F as the **truth value** of the statement.

For a much more extensive interactive treatment of logic, including discussion of proofs, rules of inference, and an introduction to the predicate calculus, go online and follow:

Online Text
→ On Line Topics in Finite Mathematics
→ Introduction to Logic

EXAMPLE 1 Statements

a. "$2 + 2 = 4$" is a statement because it can be either true or false.[1] Because it happens to be a true statement, its truth value is T.

b. "$1 = 0$" is also a statement, but its truth value is F.

c. "It will rain tomorrow" is a statement. To determine its truth value, we shall have to wait for tomorrow.

d. "Solve the following equation for x" is not a statement, because it cannot be assigned any truth value whatsoever. It is an imperative, or command, rather than a declarative sentence.

e. "The number 5" is not a statement, because it is not even a complete sentence.

[1] Is "$2 + 2 = 4$" a sentence? Read it aloud: "Two plus two equals four," is a perfectly respectable English sentence.

f. "This statement is false" gets us into a bind: If it were true, then, because it is declaring itself to be false, it must be false. On the other hand, if it were false, then its declaring itself false is a lie, so it is true! In other words, if it is true, then it is false, and if it is false, then it is true, and we go around in circles. We get out of this bind by saying that because the sentence cannot be either true or false, we refuse to call it a statement. An equivalent pseudo-statement is: "I am lying," so this sentence is known as **the liar's paradox**.

Note Sentences that refer to themselves, or *self-referential sentences*, as illustrated in Example 1(f), are not permitted to be statements. This eliminates the liar's paradox and several similar problems. ■

We shall use letters like p, q, r, and so on to stand for statements. Thus, for example, we might decide that p should stand for the statement "the moon is round." We write

p: "the moon is round" p is the statement that the moon is round

to express this.

We can form new statements from old ones in several different ways. For example, starting with p: "I am an Anchovian," we can form the **negation** of p: "It is not the case that I am an Anchovian" or simply "I am not an Anchovian."

Negation of a Statement

If p is a statement, then its **negation** is the statement "not p" and is denoted by $\sim p$. We mean by this that, if p is true, then $\sim p$ is false, and vice versa.

Quick Examples

1. If p: "$2+2=4$," then $\sim p$: "It is not the case that $2+2=4$," or, more simply, $\sim p$: "$2+2\neq 4$."
2. If q: "$1=0$," then $\sim q$: "$1\neq 0$."
3. If r: "Diamonds are a pearl's best friend," then $\sim r$: "Diamonds are not a pearl's best friend."
4. If s: "All politicians are crooks," then $\sim s$: "Not all politicians are crooks."
5. **Double Negation:** If p is any statement, then the negation of $\sim p$ is $\sim(\sim p)$: "not (not p)," or, in other words, p. Thus $\sim(\sim p)$ has the same meaning as p.

Notes

1. Notice in Quick Example 1 above that $\sim p$ is false, because p is true. However, in Quick Example 2, $\sim q$ is true, because q is false. A statement of the form $\sim q$ can very well be true; it is a common mistake to think it must be false.
2. Saying that not all politicians are crooks is not the same as saying that no politicians are crooks, but is the same as saying that some (meaning one or more) politicians are not crooks.
3. The symbol $\sim$ is our first example of a **logical operator**.
4. When we say in Quick Example 5 above that $\sim(\sim p)$ has the same meaning as p, we mean that they are *logically equivalent*—a notion we will make precise below. ■

Here is another way we can form a new statement from old ones. Starting with p: "I am wise," and q: "I am strong," we can form the statement "I am wise and I am strong." We denote this new statement by $p \wedge q$, read "*p and q*." In order for $p \wedge q$ to be true, *both p and q* must be true. Thus, for example, if I am wise but not strong, then $p \wedge q$ is false. The symbol $\wedge$ is another logical operator. The statement $p \wedge q$ is called the **conjunction** of p and q.

Conjunction

The **conjunction** of p and q is the statement $p \wedge q$, which we read "p and q." It can also be said in a number of different ways, such as "p even though q." The statement $p \wedge q$ is true when both p and q are true and false otherwise.

Quick Examples

1. If p: "This galaxy will ultimately disappear into a black hole" and q: "$2 + 2 = 4$," then $p \wedge q$ is the statement "Not only will this galaxy ultimately disappear into a black hole, but $2 + 2 = 4$!"
2. If p: "$2 + 2 = 4$" and q: "$1 = 0$," then $p \wedge q$: "$2 + 2 = 4$ and $1 = 0$." Its truth value is F because q is F.
3. With p and q as in Quick Example 1, the statement $p \wedge (\sim q)$ says: "This galaxy will ultimately disappear into a black hole and $2 + 2 \neq 4$," or, more colorfully, as "Contrary to your hopes, this galaxy is doomed to disappear into a black hole; moreover, two plus two is decidedly *not* equal to four!"

Notes

1. We sometimes use the word "but" as an emphatic form of "and." For instance, if p: "It is hot," and q: "It is not humid," then we can read $p \wedge q$ as "It is hot but not humid." There are always many ways of saying essentially the same thing in a natural language; one of the purposes of symbolic logic is to strip away the verbiage and record the underlying logical structure of a statement.
2. A **compound statement** is a statement formed from simpler statements via the use of logical operators. Examples are $\sim p$, $(\sim p) \wedge (q \wedge r)$ and $p \wedge (\sim p)$. A statement that cannot be expressed as a compound statement is called an **atomic statement**.[2] For example, "I am clever" is an atomic statement. In a compound statement such as $(\sim p) \wedge (q \wedge r)$, we refer to p, q, and r as the **variables** of the statement. Thus, for example, $\sim p$ is a compound statement in the single variable p. ■

Before discussing other logical operators, we pause for a moment to talk about **truth tables**, which give a convenient way to analyze compound statements.

Truth Table

The **truth table** for a compound statement shows, for each combination of possible truth values of its variables, the corresponding truth value of the statement.

[2]"Atomic" comes from the Greek for "not divisible." Atoms were originally thought to be the indivisible components of matter, but the march of science proved that wrong. The name stuck, though.

Quick Examples

1. The truth table for negation, that is, for $\sim p$, is:

p	$\sim p$
T	F
F	T

Each row shows a possible truth value for p and the corresponding value of $\sim p$.

2. The truth table for conjunction, that is, for $p \wedge q$, is:

p	q	$p \wedge q$
T	T	T
T	F	F
F	T	F
F	F	F

Each row shows a possible combination of truth values of p and q and the corresponding value of $p \wedge q$.

EXAMPLE 2 Construction of Truth Tables

Construct truth tables for the following compound statements.

a. $\sim(p \wedge q)$ **b.** $(\sim p) \wedge q$

Solution

a. Whenever we encounter a complex statement, we work from the inside out, just as we might do if we had to evaluate an algebraic expression like $-(a + b)$. Thus, we start with the p and q columns, then construct the $p \wedge q$ column, and finally, the $\sim(p \wedge q)$ column.

p	q	$p \wedge q$	$\sim(p \wedge q)$
T	T	T	F
T	F	F	T
F	T	F	T
F	F	F	T

Notice how we get the $\sim(p \wedge q)$ column from the $p \wedge q$ column: we reverse all the truth values.

b. Because there are two variables, p and q, we again start with the p and q columns. We then evaluate $\sim p$, and finally take the conjunction of the result with q.

p	q	$\sim p$	$(\sim p) \wedge q$
T	T	F	F
T	F	F	F
F	T	T	T
F	F	T	F

Because we are "and-ing" $\sim p$ with q, we look at the values in the $\sim p$ and q columns and combine these according to the instructions for "and." Thus, for example, in the first row we have $F \wedge T = F$ and in the third row we have $T \wedge T = T$.

Here is a third logical operator. Starting with p: "You are over 18" and q: "You are accompanied by an adult," we can form the statement "You are over 18 or are accompanied by an adult," which we write symbolically as $p \vee q$, read "*p or q*." Now in English the word "or" has several possible meanings, so we have to agree on which one we want here. Mathematicians have settled on the **inclusive or:** $p \vee q$ means p is true or q is true *or both are true.*[3] With p and q as above, $p \vee q$ stands for "You are over 18 or are accompanied by an adult, or both." We shall sometimes include the phrase "or both" for emphasis, but even if we leave it off we still interpret "or" as inclusive.

Disjunction

The **disjunction** of p and q is the statement $p \vee q$, which we read "p or q." Its truth value is defined by the following truth table.

p	q	$p \vee q$
T	T	T
T	F	T
F	T	T
F	F	F

This is the **inclusive** or, so $p \vee q$ is true when p is true or q is true *or both* are true.

Quick Examples

1. Let p: "The butler did it" and let q: "The cook did it." Then $p \vee q$: "Either the butler or the cook did it."
2. Let p: "The butler did it," and let q: "The cook did it," and let r: "The lawyer did it." Then $(p \vee q) \wedge (\sim r)$: "Either the butler or the cook did it, but not the lawyer."

Note The only way for $p \vee q$ to be false is for *both* p and q to be false. For this reason, we can say that $p \vee q$ also means "p and q are not both false." ■

To introduce our next logical operator, we ask you to consider the following statement: "If you earn an A in logic, then I'll buy you a new car." It seems to be made up out of two simpler statements,

p: "You earn an A in logic," and
q: "I will buy you a new car."

[3]There is also the **exclusive or:** "p or q *but not both.*" This can be expressed as $(p \vee q) \wedge \sim(p \wedge q)$. Do you see why?

The original statement says: *if p is true, then q is true*, or, more simply, **if** p, **then** q. We can also phrase this as p **implies** q, and we write the statement symbolically as $p \rightarrow q$.

Now let us suppose for the sake of argument that the original statement: "If you earn an A in logic, then I'll buy you a new car," is true. This does *not* mean that you *will* earn an A in logic. All it says is that *if* you do so, then I will buy you that car. Thinking of this as a promise, the only way that it can be broken is if you *do* earn an A and I do *not* buy you a new car. With this in mind, we define the logical statement $p \rightarrow q$ as follows.

Conditional

The **conditional** $p \rightarrow q$, which we read "if p, then q" or "p implies q," is defined by the following truth table:

p	q	$p \rightarrow q$
T	T	T
T	F	F
F	T	T
F	F	T

The arrow "$\rightarrow$" is the **conditional** operator, and in $p \rightarrow q$ the statement p is called the **antecedent,** or **hypothesis,** and q is called the **consequent,** or **conclusion.** A statement of the form $p \rightarrow q$ is also called an **implication.**

Quick Examples

1. "If $1 + 1 = 2$ then the sun rises in the east" has the form $p \rightarrow q$ where p: "$1 + 1 = 2$" is true and q: "the sun rises in the east." is also true. Therefore the statement is true.
2. "If the moon is made of green cheese, then I am Arnold Schwartzenegger" has the form $p \rightarrow q$ where p is false. From the truth table, we see that $p \rightarrow q$ is therefore true, regardless of whether or not I am Arnold Schwartzenegger.
3. "If $1 + 1 = 2$ then $0 = 1$" has the form $p \rightarrow q$ where this time p is true but q is false. Therefore, by the truth table, the given statement is false.

Notes

1. The only way that $p \rightarrow q$ can be false is if p is true and q is false—this is the case of the "broken promise" in the car example above.
2. If you look at the last two rows of the truth table, you see that we say that "$p \rightarrow q$" is true when p is false, *no matter what the truth value of q*. Think again about the promise—if you don't get that A, then whether or not I buy you a new car, I have not broken my promise. It may seem strange at first to say that F $\rightarrow$ T is T and F $\rightarrow$ F is also T, but, as they did in choosing to say that "or" is always inclusive, mathematicians agreed that the truth table above gives the most useful definition of the conditional. ■

It is usually misleading to think of "if p then q" as meaning that *p causes q*. For instance, tropical weather conditions cause hurricanes, but one cannot claim that if there are tropical weather conditions, then there are (always) hurricanes. Here is a list of some English phrases that *do* have the same meaning as $p \rightarrow q$.

Some Phrasings of the Conditional

We interpret each of the following as equivalent to the conditional $p \to q$.

If p then q.	p implies q.
q follows from p.	Not p unless q.
q if p.	p only if q.
Whenever p, q.	q whenever p.
p is sufficient for q.	q is necessary for p.
p is a sufficient condition for q.	q is a necessary condition for p.

Quick Example

"If it's Tuesday, this must be Belgium" can be rephrased in several ways as follows:

"Its being Tuesday implies that this is Belgium."
"This is Belgium if it's Tuesday."
"It's Tuesday only if this is Belgium."
"It can't be Tuesday unless this is Belgium."
"Its being Tuesday is sufficient for this to be Belgium."
"That this is Belgium is a necessary condition for its being Tuesday."

Notice the difference between "if" and "only if." We say that "p only if q" means $p \to q$ because, assuming that $p \to q$ is true, p can be true only if q is also. In other words, the only line of the truth table that has $p \to q$ true and p true also has q true. The phrasing "p is a sufficient condition for q" says that it suffices to know that p is true to be able to conclude that q is true. For example, it is sufficient that you get an A in logic for me to buy you a new car. Other things might induce me to buy you the car, but an A in logic would suffice. The phrasing "q is necessary for p" says that for p to be true, q must be true (just as we said for "p only if q").

Q: *Does the commutative law hold for the conditional? In other words, is $p \to q$ the same as $q \to p$?*

A: No, as we can see in the following truth table:

p	q	$p \to q$	$q \to p$
T	T	T	T
T	F	F	T
F	T	T	F
F	F	T	T

not the same

Converse and Contrapositive

The statement $q \to p$ is called the **converse** of the statement $p \to q$. A conditional and its converse are *not* the same.

The statement $\sim q \to \sim p$ is the **contrapositive** of the statement $p \to q$. A conditional and its contrapositive are logically equivalent in the sense we define below: they have the same truth value for all possible values of p and q.

EXAMPLE 3 Converse and Contrapositive

Give the converse and contrapositive of the statement "If you earn an A in logic, then I'll buy you a new car."

Solution This statement has the form $p \to q$ where p: "you earn an A" and q: "I'll buy you a new car." The converse is $q \to p$. In words, this is "If I buy you a new car then you earned an A in logic."

The contrapositive is $(\sim q) \to (\sim p)$. In words, this is "If I don't buy you a new car, then you didn't earn an A in logic."

Assuming that the original statement is true, notice that the converse is not necessarily true. There is nothing in the original promise that prevents me from buying you a new car if you do not earn the A. On the other hand, the contrapositive is true. If I don't buy you a new car, it must be that you didn't earn an A; otherwise I would be breaking my promise.

It sometimes happens that we do want both a conditional and its converse to be true. The conjunction of a conditional and its converse is called a **biconditional**.

Biconditional

The **biconditional**, written $p \leftrightarrow q$, is defined to be the statement $(p \to q) \wedge (q \to p)$. Its truth table is the following:

p	q	$p \leftrightarrow q$
T	T	T
T	F	F
F	T	F
F	F	T

Phrasings of the Biconditional

We interpret each of the following as equivalent to $p \leftrightarrow q$.

p if and only if q.
p is necessary and sufficient for q.
p is equivalent to q.

Quick Example

"I teach math if and only if I am paid a large sum of money" can be rephrased in several ways as follows:

"I am paid a large sum of money if and only if I teach math."
"My teaching math is necessary and sufficient for me to be paid a large sum of money."
"For me to teach math, it is necessary and sufficient that I be paid a large sum of money."

A.2 Logical Equivalence

We mentioned above that we say that two statements are **logically equivalent** if for all possible truth values of the variables involved the two statements always have the same truth values. If s and t are equivalent, we write $s \equiv t$. This is *not* another logical statement. It is simply the claim that the two statements s and t are logically equivalent. Here are some examples.

EXAMPLE 4 Logical Equivalence

Use truth tables to show the following:

a. $p \equiv \sim(\sim p)$. This is called **double negation**.

b. $\sim(p \wedge q) \equiv (\sim p) \vee (\sim q)$. This is one of **DeMorgan's Laws**.

Solution

a. To demonstrate the logical equivalence of these two statements, we construct a truth table with columns for both p and $\sim(\sim p)$.

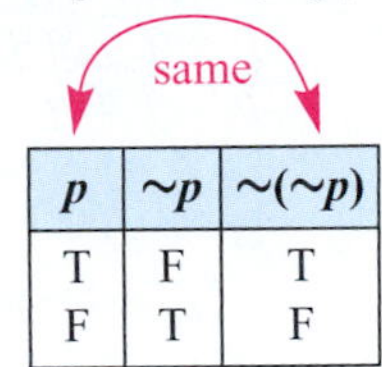

p	$\sim p$	$\sim(\sim p)$
T	F	T
F	T	F

Because the p and $\sim(\sim p)$ columns contain the same truth values in all rows, the two statements are logically equivalent.

b. We construct a truth table showing both $\sim(p \wedge q)$ and $(\sim p) \vee (\sim q)$.

same

p	q	$p \wedge q$	$\sim(p \wedge q)$	$\sim p$	$\sim q$	$(\sim p) \vee (\sim q)$
T	T	T	F	F	F	F
T	F	F	T	F	T	T
F	T	F	T	T	F	T
F	F	F	T	T	T	T

Because the $\sim(p \wedge q)$ column and $(\sim p) \vee (\sim q)$ column agree, the two statements are equivalent.

Before we go on... The statement $\sim(p \wedge q)$ can be read as "It is not the case that both p and q are true" or "p and q are not both true." We have just shown that this is equivalent to "Either p is false or q is false." ■

Here are the two equivalences known as DeMorgan's Laws.

DeMorgan's Laws

If p and q are statements, then

$$\sim(p \wedge q) \equiv (\sim p) \vee (\sim q)$$

$$\sim(p \vee q) \equiv (\sim p) \wedge (\sim q)$$

Quick Example

Let p: "the President is a Democrat," and q: "the President is a Republican." Then the following two statements say the same thing:

$\sim(p \wedge q)$: "the President is not both a Democrat and a Republican."

$(\sim p) \vee (\sim q)$: "either the President is not a Democrat, or he is not a Republican (or he is neither)."

Here is a list of some important logical equivalences, some of which we have already encountered. All of them can be verified using truth tables as in Example 4. (The verifications of some of these are in the exercise set.)

Important Logical Equivalences

$\sim(\sim p) \equiv p$	the Double Negative Law
$p \wedge q \equiv q \wedge p$	the Commutative Law for Conjunction
$p \vee q \equiv q \vee p$	the Commutative Law for Disjunction
$(p \wedge q) \wedge r \equiv p \wedge (q \wedge r)$	the Associative Law for Conjunction
$(p \vee q) \vee r \equiv p \vee (q \vee r)$	the Associative Law for Disjunction
$\sim(p \vee q) \equiv (\sim p) \wedge (\sim q)$	DeMorgan's Laws
$\sim(p \wedge q) \equiv (\sim p) \vee (\sim q)$	
$p \wedge (q \vee r) \equiv (p \wedge q) \vee (p \wedge r)$	the Distributive Laws
$p \vee (q \wedge r) \equiv (p \vee q) \wedge (p \vee r)$	
$p \wedge p \equiv p$	Absorption Laws
$p \vee p \equiv p$	
$p \rightarrow q \equiv (\sim q) \rightarrow (\sim p)$	Contrapositive Law

Note that these logical equivalences apply to *any* statement. The ps, qs, and rs can stand for atomic statements or compound statements, as we see in the next example.

EXAMPLE 5 Applying Logical Equivalences

a. Apply DeMorgan's law (once) to the statement $\sim([p \wedge (\sim q)] \wedge r)$.

b. Apply the distributive law to the statement $(\sim p) \wedge [q \vee (\sim r)]$.

c. Consider: "You will get an A if either you are clever and the sun shines, or you are clever and it rains." Rephrase the condition more simply using the distributive law.

Solution

a. We can analyze the given statement from the outside in. It is first of all a negation, but further, it is the negation $\sim(A \wedge B)$, where A is the compound statement $[p \wedge (\sim q)]$ and B is r:

$$\sim(\underbrace{[p \wedge (\sim q)]}_{A} \wedge \underbrace{r}_{B})$$

Now one of DeMorgan's laws is

$$\sim(A \wedge B) \equiv (\sim A) \vee (\sim B)$$

Applying this equivalence gives

$$\sim([p \wedge (\sim q)] \wedge r) \equiv (\sim[p \wedge (\sim q)]) \vee (\sim r)$$

b. The given statement has the form $A \wedge [B \vee C]$, where $A = (\sim p)$, $B = q$, and $C = (\sim r)$. So, we apply the distributive law $A \wedge [B \vee C], \equiv [A \wedge B] \vee [A \wedge C]$:

$$(\sim p) \wedge [q \vee (\sim r)] \equiv [(\sim p) \wedge q] \vee [(\sim p) \wedge (\sim r)]$$

(We need not stop here: The second expression on the right is just begging for an application of DeMorgan's law. . .)

c. The condition is "either you are clever and the sun shines, or you are clever and it rains." Let's analyze this symbolically: Let p: "You are clever," q: "The sun shines," and r: "It rains." The condition is then $(p \wedge q) \vee (p \wedge r)$. We can "factor out" the p using one of the distributive laws in reverse, getting

$$(p \wedge q) \vee (p \wedge r) \equiv p \wedge (q \vee r)$$

We are taking advantage of the fact that the logical equivalences we listed can be read from right to left as well as from left to right. Putting $p \wedge (q \vee r)$ back into English, we can rephrase the sentence as "You will get an A if you are clever and either the sun shines or it rains."

Before we go on... In part (a) of Example 5 we could, if we wanted, apply DeMorgan's law again, this time to the statement $\sim[p \wedge (\sim q)]$ that is part of the answer. Doing so gives

$$\sim[p \wedge (\sim q)] \equiv (\sim p) \vee \sim(\sim q) \equiv (\sim p) \vee q$$

Notice that we've also used the double negative law. Therefore, the original expression can be simplified as follows:

$$\sim([p \wedge (\sim q)] \wedge r) \equiv (\sim[p \wedge (\sim q)]) \vee (\sim r) \equiv ((\sim p) \vee q) \vee (\sim r)$$

which we can write as

$$(\sim p) \vee q \vee (\sim r)$$

because the associative law tells us that it does not matter which two expressions we "or" first. ■

A.3 Tautologies, Contradictions, and Arguments

Tautologies and Contradictions

A compound statement is a **tautology** if its truth value is always T, regardless of the truth values of its variables. It is a **contradiction** if its truth value is always F, regardless of the truth values of its variables.

Quick Examples

1. $p \vee (\sim p)$ has truth table

p	$\sim p$	$p \vee (\sim p)$
T	F	T
F	T	T

all T's

and is therefore a tautology.

2. $p \wedge (\sim p)$ has truth table

p	$\sim p$	$p \wedge (\sim p)$
T	F	F
F	T	F

and is therefore a contradiction.

When a statement is a tautology, we also say that the statement is **tautological**. In common usage this sometimes means simply that the statement is self-evident. In logic it means something stronger: that the statement is always true, under all circumstances. In contrast, a contradiction, or **contradictory** statement, is *never* true, under any circumstances.

Some of the most important tautologies are the **tautological implications**, tautologies that have the form of implications. We look at two of them: Direct Reasoning, and Indirect Reasoning:

Modus Ponens or Direct Reasoning

The following tautology is called ***modus ponens*** or **direct reasoning:**

$$[(p \to q) \wedge p] \to q$$

In Words

If an implication and its antecedent (p) are both true, then so is its consequent (q).

Quick Example

If my loving math implies that I will pass this course, and if I do love math, then I will pass this course.

Note You can check that the statement $[(p \to q) \wedge p] \to q$ is a tautology by drawing its truth table. ■

Tautological implications are useful mainly because they allow us to check the validity of **arguments**.

Argument

An **argument** is a list of statements called **premises** followed by a statement called the **conclusion.** If the premises are $P_1, P_2, \ldots, P_n$ and the conclusion is C, then we say that the argument is **valid** if the statement $(P_1 \wedge P_2 \wedge \ldots \wedge P_n) \rightarrow C$ is a tautology. In other words, an argument is valid if the truth of all its premises logically implies the truth of its conclusion.

Quick Examples

1. The following is a valid argument:

$$\begin{array}{l} p \rightarrow q \\ \underline{p \qquad\quad} \\ \therefore \quad q \end{array}$$

(This is the traditional way of writing an argument: We list the premises above a line and then put the conclusion below; the symbol "∴" stands for the word "therefore.") This argument is valid because the statement $[(p \rightarrow q) \wedge p] \rightarrow q$ is a tautology, namely *modus ponens*.

2. The following is an invalid argument:

$$\begin{array}{l} p \rightarrow q \\ \underline{q \qquad\quad} \\ \therefore \quad p \end{array}$$

The argument is invalid because the statement $[(p \rightarrow q) \wedge q] \rightarrow p$ is not a tautology. In fact, if p is F and q is T, then the whole statement is F.

The argument in Quick Example 2 above is known as the *fallacy of affirming the consequent*. It is a common invalid argument and not always obviously flawed at first sight, so is often exploited by advertisers. For example, consider the following claim: All Olympic athletes drink Boors, so you should too. The suggestion is that, if you drink Boors, you will be an Olympic athlete:

If you are an Olympic Athlete you drink Boors.	Premise (Let's pretend this is True)
You drink Boors.	Premise (True)
∴ You are an Olympic Athlete.	Conclusion (May be false!)

This is an error that Boors hopes you will make!

There is, however, a correct argument in which we *deny* the consequent:

Modus Tollens or Indirect Reasoning

The following tautology is called ***modus tollens*** or **indirect reasoning:**

$$[(p \rightarrow q) \wedge (\sim q)] \rightarrow (\sim p)$$

In Words

If an implication is true but its consequent (q) is false, then its antecedent (p) is false.

In Argument Form

$p \to q$
$\underline{\sim q}$
$\therefore \ \sim p$

Quick Example

If my loving math implies that I will pass this course, and if I do not pass the course, then it must be the case that I do not love math.
In argument form:

If I love math, then I will pass this course.
I will not pass the course.
Therefore, I do not love math.

Note This argument is not as direct as *modus ponens;* it contains a little twist: "If I loved math I would pass this course. However, I will not pass this course. Therefore, it must be that I don't love math (else I *would* pass this course)." Hence the name "indirect reasoning."

Note that, again, there is a similar, but fallacious argument to avoid, for instance: "If I were an Olympic athlete then I would drink Boors ($p \to q$). However, I am not an Olympic athlete ($\sim p$). Therefore, I won't drink Boors. ($\sim q$)." This is a mistake Boors certainly hopes you do *not* make! ■

There are other interesting tautologies that we can use to justify arguments. We mention one more and refer the interested reader to the website for more examples and further study.

For an extensive list of tautologies go online and follow:
Chapter L Logic
→ List of Tautologies and Tautological Implications

Disjunctive Syllogism or "One or the Other"

The following tautologies are both known as the **disjunctive syllogism** or **one-or-the-other:**

$$[(p \vee q) \wedge (\sim p)] \to q \qquad [(p \vee q) \wedge (\sim q)] \to p$$

In Words
If one or the other of two statements is true, but one is known to be false, then the other must be true.

In Argument Form

$p \vee q$ $p \vee q$
$\underline{\sim p}$ $\underline{\sim q}$
$\therefore \ q$ $\therefore \ p$

Quick Example

The butler or the cook did it. The butler didn't do it. Therefore, the cook did it.
In argument form:

The butler or the cook did it.
The butler did not do it.
Therefore, the cook did it.

A EXERCISES

Which of Exercises 1–10 are statements? Comment on the truth values of all the statements you encounter. If a sentence fails to be a statement, explain why. **HINT** [See Example 1.]

1. All swans are white.

2. The fat cat sat on the mat.

3. Look in thy glass and tell whose face thou viewest.[4]

4. My glass shall not persuade me I am old.[5]

5. There is no largest number.

6. 1,000,000,000 is the largest number.

7. Intelligent life abounds in the universe.

8. There may or may not be a largest number.

9. This is exercise number 9.

10. This sentence no verb.[6]

Let p: "Our mayor is trustworthy," q: "Our mayor is a good speller," and r = "Our mayor is a patriot." Express each of the statements in Exercises 11–16 in logical form: **HINT** [See Quick Examples on pages A2, A3, A5.]

11. Although our mayor is not trustworthy, he is a good speller.

12. Either our mayor is trustworthy, or he is a good speller.

13. Our mayor is a trustworthy patriot who spells well.

14. While our mayor is both trustworthy and patriotic, he is not a good speller.

15. It may or may not be the case that our mayor is trustworthy.

16. Our mayor is either not trustworthy or not a patriot, yet he is an excellent speller.

Let p: "Willis is a good teacher," q: "Carla is a good teacher," r: "Willis' students hate math," s: "Carla's students hate math." Express the statements in Exercises 17–24 in words.

17. $p \wedge (\sim r)$

18. $(\sim p) \wedge (\sim q)$

19. $q \vee (\sim q)$

20. $((\sim p) \wedge (\sim s)) \vee q$

21. $r \wedge (\sim r)$

22. $(\sim s) \vee (\sim r)$

23. $\sim(q \vee s)$

24. $\sim(p \wedge r)$

Assume that it is true that "Polly sings well," it is false that "Quentin writes well," and it is true that "Rita is good at math." Determine the truth of each of the statements in Exercises 25–32.

25. Polly sings well and Quentin writes well.

26. Polly sings well or Quentin writes well.

27. Polly sings poorly and Quentin writes well.

28. Polly sings poorly or Quentin writes poorly.

29. Either Polly sings well and Quentin writes poorly, or Rita is good at math.

30. Either Polly sings well and Quentin writes poorly, or Rita is not good at math.

31. Either Polly sings well or Quentin writes well, or Rita is good at math.

32. Either Polly sings well and Quentin writes well, or Rita is bad at math.

Find the truth value of each of the statements in Exercises 33–48. **HINT** [See Quick Examples on page A6.]

33. "If $1 = 1$, then $2 = 2$."

34. "If $1 = 1$, then $2 = 3$."

35. "If $1 \neq 0$, then $2 \neq 2$."

36. "If $1 = 0$, then $1 = 1$."

37. "A sufficient condition for 1 to equal 2 is $1 = 3$."

38. "$1 = 1$ is a sufficient condition for 1 to equal 0."

39. "$1 = 0$ is a necessary condition for 1 to equal 1."

40. "$1 = 1$ is a necessary condition for 1 to equal 2."

41. "If I pay homage to the great Den, then the sun will rise in the east."

42. "If I fail to pay homage to the great Den, then the sun will still rise in the east."

43. "In order for the sun to rise in the east, it is necessary that it sets in the west."

44. "In order for the sun to rise in the east, it is sufficient that it sets in the west."

45. "The sun rises in the west only if it sets in the west."

46. "The sun rises in the east only if it sets in the east."

47. "In order for the sun to rise in the east, it is necessary and sufficient that it sets in the west."

48. "In order for the sun to rise in the west, it is necessary and sufficient that it sets in the east."

Construct the truth tables for the statements in Exercises 49–62. **HINT** [See Example 2.]

49. $p \wedge (\sim q)$

50. $p \vee (\sim q)$

51. $\sim(\sim p) \vee p$

52. $p \wedge (\sim p)$

53. $(\sim p) \wedge (\sim q)$

54. $(\sim p) \vee (\sim q)$

55. $(p \wedge q) \wedge r$

56. $p \wedge (q \wedge r)$

57. $p \wedge (q \vee r)$

58. $(p \wedge q) \vee (p \wedge r)$

59. $p \rightarrow (q \vee p)$

60. $(p \vee q) \rightarrow \sim p$

61. $p \leftrightarrow (p \vee q)$

62. $(p \wedge q) \leftrightarrow \sim p$

Use truth tables to verify the logical equivalences given in Exercises 63–72.

63. $p \wedge p \equiv p$

64. $p \vee p \equiv p$

65. $p \vee q \equiv q \vee p$ (Commutative law for disjunction)

66. $p \wedge q \equiv q \wedge p$ (Commutative law for conjunction)

[4] William Shakespeare Sonnet 3.

[5] *Ibid.*, Sonnet 22.

[6] From *Metamagical Themas: Questing for the Essence of Mind and Pattern* by Douglas R. Hofstadter (Bantam Books, New York 1986).

67. $\sim(p \vee q) \equiv (\sim p) \wedge (\sim q)$

68. $\sim(p \wedge (\sim q)) \equiv (\sim p) \vee q$

69. $(p \wedge q) \wedge r \equiv p \wedge (q \wedge r)$
(Associative law for conjunction)

70. $(p \vee q) \vee r \equiv p \vee (q \vee r)$
(Associative law for disjunction)

71. $p \rightarrow q \equiv (\sim q) \rightarrow (\sim p)$

72. $\sim(p \rightarrow q) \equiv p \wedge (\sim q)$

In Exercises 73–78, use truth tables to check whether the given statement is a tautology, a contradiction, or neither. **HINT** [See Quick Examples on page A12.]

73. $p \wedge (\sim p)$

74. $p \wedge p$

75. $p \wedge \sim(p \vee q)$

76. $p \vee \sim(p \vee q)$

77. $p \vee \sim(p \wedge q)$

78. $q \vee \sim(p \wedge (\sim p))$

Apply the stated logical equivalence to the given statement in Exercises 79–84. **HINT** [See Example 5a, b.]

79. $p \vee (\sim p)$; the commutative law

80. $p \wedge (\sim q)$; the commutative law

81. $\sim(p \wedge (\sim q))$; DeMorgan's law

82. $\sim(q \vee (\sim q))$; DeMorgan's law

83. $p \vee ((\sim p) \wedge q)$; the distributive law

84. $(\sim q) \wedge ((\sim p) \vee q)$; the distributive law

In Exercises 85–88, use the given logical equivalence to rewrite the given sentence. **HINT** [See Example 5c.]

85. It is not true that both I am Julius Caesar and you are a fool. DeMorgan's law.

86. It is not true that either I am Julius Caesar or you are a fool. DeMorgan's law.

87. Either it is raining and I have forgotten my umbrella, or it is raining and I have forgotten my hat. The distributive law.

88. I forgot my hat or my umbrella, and I forgot my hat or my glasses. The distributive law.

Give the contrapositive and converse of each of the statements in Exercises 89 and 90, phrasing your answers in words.

89. "If I think, then I am."

90. "If these birds are of a feather, then they flock together."

Exercises 91 and 92 are multiple choice. Indicate which statement is equivalent to the given statement, and say why that statement is equivalent to the given one.

91. "In order for you to worship Den, it is necessary for you to sacrifice beasts of burden."

(A) "If you are not sacrificing beasts of burden, then you are not worshiping Den."

(B) "If you are sacrificing beasts of burden, then you are worshiping Den."

(C) "If you are not worshiping Den, then you are not sacrificing beasts of burden."

92. "In order to read the Tarot, it is necessary for you to consult the Oracle."

(A) "In order to consult the Oracle, it is necessary to read the Tarot."

(B) "In order not to consult the Oracle, it is necessary not to read the Tarot."

(C) "In order not to read the Tarot, it is necessary not to read the Oracle."

In Exercises 93–102, write the given argument in symbolic form (use the underlined letters to represent the statements containing them), then decide whether it is valid or not, If it is valid, name the validating tautology. **HINT** [See Quick Examples on pages A12, A13, A14.]

93. If I am hungry I am also thirsty. I am hungry. Therefore, I am thirsty.

94. If I am not hungry, then I certainly am not thirsty either. I am not thirsty, and so I cannot be hungry.

95. For me to bring my umbrella, it's sufficient that it rain. It is not raining. Therefore, I will not bring my umbrella.

96. For me to bring my umbrella, it's necessary that it rain. But it is not raining. Therefore, I will not bring my umbrella.

97. For me to pass math, it is sufficient that I have a good teacher. I will not pass math. Therefore, I have a bad teacher.

98. For me to pass math, it is necessary that I have a good teacher. I will pass math. Therefore, I have a good teacher.

99. I will either pass math or I have a bad teacher. I have a good teacher. Therefore, I will pass math.

100. Either roses are not red or violets are not blue. But roses are red. Therefore, violets are not blue.

101. I am either smart or athletic, and I am athletic. So I must not be smart.

102. The president is either wise or strong. She is strong. Therefore, she is not wise.

In Exercises 103–108, use the stated tautology to complete the argument.

103. If John is a swan, it is necessary that he is green. John is indeed a swan. Therefore, _____. (*Modus ponens.*)

104. If Jill had been born in Texas, then she would be able to ride horses. But Jill cannot ride horses. Therefore, ____. (*Modus tollens.*)

105. If John is a swan, it is necessary that he is green. But John is not green. Therefore, _____. (*Modus tollens.*)

106. If Jill had been born in Texas, then she would be able to ride horses. Jill was born in Texas. Therefore, ____ (*Modus ponens.*)

107. Peter is either a scholar or a gentleman. He is not, however, a scholar. Therefore, ____. (Disjunctive syllogism.)

108. Pam is either a plumber or an electrician. She is not, however, an electrician. Therefore, ____ (Disjunctive syllogism.)

COMMUNICATION AND REASONING EXERCISES

109. If two statements are logically equivalent, what can be said about their truth tables?

110. If a proposition is neither a tautology nor a contradiction, what can be said about its truth table?

111. If A and B are two compound statements such that $A \vee B$ is a contradiction, what can you say about A and B?

112. If A and B are two compound statements such that $A \wedge B$ is a tautology, what can you say about A and B?

113. Give an example of an instance where $p \to q$ means that q causes p.

114. Complete the following. If $p \to q$, then its converse, ___ , is the statement that ___ and (is/is not) logically equivalent to $p \to q$.

115. Give an instance of a biconditional $p \leftrightarrow q$ where neither one of p or q causes the other.

Appendix B Area Under a Normal Curve

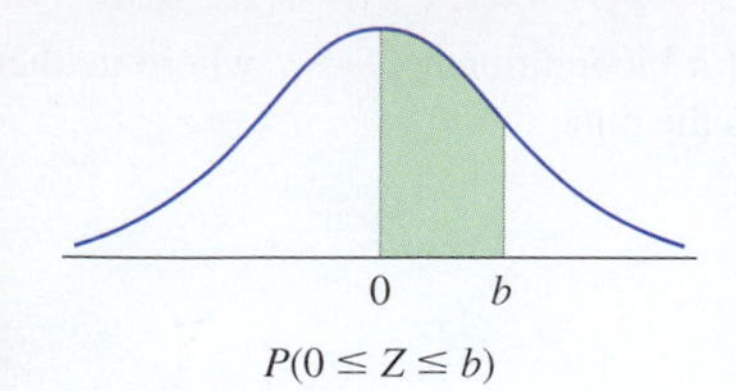

$P(0 \leq Z \leq b)$

The table below gives the probabilities $P(0 \leq Z \leq b)$ where Z is a standard normal variable. For example, to find $P(0 \leq Z \leq 2.43)$, write 2.43 as $2.4 + 0.03$, and read the entry in the row labeled 2.4 and the column labeled 0.03. From the portion of the table shown at left, you will see that $P(0 \leq Z \leq 2.43) = .4925$.

Z	0.00	0.01	0.02	0.03 ↓
2.3	.4893	.4896	.4898	.4901
→ 2.4	.4918	.4920	.4922	.4925
2.5	.4938	.4940	.4941	.4943

Z	0.00	0.01	0.02	0.03	0.04	0.05	0.06	0.07	0.08	0.09
0.0	.0000	.0040	.0080	.0120	.0160	.0199	.0239	.0279	.0319	.0359
0.1	.0398	.0438	.0478	.0517	.0557	.0596	.0636	.0675	.0714	.0753
0.2	.0793	.0832	.0871	.0910	.0948	.0987	.1026	.1064	.1103	.1141
0.3	.1179	.1217	.1255	.1293	.1331	.1368	.1406	.1443	.1480	.1517
0.4	.1554	.1591	.1628	.1664	.1700	.1736	.1772	.1808	.1844	.1879
0.5	.1915	.1950	.1985	.2019	.2054	.2088	.2123	.2157	.2190	.2224
0.6	.2257	.2291	.2324	.2357	.2389	.2422	.2454	.2486	.2517	.2549
0.7	.2580	.2611	.2642	.2673	.2704	.2734	.2764	.2794	.2823	.2852
0.8	.2881	.2910	.2939	.2967	.2995	.3023	.3051	.3078	.3106	.3133
0.9	.3159	.3186	.3212	.3238	.3264	.3289	.3315	.3340	.3365	.3389
1.0	.3413	.3438	.3461	.3485	.3508	.3531	.3554	.3577	.3599	.3621
1.1	.3643	.3665	.3686	.3708	.3729	.3749	.3770	.3790	.3810	.3830
1.2	.3849	.3869	.3888	.3907	.3925	.3944	.3962	.3980	.3997	.4015
1.3	.4032	.4049	.4066	.4082	.4099	.4115	.4131	.4147	.4162	.4177
1.4	.4192	.4207	.4222	.4236	.4251	.4265	.4279	.4292	.4306	.4319
1.5	.4332	.4345	.4357	.4370	.4382	.4394	.4406	.4418	.4429	.4441
1.6	.4452	.4463	.4474	.4484	.4495	.4505	.4515	.4525	.4535	.4545
1.7	.4554	.4564	.4573	.4582	.4591	.4599	.4608	.4616	.4625	.4633
1.8	.4641	.4649	.4656	.4664	.4671	.4678	.4686	.4693	.4699	.4706
1.9	.4713	.4719	.4726	.4732	.4738	.4744	.4750	.4756	.4761	.4767
2.0	.4772	.4778	.4783	.4788	.4793	.4798	.4803	.4808	.4812	.4817
2.1	.4821	.4826	.4830	.4834	.4838	.4842	.4846	.4850	.4854	.4857
2.2	.4861	.4864	.4868	.4871	.4875	.4878	.4881	.4884	.4887	.4890
2.3	.4893	.4896	.4898	.4901	.4904	.4906	.4909	.4911	.4913	.4916
2.4	.4918	.4920	.4922	.4925	.4927	.4929	.4931	.4932	.4934	.4936
2.5	.4938	.4940	.4941	.4943	.4945	.4946	.4948	.4949	.4951	.4952
2.6	.4953	.4955	.4956	.4957	.4959	.4960	.4961	.4962	.4963	.4964
2.7	.4965	.4966	.4967	.4968	.4969	.4970	.4971	.4972	.4973	.4974
2.8	.4974	.4975	.4976	.4977	.4977	.4978	.4979	.4979	.4980	.4981
2.9	.4981	.4982	.4982	.4983	.4984	.4984	.4985	.4985	.4986	.4986
3.0	.4987	.4987	.4987	.4988	.4988	.4989	.4989	.4989	.4990	.4990

Answers to Chapter Review Exercises

Chapter 1

Review

1. a. 1 **b.** −2 **c.** 0 **d.** −1 **3. a.** 1 **b.** 0 **c.** 0 **d.** −1

5.

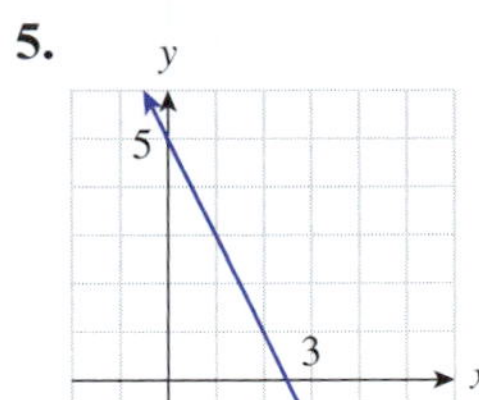

7.

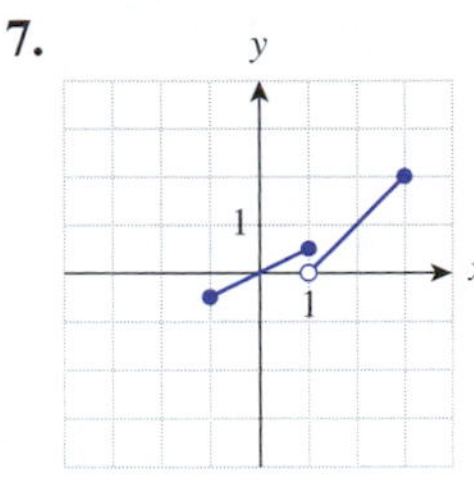

9. Absolute value **11.** Linear **13.** Quadratic
15. $y = -3x + 11$ **17.** $y = 1.25x - 4.25$
19. $y = (1/2)x + 3/2$ **21.** $y = 4x - 12$
23. $y = -x/4 + 1$ **25.** $y = -0.214x + 1.14$, $r \approx -0.33$
27. a. Exponential. Graph:

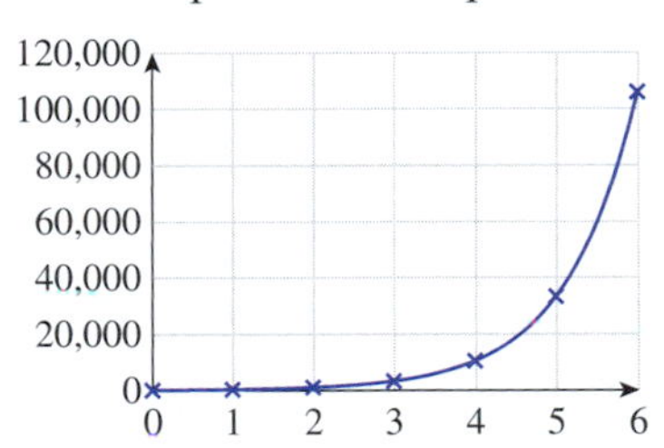

b. The ratios (rounded to 1 decimal place) are:

$V(1)/V(0)$	$V(2)/V(1)$	$V(3)/V(2)$	$V(4)/V(3)$	$V(5)/V(4)$	$V(6)/V(5)$
3	3.3	3.3	3.2	3.2	3.2

They are close to 3.2. **c.** About 343,700 visits/day
29. a. 2.3; 3.5; 6 **b.** For Web site traffic of up to 50,000 visits per day, the number of crashes is increasing by 0.03 per additional thousand visits. **c.** 140,000 **31. a.** (A) **b.** (A) Leveling off (B) Rising (C) Rising; begins to fall after 7 months (D) Rising **33. a.** The number of visits would increase by 30 per day. **b.** No; it would increase at a slower and slower rate and then begin to decrease. **c.** Probably not. This model predicts that Web site popularity will start to decrease as advertising increases beyond \$8,500 per month, and then drop toward zero. **35. a.** $v = 0.05c + 1,800$ **b.** 2,150 new visits per day **c.** \$14,000 per month **37.** $d = 0.95w + 8$; 86 kg **39. a.** Cost: $C = 5.5x + 500$; Revenue: $R = 9.5x$; Profit $P = 4x - 500$ **b.** More than 125 albums per week **c.** More than 200 albums per week
41. a. $q = -80p + 1,060$ **b.** 100 albums per week **c.** \$9.50, for a weekly profit of \$700
43. a. $q = -74p + 1,015.5$ **b.** 239 albums per week

Chapter 2

Review

1. \$7,425.00 **3.** \$7,604.88 **5.** \$6,757.41
7. \$4,848.48 **9.** \$4,733.80 **11.** \$5,331.37
13. \$177.58 **15.** \$112.54 **17.** \$187.57
19. \$9,584.17 **21.** 5.346% **23.** 14.0 years
25. 10.8 years **27.** 7.0 years **29.** 168.85%
31. 85.28% if she sold in February 2010. **33.** No. Simple interest increase is linear. We can compare slopes between successive points to see if the slope remained roughly constant: From December 2002 to August 2004 the slope was $(16.31 - 3.28)/(20/12) = 7.818$ while from August 2004 to March 2005 the slope was $(33.95 - 16.31)/(7/12) = 30.24$. These slopes are quite different. **35.** 2003

Year	2000	2001	2002	2003	2004
Revenue	\$180,000	\$216,000	\$259,200	\$311,040	\$373,248

37. At least 52,515 shares **39.** \$3,234.94
41. \$231,844 **43.** 7.75% **45.** \$420,275
47. \$140,778 **49.** \$1,453.06 **51.** \$2,239.90 per month
53. \$53,055.66 **55.** 5.99%

Chapter 3

Review

1. One solution

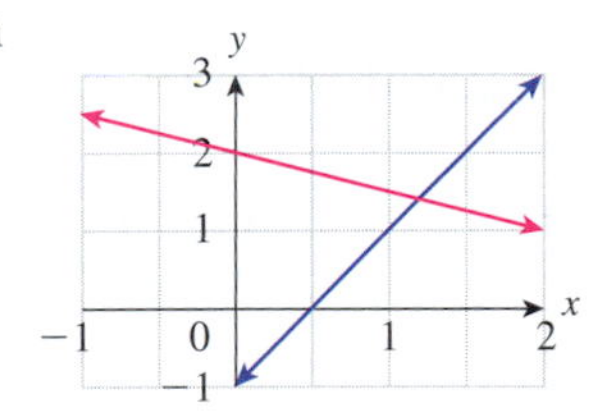

3. Infinitely many solutions

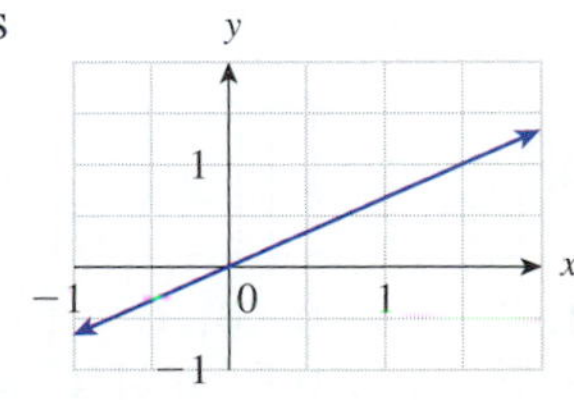

5. One solution

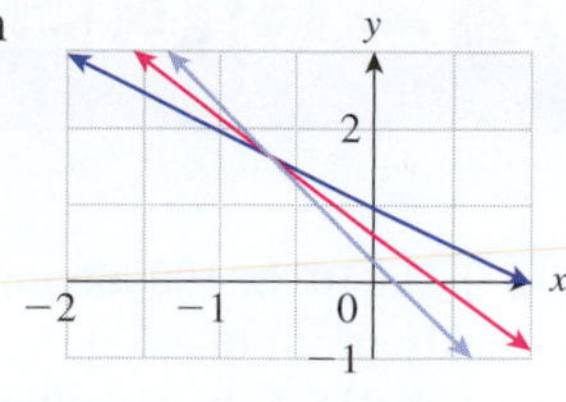

7. $(6/5, 7/5)$ **9.** $(3y/2, y)$; y arbitrary **11.** $(-0.7, 1.7)$ **13.** $(-1, -1, -1)$ **15.** $(z-2, 4(z-1), z)$; z arbitrary **17.** No solution **19.** $-40°$ **21.** It is impossible; setting $F = 1.8C$ leads to an inconsistent system of equations. **23.** $x + y + z + w = 10$; Linear **25.** $w = 0$; Linear **27.** $-1.3y + z = 0$ or $1.3y - z = 0$; Linear **29.** 550 packages from Duffin House, 350 from Higgins Press **31.** 1,200 packages from Duffin House, 200 from Higgins Press **33.** \$40 **35.** 7 of each **37.** 5,000 hits per day at OHaganBooks.com, 1,250 at JungleBooks.com, 3,750 at FarmerBooks.com **39.** 100 shares of HAL, 20 shares of POM, and 80 shares of WELL **41.** Billy-Sean is forced to take exactly the following combination: Liberal Arts: 52 credits, Sciences: 12 credits, Fine Arts: 12 credits, Mathematics: 48 credits. **43. a.** $x = 100$, $y = 100 + w$, $z = 300 - w$, w arbitrary **b.** 100 book orders per day **c.** 300 book orders per day **d.** $x = 100$, $y = 400$, $z = 0$, $w = 300$ **e.** 100 book orders per day **45.** Yes; New York to OHaganBooks.com: 450 packages, New York to FantasyBooks.com: 50 packages, Illinois to OHaganBooks.com: 150 packages, Illinois to FantasyBooks.com: 150 packages

Chapter 4

Review

1. Undefined **3.** $\begin{bmatrix} 1 & 8 \\ 5 & 11 \\ 6 & 13 \end{bmatrix}$ **5.** $\begin{bmatrix} 1 & 3 \\ 2 & 3 \\ 3 & 3 \end{bmatrix}$ **7.** $\begin{bmatrix} 1 & -2 \\ 0 & 1 \end{bmatrix}$

9. $\begin{bmatrix} 2 & 4 \\ 1 & 12 \end{bmatrix}$ **11.** $\begin{bmatrix} 1 & 1 \\ 0 & 1 \end{bmatrix}$ **13.** $\begin{bmatrix} 1 & -1/2 & -5/2 \\ 0 & 1/4 & -1/4 \\ 0 & 0 & 1 \end{bmatrix}$

15. Singular

17. $\begin{bmatrix} 1 & 2 \\ 3 & 4 \end{bmatrix}\begin{bmatrix} x \\ y \end{bmatrix} = \begin{bmatrix} 0 \\ 2 \end{bmatrix}$; $\begin{bmatrix} x \\ y \end{bmatrix} = \begin{bmatrix} 2 \\ -1 \end{bmatrix}$

19. $\begin{bmatrix} 1 & 1 & 1 \\ 1 & 2 & 1 \\ 1 & 1 & 2 \end{bmatrix}\begin{bmatrix} x \\ y \\ z \end{bmatrix} = \begin{bmatrix} 2 \\ 3 \\ 1 \end{bmatrix}$; $\begin{bmatrix} x \\ y \\ z \end{bmatrix} = \begin{bmatrix} 2 \\ 1 \\ -1 \end{bmatrix}$

21. $R = [1 \quad 0 \quad 0]$, $C = [0 \quad 1 \quad 0 \quad 0]^T$, $e = 1$
23. $R = [0 \quad 0.8 \quad 0.2]$, $C = [0.2 \quad 0 \quad 0.8]$, $e = -0.2$

25. $\begin{bmatrix} 1{,}100 \\ 700 \end{bmatrix}$ **27.** $\begin{bmatrix} 48{,}125 \\ 22{,}500 \\ 10{,}000 \end{bmatrix}$

29. Inventory − Sales + Purchases =

$$\begin{bmatrix} 2{,}500 & 4{,}000 & 3{,}000 \\ 1{,}500 & 3{,}000 & 1{,}000 \end{bmatrix} - \begin{bmatrix} 300 & 500 & 100 \\ 100 & 600 & 200 \end{bmatrix} + \begin{bmatrix} 400 & 400 & 300 \\ 200 & 400 & 300 \end{bmatrix} = \begin{bmatrix} 2{,}600 & 3{,}900 & 3{,}200 \\ 1{,}600 & 2{,}800 & 1{,}100 \end{bmatrix}$$

31. $N =$

$$\begin{bmatrix} 2{,}600 & 3{,}900 & 3{,}200 \\ 1{,}600 & 2{,}800 & 1{,}100 \end{bmatrix} - x\begin{bmatrix} 280 & 550 & 100 \\ 50 & 500 & 120 \end{bmatrix} + x\begin{bmatrix} 400 & 400 & 300 \\ 200 & 400 & 300 \end{bmatrix} = \begin{bmatrix} 2{,}600 & 3{,}900 & 3{,}200 \\ 1{,}600 & 2{,}800 & 1{,}100 \end{bmatrix} + x\begin{bmatrix} 120 & -150 & 200 \\ 150 & -100 & 180 \end{bmatrix}$$; 28 months from now (July 1)

33. Revenue = Quantity × Price =

$$\begin{bmatrix} 280 & 550 & 100 \\ 50 & 500 & 120 \end{bmatrix}\begin{bmatrix} 5 \\ 6 \\ 5.5 \end{bmatrix} = \begin{bmatrix} 5{,}250 \\ 3{,}910 \end{bmatrix} \begin{matrix} \text{Texas} \\ \text{Nevada} \end{matrix}$$

35. July 1: 1,000 shares, August 1: 2,000 shares, September 1: 2,000 shares
37. Loss = Number of shares × (Purchase price − Dividends − Selling price) = $[1{,}000 \quad 2{,}000 \quad 2{,}000]$

$$\left(\begin{bmatrix} 20 \\ 10 \\ 5 \end{bmatrix} - \begin{bmatrix} 0.10 \\ 0.10 \\ 0 \end{bmatrix} - \begin{bmatrix} 3 \\ 1 \\ 1 \end{bmatrix}\right) = [42{,}700]$$

39. $[2{,}000 \quad 4{,}000 \quad 4{,}000]\begin{bmatrix} 0.8 & 0.1 & 0.1 \\ 0.4 & 0.6 & 0 \\ 0.2 & 0 & 0.8 \end{bmatrix} =$

$[4{,}000 \quad 2{,}600 \quad 3{,}400]$ **41.** The matrix shows that no JungleBooks customers switched directly to FarmerBooks, so the only way to get to FarmerBooks is via OHaganBooks. **43.** Go with the "3 for 1" promotion and gain 20,000 customers from JungleBooks. **45.** JungleBooks will go with "3 for 2" and OHaganBooks will go with Finite Math, resulting in a gain of 15,000 customers to OHaganBooks. **47.** Choose between the "3 for 1" and Finite Math promotions with probabilities 60% and 40%, respectively. You would expect to gain 12,000 customers from JungleBooks (if you played this game many times).

49. $A = \begin{bmatrix} 0.1 & 0.5 \\ 0.01 & 0.05 \end{bmatrix}$

51. \$1,190 worth of paper, \$1,802 worth of books

Chapter 5

Review

1.

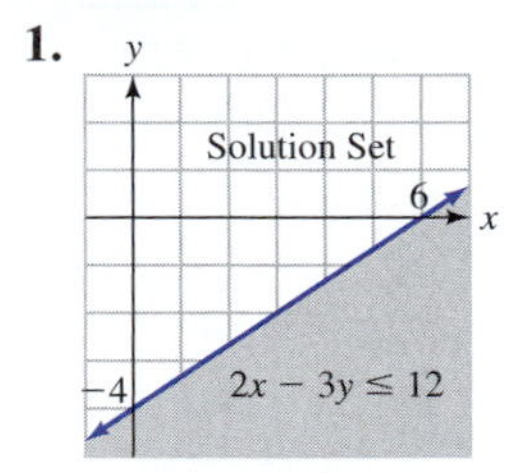

Unbounded

3.

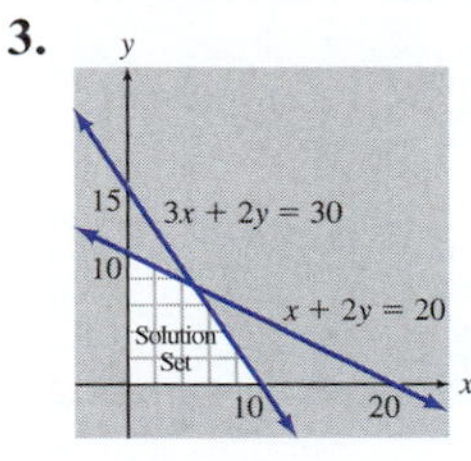

Bounded; corner points: (0, 0), (0, 10), (5, 15/2), (10, 0)

5. $p = 21; x = 9, y = 3$ **7.** $c = 22; x = 8, y = 6$ **9.** $p = 45;\ x = 0, y = 15, z = 15$ **11.** $p = 220; x = 20, y = 20, z = 60$ **13.** $c = 30; x = 30, y = 0, z = 0$ **15.** No solution; feasible region unbounded **17.** $c = 50; x = 20, y = 10, z = 0, w = 20$, OR $x = 30, y = 0, z = 0, w = 20$ **19.** $c = 60; x = 24, y = 12$ OR $x = 0, y = 60$ **21.** $c = 20; x = 0,\ y = 20$ **23.** $R = [1/2 \quad 1/2 \quad 0], C = [0 \quad 1/3 \quad 2/3]^T, e = 0$ **25.** $R = [1/27 \quad 7/9 \quad 5/27]$, $C = [8/27 \quad 5/27 \quad 14/27]^T,\ e = 8/27$ **27.** (A) **29.** 35 **31.** 400 packages from each for a minimum cost of $52,000 **33. a.** (B), (D) **b.** 450 packages from Duffin House, 375 from Higgins Press for a minimum cost of $52,500 **35.** 220 shares of EEE and 20 shares of RRR. The minimum total risk index is 500. **37.** 240 Sprinkles, 120 Storms, and no Hurricanes **39.** Order 600 packages from Higgins and none from the others, for a total cost of $90,000. **41. a.** Let $x = \#$ science credits, $y = \#$ fine arts credits, $z = \#$ liberal arts credits, and $w = \#$ math credits. Minimize $C = 300x + 300y + 200z + 200w$ subject to: $x + y + z + w \geq 120; x - y \geq 0; -2x + w \leq 0; -y + 3z - 3w \leq 0; x \geq 0, y \geq 0, z \geq 0, w \geq 0$. **b.** Billy-Sean should take the following combination: Sciences—24 credits, Fine Arts—no credits, Liberal Arts—48 credits, Mathematics—48 credits, for a total cost of $26,400. **43.** Smallest cost is $20,000; New York to OHaganBooks.com: 600 packages, New York to FantasyBooks.com: 0 packages, Illinois to OHaganbooks.com: 0 packages, Illinois to FantasyBooks.com: 200 packages.
45. FantasyBooks.com should choose between "2 for 1" and "3 for 2" with probabilities 20% and 80%, respectively. OHaganBooks.com should choose between "3 for 1" and "Finite Math" with probabilities 60% and 40%, respectively. OHaganBooks.com expects to gain 12,000 customers from FantasyBooks.com.

Chapter 6

Review

1. $N = \{-3, -2, -1\}$ **3.** $S = \{(1, 2), (1, 3), (1, 4), (1, 5), (1, 6), (2, 1), (2, 3), (2, 4), (2, 5), (2, 6), (3, 1), (3, 2), (3, 4), (3, 5), (3, 6), (4, 1), (4, 2), (4, 3), (4, 5), (4, 6), (5, 1), (5, 2), (5, 3), (5, 4), (5, 6), (6, 1), (6, 2), (6, 3), (6, 4), (6, 5)\}$ **5.** $A \cup B' = \{\text{a, b, d}\}$, $A \times B' = \{(\text{a, a}), (\text{a, d}), (\text{b, a}), (\text{b, d})\}$ **7.** $A \times B$ **9.** $(P \cap E' \cap Q)'$ or $P' \cup E \cup Q'$ **11.** $n(A \cup B) = n(A) + n(B) - n(A \cap B), n(C') = n(S) - n(C)$; 100 **13.** $n(A \times B) = n(A)n(B), n(A \cup B) = n(A) + n(B) - n(A \cap B), n(A') = n(S) - n(A)$; 21 **15.** $C(12, 1)C(4, 2)C(11, 3)C(4, 1)C(4, 1)C(4, 1)$ **17.** $C(4, 1)C(10, 1)$ **19.** 6 **21.** $C(4, 4)C(8, 1) = 8$ **23.** $C(3, 2)C(9, 3) + C(3, 3)C(9, 2) = 288$ **25.** The set of books that are either sci-fi or stored in Texas (or both); $n(S \cup T) = 112{,}000$ **27.** The set of books that are either stored in California or not sci-fi; $n(C \cup S') = 175{,}000$ **29.** The romance books that are also horror books or stored in Texas; $n(R \cap (T \cup H)) = 20{,}000$ **31.** 1,000 **33.** FarmerBooks.com; 1,800 **35.** FarmerBooks.com; 5,400 **37.** $26 \times 26 \times 26 = 17{,}576$ **39.** $26 \times 25 \times 9 \times 10 = 58{,}500$ **41.** 60,000 **43.** 19,600

Chapter 7

Review

1. $n(S) = 8,\ E = \{\text{HHT, HTH, HTT, THH, THT, TTH, TTT}\}, P(E) = 7/8$ **3.** $n(S) = 36; E = \{(1, 6), (2, 5), (3, 4), (4, 3), (5, 2), (6, 1)\}; P(E) = 1/6$ **5.** $n(S) = 6; E = \{2\}; P(E) = 1/8$ **7.** .76 **9.** .25 **11.** .5 **13.** 7/15 **15.** 8/792 **17.** 48/792 **19.** 288/792 **21.** $C(8, 5)/C(52, 5)$ **23.** $C(4, 3)C(1, 1)C(3, 1)/C(52, 5)$ **25.** $C(9, 1)C(8, 1)C(4, 3)C(4, 2)/C(52, 5)$ **27.** 1/5; dependent **29.** 1/6; independent **31.** 1; dependent **33.** $P = \begin{bmatrix} 1/2 & 1/2 \\ 1/4 & 3/4 \end{bmatrix}$ **35.** Brand A: $65/192 \approx .339$, Brand B: $127/192 \approx .661$ **37.** 14/25 **39.** 15/94 **41.** 79/167 **43.** 98% **45.** 6.9% **47.** .931 **49.** $P(H \cap C)$, since $P(H|C) > P(H)$ gives $P(H \cap C) > P(H)P(C)$ **51.** .246 **53.** 91% **55.** $[.2 \quad .4 \quad .4]\begin{bmatrix} 0.8 & 0.1 & 0.1 \\ 0.4 & 0.6 & 0 \\ 0.2 & 0 & 0.8 \end{bmatrix} = [.4 \quad .26 \quad .34]$, so 40% for OHaganBooks.com, 26% for JungleBooks.com, and 34% for FarmerBooks.com **57.** Here are three: (1) it is possible for someone to be a customer at two different enterprises; (2) some customers may stop using all three of the companies; and (3) new customers can enter the field.

Chapter 8

Review

1.

x	0	1	2
$P(X = x)$	1/4	1/2	1/4

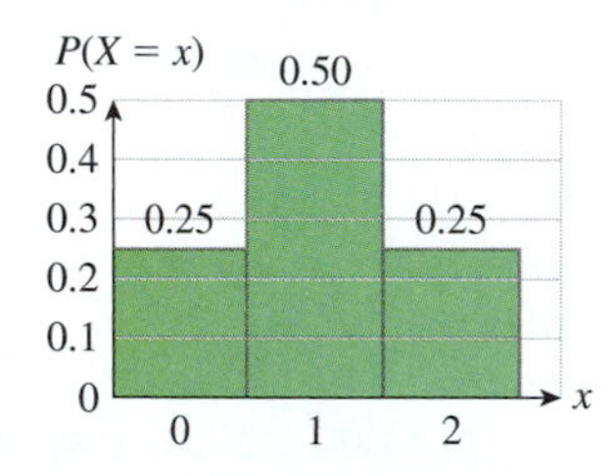

3.

x	2	3	4	5	6	7	8
$P(X = x)$	1/16	2/16	3/16	4/16	3/16	2/16	1/16

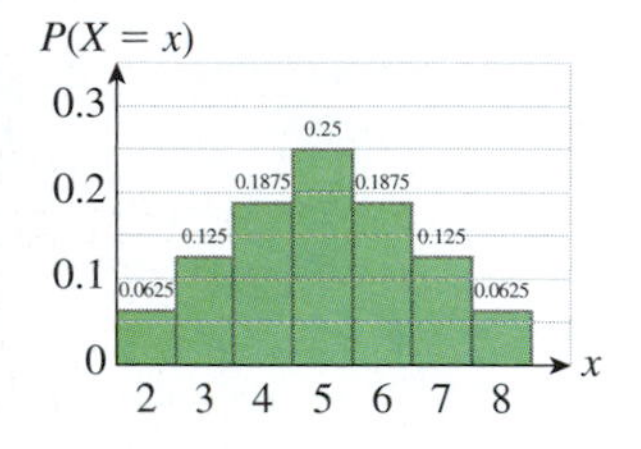

5.

x	0	1	2	3
$P(X = x)$	.2071	.4439	.2908	.0582

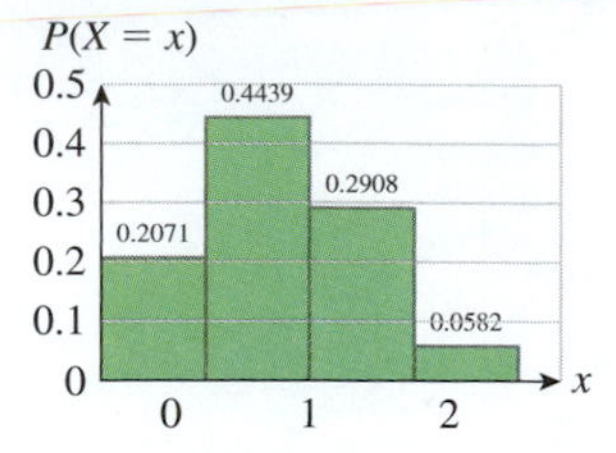

7. $\bar{x} = 2$, $m = 2$, $s \approx 2.7386$ **9.** Two examples are: 0, 0, 0, 4 and $-1, -1, 1, 5$ **11.** An example is $-1, -1, -1, 1, 1, 1$ **13.** .4165 **15.** .9267 **17.** .0067 **19.** .7330 **21.** $\mu = 1.5$, $\sigma = 0.8660$; 2

23.

x	−3	−2	−1	0	1	2	3
$P(X = x)$	1/16	2/16	3/16	4/16	3/16	2/16	1/16

$\mu = 0$, $\sigma = 1.5811$; within 1.3 standard deviations of the mean **25.** [49.4, 150.6] **27.** 260 **29.** .4332 **31.** .0358 **33.** .3085 **35.** $12.15 **37.** $27,210 **39.** False; let $X =$ price and $Y =$ weekly sales. Then weekly Revenue $= XY$. However, $27{,}210 \neq 12.15 \times 2{,}620$. In other words, $E(XY) \neq E(X)E(Y)$.

41. a.

x	2	4	6	8	10
$P(X = x)$	.25	.35	.15	.15	.10

$\mu = 5$, $\sigma = 2.5690$ **b.** Between 2.431 and 7.569 orders per million residents; the empirical rule does not apply because the distribution is not symmetric. **c.** (A) **43.** .190 **45.** .060 **47.** 0.5 **49.** .284 **51.** .108 **53.** Using normal distribution table: 407,000 people; more accurate answer: 423,000 people **55.** 148

Index

Index

Index of Applications

Business and Economics

General Interest